# Biosensor Based Advanced Cancer Diagnostics

# Biosensor Based Advanced Cancer Diagnostics

## From Lab to Clinics


Edited by

**Raju Khan**
Microfluidics & MEMS Centre, CSIR-Advanced Materials and Processes Research Institute (AMPRI), Bhopal, India;
Academy of Scientific and Innovative Research (AcSIR), Ghaziabad, India

**Arpana Parihar**
Microfluidics & MEMS Centre, CSIR-Advanced Materials and Processes Research Institute (AMPRI), Bhopal, India

**Sunil K. Sanghi**
Microfluidics & MEMS Centre, CSIR-Advanced Materials and Processes Research Institute (AMPRI), Bhopal, India



ELSEVIER

ACADEMIC PRESS
An imprint of Elsevier

Academic Press is an imprint of Elsevier
125 London Wall, London EC2Y 5AS, United Kingdom
525 B Street, Suite 1650, San Diego, CA 92101, United States
50 Hampshire Street, 5th Floor, Cambridge, MA 02139, United States
The Boulevard, Langford Lane, Kidlington, Oxford OX5 1GB, United Kingdom



**Notices**

Knowledge and best practice in this field are constantly changing. As new research and experience broaden our understanding, changes in research methods, professional practices, or medical treatment may become necessary.

Practitioners and researchers must always rely on their own experience and knowledge in evaluating and using any information, methods, compounds, or experiments described herein. In using such information or methods they should be mindful of their own safety and the safety of others, including parties for whom they have a professional responsibility.

To the fullest extent of the law, neither the Publisher nor the authors, contributors, or editors, assume any liability for any injury and/or damage to persons or property as a matter of products liability, negligence or otherwise, or from any use or operation of any methods, products, instructions, or ideas contained in the material herein.

**British Library Cataloguing-in-Publication Data**
A catalogue record for this book is available from the British Library

**Library of Congress Cataloging-in-Publication Data**
A catalog record for this book is available from the Library of Congress

ISBN: 978-0-12-823424-2

For Information on all Academic Press publications
visit our website at https://www.elsevier.com/books-and-journals

*Publisher:* Mara Conner
*Acquisitions Editor:* Carrie Bolger
*Editorial Project Manager:* Sara Valentino
*Production Project Manager:* Prem Kumar Kaliamoorthi
*Cover Designer:* Mark Rogers

Typeset by MPS Limited, Chennai, India

# Contents

# List of contributors


**Aida Alaei** Department of Life Science Engineering, Faculty of New Sciences & Technologies, University of Tehran, Tehran, Iran

**Kavita Arora** Advanced Instrumentation & Research Facility (AIRF) and School of Computational & Integrative Sciences (SCIS), Jawaharlal Nehru University, New Delhi, India

**Neha Arya** Department of Medical Devices, National Institute of Pharmaceutical Education and Research, Ahmedabad, India; Department of Translational Medicine Centre, All India Institute of Medical Sciences, Bhopal, Bhopal, India

**Rinti Banerjee** Nanomedicine Laboratory, Department of Biosciences and Bioengineering, Indian Institute of Technology Bombay, Powai, Mumbai, India

**Saptaka Baruah** Department of Physics, Rajiv Gandhi University, Itanagar, India

**Niloy Chatterjee** Food and Nutrition Division, University of Calcutta, Kolkata, India

**Meenakshi Choudhary** Centre for Biomedical Engineering, Indian Institute of Technology Delhi, New Delhi, India

**Samraggi Coudhury** DBT-National Institute of Animal Biotechnology (DBT-NIAB), Hyderabad, Telangana, India

**Mehdi Dadmehr** Department of Biology, Payame Noor University, Tehran, Iran

**Surojeet Das** European Molecular Biology Laboratory Australia, Australian Regenerative Medicine Institute, Monash University, Melbourne, VIC, Australia

**Krishna Das Saha** Cancer Biology and Inflammatory Disorder Division, CSIR-Indian Institute of Chemical Biology, Kolkata, India

**Nitu Dogra** Proteomic & Translational Research Lab, Centre for Medical Biotechnology, Amity Institute of Biotechnology, Amity University Noida, Noida, India

**Manuela F. Frasco** BioMark Sensor Research/UC, Faculty of Sciences and Technology, Coimbra University, Coimbra, Portugal; BioMark Sensor Research/ISEP, School of Engineering, Polytechnic Institute of Porto, Porto, Portugal; CEB - Centre of Biological Engineering, Minho University, Braga, Portugal

**Sonu Gandhi** DBT-National Institute of Animal Biotechnology (DBT-NIAB), Hyderabad, Telangana, India

**Shagun Gupta** Shoolini University, Solan, India

**Morteza Hosseini** Department of Life Science Engineering, Faculty of New Sciences & Technologies, University of Tehran, Tehran, Iran

**Pouria Jafari** Department of Life Science Engineering, Faculty of New Sciences & Technologies, University of Tehran, Tehran, Iran

**Surbhi Jain** Department of Biochemistry and Genetics, Barkatullah University, Bhopal, India

**Deepak Kala** Amity Centre of Nanotechnology, Amity University, Gurugram, India

**Deepshikha Pande Katare** Proteomic & Translational Research Lab, Centre for Medical Biotechnology, Amity Institute of Biotechnology, Amity University Noida, Noida, India

**Ankur Kaushal** Amity Centre of Nanotechnology, Amity University, Gurugram, India

**Raju Khan** Microfluidics & MEMS Centre, CSIR-Advanced Materials and Processes Research Institute (AMPRI), Bhopal, India; Academy of Scientific and Innovative Research (AcSIR), Ghaziabad, India

**Lucky Krishnia** Amity Centre of Nanotechnology, Amity University, Gurugram, India

**Ashok Kumar** Department of Biochemistry, All India Institute of Medical Sciences, Bhopal, Bhopal, India

**Avinash Kumar** Department of Mechanical Engineering, Indian Institute of Information Technology Design & Manufacturing Kancheepuram, Chennai, India

**Pradip Kumar** Integrated Approach for Design and Product Development Division, CSIR-Advanced Materials and Processes Research Institute (CSIR-AMPRI), Bhopal, India

**Sanjeev Kumar** Department of Physics, Rajiv Gandhi University, Itanagar, India

**Bidyarani Maibam** Department of Physics, Rajiv Gandhi University, Itanagar, India

**Ruchi Jakhmola Mani** Proteomic & Translational Research Lab, Centre for Medical Biotechnology, Amity Institute of Biotechnology, Amity University Noida, Noida, India

**Krishnendu Manna** Cancer Biology and Inflammatory Disorder Division, CSIR-Indian Institute of Chemical Biology, Kolkata, India; Department of Food and Nutrition, University of Kalyani, Kalyani, India

**Shubhangi Mhaske** Oral and Maxillofacial Pathology and Microbiology, People's College of Dental Sciences & Research Centre, People's University, Bhopal, India

**Maryam Mousavizadegan** Department of Life Science Engineering, Faculty of New Sciences & Technologies, University of Tehran, Tehran, Iran

**Niladri Mukherjee** Cancer Biology and Inflammatory Disorder Division, CSIR-Indian Institute of Chemical Biology, Kolkata, India

**Sayali Mukherjee** Amity Institute of Biotechnology, Amity University Uttar Pradesh, Lucknow, India

**Sagar Narlawar** DBT-National Institute of Animal Biotechnology (DBT-NIAB), Hyderabad, Telangana, India

**Fatemeh Nemati** Department of Life Science Engineering, Faculty of New Sciences & Technologies, University of Tehran, Tehran, Iran

**Riyaz Ali M. Osmani** Nanomedicine Laboratory, Department of Biosciences and Bioengineering, Indian Institute of Technology Bombay, Powai, Mumbai, India

**Udwesh Panda** Department of Mechanical Engineering, Indian Institute of Information Technology Design & Manufacturing Kancheepuram, Chennai, India

**Ritu Pandey** Department of Biochemistry, All India Institute of Medical Sciences, Bhopal, Bhopal, India

**Arpana Parihar** Microfluidics & MEMS Centre, CSIR-Advanced Materials and Processes Research Institute (AMPRI), Bhopal, India; Department of Biochemistry and Genetics, Barkatullah University, Bhopal, India

**Dipesh Singh Parihar** Engineering College Tuwa, Godhra, India

**Azam Bagheri Pebdeni** Department of Life Science Engineering, Faculty of New Sciences & Technologies, University of Tehran, Tehran, Iran

**Amarjitsing Rajput** Nanomedicine Laboratory, Department of Biosciences and Bioengineering, Indian Institute of Technology Bombay, Powai, Mumbai, India; Department of Pharmaceutics, Poona College of Pharmacy, Bharti Vidyapeeth Deemed University, Erandwane, Pune, India

**Pushpesh Ranjan** Microfluidics & MEMS Centre, CSIR-Advanced Materials and Processes Research Institute (AMPRI), Bhopal, India; Academy of Scientific and Innovative Research (AcSIR), Ghaziabad, India

**Tripti Rimza** Integrated Approach for Design and Product Development Division, CSIR-Advanced Materials and Processes Research Institute (CSIR-AMPRI), Bhopal, India

**Amirreza Roshani** Department of Life Science Engineering, Faculty of New Sciences & Technologies, University of Tehran, Tehran, Iran

**Mohd Abubakar Sadique** Microfluidics & MEMS Centre, CSIR-Advanced Materials and Processes Research Institute (AMPRI), Bhopal, India

**M. Goreti F. Sales** BioMark Sensor Research/UC, Faculty of Sciences and Technology, Coimbra University, Coimbra, Portugal; BioMark Sensor Research/ISEP, School of Engineering, Polytechnic Institute of Porto, Porto, Portugal; CEB - Centre of Biological Engineering, Minho University, Braga, Portugal

**Shikha Saxena** Amity Institute of Pharmacy, Amity University, Noida, India

**Nikita Sehgal** Proteomic & Translational Research Lab, Centre for Medical Biotechnology, Amity Institute of Biotechnology, Amity University Noida, Noida, India

**Amit Seth** School of Life Science, Manipur University, Imphal, India

**Deepshikha Shahdeo** DBT-National Institute of Animal Biotechnology (DBT-NIAB), Hyderabad, Telangana, India

**Kosar Shahsavar** Department of Life Science Engineering, Faculty of New Sciences & Technologies, University of Tehran, Tehran, Iran

**Ekta Singh** Nanomedicine Laboratory, Department of Biosciences and Bioengineering, Indian Institute of Technology Bombay, Powai, Mumbai, India

**Shiv Singh** Lightweight Metallic Materials Division, CSIR-Advanced Materials and Processes Research Institute (CSIR-AMPRI), Bhopal, India

**Ayushi Singhal** Microfluidics & MEMS Centre, CSIR-Advanced Materials and Processes Research Institute (AMPRI), Bhopal, India; Academy of Scientific and Innovative Research (AcSIR), Ghaziabad, India

**Raquel Vaz** BioMark Sensor Research/UC, Faculty of Sciences and Technology, Coimbra University, Coimbra, Portugal; BioMark Sensor Research/ISEP, School of Engineering, Polytechnic Institute of Porto, Porto, Portugal; CEB - Centre of Biological Engineering, Minho University, Braga, Portugal

**Vivek Verma** Shoolini University, Solan, India

**Shalu Yadav** Microfluidics & MEMS Centre, CSIR-Advanced Materials and Processes Research Institute (AMPRI), Bhopal, India; Academy of Scientific and Innovative Research (AcSIR), Ghaziabad, India

**Monal Yuwanati** Department Of Oral Pathology and Microbiology, Saveetha Dental College and Hospitals, Saveetha Institute of Medical and Technical Sciences, Saveetha University, Chennai, India

# About the editors

**Raju Khan** is currently working as the principal scientist and associate professor at CSIR-Advanced Materials and Processes Research Institute (AMPRI), Bhopal, MP, India. Dr. Khan received his PhD & MSc in chemistry from the Jamia Millia Islamia (Central University), New Delhi, India. Dr. Khan has published several refereed papers in national and international journals, has filed patents, and has edited as well as coedited several books on biosensors and antimicrobial applications. He has completed several national and international collaborative projects such as Indo-Czech Republic, Indo-Russia, and United States. He is a recipient of the reputed BOYSCAST fellowship from the Department of Science & Technology (DST) within the Ministry Government of India. During the fellowship, he has worked as a visiting scientist at the University of Texas at San Antonio (UTSA), United States. Since then, Dr. Khan is continuously very productive with more than 15 years of R&D and teaching experiences, producing high-quality research, mentoring students, and supporting the analytical and microfluidics division as outsource facility. His current research activities include nano-biomaterials, biosensors, point-of-care diagnostics, nano-biotechnology, antimicrobials, and biomedical engineering.

**Arpana Parihar** is currently working as a Women Scientist B at CSIR-Advanced Materials and Processes Research Institute (AMPRI), Bhopal, MP, India, under the scheme of DST-WoS-B awarded from the Department of Science and Technology, Government of India. She did her PhD from Raja Rammana Centre for Advanced Technology, Indore. Her doctoral research work involves the evaluation of tumor selectivity and photodynamic therapy (PDT) efficacy of chlorin p6 through receptor-mediated targeted delivery in oral cancer. After PhD, her postdoctoral research work at the Centre for Biomedical Engineering (CBME), Indian Institute of Technology (IIT) Delhi involves the enhancement of osteoinductive and osteoconductive properties of various implants made up of metals, ceramics, and polymers. Dr. Parihar is awarded prestigious GATE, CSIR-NET, DST-WoS A, and WoS B fellowship. She has more than 7 years of research and teaching experience at various prestigious institutes that fetched several peer-reviewed papers in national and international journals of repute. Her current research activity includes fabrication of biosensors for early diagnosis of cancer, molecular docking and simulation for drug designing, tissue engineering, targeted cancer therapy, and 3D cell culture.

**Sunil K. Sanghi** was working as chief scientist, professor, and Head of Department at Microfluidics & MEMS Centre, CSIR-Advanced Materials and Processes Research Institute (AMPRI), Bhopal, India. His past research areas were on development of manual and automated procedures for all kinds of analytes in biomedical, pharmaceutical, and environmental samples using micro liquid, capillary gas chromatographic, and capillary electrophoretic separation techniques in combination with sample preparation, derivatization and reaction-detection systems, micro-chip-based separation under the concept of lab-on-a-chip. Dr. Sanghi has successfully completed several international and national collaborative R&D projects—Indo-European Union, Indo–French, New Millennium Indian Technology Leadership Initiative (NMITLI).
He was awarded the reputed Marie Curie Fellowship of the European Union, and worked as a visiting scientist for 3 years at the University of Amsterdam and Institute Curie, Paris. He holds an experience of 35 years in R&D and teaching. Recently, Dr. Sanghi has received the 2021 National Meritorious Innovation Award from Government of India.

Chapter 1

# Cancer: A *sui generis* threat and its global impact

Amarjitsing Rajput[1,2], Riyaz Ali M. Osmani[1], Ekta Singh[1] and Rinti Banerjee[1]
[1]*Nanomedicine Laboratory, Department of Biosciences and Bioengineering, Indian Institute of Technology Bombay, Powai, Mumbai, India*
[2]*Department of Pharmaceutics, Poona College of Pharmacy, Bharti Vidyapeeth Deemed University, Erandwane, Pune, India*

## 1.1 Introduction

### 1.1.1 Cancer

The history of cancer can be traced back to the ever existence of human civilization. The first-ever written record was the one that described breast cancer and obtained in 1600 BCE from the Egyptian Edwin Smith Papyrus (Hajdu, 2011). The term cancer is used generically and includes multiple disorders affecting different organs in the body (World Health Organization, 2018; National Cancer Institute, 2007). Cancer is also referred to as neoplasms or malignant tumors in specific cases. An expounding feature of cancer is the rapid transformation of normal cells into cells with abnormal and uncontrolled proliferative capacity, loss of characteristic properties like contact inhibition, and invasive behavior. These cells tend to spread into other neighboring tissues and organs in a phenomenon termed metastasis. The process of metastasis is a critical factor responsible for lethality in cancer cases.

The active cases as well cancer survivors experience many mental health disturbances including depression, anxiety, panic, and posttraumatic stress disorders which lead to complications and further health deterioration (Fox et al., 2013). Some reports suggest a strong association of mental health with impunity among cancer patients and survivors resulting in high morbidity and mortality rate (Andersen et al., 2007; Lutgendorf, 2005; Sephton et al., 2009), as well as decreased congruity to treatment and (DiMatteo, Lepper, & Croghan, 2000; Looper, 2007), consequently increase in their healthcare costs. It has also been suggested that mental health disorders in cancer patients can lead to high-risk comorbidities owing to the interconnectivity of metabolic pathways and hormonal controls and switches in the body. Several high-risk comorbidities can include diseases like cardiovascular, musculoskeletal conditions, and diabetes (Proctor et al., 2003; Talbot & Nouwen, 2000). Moreover, apart from comorbidities, mental health disorder is accountable for the highest long-term care, hospital, and ambulatory costs itself compared to other chronic conditions among cancer survivors (Anguiano, Mayer, & Piven, 2012). The current understanding of the social, psychological, and economic burden of cancer is quite restricted due to limited reports and studies at the intersection of cancer, mental health, and medical costs involved in treatment and survivorship phases. Loss of physical productivity in patients and limited scope of employment impacts the patients or survivors and their families financially and psychologically. Moreover, this factor affects the most when patients are in their earning age group where insurance coverage is endangered. Apart from direct costs, there are certain indirect costs due to mortality and morbidity incurred by the patient or survivor and the family. In addition to the above, it has been suggested that the tendency to commit suicide and the suicide rates associated with cancers are mostly attributed to mental health issues faced by the patients and survivors (Misono, Weiss, Fann, Redman, & Yueh, 2008; Rim, Guy, Yabroff, McGraw, & Ekwueme, 2016). On-going evaluation of different payment models, care coordination, and disease management programs for cancer survivors with comorbidities will be important in monitoring the impact on healthcare costs (Khushalani et al., 2018).

### 1.1.2 Pathophysiology of cancer

Cancer development is a multistage phenomenon that originates from normal cells, leading to their transformation into the precancerous lesion and, subsequently, a malignant tumor, as depicted in Fig. 1.1. The trigger to such a process

Biosensor Based Advanced Cancer Diagnostics. DOI: https://doi.org/10.1016/B978-0-12-823424-2.00019-3

arises from exposure to physical, chemical, and biological agents termed as "carcinogens" and genetic factors. International Agency for Research on Cancer (IARC) is a World Health Organization (WHO) agency that maintains a record of cancer-causing agents and their classification into different categories based on their properties and mode of action, etc. (Ferlay et al., 2013). The characteristic features of cancerous cells include by-passing apoptosis, competent growth, resistance to antigrowth signals, neoangiogenesis within tumor tissue, higher replicative potential, reprogrammed metabolic flux, and immune system bypass (Gutschner & Diederichs, 2000; Hanahan & Weinberg, 2011). A classic hallmark of cancer is the rapid transformation of normal cells into rapidly and uncontrollably dividing abnormal cells that invade adjoining tissues and organs in a process termed metastasis. The tumor at the origin of cancer is called the primary tumor, and dispersed tumors are called metastatic tumors. The metastasis phenomenon commonly occurs at later stages of tumor development via the bloodstream or lymph circulation. The key events involve local tissue invasion, intravasation in blood-vascular or lymphatics, circulation, extravasation in other tissues, consequently, proliferation and angiogenesis. Metastasis mostly occurs in pulm nary, hepatic, nervous, and skeletal tissues (National Cancer Institute, 2015). In almost all cases of cancer patients, metastases are a significant cause of fatality (National Comprehensive Cancer Network, 2013).

### 1.1.3 Genetics and epigenetics of cancer

The underlying cancer concept is a disease occurring due to a defect in cellular growth regulation and differentiation. A change in two particular sets of genes results in transforming a normal cell into a cancerous state, i.e., oncogenes and tumor suppressor genes, which are involved in promoting and inhibiting cellular growth and differentiation process, respectively (Croce, 2008). Overexpression and underexpression of such multiple genes lead to the transformation process (Knudson, 2001). Genetic alteration occurs at different levels in these genes via various mechanisms. Additionally, the tumor microenvironment is a critical determinant of whether such genetic alterations are favored for perpetuation. The presence of carcinogens, oxygen deficit, physical cellular injuries, and thermal conditions are some of the factors that constitute the tumor microenvironment (Nelson et al., 2004).

Cancer progression to invasive stages is driven by clonal evolution resulting from genetic mutations and chromosomal aberrations, leading to cellular heterogeneity within tumor cell subpopulation. This is one of the reasons for complications in the disease and designing its treatment (Merlo, Pepper, Reid, & Maley, 2006). Apart from genetic factors, epigenetic alterations have also been associated strongly and more frequently with cancer development and progression (Baylin & Ohm, 2006). In a study, variations in the pattern of methylation of several protein-coding genes, including 147 hypermethylated and 27 hypomethylated, have been associated with colon cancer cases. In 100% of cases, ten genes were directly hypermethylated (Schnekenburger & Diederich, 2012).

Moreover, epigenetic alterations in deoxyribonucleic acid (DNA) repair proteins are likely reported to be responsible for genetic instability during early progression to malignancy (Bernstein, Nfonsam, Prasad, & Bernstein, 2013; Jacinto & Esteller, 2007; Lahtz & Pfeifer, 2011). Altogether, cancer's development and progression results from a conglomeration of multiple mutations and favorable epimutations, resulting in clonal expansion. For example, an average of 60–70 protein-altering mutations are present in breast cancer, of which only 3%–4% are "driver" mutations, and the remainder may be "passenger" mutations (Vogelstein et al., 2013).

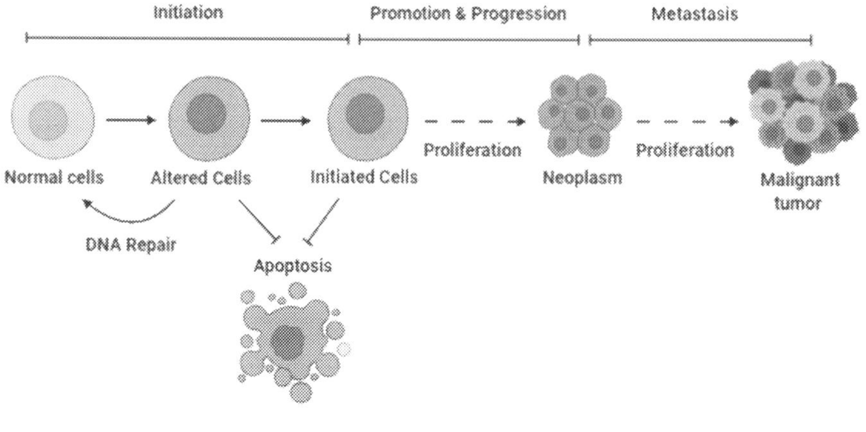


**FIGURE 1.1** Diagrammatic illustration representing various stages leading to cancer.

Cancer is a complex and highly emerging disease. Cancer's complex nature is due to evolutionary processes of mutation, genetic drift, and selection with many factors and agents interacting in the tumor microenvironment (Turajlic, Sottoriva, Graham, & Swanton, 2019). Aging is another aspect that plays an important role in cancer growth and progression. The risk and incidence of cancer's increase with age is due to the decrease in the cellular repair mechanism's efficiency.

### 1.1.4 Classification and nomenclature of cancers

Cancers can be classified in two ways; by their histology or by their site of origin. According to the International Classification of Diseases for Oncology (ICD-O), six major groups have been identified based on histology, namely; carcinoma, sarcoma, myeloma, leukemia, lymphoma, and mixed types (Fritz et al., 2000; Table 1.1). When classified by their site of origin, the most commonly reported cancers include lung, breast, colorectal, skin, stomach, cervical, and uterine cancers. Fig. 1.2 depicts the percentage of prevalent cases of some of the most common types of cancers

The nomenclature of cancers usually involved using the Latin or Greek word "-oma" as a suffix in histological type and the name of the organ or tissue of origin as the root. For instance, cancer affecting the parenchymatous liver tissue developing from malignant epithelial cells is called hepatocarcinoma. A malignant tumor developing from primitive hepatic precursor cells is called a hepatoblastoma, and when developing from fat cells is called a liposarcoma. As an example of breast cancer named "ductal carcinoma" referring to cancer from lactation ducts in the breast, a different convention can be clarified. Similarly, nomenclature means using -oma or sometimes -noma as a suffix with the organ's name as the source of noncancerous benign tumors. For example, leiomyoma, melanoma, and seminoma are considered benign tumors of smooth muscle cells, skin, and testicles. In addition, the nomenclature of such cancers takes into account the size and shape of cancer cells as seen under a microscope used in the "root" name, such as spindle cell carcinoma, small cell carcinoma, and giant cell carcinoma.

### 1.1.5 Epidemiology and demographics

According to WHO reports, as of 2018, the leading cause of death in cancer with a global estimate of 9.6 million, one out of every six cases of deaths is due to cancer worldwide (International Agency for Research on Cancer IARC, 2018). The risks associated with cancer are 20% in males and 17% in females. Moreover, the fatality in male and female cancer patients is 13% and 9%, respectively. It is the leading cause of fatality in countries with populations falling below all income levels.

Cancer causes around 70% of fatalities in low- and middle-income countries (LMICs). About one third of these deaths are associated with five main risk factors: high body mass index (BMI), sedentary lifestyle, consumption of alcohol, smoking, and low intake of nutritional greens. With the increasing population towards a westernized and urbanized lifestyle, economic transition, and exposure to risk factors prevalent in high-income countries (HICs), cancer risk is expected to increase drastically in LMICs.

One of the most critical risk factors in cancer is tobacco exposure that accounts for 22% of deaths (GBD 2015 Risk Factors Collaborators, 2016). About 25% of cancer cases in LMICs are hepatitis and human papillomavirus (HPV), causing infection-related cancer (Plummer et al., 2016). Other common factors are the delayed diagnosis, late-stage presentation of the disease, and inaccessibility to proper treatments responsible for spiking the number of cancer cases observed in LMICs. It was reported in the year 2017 that only 26% of LMICs had pathology services provided by their public sector relative to 90% of HICs. As a result, cancer has had a rising economical impact. An approximate USD $1.6 trillion dollars was reported as the total economic cost of cancer in 2010 (Stewart & Kleihues, 2003; Stewart &

**TABLE 1.1** Chemical with carcinogenic danger to humans (Centers for Disease Control and Prevention CDC, 2010a,b).

| Group | Chemical |
|---|---|
| Group 1 | Cancerous to human beings |
| Group 2A | Perhaps cancerous to people |
| Group 2B | Probably cancerous to people |
| Group 3 | Not classifiable as to its carcinogenicity to people |
| Group 4 | Not classifiable as to its carcinogenicity to people |

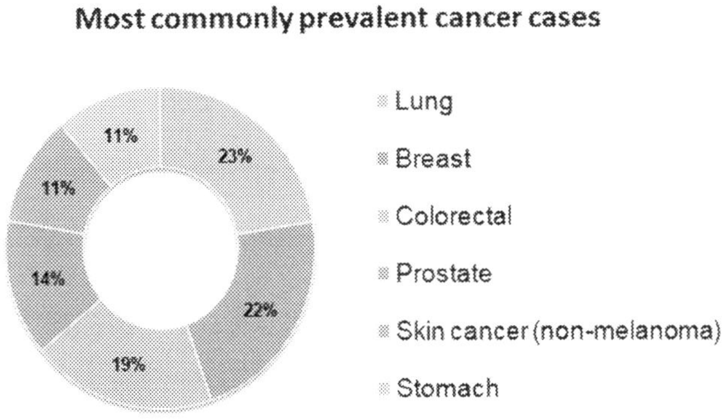


**FIGURE 1.2** A statistical representation of the percentage of cases of some of the most commonly prevalent cancers.

Wild, 2014). Lastly, it is also observed that merely one out of five LMICs have accessible essential data to drive cancer policies (International Agency for Research on Cancer IARC, n.d.).

Cancer has considerably burdened countries at all income levels around the world. With the knowledge of risk factors and technology-driven diagnosis and treatments, cancer incidence rates are controlled in many western countries. However, cases of lung, breast, and colorectal cancer, which were more prevalent in HICs, are rising in LMICs due to a shift in lifestyle patterns under the influence of the more developed HICs. Many risk factors typically observed in HICs are now frequently rising in LMICs, such as smoking, obesity, sedentary routines, and reproductive behaviors. Additionally, the burden of comorbidities and infection-related cancers, including stomach, hepatic, and cervical cancers, continues to be out of proportion in these countries.

Most cancers can be prevented by awareness, controlling predisposition factors such as tobacco exposure and unhealthy lifestyle, proper vaccination, and early diagnosis. Additionally, adequate treatment and palliative care assist in overcoming the complications of disease manifestations. A collaborative effort in this direction by nations and global agencies is indispensable to implicate cancer control measures effectively and equitably across all diasporas and economic strata.

# 1.2 Causes of cancer

## 1.2.1 Physical carcinogens

### *1.2.1.1 Ultraviolet rays*

Ultraviolet (UV) rays are electromagnetic rays that emerge from the sun and artificial materials such as welding torches and tanning beds. The ray is nothing but the discharge of energy from any material. The energy rays are classified into highest intensity rays (e.g., gamma rays, X-rays) to lowest intensity rays (e.g., radio waves). The UV ray is found in the central position of the energy spectrum.

#### 1.2.1.1.1 Classification of ultraviolet radiation

It is mainly classified into three types:

1. UVA rays (320–400 nm): It consists of minimum energy than other UV rays. These rays can bring out the aging of the skin cells and result in some implied damage to cell DNA. They are mainly associated with eternal skin damage such as crinkles and are often recognized to play a crucial role in a few skin-related cancers.
2. UVB rays (280–320 nm): It possesses somewhat more energy compared to UVA rays. They resulted in DNA damage in skin cells and considered the principal rays responsible for sunburn and most skin cancers.
3. UVC rays (200–280 nm): It consists of maximum energy than other UV rays. They mainly attack the atmosphere's ozone sheet and cannot reach the earth and cannot cause skin cancer.

#### 1.2.1.1.2 Artificial sources of ultraviolet rays

The artificial UV ray sources consist of:

- Lamps (blacklight).
- Lamps (Mercury vapor).

- Sunlamps and sunbeds (tanning beds and booths).
- Phototherapy (UV therapy).
- High-pressure lights, plasma torches, and welding flashes (xenon and xenon-mercury arc) (American Cancer Society, 2020b).

The exposure to UV radiation results in hazardous effects on human health, including skin diseases, immunosuppression, cataract, photoaging, and skin cancer (Armstrong & Kricker, 2001). The danger of UV-induced cancer progression is based on the type of UV rays to which one is exposed, the intensity of the exposure, and the amount of protection offered by the skin's natural pigment (melanin). People with fair skin are more prone to skin melanoma due to less melanin in their skin (Costa, 2020).

##### 1.2.1.1.3 Ultraviolet radiation and skin cancer pathogenesis

UV rays found in sunlight results in skin carcinoma of the squamous cells, basal cell carcinoma, and malignant melanoma. The anticancer activity of UV radiation is due to the formation of pyrimidine dimers in DNA. Pyrimidine dimers are nothing but the structure formed between the base pairs of the DNA, namely cytosine and thymine, which are members of the chemical family called pyrimidines (Costa, 2020). The mechanism responsible for carcinoma includes impairment in the functioning of the pyrimidine dimer growth regulatory gene. The UV radiation causes mutations to p53 tumor suppression genes. These genes play a vital role in the repair of DNA or the apoptosis of cells suffering from multiple DNA damage. Therefore, if p53 genes are mutated, they can no longer restore the DNA system, resulting in dysregulation of apoptosis, the proliferation of mutated keratinocytes, and the beginning of skin cancer (Benjamin & Ananthaswamy, 2007).

#### *1.2.1.2 Ionizing rays*

An ionizing ray is a form of energy discharged by the molecules which transit as electromagnetic waves (gamma or X-rays) or fragments (neutrons, alpha, or beta). The impetuous breakdown of an atom is known as radioactivity, and the extra energy emitted during this process is called ionizing radiation. The hazardous items after breakdown resulting in radiation (ionizing) are called radionucleotide. All radionuclides are simply recognized by pollution type, the radiation energy they emit, and half-life.

##### 1.2.1.2.1 Radiation source

The human population is exposed to different ionizing radiation from natural and artificial origins daily. Natural radiation comes from many sources; 60% is present in the soil, water, and air. People inhale and consume the radionucleotide daily from air, water, and food.

In addition to this, people are also exposed to radiation from a natural source such as cosmic rays, mainly of a high peak. Around 80% of the annual quantity of background radiation is received from terrestrial and cosmic radiation sources. The background radiation levels vary from one geographical location to another due to geological differences.

Human exposure to radiation arises from the artificial resources from the nuclear power plant to devices used for diagnosis or treatment in the medical field. The most common medical devices producing ionizing radiation are X-ray machines (World Health Organization WHO, 2020a).

Ionizing radiation, including electromagnetic and particulate, is considered a potent cancer-causing agent, though many years are required for exposure and tumor appearance. Radiation is responsible for a small number of cancers in humans compared to chemicals. Still, the prolonged latency and the overall effect of the repeated small doses of radiation-induced tumors make it significantly difficult (Costa, 2020). The long time exposure of this ionizing thus results in an increase in the risk of the ensuing cancer. This risk is associated with the dose-response relationship, and there are no safe limits for radiation exposure (Leuraud et al., 2015). This relationship was first noted by March; increasing leukemia cases among the different radiological professionals (March, 1944).

One of the prominent reasons for the radiation-linked cancers appears to be exposure to the radiation used in the medical field, particularly in the form of radiotherapy (unrelated malignancy) (Travis et al., 2012) or due to diagnostic radiography (Preston et al., 2002; Ronckers, Doody, Lonstein, Stovall, & Land, 2008). These iatrogenic tumors develop as de novo neoplasm in the therapeutic radiation field postlatency period, which remains for a long time (Cahan, Woodard, Higinbotham, Stewart, & Coley, 1948). They are not returning to the original cancer state (Allan & Travis, 2005).

The cancer-causing effect of ionizing radiation was first known in the 20th century with reports of skin cancer due to X-rays and radium in scientists and physicians. The various medical practices that used X-rays as therapeutic agents were prohibited due to increased leukemia incidences. The best example to explain ionizing radiation-induced carcinoma is Japan's atomic bomb attack at Hiroshima and Nagasaki in 1945. It leads to an increase in solid tumors of the lung, breast, and thyroid. Similar effects were observed due to the high radiation level in the Chernobyl disaster in Ukraine in 1986. In the case of electromagnetic radiation, it results in lung cancer in uranium mine workers in central Europe and Europe, and the Rocky Mountains of North America (Costa, 2020).

#### 1.2.1.2.2 Ionizing radiation and cancer pathogenesis

Ionizing radiation results in the modification of the chemical equilibrium of the cells. A few of these modifications may lead to cancer. Further, ionizing radiation causes damage to genetic material (DNA) and causes harmful genetic mutations that can carry forward from one generation to another. A large amount of exposure to such kind of radiation in rare cases leads to illness in a short period of time (hours or days) and death in cases of prolonged exposure (60 days). In exceptional cases, a person dies after a few hours of exposure (Ionizing Radiation Fact Sheet, 2017).

## 1.2.2 Chemical carcinogens

### *1.2.2.1 Asbestos*

Asbestos is a naturally occurring mineral in the environment as bundles of fibers are divided into thin, long-lasting threads having commercial and industrial applications. These fibers are resistant to fire, heat, and chemicals and are a nonconductor of electricity. Because of these properties, asbestos is used extensively in several industries. Chemically, asbestos is a silicate compound, which means it consists of atoms of silicon and oxygen in their molecular structure (American Cancer Society, 2020a). The fate and metabolic pathways of chemical carcinogen triggered by genotoxic and nongenotoxic effects are as shown in Fig. 1.3.

#### 1.2.2.1.1 Types of asbestos exposure

1. Occupational;
2. Secondary;
3. Environment (Pleural Mesothelioma Center, 2020).

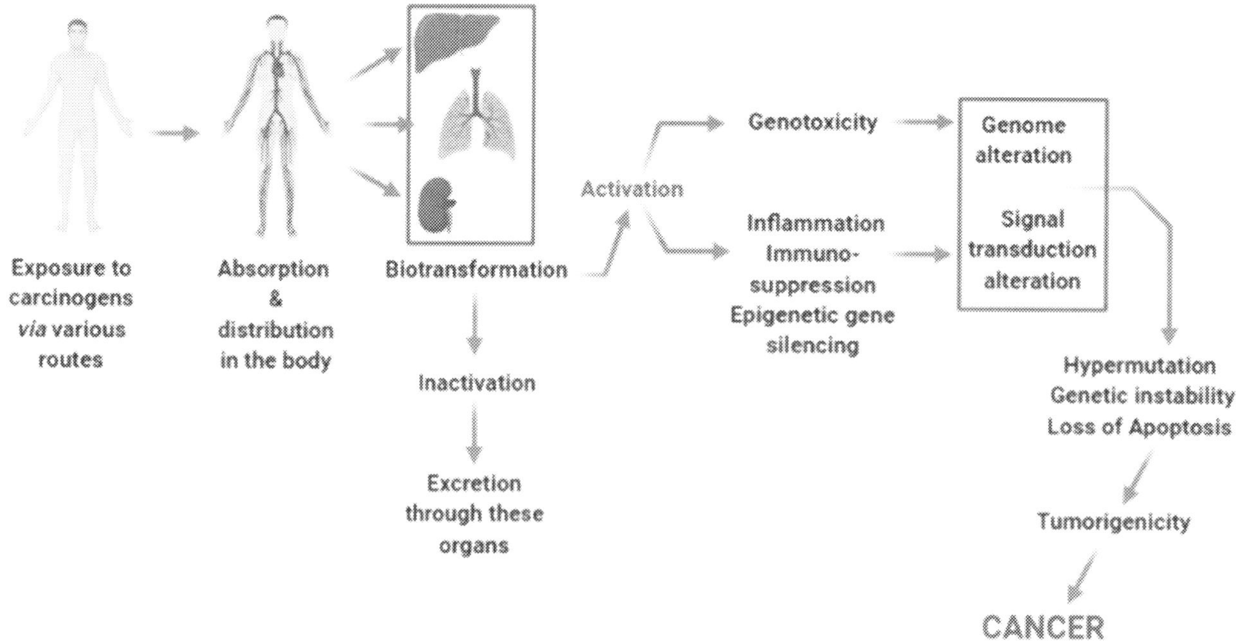


**FIGURE 1.3** Diagrammatic illustration of the fates of chemical carcinogens in the human body and metabolic pathways triggered due to their genotoxic and nongenotoxic effects.

#### 1.2.2.1.2 Asbestos and cancer

The airborne asbestos fibers are entering into the body through inhalation or swallowing and retained in the lungs and abdomen's delicate tissues. The human body cannot expel these fibers, which leads to more severe health complications, including cancer (National Cancer Institute, 2020a).

Asbestos fibers are also responsible for lung cancer, asbestosis, and pleural mesothelioma, which usually take a long time. These asbestos-related diseases lead to around 10,000 deaths per year in the United States. In 2000 to 3000 of these deaths per year, one dies on average every 3.4 h due to mesothelioma. Pleural mesothelioma is one of the most common forms of mesothelioma that occurs as a contributing factor in lung and lung cancer; pleural mesothelioma is extensively triggered after asbestos exposure (American Cancer Society, 2020a).

Asbestos disease latency times vary from 10 to 40 years or even longer. All kinds of asbestos fibers are known to cause health risks, the most potent of which is considered blue asbestos. Hundreds of thousands of workers in developed countries suffered from asbestos-related diseases in the 20th century after widespread asbestos. According to WHO, they predict asbestos-associated lung cancer, asbestosis, and mesothelioma are caused by occupational exposure, which leads to the death of >107,000 people per year (World Health Organization WHO, 2010).

### *1.2.2.2 Components of tobacco smoke*

The tobacco and components of tobacco smoke are considered a major public health threat that the world faces. It results in greater than 8 million deaths a year around the globe. The direct use of tobacco causes more than 7 million deaths and around 1.2 million deaths of the non-smoking population from exposure to second-hand smoke.

Nearly 80% of the smokers out of 1.1 billion smokers belong to poor and middle-income nations, which increases tobacco-linked ailments and deaths. The consumption of tobacco in daily routine supports poverty by spending more money on it instead of the basic needs like food and shelter. This spending behavior and chewing tobacco habit are challenging to overcome, as chewing tobacco is also addictive (World Health Organization WHO, 2019).

In tobacco smoke, it consists of many chemicals and toxic components that are dangerous to smokers and nonsmokers only. The breathing of a very little quantity of tobacco smoke is very harmful (Centers for Disease Control and Prevention CDC, 2010a,b; 2014). Tobacco smoke consists of more than 7,000 chemicals out of which 250 are responsible for causing cancer, such as carbon monoxide, ammonia, and hydrogen cyanide (National Institute of Environmental Health Sciences, 2016; Centers for Disease Control and Prevention CDC, 2010a,b; 2014).

As per the IARC, a carcinogen is a compound (e.g., chemical) which results and enhances the danger of cancer (Cahan et al., 1948). The threat to humans of these carcinogenic chemicals based on the degree of evidence from the scientific studies, IARC classifies them into different groups (Table 1.1; Centers for Disease Control and Prevention CDC, 2010a,b).

#### 1.2.2.2.1 Tobacco smoke components: As carcinogens

The IARC has classified the components of tobacco smoke as a Group 1 carcinogen (World Health Organization WHO and International Agency for Research on Cancer, IARC., 2002). The examples of such components responsible for cancer's development are as shown in Table 1.2. The exposure to such components present in tobacco smoke enhances the danger of developing cancer.

**TABLE 1.2** IARC classification of tobacco smoke carcinogens (World Health Organization WHO and International Agency for Research on Cancer, IARC., 2002).

| Group | Examples |
|---|---|
| Group1 | Benzene, arsenic, cadmium, nickel, formaldehyde |
| Group 2 A | Lead (inorganic) |
| Group 2B | Styrene, isoprene, acetaldehyde |

##### 1.2.2.2.2 Cancer pathogenesis: tobacco smoke

The cancer pathogenesis associated with tobacco smoke involves DNA damage due to a combination of harmful chemicals present in tobacco smoke. Therefore, DNA damage caused by each cigarette is a build-up of cancer damage in the same cell (Cancer Research UK, 2019).

### 1.2.2.3 *Aflatoxin (food contaminant)*

Cancer is currently known to be a leading cause of death, and in the etiology of carcinoma, diet plays a key role. The harmful effects of various foods, food components, and food contaminants have been extensively revealed from laboratory and epidemiological studies. Different governmental and international bodies have examined the relationship between food pollutants in diet and cancer (Groopman, Kensler, & Wild, 2008).

Aflatoxins are a family of toxins prepared by a specific type of fungi observed on different crops like peanuts, cottonseed, maize (corn), and tee nuts. *Aspergillus flavus* and *Aspergillus parasiticus* are the major fungi that produce aflatoxin, widely obtained from a warm and humid part of the world. This fungus is also responsible for crop contamination at harvest, in the field, and during storage.

The human population is exposed to aflatoxin through various sources such as ingestion of contaminated plant products (e.g., peanut) or meat or animal-derived dairy products that ate contaminated feed. In addition to this, farmers and different workers involved in agriculture activities may be exposed to it by inhaling dust particles produced while dealing with contaminated crops and feeds (National Cancer Institute, 2020b).

The various research activities in different laboratories have shown carcinogenic activity in primates, rodents, and fish (Abnet, 2007). The primary target organ of aflatoxin is the liver and resulted in hepatocellular carcinoma (HCC). A number of species have seen the liver as the primary target organ for HCC. Hence, consumption of the even microgram per day quantity of the aflatoxin could also cause liver cancer, and this study served as a link between aflatoxin and HCC. HCC served as the fundamental reason for cancer (National Cancer Institute, 2020b).

In different parts of the world, morbidity and mortality affect Asia and Sub-Saharan Africa (up to 600,000 new cases per year), and more than 200,000 deaths per year in the China alone (Kew, 2002; Wang et al., 2002). Inn a few species it is also served as a tumor inducer in the colon and kidney.

### 1.2.2.4 *Arsenic (drinking water contaminant)*

Arsenic, for a human being, is considered a toxic and carcinogenic component. This is affected in different organs, such as the brain, lung, liver, kidney, heart, skin, and blood. The non-carcinogenic effect can be seen in these organs on exposure to arsenic (Saint-Jacques et al., 2018).

##### 1.2.2.4.1 Sources of arsenic exposure

Arsenic is a component obtained naturally from the earth's crust and is extensively scattered all over the environment in water, air, and land. People are exposed via drinking contaminated water to high arsenic levels, which is highly toxic (inorganic form). The contaminated water source includes water used to prepare water, eat uncontaminated water, industrial processes, and smoke tobacco (World Health Organization, 2020b).

##### 1.2.2.4.2 Arsenic and oncogenesis

The Division of Cancer Epidemiology and Genetics (DCEG), an initiative of the National Institute of Health (NIH), studied the different water contaminants responsible for cancer. These consist of natural contaminants such as by-products of fertilizer—nitrate, arsenic—and components generated during the disinfection of water using chlorine. More arsenic quantity consumption leads to bladder cancer, but the danger at a lower level is not clear. The drinking water contaminated with arsenic (small quantity) leads to a risk of bladder cancer (National Cancer Institute, 2020c).

The metabolism stage for deporting the arsenic from the body causes genomic and epigenetic changes in its level. Arsenic inhibits DNA damage recovery and chromosome structure, which interferes with the genomic material's stability. Arsenic also resulted in the hyper- and hypomethylation of the specific promoter by reducing the cellular methyl groups. Hence, arsenic can lead to changes in tumor suppression and oncogenes (Ghosh, Banerjee, Giri, & Ray, 2008). The inorganic arsenic does not show any effect as it is not able to bind with DNA, but the modified arsenic (i.e., methylated arsenic) is able to bind DNA and results in the production of reactive oxygen species (ROS) and ultimately leads to the breakage of both single and double strands of the DNA (Kligerman & Tennant, 2007).

## 1.2.3 Biological carcinogens

### *1.2.3.1 Viruses*

Along with chemicals and radiation, viruses is also responsible for cancer. Viruses are nothing but the small organism which can infect cells of animals or plants. People are prone to a broad range of viruses. Viruses differ from bacteria, though both bacteria and viruses can cause human diseases. A few examples of viruses result in acquired immune deficiency syndrome (AIDS) are human immunodeficiency virus (HIV) and flu by the influenza virus.

#### 1.2.3.1.1 Viruses and carcinogenesis

**Mechanism**: Viruses can cause the disturbance in cell behavior via various pathways, By inserting their genome into the host cell DNA, viruses may directly spread lead to DNA damage (mutations). Integration can interfere with essential regulatory genes' activities. The viruses consist of their genes which affect cell regulation. This method may be useful to the virus if it permits for faster production of progeny but can be harmful to the host. Few of the viruses bear the different versions of genes collected for the earlier host cells. These modified genes are no longer working well after insertion into a new host cell which results in dysregulation and ultimately cause cancer (Fig. 1.4. van Tong, Brindley, Meyer, & Velavan, 2016; Cancer Quest, 2020).

Viruses are considered the responsible factors for 10%–15% of carcinomas throughout the world. The cancer-causing DNA viruses include Merkel cell polyomavirus (MCV), Epstein-Barr virus (EBV), Kaposi sarcoma herpesvirus, HPV, hepatitis B virus (HBV), and simian virus 40, and two RNA viruses: hepatitis C (HCV) and human T-cell lymphotropic virus type 1 (HTLV-1) as shown in Table 1.3 (Chen, Williams, Filippova, Filippov, & Duerksen-Hughes, 2014).

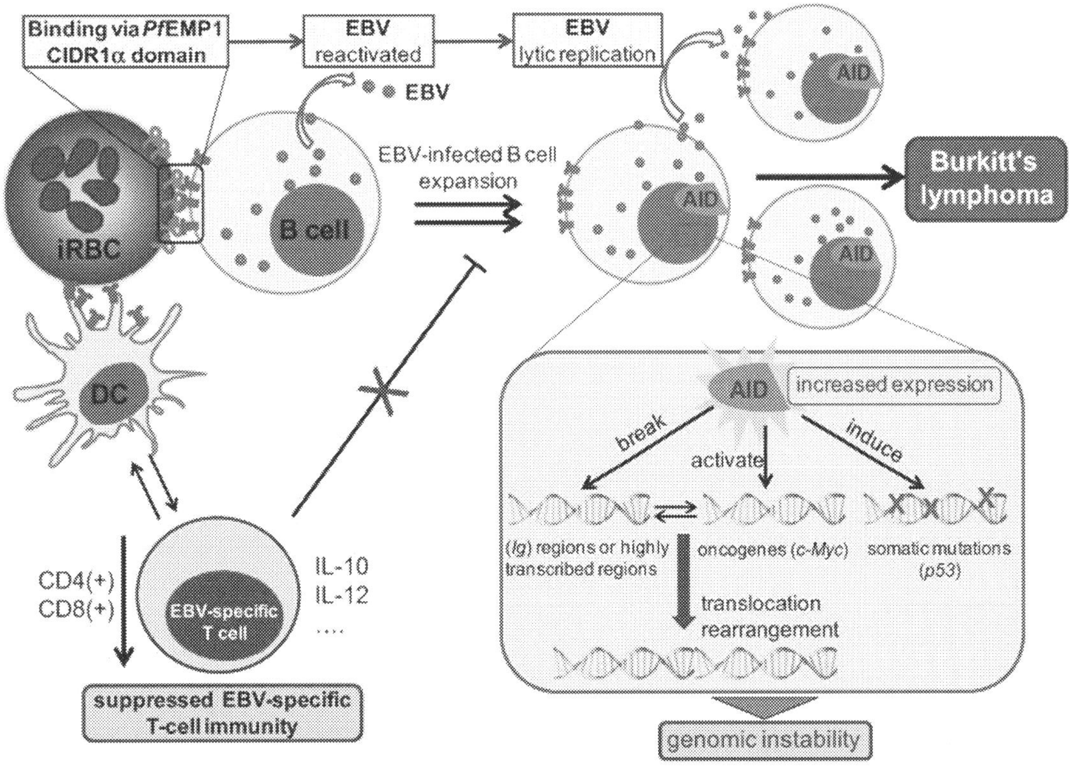


**FIGURE 1.4** Proposed mechanisms of induction of Epstein-Barr virus driven Burkitt lymphoma by falciparum malaria. *Reproduced with permission from Tong et al. (2017); van Tong, H., Brindley, P., Meyer, C., & van Tong et al. (2016). Parasite infection, carcinogenesis and human malignancy.* EBioMedicine, 15, *12–23. https://doi.org/10.1016/j.ebiom.2016.11.034 (Original work published 2016).*

**TABLE 1.3 List of viruses and associated cancers (Chen et al., 2014).**

| Virus | Associated Cancer |
|---|---|
| HCV | Hepatocellular carcinoma (HCC) (Choo, Kuo et al., 1989; Marcucci and Mele 2011) |
| HTLV-I | T-cell leukemia (ATL) (Poiesz, Ruscetti et al., 1980) |
| MCV | Merkel cell carcinoma (MCC) (Feng, Shuda et al., 2008) |
| HBV | Hepatocellular carcinoma (HCC) (Blumberg and Alter 1965; Di Bisceglie 2009) |
| HPVs | High-risk human papillomaviruses (HPV) 16 & HPV 18 (Dürst, Gissmann et al., 1983; Boshart, Gissmann et al., 1984) |
| SV40 | Human mesothelioma (Carbone, Pass et al., 1994) |
| KSHV | Kaposi's sarcoma (Chang, Cesarman et al., 1994) |
| EBV | Burkitt's lymphoma and nasopharyngeal carcinoma (Epstein 1964) |

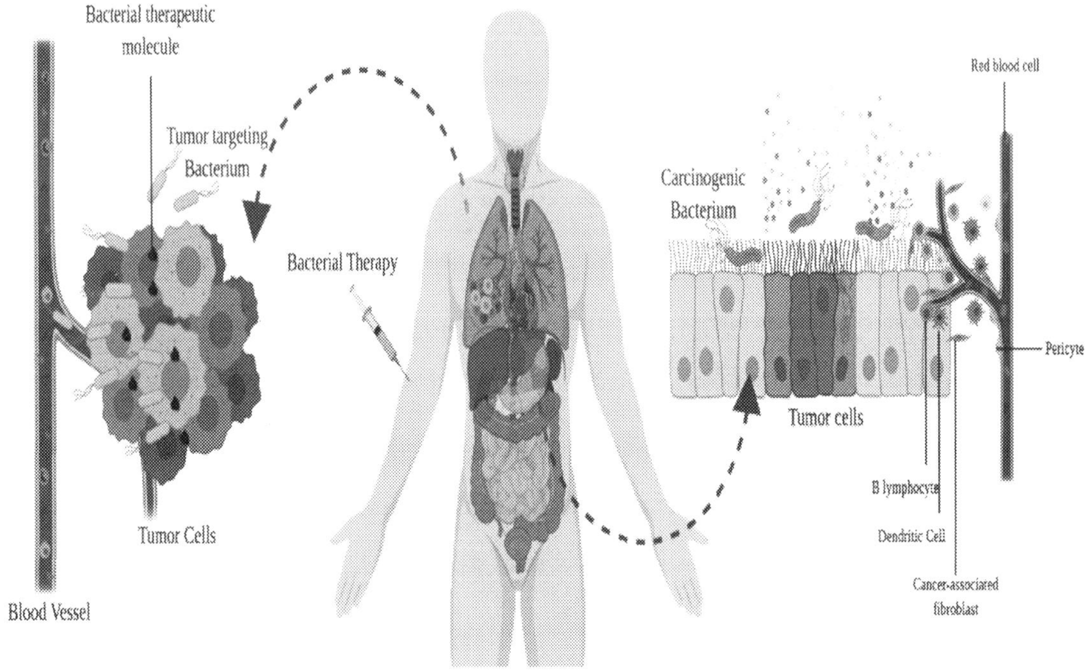


FIGURE 1.5 Role of bacteria in cancer. Some bacteria are notorious as carcinogenic bacteria, which cause tumorigenesis, while some bacteria have the potential of cancer therapy, targeted to the hypoxic tumor region and leading to the eradication of tumor cells by releasing of therapeutic molecules. *Reproduced with permission from Laliani, G., Ghasemian Sorboni, S., Lari, R., Yaghoubi, A., Soleimanpour, S., Khazaei, M., Hasanian, S., & Laliani et al. (2019). Bacteria and cancer: Different sides of the same coin.* Life Sciences, 246. *https://doi.org/10.1016/j.lfs.2020.117398 (Original work published 2020).*

### 1.2.3.2 Bacteria

It is calculated that every year around 15% of cancer worldwide, or about 1.2 million population, are due to infections. The various infections associated with bacteria, viruses, and schistosomes, lead to a higher risk of cancer, as proposed by a research group (Pisani, Parkin, Muñoz, & Ferlay, 1997).

Considering the inflammation caused by bacteria and human cancer, it is assumed that bacteria remain in the host's body for a long period and produce an atmosphere of persistent inflammation. There are some specific bacteria responsible for chronic inflammation and increased cancer risk, such as *Salmonella enterica*, serovar *Typhimurium*, or *Paratyphi* (*Helicobacter pylori*) and cannot recognize the type of bacteria (Fig. 1.5). For example, chronic urinary tract infections (UTIs) (bladder cancer), chronic ulcer and sinus osteomyelitis (squamous cell carcinoma), and chronic prostatitis (prostate cancer) are associated with this disease (Chang & Parsonnet, 2010; Laliani et al., 2019).

The pathogenic bacteria promote an attachment and encapsulation of the surface proteins. For example, different species of the pathogens belong to the *Neisseria* family favor the surface adhesins, which acts as the selective carrier with specific cell types to permit the use of specialized host cell niches (Popp, Billker, & Rudel, 2001). Similarly, fibronectin-binding proteins of *Staphylococcus aureus* and *Borrelia burgdorferi* intervene in the interaction among host cells and bacterium by means of forming a tandem of β-zippers that encourage non-phagocytic cell's engulfment of bacteria (Meenan et al., 2007; Raibaud et al., 2005).

#### 1.2.3.2.1 *Helicobacter pyroli* and gastric ulcer

The population (80%) suffering from the *H. pyroli* infection do not show any symptoms, and around 1%−3% of individuals will develop a gastric ulcer. In a few individuals, *H. pyroli* infection brings out a pathological advancement in the gastric mucosa, which initiates chronic gastritis and may advance to other gastric problems such as dysplasia, atrophic gastritis, intestinal metaplasia, and gastric ulcer (Park & Kim, 2015). *Helicobacter pyroli* infection also causes changes at the molecular level like varied immune response (Kang, 2012; Moyat & Velin, 2014), hypermethylation of promoters (Cover, Krishna, Israel, & Peek, 2003), and dysregulation of apoptosis (Kalisperati et al., 2017). The primary mechanism responsible for *H. pyroli* induced carcinogenesis is genetic damage, studied by various researchers (Jemal et al., 2011; Shi et al., 2019). The enhanced production of ROS in human gastric cells post−*H. pyroli* infection induces chronic gastritis, exclusively when the infection occurs due to cagA-positive *H. pylori* (Jemal et al., 2011).

#### 1.2.3.2.2 *Chlamydia trachomatis* and cervical cancer

The third most common cancer and the fourth most common cause of death is cervical cancer. It is found in women worldwide, and occurs in about 10% of the recently diagnosed cancer patients; about 8% of the total deaths occur due to cancer (Stone et al., 1995). One of the most common causes of cervical cancer is *Chlamydia trachomatis*. It is the most prevalent sexually transmitted pathogen in women. The studies related to epidemiology have shown a patient suffering from cervical cancer is due to *C. trachomatis* infection (Solnick & Schauer, 2001).

#### 1.2.3.2.3 *Helicobacter* species and biliary tract cancer

The different species belong to the enterohepatic family. *Helicobacter* includes *Helicobacter bilis* and *Helicobacter hepaticus*, which are bile-resistant organisms found in the gallbladder (De Martel, Plummer, Parsonnet, Van Doorn, & Franceschi, 2009). Human research has been performed to describe *Helicobacter* species' presence in the biliary tract of cancer patients compared to the control group. These studies consist of bacterial identification techniques like DNA amplication, immunohistochemistry, and culture, and they differ significantly from each other (Dove et al., 1997; Hamada et al., 2009).

#### 1.2.3.2.4 Colonic microflora and colon cancer

The key function of the colonic microflora, which results in colon cancer, is actively recommended as per the different animal model studies. The study conducted in the mice model (familial adenomatous polyposis) suggested that germ-free mice produce colon cancer at a half rate compared to control mice (El-Omar et al., 2000). Similarly, studies of inflammatory bowel disease mouse models have been developed. It was observed that germ-free mice do not develop cancers, the same as that of mice grown in normal conditions (Rakoff-Nahoum & Medzhitov, 2007; Uronis et al., 2009). So experimental results suggested how colony-forming bacteria can promote tumors.

It can also involve Toll-like receptors located on the surface of the macrophage and some cells, particularly dendritic cells. For example, in experimental colon cancer models, mice without the myeloid differentiation factor 88 (MyD88) adapter protein that is required for signal transmission of toll-like receptors and consists of fewer and smaller adenomas compared to wild mice (Botelho & Richter, 2019). Further research on colon cancer indicated that colonic bacteria are involved in a complicated interplay with the host's innate, adaptive, and regulatory immune responses, leading to carcinogenesis through these interactions (El-Omar et al., 2000).

### *1.2.3.3 Parasite*

Recent studies related to parasites suggested that few parasites, like the small parasite trematode (liver fluke) *Opisthorchis viverrini* and *Clonorchis*, are responsible for the bladder cancer that occurred due to schistosomes and cholangiocarcinoma by a group of parasite trematodes. In a large number of pandemic areas, these helminths are culpable for most of the carcinoma cases. Apart from the helminths, the parasite is linked with cancer, like theileria, which is an intracellular eukaryotic parasite. On the other hand, few parasite infection or molecules show a protective action on a few cancers like *Echinococcus's* case (Bhagwandeen, 1976).

#### 1.2.3.3.1 Parasite and cancers

***1.2.3.3.1.1* Schistosoma haematobium *and urinary bladder cancer*** The causative agent for urogenital schistosomiasis, which has been reliably responsible for bladder cancer, is *Schistosoma haematobium*. Earlier findings suggested that 65% of Zambia patients have suffered from bladder cancer, and 75% of the population suffered from squamous cell carcinoma (Bouvard et al., 2009). The various mechanisms may be responsible for urinary bladder cancer because *S. haematobium* includes oxidative stress, damage to epithelial cells and chronic inflammatory reactions (Fig. 1.6; Brindley, da Costa, & Sripa, 2015; Honeycutt, Hammam, Fu, & Hsieh, 2014; Matsuda et al., 1999; van Tong et al., 2016).

***1.2.3.3.1.2* Schistosoma japonicum *and colorectal and hepatocellular carcinoma*** It has been evidenced that *Schistosoma japonicum* has been involved in the diagnosis of colorectal cancer (Ishii et al., 1994; Qiu, Hubbard, Zhong, Zhang, & Spear, 2005). One of the studies carried out in Japan showed that 19% of the patients suffered from liver disease (chronic), and 51% of the persons infected with *S. japonicum* suffered from HCC. The immunogenic component soluble egg antigen of *S. japonicum* may be associated with carcinogenesis via chronic inflammation stimulation (Basílio-de-Oliveira, Aquino, Simon, & Eyer-Silva, 2002).

***1.2.3.3.1.3* Schistosoma mansoni *and cancer*** *Schistosomiasis mansoni* infection may be involved in the progression of HCC with the hepatitis C virus coinfection. The different case reports also suggested that *S. mansoni* is also linked with sigmoid colonic cancer and prostatic adenocarcinoma (Kim, Bae, Choi, & Hong, 2012; Salim, Hamid, Mekki, Suleiman, & Ibrahim, 2010). The existing Turkish report showed a correlation between *S. mansoni* and causes of bladder cancer (Madbouly et al., 2007). In addition, prior to the onset of colorectal cancer, schistosomal colitis can be associated. Tumor protein 53 transformed appearances in persons with colorectal cancer associated with mansoni colitis, indicating that freshwater parasitic worm disease can activate carcinogenesis by attacking oncogenes (Lopez, Mathers, Ezzati, Jamison, & Murray, 2006).

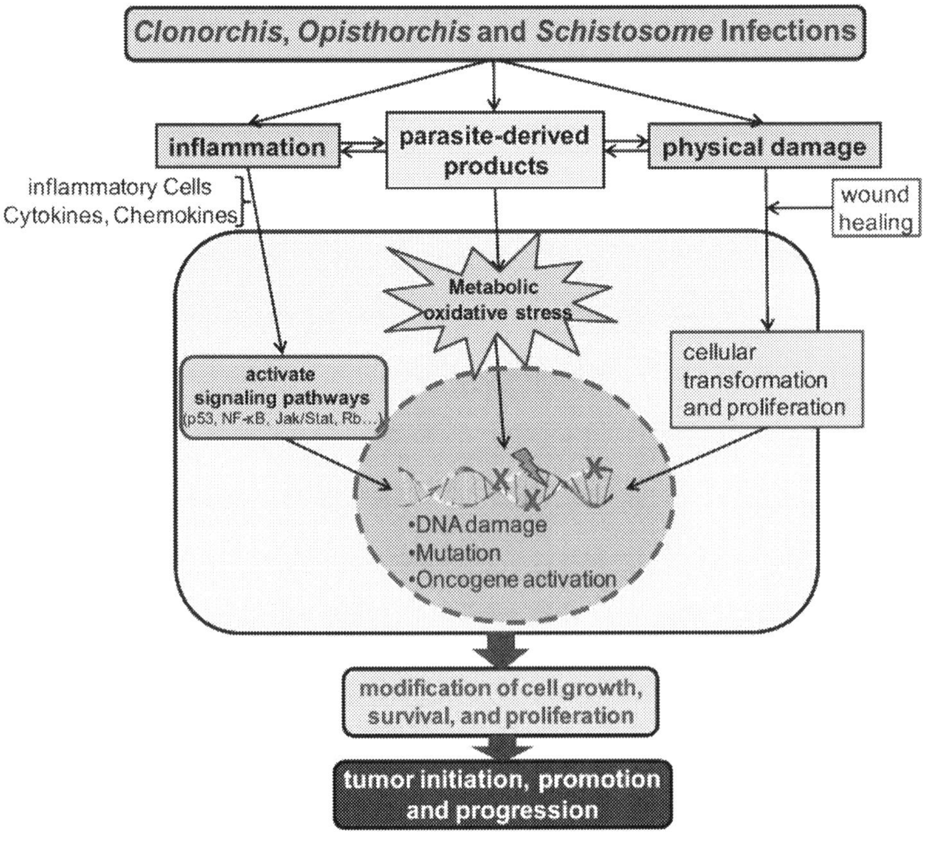


**FIGURE 1.6** Proposed mechanisms of carcinogenicity induced by infection with the liver and blood flukes *Clonorchis*, *Opisthorchis*, and *Schistosoma* species. *Reproduced with permission from Tong et al. (2017); van Tong, H., Brindley, P., Meyer, C., & van Tong et al. (2016). Parasite infection, carcinogenesis and human malignancy.* EBioMedicine, 15, *12–23. https://doi.org/10.1016/j.ebiom.2016.11.034 (Original work published 2016).*

## 1.3 Causes and risk factors of cancer

### 1.3.1 Causes

It is a matter of simple logical thinking that, when we are interested in reducing cancer incidences, we need to analyze what we already know about it and note all the probable causes and risk factors. This helps in estimating how difficult or easy it is to accomplish the reduced occurrence of the disease. In cases where the causes and risk factors cannot be well established or are merely poor or cannot be easily changed, the primary emphasis remains on potential solutions for prevention and proper care for incurable cases. Knowing the cause is incredibly beneficial so that one can decides whether it is avoidable (e.g., tobacco use) or unavoidable (e.g., atmospheric ionizing radiations) (Epping-Jordan, Galea, Tukuitonga, & Beaglehole, 2005). Chronic respiratory diseases, stroke, cancer, diabetes, and heart disease are a few prominent chronic diseases that share many unified risk factors (Parkin, 2006; World Health Organization WHO, 2005a). Notable causes and risk factors for cancer are discussed in the coming sections.

#### *1.3.1.1 Tobacco use*

The foremost and single largest contributor to the total mortality by cancer is tobacco use. It was noted to result in about one-fifth of the total cancer deaths in LMICs. It is even the principal reason for fatalities from the total number of deaths by major noncommunicable diseases. In most countries where tobacco use is extreme, it is one major issue to manage its excessive use. Usually, two basic approaches are implied:

1. Short-term approach—where adults are encouraged and aided in quitting smoking and tobacco and
2. Long-term approach—where youngsters are discouraged from starting tobacco consumption.

#### *1.3.1.2 Infections*

The highest mortality causing risk factor after tobacco in developing countries is infections chiefly by four microbes viz. HPV, hepatitis B (HBV), and hepatitis C (HCV) virus, and *H. pylori* (International Agency for Research on Cancer IARC, 1997). The prevalence of infections by the microbes above can be significantly abridged by varied means. Immunization is effective in HBV, whereas safer injection practices and rigorous blood bank screenings can be implemented for hepatitis C. In the majority of nations, HPV is prevented thorough screening and precancerous lesions treatment; however, nowadays vaccination is also adopted. HPV and HBV-allied cancers symbolize instantaneous opportunities for cancer prevention in present generations.

*H. pylori* infection leads to stomach cancer, which is an inadequate response to a therapy type of cancer. Surprisingly, the *H. pylori* and thereby allied stomach cancer have dramatically declined in most parts of the world without implementing any targeted measures. Currently, a quest for search ix whether to use antibiotics that eradicate bacteria in infected persons. Even ample research is ongoing for vaccine development. Apart from these, other local infections are equally important and hold a share of about 4% of all cancers (International Agency for Research on Cancer IARC, 1997). One excellent example is the EBV, which is well-known and recognized for its allied risk for nasopharyngeal cancer and Burkitt's lymphoma. Both the cancers above are predominant in developing nations is concerning for those nations. HIV stands next in line, which accounts for around >62,500 Kaposi's sarcoma cases annually amid developing nations, i.e., nearly 1% of total cancers. Comparatively, a smaller percentage is contributed by helminth infections (worm type), leading to colorectal and bladder cancer. The international agency has proposed the probability of liver cancer and colorectal cancer postinfection by liver flukes and schistosomes for cancer research, but with the scarcity of evidence. Additionally, HTLV-1 has proven to be a causative agent for a few non-Hodgkin's lymphomas (Institute of Medicine IOM, 2003).

#### *1.3.1.3 Diet, overweight/obesity, physical inactivity*

The activity levels, body weight, and diet of a person are very intertwined and work in different ways to decrease or encourage cancer risk. It has been evidenced that adequate intake of vegetables and fruits chiefly lowers the risk associated with lung, esophagus, pharynx, mouth, stomach, rectum, and colon (Stewart & Kleihues, 2003; World Cancer Research Fund, 1997). Other cancer risks might also be lowered significantly, but strong evidence is still lagging. The research literature totality estimates that around 6% of cancer deaths can be corroborated by low vegetable and fruit intake (Epping-Jordan et al., 2005). Another estimate allied to "diet and nutrition" relativity depicts higher counts. As per IARC's World Cancer Report, >30% of human cancers can be correlated with their nutrition and diet (Stewart & Kleihues, 2003). The report also referred to the world as one nation devoid of each nation's individuality and quoted the estimates globally. They add up to the body weight, vegetable-fruit consumption, salt, and preserved foods associated with a higher risk of nasopharyngeal and stomach cancers. In most high-income nations, red meat consumption has also appeared as a major risk factor for colorectal cancer. Physical activity and diet are crucial cancer determinants. However, the association's specificities are still unclear and undefined, but their significance in controlling cancer and other chronic diseases is well established (World Cancer Research Fund, 1997).

#### *1.3.1.4 Alcohol use*

Heavy alcohol drinking can lead to varied kinds of cancers like the liver, breast, esophagus, larynx, pharynx, and oral cavity. The same risks vary based on the cancer site but are elevated in alcoholics and account for around 5% of total deaths by cancer in LMICs (Epping-Jordan et al., 2005).

In line with the physical activity and diet, heavy alcohol consumption in certain specific medical conditions (neuropsychiatric illness, cardiovascular disease, unintentional and intentional injuries, etc.) precipitates a greater disease burden for cancer. The argument worth noting here is that the advantage of reduced cardiovascular risk is a limited intake of alcohol (documented in many studies from HICs). However, chronic drinking is still not recommended and is associated with psychological and diverse acute social tribulations (Jamison et al., 2006 Apr). A few cost-effective and effective interventions for control of chronic alcoholism are restricted access to alcoholic beverage stores selling, higher excise taxes, random breath testing of drivers, advertisement bans/restrictions, and brief counseling and advice sessions to chronic drinkers (Jamison et al., 2006).

#### *1.3.1.5 Occupational exposures*

Exposure to hazardous chemicals can occur at workplaces, be it city streets, family farms, factories, or any informal and formal workplace. As per an estimation by the International Labour Organization (ILO), >2,000,000 deaths per annum among a population group of 2,700,000 workers are owing to workplace exposures. Disability incurred and disease burden caused by this risk factor is enormous. However, we lack the quantitative estimates for most parts of the world (Bruce, Rehfuess, Mehta, Hutton, & Smith, 2006). About 25 chemicals, or mix of chemicals, have been identified and listed by the IARC as extremely carcinogenic. The list expands every year with an equal number of already existing carcinogens (Stewart & Kleihues, 2003).

#### *1.3.1.6 Pollution*

Water, soil, and air (outdoor and indoor) pollution account for 1%–4% of cancer cases globally (Institute of Medicine IOM, 2003; Stewart & Kleihues, 2003). Though the per person risk levels are pretty low, the total involuntarily population exposure is relatively high and forms a significant share in cancer cases. Lung cancer is the main type implicated through outdoor and indoor air pollution. In addition to that, vehicular exhaust (complex chemicals mixture) causes ambient air pollution. In the recent past decades, vehicular emission levels have declined significantly HICs, whereas they have been leveled or increased in LMICs. In underprivileged areas of East Asia, South Asia, the Pacific, and sub-Saharan Africa, heating, cooking, and cooked food fumes contribute to indoor air pollution. In the places above, the smokiest fuels, biomass, and coal, are often used. Indoor tobacco smoke exposure impacts are relatively less to the person exposed. Due to outdoor/indoor air pollution, cancer deaths greatly outweigh the overall deaths in adults and children due to cardiovascular, infectious problems and acute respiratory infections. The water of low or deteriorated quality is often related to many infectious diseases. It is reported that a few cancers result from contamination of water via chlorinated by-products and because of excessive arsenic in the soil of particular localities. Getting exposed to a range of harmful substances from industrial wastes could also cause cancer, but sorting out such chemicals and further quantifying them is a challenge in itself (Darby, Hill, & Doll, 2001; Darby et al., 2005).

### 1.3.2 Risk factors

#### *1.3.2.1 Contaminants from food*

Apart from the energy balance and macro/micronutrient role in altering cancer risks, food contaminants can also be manmade or naturally occurring carcinogens. Although the degree of cancer burden from food pollutants has not been predominantly calculated, several studies note that it has a relatively lower effect (Stewart & Kleihues, 2003).

#### 1.3.2.2 *Medical drugs*

As per the very old saying, even the drug at elevated and higher doses turns to be a poison, arrays of potential drugs have been noted to be carcinogenic. Ironically, the major group of such drugs is comprised of anticancer drugs themselves. But the curing potential or efficacy of these drugs outweighs their risks of causing cancer; thereby, these are mostly knowingly accepted by the patients. However, for certain medications with proven carcinogenicity and that is used in comparatively less severe conditions, restriction (e.g., phenacetin) and ban or withdrawal from the market (e.g., diethylstilbestrol) is implicated (Stewart & Kleihues, 2003).

#### 1.3.2.3 *Reproductive and hormonal factors*

Steroidal sex hormones (like estrogens, androgens, progestogens) significantly affect the growth and development of a few cancers, especially in women, and are considered prominent risk factors; the predominantly led cases are of ovarian, breast, and endometrial cancer. Additionally, it is also a well-recognized fact that child delivery/birth protects women against a range of cancers (like ovarian, endometrial, breast). Those having no child are at higher risk. Also, obesity that alters various circulating hormones enhances breast cancer risk in postmenopausal life. In men's case, the hormone testosterone governs the prostate gland; hence, altered testosterone levels also elevate the degree of prostate cancer risk. Even a significant impact and correlation between obesity and prostate cancer does exist (Stewart & Kleihues, 2003).

#### 1.3.2.4 *Natural, medical, and industrial sources-based ionizing radiations*

Knowingly or unknowingly, everyone is exposed to varied ionizing radiations now and then, in the form of radiations from outer space (e.g., $\gamma$-rays), and human-made and globally distributed X-rays and $\gamma$-rays radiating from nuclear reactors and nuclear weapon testing centers. Another noteworthy example is radon gas (natural uranium decay product), ubiquitous in the atmosphere. While dispersed in the atmosphere, it is of less consequence, whereas, in radon-concentrated airtight closed buildings, chances of substantial exposure are there. Though radon in its original form is inert, decaying it generates radioactive alpha particles that get lodged in the lung's air sacs, resulting in local tissue damage and, consequently, lung cancer (Darby et al., 2001; International Agency for Research on Cancer IARC, 2004). Moreover, medical radiation (like X-rays) from diagnostic centers and radiotherapy centers contributes to almost 40% of the total global human exposure to ionizing radiation. Still, it is probable to be considerably less in the regions where such technologies are scarcely used.

#### 1.3.2.5 *Ultraviolet radiation*

The nature-derived UV radiation from the sun holds a significant share of world's melanoma cases (World Health Organization WHO, 2003). Annually, around 160,000 cases of melanoma are reported in Australia, North America, New Zealand, and Europe (Stewart & Kleihues, 2003). It also accounts for several nonmelanoma skin cancers in light complexion populations that are usually decidedly excised and curable. The interventions for preventing UV-allied cancers are avoiding direct sun ray exposure, using sunscreen, sunglasses, protective clothing, and undertaking all requisite measures to access physical items and create public awareness. Comprehended guidelines and information for diverse countries for protection against UV rays are also given as a part of the United Nation's INTERSUN Programme (Fleming, Brady, Mieszkalski, & Cooper, 1995; World Health Assembly, 2005).

#### 1.3.2.6 *Immunosuppression*

Now it is a well-established fact that the immune system of the body protects itself strongly against cancer. Hence, individuals with compromised or suppressed immunity are at a higher risk and prevalence of cancer. Drugs used in certain medical conditions are immunosuppressants, and their use is common. One such disease is HIV, which is often associated with rare cancers such as Kaposi sarcoma and some non-Hodgkin lymphomas.

#### 1.3.2.7 *Genetic susceptibility*

The risk factors mentioned above can also cause cancer, and genetic materials inherited from generation to generation could lead to specific cancerous characteristics and cases (Institute of Medicine IOM, 2003). Mutation occurring in a particular gene sometimes ends in a too high-risk level for cancer. For instance, BRCA1 gene mutations in women result in around 70% higher risks of breast and ovarian cancer. However, luckily higher penetrance and inheritance of such genes are quite rare. Moreover, an enhanced degree of mutation results in post environmental interactions, which is a critical factor in cancer development in people with certain genetic traits.

## 1.4 Early detection and management

### 1.4.1 Diagnosis and staging

Accurate diagnosis is key to providing cancer patients with adequate treatment and services. Pertinent care for cancer is the right way for a precise recovery. Not only physical examination but also imaging, laboratory, and pathology techniques are useful for cancer diagnostics. For initial diagnosis and staging, similar methods are applied, and then re-evaluations are required to assess a patient's state at further stages.

New and more advanced (highly expensive) procedures and diagnostic methods have been available for cancer management than those used before, in line with the requirements. The measures are small and restricted. Mostly more than one test is to be done in a line when detected with cancer. Ultrasonography, computed tomography (CT), conventional X-rays, scanning, and magnetic resonance imaging, all fall under imaging techniques. These are implied for visualizing cancer, more precisely, the anatomy of cancer. Positron emission tomography and single-photon emission CT are the two often used nuclear imaging techniques for cancer imaging in recent years. These methods help differentiate and detect the metabolic activities inside cells, and by assessing the level of activities; they can be categorized as cancerous or noncancerous cells.

Tests on tissues, blood, urine, and other body fluids are the pathological tests for detecting cancer. Vein puncture (blood drawing) or phlebotomy, fine needle aspiration cytology, or fine needle biopsy and surgical methods are the ways of collecting samples from suspects. To have a more detailed report on cancerous cells, tests like tumor markers or liver function tests (an increase of biological or chemical elements/compounds in the cancerous state) can be done to have in-depth knowledge of the condition of cancer.

For laboratory examination of certain hard and solid tumors, surgical excision of a sample tissue specimen of the cancer is needed, known as a biopsy. Moreover, the cells which are seen in body fluids are also examined by pathological methods. Detailed examinations like microscopic methods are used to assess the growth into other tissues, the size of the cancerous tumor, and the level of activity and grade of cancer (by scoring the difference between the normal tissue and cancerous tissue). When the surgical method is adopted, more knowledge and detailed information about cancers can be known. The surgical method and its results provide detailed information about the appearance, size, and advent of tumors and gives ideas about the condition of lymph nodes and relative organs.

### 1.4.2 Management

The primary prevention and interventions for it at an early stage are the most specific way to mitigate the cancer burden. This preliminary preventive approach includes the awareness of cancer-related symptoms or risk factors. Suppose they eliminate or reduce the symptoms or risk factors. In that case, prevention is possible by limitation of action, restriction of the environment, or sufficient vaccination or therapy of infectious agents. Not to our surprise, proofs about cancer prevention and early observation have been seen in lucrative countries. These prevention plans are restricted to fit the conditions in those countries where more of this knowledge relies directly on cancer severity.

Tobacco usage is the most prevalent cancer cause that dominates all. Luckily, some researchers and mediations are working on reducing the use of tobacco. Prevention through vaccination or therapy is necessary for LMICs due to the rise in cancer from infectious sources. In lucrative countries, they are still trying to inspire people with other known prevention factors like maintaining a healthy diet, eating patterns, weight management, exercising, and alcohol restriction, to list a few.

#### *1.4.2.1 Surgery*

The main centerpiece of cancer treatment was surgery, where there was a different range of cancer treatments available. Later on, chemotherapy and radiotherapy have been established in recent treatment methodologies. Until the mid-20th century, surgical resection was the only possible and available procedure. For patients to survive longer-term, they usually rely on the removal of the necessary tumor via surgery. Surgical intervention principally includes removing solid tumors (along with marginal normal tissue) and regional lymph nodes, and is mostly accompanied by newer therapeutic modalities. As per a recent report, around 90% of solid tumors in the United States population are productively healed by surgical methods solely or with other treatments combined.

Cancers of the thyroid, rectum, breast, lung, and stomach are the types of cancers mostly cured by surgical resection, most of which are melanomas (Roos et al., 2005). These surgeries are performed on a scale from primary to the most complex levels, which involves all the settings required to carry out surgeries. The most preferred trend is to promote lower radical surgery and then use chemotherapy or radiotherapy to facilitate surgeries. For instance, cutting the limb in scientific terms is called surgical amputation of the limb and is done to treat the bone or soft tissue sarcomas at its extremes. Adding the new treatment methods, nowadays, limbs are usually saved by performing tumor-specific excision. More advanced surgical interventions are generally adopted in common (like laparoscopy, imaging techniques, ultrasound, etc.), where resources exist.

It's not the case that all surgeries are performed with the intent of complete cure; in contrast, most of the case's intent was unknown until the actual surgery began. On doing surgery, if it's not possible to eradicate the tumor, it is removed to the extent possible to debulk the tumor and reduce allied symptoms. This approach could be life-saving when the cancer is associated with vitals and leads to intervention in normal functioning.

### *1.4.2.2 Radiotherapy*

For cancer treatment, even ionizing radiations (such as X-rays, radioactive particles, and $\gamma$-rays) are used and referred to as radiotherapy. The medical discipline that involves the treatment of cancer with radiation is known as "radiation oncology." Radiotherapy is also used to set free or relieve the pain of patients suffering from incurable cancers. Radiotherapy is an immensely used cancer treatment method, but its use is restricted in some medical cases which are not allied to cancer (e.g., keloid or heaped-up scars).

#### 1.4.2.2.1 Curative radiotherapy

Many cancer patients are cured and treated in lucrative countries with treatments such as radiotherapy alone or surgery, chemotherapy, or both. One of the majorly used treatments is radiotherapy, which is considered to have less side effects and a higher number of curable rates among cancer patients. This involves treating Hodgkin's lymphoma, low-grade lymphomas, nasopharyngeal cancer, pituitary tumors, cervical cancer, deep-seated gliomas, low-grade, and early-stage lymphomas. Organs are at risk in surgery, and owing to that, radiotherapy is mostly chosen over it, whereas control over the tumor is to the same degree. Laryngeal cancer and prostate cancer are examples of the same preference. For the removal of a small tumor, surgery alone can be efficacious. Radiotherapy is combined with surgery in most cases to minimize tumor size or reduce the chances of the reappearance of cancer so that the patient gets treated well with fewer side effects.

Chemotherapy proves to be effective in disseminated micrometastases, while radiotherapy is operative in treating large tumors. There are various dose-limiting toxicities of chemotherapy and radiotherapy, implying the possibility of delivering a higher level treatment dose with two techniques rather than either of them. Using a higher radiation dose or using a chemical medium to safeguard normal tissues are the two most promising treatment strategies that are further intensified with other techniques.

#### 1.4.2.2.2 Palliative radiotherapy

An increase in tumor size shows symptoms of their presence physically, for instance, orifice blockage or pressing effect on the nearby organs and blocking effect on passages. Shrinking of tumors directly via radiotherapy can show prominent symptoms. In most of the issues, it may minimize or abolish the need for analgesics like opioids.

For people suffering from hemoptysis (bloody cough), reduced breathing patterns, cough, incurable lung cancer, etc., radiotherapy has been proven the most useful way. Radiotherapy is used to control fungating masses (rapid and increased number of tissues) and breast cancer, prostate cancer, and to get rid of urinary obstruction. Short-course radiotherapy is more effective in treating and alleviating bone and brain metastases, spinal cord contraction, and other nerves (Laperriere & Bernstein, 1994). Prevention of paraplegia and compression of the spinal cord can be accomplished with the use of precise radiotherapy. Radiotherapy can also help extend people's lives from incurable and dreadful cancer, such as high-grade gliomas (World Health Organization WHO, 2006).

### *1.4.2.3 Chemotherapy*

One of the chief approaches of cancer management therapy is known as chemotherapy, wherewith the use of medications and beneficial drugs it is attempted to treat cancer to give a benefit of higher chances of survival (or cure) and at least a better quality of life and the considerable increase in the life span. The medical experts are educated in this for use during chemotherapy and are known as medical oncologists (who are especially well trained for handling cancer-related issues). Various techniques are available to kill the cancer cells, one of which is "cytotoxic" drugs. Hormonal therapy is another therapy specially undertaken to abolish, minimize, or block some cancers (majorly in prostate and breast cancer). Generally, drugs are combined; rarely combination involves three or four drugs, designed as a regime used for a longer period. The major classes of chemotherapeutic agents can be found in Table 1.4.

#### 1.4.2.4 WHO model list of essential medicines

Considering each allied term and condition, the WHO model list of necessary medicines is designed and revised from time to time. This is used as a guiding script for national and crucial/essential medicine lists that "mainly look after the people's healthcare needs" (Institute of Medicine IOM, 2004), altered as per the circumstances precisely. There are two types of lists categorized and formalized as a core list and complimentary list.

The core list involves "minimal medicinal needs for a primary healthcare system." The best and successful, effective, secured, and avaricious medicines for "primary conditions" are selected considering the present necessity and evaluated for future healthcare relevance, healthcare importance, and probability for safe and most effective treatment. They know that cancer is one such utmost condition that must be on the priority list and demands observation and special services for diagnosis. Thus cancer chemotherapeutic agents can be seen on the list (Table 1.5).

#### 1.4.2.5 Psychosocial services

Psychosocial stress may be seen in people suffering from cancer at any stage from their initial diagnosis until they reach the advanced cancer stage and also amid the survival point of their illness. This type of stress, or call it distress, is defined as the behavioral impact on one's social, spiritual, or psychological conditions. These behavioral modifications can interfere with the person's diagnostic stage that can ultimately result in depression. Step by step, these make it more worse (from the start of diagnosis, between treatment, and at the end of the treatment) (Institute of Medicine IOM, 2004).

The patients suffering from cancer get most of the sustenance from psychosocial support to deal with and handle the disease, its effect, and the pain of the last stage of cancer plus the life after recovering from cancer. At all stages of cancer, this kind of care and special host services are taken from the beginning of diagnosis and the treatment or procedure. Major emotional help is also needed by people who are close to the patient, or their relatives and family members. These psychosocial methods are considered crucial stages of cancer as they are still under process but are considered important factors in cancer therapy. In LMICs very little share of knowledge is given. At the primary level, people suffering from cancer seek help to cope with their disease through exceptional care and talking to people. Understanding the treatment and considering the social and emotional state, and making adjustments per the diagnosis, treatment, and behavior are other chief concerns (Institute of Medicine IOM, 2004).

**TABLE 1.4** Major classes of chemotherapeutic drugs for cancer therapy.

| Drug type | Mode of action and examples |
|---|---|
| Antimetabolites | (a) Block cell growth by interfering with certain activities, usually DNA synthesis, halting normal development and reproduction; (b) Used to treat acute and chronic leukemias, choriocarcinoma, and some tumors of the gastrointestinal tract, breast, and ovary; (c) 6-mercaptopurine and 5-fluorouracil (5-FU) are commonly used. |
| Alkylating agents | (a) Kill cells by directly attacking DNA; (b) Used to treat chronic leukemias, Hodgkin's disease, lymphomas, and certain carcinomas of the lung, breast, prostate, and ovary; (c) Cyclophosphamide is a commonly used alkylating agent. |
| Nitrosoureas | (a) Act similarly to alkylating agents and also inhibit changes necessary for DNA repair; (b) Cross the blood-brain barrier and are therefore used to treat brain tumors, lymphomas, multiple myeloma, and malignant melanoma; (c) Carmustine (BCNU) and lomustine (CCNU) are the major nitrosourea drugs. |
| Antitumor antibiotics | (a) Diverse group of compounds that generally act by binding with DNA and preventing RNA synthesis; (b) Widely used to treat a variety of cancers; (c) Doxorubicin, adriamycin, mitomycin-C, and bleomycin are the most common drugs in this category. |
| Hormonal agents | (a) Includes adrenocorticosteroids, estrogens, antiestrogens, progesterone, and androgens that modify the growth of certain hormone-dependent cancers; (b) Tamoxifen, used for estrogen-dependent breast cancer, is an example. |
| Plant (vinca) alkaloids | (a) Act by blocking cell division during mitosis; (b) Commonly used to treat acute lymphocytic leukemia (ALL), Hodgkin's and non-Hodgkin's lymphomas, neuroblastomas, Wilms' tumor, and cancers of the lung, breast, and testes; (c) Vincristine and vinblastine are commonly used agents in this group. |

**TABLE 1.5** WHO model list of essential medicines: Antineoplastic drugs.

| | |
|---|---|
| **Cytotoxic medicines (complementary list)** | |
| Asparaginase | Powder for injection |
| Bleomycin | Powder for injection |
| Calcium folinate | Tablet; injectable liquid |
| Chlorambucil | Tablet |
| Chlormethine | Powder for injection |
| Cisplatin | Powder for injection |
| Cyclophosphamide | Tablet; powder for injection |
| Cytarabine | Powder for injection |
| Dacarbazine | Powder for injection |
| Dactinomycin | Powder for injection |
| Daunorubicin | Powder for injection |
| Doxorubicin | Powder for injection |
| Etoposide | Capsule; injectable liquid |
| Fluorouracil | Injectable liquid |
| Levamisole | Tablet |
| Mercaptopurine | Tablet |
| Methotrexate | Tablet; powder for injection |
| Procarbazine | Capsule |
| Vinblastine | Powder for injection |
| Vincristine | Powder for injection |
| **Hormones and antihormones (complementary list[a])** | |
| Dexamethasone | Injectable liquid |
| Hydrocortisone | Powder for injection |
| Prednisolone | Tablet |
| Tamoxifen | Tablet |
| **Medicines used in palliative care** | |

[a]*The complementary list includes essential medicines for priority diseases for which specialized diagnostic or monitoring facilities, and specialist medical care and specialist training are needed.*

Additionally, to the treatment mentioned above, a wide range of important and particular interventions can be listed as follows:

1. Counseling in different attributes, such as psychotherapeutic therapies, crisis, counseling for family and community therapy, grief therapy, and sexual counseling.
2. Psychopharmacological intrusions.
3. Complementary approaches, such as dance therapy, massage therapy, exercise, music, treatment of art and acupuncture, and many others.

A number of suppliers offer psychosocial facilities. In psychosocial support, nurses stand in the main front stream for cancer patients to give them exceptional support. Other practitioners like physicians can also provide care if the cancer patient has an approach toward them. Other than this, more respect can also be expected from spiritual sources, counselors, social helpers, psychologists who can offer extra support. The most known groups of help are the sufferers who are the patient and the cancer survivors themselves. Other means which serve are the old age groups and communities where they give psychosocial help to patients and their relatives. For instance, psychosocial support offered by Reach to Recovery International (RRI) under the guidance of the International Union Against Cancer is a unique example of service offering centers by breast cancer survivors. Traditionally, RRI centers can be found in every corner of the world and are run by local people. This mainly stands on the pillar that if one female has gone through breast cancer, she will share her tough goings, time, and happenings to help another female undergoing the same situation.

## 1.5 Current management

The other way to reduce the cancer mortality rate is to detect the early stage of cancer or premalignant phases, considering that effective management is seen in cancer treatment. Highly detectable cancer widely seen in HICs are breast, cervical, and colon cancers. Controversies arise upon the screening value when prostate cancer detection is done using the prostate-specific antigen test. Detection of cancer needs a specific base for screening when it reaches a considerable population, and the same is for its treatment capacity. When a country undertakes the detection programs, it principally depends on whether it is ready to deal with both aspects: management and screening.

The individual seeks care and other special services if they have been detected with cancer, either through screening tests or observed with cancer signs and symptoms. Depending on the form and stage of cancer, individuals who have been diagnosed with cancer should receive the required services. For specific individuals who treat cancer, this involves using multiple procedures such as surgery, radiotherapy, chemotherapy, or, more frequently, a combination of all of these with an integrated medical team working together. In brief, if it is detected at early stages, soothing care for symptoms control and appropriate treatment can be useful. If cancers reach beyond a particular stage or maybe advanced cancer or other types of cancer that are difficult to treat and beyond a possible cure, gentle care alone may be very useful. Psychologists deal with people and with their psychological and social needs. Individuals suffering from cancer and who are dealing with cancer at their last stages need psychological services. These perspectives are categorized and come under the terms of "cancer management."

## 1.6 Conclusions and future prospects

Cancer is the leading cause of death worldwide and affecting most of the population. Cancer development is a multistage process that arises from normal cells, leading to their transformation into the precancerous lesion and, subsequently, a malignant tumor. It occurs due to certain factors such as physical, chemical, and biological agents termed as "carcinogens" and genetic factors. The risks associated with cancer are 20% in males and 17% in females. Moreover, the fatality in male and female cancer patients is 13% and 9%, respectively. Cancer occurs because of several factors such as physical (UV and ionizing rays), chemical compounds (asbestos, aflatoxin, arsenic), and biological agents (bacteria, viruses, parasites). Other factors responsible for cancer are tobacco, infections, physical inactivity, pollution, etc. The diagnosis of cancer in the early stages is crucial to treat the disease. Hence, various techniques were studied for the diagnosis of different types of cancers. Finally, for cancer management, surgery, radiotherapy, chemotherapy are adopted along with advanced techniques.

## Conflict of interest

The author(s) express no conflict of interest for the content of this book chapter.

## List of abbreviations

| | |
|---|---|
| **AIDS** | Acquired immune deficiency syndrome |
| **BMI** | Body mass index |
| **CT** | Computed tomography |
| **DCEG** | Division of Cancer Epidemiology and Genetics |
| **DNA** | Deoxyribonucleic acid |
| **EBV** | Epstein-Barr virus |
| **h** | hour |
| **HBV** | Hepatitis B virus |
| **HCC** | Hepatocellular carcinoma |
| **HCV** | Hepatitis C virus |
| **HICs** | High-income countries |
| **HIV** | Human immunodeficiency virus |
| **HPV** | Human papillomavirus |
| **HTLV-1** | Human T-cell lymphotropic virus type 1 |
| **IARC** | International Agency for Research on Cancer |
| **ILO** | International Labour Organization |
| **KSHV** | Kaposi sarcoma herpesvirus |
| **LMICs** | Low- and middle-income countries |
| **MCC** | Merkel cell carcinoma |
| **MCV** | Merkel cell polyomavirus |
| **NIH** | National Institute of Health |

| | |
|---|---|
| **ROS** | Reactive oxygen species |
| **RRI** | Reach to Recovery International |
| **UV** | Ultraviolet |
| **WHO** | World Health Organization |

## References


Abnet, C. C. (2007). Carcinogenic food contaminants. *Cancer Investigation*, *25*(3), 189–196. Available from https://doi.org/10.1080/07357900701208733.

Allan, J. M., & Travis, L. B. (2005). Mechanisms of therapy-related carcinogenesis. *Nature Reviews. Cancer*, *5*(12), 943–955. Available from https://doi.org/10.1038/nrc1749.

American Cancer Society. *Asbestos and cancer risk*. (2020a). <https://www.cancer.org/cancer/cancer-causes/asbestos.html> Accessed 17.04.20.

American Cancer Society. *Ultraviolet (UV) radiation*. (2020b). <https://www.cancer.org/cancer/cancer-causes/radiation-exposure/uv-radiation.html> Accessed 15.04.20.

Andersen, B. L., Farrar, W. B., Golden-Kreutz, D., Emery, C. F., Glaser, R., Crespin, T., & Carson, W. E. (2007). Distress reduction from a psychological intervention contributes to improved health for cancer patients. *Brain, Behavior, and Immunity*, *21*(7), 953–961. Available from https://doi.org/10.1016/j.bbi.2007.03.005.

Anguiano, l, Mayer, D., Piven, M., et al. (2012). A literature review of suicide in cancer patients. *Cancer Nursing*, *35*(4), 14–26.

Armstrong, B. K., & Kricker, A. (2001). The epidemiology of UV induced skin cancer. *Journal of Photochemistry and Photobiology B: Biology*, *63* (1–3), 8–18. Available from https://doi.org/10.1016/S1011-1344(01)00198-1.

Basílio-de-Oliveira, C. A., Aquino, A., Simon, E. F., & Eyer-Silva, W. A. (2002). Concomitant prostatic schistosomiasis and adenocarcinoma: case report and review. *The Brazilian Journal of Infectious Diseases: An Official Publication of the Brazilian Society of Infectious Diseases*, *6*(1), 45–49. Available from https://doi.org/10.1590/S1413-86702002000100007.

Baylin, S. B., & Ohm, J. E. (2006). Epigenetic gene silencing in cancer—A mechanism for early oncogenic pathway addiction? *Nature Reviews. Cancer*, *6*(2), 107–116. Available from https://doi.org/10.1038/nrc1799.

Benjamin, C. L., & Ananthaswamy, H. N. (2007). p53 and the pathogenesis of skin cancer. *Toxicology and Applied Pharmacology*, *224*(3), 241–248. Available from https://doi.org/10.1016/j.taap.2006.12.006.

Bernstein, C., Nfonsam, V., Prasad, A. R., & Bernstein, H. (2013). Epigenetic field defects in progression to cancer. *World Journal of Gastrointestinal Oncology*, *5*(3), 43–49. Available from https://doi.org/10.4251/wjgo.v5.i3.43.

Bhagwandeen, S. B. (1976). Schistosomiasis and carcinoma of the bladder in Zambia. *South African Medical Journal*, *50*(41), 1616–1620.

Blumberg, B. S., & Alter, H. J. (1965). A new antigen in leukemia sera. *Jama*, *191*(7), 541–546.

Boshart, M., Gissmann, L., Ikenberg, H., Kleinheinz, A., Scheurlen, W., & zur Hausen, H. (1984). A new type of papillomavirus DNA, its presence in genital cancer biopsies and in cell lines derived from cervical cancer. *The EMBO Journal*, *3*(5), 1151–1157.

Botelho, M. C., & Richter, J. (2019). Parasites and cancer. *Frontiers in Medicine*, *6*, 55.

Bouvard, V., Baan, R., Straif, K., Grosse, Y., Secretan, B., El Ghissassi, F., ... Cogliano, V. (2009). A review of human carcinogens–part B: Biological agents. *The Lancet Oncology*, *10*(4), 321–322. Available from https://doi.org/10.1016/s1470-2045(09)70096-8.

Brindley, P. J., da Costa, J. M. C., & Sripa, B. (2015). Why does infection with some helminths cause cancer? *Trends in Cancer*, *1*(3), 174–182. Available from https://doi.org/10.1016/j.trecan.2015.08.011.

Bruce, N., Rehfuess, E., Mehta, S., Hutton, G., & Smith, K. (2006). Indoor air pollution. In D. T. Jamison, J. G. Breman, A. R. Measham, G. Alleyne, M. Claeson, D. B. Evans, P. Jha, A. Mills, & P. Musgrove (Eds.), *Disease control priorities in developing countries* (2nd (ed.), pp. 793–815). New York: Oxford University Press.

Cahan, W. G., Woodard, H. Q., Higinbotham, N. L., Stewart, F. W., & Coley, B. L. (1948). Sarcoma in irradiated bone. Report of eleven cases. *Cancer*, *1*(1), 3–29. https://doi.org/10.1002/1097-0142(194805)1:1 < 3::aid-cncr2820010103 > 3.0.co;2-7.

Cancer Quest. *Viruses and cancer*. (2020). <https://www.cancerquest.org/cancer-biology/viruses-and-cancer> Accessed 24.04.20.

Cancer Research UK (2019). How does smoking cause cancer? https://www.cancerresearchuk.org/about-cancer/causes-of-cancer/smoking-and-cancer/how-does-smoking-cause-cancer#tobacco10 (accessed April 21, 2020).

Carbone, M., Pass, H. I., Rizzo, P., Marinetti, M., Di Muzio, M., Mew, D., Levine, A. S., & Procopio, A. (1994). Simian virus 40-like DNA sequences in human pleural mesothelioma. *Oncogene*, *9*(6), 1781–1790.

Centers for Disease Control and Prevention (CDC). *Surgeon general's report. How tobacco smoke causes disease*. (2010a). <https://www.cdc.gov/tobacco/data_statistics/sgr/2010/index.htm> Accessed 21.04.20.

Centers for Disease Control and Prevention (CDC). *How tobacco smoke causes disease: The biology and behavioral basis for smoking-attributable disease: A report of the surgeon general*. (2010b). <https://www.cdc.gov/tobacco/data_statistics/sgr/2010/index.htm> Accessed 21.04.20.

Centers for Disease Control and Prevention (CDC). *Surgeon general's report: The health consequences of smoking-50 years of progress*. (2014). <https://www.cdc.gov/tobacco/data_statistics/sgr/50th-anniversary/index.htm> Accessed 21.04.20.

Chang, A. H., & Parsonnet, J. (2010). Role of bacteria in oncogenesis. *Clinical Microbiology Reviews*, *23*(4), 837–857. Available from https://doi.org/10.1128/CMR.00012-10.

Chang, Y., Cesarman, E., Pessin, M. S., Lee, F., Culpepper, J., Knowles, D. M., & Moore, P. S. (1994). Identification of herpesvirus-like DNA sequences in AIDS-associated Kaposi's sarcoma. *Science*, *266*(5192), 1865–1869.

Chen, Y., Williams, V., Filippova, M., Filippov, V., & Duerksen-Hughes, P. (2014). Viral carcinogenesis: Factors inducing DNA damage and virus integration. *Cancers*, *6*(4), 2155–2186. Available from https://doi.org/10.3390/cancers6042155.

Choo, Q.-L., Kuo, G., Weiner, A. J., Overby, L. R., Bradley, D. W., & Houghton, M. (1989). Isolation of a cDNA clone derived from a blood-borne non-A, non-B viral hepatitis genome. *Science*, *244*(4902), 359–362.

Costa, J. *Cancer. Encyclopædia britannic*. (2020). <https://www.britannica.com/science/cancer-disease/Radiation> Accessed 15.04.20.

Cover, T. L., Krishna, United States, Israel, D. A., & Peek, R. M. (2003). Induction of gastric epithelial cell apoptosis by *Helicobacter pylori* vacuolating cytotoxin. *Cancer Research*, *63*(5), 951–957.

Croce, C. M. (2008). Oncogenes and cancer. *New England Journal of Medicine*, *358*(5), 502–511. Available from https://doi.org/10.1056/NEJMra072367.

Darby, S., Hill, D., Auvinen, A., Barros-Dios, J. M., Baysson, H., Bochicchio, F., ... Doll, R. (2005). Radon in homes and risk of lung cancer: Collaborative analysis of individual data from 13 European case-control studies. *BMJ (Clinical Research ed.)*, *330*(7485), 223.

Darby, S., Hill, D., & Doll, R. (2001). Radon: A likely carcinogen at all exposures. *Annals of Oncology*, *12*(10), 1341–1351. Available from https://doi.org/10.1023/A:1012518223463.

De Martel, C., Plummer, M., Parsonnet, J., Van Doorn, L. J., & Franceschi, S. (2009). Helicobacter species in cancers of the gallbladder and extrahepatic biliary tract. *British Journal of Cancer*, *100*(1), 194–199. Available from https://doi.org/10.1038/sj.bjc.6604780.

Di Bisceglie, A. M. (2009). Hepatitis B and hepatocellular carcinoma. *Hepatology*, *49*(S5), S56–S60.

DiMatteo, M. R., Lepper, H. S., & Croghan, T. W. (2000). Depression is a risk factor for noncompliance with medical treatment *meta*-analysis of the effects of anxiety and depression on patient adherence. *Archives of Internal Medicine*, *160*(14), 2101–2107. Available from https://doi.org/10.1001/archinte.160.14.2101.

Dove, W. F., Clipson, L., Gould, K. A., Luongo, C., Marshall, D. J., Moser, A. R., ... Jacoby, R. F. (1997). Intestinal neoplasia in the Apc(Min) mouse: Independence from the microbial and natural killer (beige locus) status. *Cancer Research*, *57*(5), 812–814.

Dürst, M., Gissmann, L., Ikenberg, H., & Zur Hausen, H. (1983). A papillomavirus DNA from a cervical carcinoma and its prevalence in cancer biopsy samples from different geographic regions. *Proceedings of the National Academy of Sciences*, *80*(12), 3812–3815.

El-Omar, E. M., Carrington, M., Chow, W. H., McColl, K. E. L., Bream, J. H., Young, H. A., ... Rabkin, C. S. (2000). Interleukin-1 polymorphisms associated with increased risk of gastric cancer. *Nature*, *404*(6776), 398–402. Available from https://doi.org/10.1038/35006081.

Epping-Jordan, J. A. E., Galea, G., Tukuitonga, C., & Beaglehole, R. (2005). Preventing chronic diseases: Taking stepwise action. *Lancet*, *366*(9497), 1667–1671. Available from https://doi.org/10.1016/S0140-6736(05)67342-4.

Epstein, M. A. (1964). Virus particles in cultured lymphoblasts from Burkitt's lymphoma. *Lancet*, *1*, 702–703.

Feng, H., Shuda, M., Chang, Y., & Moore, P. S. (2008). Clonal integration of a polyomavirus in human Merkel cell carcinoma. *Science*, *319*(5866), 1096–1100.

Ferlay, J., Soerjomataram, I., Ervik, M., Dikshit, R., Eser, S., Mathers, C., ... Bray, F. (2013). *Cancer incidence and mortality worldwide: IARC cancer base*. 2012(11). Lyon: International Agency for Research on Cancer.

Fleming, I. D., Brady, L. W., Mieszkalski, G. B., & Cooper, M. R. (1995). Basis for major current therapies for cancer. In G. P. Murphy, W. Lawrence, & R. E. Lenhard (Eds.), *American cancer society textbook of clinical oncology* (2nd ed., p. 96). Atlanta, GA: American Cancer Society.

Fox, J. P., Philip, E. J., Gross, C. P., Desai, R. A., Killelea, B., & Desai, M. M. (2013). Associations between mental health and surgical outcomes among women undergoing mastectomy for cancer. *Breast Journal*, *19*(3), 276–284. Available from https://doi.org/10.1111/tbj.12096.

Fritz, A., Percy, C., Jack, A., et al., (2000). eds. WHO International Classification of Diseases for Oncology ICD-O, 3rd ed Geneva: WHO.

GBD 2015 Risk Factors Collaborators. (2016). Global, regional, and national comparative risk assessment of 79 behavioural, environmental and occupational, and metabolic risks or clusters of risks, 1990–2015: A systematic analysis for the Global Burden of Disease Study 2015. *Lancet (London, England)*, *388*(10053), 1659–1725.

Ghosh, P., Banerjee, M., Giri, A. K., & Ray, K. (2008). Toxicogenomics of arsenic: Classical ideas and recent advances. *Mutation Research – Reviews in Mutation Research*, *659*(3), 293–301. Available from https://doi.org/10.1016/j.mrrev.2008.06.003.

Groopman, J. D., Kensler, T. W., & Wild, C. P. (2008). Protective interventions to prevent aflatoxin-induced carcinogenesis in developing countries. *Annual review of public health*, *29*, 187–203. Available from https://doi.org/10.1146/annurev.publhealth.29.020907.090859.

Gutschner, T., & Diederichs, S. (2000). The hallmarks of cancer. *Cell*, *100*, 57–70.

Hajdu, S. I. (2011). A note from history: Landmarks in history of cancer, part 1. *Cancer*, *117*(5), 1097–1102. Available from https://doi.org/10.1002/cncr.25553.

Hamada, T., Yokota, K., Ayada, K., Hirai, K., Kamada, T., Haruma, K., ... Oguma, K. (2009). Detection of helicobacter hepaticus in human bile samples of patients with biliary disease. *Helicobacter*, *14*(6), 545–551. Available from https://doi.org/10.1111/j.1523-5378.2009.00729.x.

Hanahan, D., & Weinberg, R. (2011). Hallmarks of cancer: The next generation. *Cell.*, *144*(5), 646–674.

Honeycutt, J., Hammam, O., Fu, C. L., & Hsieh, M. H. (2014). Controversies and challenges in research on urogenital schistosomiasis-associated bladder cancer. *Trends in Parasitology*, *30*(7), 324–332. Available from https://doi.org/10.1016/j.pt.2014.05.004.

Institute of Medicine (IOM). (2003). In S. J. Curry, T. Byers, & M. Hewitt (Eds.), *Fulfilling the potential of cancer prevention and early detection*. Washington, DC: The National Academies Press.

Institute of Medicine (IOM). (2004). In M. Hewitt, R. Herdman, & J. Holland (Eds.), *Meeting psychosocial needs of women with breast cancer*. Washington, DC: The National Academies Press.

International Agency for Research on Cancer (IARC). (n.d.). *Global initiative for cancer registry development*. Lyon: IARC.

International Agency for Research on Cancer (IARC). (1997). *IARC monographs on the evaluation of carcinogenic risks to humans: Schistosomes. Liver flukes and Helicobacter pylori*. Lyon, France: IARC.

International Agency for Research on Cancer (IARC). (2004). *GLOBOCAN 2002*. Lyon, France: IARC.

International Agency for Research on Cancer (IARC). *Latest global cancer data: Cancer burden rises to 18.1 million new cases and 9.6 million cancer deaths in 2018" (PDF)*. (2018). <http://www.iarc.fr> Retrieved 05.12.18.

Ionizing Radiation Fact Sheet. On Health 2017. https://www.onhealth.com/content/1/ionizing_radiation_fact_sheet (accessed April 16, 2020).

Ishii, A., Matsuoka, H., Aji, T., Ohta, N., Arimoto, S., Wataya, Y., & Hayatsu, H. (1994). Parasite infection and cancer: With special emphasis on *Schistosoma japonicum* infections (Trematoda). A review. *Mutation Research - Fundamental and Molecular Mechanisms of Mutagenesis*, *305*(2), 273–281. Available from https://doi.org/10.1016/0027-5107(94)90247-X.

Jacinto, F. V., & Esteller, M. (2007). Mutator pathways unleashed by epigenetic silencing in human cancer. *Mutagenesis*, *22*(4), 247–253. Available from https://doi.org/10.1093/mutage/gem009.

Jamison, D. T., Breman, J. G., Measham, A. R., Alleyne, G., Claeson, M., Evans, D. B., & ... Musgrove, P. (Eds.), (2006). *Disease conrol priorities in developing countries*. The World Bank, Apr 2.

Jemal, A., Bray, F., Center, M. M., Ferlay, J., Ward, E., & Forman, D. (2011). Global cancer statistics. *CA Cancer Journal for Clinicians*, *61*(2), 69–90. Available from https://doi.org/10.3322/caac.20107.

Kalisperati, P., Spanou, E., Pateras, I. S., Korkolopoulou, P., Varvarigou, A., Karavokyros, I., ... Sougioultzis, S. (2017). Inflammation, DNA damage, *Helicobacter pylori* and gastric tumorigenesis. *Frontiers in Genetics*, *8*, 20. Available from https://doi.org/10.3389/fgene.2017.00020.

Kang, G. H. (2012). CpG island hypermethylation in gastric carcinoma and its premalignant lesions. *Korean Journal of Pathology*, *46*(1), 1. Available from https://doi.org/10.4132/koreanjpathol.2012.46.1.1.

Kew, M. C. (2002). Epidemiology of hepatocellular carcinoma. *Toxicology*, *181–182*, 35–38. Available from https://doi.org/10.1016/S0300-483X (02)00251-2.

Khushalani, J. S., Qin, J., Cyrus, J., Buchanan Lunsford, N., Rim, S. H., Han, X., ... Ekwueme, D. U. (2018). Systematic review of healthcare costs related to mental health conditions among cancer survivors. *Expert Review of Pharmacoeconomics and Outcomes Research*, *18*(5), 505–517. Available from https://doi.org/10.1080/14737167.2018.1485097.

Kim, E. M., Bae, Y. M., Choi, M. H., & Hong, S. T. (2012). Cyst formation, increased anti-inflammatory cytokines and expression of chemokines support for *Clonorchis sinensis* infection in FVB mice. *Parasitology International*, *61*(1), 124–129. Available from https://doi.org/10.1016/j.parint.2011.07.001.

Kligerman, A. D., & Tennant, A. H. (2007). Insights into the carcinogenic mode of action of arsenic. *Toxicology and Applied Pharmacology*, *222*(3), 281–288. Available from https://doi.org/10.1016/j.taap.2006.10.006.

Knudson, A. G. (2001). Two genetic hits (more or less) to cancer. *Nature Reviews. Cancer*, *1*(2), 157–162. Available from https://doi.org/10.1038/35101031.

Lahtz, C., & Pfeifer, G. P. (2011). Epigenetic changes of DNA repair genes in cancer. *Journal of Molecular Cell Biology*, *3*(1), 51–58. Available from https://doi.org/10.1093/jmcb/mjq053.

Laliani, G., Ghasemian Sorboni, S., Lari, R., Yaghoubi, A., Soleimanpour, S., Khazaei, M., ... Laliani. (2019). Bacteria and cancer: Different sides of the same coin. *Life Sciences*, *246*, 117398. Available from https://doi.org/10.1016/j.lfs.2020.117398, et al. 2020 Apr 1 (Original work published 2020).

Laperriere, N. J., & Bernstein, M. (1994). Radiotherapy for brain tumors. *CA: A Cancer Journal for Clinicians*, *44*(2), 96–108. Available from https://doi.org/10.3322/canjclin.44.2.96.

Leuraud, K., Richardson, D. B., Cardis, E., Daniels, R. D., Gillies, M., O'Hagan, J. A., ... Kesminiene, A. (2015). Ionising radiation and risk of death from leukaemia and lymphoma in radiation-monitored workers (INWORKS): An international cohort study. *The Lancet Haematology*, *2*(7), e276–e281. Available from https://doi.org/10.1016/S2352-3026(15)00094-0.

Looper, K. J. (2007). Potential medical and surgical complications of serotonergic antidepressant medications. *Psychosomatics*, *48*(1), 1–9. Available from https://doi.org/10.1176/appi.psy.48.1.1.

Lopez, A. D., Mathers, C. D., Ezzati, M., Jamison, D. T., & Murray, C. J. L. (Eds.), (2006). *Global burden of disease and risk factors*. New York: Oxford University Press.

Lutgendorf, S. K. (2005). Stress, spirituality, and cytokines in aging and cancer. *Gynecologic oncology*, *99*(3Suppl 1), S139–S140. Available from https://doi.org/10.1016/j.ygyno.2005.07.065.

Madbouly, K. M., Senagore, A. J., Mukerjee, A., Hussien, A. M., Shehata, M. A., Navine, P., ... Fazio, V. W. (2007). Colorectal cancer in a population with endemic *Schistosoma mansoni*: Is this an at-risk population? *International Journal of Colorectal Disease*, *22*(2), 175–181. Available from https://doi.org/10.1007/s00384-006-0144-3.

March, H. C. (1944). Leukemia in radiologists. *Radiology*, *43*, 275–278. Available from https://doi.org/10.1148/43.3.275.

Marcucci, F., & Mele, A. (2011). Hepatitis viruses and non-Hodgkin lymphoma: epidemiology, mechanisms of tumorigenesis, and therapeutic opportunities. *Blood, The Journal of the American Society of Hematology*, *117*(6), 1792–1798.

Matsuda, K., Masaki, T., Ishii, S., Yamashita, H., Watanabe, T., Nagawa, H., ... Kojima, S. (1999). Possible associations of rectal carcinoma with *Schistosoma japonicum* infection and membranous nephropathy: A case report with a review. *Japanese Journal of Clinical Oncology*, *29*(11), 576–581. Available from https://doi.org/10.1093/jjco/29.11.576.

Meenan, N. A. G., Visai, L., Valtulina, V., Schwarz-Linek, U., Norris, N. C., Gurusiddappa, S., ... Potts, J. R. (2007). The tandem β-zipper model defines high affinity fibronectin-binding repeats within *Staphylococcus aureus* FnBPA. *Journal of Biological Chemistry*, *282*(35), 25893–25902. Available from https://doi.org/10.1074/jbc.M703063200.

Merlo, L. M. F., Pepper, J. W., Reid, B. J., & Maley, C. C. (2006). Cancer as an evolutionary and ecological process. *Nature Reviews. Cancer*, *6*(12), 924–935. Available from https://doi.org/10.1038/nrc2013.

Misono, S., Weiss, N. S., Fann, J. R., Redman, M., & Yueh, B. (2008). Incidence of suicide in persons with cancer. *Journal of Clinical Oncology*, *26* (29), 4731–4738. Available from https://doi.org/10.1200/JCO.2007.13.8941.

Moyat, M., & Velin, D. (2014). Immune responses to *Helicobacter pylori* infection. *World Journal of Gastroenterology: WJG*, *20*(19), 5583–5593. Available from https://doi.org/10.3748/wjg.v20.i19.5583.

National Cancer Institute. *Defining cancer* (Vol. 17). (2007). Retrieved 28.03.18.

National Cancer Institute. (2015). *Metastatic cancer: Questions and answers*. https://www.cancer.gov/Metastatic cancer: Questions and answers. Retrieved 28.03.18.

National Cancer Institute. *Abo t cancer, causes, prevention, risk, substances and asbestos fact sheet*. (2020a). <http://www.cancer.gov/aboutcancer/causes> Accessed 17.04.20.

National Cancer Institute. *Aflatoxins – Cancer-causing substances*. (2020b). <https://www.cancer.gov/about-cancer/causes-prevention/risk/substances/aflatoxins> Accessed 22.04.20.

National Cancer Institute. *Water contaminants and cancer risk: Arsenic, disinfection by products, and nitrate*. (2020c). <https://dceg.cancer.gov/research/what-we-study/drinking-water-contaminants> Accessed 23.04.20.

National Comprehensive Cancer Network. What is metastasized cancer?. (2013). Archived from the original on 07.07.13. https://www.nccn.org/. Retrieved 18.07.13.

National Institute of Environmental Health Sciences. *Sur.report/toba. smoke causes, 2016 National Toxicology Program: 14th Report on carcinogens*. (2016). <https://ntp.niehs.nih.gov/whatwestudy/assessments/cancer/roc/2016/index.html> Accessed 21.04.20.

Nelson, D. A., Tan, T. T., Rabson, A. B., Anderson, D., Degenhardt, K., & White, E. (2004). Hypoxia and defective apoptosis drive genomic instability and tumorigenesis. *Genes and Development*, *18*(17), 2095–2107. Available from https://doi.org/10.1101/gad.1204904.

Park, Y. H., & Kim, N. (2015). Review of atrophic gastritis and intestinal metaplasia as a premalignant lesion of gastric cancer. *Journal of Cancer Prevention*, *20*(1), 25–40. Available from https://doi.org/10.15430/JCP.2015.20.1.25.

Parkin, D. M. (2006). The global health burden of infection-associated cancers in the year 2002. *International Journal of Cancer*, *118*(12), 3030–3044. Available from https://doi.org/10.1002/ijc.21731.

Pisani, P., Parkin, D. M., Muñoz, N., & Ferlay, J. (1997). Cancer and infection: Estimates of the attributable fraction in 1990. *Cancer Epidemiology Biomarkers and Prevention*, *6*(6), 387–400.

Pleural Mesothelioma Center. *What is asbestos: Types & potential risks after exposure: Giving mesothelioma cancer patients hope* (Vol. 17). (2020). <https://www.pleuralmesothelioma.com/asbestos> Accessed 17.04.20.

Plummer, M., de Martel, C., Vignat, J., Ferlay, J., Bray, F., & Franceschi, S. (2016). Global burden of cancers attributable to infections in 2012: A synthetic analysis. *The Lancet Global Health*, *4*(9), e609–e616. Available from https://doi.org/10.1016/S2214-109X(16)30143-7.

Poiesz, B. J., Ruscetti, F. W., Gazdar, A. F., Bunn, P. A., Minna, J. D., & Gallo, R. C. (1980). "Detection and isolation of type C retrovirus particles from fresh and cultured lymphocytes of a patient with cutaneous T-cell lymphoma.". *Proceedings of the National Academy of Sciences*, *77*(12), 7415–7419.

Popp, A., Billker, O., & Rudel, T. (2001). Signal transduction pathways induced by virulence factors of *Neisseria gonorrhoeae*. *International Journal of Medical Microbiology*, *291*(4), 307–314. Available from https://doi.org/10.1078/1438-4221-00134.

Preston, D. L., Mattsson, A., Holmberg, E., Shore, R., Hildreth, N. G., & Boice, J. D., Jr (2002). Radiation effects on breast cancer risk: A pooled analysis of eight cohorts. *Radiation Research*, *158*(2), 220–235. Available from https://doi.org/10.1667/0033-7587(2002)158[0220:REOBCR]2.0.CO;2.

Proctor, E. K., Morrow-Howell, N. L., Doré, P., Wentz, J., Rubin, E. H., Thompson, S., & Li, H. (2003). Comorbid medical conditions among depressed elderly patients discharged home after acute psychiatric care. *American Journal of Geriatric Psychiatry*, *11*(3), 329–338. Available from https://doi.org/10.1097/00019442-200305000-00010.

Qiu, D. C., Hubbard, A. E., Zhong, B., Zhang, Y., & Spear, R. C. (2005). A matched, case-control study of the association between *Schistosoma japonicum* and liver and colon cancers, in rural China. *Annals of Tropical Medicine and Parasitology*, *99*(1), 47–52. Available from https://doi.org/10.1179/136485905X19883.

Raibaud, S., Schwarz-Linek, U., Kim, J. H., Jenkins, H. T., Baines, E. R., Gurasiddappa, S., ... Potts, J. R. (2005). Borrelia burgdorferi binds fibronectin through a tandem β-zipper, a common mechanism of fibronectin binding in *Staphylococci*, *Streptococci*, and *Spirochetes*. *Journal of Biological Chemistry*, *280*(19), 18803–18809. Available from https://doi.org/10.1074/jbc.M501731200.

Rakoff-Nahoum, S., & Medzhitov, R. (2007). Regulation of spontaneous intestinal tumorigenesis through the adaptor protein MyD88. *Science (New York, N.Y.)*, *317*(5834), 124–127. Available from https://doi.org/10.1126/science.1140488.

Rim, S. H., Guy, G. P., Jr, Yabroff, K. R., McGraw, K. A., & Ekwueme, D. U. (2016). The impact of chronic conditions on the economic burden of cancer survivorship: A systematic review. *Expert Review of Pharmacoeconomics and Outcomes Research*, *16*(5), 579–589. Available from https://doi.org/10.1080/14737167.2016.1239533.

Ronckers, C. M., Doody, M. M., Lonstein, J. E., Stovall, M., & Land, C. E. (2008). Multiple diagnostic X-rays for spine deformities and risk of breast cancer. *Cancer Epidemiology Biomarkers and Prevention*, *17*(3), 605–613. Available from https://doi.org/10.1158/1055-9965.EPI-07-2628.

Roos, D. E., Turner, S. L., O'Brien, P. C., Smith, J. G., Spry, N. A., Burmeister, B. H., … Ball, D. L. (2005). Randomized trial of 8Gy in 1 versus 20Gy in 5 fractions of radiotherapy for neuropathic pain due to bone metastases (Trans-Tasman Radiation Oncology Group, TROG 96.05). *Radiotherapy and Oncology*, *75*(1), 54–63. Available from https://doi.org/10.1016/j.radonc.2004.09.017.

Saint-Jacques, N., Brown, P., Nauta, L., Boxall, J., Parker, L., & Dummer, T. J. B. (2018). Estimating the risk of bladder and kidney cancer from exposure to low-levels of arsenic in drinking water, Nova Scotia, Canada. *Environment International*, *110*, 95–104. Available from https://doi.org/10.1016/j.envint.2017.10.014.

Salim, O. E., Hamid, H. K., Mekki, S. O., Suleiman, S. H., & Ibrahim, S. Z. (2010). Colorectal carcinoma associated with chistosomiasis: A possible causal relationship. *World Journal of Surgical Oncology*, *8*(1), 68. Available from https://doi.org/10.1186/1477-7819-8-68.

Schnekenburger, M., & Diederich, M. (2012). Epigenetics offer new horizons for colorectal cancer prevention. *Current Colorectal Cancer Reports*, *8* (1), 66–81. Available from https://doi.org/10.1007/s11888-011-0116-z.

Sephton, S. E., Dhabhar, F. S., Keuroghlian, A. S., Giese-Davis, J., McEwen, B. S., Ionan, A. C., & Spiegel, D. (2009). Depression, cortisol, and suppressed cell-mediated immunity in metastatic breast cancer. *Brain, Behavior, and Immunity*, *23*(8), 1148–1155. Available from https://doi.org/10.1016/j.bbi.2009.07.007.

Shi, Y., Wang, P., Guo, Y., Liang, X., Li, Y., & Ding, S. (2019). *Helicobacter pylori*-induced DNA damage is a potential driver for human gastric cancer AGS cells. *DNA and Cell Biology*, *38*(3), 272–280. Available from https://doi.org/10.1089/dna.2018.4487.

Solnick, J. V., & Schauer, D. B. (2001). Emergence of diverse *Helicobacter* species in the pathogenesis of gastric and enterohepatic diseases. *Clinical Microbiology Reviews*, *14*(1), 59–97. Available from https://doi.org/10.1128/CMR.14.1.59-97.2001.

Stewart, B. W., & Kleihues, P. (2003). *World cancer report*. Lyon, France: IARC Press.

Stewart, B. W., & Wild, C. P. (Eds.), (2014). *World cancer report 2014*. Lyon: International Agency for Research on Cancer.

Stone, K. M., Zaidi, A., Rosero-Bixby, L., Oberle, M. W., Reynolds, G., Larsen, S., … Guinan, M. E. (1995). Sexual behavior, sexually transmitted diseases, and risk of cervical cancer. *Epidemiology (Cambridge, Mass.)*, *6*(4), 409–414. Available from https://doi.org/10.1097/00001648-199507000-00014.

Talbot, F., & Nouwen, A. (2000). A review of the relationship between depression and diabetes in adults: Is there a link? *Diabetes Care*, *23*(10), 1556–1562. Available from https://doi.org/10.2337/diacare.23.10.1556.

Travis, L. B., Ng, A. K., Allan, J. M., Pui, C. H., Kennedy, A. R., Xu, X. G., … Boice, J. D. (2012). Second malignant neoplasms and cardiovascular disease following radiotherapy. *Journal of the National Cancer Institute*, *104*(5), 357–370. Available from https://doi.org/10.1093/jnci/djr533.

Turajlic, S., Sottoriva, A., Graham, T., & Swanton, C. (2019). Resolving genetic heterogeneity in cancer. *Nature Reviews. Genetics*, *20*(7), 404–416. Available from https://doi.org/10.1038/s41576-019-0114-6.

Uronis, J. M., Mühlbauer, M., Herfarth, H. H., Rubinas, T. C., Jones, G. S., & Jobin, C. (2009). Modulation of the intestinal microbiota alters colitis-associated colorectal cancer susceptibility. *PLoS One*, *4*(6), e6026. Available from https://doi.org/10.1371/journal.pone.0006026.

van Tong, H., Brindley, P. J., Meyer, C. G., & Velavan, T. P. (2016). Parasite infection, carcinogenesis and human malignancy. *EBioMedicine*, *15*, 12–23. Available from https://doi.org/10.1016/j.ebiom.2016.11.034.

Vogelstein, B., Papadopoulos, N., Velculescu, V. E., Zhou, S., Diaz, L. A., & Kinzler, K. W. (2013). Cancer genome landscapes. *Science (New York, N.Y.)*, *340*(6127), 1546–1558. Available from https://doi.org/10.1126/science.1235122.

Wang, X. W., Hussain, S. P., Huo, T. I., Wu, C. G., Forgues, M., Hofseth, L. J., … Harris, C. C. (2002). Molecular pathogenesis of human hepatocellular carcinoma. *Toxicology*, *181–182*, 43–47. Available from https://doi.org/10.1016/S0300-483X(02)00253-6.

World Cancer Research Fund. (1997). *Food, nutrition, and the prevention of cancer: A global perspective*. Washington, DC: American Institute for Cancer Research.

World Health Assembly. (2005). *Cancer prevention and control*. Geneva, Switzerland: WHO. https://www.who.int/WorldHealthAssembly. accessed August 16, 2020.

World Health Organization (WHO). (2002). Tobacco smoke and involuntary smoking. In *Monographs on the evaluation of carcinogenic risks to humans* (Vol. 83, p. 1187). International Agency for Research on Cancer, IARC.

World Health Organization (WHO). (2003). *INTERSUN: The global UV project, a guide and compendium*. Geneva, Switzerland: WHO.

World Health Organization (WHO). (2005a). *Preventing chronic diseases: A vital investment*. Geneva, Switzerland: WHO.

World Health Organization (WHO). *Essential medicines?*. (2006). <http://www.who.int/medicines/services/essmedicines_def/en/index.html> Accessed 05.08.20.

World Health Organization (WHO). *Asbestos: Elimination of asbestos-related diseases. Fact sheet no. 343*. (2010). <http://www.who.int/mediacenter/factsheets/fs343/en/index.html>.

World Health Organization. *Cancer*. (2018). https://www.who.int/WorldHealthOrganization. Cancer. Retrieved 19.12.18.

World Health Organization (WHO). *Tobacco*. (2019). <http://https://www.who.int/news-room/fact-sheets/detail/tobacco> Accessed 21.04.20.

World Health Organization (WHO). *Ionizing radiation health effects and protective measures*. (2020a). <http://www.who.int/news-room/fact-sheets/detail> Accessed 16.04.20.

World Health Organization. *Arsenic*. (2020b). <https://www.who.int/news-room/fact-sheets/detail/arsenic> Accessed 23.04.20.

Chapter 2

# Types of cancer diagnostics, the current achievements, and challenges

Niladri Mukherjee[1,*], Niloy Chatterjee[2,*], Krishnendu Manna[1,3,*] and Krishna Das Saha[1]
[1]*Cancer Biology and Inflammatory Disorder Division, CSIR-Indian Institute of Chemical Biology, Kolkata, India*
[2]*Food and Nutrition Division, University of Calcutta, Kolkata, India*
[3]*Department of Food and Nutrition, University of Kalyani, Kalyani, India*

## 2.1 Introduction

Cancer is accounted as the second most common cause of death in humans (WHO, n.d.) and an estimated 13% of all the deaths worldwide are cancer-related annually with an alarming 70% of these cancer-related demises from low and middle-income countries (WHO, n.d.). Cancer diagnostics and surveillance methods are comprised of different methods that include histology or biopsy, ultrasound, mammography, digital mammography, magnetic resonance imaging (MRI), computed tomography, positron emission tomography (PET), magnetic resonance spectroscopy (MRS), etc. More recent diagnostic techniques are immunohistochemistry (IHC), in situ hybridization [fluorescence in situ hybridization (FISH), CSH], polymerase chain reaction (PCR), RT-PCR (real-time-PCR), flow cytometry, and microarray.

In practice, a cancer diagnosis has become easier in the past decade with the inclusion of recent techniques as well as further development of relatively old diagnostic procedures. The major concern of any cancer diagnostic is its accuracy, with the least false-positive or false-negative results, as well as it should be timely with early presentation of any symptoms (Maxim et al., 2014). Another fact for better survival chances remained with the patients to pinpoint warning signs when there is any doubt of the disease and initially report to primary care (2005); although, at the earlier stages cancers may be far from apparent to the physicians to connect them with malignancy. The new diagnostic tests like microarray or RT-PCR can play a major role.

This present chapter provides an overall idea about the different diagnosis measures for the detection of cancer. This chapter comprises both the common as well as relatively new diagnostic measures with equal importance. We also present a separate segment for describing diagnostic arrangements taken for a few most commonly observed cancers. One should always keep in mind that when a diagnosis is accurate and timely, a patient can benefit from longer survival chances, be completely cured of the illness, and thus represent the best opportunity for a positive health outcome (Kruk et al., 2018; National Cancer Institute, n.d.).

## 2.2 What is cancer

It can be annoyingly complicated for one to define and intersperse "cancer" in a sentence or more. Cancer is a group of more than 100 different diseases arising from varying initial triggers and can develop almost anywhere in the body (National Cancer Institute, n.d.). First mentioned in Nathan Bailey's *Universal Etymological English Dictionary* in 1721, cancer however evaded its precise revealing by employing its origin, diagnosis, and its various stages, primarily challenging the researchers and medical practitioners to define and differentiate between "real" and false cancers (2018). Interestingly, different common names for the disease—"canker," "cancer," "chancre," or "Kanker" were drawn from the identical etymological Greek name *karkinos* representing "crab."

Briefly, the origin of cancer starts from an anomaly of the normal cellular process that implicates that cells, the most basic structural and functional of the human body, grow and divide in a well maintained and very precise

*Niladri Mukherjee, Krishnendu Manna, and Niloy Chatterjee contributed equally.

Biosensor Based Advanced Cancer Diagnostics. DOI: https://doi.org/10.1016/B978-0-12-823424-2.00022-3

calculative process to generate new cells. And as a rule, aged, damaged, and senescence cells die for new cells to take their place. The dying of the cells is normally called programmed cell deaths, having apoptosis is the most pronounced and discussed one (Elmore, 2007). Cancer originates with abnormal genetic alterations that can hinder this systematic cellular arrangement resulting in uncontrolled growth, division, and life-span of the altered and affected cells with all these abnormalities continuing to their daughter cells (Fouad & Aanei, 2017). As a result, the cells started to grow frenziedly; in most cases, a visual modification or enlargement results in a mass called a tumor. However, the tumor can be either benign (specified nonspreading neoplastic body) or cancerous, alternatively called malignant, when it can grow and extend to other nonspecified places of the body. In most instances, cancerous cells originate with unfavorable mutations in their DNA; however, some inherited predisposed genetic flaw (e.g., BRCA1 and BRCA2 mutations), environmental factors (pollution, etc.), infections, exposure to carcinogenic chemicals (e.g., heavy metals, dichlorodiphenyl-trichloroethane (DDT), poor lifestyle choices, and addiction (e.g., smoking, alcohol abuse) potentially mutate DNA and augment cancer possibility (Parsa, 2012). Though, in most of the cases as a protective strategy, mutated cells with inherent cellular machinery can detect the DNA damage and repair it. If not able to do so usually perform apoptosis to get rid of the potentially cancerous cell out from the system. If not able to do so, cancer will transpire by the growth, division, and spreading of it abnormally (Borges et al., 2008).

Generally, cancer is associated with neoplastic growth, commonly called a tumor, however, some types like leukemia, lymphoma (most cases), and myeloma are not associated with the tumor. Based on its origin, cancer usually is differentiated into four major types. First and most elaborate as well as most common is (1) *Carcinoma*: Carcinoma is the cancer of epithelial tissue that includes the skin as well as surfaces of internal organs and glands. Neoplastic bodies connected with carcinoma usually are solid tumors. Some of the most common cancers, namely, breast cancer, lung cancer, colorectal cancer, and prostate cancer are carcinoma. (2) *Sarcoma*: Sarcoma is the cancer of connective tissue that supports and connects the body. Sarcoma can be of elaborate and diversified spaces of origin that includes fat, muscles, nerves, tendons, joints, blood vessels, lymph vessels, cartilage, or bone. (3) *Leukemia*: Leukemia is commonly known as blood cancer and appears when blood cells become cancerous and grow uncontrollably. There are four major types of leukemia namely chronic lymphocytic leukemia, acute lymphocytic leukemia, chronic myeloid leukemia, and acute myeloid leukemia. (4) *Lymphomas*: Lymphoma is the cancer of the lymphatic system with the two major types being Hodgkin lymphoma and non-Hodgkin lymphoma.

The major concerns, worries, and difficulties associated with cancer are that the tumor or cancerous cells can invade other organs, a process known as metastasis (Van Zijl et al., 2011). Lymphatic system and bloodstream work are the path of metastasis (Van Zijl et al., 2011). The primary tumor is the place of origin of the malignant neoplast and after invading other body parts the metastatic tumors that grow are called "secondary cancers." For example, if breast cancer spreads to the lungs, it is called metastatic breast cancer, not lung cancer. Metastasis is very important and a deciding factor for the detection and treatment and, with the vast improvement of diagnosis in modern medical science nowadays, cancer can be identified a lot earlier; arguably the single most important factor for the survival and cure of the patients (Anand et al., 2008). Some types of metastatic cancer are curable, but many are not. That opens the scope for the discussion in this chapter for the various types of cancer diagnostics and the associated challenges and achievements.

## 2.3 What is diagnostics

*Diagnosis* can be defined as the means of pronouncement of a specific disease to the patient, and often regarded as the preliminary course of the clinical arrangement after the medical practitioner suspects a disease. Affirmation that is followed by appropriate treatment for the diagnosed disease and, if not confirmed for a suspected one, further diagnostic measures will be pursued. Even for some diseases like cancer, not only confirmation is essential but also the degree of development of the malignancy is important to determine a course of treatment. Even thereafter constant and periodic diagnostic evaluation is necessary to be aware of the stage of the disease, its signs of progress, stability, or regression (Ryu, 2015). Thus, at the same time, diagnostic tests are essential for the assessment of the efficacy of chosen treatment and, therefore, widens the scope and importance of diagnosis in cancer. It is also important to note in this pretext the closely associated, though different, medical observatory terms. *Monitoring* can be termed as constant observation of the disease status to notice if it is in a controlled situation. Monitoring is very common in chronic diseases like diabetes. Monitoring is essential for such chronic diseases as the patient's condition can be controlled through the use of proper medications, hormones, or lifestyle changes. *Screening* for any particular disease comprises a blind evaluation within a community or for an individual that may or may not present any clinical signs or symptoms. Screening is implemented for the application of treatment as soon as possible. It is also important concerning cancer, as persons genetically

**TABLE 2.1 Role and related phases of cancer diagnostics. (Bolboacă, 2019)**

| Terms | Specific example | Explainable criteria |
|---|---|---|
| Confirmation or exclusion | Brain natriuretic peptide can diagnose left ventricular dysfunction | Confirm (rule-in) or exclude (rule-out) the disease |
| Triage | Renal Doppler resistive index can diagnose hemorrhagic shock in polytrauma patients | An initial test for rapid application and should represent a small number of false-positive results |
| Monitoring | Glycohemoglobin (A1c Hb) for the diagnose of overall glycemic control of patients | Repeated test for the appraisal of a therapeutic application |
| Prognosis | PET/CT scan to diagnose distant metastasis in cervical and endometrial cancer | Estimation of the disease progression |
| Screening | Cytology test for the screening of cervical uterine cancer | To confirm the disease in asymptomatic persons |

Adapted from Bolboacă SD. Medical Diagnostic Tests: A Review of Test Anatomy, Phases, and Statistical Treatment of Data. Comput Math Methods Med. 2019;2019:1891569.

predisposed to any scrupulous cancer can avoid potential fatality if considered under treatment in the earliest stage before any complications may arise (Anand et al., 2008). Screening tests are mandatorally simple and cheap for it to be applied to masses and in large scales; it's additionally preferable if applied rapidly on field (Maxim et al., 2014). Lastly, *Prognosis* is the measure for the assessment of the likelihood of a disease to be developed in the future within a person and a way to get rid of the potential disease by applying precautions at the earliest. Genetic studies are the key in prognostic tests and can unfold a person's probability and possibilities for developing a disease, more importantly for different types of cancers, and thus he or she can take preventive measures (Jackson et al., 2018). Different phases of cancer diagnostics are presented in Table 2.1.

Though the scope of diagnosis is vast and the importance is immense about cancer, it is essential to be acquainted with the fact that diagnostics are not meant to treat patients or cure illnesses. The major impact of these results is associated with healthcare decisions and, therefore, significant and vital. However, diagnostics are not self exclamatory and are closely dependent upon a medical professional's precise and correct judgment and apposite choice of treatment.

## 2.4 Importance of diagnostics

Correct and judicious diagnosis is crucial in the management of any disease and the possibility of misdiagnosis, no-diagnosis, and delayed diagnosis can be just as crucial; this may be more critical of some malignancies where time can be the ultimate factor for survival of the patient (Mayo Clinic, n.d.). Diagnostics, however, is an evolving process since both the understanding of disease mechanisms and novelties of diagnostic measures are continuously upgrading (Croft et al., 2015). A diagnostic test could be used in clinical settings for confirmation/exclusion, triage, monitoring, prognosis, or screening (Cohen et al., 2016). Some important features or acceptability of any diagnostic test rely on its accuracy (the result should be the same if conducted multiple times), reproducibility (the result should be the same irrespective of persons conducting it), and feasibility (diagnostic method should be reachable and inexpensive) and also should be easy to comprehend for taking clinical decisions. The timing of cancer symptom presentation to diagnosis is presented in Fig. 2.1.

## 2.5 Different types of cancer diagnostics

Cancer or malignant neoplasts can sometimes be very conspicuous based on the fact that the majority of them originate and grow on or near the facade of the body. Visual diagnosis can most of the time hold significance in medical textbook descriptions of this disease, which can sometimes be unworthy if not appraised by a person with ample knowledge or experience, leading to false-positive or false-negative avowal. This also led to open the scope of modern diagnostics to authenticate malignant cases. If we consider past practices from the 1580s to the first years of the 18th century, the color of malignant tumors was of major apprehension over malignancy with tumors having unspecified bruised shade to blackish or bluish shade. Additionally, importance was also given towards the rough, un-uniform, or circular nature of the tumors with an uneven surface appearance in confirming the tumors as cancerous. The German physician Christof

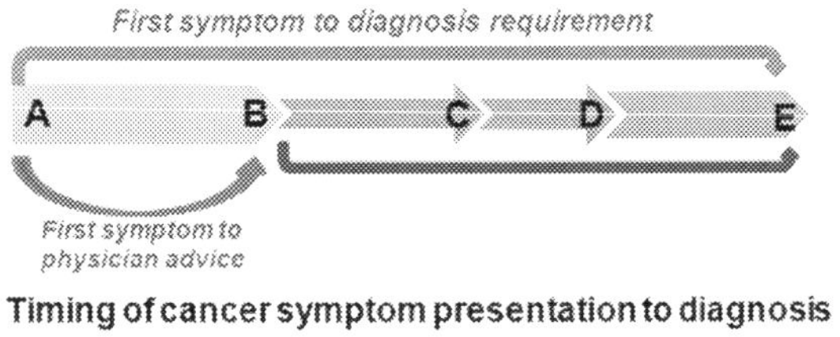


**FIGURE 2.1 Timing of cancer symptom presentation to diagnosis.** (A) First symptom presentation; (B) First requirement for medical attention; (C) Recognition of cancer as a possibility for the first time; (D) Time for definitive diagnostics announcement; (E) Start of treatment.

Wirsung described in his text that "the Canker causeth . . . great pain and beating, whereof *Schirrhus* is free" (Skuse, 2015). Furthermore, the pains associated with cancer could be aligned with the crab. Even as late as 1597, physician Peter Lowe asserted that not only did cancers look like crabs, they "gnaweth, eateth, and goeth like this fish" (Skuse, 2015). The major problem thus raised was that these were characteristics that could also be associated with more benign tumors and resulted in false-positive conclusions. For example, in the 1698 edition of *The Compleat Midwife's Practice* it was written that breast cancer can be confirmed by the association of "the crooked windings" and "retorted veins" around the neoplasts (Skuse, 2015).

Diagnosis of cancer malignancy primarily depends upon the choices, practices, and reliability of tests and varied greatly on the choices of the practitioners and surely on the varied complications arising due to the neoplast. However, more modern and up-to-date diagnostics are being constantly advised and practiced, though some of the relatively old techniques like biopsy and imaging are still considered as reliable and most practiced with constant upgradations and modifications (Ulrich et al., 2020). The diagnosis of cancer principally dependent on the analysis of tissue and cell specimens acquired by different techniques like a surgical biopsy, core or aspirational needle biopsy, venipuncture, pleural or ascitic sap collection, scraping of tissue surfaces, and collection of exfoliative cells from urine and sputum (Al-Abbadi, 2011). The conventional histopathology based on light microscopy remained the typical diagnostic method for years, however, recently been complemented with ultrastructure, IHC, and molecular diagnostics (Makki, 2016). Simultaneously, the advancement of imaging techniques also is helping the physicians to furthermore characterize the neoplasts with higher resolution images and with anatomic and spatial 2D and 3D images to a focus on molecular, functional, biologic, and genetic imaging (Bayer et al., 2013). Additionally, the development of advanced and refined technologies like mass spectrometry, microarray, and automated DNA sequencing has also nowadays been regularly opted, advised, and relied upon by clinicians in cancer diagnosis (Gray et al., 2015). The use of enzyme histochemistry and electron microscopy manifold increased the possibilities of the primary microanatomic evaluation to include biochemical and subcellular ultra-structural features. Some of them are discussed in the following paragraph.

### 2.5.1 Clinical symptoms

Doubt over the origin or presence of malignant neoplasts or cancerous growth begins with varied clinical symptoms. Symptoms of cancer can be diverse according to the character of cancer and its location. Tumors that occupied internal or external places of organs like the stomach, small intestine, large intestine, or colon generally can lead to gastrointestinal obstruction accompanied by occasional, continuous, or profuse bleeding in the stool and/or vomiting (Yang & Pan, 2014). Tumors of the kidney or bladder resulted in hematuria, and tumors in endocrine glands can cause Cushing's disease, hypoglycemia, etc.; whereas, tumors of the brain or spinal cord can often result in neurologic symptoms such as loss of coordination or seizures (Pelosof & Gerber, 2010). However, the more challenging scenario for the diagnosis occurs when cancer presents nonspecific symptoms and is raised in abnormal locations related to its primary nature and referred to as paraneoplastic disorders (Pelosof & Gerber, 2010). In these cases, varied symptoms are observed like weight loss, low-grade fever, seizures, lethargy and loss of appetite, diarrhea, skin rash, hair loss, and general arthritic-like symptoms. In these cases, specialized diagnostic techniques like X-rays, CT scans, MRIs, etc. are needed and advised (Fass, 2008).

### 2.5.2 Physical examination

Physical examination is the foremost diagnostic measure applied by the physician to predict any kind of neoplast. Physicians generally access any abnormalities like changes in skin color or enlargement of an organ that may be due to malignancy.

### 2.5.3 Laboratory tests

Laboratory tests can help physicians to identify some types of cancers. Alterations in the levels, parameters, or markers of certain substances can be a sign of cancer. However, abnormal lab results are not conclusive of cancer. Leukemia can be primarily accessed by complete blood count to notice an unusual number or type of white blood cells. There are different lab tests currently available with blood or tissue samples for tumor markers which are substances produced by cancer cells or other cells in response to cancer.

### 2.5.4 Ultrasound

Ultrasound is extensively used for cancer diagnosis of abdominal organs, heart, breast, muscles, tendons, and even arteries and veins. Additionally, ultrasound can also characterize cancerous lesions based on parameters like shape, size, and density and can be extremely beneficial during screening mammograms in women (Thigpen et al., 2018). Though the detailing neoplasts obtain is inferior compared to CT or MRI, there are a few advantages over them that cannot be overlooked and may even be a preferred diagnostic measure. Especially in studies of moving structures in real-time; additionally, ultrasound does not emit ionizing radiation, contains speckle which may be used in elastography, and is also relatively cheap and less time consuming (Chen et al., 2020; Wu & Shu, 2018). Diagnosis of neoplasts of hollow organs like the urinary bladder is preferably done by ultrasound.

### 2.5.5 Imaging tests

#### *2.5.5.1 Normal X-rays*

Normal X-ray is used to create pictures inside of the body. Abnormalities or neoplasts can be visible and after close examination, physicians can conclude of it's of a malignant nature.

#### *2.5.5.2 Ultrasound examination*

Ultrasound uses high-frequency broadband sound waves in the megahertz range (Miller et al., 2012). The sound waves echo off tissues inside. A computer program utilizes these echoes to create pictures of inside the body called a sonogram (3D) (Miller et al., 2012).

#### *2.5.5.3 Positron emission tomography*

PET scan is also a type of nuclear scan that makes detailed 3D pictures of areas inside the body where glucose is taken up (Zhu et al., 2011). As cancer cells have high energy demands, they often accumulate higher glucose than healthy cells. Before the scan, the patient is provided with a radioactive material that acts as a tracer. PET is useful in diagnosing cancerous growth with the aid of radio-labeled tracers in vivo (Chen & Chen, 2011). PET utilizes, and is directly linked with, the metabolic activity of underlying tissue based upon measuring glucose, oxygen, and amino acid metabolism or receptor density (Zhu et al., 2011) Fluorine-18 fluorodeoxyglucose (18FFDG) is the most used molecule which enters cells and gets phosphorylated to FDG-6-phosphate and trapped within malignant tumor cells having high energy demand and thus augmented glucose metabolism (Chen & Chen, 2011). PET is considered the most accurate noninvasive diagnosing technique for the detection of lung cancer (Dabbagh Kakhki, 2007) and also preferred over CT during diagnosing of intrahepatic metastases in colorectal cancers and metastatic deposits in lymph nodes very small in size ($<1$ cm) which can be diagnosed false negative by CT (O'Connor et al., 2011). However, during observations of large masses, it can sometimes wrongly diagnose as benign if FDG uptake is low (Takalkar et al., 2008) and additionally not advised for diagnosing malignancy for normal inflammatory tissue that can have a higher and irregular FDG uptake (Chen & Chen, 2011; Purohit et al., 2014).

#### *2.5.5.4 Computed tomography*

The CT scan is connected with an X-ray machine linked with a computer to take a series of pictures of organs from different angles and create detailed 3D images (Bercovich & Javitt, 2018). At times, the patient is given a dye or other contrast material orally or injected into a vein. Images from CT scans are then used to diagnose cancer by learned professionals or physicians. Recent innovations include spiral (helical) CT, multiphase imaging, and multidetector scanning (Pelc, 2014). The major advantages of CT as a diagnostic measure are fast data acquisition with superior detection and characterization of neoplasts. Spiral CT is extensively used for the detection of cancers of pulmonary organs and liver and also before pancreatic or renal cancer treatment with improved modern detection of pulmonary emboli with the aid of CT angiography and endoscopic viewing of hollow organs (Pelc, 2014).

#### 2.5.5.5 Magnetic resonance imaging

MRI uses a powerful magnet and radiowaves for taking pictures of the body or organs in slices which then are used to create detailed images (Berger, 2002). Sometimes a special dye called a contrast agent is injected before or during the MRI, making tumors brighter in the pictures (Berger, 2002). The benefits of MRI primarily consist of excellent soft-tissue contrast, multiplanar and 3D image acquisition, free from ionizing radiation, and bony noises. In MRI, powerful magnets are utilized for polarizing and excite hydrogen nuclei present in water molecules inside tissue and can produce a measurable signal, resulting in images of the body. Presently the use of supercoils enabled augmented sensitivity and specificity in the image output (Chambers et al., 2014). A contrast agent is required during image acquisition in MRIs and chelates of gadolinium with three unpaired electrons are normally used for generating an enhanced magnetic field. MRI is preferably used for diagnosing neoplastic growth in the brain, head and neck, spine, liver, adrenal glands, breast (due to dense breast, silicone implants and additionally when aided with mental hindrance for scarring and distress) (Berger, 2002; Wang et al., 2018). MRI can acquire images very promptly and also can visualize superimposed anatomical changes due to the cancerous growth (Chambers et al., 2014). Even for breast cancers, MRI can be very useful for its ability to diagnose early breast cancer given higher sensitivity (95%–100%) and with very low false-negative results (5%–10%) (Wernli et al., 2019). Dynamic contrast-enhanced MRI (DCE-MRI) is an enhanced technique for the assessment of parameters that are flowing inside the body like blood flow, perfusion, and vascular permeability, and additionally can evaluate interstitial space within a tumor (Gordon et al., 2014).

#### 2.5.5.6 Magnetic resonance spectroscopy

MRS is a noninvasive method for studying tumor biochemistry and physiology. P31 MRS is proportional to tissue energetics and pH, and H1-MRS provides information on cell membrane synthesis and degradation (Yang & Hinner, 2015). MRS is diagnostically very important to verify tumor grade and very useful during therapy as diagnostic sensitivity and specificity can reach as high as 92% in distinguishing malignant from benign tumors (Zhu et al., 2011), even in a subgroup of 20 younger women, the sensitivity and specificity approached 100% (Yankaskas et al., 2010).

#### 2.5.5.7 Nuclear scan or radionuclide scan

A nuclear or radionuclide scan uses radioactive material for capturing images inside of the body. Before the nuclear scan, the patient is injected with a small amount of radioactive material, or tracer which flows through the bloodstream and deposits in certain bones or organs for marking known as "hot spots" [National Research Council (United States) and Institute of Medicine (United States) Committee on State of the Science of Nuclear Medicine), 2007]. The radioactive materials lose their radioactivity over time as it passes through urine or stool. Bone scans are a type of nuclear scan that checks for abnormal areas or damage in the bones for diagnosis of metastatic bone tumors (O'Sullivan et al., 2015).

### 2.5.6 Cytologic and histopathological technique (biopsy)

In many instances, a biopsy can be the only way to definitively diagnose cancer. A biopsy is a procedure where practitioners remove a sample of tissue and prepare it for visual observation using types of dyes and stains. Histopathology remained as the most trusted diagnostic measure for tumors and for differentiating them as benign or malignant (Thway et al., 2014) however, it alone insufficient for providing details of the cellular changes associated with the clinical nature of the tumor (Gurcan et al., 2009). Additionally, serous effusions from the pleural, peritoneal, or pelvic cavity can be used as biopsy material for the diagnosis of malignant tumors (Dixit et al., 2017) Major diagnosing features of malignancy during histopathology are the presence of hyperchromatic nuclei, higher nucleus to cytoplasm ratio, the bewilderment of the cells, etc. (Malhotra et al., 2013). Higher numbers of mitotic cells are also observed in the malignant tumors compared to benign tumors (Malhotra et al., 2013). Additionally, the use of special stains can diagnose types of tumors, e.g., toluidine blue stains the metachromatic granules of mast cells, and differentiate them from other tumors (Sridharan & Shankar, 2012). A pathologist then examines the tissue under a microscope and runs other tests to confirm the tissue as cancerous. Normal cells look uniform, with similar sizes and orderly organization whereas, cancer cells look less orderly, with varying sizes and without apparent organization. The biopsy sample may be obtained in several ways. The doctors can surgically remove some tissue from the tumor or can use a needle to withdraw tissue or fluid. This method is used for bone marrow aspirations, spinal taps, and some breast, prostate, and liver biopsies. In an excisional biopsy, the surgeon removes the entire area of abnormal cells with some of the normal surrounding tissue, and during incisional biopsy the surgeon removes just part of the abnormal area.

### 2.5.7 Endoscopy

A doctor uses a thin, lighted tube called an endoscope to examine areas inside the body. Endoscopes go into natural body openings and after observing abnormal tissue during the exam, can remove part of the abnormal tissue along with some of the surrounding normal tissue through the endoscope (McGill et al., 2009). There are a couple variations of endoscopy namely colonoscopy (examination of the colon and rectum), bronchoscopy (examination of the trachea, bronchi, and lungs).

### 2.5.8 Tumor markers

Gold et al. (1978) was first successful in isolating a glycoprotein from human colonic cancer specimens and is the first "tumor antigen," later named carcinoembryonic antigen (CEA) (Gold et al., 1978). Now there are hundreds of tumor markers available, however, their usefulness in can cancer diagnosis is a continuous process and is being utilized more in recent days (Sharma, 2009). Therefore, tumor markers can be differentiated as biological or biochemical substances produced by tumors and can be detected from body fluids, body tissues, and even in urine, sweat, or excreta in higher or sometimes even in lower amounts than normal (Sharma, 2009). Detection or diagnostics of these tumor markers are varied and some of the most trusted methods are antigen-antibody based techniques like enzyme-linked immunosorbent assay (ELISA), RIA (radio-immunoassay), precipitin tests, flow cytometry, IHC, immunoscintigraphy, etc. Though potential is enormous, in most cases tumor marker levels alone are still not regarded a sufficient to diagnose cancer and, therefore, the scope of further development of this diagnostic technique is very high (National Cancer Institute, n.d., 2016). Besides diagnosis, tumor marker levels can also reflect the stage of the disease and possible prognosis (Sharma, 2009). For example, SC6-Ag is a valuable tumor marker for the diagnosis of pancreatic cancer before and after surgery (Liu et al., 2005). A few more examples are adhesion molecule L1 for esophageal adenocarcinoma (Rawnaq et al., 2009), Y-Box-binding Protein-1 for neuroblastoma (Wachowiak et al., 2010), etc.

### 2.5.9 Serological methods

Serum tumor markers are utilized for serological diagnostic methods through immunological procedures of ELISA and RIA (Sharma, 2009).ELISA is used for the detection and quantification of cancer markers used as an antigen present in biological fluids (Yin et al., 2010). Among different ELISA methods, DualAntibody Sandwich ELISA is mostly used with as high as 80% among all ELISA methods (Sakamoto et al., 2018). Whereas, RIA is used to measure antigen or antibody by competitive binding of radio-labeled and unlabeled antigen to a high-affinity antibody using gamma-emitting isotopes of iodine and beta-emitting isotopes of tritium mostly (Shan et al., 2000). The presence of CEA, Alpha Fetoprotein (AFP), prostate-specific antigen (PSA), and other markers in the serum of the cancer patients can also be diagnosed through ELFA (enzyme-linked fluorescent assay) (Vaidyanathan & Vasudevan, 2012).

### 2.5.10 Immunohistochemistry

IHC is an enhanced method comprising both histology as well as immunologic diagnostic methods and based on detection of specific antigenic determinants present in the cells of the isolated tissues from the neoplasts with the aid of polyclonal or monoclonal antibodies (Karunakaran et al., 2012). IHC can be very useful in diagnosing metastatic tumors of an unknown primary site and also in the diagnosis of undifferentiated tumors where light microscopy is of little use (Karunakaran et al., 2012). Various important features of the different malignant neoplasts namely, poorly differentiated carcinoma, anaplastic large cell lymphoma, amelanotic melanoma, and less common sarcoma can be decided through IHC (Hudacko et al., 2011). A few diagnostic examples of IHC are the detection of cytokeratins for an epithelial origin (Selves et al., 2018); leukocyte common antigen in neoplasts of lymphoid origin (Kurtin & Pinkus, 1985); and S100, HMB 45 protein expression is characteristic of malignant melanoma (Viray et al., 2013); estrogen, progesterone, and Her-2 (c-erbB2) receptor status for breast cancer for determining disease prognosis during treatment (Mouttet et al., 2016). Detection of proteins involved in the regulation of cell cycle like cyclin D1 and E can also be of diagnostic importance in breast cancer and squamous cell carcinoma of the head and neck (Casimiro et al., 2012). Additionally, expression evaluations of oncoproteins like p53, c-myc, c-met, LKB1 are diagnostically found in human lung cancer (Liang et al., 2009), bladder cancer (Sauter et al., 1995), and head and neck cancers (Waitzberg et al., 2004). A few more markers used for diagnostic purposes in IHC are neural markers like neuron-specific enolase and synaptophysin for neuroectodermal tumors (Mjønes et al., 2017), desmin, and myoglobin, which are indicative of rhabdomyosarcoma (Parham et al., 1991).

## 2.5.11 Flow cytometry

Diagnostic purposes of flow cytometry mainly classified for hematologic malignancies have been variedly evolved during the past few decades (Cools & Vandenberghe, 2009). Availability of superior antibodies and improved gating tactics enhanced the diagnostic application of flow cytometry. Typically, light scatter is considered as a better technology to fix cells for the permeabilization of intracellular antigens which in turn can be detected by flow cytometry (Ciáurriz et al., 2017). An example of the diagnostic strategy with flow cytometry is that TdT can only be expressed in T cells residing in the thymus and limitedly in a few bone marrow cells, whereas, the majority of lymphoblastic lymphoma expresses TdT; thus, if TdT expression is found in the peripheral blood or cerebrospinal fluid it signifies malignancy (Hooijkaas et al., 1989). Another example is that majority of B-lineage cells express a combination of TdT, CD19, and CD10, therefore, the combination of these markers with the addition of certain aberrant markers such as CD13, CD33, or CD15 is the identification of ALL cells from normal bone marrow or peripheral blood cells (Bento et al., 2017). Additionally, CLL can also be identified in the peripheral blood and bone marrow if coexpression of CD5 with either CD20,CD19, or CD5 with kappa and lambda light chain ids detected (Strati & Shanafelt, 2015). Different markers of some cancers are presented in Table 2.2.

**TABLE 2.2** Some common cancers and associated tumor markers. (Kumar & Pawaiya, 2010)

| Name of some common cancer types | Name of the tumor markers | Type of tumor markers |
|---|---|---|
| Cancer of colon, breast, lung, pancreas, stomach, and ovary | Carcinoembryonic antigen | Tumor antigen |
| Hepatocellular carcinoma, testicular germ cell tumors | Alpha-fetoprotein | |
| Prostate cancer | Prostate-specific antigen | |
| Non mucinous ovarian carcinomas | CA125 | |
| Gastrointestinal adenocarcinoma, pancreatic tumors | CA19−9 | |
| Melanoma | Granules of melanin | Cytoplasmic proteins |
| Sarcoma | Actin | |
| Epithelial tumors | Cytokeratin | |
| An endothelial cell in the vascular tumor | Factor III | |
| Astrocytoma | Glial fibrillary protein | |
| Medullary carcinoma of the thyroid | Human chorionic gonadotropin | Hormones |
| Medullary carcinoma of the thyroid | Calcitonin | |
| Pheochromocytoma | Catecholamines and metabolites | |
| Islet cell tumor | Insulin production | |
| Islet cell tumor | Insulin production | Enzymes |
| Prostate cancer | Prostatic acid phosphatases | |
| Small cell cancer of lung, neuroblastoma | Neuron-specific enolase | |
| Colon cancer, Pancreatic cancer | Galactosyl transferase II | |

Adapted from Kumar and Pawaiya; Advances in Cancer Diagnostics. Braz J Vet Pathol; 2010, 3(2),142−153.

### 2.5.12 Fluorescence in situ hybridization technique

FISH is the technique for specific hybridization of a labeled nucleic acid probe to complementary gene sequence and subsequent visualization by the autoradiographic or immunocytochemical method in tissue section, smears, or cytocentrifuged cell suspensions (Shakoori, 2017). As chromosomal abnormalities can frequently be found in malignant cells, FISH can be utilized for diagnostic purposes. This technique can detect chromosomal duplications, deletions, segmental amplifications, translocations, and inversions in malignant cells (Cui et al., 2016). Additionally, comparative genomic hybridization (CGH) a newly evolved method used for studies of chromosomal gains and losses in genomic complement (Weiss et al., 1999). Several investigators have found this method to be useful in cancer studies with the affirmation that different tumor types or stages have distinct CGH patterns (Grade et al., 2015).

### 2.5.13 Polymerase chain reaction

PCR can be very helpful in molecular level detection of malignancy by diagnosing complex profiles ("finger-prints") or unique molecular alteration specific for tumor types (Samuelsson et al., 2010). PCR detection can be done on wide and varied bio-molecules like chromosomes, DNA, or RNA. In carcinogenesis, microsatellites, or simple sequence repeats or SSRs are associated with loss of heterozygosity (LOH) as tumor suppressor genes and can be detected by PCR (Zheng et al., 2005). PCR can be useful for the early diagnosis of malignancy of leukemia and lymphomas (Drexler et al., 1995). Quantitative PCR also allows the estimation of the amount of alteration or modification of gene expression. Real-time quantitative PCR can a diversified and accurate diagnostic technique for the detection of oncogenes and tumor suppressor genes in different malignancies (Ståhlberg et al., 2005).

### 2.5.14 Microarray

DNA microarray technology is a very promising diagnostic approach for both qualitative and quantitative screening for sequence variations in the genomic DNA of cancer cells (Miller & Tang, 2009). DNA microarray-based sequence analysis diagnoses mutational detection to polymorphism genotyping (Miller & Tang, 2009). Labeled DNA for analysis binds strongly only to those targets that are fully complementary to one of its subsequences thereafter, the specific binding profile additionally checked for the whole original sequence (Miller & Tang, 2009). Microarray is useful for diagnosing mutational analysis for the understanding of the disease process and clinical practice and high-density oligonucleotide arrays are implied (Miller & Tang, 2009) and were very much successful for detecting alterations of breast cancer-associated genes *BRCA1* and *BRCA2* (Wiggins et al., 2020). Microarray is also very much useful in detecting polymorphism genotyping for the prediction of how to sequence polymorphism may impact biological functions and be associated with heritable phenotypes (Mao et al., 2007). Single nucleotide polymorphisms (SNPs) microarray is an oligo array in which SNPs are screened by a set of oligonucleotide probes (LaFramboise, 2009) and also can detect LOH (Zheng et al., 2005). Microarray-based expression comparisons also diagnose a panel of up or downregulated genes considered molecular markers for cancer (Narrandes & Xu, 2018), because the expression of genes differ variedly in different types of tumors and may explain some traits (Tarca et al., 2006). cDNA microarray screening was successful in the identification of 176 genes that share a distinct expression pattern between mutated *BRCA1* and *BRCA2* hereditary breast cancers (Inês & Daniele, 2017). There are very new developments of diagnostics based on analyses of circulating DNA (ctDNA) and circulating tumor cells (CTCs) (Neumann et al., 2018) and released new possibilities for cancer detection and prognosis (Calabuig-Fariñas et al., 2016). CTCs and ctDNA from patients can be analyzed for the detection of different tumor markers and can also detect mutations, microsatellite instability, hypermethylation, and varied gene expression (Han et al., 2017). This method can also be utilized to detect malignancy from other body fluids like saliva, urine, bronchoalveolar lavage, sputum, and ductal lavage (Alena et al., 2020).

### 2.5.15 Alternative and new diagnostic measures for cancers

Alternative biopsies recently attained interest in diagnostic oncology for their potential to isolate and analyze biomarkers during the personalization of cancer therapy. A key advantage of alternative biopsies is that they don't require invasive tissue collection methods that are often associated with mental and physical trauma (Kettritz, 2011). A recent example of this is the development of an at-home urine test for prostate cancer (Bax et al., 2018). Although urine collected by patients in their own homes is not assisted by digital rectal examination (DRE) to boost the levels of prostatic secretions, it is comparable in quality to the traditional method, however, without the time, inconvenience, cost, discomfort, and expense of visiting the clinic (Bax et al., 2018). A noninvasive and pain-free breathalyzer was also developed to collect exhaled breath samples with potential biomarker volatile organic compounds for the diagnose of certain cancers at the earlier stages (Xiaohua et al., 2016). For instance, diagnosis trials also aim to assess whether the breathalyzer can differentiate between patients with and without different cancer types by comparing the breath biomarkers of patients with gastric, esophageal, pancreatic, renal, prostate, and bladder cancer from matched controls (Bax et al., 2018). Liquid biopsies were also developed over the past years, and it is of immense importance for blood tests in respect to cancer diagnostics. A pilot study was conducted at the University of Nottingham (UK) for the development of a blood test aimed at the early detection of breast cancer (Sullivan et al., 2017) through detection of tumor-associated antigens and, can be indicative of breast cancer long before 5 years of the appearance of clinical symptoms (Sharma et al., 2010).

Ensuring accuracy in diagnosis cancer is not to differentiate it by cancer type; rather, focus can be concentrated on tumor mutation and genetic factors. Utilizing the latest minimally invasive biopsy procedures diagnosis, one is able to extract genetic material from patients and attain insight into their individual cancer history. Recently, a multicenter research group developed a new method: single-cell tri-channel processing (strip) (Sanders et al., 2020) that allows researchers to study genetic variation within the DNA of a single cell as well as identify new ones. Interestingly, by applying this method in patient leukemia cells, they found four times more variants than could be detected by normal diagnostics.

Artificial intelligence (AI) has an enormous potential in medicine and mostly done during diagnostics. Presently, AI is becoming more precise in identifying cancers and can complement current conventional diagnostics (Hosny et al., 2018). AI can aid the current practice of radiography during cancer diagnostics. Radiography, despite being highly effective, can sometimes be prone to human error. AI would emerge as a powerful tool for image analyses. A recent example of this is the development of AI models for breast cancer detection using X-ray images (Sadoughi et al., 2018). One such study, from Karolinska Institute (Stockholm, Sweden) and Tampere University (Finland), used over 8,000 digitized biopsies to train an AI system to discern which biopsies exhibited cancer and provided a comparable level of accuracy with expert neuropathologists (Robertson et al., 2018). A similar study from the RIKEN Center for Advanced Intelligence Projects (Tokyo, Japan) demonstrated the efficacy of unsupervised learning by the AI technology using deep neural networks to identify predictive features in pathology images from cancer patients (Yamamoto et al., 2019).

### 2.5.16 Nanoparticles in cancer diagnosis

Although nanotechnology has still not directly applied clinically for cancer diagnosis, application of it is available in a variety of medical tests and screenings. One example is gold nanoparticles in home pregnancy tests (Rojanathanes et al., 2008). For diagnostic purposes for malignancy, nanoparticles can be useful to capture cancer biomarkers like cancer-associated proteins, circulating tumor DNA, CTCs, and exosomes (Zhang et al., 2019). Nanoparticle surfaces can also be densely covered with antibodies, small molecules, peptides, aptamers, and other moieties and can be used as a diagnostic tool to recognize specific cancer molecules. With the aid of nanoparticles, one also can join binding ligands to cancer cells therefore multivalent effects can be achieved to improve the specificity and sensitivity of a diagnostic assay.

## 2.6 Factors that can amend cancer diagnostics

There are a few troubleshoots during cancer diagnosis; first of them is sample preparation which is essential for every diagnostic measurement and a key pre-analytical step. One should always be careful about applying definite methods for the collection of the appropriate type of specimen, abiding standard procedures, and be very careful to preserve specimen integrity. For diagnosing cancer by applying different molecular assays, the obtained sample should be aseptic and the DNA, RNA, and proteins of the sample should be stable, having unaltered integrity during collection and transport. There are various factors during the transport and storage and, additionally, time for evaluation can affect specimen quality and can even produce false negative or false-positive results (Maxim et al., 2014). Analytical assay development followed by clinical validation of cancer biomarkers is extensively dependent on test accuracy, reproducibility, and interpretation, therefore, should be critically validated by a specific expert, and should be estimated before diagnostic deliberation. Therefore, data interpretation is an integral part of a diagnostic assay (Bruno, 2011). It should always be considered and taken care of, dependent on the accurate information regarding the analytical and clinical performance of a diagnostic test, so the doctor more likely can take the appropriate clinical decisions.

## 2.7 Diagnostics for some typical and mostly observed cancer types

### 2.7.1 Breast cancer

Breast cancer is a very significant and sensitive cancer, however, primary care evidence is surprisingly very low in referrals of women with breast symptoms; although, there are several ways for the screening on genetics for a possible breast cancer appearance. As a result, most breast cancers are not recognized due to screening; on the contrary, three fourths of all breast cancer cases are diagnosed after presentation with symptoms (Sharma et al., 2010) and sometimes way after the first appearance, which is critical for treatment as a result of most of the deaths due to breast cancer is because of the late diagnosis when the cancer metastases. This scenario may be the total opposite if they were diagnosed earlier and lead to full remission. It is estimated that overall, 8% of women with a breast lump reported to primary care are then diagnosed with cancer (Haas et al., 2005). This is very much associated with the age of patient; it is 15 times higher in women aged 45–64 years than in those aged below 25 years (Berg et al., 2010). Breast cancer is of heterogeneous types and presents a wide variety of clinical presentations, histological types, and growth rates. Therefore, initial diagnosis requires careful assessment of various clinical and pathological parameters; traditional diagnostic features may not always be sufficient to predict the exact type of breast cancer (Wang, 2017). In primary breast cancer, metastasis to axillary lymph nodes are most commonly observed, and about 60%–70% of lymph-node negative patients can be cured by local–regional treatment alone (Al-Mahmood et al., 2018). However, a relapse can eventually be fatal and, therefore, patients having a high risk need special care with adjuvant systemic therapy (Rueda et al., 2019). Genomic Health (CA, USA) has commercialized the Oncotype Dx assay, a set of 16 signature genes in their RNA expression and five control genes as an early diagnostic measurement (Vieira & Schmitt, 2018). Agendia (Amsterdam, The Netherlands) also has commercialized Mammaprint assay, a 70-gene signature which is valid for women under the age of 55 (Slodkowska & Ross, 2009).

### 2.7.2 Lung cancer

Lung cancer diagnosis and the early prognosis is a generally poor and delayed presentation of symptoms restricts early diagnosis and curative treatment is often impossible. Additionally, lung cancer symptoms are nonspecific making it difficult to specifically identify the symptom is due to malignancy. These factors, combined, statistically are only 20% of UK lung cancer patients eligible for surgery, 17% was performed, and only half of the opted patients survived 5 years (Jones & Baldwin, 2018). Earlier diagnosis is therefore very important for lung cancer, however, conventional sputum cytology is not advised because of its low sensitivity. Annual cytology with improved computer-assisted image analysis, in smokers with chronic obstructive pulmonary disease, is currently implied in the UK Lung-SEARCH trial (Spiro et al., 2019). Spiral CT can detect lung nodules as small as 0.5 cm, although the problem with such small nodules is the high false-positive rate (21%–33%) (Gould et al., 2013). One silver lining for lung cancer diagnostics is that assessment of it is easier than most other cancers; as the main test, chest X-ray is easily available, convincingly cheap, and fairly accurate. False-positive chest X-rays for lung cancers are relatively few; however, the threshold limit for the cancerous nodule must be over 3 cm (Rubin, 2015) and may sometimes require a repeat test. False-negative X-rays can occur in a quarter of cancers when some lesions are misdiagnosed by the radiologist and some are not visible.

### 2.7.3 Colorectal cancer

Contemporary UK recommendation advised that 6 weeks of rectal bleeding or bleeding with diarrhea as an urgent referral for colorectal cancer (Walsh et al., 2018) and the risk is even higher for people over 60 years of age even with or without accompanying symptoms. The threshold value of hemoglobin plus iron-deficiency at which rapid investigation for colorectal cancer is recommended is 10 g/dL in women and 11 g/dL in men (Hamilton et al., 2008). The statistics for possible colorectal cancers for these people is about 13.3% and 7.7%, respectively (Rawla et al., 2019). Alarmingly, patients with milder levels of anemia may need immediate investigation for possible colorectal cancer, as this associated milder anemia presents a increased risk of mortality (Muñoz et al., 2014). Colonoscopy represents good performance characteristics although a relatively costly and uncomfortable diagnostic measure. A relatively comfortable CT colonography can be an alternate although also with lower specificity and accuracy as it can fail to spot one tenth of the lesions when larger than 1 cm (Wylie & Burling, 2011). Examination of various biomarkers, by molecular biology and immunology techniques, correlate with the presence of colorectal cancer and can be sought after diagnostic measures.

### 2.7.4 Prostate cancer

Although prostate cancer is alikened to lung cancers, for most of the patients are not diagnosed at the earlier stages, the major relief is that the survivability rate of prostate cancer patients is much higher. Even considering the fact, prostate cancer is the second leading cause of male cancer-related deaths in the United States, and its prevalence increases with age. Also, statistics showed that screening for prostate cancer represents very little evidence of mortality benefit (Catalona, 2018). As the benefit for the treatment of symptomatic prostate cancer denotes survivability, it is very much advisable to search for prostate cancer when a man presents with lower urinary tract symptoms which is often connected with prostate gland enlargement; however, one should also diagnose whether the enlargement is benign or malignant. The percentage of malignancy in those cases can be as high as 3% (Hajmanoochehri & Rabiee, 2015). Measurement of the PSA is especially regarded as the screening tool; although, a small number of men can have prostate cancer with a PSA as low as 1 ng/mL (Thompson & Ankerst, 2007). Urologists consider the level of free PSA or abnormal DRE findings in the selection of patients for biopsy; however, there is no published evidence to support such measurements in primary care. The standard for prostate cancer detection is trans-rectal ultrasound-guided sextant needle biopsy, a method introduced in 1989 (Thompson & Ankerst, 2007) however, with false-negative rates as high as about 20% (Thompson & Ankerst, 2007). As a molecular marker, high sensitivity is associated with *GSTP1* to detect the presence of both prostatic intraepithelial neoplasia and prostate cancer, and ability to distinguish these from benign prostatic hyperplasia, and a prevalence of methylation in the range of 70%–90% in prostate cancer (Dumache et al., 2014). Also, molecular evaluation of two genes, prostate cancer antigen 3 (*PCA3*) and *PSA* by RT-PCR can be very much useful in diagnosing prostate cancer.

### 2.7.5 Ovarian cancer

Ovarian cancer is a complicated cancer in regards to the diagnosis as well as treatments. Ovarian cancer is associated with a lot of nonspecific symptoms like fatigue, abdominal pain, and urinary frequency. Rapid and abnormal abdominal distension is associated with a pretty high risk of ovarian cancer and can have a predictive value as high as 2.5% (Hamilton et al., 2009). The molecular marker for ovarian cancer is serum CA125 having reasonable specificity. Ultrasound is the advised diagnostic procedure and is executed trans-vaginally which is extensively tested in screening studies.

### 2.7.6 Biopsy and diagnosis of carcinoma of unknown primary origin

Carcinoma of unknown primary (CUP) refers to the metastatic representation of malignancy when represented without an identifiable primary tumor site and comprises about 3%–5% of all cancers (Varadhachary, 2007). The major problem of these cancers is, for the treatment as diagnosed and the cancer therapy are extensively dependent on the origin of the primary tumor and type of cancer. Therefore, diagnosing CUP and finding out its origin is a major and very important diagnostic task. Different methods are currently used to resolve this problem with special importance is given to molecular and immunologic assessments. IHC markers, using panels of 4–14 tissue-specific markers, are used to identify the origin of the tumor, with sensitivity and specificity over 85% (Selves et al., 2018). Some other diagnostic tools that are applied are chest X-rays, CT scans, and PET scans; however, these methods are not advised due to poor accuracy. microRNA profiling to identify the origin of tumors can also be useful in the metastatic tumor, since fixed tissue samples are the standard material in current practice, qRT-PCR can produce reliable results.

### 2.7.7 Circulating tumor cells

CTCs are extremely rare and the frequency is estimated at one tumor cell for every million peripheral blood mononuclear cells (Micalizzi et al., 2017). CellSearch is a technique that can detect, enumerate, and characterize CTCs, defined as nucleic acid-positive/CD45-negative/cytokeratin-positive, in the blood (Huebner et al., 2018). This method can also be applied for the clinical assessment of CTCs in metastatic breast, colorectal, and prostate cancer (Huebner et al., 2018). Among the most observed cancers, a larger percentage of the prostate (57%) and breast (37%) cancer patients exhibited two or more CTCs (Cieślikowski et al., 2020). Circulating endothelial cells (CECs) considered surrogate markers as CEC levels correlate with disease progression and reflect changes in the vascular endothelial growth factor (VEGF) pathway (Goon et al., 2009). Transcription analyses with multigene RT-PCR were also conducted to analyze the expression of 37 genes in CTCs (Park et al., 2016), FISH on CECs also demonstrated that patients who had CECs showed abnormal copy numbers (Katz et al., 2010).

### 2.7.8 Other cancers that need early diagnosing

The risk of esophageal cancer with dysphagia is 5.7% in men and 2.4% in women, and the risk of urinary tract malignancy with hematuria 7.4% and 3.4%; therefore, these cancers need early investigations. Similarly, patients with enlarged cervical lymph nodes can be associated with malignancy; however, less than 2% of cervical lymphadenopathy presented as malignant (Mohseni et al., 2014). The risk of a brain tumor with headache is 1 in 1000, therefore, it may not always require scanning. However, in exceptionally acute pain and if the condition is prolonged for a considerable period, it may require diagnosing for brain tumors and possible malignancy.

## 2.8 Achievements, challenges, and future aim of cancer diagnostics

The major achievement of cancer diagnostics is the early, easy, and accurate detection of the type and stages of cancer, which directly influences patient survival and treatment prognosis. Because of the rapid evolution of both diagnostic techniques and therapeutic interventions, a greater collaboration and teamwork have become achievable between clinicians, laboratories, researchers, and regulatory authorities for providing better care and services towards the patients. However, the scope of further defining the analytical and clinical performance of diagnostic tests will always be there.

The major challenge for cancer diagnostics is to develop newer and more sensitive techniques to screen and diagnose the patients at the earliest as time, possibly the most defining and required factor for some cancer patients. Still, biopsy followed by histopathology as the most practiced diagnostic techniques with recent advancement of various imaging techniques (MRI, CT, MRS). IHC, PCR, flow cytometry, FISH, CSH, and microarray are also considered as important with constant upgrading on their reliability for their participation in early, rapid and decisive diagnosis, prognosis, and therapeutics of cancer. MRI, is currently applied as the next most widely used adjunct imaging modality in evaluating the extent of disease and in discriminating between benign from malignant lesions (Rahbar & Partridge, 2016). IHC is also useful in the detection of tumor markers and oncoproteins expressions. Molecular techniques like PCR, RT-PCR, in situ hybridization are used to detect cellular DNA mutations, genetic alterations, and abnormal expressions of certain genes; and they are highly applicable for screening and early detection of possible malignancies in high-risk populations. With the enhancement of diagnostic accuracy and lowering the current threshold for detection, we are hopeful for the least loss of lives due to cancer.

Any diagnostic test falling between perfect and useless test, and no diagnostic test, can tell us with certainty if a patient has a particular disease or not, cancer diagnostics are no exception, however, a mere error potentially can lead to the death of a person. In addition to the high sensitivity and specificity, cancer diagnostic assays will have to be accepted in clinical practices owing to the ease of application and less time required to master the new diagnostic test for the pathologists. The diagnostic test should be configured to detect almost any target, their requirement for minimal quantities of sample, and their ability to be automated. Additionally, the utilization of molecular markers for diagnostic purposes requires them to be stable for a considerable time. It is also immensely significant to determine the particular application for a diagnostic test, and specificity and sensitivity should be of higher value during application. There is no doubt that constant improvisation, discovery, validation, commercialization, and clinical adoption of novel cancer diagnostics can be of immense importance and value which directly reflects inpatient survivability.

## Acknowledgments

Space limitation constrained us to include only limited and selected publications, but we are grateful to all the uncited related articles/studies which are also important equally for the advancement of cancer diagnostics.

## Conflict of interest

None.

## References


Alena, L., Marek, S., Lenka, K., Giordano, F. A., Peter, K., & Olga, G. (2020). Liquid biopsy is instrumental for 3 p.m. dimensional solutions in cancer management. *Journal of Clinical Medicine*, 2749. Available from https://doi.org/10.3390/jcm9092749.

Al-Mahmood, S., Sapiezynski, J., Garbuzenko, O. B., & Minko, T. (2018). Metastatic and triple-negative breast cancer: Challenges and treatment options. *Drug Delivery and Translational Research*, *8*(5), 1483–1507. Available from https://doi.org/10.1007/s13346-018-0551-3.

Anand, P., Kunnumakara, A. B., Sundaram, C., Harikumar, K. B., Tharakan, S. T., Lai, O. S., ... Aggarwal, B. B. (2008). Cancer is a preventable disease that requires major lifestyle changes. *Pharmaceutical Research*, *25*(9), 2097–2116. Available from https://doi.org/10.1007/s11095-008-9661-9.

Bax, C., Taverna, G., Eusebio, L., Sironi, S., Grizzi, F., Guazzoni, G., & Capelli, L. (2018). Innovative diagnostic methods for early prostate cancer detection through urine analysis: A review. *Cancers (Basel)*, *10*(4), 123. Available from https://doi.org/10.3390/cancers10040123.

Bayer, C. L., Joshi, P. P., & Emelianov, S. Y. (2013). Photoacoustic imaging: A potential tool to detect early indicators of metastasis. *Expert Review of Medical Devices*, *10*(1), 125–134. Available from https://doi.org/10.1586/erd.12.62.

Bento, L. C., Correia, R. P., Mangueira, C. L. P., Barroso, R. D. S., Rocha, F. A., Bacal, N. S., & Marti, L. C. (2017). The use of flow cytometry in myelodysplastic syndromes: A review. *Frontiers in Oncology*, *7*, 270. Available from https://doi.org/10.3389/fonc.2017.00270.

Bercovich, E., & Javitt, M. C. (2018). Medical imaging: From roentgen to the digital revolution, and beyond. *Rambam Maimonides Medical journal*, *9*(4). Available from https://doi.org/10.5041/RMMJ.10355.

Berg, W. A., Sechtin, A. G., Marques, H., & Zhang, Z. (2010). Cystic breast masses and the ACRIN 6666 experience. *Radiologic Clinics of North America*, *48*(5), 931–987. Available from https://doi.org/10.1016/j.rcl.2010.06.007.

Berger, A. (2002). Magnetic resonance imaging. *BMJ (Clinical Research ed.)*, *324*(7328), 35. Available from https://doi.org/10.1136/bmj.324.7328.35.

Bolboacă, S. D. (2019). Medical diagnostic tests: A review of test anatomy, phases, and statistical treatment of data. *Comput Math Methods Med*, *2019*(1891569). Available from https://doi.org/10.1155/2019/1891569.

Borges, H. L., Linden, R., & Wang, J. Y. J. (2008). DNA damage-induced cell death: Lessons from the central nervous system. *Cell Research*, *18*(1), 17–26. Available from https://doi.org/10.1038/cr.2007.110.

Bruno, P. (2011). The importance of diagnostic test parameters in the interpretation of clinical test findings: The prone hip extension test as an example. *The Journal of Canadian Chiropractic Association*, *55*(2), 69–75.

Calabuig-Fariñas, S., Jantus-Lewintre, E., Herreros-Pomares, A., & Camps, C. (2016). Circulating tumor cells versus circulating tumor DNA in lung cancer-which one will win? *Translational Lung Cancer Research*, *5*(5), 466–482. Available from https://doi.org/10.21037/tlcr.2016.10.02.

Casimiro, M. C., Crosariol, M., Loro, E., Li, Z., & Pestell, R. G. (2012). Cyclins and cell cycle control in cancer and disease. *Genes and Cancer*, *3* (11–12), 649–657. Available from https://doi.org/10.1177/1947601913479022.

Catalona, W. J. (2018). Prostate cancer screening. *Medical Clinics of North America*, *102*(2), 199–214. Available from https://doi.org/10.1016/j.mcna.2017.11.001.

Chambers, S., Cooney, A., Caplan, N., Dowen, D., & Kader, D. (2014). The accuracy of magnetic resonance imaging (MRI) in detecting meniscal pathology. *Journal of the Royal Naval Medical Service*, *100*(2), 157–160.

Chen, K., & Chen, X. (2011). Positron emission tomography imaging of cancer biology: Current status and future prospects. *Seminars in Oncology*, *38*(1), 70–86. Available from https://doi.org/10.1053/j.seminoncol.2010.11.005.

Chen, M., Ma, Z., & Cao. (2020). Retroperitoneal metastasis synchronous with brain and mediastinal lymph nodes metastasis from breast invasive ductal carcinoma as the first site of distant metastasis: A case report and review of literature. *International Journal of Clinical and Experimental Pathology*, *13*(7), 1693–1697.

Ciáurriz, M., Beloki, L., Bandrés, E., Mansilla, C., Zabalza, A., Pérez-Valderrama, E., ... Ramírez, N. (2017). Streptamer technology allows accurate and specific detection of CMV-specific HLA-A*02 CD8 + T cells by flow cytometry. *Cytometry Part B - Clinical Cytometry*, *92*(2), 153–160. Available from https://doi.org/10.1002/cyto.b.21367.

Cieślikowski, W. A., Budna-Tukan, J., Świerczewska, M., Ida, A., Hrab, M., Jankowiak, A., ... Antczak, A. (2020). Circulating tumor cells as a marker of disseminated disease in patients with newly diagnosed high-risk prostate cancer. *Cancers (Basel)*, *12*(1), 160. Available from https://doi.org/10.3390/cancers12010160.

Cohen, J. F., Korevaar, D. A., Altman, D. G., Bruns, D. E., Gatsonis, C. A., Hooft, L., ... Bossuyt, P. M. M. (2016). STARD 2015 guidelines for reporting diagnostic accuracy studies: Explanation and elaboration. *BMJ Open*, *6*(11), e012799. Available from https://doi.org/10.1136/bmjopen-2016-012799.

Cools, J., & Vandenberghe, P. (2009). New flow cytometry in hematologic malignancies. *Haematologica*, *94*(12), 1639–1641. Available from https://doi.org/10.3324/haematol.2009.013482.

Croft, P., Altman, D. G., Deeks, J. J., Dunn, K. M., Hay, A. D., Hemingway, H., ... Timmis, A. (2015). The science of clinical practice: Disease diagnosis or patient prognosis? Evidence about "what is likely to happen" should shape clinical practice. *BMC Medicine*, *13*(1), 20. Available from https://doi.org/10.1186/s12916-014-0265-4.

Cui, C., Shu, W., & Li, P. (2016). Fluorescence in situ hybridization: Cell-based genetic diagnostic and research applications. *Frontiers in Cell and Developmental Biology*, *4*, 89. Available from https://doi.org/10.3389/fcell.2016.00089.

Dabbagh Kakhki, V. (2007). Positron emission tomography in the management of lung cancer. *Annals of Thoracic Medicine*, *2*(2), 69–76. Available from https://doi.org/10.4103/1817-1737.32235.

Dixit, R., Agarwal, K., Gokhroo, A., Patil, C., Meena, M., Shah, N., & Arora, P. (2017). Diagnosis and management options in malignant pleural effusions. *Lung India*, *34*(2), 160–166. Available from https://doi.org/10.4103/0970-2113.201305.

Drexler, H. G., Borkhardt, A., & Janssen, J. W. G. (1995). Detection of chromosomal translocations: In leukemia-lymphoma cells by polymerase chain reaction. *Leukemia and Lymphoma*, *19*(5–6), 359–380. Available from https://doi.org/10.3109/10428199509112194.

Dumache, R., Puiu, M., Motoc, M., Vernic, C., & Dumitrascu, V. (2014). Prostate cancer molecular detection in plasma samples by glutathione S-transferase P1 (GSTP1) methylation analysis. *Clinical Laboratory*, *60*(5), 847–852. Available from https://doi.org/10.7754/Clin.Lab.2013.130701.

Elmore, S. (2007). Apoptosis: A review of programmed cell death. *Toxicologic Pathology*, *35*(4), 495–516. Available from https://doi.org/10.1080/01926230701320337.

Fass, L. (2008). Imaging and cancer: A review. *Molecular Oncology*, *2*(2), 115–152. Available from https://doi.org/10.1016/j.molonc.2008.04.001.

Fouad, Y. A., & Aanei, C. (2017). Revisiting the hallmarks of cancer. *American Journal of Cancer Research*, *7*(5), 1016–1036. Available from http://www.ajcr.us/files/ajcr0053932.pdf.

Gold, P., Shuster, J., & Freedman, S. O. (1978). Carcinoembryonic antigen (CEA) in clinical medicine. Historical perspectives, pitfalls and projections. *Cancer*, *42*(3 S), 1399–1405. Available from https://doi.org/10.1002/1097-0142(197809)42:3 + < 1399::AID-CNCR2820420803 > 3.0.CO;2-P.

Goon, P. K. Y., Lip, G. Y. H., Stonelake, P. S., & Blann, A. D. (2009). Circulating endothelial cells and circulating progenitor cells in breast cancer: Relationship to endothelial damage/dysfunction/apoptosis, clinicopathologic factors, and the Nottingham prognostic index. *Neoplasia (New York, N.Y.)*, *11*(8), 771–779. Available from https://doi.org/10.1593/neo.09490.

Gordon, Y., Partovi, S., Müller-Eschner, M., et al. (2014). Dynamic contrast-enhanced magnetic resonance imaging: fundamentals and application to the evaluation of the peripheral perfusion. *Cardiovascular Diagnosis Therapy*, *4*(2), 147–164. Available from https://doi.org/10.3978/j.issn.2223-3652.2014.03.01.

Gould, M. K., Donington, J., Lynch, W. R., Mazzone, P. J., Midthun, D. E., Naidich, D. P., & Wiener, R. S. (2013). Evaluation of individuals with pulmonary nodules: When is it lung cancer? Diagnosis and management of lung cancer, 3rd ed: American college of chest physicians evidence-based clinical practice guidelines. *Chest*, *143*(5), e93–e120. Available from https://doi.org/10.1378/chest.12-2351.

Grade, M., Difilippantonio, M. J., & Camps, J. (2015). *Patterns of chromosomal aberrations in solid tumors. Recent results in cancer research* (200, pp. 115–142). Springer New York LLC. Available from https://doi.org/10.1007/978-3-319-20291-4_6.

Gray, P. N., Dunlop, C. L. M., & Elliott, A. M. (2015). Not all next generation sequencing diagnostics are created equal: Understanding the nuances of solid tumor assay design for somatic mutation detection. *Cancers (Basel)*, *7*(3), 1313–1332. Available from https://doi.org/10.3390/cancers7030837.

Gurcan, M. N., Boucheron, L. E., Can, A., Madabhushi, A., Rajpoot, N. M., & Yener, B. (2009). Histopathological image analysis: A review. *IEEE Reviews in Biomedical Engineering*, *2*, 147–171. Available from https://doi.org/10.1109/RBME.2009.2034865.

Haas, J. S., Kaplan, C. P., Brawarsky, P., & Kerlikowske, K. (2005). Evaluation and outcomes of women with a breast lump and a normal mammogram result. *Journal of General Internal Medicine*, *20*(8), 692–696. Available from https://doi.org/10.1111/j.1525-1497.2005.0149.x.

Hajmanoochehri, F., & Rabiee, E. (2015). FNAC accuracy in diagnosis of thyroid neoplasms considering all diagnostic categories of the Bethesda reporting system: A single-institute experience. *Journal of Cytology*, *32*(4), 238–243. Available from https://doi.org/10.4103/0970-9371.171234.

Hamilton, W., Lancashire, R., Sharp, D., Peters, T. J., Cheng, K. K., & Marshall, T. (2008). The importance of anaemia in diagnosing colorectal cancer: A case-control study using electronic primary care records. *British Journal of Cancer*, *98*(2), 323–327. Available from https://doi.org/10.1038/sj.bjc.6604165.

Hamilton, W., Peters, T. J., Bankhead, C., & Sharp, D. (2009). Risk of ovarian cancer in women with symptoms in primary care: Population based case-control study. *BMJ (Online)*, *339*(7721), b2998. Available from https://doi.org/10.1136/bmj.b2998, 616.

Han, X., Wang, J., & Sun, Y. (2017). Circulating tumor DNA as biomarkers for cancer detection. *Genomics, Proteomics and Bioinformatics*, *15*(2), 59–72. Available from https://doi.org/10.1016/j.gpb.2016.12.004.

Hooijkaas, H., Hahlen, K., Adriaansen, H. J., Dekker, I., Van Zanen, G. E., & Van Dongen, J. J. M. (1989). Terminal deoxynucleotidyl transferase (TdT)-positive cells in cerebrospinal fluid and development of overt CNS leukemia: A 5-year follow-up study in 113 children with a TdT-positive leukemia or non-Hodgkin's lymphoma. *Blood*, *74*(1), 416–422. Available from https://doi.org/10.1182/blood.v74.1.416.416.

Hosny, A., Parmar, C., Quackenbush, J., Schwartz, L. H., & Aerts, H. J. W. L. (2018). Artificial intelligence in radiology. *Nature Reviews. Cancer*, *18*(8), 500–510. Available from https://doi.org/10.1038/s41568-018-0016-5.

Hudacko, R., Rapkiewicz, A., Berman, R. S., & Simsir, A. (2011). ALK-negative anaplastic large cell lymphoma mimicking a soft tissue sarcoma. *Journal of Cytology*, *28*(4), 230–233. Available from https://doi.org/10.4103/0970-9371.86362.

Huebner, H., Fasching, P. A., Gumbrecht, W., Jud, S., Rauh, C., Matzas, M., … Ruebner, M. (2018). Filtration based assessment of CTCs and CellSearch® based assessment are both powerful predictors of prognosis for metastatic breast cancer patients. *BMC Cancer*, *18*(1), 204. Available from https://doi.org/10.1186/s12885-018-4115-1.

Inês, G., & Daniele, M. G. (2017). BRCA1 and BRCA2 mutations and treatment strategies for breast cancer. *Integrative Cancer Science and Therapeutics*. Available from https://doi.org/10.15761/ICST0.1000228.

Jackson, M., Marks, L., May, G. H. W., & Wilson, J. B. (2018). The genetic basis of disease, [published correction appears in *Essays Biochem*. 2020 Oct 8;64(4):681]*Essays in Biochemistry*, *62*(5), 643–723. Available from https://doi.org/10.1042/EBC20170053.

Jones, G. S., & Baldwin, D. R. (2018). Recent advances in the management of lung cancer. *Clinical Medicine, Journal of the Royal College of Physicians of London*, *18*(Suppl 2), s41–s46. Available from https://doi.org/10.7861/clinmedicine.18-2-s41.

Karunakaran, K., Murugesan, P., Jeyapradha, D., & Rajeshwar, G. (2012). Applications of immunohistochemistry. *Journal of Pharmacy and Bioallied Sciences, 307.* Available from https://doi.org/10.4103/0975-7406.100281.

Katz, R. L., He, W., Khanna, A., Fernandez, R. L., Zaidi, T. M., Krebs, M., ... El-Zein, R. (2010). Genetically abnormal circulating cells in lung cancer patients: An antigen-independent fluorescence in situ hybridization-based case-control study. *Clinical Cancer Research, 16*(15), 3976–3987. Available from https://doi.org/10.1158/1078-0432.CCR-09-3358.

Kettritz, U. (2011). Minimally invasive biopsy methods - Diagnostics or therapy? Personal opinion and review of the literature. *Breast Care (Basel), 6* (2), 94–97. Available from https://doi.org/10.1159/000327889.

Kruk, M. E., Gage, A. D., Arsenault, C., Jordan, K., Leslie, H. H., Roder-DeWan, S., ... Pate, M. (2018). High-quality health systems in the sustainable development goals era: Time for a revolution, [published correction appears in The Lancet Global Health 2018 Sep 18; 2018 Nov;6(11): e1162]*The Lancet Global Health, 6*(11), e1196–e1252. Available from https://doi.org/10.1016/S2214-109X(18)30386-3.

Kumar, P., & Pawaiya, R. V. S. (2010). Advances in cancer diagnostics. *Braz J Vet Pathol, 3*(2), 142–153.

Kurtin, P., & Pinkus, G. S. (1985). Leukocyte common antigen—A diagnostic discriminant between hematopoietic and nonhematopoietic neoplasms in paraffin sections using monoclonal antibodies: Correlation with immunologic studies and ultrastructural localization. *Human Pathology, 16*(4), 353–365. Available from https://doi.org/10.1016/s0046-8177(85)80229-x.

LaFramboise, T. (2009). Single nucleotide polymorphism arrays: A decade of biological, computational and technological advances. *Nucleic Acids Research, 37*(13), 4181–4193. Available from https://doi.org/10.1093/nar/gkp552.

Liang, X., Nan, K. J., Li, Z. L., & Xu, Q. Z. (2009). Overexpression of the LKB1 gene inhibits lung carcinoma cell proliferation partly through degradation of c-myc protein. *Oncology Reports, 21*(4), 925–931. Available from https://doi.org/10.3892/or_00000305.

Liu, M. P., Guo, X. Z., Xu, J. H., Wang, D., Li, H. Y., Cui, Z. M., ... Ren, L. N. (2005). New tumor-associated antigen SC6 in pancreatic cancer. *World Journal of Gastroenterology, 11*(48), 7671–7675. Available from https://doi.org/10.3748/wjg.v11.i48.7671.

Makki, J. S. (2016). Diagnostic implication and clinical relevance of ancillary techniques in clinical pathology practice. *Clinical Medicine Insights: Pathology, 9*(1), 5–11. Available from https://doi.org/10.4137/CPath.s32784.

Malhotra, S., Kazlouskaya, V., Andres, C., Gui, J., & Elston, D. (2013). Diagnostic cellular abnormalities in neoplastic and non-neoplastic lesions of the epidermis: A morphological and statistical study. *Journal of Cutaneous Pathology, 40*(4), 371–378. Available from https://doi.org/10.1111/cup.12090.

Mao, X., Young, B. D., & Lu, Y. J. (2007). The application of single nucleotide polymorphism microarrays in cancer research. *Current Genomics, 8* (4), 219–228. Available from https://doi.org/10.2174/138920207781386924.

Maxim, L. D., Niebo, R., & Utell, M. J. (2014). Screening tests: A review with examples, [published correction appears in *Inhalation Toxicology* 2019 Jun;*31*(7):298]*Inhalation Toxicology, 26*(13), 811–828. Available from https://doi.org/10.3109/08958378.2014.955932.

Mayo Clinic. (n.d.). Cancer – Symptoms and causes. Available from: https://www.mayoclinic.org/diseases-conditions/cancer/symptoms-causes/syc-20370588

McGill, S., Soetikno, R., & Kaltenbach, T. (2009). Image-enhanced endoscopy in practice. *Canadian Journal of Gastroenterology, 23*(11), 741–746. Available from https://doi.org/10.1155/2009/143949.

Micalizzi, D. S., Maheswaran, S., & Haber, D. A. (2017). A conduit to metastasis: Circulating tumor cell biology. *Genes and Development, 31*(18), 1827–1840. Available from https://doi.org/10.1101/gad.305805.117.

Miller, D. L., Smith, N. B., Bailey, M. R., Czarnota, G. J., Hynynen, K., & Makin, I. R. S. (2012). Overview of therapeutic ultrasound applications and safety considerations. *Journal of Ultrasound in Medicine, 31*(4), 623–634. Available from https://doi.org/10.7863/jum.2012.31.4.623.

Miller, M. B., & Tang, Y. W. (2009). Basic concepts of microarrays and potential applications in clinical microbiology. *Clinical Microbiology Reviews, 22*(4), 611–633. Available from https://doi.org/10.1128/CMR.00019-09.

Mjønes, P., Sagatun, L., Nordrum, I. S., & Waldum, H. L. (2017). Neuron-Specific enolase as an immunohistochemical marker is better than its reputation. *Journal of Histochemistry and Cytochemistry, 65*(12), 687–703. Available from https://doi.org/10.1369/0022155417733676.

Mohseni, S., Shojaiefard, A., Khorgami, Z., Alinejad, S., Ghorbani, A., & Ghafouri, A. (2014). Peripheral lymphadenopathy: Approach and diagnostic tools. *Iranian Journal of Medical Sciences, 39*(2 Suppl), 158–170. Available from http://ijms.sums.ac.ir/index.php/IJMS/article/download/638/145.

Mouttet, D., Laé, M., Caly, M., Gentien, D., Carpentier, S., Peyro-Saint-Paul, H., ... Reyal, F. (2016). Estrogen-receptor, progesterone-receptor and HER2 status determination in invasive breast cancer. concordance between immuno-histochemistry and MapQuant™ microarray based assay. *PLoS One, 11*(2), e0146474. Available from https://doi.org/10.1371/journal.pone.0146474.

Muñoz, M., Gómez-Ramírez, S., Martín-Montañez, E., & Auerbach, M. (2014). Perioperative anemia management in colorectal cancer patients: A pragmatic approach. *World Journal of Gastroenterology, 20*(8), 1972–1985. Available from https://doi.org/10.3748/wjg.v20.i8.1972.

Narrandes, S., & Xu, W. (2018). Gene expression detection assay for cancer clinical use. *Journal of Cancer, 9*(13), 2249–2265. Available from https://doi.org/10.7150/jc.24744.

National Cancer Institute. Cancer. (n.d.). Available from: https://www.cancer.gov/about-cancer/understanding/what-is-cancer.

National Research Council (United States) and Institute of Medicine (United States) Committee on State of the Science of Nuclear Medicine. (2007). *Advancing Nuclear Medicine Through Innovation.* Washington (DC): National Academies Press (United States) 2, Nuclear Medicine. (n.d.). Available from. Available from https://www.ncbi.nlm.nih.gov/books/NBK11471/.

Neumann, M. H. D., Bender, S., Krahn, T., & Schlange, T. (2018). ctDNA and CTCs in liquid biopsy – Current status and where we need to progress. *Computational and Structural Biotechnology Journal, 16*, 190–195. Available from https://doi.org/10.1016/j.csbj.2018.05.002.

O'Connor, O. J., McDermott, S., Slattery, J., Sahani, D., & Blake, M. A. (2011). The use of PET-CT in the assessment of patients with colorectal carcinoma. *International Journal of Surgical Oncology, 2011*, 846512. Available from https://doi.org/10.1155/2011/846512, 1–14.

O'Sullivan, G. J., Carty, F. L., & Cronin, C. G. (2015). Imaging of bone metastasis: An update. *World Journal of Radiology*, *7*(8), 202–211. Available from https://doi.org/10.4329/wjr.v7.i8.202.

Parham, D. M., Holt, H., Kent Williams, W., Webber, B., & Maurer, H. (1991). Immunohistochemical study of childhood rhabdomyosarcomas and related neoplasms. Results of an intergroup rhabdomyosarcoma study project. *Cancer*, *67*(12), 3072–3080, https://doi.org/10.1002/1097-0142(19910615)67:12 < 3072::AID-CNCR2820671223 > 3.0.CO;2-Z.

Park, S. M., Wong, D. J., Ooi, C. C., Kurtz, D. M., Vermesh, O., Aalipour, A., ... Gambhir, S. S. (2016). Molecular profiling of single circulating tumor cells from lung cancer patients. *Proceedings of the National Academy of Sciences of the United States of America*, *113*(52), E8379–E8386. Available from https://doi.org/10.1073/pnas.1608461113.

Parsa, N. (2012). Environmental factors inducing human cancers. *Iranian Journal of Public Health*, *41*(11), 1–9. Available from http://ijph.ir/pdfs/1-%20IJPH%20Dr%20Parsa%2012403%20RA%20RTG%2091.8.12.pdf.

Pelc, N. J. (2014). Recent and future directions in CT imaging. *Annals of Biomedical Engineering*, *42*(2), 260–268. Available from https://doi.org/10.1007/s10439-014-0974-z.

Pelosof, L. C., & Gerber, D. E. (2010). Paraneoplastic syndromes: an approach to diagnosis and treatment, [published correction appears in *Mayo Clinic Proceedings* 2011 Apr;*86*(4):364. Dosage error in article text]*Mayo Clinic Proceedings.*, *85*(9), 838–854. Available from https://doi.org/10.4065/mcp.2010.0099.

Purohit, B. S., Ailianou, A., Dulguerov, N., Becker, C. D., Ratib, O., & Becker, M. (2014). FDG-PET/CT pitfalls in oncological head and neck imaging. *Macromolecular Research*, *5*(5), 585–602. Available from https://doi.org/10.1007/s13244-014-0349-x.

Rahbar, H., & Partridge, S. C. (2016). Multiparametric MR imaging of breast cancer. *Magnetic Resonance Imaging Clinics of North America*, *24*(1), 223–238. Available from https://doi.org/10.1016/j.mric.2015.08.012.

Rawla, P., Sunkara, T., & Barsouk, A. (2019). Epidemiology of colorectal cancer: Incidence, mortality, survival, and risk factors. *Przeglad Gastroenterologiczny*, *14*(2), 89–103. Available from https://doi.org/10.5114/pg.2018.81072.

Rawnaq, T., Kleinhans, H., Uto, M., Schurr, P. G., Reichelt, U., Cataldegirmen, G., ... Kaifi, J. T. (2009). Subset of esophageal adenocarcinoma expresses adhesion molecule l1 in contrast to squamous cell carcinoma. *Anticancer Research*, *29*(4), 1195–1199.

Robertson, S., Azizpour, H., Smith, K., & Hartman, J. (2018). Digital image analysis in breast pathology-from image processing techniques to artificial intelligence. *Translational Research*, *194*, 19–35. Available from https://doi.org/10.1016/j.trsl.2017.10.010.

Rojanathanes, R., Sereemaspun, A., Pimpha, N., Buasorn, V., Ekawong, P., & Wiwanitkit, V. (2008). Gold nanoparticle as an alternative tool for a urine pregnancy test. *Taiwanese Journal of Obstetrics and Gynecology*, *47*(3), 296–299. Available from https://doi.org/10.1016/S1028-4559(08)60127-8.

Rubin, G. D. (2015). Lung nodule and cancer detection in computed tomography screening. *Journal of Thoracic Imaging*, *30*(2), 130–138. Available from https://doi.org/10.1097/RTI.0000000000000140.

Rueda, O. M., Sammut, S. J., Seoane, J. A., Chin, S. F., Caswell-Jin, J. L., Callari, M., ... Curtis, C. (2019). Dynamics of breast-cancer relapse reveal late-recurring ER-positive genomic subgroups. *Nature*, *567*(7748), 399–404. Available from https://doi.org/10.1038/s41586-019-1007-8.

Ryu, Y. J. (2015). Diagnosis of pulmonary tuberculosis: Recent advances and diagnostic algorithms. *Tuberculosis and Respiratory Diseases*, *78*(2), 64–71. Available from https://doi.org/10.4046/trd.2015.78.2.64.

Sadoughi, F., Kazemy, Z., Hamedan, F., Owji, L., Rahmanikatigari, M., & Azadboni, T. T. (2018). Artificial intelligence methods for the diagnosis of breast cancer by image processing: A review. *Breast Cancer: Targets and Therapy*, *10*, 219–230. Available from https://doi.org/10.2147/BCTT.S175311.

Sakamoto, S., Putalun, W., Vimolmangkang, S., Phoolcharoen, W., Shoyama, Y., Tanaka, H., & Morimoto, S. (2018). Enzyme-linked immunosorbent assay for the quantitative/qualitative analysis of plant secondary metabolites, [published correction appears in *Journal of Natural Medicines* 2018 Jan 5]*Journal of Natural Medicines*, *72*(1), 32–42. Available from https://doi.org/10.1007/s11418-017-1144-z.

Samuelsson, J. K., Alonso, S., Yamamoto, F., & Perucho, M. (2010). DNA fingerprinting techniques for the analysis of genetic and epigenetic alterations in colorectal cancer. *Mutation Research – Fundamental and Molecular Mechanisms of Mutagenesis*, *693*(1–2), 61–76. Available from https://doi.org/10.1016/j.mrfmmm.2010.08.010.

Sanders, A. D., Meiers, S., Ghareghani, M., Porubsky, D., Jeong, H., van Vliet, M. A. C. C., ... Korbel, J. O. (2020). Single-cell analysis of structural variations and complex rearrangements with tri-channel processing. *Nature Biotechnology*, *38*(3), 343–354. Available from https://doi.org/10.1038/s41587-019-0366-x.

Sauter, G., Carroll, P., Moch, H., Kallioniemi, A., Kerschmann, R., Narayan, P., ... Waldman, F. M. (1995). c-myc Copy number gains in bladder cancer detected by fluorescence in situ hybridization. *American Journal of Pathology*, *146*(5), 1131–1139.

Selves, J., Long-Mira, E., Mathieu, M. C., Rochaix, P., & Ilié, M. (2018). Immunohistochemistry for diagnosis of metastatic carcinomas of unknown primary site. *Cancers*, *10*(4), 108. Available from https://doi.org/10.3390/cancers10040108.

Shakoori, A. R. (2017). *Fluorescence in situ hybridization (FISH) and its applications. Chromosome structure and aberrations* (pp. 343–367). Springer India. Available from https://doi.org/10.1007/978-81-322-3673-3_16.

Shan, G., Huang, W., Gee, S. J., Buchholz, B. A., Vogel, J. S., & Hammock, B. D. (2000). Isotope-labeled immunoassays without radiation waste. *Proceedings of the National Academy of Sciences of the United States of America*, *97*(6), 2445–2449. Available from https://doi.org/10.1073/pnas.040575997.

Sharma, G. N., Dave, R., Sanadya, J., Sharma, P., & Sharma, K. K. (2010). Various types and management of breast cancer: An overview. *Journal of Advanced Pharmaceutical Technology and Research*, *1*(2), 109–126.

Sharma, S. (2009). Tumor markers in clinical practice: General principles and guidelines. *Indian Journal of Medical and Paediatric Oncology*, *30*(1), 1–8. Available from https://doi.org/10.4103/0971-5851.56328, 1.

Skuse, A. (2015). *Constructions of cancer in Early Modern England, 1580–1720: Ravenous natures*. London: Palgrave Macmillan ISBN 9781137487537 doi. Available from https://doi.org/10.1057/9781137487537.

Slodkowska, E. A., & Ross, J. S. (2009). MammaPrint™ 70-gene signature: Another milestone in personalized medical care for breast cancer patients. *Expert Review of Molecular Diagnostics*, *9*(5), 417–422. Available from https://doi.org/10.1586/erm.09.32.

Spiro, S. G., Shah, P. L., Rintoul, R. C., George, J., Janes, S., Callister, M., ... Taylor, M. N. (2019). Sequential screening for lung cancer in a high-risk group: Randomised controlled trial: LungSEARCH: A randomised controlled trial of surveillance using sputum and imaging for the early detection of lung cancer in a high-risk group. *European Respiratory Journal*, *54*(4), 1900581. Available from https://doi.org/10.1183/13993003.00581-2019.

Sridharan, G., & Shankar, A. A. (2012). Toluidine blue: A review of its chemistry and clinical utility. *Journal of Oral and Maxillofacial Pathology*, *16*(2), 251–255. Available from https://doi.org/10.4103/0973-029X.99081.

Ståhlberg, A., Zoric, N., Åman, P., & Kubista, M. (2005). Quantitative real-time PCR for cancer detection: The lymphoma case. *Expert Review of Molecular Diagnostics*, *5*(2), 221–230. Available from https://doi.org/10.1586/14737159.5.2.221.

Strati, P., & Shanafelt, T. D. (2015). Monoclonal B-cell lymphocytosis and early-stage chronic lymphocytic leukemia: Diagnosis, natural history, and risk stratification. *Blood*, *126*(4), 454–462. Available from https://doi.org/10.1182/blood-2015-02-585059.

Sullivan, F. M., Farmer, E., Mair, F. S., Treweek, S., Kendrick, D., Jackson, C., ... Schembri, S. (2017). Detection in blood of autoantibodies to tumour antigens as a case-finding method in lung cancer using the EarlyCDT®-Lung test (ECLS): Study protocol for a randomized controlled trial. *BMC Cancer*, *17*(1), 187. Available from https://doi.org/10.1186/s12885-017-3175-y.

Takalkar, A. M., El-Haddad, G., & Lilien, D. L. (2008). FDG-PET and PET/CT - Part II. *Indian Journal of Radiology and Imaging*, *18*(1), 17–36. Available from https://doi.org/10.4103/0971-3026.38504.

Tarca, A. L., Romero, R., & Draghici, S. (2006). Analysis of microarray experiments of gene expression profiling. *American Journal of Obstetrics and Gynecology*, *195*(2), 373–388. Available from https://doi.org/10.1016/j.ajog.2006.07.001.

Thigpen, D., Kappler, A., & Brem, R. (2018). The role of ultrasound in screening dense breasts – A review of the literature and practical solutions for implementation. *Diagnostics (Basel)*, *8*(1), 20. Available from https://doi.org/10.3390/diagnostics8010020.

Thompson, I. M., & Ankerst, D. P. (2007). Prostate-specific antigen in the early detection of prostate cancer. *CMAJ: Canadian Medical Association Journal (Journal de l'Association Medicale Canadienne)*, *176*(13), 1853–1858. Available from https://doi.org/10.1503/cmaj.060955.

Thway, K., Wang, J., Mubako, T., & Fisher, C. (2014). Histopathological diagnostic discrepancies in soft tissue tumours referred to a specialist centre: Reassessment in the era of ancillary molecular diagnosis. *Sarcoma*, *2014*, 686902. Available from https://doi.org/10.1155/2014/686902.

Ulrich, B., Trimboli, R. M., Alexandra, A., Corinne, B., Baltzer, P. A. T., Maria, B., ... Francesco, S. (2020). Image-guided breast biopsy and localisation: Recommendations for information to women and referring physicians by the European Society of Breast Imaging. *Insights into Imaging*. Available from https://doi.org/10.1186/s13244-019-0803-x.

Vaidyanathan, K., & Vasudevan, D. M. (2012). Organ specific tumor markers: What's new? *Indian Journal of Clinical Biochemistry*, *27*(2), 110–120. Available from https://doi.org/10.1007/s12291-011-0173-8.

Van Zijl, F., Krupitza, G., & Mikulits, W. (2011). Initial steps of metastasis: Cell invasion and endothelial transmigration. *Mutation Research - Reviews in Mutation Research*, *728*(1–2), 23–34. Available from https://doi.org/10.1016/j.mrrev.2011.05.002.

Varadhachary, G. R. (2007). Carcinoma of Unknown Primary Origin. *Gastrointest Cancer Res*, *1*(6), 229–235.

Vieira, A. F., & Schmitt, F. (2018). An update on breast cancer multigene prognostic tests-emergent clinical biomarkers. *Frontiers in Medicine (Lausanne)*, *5*, 248. Available from https://doi.org/10.3389/fmed.2018.00248.

Viray, H., Bradley, W. R., Schalper, K. A., Rimm, D. L., & Rothberg, B. E. G. (2013). Marginal and joint distributions of S100, HMB-45, and Melan-A across a large series of cutaneous melanomas. *Archives of Pathology and Laboratory Medicine*, *137*(8), 1063–1073. Available from https://doi.org/10.5858/arpa.2012-0284-OA.

Wachowiak, R., Thieltges, S., Rawnaq, T., Kaifi, J. T., Fiegel, H., Metzger, R., ... Izbicki, J. R. (2010). Y-box-binding protein-1 is a potential novel tumour marker for neuroblastoma. *Anticancer Research*, *30*(4), 1239–1242, PMID:. Available from 20530434.

Waitzberg, A. F. L., Nonogaki, S., Nishimoto, I. N., Kowalski, L. P., Miguel, R. E. V., Brentani, R. R., & Brentani, M. M. (2004). Clinical significance of c-myc and p53 expression in head and neck squamous cell carcinomas. *Cancer Detection and Prevention*, *28*(3), 178–186. Available from https://doi.org/10.1016/j.cdp.2004.02.003.

Walsh, C. J., Delaney, S., & Rowlands, A. (2018). Rectal bleeding in general practice: New guidance on commissioning. *British Journal of General Practice*, *68*(676), 514–515. Available from https://doi.org/10.3399/bjgp18X699485.

Wang, F., Liu, J., Zhang, R., Bai, Y., Li, C., Li, B., ... Zhang, T. (2018). CT and MRI of adrenal gland pathologies. *Quantitative Imaging in Medicine and Surgery*, *8*(8), 853–875. Available from https://doi.org/10.21037/qims.2018.09.13.

Wang, L. (2017). Early diagnosis of breast cancer. *Sensors (Basel) (Switzerland)*, *17*(7), 1572. Available from https://doi.org/10.3390/s17071572.

Weiss, M. M., Hermsen, M. A., Meijer, G. A., van Grieken, N. C., Baak, J. P., Kuipers, E. J., & van Diest, P. J. (1999). Comparative genomic hybridisation. *Molecular Pathology*, *52*(5), 243–251. Available from https://doi.org/10.1136/mp.52.5.243.

Wernli K., Brandzel S., Buist D., et al. Is breast MRI better at finding second breast cancers than mammograms alone for breast cancer survivors? [Internet] Washington (DC): Patient-Centered Outcomes Research Institute (PCORI); 2019 May. Available from: https://www.ncbi.nlm.nih.gov/books/NBK554228/ doi: 10.25302/5.2019.CE.13046656

Wiggins, G. A. R., Walker, L. C., & Pearson, J. F. (2020). Genome-wide gene expression analyses of brc.1-and brc.2-associated breast and ovarian tumours. *Cancers (Basel)*, *12*(10), 3015. Available from https://doi.org/10.3390/cancers12103015, 1–17.

World Health Organisation (WHO). (n.d.). <https://www.who.int/news-room/fact-sheets/detail/cancer>.

Wu, M., & Shu, J. (2018). Multimodal molecular imaging: Current status and future directions. *Contrast Media and Molecular Imaging, 2018*, 1382183. Available from https://doi.org/10.1155/2018/1382183.

Wylie, P. N., & Burling, D. (2011). CT colonography: What the gastroenterologist needs to know. *Frontline Gastroenterology*, 2(2), 96–104. Available from https://doi.org/10.1136/fg.2009.000380.

Xiaohua, S., Kang, S., & Tie, W. (2016). Detection of volatile organic compounds (VOCs) from exhaled breath as noninvasive methods for cancer diagnosis. *Analytical and Bioanalytical Chemistry*, 2759–2780. Available from https://doi.org/10.1007/s00216-015-9200-6.

Yamamoto, Y., Tsuzuki, T., Akatsuka, J., Ueki, M., Morikawa, H., Numata, Y., ... Kimura, G. (2019). Automated acquisition of explainable knowledge from unannotated histopathology images. *Nature Communications*, *10*(1), 5642. Available from https://doi.org/10.1038/s41467-019-13647-8.

Yang, N. J., & Hinner, M. J. (2015). Getting across the cell membrane: An overview for small molecules, peptides, and proteins. *Methods in Molecular Biology*, *1266*, 29–53. Available from https://doi.org/10.1007/978-1-4939-2272-7_3.

Yang, X. F., & Pan, K. (2014). Diagnosis and management of acute complications in patients with colon cancer: Bleeding, obstruction, and perforation. *Chinese Journal of Cancer Research*, *26*(3), 331–340. Available from https://doi.org/10.3978/j.issn.1000-9604.2014.06.11.

Yankaskas, B. C., Haneuse, S., Kapp, J. M., Kerlikowske, K., Geller, B., & Buist, D. S. M. (2010). Performance of first mammography examination in women younger than 40 years. *Journal of the National Cancer Institute*, *102*(10), 692–701. Available from https://doi.org/10.1093/jnci/djq090.

Yin, Y., Cao, Y., Xu, Y., & Li, G. (2010). Colorimetric immunoassay for detection of tumor markers. *International Journal of Molecular Sciences*, *11*(12), 5077–5094. Available from https://doi.org/10.3390/ijms11125077.

Zhang, Y., Li, M., Gao, X., Chen, Y., & Liu, T. (2019). Nanotechnology in cancer diagnosis: Progress, challenges and opportunities. *Journal of Hematology and Oncology*, *12*(1), 137. Available from https://doi.org/10.1186/s13045-019-0833-3.

Zheng, H. T., Peng, Z. H., Li, S., & He, L. (2005). Loss of heterozygosity analyzed by single nucleoside polymorphism array in cancer. *World Journal of Gastroenterology*, *11*(43), 6740–6744. Available from https://doi.org/10.3748/wjg.v11.i43.6740.

Zhu, A., Lee, D., & Shim, H. (2011). Metabolic positron emission tomography imaging in cancer detection and therapy response. *Seminars in Oncology*, *38*(1), 55–69. Available from https://doi.org/10.1053/j.seminoncol.2010.11.012.

Al-Abbadi, M. A. (2011). Basics of cytology. *Avicenna Journal of Medicine*, *1*(1), 18–28. Available from https://doi.org/10.4103/2231-0770.83719.

Chapter 3

# Biomarkers associated with different types of cancer as a potential candidate for early diagnosis of oncological disorders

Arpana Parihar[1,2], Surbhi Jain[1], Dipesh Singh Parihar[3], Pushpesh Ranjan[2,4] and Raju Khan[2,4]

[1]*Department of Biochemistry and Genetics, Barkatullah University, Bhopal, India*

[2]*Microfluidics & MEMS Centre, CSIR-Advanced Materials and Processes Research Institute (AMPRI), Bhopal, India*

[3]*Engineering College Tuwa, Godhra, India*

[4]*Academy of Scientific and Innovative Research (AcSIR), Ghaziabad, India*

## 3.1 Introduction

The deadliest disease in which cells divide uncontrollably and at a later stage start invading nearby tissue is known as cancer (Bray et al., 2018). Generally, hereditary mutations, consumption of carcinogens, radiation exposures, alcohol usage, smoking, and extreme lifestyles may cause cancer. Lung, gastric, liver, breast and colorectal cancer are among the topmost cancers in the world which cause higher mortality. The number of new cases of different types of cancers reported and number of cancer-related deaths in year 2018 are shown in Table 3.1. Cancer diagnosis at an early stage leads to its successful treatment. The conventional diagnostic approaches—x-ray, CT-scan and tissue biopsy—are unable to detect it at initial stages, thereby, the delay in treatment claimed several lives due to cancer worldwide (Bray et al., 2018; Cohen et al., 2018; Newman et al., 2014). Recent advances in the field of cancer biology led to the discovery of several biomolecules which are specifically associated with cancer progression and development, hence termed as "biomarkers." These biomarkers can be classified on the basis of their chemical nature and function as: genomic, transcriptomic, proteomic, and metabolomic. Cancer-associated biomarkers play a crucial role in therapeutic designing, disease diagnostic Fig. 3.1 and prognostic Fig. 3.2.

Normally, a living cell has a finite life span and its genome deoxy ribonucleic acid (DNA) transcribes into ribonucleic acid (RNA) which leads to synthesis of proteins that take part in various physiological and metabolic process necessary for

**TABLE 3.1** The number of new cases of different types of cancers reported and number of cancer-related deaths in the year 2018.

| S. no. | Cancer | No. of new cases reported (% of all cancers) | No. of deaths caused (% of all cancer) |
|---|---|---|---|
| 1 | Lung | 2,093,876 (11.6) | 1,761,007 (18.4) |
| 2 | Gastric | 1,033,701 (5.7) | 782,685 (8.2) |
| 3 | Liver | 841,080 (4.7) | 781,631 (8.2) |
| 4 | Breast | 2,088,849 (11.6) | 626,679 (6.6) |
| 5 | Colo-rectal | 1,096,601 (6.1) | 551,269 (5.8) |

Biosensor Based Advanced Cancer Diagnostics. DOI: https://doi.org/10.1016/B978-0-12-823424-2.00007-7

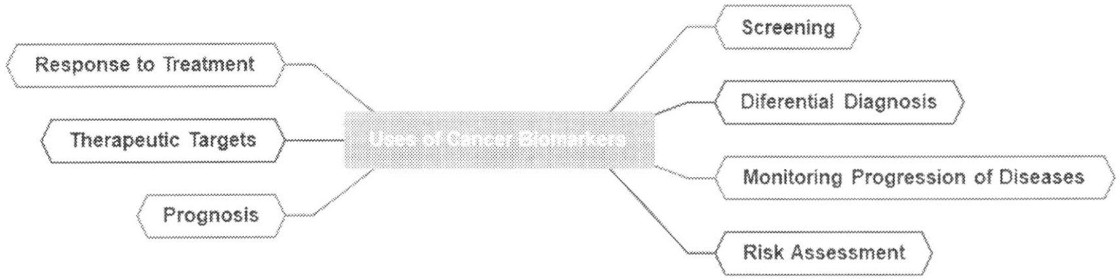


**FIGURE 3.1** Application of cancer Biomarkers.

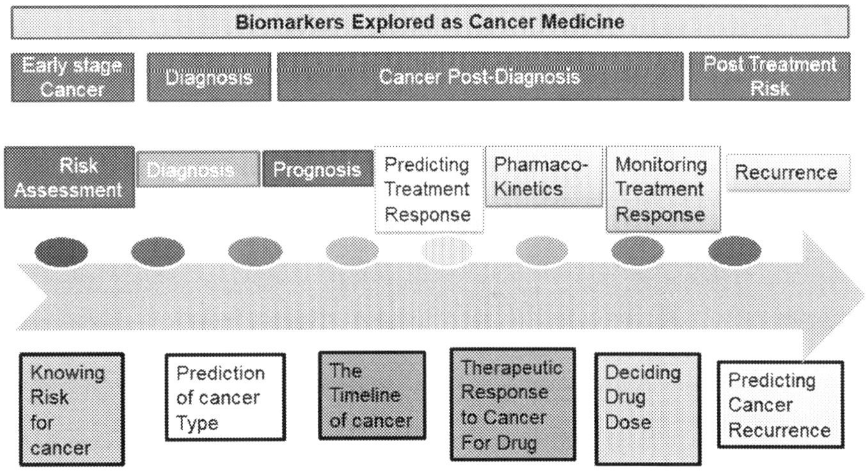


**FIGURE 3.2** Application of cancer Biomarkers in designing cancer therapeutics and diagnostics.

the body. Any alteration such as mutation in DNA in these processes leads to disturbance and causes cancer. Detection of mutation in DNA can be exploited as a predictive marker for cancer risk. Similarly, detection of the expression level of RNA, proteins, and metabolites can provide significant information with respect to disease progression and profiling. In this chapter, we gathered and presented the information about various biomarkers associated with the top five types of cancer in world, which can be exploited in designing sensitive and effective diagnostic technology for early detection of cancer.

## 3.2 Cancer biomarkers

### 3.2.1 Lung cancer biomarkers

Cancer is the second leading cause of death worldwide. Among different types of cancers, lung cancer is the most common cause of cancer deaths, according to the fact-sheet published by Global Cancer Statistics 2018. More than 2 million new cases of lung cancer were reported in 2018. Lung cancer alone caused 1,761,007 deaths (18.4% of all cancers) worldwide in 2018 (Bray et al., 2018). At initial stages of lung cancer, nodules in lungs are neither determined by CT scans nor accessible by tissue biopsy are observed. Liquid biopsy to trace molecular biomarkers allows the diagnosis of lung cancer at very early stages when a CT scan and biopsy process unable to do it. To detect lung cancer, biomarkers can be detected in different clinical samples such as blood, sputum, saliva, urine, and even exhaled breath. Genetic biomarkers that can be used to detect lung cancer include DNA mutations, circulating tumor DNA in genes such as *CA-125, CEA, CA 19–9, PRL, HGF, OPN, MPO, and TIMP-1* (Aravanis, Lee, & Klausner, 2017; Cohen et al., 2018; Guo et al., 2016; Newman et al., 2014; Phallen et al., 2017; Showe et al., 2009). Other than genetic biomarkers, epigenetic biomarkers like SNP, NER pathway mutations, DNA methylation in genes such as *SAX2, RASSF1A, and PTGER4* are also discovered for lung cancer diagnosis (Weiss, Schlegel, Kottwitz, König, & Tetzner, 2017; Zhang et al., 2017). Besides DNA, mRNA mutations and miRNA expressions are also promising candidates as lung cancer biomarkers. (Xing et al., 2015) The protein biomarkers which can be detected in blood are *CEA, A-125, CYFRA21–1,* (Doseeva, Colpitts, Gao, Woodcock, & Knezevic, 2015) *LG3BP and C163A* (Silvestri et al., 2018) *SCC, NSE, proGRP* (Liu, 2017) and complement fragment C4d (Ajona et al., 2013, 2018). Autoantibodies against lung tissues can also be detected as biomarker like *NY-ESO-1,* (Doseeva et al., 2015) *p53, CAGE, GBU4−5, SOX2, HuD, and MAGE A4* (Chapman et al., 2012; Healey et al., 2013; John et al., 2018). Vascular endothelial growth factor (*VEGF*) is a protein responsible for angiogenesis and blood vessel permeability. Researchers have found a potential role of *VEGF* in the diagnosis of lung cancer ((Banys-Paluchowski et al., 2018)Abdallah, Belal, El Bastawisy, & Gaafar, 2014; Lai et al., 2018; Lumachi et al., 2017). Abdallah et al. studied levels of *VEGF 165* in serum samples of advanced non small cell lung cancer (NSCLC) patients and healthy controls (Abdallah et al., 2014). The author found VEGFto be significantly elevated in NSCLC patients (707 pg/mL) than healthy controls (48 pg/mL). This study, proves the diagnostic potential of VEGF in NSCLC. Similarly, Lai et al. compared the diagnostic ability of VEGF in NSCLC to *CEA, CA-125 and CYFRA21−1* and found that *VEGF* is more sensitive and specific than other biomarkers *CEA, CA-125, and CYFRA21−1* (Lai et al., 2018). Moreover, the author suggested that *VEGF* combined with *CEA, CA-125, and CYFRA21−1* increases the diagnostic sensitivity and specificity to 85% and 90%, respectively. Mathe et al. identified two small molecules—creatine riboside and N-acetylneuraminic acid (NANA)—to be highly elevated in urine samples of NSCLC (Mathé et al., 2014). In this large cohort study, creatine riboside and NANA were identified as potential urinary metabolomic diagnostic biomarkers for NSCLC. Urine, sputum, and exhaled breath samples can be used to detect volatile metabolomics and metagenomics biomarkers. Further, with the help of the integration of bioinformatic, advanced metabolomic and metagenomic analysis tools the assement of NSCLC can be easily done with high accuracy (Mathé et al., 2014; Nakhleh et al., 2017; Newman et al., 2014; Xing et al., 2015).

### 3.2.2 Gastric cancer biomarkers

Gastric cancer (GC) has the second highest mortality rates worldwide (Bray et al., 2018). It is due the fact that GC is diagnosed at late stages where prognosis is worse. Early detection may help in decreasing the mortality and morbidity rates due to GC. There are several molecular biomarkers discovered till date to detect and diagnose GC early, but none are used for GC diagnostic purpose in clinics (Nick, Jenna-Lynn, Shahid, Kanthan, & Kanthan, 2016). In a study conducted by Kania et al. higher survivin, caspase mRNA and protein expression were detected in GC compared to normal gastric mucosa (Kania, Konturek, Marlicz, Hahn, & Konturek, 2003). In another study conducted by Kim et al., using 2 K cDNA microarray showed that six genes (*SKB1*, *NT5C3*, *ZNF9*, *p30*, *CDC20*, and *FEN1*) were found to be upregulated where two genes (*MT2A* and *CXX1*) were found to be downregulated in GC. Among them *CDC20 and MT2A* were recognized as a potent GC biomarkers (Kim et al., 2005). *HER2* gene upregulation was a known biomarker for breast cancer (BC; Gallardo et al., 2012) and its role in gastric cancer is ambiguous. Researchers found that the *HER2* gene is overexpressed in some cases of GC but not in all (Grabsch, Sivakumar, Gray, Gabbert, & Müller, 2010). E-cadherins are one of the most important biomarkers of GC. Gene *CDH-1* is responsible to prevent tumor formation in the stomach. Any kind of mutation in the *CDH-1* gene causes increased cell proliferation and tumor formation in the stomach (Corso et al., 2013). Proteinaceous receptor biomarkers such as *FGFR, mTOR, MET, and PD-1* can be detected in blood stream or serum to diagnose GC (Deng et al., 2012; Lee et al., 2011; Muro et al., 2016; Ohtsu et al., 2013). Wang D. et al. identified low serum levels of 2,4-hexadienoic acid, 4-methylphenyl dodecanoate, and glycerol tributanoate as a potential metabolomic marker in GC prognosis and diagnosis (Wang et al., 2017). Lario et al. identified histidine, tryprophan, and phenylacetylglutamine as metabolomic biomarkers to distinguish GC and non-GC patients (Lario et al., 2017). Rodriguez et al. suggested that there occurs an enrichment of pro-inflammatory oral bacterial species, increased abundance of lactic acid producing bacteria, and enrichment of short chain fatty acid production in GC which could be helpful in risk assessment (Rodríguez, 2017). The author also found *Lactococcus*, *Veilonella*, and *Fusobacteriaceae* (*Fusobacterium* and *Leptotrichia*) to be enriched in gastric microbiota. Hu et al. identified alterations in gastric microbiome of GC patients compared to gastritis patients. Hu (2018) opportunitistic pathogen found in oral microbiome such as genera *Neisseria*, *Alloprevotella*, and *Aggregatibacter*, species *Streptococcus_mitis_oralis_pneumoniae* and strain *Porphyromonas_endodontalis.t*_GCF_000174815 were abundant in GC patients gastric microbiomes. Moreover, *Sphingobium yanoikuyae* were found to be depleted in gastric microbiomes of GC. These *Sphingobium* are known to be responsible for carcinogenic compounds (Grabsch et al., 2010).

### 3.2.3 Liver cancer biomarkers

Liver cancer is the third leading cause of cancer death worldwide. It caused 781,631 deaths in the year 2018 that is, 8.2% of all types of cancer deaths around the globe (Bray et al., 2018). The high mortality of liver cancer is due to late diagnosis. Among different types of liver cancers, hepatocellular carcinoma (HCC) is the most lethal type of liver cancer (Mittal et al., 2016). Scientists around the world are working to develop biomarkers which can detect liver cancer early (Park et al., 2012; Seregni, Botti, & Bombardieri, 1995; Yuen et al., 2000). Alpha-fetoprotein (*AFP*) is the most widely used to detect liver cancer but its sensitivity and specificity is poor. Since, AFP levels are elevated in liver infections and inflammations also in othertype of liver cancer (Sterling et al., 2011). Studies suggest that AFP in combination with other protien biomarkers (such as *DCP, AFP-L3, and HCCR-1*) can be used as a more potent biomarker of HCC (Weiss et al., 2017; Zhang et al., 2017). Reactive fraction of alpha-fetoprotein (AFP-L3), lens culinaris agglutinin (LCA) and des-gammacarboxy prothrombin (DCP) are being used to detect hepatic cancer in Japan since the 1990s (Marrero & Lok, 2004; Spangenberg et al., 2006). However, biomarkers such as AFP, AFP-L3, glypican, osteopontin, and DCP are also found elevated in diseases and liver pathologies other than liver cancer (Carr et al., 2011; Behnes et al., 2013; Komine-Aizawa et al., 2015; Lim et al., 2012; Nakamura et al., 2018; Özkan et al., 2011; Simao, 2015; Sterling et al., 2009). HCCR-1, a protein encoded by human cervical cancer oncogene is identified in cervical cancer patient, but also found to be a biomarker of HCC along with cervical cancer (Yoon et al., 2004). In a study conducted by Zhang et al., HCCR-1 in combination with AFP is found to be a promising serological biomarker for small HCC (Zhang et al., 2012). In an official, large scale and multicenter clinical trial conducted by Yan Fu, heat shock protein 90alpha (Hsp90α) is found to be a promising biomarker in HCC diagnosis and discriminating HCC and other liver pathologies in patients (Fu et al., 2017). Hsp90α is a major molecular chaperone which is highly conserved (Frydman, 2001). Hsp90α can be secreted into the extracellular space by cancer cells after being translocated to the cell surface (Eustace et al., 2004). Moreover, plasma levels of secreted Hsp90α increase radically in cancer patients, and show a relationship confidently with tumor malignancy and metastatic ability (Xiaofeng et al., 2009). Circulating cell-free DNA (CCF) in blood released by cancerous or necrotic cells is a novel and promising biomarker in cancer diagnosis. Zhao et al. identified three major differentially methylated genes (DMGs) in genome-wide CCF-DNA profiles including *ZNF300, SLC22A20, and SHISA7* using methylcap-seq technique. These DMGs can differentiate chronic hepatitis B infection, Liver cirrhosis, and HCC (Zhao et al., 2014). Wen et al. also developed a methylated CpG tandems amplification and sequencing (MCTA-Seq) method which can detect thousands of hypermethylated DNA segments simultaneously in CCF-DNA (Wen, Li, et al., 2015; Wen, Han, et al., 2015). This is a highly sensitive method which can detect DNA hypermethylation in DNA samples as small as 2.5 copies of haploid genome (i.e., 7.5 pg). Hypermethylated CpG islands—RGS10, ST8SIA6, RUNX2, and VIM—are the four major biomarkers identified in cancer patients. Yan et al. suggested that elevated CCF-DNA levels in combination with AFP levels and age can be used potentially to identify HCC patients (Yan et al., 2018). The study also revealed that CCF-DNA levels are associated with BMI in liver fibrosis and are less than CCF-DNA levels in HCC. Wen et al. identified a panel of four blood-based miRNAs to early diagnose HCC in preclinical settings which includes miR-20a-5p, miR-320a, miR-324−3p, and miR-375. In the same study, a panel of eight miRNAs was also identified which can differentiate HCC and normal patients. Poto et al. identified plasma metabolites that may be correlated with HCC in a race-specific manner in view of AA (African Americans) and EA (European Americans) (Poto, 2018). The levels of selected plasma metabolites were quantified by GC-SIM-Ms and analyzed using a least absolute shrinkage and selection operator (LASSO) regression model. Several metabolites including alpha tocopherol for AA and EA combined, valine for AA only, and glycine for EA only, exhibited better performance than AFP as HCC diagnostic biomarker. However, study on larger samples is required to confirm the role of alpha tocopherol, valine, and glycine as cancer biomarkers (Marrero & Lok, 2004).

### 3.2.4 Breast cancer biomarkers

Breast Cancer (BC) is the most prevalent cancer in women in the world. It caused (626,679) 6.6% of cancer deaths in the year 2018, being the fourth leading cause of cancer deaths in the world (Bray et al., 2018). Early diagnosis is the key to better prognosis of BC (Ranjan et al., 2020). There are several molecular biomarkers developed to detect BC early. Certain elevated proteins or auto-antibodies against them (HER2, EGFR, CA-15−9, CA27.29, p53, VEGF, EGF, cathepsin D, cyclin E, human epidydimal protein 4) serve as potent biomarkers to detect BC (Streckfus et al., 2000; Chapman et al., 2012; Healey et al., 2013; John et al., 2018). HER2 is a tyrosine kinase receptor which is encoded by the *HER2* gene. The elevated HER2 protein can be detected in serum and saliva of HER2 positive BC patients, which serves as a diagnostic tool (Chapman et al., 2007; Laidi, Bouziane, Errachid, & Zaoui, 2016; Ludovini et al., 2008; Streckfus, Arreola, Edwards, & Bigler, 2012). Trastuzumab, a humanized monoclonal antibody which targets human epidermal growth factor receptor 2, found to be an efficient treatment for HER2 positive BC cells (Hudis, 2007). However, some of the HER2 positive BC patients show resistance to trastuzumab treatment. The number of gene expression via metagenes analysis were identified that mirror the level of pathway activation according to the expression of proteins and/or phosphorylated-proteins pertinent in cancer. Identification of metagenes is done by integrating RNA and protein expression data from RPPA (reverse phase protein assay) analysis of HER2 positive BC patients from

TCGA (The Tumor Cancer Genomic Atlas project). Among the identified metagenes, AnnexinA1 metagene is found to be responsible for trastuzumab resistance in HER2 positive BC patients (Sonnenblick et al., 2015). Another study in 2017 using reverse phase protein arrays RPPA, discovered two metabolism related proteins: serine hydroxymethyltransferase 2 (*SHMT2*) and amino acid transporter ASCT2, to be poor prognostic biomarkers of BC (Bernhardt et al., 2017). Inflammatory proteins such as fibrinogen and soluble intracellular adhesion molecule 1 are associated directly and inversely, respectively, with BC risk (Tobias et al., 2018). In triple negative BC, MicroRNAs can be detected to diagnose and monitor BC, miR-126–5p, and miR-34a quantification in blood using RT-qPCR with 84% sensitivity (Kahraman et al., 2018). Furthermore, exosomal miR-101 and miR-372 are found considerably higher in the serum exosomes of patients with BC than in healthy controls. MiR-373 is found to be significantly high in triple-negative patients when compared with luminal cancers or healthy controls. Thus, miR-372 can be targeted to treat triple negative BC (Eichelser et al., 2014). BC risk can also be estimated in patients by genome screening for *BRCA1/BRCA2* genes on chromosome 13 and 17, respectively. Mutations in these tumor suppressor genes indicate a risk of hereditary breast and ovarian cancer. These genes were first discovered by Mary-Claire King, PhD, who was professor of genomic science and medicine at University of Washington (Miki et al., 1994; Wooster et al., 1995). Other genetic mutations which are being used to predict BC risk and thus used in BC diagnosis are PTEN, TP53, *CHEK2, ATM, PALB* and *BRIP1 2, FGFR2, MAP3K1*, and TGFB1 (Hirshfield, Rebbeck, & Levine, 2010; Nathanson, Wooster, & Weber, 2001; Stratton & Rahman, 2008). Models to distinguish early and metastatic BC, and to predict relapse is made using metabolites in biological samples such as serum, blood, etc., Tenori et al. confirmed that the serum metabolomic prognostic model which predicts relapse with 90% sensitivity, 67% specificity, and 73% predictive accuracy. Lower levels of histidine and higher levels of glucose and lipids are found strongly associated to BC relapse (Tenori et al., 2015). Studies have proved a direct relationship between estrogen levels, gut microbiota, and a risk of BC in postmenopausal women. Microbiome of post-menopausal cancer patients was found to be more diverse at species level and abundance (Flores et al., 2012; Key et al., 2011; Zhu et al., 2018). Zhu et al. recently using shotgun metagenomics analysis identified several bacterial species (some elevated and some decreased in number) which are related to post-menopausal BC in women. These species and their abundance in fecus samples could be used to diagnose BC cancer early (Zhu et al., 2018). Another similar clinical trial is going on at the University of Granada, Spain. This is the first study which evaluates the relationship between bacteria, archaea, viruses, and fungi in the gut and the risk of BC (Diaz, 2019).

### 3.2.5 Colorectal cancer biomarkers

Colorectal cancer/bowel cancer (CRC) is a type of cancer which occurs in (either or both) parts of large intestine from colon to rectum. Colon cancer alone took 551,269 lives in 2018. Together, colorectal cancer caused 9% of total cancer deaths last year (Bray et al., 2018). Liquid biopsy-based biomarker detection is an easy, quick, and noninvasive method for diagnosis, prognosis, and monitoring of colorectal cancer. It is suggested that mutations in *KRAS/BRAF*, *TP53* genes, DNA repair genes, and microsatellite instability can be explored for risk assessment of CRC. The *KRAS/BRAF* mutations cause cituximab resistance in EGFR-related CRC. Sclafani et al. detected ctDNA mutations in KRAS/BRAF genes with high sensitivity by using ddPCR in a sample size of 97 patients, however this study needs further trials for assessment of locally advanced rectal cancer (Sclafani et al., 2018). Mutant KRAS/BRAF detection allows CRC diagnosis for targeted therapy. *TP53* is a tumor suppressor gene which regulates the cell cycle, cell proliferation, and apoptosis. Mutation in *TP53* besides DNA repair genes causes mutation in proto-oncogenes. The proto-oncogenes turn into oncogenes and cause cancer (Croce, 2008). Apoptosis-stimulating proteins of *p53* (ASPP) are a group of three proteins ASPP-1, ASPP-2 and inhibitor of apoptosis stimulating proteins of *p53* (iASPP). These protein levels are associated with different types of cancers (Mak et al., 2013; Schittenhelm et al., 2013; Shi et al., 2015; Song, 2015). In a study conducted by Lin et al. reported significantly lower levels of the protein and mRNA expression of *ASPP1 and ASPP2*, and drastically increased levels of iASPP in human CRC specimens compared with the noncancerous controls (Yin et al., 2018). The study also suggests that increased iASPP levels can be combined with elevated AFP in blood which could be used as a potential biomarker for diagnosis and prognosis of CRC. Studies have identified methylated DNA panels which can aid in prognosis and monitoring of CRC which include *EYA4, GRIA4, ITGA4, MAP3K14-AS1, MSC*, (Barault et al., 2018) *BCAT1* and *IKZF1* (Symonds et al., 2018). Different types of protein are identified as potential CRC diagnostic biomarker in several studies such as serum nucleoside diphosphate kinase A, (Otero-Estévez et al., 2016) inflammatory cytokines IL-1, IL-6, IL-10, TNFalpha, CCL2, CXCL8, VEGF (Kim et al., 2016) and the acute phase protein pentraxin-3, (Di Caro et al., 2016) plasma C-reactive protein (CRP), interleukin-6 (IL-6), and tumor necrosis factor receptor 2 (TNFR-2), (Kim et al., 2005) neurotensin, (Kontovounisios, Qiu, Rasheed, Darzi, & Tekkis, 2017) and

Galectin-1 and 90 K/Mac-2BP (Wu et al., 2015). Carcinoembryonic antigen (CEA) is the major oncofetal glycoprotein biomarker found elevated in CRC patients blood. CEA is an antigen produced by normal cells and tumor cells (at higher levels) in the intestine (Gold & Freedman, 1965). Besides imaging techniques, CEA has been clinically used since a long time for diagnosing, and monitoring colorectal cancer (Duffy et al., 2007; Locker et al., 2006). CEA as a biomarker is not 100% accurate in diagnosing tumor growth. Therefore, researchers are trying to use it in combination with other biomarkers. Carbohydrate antigen 19–9 (CA 19–9) increases sensitivity and specificity of CRC diagnosis with CEA. Polat et al. estimated lymph node invasion and distant metastasis in colorectal cancer by measuring levels of CEA and CA 19–9 preoperatively (Polat et al., 2014). It is reconfirmed in several studies that CEA in a panel of biomarkers such as CEA and CA 19–9, (Zhang, Lin, & Zhang, 2015) *CEA, CA 19–9, TK-1, and CA 72–4* (Ning et al., 2018) can play a pivoted role in GC and CRC diagnosis. CA 19–9 cannot be used as a diagnostic biomarker (Ning et al., 2018; Polat et al., 2014; Zhang et al., 2015), but it is proven to be a better prognosis biomarker than CEA (Diez et al., 1994; Filella et al., 1992; Reiter et al., 2000). A recent study by Ma et al. reports that methylated septin9 gene with CEA can be used in more accurate diagnosis (Ma et al., 2019). mSEPT9 is proved as a more sensitive biomarker than CEA but mSEPT9 and CEA together are even more sensitive. The mSEPT9 is also proven to predict recurrence of CRC in the same study. A critical role in biological/cellular processes of cancer such as proliferation, apoptosis, differentiation, invasion, and metastasis is played by miRNAs (Bartel, 2009). MiRNAs such as miR-21, miR-221, miR-222, mir-31, miR-92a, miR-181b, and miR-203, etc. can also be used forcolorectal cancer diagnosis and prognosis (Du et al., 2014; Basati et al., 2014; Wang et al., 2017). Pu et al. by using a quantitative reverse transcription-polymerase chain reaction, determined plasma levels of miR-21, miR-221, and miR-222 in CRC patients and control. The research suggested direct amplification of miR-21 as a potential diagnostic and prognostic biomarker. It also indicates association of miR-21 to p53 (Pu et al., 2010). Later, Wang et al. also discovered a six miRNAs signature to detect CRC with sensitivity and specificity of 93% and 91%, respectively (Wang et al., 2014). The miRNA signature includes miR-21, let-7g, miR-31, miR-92a, miR-181b, and miR-203. The signature can be used in diagnosis and monitoring of CRC. Colorectal adenomatous polyps are considered as precursors of CRC. Gu et al. differentiated colorectal polyps, CRC, and controls using serum based metabolic profiles (Gu et al., 2019). The study discovered abnormal metabolic pathways such as: glutamine and glutamate metabolism pathways, alanine, aspartate, and glutamate metabolism pathways (in colorectal polyps and CRC), pyruvate and glycerolipid metabolism pathways (in colorectal polyps), glycolysis and glycine, serine, and threonine metabolism (in CRC). Cellular proliferation might be due to the changed metabolism. For population screening of colorectal cancer, colonoscopy, blood cells/mutated, DNA detection in fecal samples and blood samples are utilized (Church, 2013; Schreuders et al., 2015; Singal et al., 2018). These methods are inconvenient, costly and tedious option for early diagnose CRC. A novel method for population screening of CRC in early stages using urine samples is developed by (Deng et al. (2019), Deng et al. (2012), Lee et al. (2011), Muro et al. (2016), and Ohtsu et al. (2013). The study identifies diacetylspermine and kynurenine as metabolomic predictors for CRC with the specificity and sensitivity values, 90.6% and 74.3%, respectively.

## 3.3 Concluding remarks

Research in the field of cancer-specific biomarkers have provided a promising source of novel diagnostic tools. Various groups have reported that altered cancer-associated biomarkers can be exploited to diagnose and monitor various cancers with greater sensitivity and specificity Assessment of genomic and transcriptomic biomarkers found to be potentially very sensitive approaches for discriminating between cancerous non-cancerous (benign) conditions. Besides, this one could detect cancers at a much earlier stage by quantitative analysis of potential biomarker associated with specific cancer. Given the possible diagnostic power of genomic, transcriptomic and proteomic, and metabolomic and biomarkers, these are currently one of the most promising areas of research in the field of development of cancer prognostic and diagnostics devices.

## References

Abdallah, A., Belal, M., El Bastawisy, A., & Gaafar, R. (2014). Plasma vascular endothelial growth factor 165 in advanced non-small cell lung cancer. *Oncology Letters*, 7(6), 2121–2129. Available from https://doi.org/10.3892/ol.2014.2016.

Ajona, D., Okrój, M., Pajares, M. J., Agorreta, J., Lozano, M. D., Zulueta, J. J., ... Pio, R. (2018). Complement C4d-specific antibodies for the diagnosis of lung cancer. *Oncotarget*, 9(5), 6346–6355. Available from https://doi.org/10.18632/oncotarget.23690.

Ajona, D., Pajares, M. J., Corrales, L., Perez-Gracia, J. L., Agorreta, J., Lozano, M. D., … Pio, R. (2013). Investigation of complement activation product C4d as a diagnostic and prognostic biomarker for lung cancer. *Journal of the National Cancer Institute*, *105*(18), 1385–1393. Available from https://doi.org/10.1093/jnci/djt205.

Aravanis, A. M., Lee, M., & Klausner, R. D. (2017). Next-generation sequencing of circulating tumor DNA for early cancer detection. *Cell*, *168*(4), 571–574. Available from https://doi.org/10.1016/j.cell.2017.01.030.

Banys-Paluchowski, M., Witzel, I., Riethdorf, S., Pantel, K., Rack, B., Janni, W., … Müller, V. (2018). The clinical relevance of serum vascular endothelial growth factor (VEGF) in correlation to circulating tumor cells and other serum biomarkers in patients with metastatic breast cancer. *Breast Cancer Research and Treatment*, *172*(1), 93–104. Available from https://doi.org/10.1007/s10549-018-4882-z.

Barault, L., Amatu, A., Siravegna, G., Ponzetti, A., Moran, S., Cassingena, A., … Di Nicolantonio, F. (2018). Discovery of methylated circulating DNA biomarkers for comprehensive non-invasive monitoring of treatment response in metastatic colorectal cancer. *Gut*, *67*(11), 1995–2005. Available from https://doi.org/10.1136/gutjnl-2016-313372.

Bartel, D. P. (2009). MicroRNAs: Target recognition and regulatory functions. *Cell*, *136*(2), 215–233. Available from https://doi.org/10.1016/j.cell.2009.01.002.

Basati, G., Emami Razavi, A., Abdi, S., & Mirzaei, A. (2014). Elevated level of microRNA-21 in the serum of patients with colorectal cancer. *Medical Oncology*, *31*(10), 1–5. Available from https://doi.org/10.1007/s12032-014-0205-3.

Behnes, M., Brueckmann, M., Lang, S., Espeter, F., Weiss, C., Neumaier, M., … Hoffmann, U. (2013). Diagnostic and prognostic value of osteopontin in patients with acute congestive heart failure. *European Journal of Heart Failure*, *15*(12), 1390–1400. Available from https://doi.org/10.1093/eurjhf/hft112.

Bernhardt, S., Bayerlová, M., Vetter, M., Wachter, A., Mitra, D., Hanf, V., … Kantelhardt, E. J. (2017). Proteomic profiling of breast cancer metabolism identifies SHMT2 and ASCT2 as prognostic factors. *Breast Cancer Research*, *19*(1). Available from https://doi.org/10.1186/s13058-017-0905-7.

Bray, F., Ferlay, J., Soerjomataram, I., Siegel, R. L., Torre, L. A., & Jemal, A. (2018). Global cancer statistics 2018: GLOBOCAN estimates of incidence and mortality worldwide for 36 cancers in 185 countries. *CA Cancer Journal for Clinicians*, *68*(6), 394–424. Available from https://doi.org/10.3322/caac.21492.

Carr, B. I., Wang, Z., Wang, M., & Wei, G. (2011). Differential effects of vitamin K1 on AFP and DCP levels in patients with unresectable HCC and in HCC cell lines. *Digestive Diseases and Sciences*, *56*(6), 1876–1883. Available from https://doi.org/10.1007/s10620-010-1521-x.

Chapman, C. J., Healey, G. F., Murray, A., Boyle, P., Robertson, C., Peek, L. J., … Robertson, J. F. R. (2012). EarlyCDT®-Lung test: Improved clinical utility through additional autoantibody assays. *Tumor Biology*, *33*(5), 1319–1326. Available from https://doi.org/10.1007/s13277-012-0379-2.

Chapman, C., Murray, A., Chakrabarti, J., Thorpe, A., Woolston, C., Sahin, U., … Robertson, J. (2007). Autoantibodies in breast cancer: Their use as an aid to early diagnosis. *Annals of Oncology*, *18*(5), 868–873. Available from https://doi.org/10.1093/annonc/mdm007.

Church, J. (2013). Complications of colonoscopy. *Gastroenterology Clinics of North America*, *42*(3), 639–657. Available from https://doi.org/10.1016/j.gtc.2013.05.003.

Cohen, J. D., Li, L., Wang, Y., Thoburn, C., Afsari, B., Danilova, L., … Papadopoulos, N. (2018). Detection and localization of surgically resectable cancers with a multi-analyte blood test. *Science*, *359*(6378), 926–930. Available from https://doi.org/10.1126/science.aar3247.

Corso, G., Carvalho, J., Marrelli, D., Vindigni, C., Carvalho, B., Seruca, R., … Oliveira, C. (2013). Somatic mutations and deletions of the e-cadherin gene predict poor survival of patients with gastric cancer. *Journal of Clinical Oncology*, *31*(7), 868–875. Available from https://doi.org/10.1200/JCO.2012.44.4612.

Croce, C. M. (2008). Oncogenes and cancer. *New England Journal of Medicine*, *358*(5), 502–511. Available from https://doi.org/10.1056/NEJMra072367.

Deng, N., Goh, L. K., Wang, H., Das, K., Tao, J., Tan, I. B., … Tan, P. (2012). A comprehensive survey of genomic alterations in gastric cancer reveals systematic patterns of molecular exclusivity and co-occurrence among distinct therapeutic targets. *Gut*, *61*(5), 673–684. Available from https://doi.org/10.1136/gutjnl-2011-301839.

Deng, L., Ismond, K., Liu, Z., Constable, J., Wang, H., Alatise, O. I., … Chang, D. (2019). Urinary metabolomics to identify a unique biomarker panel for detecting colorectal cancer: A multicenter study. *Cancer Epidemiology Biomarkers and Prevention*, *28*(8), 1283–1292. Available from https://doi.org/10.1158/1055-9965.EPI-18-1291.

Di Caro, G., Carvello, M., Pesce, S., Erreni, M., Marchesi, F., Todoric, J., … Spinelli, A. (2016). Circulating inflammatory mediators as potential prognostic markers of human colorectal cancer. *PLoS ONE*, *11*(2). Available from https://doi.org/10.1371/journal.pone.0148186.

Diaz, P. (2019). Association of breast and gut microbiota dysbiosis and the risk of breast cancer: A case-control clinical study. *BMC Cancer*, *19*.

Diez, M., Granell, J., Torres, A., Gomez, A., Balibrea, J. L., Pollán, M., … Maestro, M. L. (1994). Prognostic significance of serum c.125 antigen assay in patients with non-small cell lung cancer. *Cancer*, *73*(5), 1368–1376. <https://doi.org/10.1002/1097-0142(19940301)73:5 < 1368::AID-CNCR2820730510 > 3.0.CO;2-O>.

Doseeva, V., Colpitts, T., Gao, G., Woodcock, J., & Knezevic, V. (2015). Performance of a multiplexed dual analyte immunoassay for the early detection of non-small cell lung cancer. *Journal of Translational Medicine*, *13*(1). Available from https://doi.org/10.1186/s12967-015-0419-y.

Du, M., Liu, S., Gu, D., Wang, Q., Zhu, L., Kang, M., … Wang, M. (2014). Clinical potential role of circulating microRNAs in early diagnosis of colorectal cancer patients. *Carcinogenesis*, *35*(12), 2723–2730. Available from https://doi.org/10.1093/carcin/bgu189.

Duffy, M. J., van Dalen, A., Haglund, C., Hansson, L., Holinski-Feder, E., Klapdor, R., … Topolcan, O. (2007). Tumour markers in colorectal cancer: European Group on Tumour Markers (EGTM) guidelines for clinical use. *European Journal of Cancer*, *43*(9), 1348–1360. Available from https://doi.org/10.1016/j.ejc.2007.03.021.

Eichelser, C., Stückrath, I., Müller, V., Milde-Langosch, K., Wikman, H., Pantel, K., & Schwarzenbach, H. (2014). Increased serum levels of circulating exosomal microRNA-373 in receptor-negative breast cancer patients. *Oncotarget*, *5*(20), 9650–9663. Available from https://doi.org/10.18632/oncotarget.2520.

Eustace, B. K., Sakurai, T., Stewart, J. K., Yimlamai, D., Unger, C., Zehetmeier, C., ... Jay, D. G. (2004). Functional proteomic screens reveal an essential extracellular role for hsp90α in cancer cell invasiveness. *Nature Cell Biology*, *6*(6), 507–514. Available from https://doi.org/10.1038/ncb1131.

Filella, X., Molina, R., Grau, J. J., Piqué, J. M., Garcia-Valdecasas, J. C., Astudillo, E., ... Ballesta, A. M. (1992). Prognostic value of CA 19.9 levels in colorectal cancer. *Annals of Surgery*, *216*(1), 55–59. Available from https://doi.org/10.1097/00000658-199207000-00008.

Flores, R., Shi, J., Fuhrman, B., Xu, X., Veenstra, T. D., Gail, M. H., ... Goedert, J. J. (2012). Fecal microbial determinants of fecal and systemic estrogens and estrogen metabolites: A cross-sectional study. *Journal of Translational Medicine*, *10*(1). Available from https://doi.org/10.1186/1479-5876-10-253.

Frydman, J. (2001). Folding of newly translated proteins in vivo: The role of molecular chaperones. *Annual Review of Biochemistry*, *70*, 603–648. Available from https://doi.org/10.1146/annurev.biochem.70.1.603.

Fu, Y., Xu, X., Huang, D., Cui, D., Liu, L., Liu, J., ... Luo, Y. (2017). Plasma heat shock protein 90alpha as a biomarker for the diagnosis of liver cancer: An official, large-scale, and multicenter clinical trial. *EBioMedicine*, *24*, 56–63. Available from https://doi.org/10.1016/j.ebiom.2017.09.007.

Gallardo, A., Lerma, E., Escuin, D., Tibau, A., Muñoz, J., Ojeda, B., ... Peiró, G. (2012). Increased signalling of EGFR and IGF1R, and deregulation of PTEN/PI3K/Akt pathway are related with trastuzumab resistance in HER2 breast carcinomas. *British Journal of Cancer*, *106*(8), 1367–1373. Available from https://doi.org/10.1038/bjc.2012.85.

Gold, P., & Freedman, S. O. (1965). Specific carcinoembryonic antigens of the human digestive system. *The Journal of Experimental Medicine*, *122*(3), 467–481. Available from https://doi.org/10.1084/jem.122.3.467.

Grabsch, H., Sivakumar, S., Gray, S., Gabbert, H. E., & Müller, W. (2010). HER2 expression in gastric cancer: Rare, heterogeneous and of no prognostic value-conclusions from 924 cases of two independent series. *Cellular Oncology*, *32*(1–2), 57–65. Available from https://doi.org/10.3233/CLO-2009-0497.

Gu, J., Xiao, Y., Shu, D., Liang, X., Hu, X., Xie, Y., ... Li, H. (2019). Metabolomics analysis in serum from patients with colorectal polyp and colorectal cancer by 1H-NMR spectrometry. *Disease Markers*, *2019*. Available from https://doi.org/10.1155/2019/3491852.

Guo, N., Lou, F., Ma, Y., Li, J., Yang, B., Chen, W., ... Liu, Y. (2016). Circulating tumor DNA detection in lung cancer patients before and after surgery. *Scientific Reports*, *6*. Available from https://doi.org/10.1038/srep33519.

Healey, G. F., Lam, S., Boyle, P., Hamilton-fairley, G., Peek, L. J., & Robertson, J. F. R. (2013). Signal stratification of autoantibody levels in serum samples and its application to the early detection of lung cancer. *Journal of Thoracic Disease*, *5*(5), 618–625. Available from https://doi.org/10.3978/j.issn.2072-1439.2013.08.65.

Hirshfield, K. M., Rebbeck, T. R., & Levine, A. J. (2010). Germline mutations and polymorphisms in the origins of cancers in women. *Journal of Oncology*, 1–11. Available from https://doi.org/10.1155/2010/297671.

Hu, Y. (2018). The gastric microbiome is perturbed in advanced gastric adenocarcinoma identified through shotgun metagenomics. *Frontiers in Cellular and Infection Microbiology*, *8*.

Hudis, C. A. (2007). Trastuzumab – Mechanism of action and use in clinical practice. *New England Journal of Medicine*, *357*(1), 39–51. Available from https://doi.org/10.1056/NEJMra043186.

John, E., Derek, W., Mark, A., Geoffrey, H.-F., Jett, J. R., & Gorlova, O. Y. (2018). Cost-effectiveness of an autoantibody test (EarlyCDT-Lung) as an aid to early diagnosis of lung cancer in patients with incidentally detected pulmonary nodules. *PLOS ONE*, e0197826. Available from https://doi.org/10.1371/journal.pone.0197826.

Kahraman, M., Röske, A., Laufer, T., Fehlmann, T., Backes, C., Kern, F., ... Schrauder, M. G. (2018). MicroRNA in diagnosis and therapy monitoring of early-stage triple-negative breast cancer. *Scientific Reports*, *8*(1). Available from https://doi.org/10.1038/s41598-018-29917-2.

Kania, J., Konturek, S. J., Marlicz, K., Hahn, E. G., & Konturek, P. C. (2003). Expression of survivin and caspase-3 in gastric cancer. *Digestive Diseases and Sciences*, *48*(2), 266–271. Available from https://doi.org/10.1023/A:1021915124064.

Key, T. J., Appleby, P. N., Reeves, G. K., Roddam, A. W., Helzlsouer, K. J., Alberg, A. J., ... Strickler, H. D. (2011). Circulating sex hormones and breast cancer risk factors in postmenopausal women: Reanalysis of 13 studies. *British Journal of Cancer*, *105*(5), 709–722. Available from https://doi.org/10.1038/bjc.2011.254.

Kim, J. M., Sohn, H. Y., Yoon, S. Y., Oh, J. H., Yang, J. O., Kim, J. H., ... Kim, N. S. (2005). Identification of gastric cancer-related genes using a cDNA microarray containing novel expressed sequence tags expressed in gastric cancer cells. *Clinical Cancer Research*, *11*(2), 473–482.

Kim, C., Zhang, X., Chan, A. T., Sesso, H. D., Rifai, N., Stampfer, M. J., & Ma, J. (2016). Inflammatory biomarkers, aspirin, and risk of colorectal cancer: Findings from the physicians' health study. *Cancer Epidemiology*, *44*, 65–70. Available from https://doi.org/10.1016/j.canep.2016.07.012.

Komine-Aizawa, S., Masuda, H., Mazaki, T., Shiono, M., Hayakawa, S., & Takayama, T. (2015). Plasma osteopontin predicts inflammatory bowel disease activities. *International Surgery*, *100*(1), 38–43. Available from https://doi.org/10.9738/INTSURG-D-13-00160.1.

Kontovounisios, C., Qiu, S., Rasheed, S., Darzi, A., & Tekkis, P. (2017). The role of neurotensin as a novel biomarker in the endoscopic screening of high-risk population for developing colorectal neoplasia. *Updates in Surgery*, *69*(3), 397–402. Available from https://doi.org/10.1007/s13304-017-0464-6.

Lai, Y., Wang, X., Zeng, T., Xing, S., Dai, S., Wang, J., ... Liu, W. (2018). Serum VEGF levels in the early diagnosis and severity assessment of non-small cell lung cancer. *Journal of Cancer*, *9*(9), 1538–1547. Available from https://doi.org/10.7150/jc.23973.

Laidi, F., Bouziane, A., Errachid, A., & Zaoui, F. (2016). Usefulness of salivary and serum auto-antibodies against tumor biomarkers HER2 and MUC1 in breast cancer screening. *Asian Pacific Journal of Cancer Prevention*, *17*(1), 335–339. Available from https://doi.org/10.7314/APJCP.2016.17.1.335.

Lario, S., Ramírez-Lázaro, M. J., Sanjuan-Herráez, D., Brunet-Vega, A., Pericay, C., Gombau, L., . . . Calvet, X. (2017). Plasma sample based analysis of gastric cancer progression using targeted metabolomics. *Scientific Reports*, *7*(1). Available from https://doi.org/10.1038/s41598-017-17921-x.

Lee, J., Seo, J. W., Jun, H. J., Ki, C. S., Park, S. H., Park, Y. S., . . . Park, J. O. (2011). Impact of MET amplification on gastric cancer: Possible roles as a novel prognostic marker and a potential therapeutic target. *Oncology Reports*, *25*(6), 1517–1524. Available from https://doi.org/10.3892/or.2011.1219.

Lim, A. M., Rischin, D., Fisher, R., Cao, H., Kwok, K., Truong, D., . . . Le, Q. T. (2012). Prognostic significance of plasma osteopontin in patients with locoregionally advanced head and neck squamous cell carcinoma treated on TROG 02.02 phase III trial. *Clinical Cancer Research*, *18*(1), 301–307. Available from https://doi.org/10.1158/1078-0432.CCR-11-2295.

Liu, L., et al. (2017). The combination of the tumor markers suggests the histological diagnosis of lung cancer. *Biomed Res Int*.

Locker, G. Y., Hamilton, S., Harris, J., Jessup, J. M., Kemeny, N., Macdonald, J. S., . . . Bast, R. C. (2006). ASCO 2006 update of recommendations for the use of tumor markers in gastrointestinal cancer. *Journal of Clinical Oncology*, *24*(33), 5313–5327. Available from https://doi.org/10.1200/JCO.2006.08.2644.

Ludovini, V., Gori, S., Colozza, M., Pistola, L., Rulli, E., Floriani, I., . . . Crinò, L. (2008). Evaluation of serum HER2 extracellular domain in early breast cancer patients: Correlation with clinicopathological parameters and survival. *Annals of Oncology*, *19*(5), 883–890. Available from https://doi.org/10.1093/annonc/mdm585.

Lumachi, F., et al. (2017). Diagnostic value of pleural cytology together with pleural CEA and VEGF in patients with NSCLC and lung metastases from breast cancer. *Journal of Thoracic Oncology*, *12*(1), S975–S976.

Ma, Z. Y., Law, W. L., Ng, E. K. O., Chan, C. S. Y., Lau, K. S., Cheng, Y. Y., . . . Leung, W. K. (2019). Methylated septin 9 and carcinoembryonic antigen for serological diagnosis and monitoring of patients with colorectal cancer after surgery. *Scientific Reports*, *9*(1). Available from https://doi.org/10.1038/s41598-019-46876-4.

Mak, V. C. Y., Lee, L., Siu, M. K. Y., Wong, O. G. W., Lu, X., Ngan, H. Y. S., . . . Cheung, A. N. Y. (2013). Downregulation of ASPP2 in choriocarcinoma contributes to increased migratory potential through Src signaling pathway activation. *Carcinogenesis*, 2170–2177. Available from https://doi.org/10.1093/carcin/bgt161.

Marrero, J. A., & Lok, A. S. F. (2004). *Newer markers for hepatocellular carcinoma*, . *Gastroenterology* (127, 5, pp. S113–S119). W.B. Saunders. Available from https://doi.org/10.1053/j.gastro.2004.09.024.

Mathé, E. A., Patterson, A. D., Haznadar, M., Manna, S. K., Krausz, K. W., Bowman, E. D., . . . Harris, C. C. (2014). Noninvasive urinary metabolomic profiling identifies diagnostic and prognostic markers in lung cancer. *Cancer Research*, *74*(12), 3259–3270. Available from https://doi.org/10.1158/0008-5472.CAN-14-0109.

Miki, Y., Swensen, J., Shattuck-Eidens, D., Futreal, P. A., Harshman, K., Tavtigian, S., . . . Skolnick, M. H. (1994). A strong candidate for the breast and ovarian cancer susceptibility gene BRCA1. *Science*, *266*(5182), 66–71. Available from https://doi.org/10.1126/science.7545954.

Mittal, S., Kanwal, F., Ying, J., Chung, R., Sada, Y. H., Temple, S., . . . El-Serag, H. B. (2016). Effectiveness of surveillance for hepatocellular carcinoma in clinical practice: A United States cohort. *Journal of Hepatology*, *65*(6), 1148–1154. Available from https://doi.org/10.1016/j.jhep.2016.07.025.

Muro, K., Chung, H. C., Shankaran, V., Geva, R., Catenacci, D., Gupta, S., . . . Bang, Y. J. (2016). Pembrolizumab for patients with PD-L1-positive advanced gastric cancer (KEYNOTE-012): A multicentre, open-label, phase 1b trial. *The Lancet Oncology*, *17*(6), 717–726. Available from https://doi.org/10.1016/S1470-2045(16)00175-3.

Nakamura, M., Xu, C., Diack, C., Ohishi, N., Lee, R. M., Iida, S., . . . Chen, Y. C. (2018). Time-to-event modelling of effect of codrituzumab on overall survival in patients with hepatocellular carcinoma. *British Journal of Clinical Pharmacology*, *84*(5), 944–951. Available from https://doi.org/10.1111/bcp.13530.

Nakhleh, M. K., Amal, H., Jeries, R., Broza, Y. Y., Aboud, M., Gharra, A., . . . Haick, H. (2017). Diagnosis and classification of 17 diseases from 1404 subjects via pattern analysis of exhaled molecules. *ACS Nano*, *11*(1), 112–125. Available from https://doi.org/10.1021/acsnano.6b04930.

Nathanson, K. N., Wooster, R., & Weber, B. L. (2001). Breast cancer genetics: What we know and what we need. *Nature Medicine*, *7*(5), 552–556. Available from https://doi.org/10.1038/87876.

Newman, A. M., Bratman, S. V., To, J., Wynne, J. F., Eclov, N. C. W., Modlin, L. A., . . . Diehn, M. (2014). An ultrasensitive method for quantitating circulating tumor DNA with broad patient coverage. *Nature Medicine*, *20*(5), 548–554. Available from https://doi.org/10.1038/nm.3519.

Nick, B., Jenna-Lynn, S., Shahid, A., Kanthan, S. C., & Kanthan, R. (2016). Gastric biomarkers: A global review. *World Journal of Surgical Oncology*. Available from https://doi.org/10.1186/s12957-016-0969-3.

Ning, S., Wei, W., Li, J., Hou, B., Zhong, J., Xie, Y., . . . Zhang, L. (2018). Clinical significance and diagnostic capacity of serum TK1, CEA, CA 19–9 and CA 72–4 levels in gastric and colorectal cancer patients. *Journal of Cancer*, *9*(3), 494–501. Available from https://doi.org/10.7150/jc.21562.

Ohtsu, A., Ajani, J. A., Bai, Y. X., Bang, Y. J., Chung, H. C., Pan, H. M., . . . Van Cutsem, E. (2013). Everolimus for previously treated advanced gastric cancer: Results of the randomized, double-blind, phase III GRANITE-1 study. *Journal of Clinical Oncology*, *31*(31), 3935–3943. Available from https://doi.org/10.1200/JCO.2012.48.3552.

Otero-Estévez, O., De Chiara, L., Barcia-Castro, L., Páez De La Cadena, M., Rodríguez-Berrocal, F. J., Cubiella, J., . . . Martínez-Zorzano, V. S. (2016). Evaluation of serum nucleoside diphosphate kinase A for the detection of colorectal cancer. *Scientific Reports*, 6. Available from https://doi.org/10.1038/srep26703.

Özkan, H., Erdal, H., Koçak, E., Tutkak, H., Karaeren, Z., Yakut, M., & Köklü, S. (2011). Diagnostic and prognostic role of serum glypican 3 in patients with hepatocellular carcinoma. *Journal of Clinical Laboratory Analysis*, *25*(5), 350–353. Available from https://doi.org/10.1002/jcla.20484.

Park, W. H., Shim, J. H., Han, S. B., Won, H. J., Shin, Y. M., Kim, K. M., … Lee, H. C. (2012). Clinical utility of des-γ-carboxyprothrombin kinetics as a complement to radiologic response in patients with hepatocellular carcinoma undergoing transarterial chemoembolization. *Journal of Vascular and Interventional Radiology*, *23*(7), 927–936. Available from https://doi.org/10.1016/j.jvir.2012.04.021.

Phallen, J., Sausen, M., Adleff, V., Leal, A., Hruban, C., White, J., … Velculescu, V. E. (2017). Direct detection of early-stage cancers using circulating tumor DNA. *Science Translational Medicine*, *9*(403). Available from https://doi.org/10.1126/scitranslmed.aan2415.

Polat, E., Duman, U., Duman, M., Atici, A. E., Reyhan, E., Dalgic, T., … Yol, S. (2014). Diagnostic value of preoperative serum carcinoembryonic antigen and carbohydrate antigen 19–9 in colorectal cancer. *Current Oncology*, *1*. Available from https://doi.org/10.3747/co.21.1711.

Poto, C. (2018). Identification of race-associated metabolite biomarkers for hepatocellular carcinoma in patients with liver cirrhosis and hepatitis C virus infection. *PLoS ONE*, *13*(3).

Pu, X. X., Huang, G. L., Guo, H. Q., Guo, C. C., Li, H., Ye, S., … Lin, T. Y. (2010). Circulating miR-221 directly amplified from plasma is a potential diagnostic and prognostic marker of colorectal cancer and is correlated with p53 expression. *Journal of Gastroenterology and Hepatology (Australia)*, *25*(10), 1674–1680. Available from https://doi.org/10.1111/j.1440-1746.2010.06417.x.

Ranjan P, Parihar A, Jain S, Kumar N, Dhand C, Murali S., … Khan, R. (2020). Biosensor-based diagnostic approaches for various cellular biomarkers of breast cancer: A comprehensive review. *Analytical Biochemistry*. *610*, 113996. Available from https://doi.org/10.1016/j.ab.2020.113996.

Reiter, W., Stieber, P., Reuter, C., Nagel, D., Lau-Werner, U., & Lamerz, R. (2000). Multivariate analysis of the prognostic value of CEA and CA 19–9 serum levels in colorectal cancer. *Anticancer Research*, *20*(6 D), 5195–5198.

Rodríguez, N. (2017). Dysbiosis of the microbiome in gastric carcinogenesis. *Scientific Reports*, 7.

Schittenhelm, M. M., Illing, B., Ahmut, F., Rasp, K. H., Blumenstock, G., Döhner, K., … Kampa-Schittenhelm, K. M. (2013). Attenuated expression of apoptosis stimulating protein of p53–2 (ASPP2) in human acute leukemia is associated with therapy failure. *PLoS ONE*, *8*(11). Available from https://doi.org/10.1371/journal.pone.0080193.

Schreuders, E. H., Ruco, A., Rabeneck, L., Schoen, R. E., Sung, J. J. Y., Young, G. P., & Kuipers, E. J. (2015). Colorectal cancer screening: A global overview of existing programmes. *Gut*, *64*(10), 1637–1649. Available from https://doi.org/10.1136/gutjnl-2014-309086.

Sclafani, F., Chau, I., Cunningham, D., Hahne, J. C., Vlachogiannis, G., Eltahir, Z., … Valeri, N. (2018). KRAS and BRAF mutations in circulating tumour DNA from locally advanced rectal cancer. *Scientific Reports*, *8*(1). Available from https://doi.org/10.1038/s41598-018-19212-5.

Seregni, E., Botti, C., & Bombardieri, E. (1995). Biochemical characteristics and clinical applications of alpha-fetoprotein isoforms. *Anticancer Research*, *15*(4), 1491–1499.

Shi, Y., Han, Y., Xie, F., Wang, A., Feng, X., Li, N., … Chen, D. (2015). ASPP2 enhances oxaliplatin (L-OHP)-induced colorectal cancer cell apoptosis in a p53-independent manner by inhibiting cell autophagy. *Journal of Cellular and Molecular Medicine*, *19*(3), 535–543. Available from https://doi.org/10.1111/jcmm.12435.

Showe, M. K., Vachani, A., Kossenkov, A. V., Yousef, M., Nichols, C., Nikonova, E. V., … Showe, L. C. (2009). Gene expression profiles in peripheral blood mononuclear cells can distinguish patients with non-small cell lung cancer from patients with nonmalignant lung disease. *Cancer Research*, *69*(24), 9202–9210. Available from https://doi.org/10.1158/0008-5472.CAN-09-1378.

Silvestri, G. A., Tanner, N. T., Kearney, P., Vachani, A., Massion, P. P., Porter, A., … Mazzone, P. J. (2018). Assessment of plasma proteomics biomarker's ability to distinguish benign from malignant lung nodules: Results of the PANOPTIC (Pulmonary Nodule Plasma Proteomic Classifier) trial. *Chest*, *154*(3), 491–500. Available from https://doi.org/10.1016/j.chest.2018.02.012.

Simao, A. (2015). Plasma osteopontin is a biomarker for the severity of alcoholic liver cirrhosis, not for hepatocellular carcinoma screening. *BMC Gastroenterology*, 15.

Singal, A. G., Corley, D. A., Kamineni, A., Garcia, M., Zheng, Y., Doria-Rose, P. V., … Halm, E. A. (2018). Patterns and predictors of repeat fecal immunochemical and occult blood test screening in four large health care systems in the United States. *American Journal of Gastroenterology*, *113*(5), 746–754. Available from https://doi.org/10.1038/s41395-018-0023-x.

Song, B. (2015). Downregulation of ASPP2 in pancre- atic cancer cells contributes to increased resistance to gemcitabine through autophagy activation. *Mol Cancer*, 14.

Sonnenblick, A., Brohée, S., Fumagalli, D., Rothé, F., Vincent, D., Ignatiadis, M., … Sotiriou, C. (2015). Integrative proteomic and gene expression analysis identify potential biomarkers for adjuvant trastuzumab resistance: Analysis from the Fin-her phase III randomized trial. *Oncotarget*, *6* (30), 30306–30316. Available from https://doi.org/10.18632/oncotarget.5080.

Spangenberg, H. C., Thimme, R., & Blum, H. E. (2006). Serum markers of hepatocellular carcinoma. *Seminars in Liver Disease*, *26*(4), 385–390. Available from https://doi.org/10.1055/s-2006-951606.

Sterling, R. K., Jeffers, L., Gordon, F., Venook, A. P., Reddy, K. R., Satomura, S., … Sherman, M. (2009). Utility of lens culinaris agglutinin-reactive fraction of α-fetoprotein and des-gamma-carboxy prothrombin, alone or in combination, as biomarkers for hepatocellular carcinoma. *Clinical Gastroenterology and Hepatology*, *7*(1), 104–113. Available from https://doi.org/10.1016/j.cgh.2008.08.041.

Sterling, R. K., Wright, E. C., Morgan, T. R., Seeff, L. B., Hoefs, J. C., Di Bisceglie, A. M., … Lok, A. S. (2011). Frequency of elevated hepatocellular carcinoma (HCC) biomarkers in patients with advanced hepatitis C. *American Journal of Gastroenterol*, *107*(1). Available from https://doi.org/10.1038/ajg.2011.312, Epub.

Stratton, M. R., & Rahman, N. (2008). The emerging landscape of breast cancer susceptibility. *Nature Genetics*, *40*(1), 17–22. Available from https://doi.org/10.1038/ng.2007.53.

Streckfus, C. F., Arreola, D., Edwards, C., & Bigler, L. (2012). Salivary protein profiles among her2/neu-receptor-positive and -negative breast cancer patients: Support for using salivary protein profiles for modeling breast cancer progression. *Journal of Oncology*. Available from https://doi.org/10.1155/2012/413256.

Streckfus, C., Bigler, L., Tucci, M., & Thigpen, J. T. (2000). A preliminary study of CA15–3, c-erbB-2, epidermal growth factor receptor, cathepsin-D, and p53 in saliva among women with breast carcinoma. *Cancer Investigation*, *18*(2), 101–109. Available from https://doi.org/10.3109/07357900009038240.

Symonds, E. L., Pedersen, S. K., Murray, D. H., Jedi, M., Byrne, S. E., Rabbitt, P., ... Young, G. P. (2018). Circulating tumour DNA for monitoring colorectal cancer—A prospective cohort study to assess relationship to tissue methylation, cancer characteristics and surgical resection. *Clinical Epigenetics*, *10*(1). Available from https://doi.org/10.1186/s13148-018-0500-5.

Tenori, L., Oakman, C., Morris, P. G., Gralka, E., Turner, N., Cappadona, S., ... Di Leo, A. (2015). Serum metabolomic profiles evaluated after surgery may identify patients with oestrogen receptor negative early breast cancer at increased risk of disease recurrence. Results from a retrospective study. *Molecular Oncology*, *9*(1), 128–139. Available from https://doi.org/10.1016/j.molonc.2014.07.012.

Tobias, D. K., Akinkuolie, A. O., Chandler, P. D., Lawler, P. R., Manson, J. E., Buring, J. E., ... Mora, S. (2018). Markers of inflammation and incident breast cancer risk in the women's health study. *American Journal of Epidemiology*, *187*(4), 705–716. Available from https://doi.org/10.1093/aje/kwx250.

Wang, D., Li, W., Zou, Q., Yin, L., Du, Y., Gu, J., & Suo, J. (2017). Serum metabolomic profiling of human gastric cancer and its relationship with the prognosis. *Oncotarget*, *8*(66), 110000–110015. Available from https://doi.org/10.18632/oncotarget.21314.

Wang, J., Huang, S. K., Zhao, M., Yang, M., Zhong, J. L., Gu, Y. Y., ... Huang, C. Z. (2014). Identification of a circulating microRNA signature for colorectal cancer detection. *PLoS ONE*, *9*(4). Available from https://doi.org/10.1371/journal.pone.0087451.

Weiss, G., Schlegel, A., Kottwitz, D., König, T., & Tetzner, R. (2017). Validation of the SHOX2/PTGER4 DNA methylation marker panel for plasma-based discrimination between patients with malignant and nonmalignant lung disease. *Journal of Thoracic Oncology*, *12*(1), 77–84. Available from https://doi.org/10.1016/j.jtho.2016.08.123.

Wen, L., Li, J., Guo, H., Liu, X., Zheng, S., Zhang, D., ... Peng, J. (2015). Genome-scale detection of hypermethylated CpG islands in circulating cell-free DNA of hepatocellular carcinoma patients. *Cell Research*, *25*(11), 1250–1264. Available from https://doi.org/10.1038/cr.2015.126.

Wen, Y., Han, J., Chen, J., Dong, J., Xia, Y., Liu, J., ... Hu, Z. (2015). Plasma miRNAs as early biomarkers for detecting hepatocellular carcinoma. *International Journal of Cancer*, *137*(7), 1679–1690. Available from https://doi.org/10.1002/ijc.29544.

Wooster, R., Bignell, G., Lancaster, J., Swift, S., Seal, S., Mangion, J., ... Stratton, M. R. (1995). Identification of the breast cancer susceptibility gene BRCA2. *Nature*, *378*(6559), 789–792. Available from https://doi.org/10.1038/378789a0.

Wu, K. L., Chen, H. H., Pen, C. T., Yeh, W. L., Huang, E. Y., Hsiao, C. C., & Yang, K. D. (2015). Circulating Galectin-1 and 90K/Mac-2BP correlated with the tumor stages of patients with colorectal cancer. *BioMed Research International*, 2015. Available from https://doi.org/10.1155/2015/306964.

Xiaofeng, W., Xiaomin, S., Wei, Z., Yan, F., Hubing, S., Yun, L., ... Yongzhang, L. (2009). The regulatory mechanism of Hsp90α secretion and its function in tumor malignancy. *Proceedings of the National Academy of Sciences*, 21288–21293. Available from https://doi.org/10.1073/pnas.0908151106.

Xing, L., Su, J., Guarnera, M. A., Zhang, H., Cai, L., Zhou, R., ... Jiang, F. (2015). Sputum microRNA biomarkers for identifying lung cancer in indeterminate solitary pulmonary nodules. *Clinical Cancer Research*, *21*(2), 484–489. Available from https://doi.org/10.1158/1078-0432.CCR-14-1873.

Yan, L., Chen, Y., Zhou, J., Zhao, H., Zhang, H., & Wang, G. (2018). Diagnostic value of circulating cell-free DNA levels for hepatocellular carcinoma. *International Journal of Infectious Diseases*, *67*, 92–97. Available from https://doi.org/10.1016/j.ijid.2017.12.002.

Yin, L., Lin, Y., Wang, X., Su, Y., Hu, H., Li, C., ... Jiang, Y. (2018). The family of apoptosis-stimulating proteins of p53 is dysregulated in colorectal cancer patients. *Oncology Letters*, *15*(5), 6409–6417. Available from https://doi.org/10.3892/ol.2018.8151.

Yoon, S. K., Lim, N. K., Ha, S. A., Park, Y. G., Choi, J. Y., Chung, K. W., ... Kim, J. W. (2004). The human cervical cancer oncogene protein is a biomarker for human hepatocellular carcinoma. *Cancer Research*, *64*(15), 5434–5441. Available from https://doi.org/10.1158/0008-5472.CAN-03-3665.

Yuen, M. F., Cheng, C. C., Lauder, I. J., Lam, S. K., Ooi, C. G. C., & Lai, C. L. (2000). Early detection of hepatocellular carcinoma increases the chance of treatment: Hong kong experience. *Hepatology*, *31*(2), 330–335. Available from https://doi.org/10.1002/hep.510310211.

Zhang, C., Yu, W., Wang, L., Zhao, M., Guo, Q., Lv, S., ... Lou, J. (2017). DNA methylation analysis of the SHOX2 and RASSF1A panel in bronchoalveolar lavage fluid for lung cancer diagnosis. *Journal of Cancer*, *8*(17). Available from https://doi.org/10.7150/jc.21368.

Zhang, G., Ha, S. A., Kim, H. K., Yoo, J., Kim, S., Lee, Y. S., ... Kim, J. W. (2012). Combined analysis of AFP and HCCR-1 as an useful serological marker for small hepatocellular carcinoma: A prospective cohort study. *Disease Markers*, *32*(4), 265–271. Available from https://doi.org/10.3233/DMA-2011-0878.

Zhang, S. Y., Lin, M., & Zhang, H. B. (2015). Diagnostic value of carcinoembryonic antigen and carcinoma antigen 19–9 for colorectal carcinoma. *International Journal of Clinical and Experimental Pathology*, *8*(8), 9404–9409. Available from http://www.ijcep.com/files/ijcep0009775.pdf.

Zhao, Y., Xue, F., Sun, J., Guo, S., Zhang, H., Qiu, B., ... Xia, Q. (2014). Genome-wide methylation profiling of the different stages of hepatitis B virus-related hepatocellular carcinoma development in plasma cell-free DNA reveals potential biomarkers for early detection and high-risk monitoring of hepatocellular carcinoma. *Clinical Epigenetics*, *6*(1). Available from https://doi.org/10.1186/1868-7083-6-30.

Zhu, J., Liao, M., Yao, Z., Liang, W., Li, Q., Liu, J., ... Mo, Z. (2018). Breast cancer in postmenopausal women is associated with an altered gut metagenome. *Microbiome*, *6*(1). Available from https://doi.org/10.1186/s40168-018-0515-3.

Chapter 4

# Biosensors: concept and importance in point-of-care disease diagnosis

Raquel Vaz[1,2,3], Manuela F. Frasco[1,2,3] and M. Goreti F. Sales[1,2,3]

[1]*BioMark Sensor Research/UC, Faculty of Sciences and Technology, Coimbra University, Coimbra, Portugal*

[2]*BioMark Sensor Research/ISEP, School of Engineering, Polytechnic Institute of Porto, Porto, Portugal*

[3]*CEB - Centre of Biological Engineering, Minho University, Braga, Portugal*

## 4.1 Introduction

Point-of-care (POC) disease diagnosis is of extreme importance because on-spot diagnosis alleviates patient stress, increases the chance of early diagnosis, survival rates, and facilitates quick medical decisions. According to the World Health Organization, cancer is the second leading cause of global morbidity and mortality. The number of new patients diagnosed with cancer is expected to increase to 29.5 million by 2040, due to pollution exposure, obesity, smoking, alcohol consumption, and increase of life expectancy (World Health Organization, 2020). One of cancer's primary reasons for high mortality is the lack of early diagnosis. Moreover, its major diagnostic techniques rely on imaging, which has associated drawbacks, such as inability to differentiate between malignant and benign tumors, exposure to high doses of radiation, and hardships in data analysis. Therefore, the scientific community has been focusing into the development of POC technologies to specifically recognize cancer biomarkers, such as proteins, circulating tumor cells (CTCs), microRNAs (miRs) and exosomes (Shandilya et al., 2019). Those technologies are meant to be simple and avoid pre-processing of samples, so that any health professional can efficiently manipulate them.

Such demand for POC devices can be fulfilled by biosensors whose implementation in our everyday lives has been progressing fast, due to their versatility and easy application in disease diagnosis and monitoring, drug delivery, food quality control, detection of pollutants and microenvironment sensing. A biosensor is a sensing device that converts the chemical or biological reaction originating from the recognition of the analyte into a measurable signal. Therefore, a biosensor is constituted by the recognition element (e.g., enzymes, nucleic acid probes, antibodies, bioaffinity groups or tissues) that is immobilized on the sensor substrate and targets the analyte; the transducer, which converts the previous linkage into a measurable signal (e.g., electrochemical, optical, thermometric, magnetic or piezoelectric); and the output system that amplifies and displays the signal (Fig. 4.1) (Asal, Özen, Şahinler, Baysal, & Polatoğlu, 2019; Kawamura & Miyata, 2016). Unlike conventional techniques, such as enzyme linked immunosorbent assays (ELISA) or polymerase chain reaction (PCR), biosensors bring automation with enhanced reproducibility, real-time and fast analysis, can usually be reused, are amenable to functionalization, fit many different structural designs and are cost-efficient (Pirzada & Altintas, 2019).

Selectivity, sensitivity, limit of detection (LOD), working range, and reusability remain crucial characteristics of biosensors, especially for disease diagnosis (Asal et al., 2019). Furthermore, addressing power consumption, size, ease of use, robustness, and cost is mandatory for POC analysis and entrance on the market. Miniaturization reduces cost and power intake, while the materials used to fabricate the sensors impact not only the cost but also sensitivity and biocompatibility (Sadana & Sadana, 2015). In recent years, nanomaterials have been incorporated into biomedical sensors, due to their remarkable physicochemical properties, such as lower binding energy than bulk materials and high surface area, which increases their chemical reactivity. Thus, nanomaterials facilitate the immobilization of recognition biomolecules at the sensor surface, increase sensitivity, besides being easily engineered and adaptable to different detection mechanisms (Pirzada & Altintas, 2019).

Biosensor Based Advanced Cancer Diagnostics. DOI: https://doi.org/10.1016/B978-0-12-823424-2.00001-6

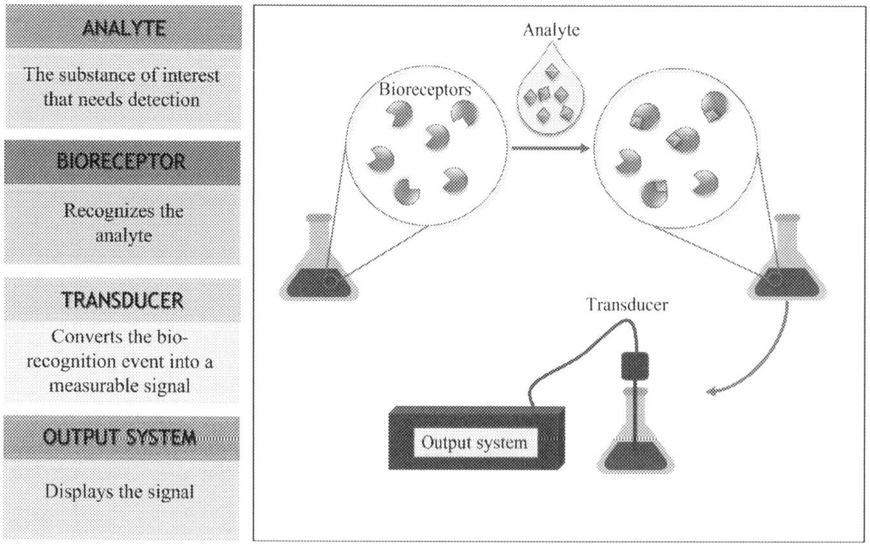


**FIGURE 4.1** Schematic representation of a biosensor.

### 4.1.1 Historical perspective of biosensors

The first biosensor dates from 1962, developed by Professor Leland C. Clark Jr., and corresponds to an oxygen electrode modified with the immobilization of the enzyme glucose oxidase (GOx). The device relied on applying a potential between the Ag/AgCl reference and the platinum working electrode; thus, the oxygen in solution between the two electrodes is reduced, creating an electric current. When glucose is in the system, it is oxidized by the enzyme in the presence of oxygen, this one being reduced to hydrogen peroxide. The generated current is proportional to oxygen levels, which in turn enables to determine glucose concentration (Newman & Turner, 2008). Thus, the first biosensor was an enzyme-based amperometric sensor for glucose. Devices based on the Clark electrode and developed with other enzymes have been extensively studied. In 1969 the first enzyme-based potentiometric sensor was described by Guilbault and Montalvo, that immobilized urease in an ammonium-selective liquid membrane electrode to sense urea (Guilbault & Montalvo, 1969), and the first ion-sensitive field-effect transistor appeared in 1970 (Bergveld, 1970), leading to much research in the following decades.

Only in the 1970s other bioreceptors, such as organelles, bacteria and antibodies were used, being coupled to different transduction methods. For example, in 1972, a piezoelectric device was fabricated for the determination of bovine serum albumin antibodies due to mass changes on the biosensor surface (Shons, Dorman, & Najarian, 1972); and in 1974 it was invented the first thermal device, which had the disadvantage of being affected by ambient thermal fluctuations (Cooney et al., 1974). In 1975 there was a boost of new ideas, such as the first optical biosensor for alcohol recognition described by Lubbers and Opitz, using a fiber-optic sensor with immobilized alcohol oxidase (Lübbers & Opitz, 1975); the first microbial electrode also to measure alcohol (Diviés, 1975); and Yellow Springs Instruments created the first commercial biosensor for glucose sensing (Bhalla, Jolly, Formisano, & Estrela, 2016). The 1980s continued to bring high innovation in optical sensors, as is the case of the fiber-optic biosensor developed by Schultz for glucose detection (Schultz, 1982) and the first surface plasmon resonance (SPR) immunosensor by Liedberg, Nylander, and Lunström (1983).

Although biosensors started being developed in the 1960s, almost all of them only entered the commercial stage more than twenty years later (Fig. 4.2). An exception happened in 1976, with the electrochemical lactate analyzer LA 640 by La Roche (Switzerland), which incorporated potassium hexacyanoferrate for transfer of electrons between lactate dehydrogenase and an electrode, and could be applied not only at clinics but also for sports. Further on, other electrochemical biosensors based on the same principle of using a mediator to shuttle electrons came onto the market: the MediSense ExactTech (1987), the Abbott (1996), the Boehringer Mannheim, the Bayer and the LifeScan (1998) were devices for home blood glucose detection that dominated the market (Newman & Turner, 2008).

The Pharmacia BIAcore instrument was launched in 1990. It was based on SPR technology for affinity measurements with proteins, DNA, RNA and membranes, as well as for drug screening and immunoreactivity (Jönsson et al., 1991). In 1992 the first multianalyte analyzer, now commercialized by Abbott Laboratories, was the i-STAT instrument, containing thin-film electrodes microfabricated on silicon chips, enabling the simultaneous measurement of electrolytes, glucose, urea nitrogen (Erickson & Wilding, 1993). In the 21st century, the scientific community has been focusing on

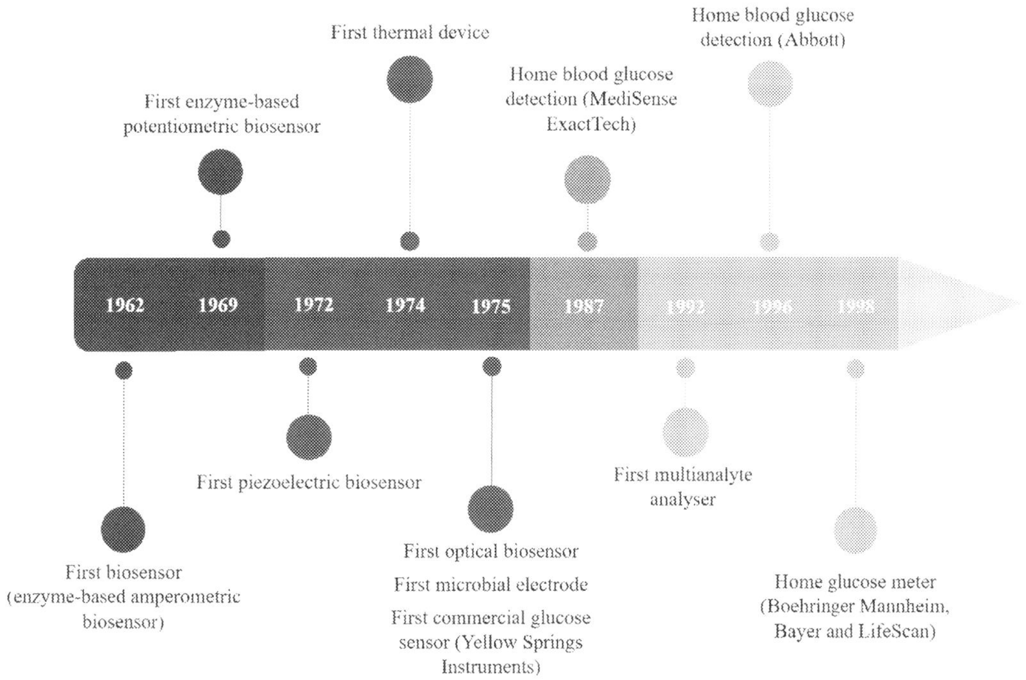


**FIGURE 4.2** Historical perspective of biosensors development in the 20th century.

enhancing the receptor/transducer interface, improving the sensitivity and selectivity of mainly electrochemical, optical, and piezoelectric biosensors. For that, nanotechnology and mimicry of nature to produce synthetic receptors gained strong popularity. These biosensors will be discussed in detail in Section 4.2.

## 4.1.2 Classification

### *4.1.2.1 Support materials*

The concept of a biosensor has evolved enormously through the years, and they can be analyzed taking into account the role of the materials involved in the construction of the device. The support materials of the biosensors must display good chemical and mechanical stability, have functional groups, be cheap, and biocompatible. Those materials are used to preserve the structure of the immobilized biomolecules, thus contributing to their reactivity and stability, while being fairly inert to the reaction system. For that purpose, porosity, hydrophilicity, and surface functional groups show extreme importance (Elnashar, 2010). For example, a large surface area and pore size improves the loading capacity of the recognition receptor and facilitates the transport of samples within the sensor, as it is the case of mesoporous materials (Chong & Zhao, 2004). However, if the pores are too wide, the surface area decreases, and if the pores are too small, they will exclude most biomolecules. Moreover, hydrophobicity should be minimized to prevent denaturation of proteins and nonspecific adsorption (Sirisha et al., 2016). The use of nanomaterials as particles, fibers or tubes instead of direct adsorption of biomolecules onto bulk materials prevents their denaturation and loss of bioactivity. Nanomaterials can also be functionalized to increase the surface area of the support material and the sensitivity of the sensor (Farzin, Shamsipur, Samandari, & Sheibani, 2018).

On the basis of the chemical composition of the material, there are inorganic and organic supports. Inorganic materials include silica, alumina, metal oxides, zirconia, and glass, being usually chosen due to their thermomechanical properties and antimicrobial effect. Organic supports can be divided into natural polymers or synthetic polymers; whereas, the first comprises polysaccharides (e.g., chitosan, alginate, cellulose, carrageenans, starch, pectin and agarose) or proteins (e.g., silk, collagen and zein), the second includes polystyrene, polyethylene glycol, polyacrylate, and polyvinyl chloride (Sirisha et al., 2016). It is also possible to construct organic-inorganic hybrids, which possess chemical inertness of the inorganic precursor while holding higher bioaffinity due to the organic precursor (Zdarta, Meyer, Jesionowski, & Pinelo, 2018).

#### 4.1.2.2 Recognition elements

The recognition system of the biosensor is chosen according to the analyte aimed to detect, besides taking into account other parameters such as the pH environment and temperature. It can be constituted by enzymes, aptamers, nucleic acids, antigens or antibodies, tissues, or cells (Asal et al., 2019).

Enzymes constitute a classical recognition element, which show specificity and are simple to use. Their high selectivity allows having a great amount of activity in small areas. However, it is necessary to purify them and enzymes show poor stability, besides being dependent on pH and temperature (Justino, Freitas, Pereira, Duarte, & Rocha Santos, 2015). Nanoparticles are a promising tool to enhance the detection sensitivity when using enzymes as a receptor system, because their high surface area enables a greater amount of enzymes immobilized on the surface of the biosensor (Du et al., 2011).

Other well-expanded methods rely on antibodies and antigens, whose main advantage is their high affinity and specificity, as well as the possibility of designing recombinant antibodies (Alhadrami, 2018). Despite the huge number of applications and biosensor strategies that have been developed using antibodies, other biomolecules present clear advantages. Such is the case of aptamers, whose structure can be easily designed, while being thermally stable and reusable, amenable to functionalization with reporter molecules (e.g., fluorophores) or other functional groups (e.g., thiols, disulfides, amines, or biotin) and do not require the use of animals to produce them (Justino et al., 2015). Biosensors based on nucleic acid systems (natural or synthetic analogs) work through the hybridization of oligonucleotides with complementary strands, creating selective and stable complexes. They are also easily modified, with incorporated end-labels, such as thiols, disulfides, amines, or biotin (Alhadrami, 2018; Rastislav, Miroslav, & Ernest, 2012).

The immobilization of the biorecognition element can be performed directly on the transducer or on supports attached to the transducer. It is mandatory that the active part of the molecules remain accessible to the target analyte, which is possible through irreversible immobilization methods, such as covalent binding, cross-linking, and entrapment, or through reversible methods, as it is the case of adsorption and affinity binding (Asal et al., 2019).

Tissue-based biosensors started being developed in 1978, where the tissue was used as a source of enzymes with high stability. They were not removed from their natural environment, which also represented affordable costs (Rechnitz, 1978). Also, they were already used to sense cancer biomarkers, pathogens, and extracellular proteins (Anwarul et al., 2014). Using whole cells as a source of stable enzymes has also been shown to be an efficient strategy, with low-cost of preparation, and without the need of extensive purification protocols. Also, measuring the metabolic status gives insight into a more relevant biological response regarding chemicals sublethal concentration, composition, carcinogenicity, and mutagenicity (Hassan, Van Ginkel, Hussein, Abskharon, & Oh, 2016; He, Yuan, Zhong, Siddikee, & Dai, 2016).

Phage-based biosensors also have some advantages since phages infect specific bacteria, displaying certain peptides or proteins in their surface, which enables easy detection of targets. Viruses are also more robust than antibodies and stable over a wide range of pH values (Brown et al., 2020; Liu et al., 2020).

One of the most successful synthetic routes to mimic biorecognition elements is the molecular imprinting technology. Molecularly imprinted polymers (MIPs) are created through the mixing of monomers, cross-linker, and the template molecule. After polymerization, the template is removed, leaving behind a polymeric matrix with size, shape, and functional groups complementary to the template molecule. Therefore, MIPs are also considered plastic antibodies, with the leading advantage of having higher thermal, chemical, and mechanical tolerance than antibodies and proteins, as well as cheaper production (Martins, Marques, Fortunato, & Sales, 2020; Piloto, Ribeiro, Rodrigues, Santos, & Ferreira Sales, 2020; Rebelo et al., 2019).

#### 4.1.2.3 Transduction mechanisms

Biosensors can also be categorized according to the physical-chemical signal transduction, being electrochemical, thermal, optical, and piezoelectric the most common (Fig. 4.3). In electrochemical biosensors, the recognition reaction produces or consumes ions or electrons, which leads to measurable electrical variations. It is the most widespread approach, due to high sensitivity, ability of inclusion in portable devices and cheap fabrication (Rastislav et al., 2012). Optical transducers include colorimetric, fluorescent, SPR, surface-enhanced raman spectroscopy (SERS) and photonic structures, with the common aspect of using light as the physical parameter changing with the binding of the target analyte. These biosensors are characterized by rapid and sensitive signals, besides allowing easy expansion towards simultaneous detection of multiple analytes (Alhadrami, 2018; Fei et al., 2013; Kadhem, Xiang, Nagel, Lin, & de Cortalezzi, 2018; Rastislav et al., 2012). Piezoelectric biosensors are mass-sensitive sensors, which allow label-free detection and real-time operation with high sensitivity (Alhadrami, 2018). Thermometric biosensors measure the variation of

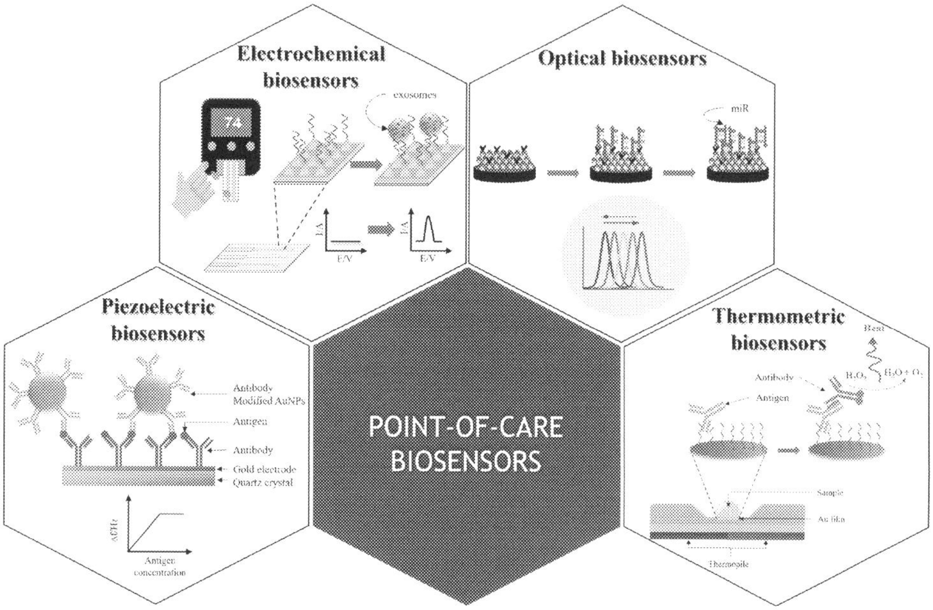


**FIGURE 4.3** Illustration of point-of-care devices based on common transduction methods of the biosensor.

temperature correlated with the binding of the analyte. Their main advantage is the insensitivity to the electrochemical and optical properties of the sample (Rastislav et al., 2012).

With all the information stated above, we can define a most complete concept of a biosensor as an analytical device that uses biochemical reactions mediated by biological (e.g., enzymes, oligonucleotides, antibodies, cells, tissues) or biomimetic receptors as the recognition element either integrated within or intimately associated with a physicochemical transducer (Bettazzi, Marrazza, Minunni, Palchetti, & Scarano, 2017; Labuda et al., 2010).

In Section 4.2, new POC biosensors developed for cancer diagnosis in the past decades, based on their transduction method, will be thoroughly explored. Next, the inclusion of biomaterials in biosensors is reviewed, due to the increase awareness of the environmental impact of biosensors research and fabrication. The discussion is then expanded to give emphasis on new trends for POC biosensor design, and those which are already commercially available.

## 4.2 POC biosensors for cancer diagnosis

Among the deadliest diseases, cancer is definitely one to be feared, attacking any organ of the human body. From the first cellular malignant modification until symptoms appear, a long time may pass. Therefore, early diagnosis is essential to prevent proliferation and metastasis, increasing the chances of patient survival (Mahato et al., 2018).

Recently, POC devices have been extensively studied, regarding their design, materials, sensitivity, and specificity, to meet the demands of cancer diagnosis, such as simplicity, low-cost, requirement of low sample volume, and accuracy (Fig. 4.4) (Mohammadniaei, Nguyen, Tieu, & Lee, 2019). Biomarkers for cancer diagnosis vary according to the type of cancer, although exosomes, miRs, proteins and CTCs are the most common (Mahato, Kumar, Maurya, & Chandra, 2018).

CTCs are cells that migrate from malignant tissues to form malignant tumors in other parts of the body (Ankeny et al., 2016; Piñeiro, Martínez-Pena, & López-López, 2020). They are extremely rare during early metastasis and detecting them is very challenging (Piñeiro et al., 2020). Many studies have been also devoted to exosomes, which are a type of extracellular vesicles originated from multivesicular bodies, with around 40–160 nm, whose cargo includes genetic material, proteins, lipids, and metabolites (Garcia-Cordero & Maerkl, 2020). They are known to be released at physiological healthy conditions, but are also involved in tumor growth, progression, and metastasis (Ge et al., 2020; Gener Lahav et al., 2019; Gowda et al., 2020; Kalluri & LeBleu, 2020; Sagredo, Sepulveda, Roa, & Oróstica, 2017). The miRs, which may be contained in exosomes or directly secreted to the blood current by cancerous cells, are single-stranded non-coding RNAs with 20–24 nucleotides, which influence the phenotypic alteration of cells and the creation of tumorous niches. For example, miR-145 is a tumor suppressor found in low levels in colon, ovary, breast, and prostate cancer (Larrea et al., 2016), whereas miR-21–5p is overexpressed in the same pathologies (Cai et al., 2018; Danarto, Astuti, Umbas, & Haryana, 2020; Xu et al., 2018).

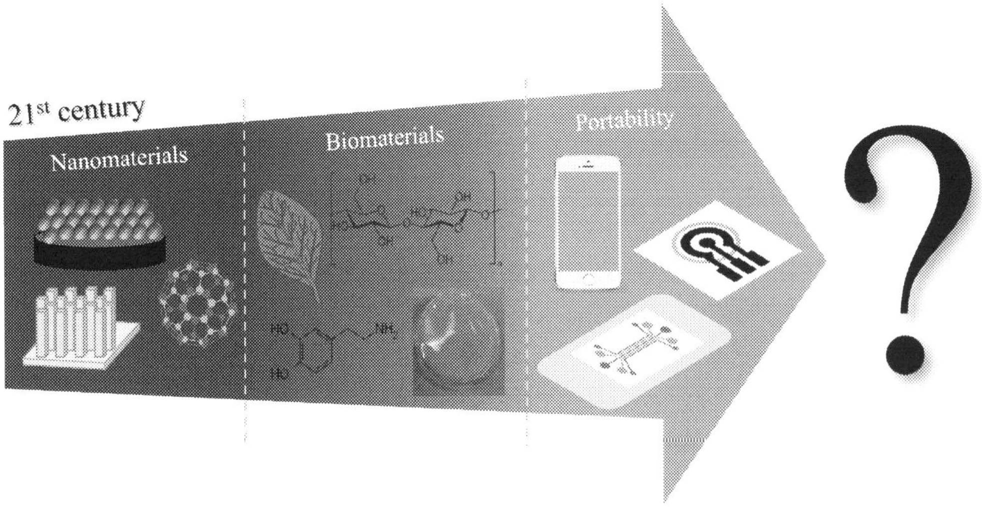


**FIGURE 4.4** Evolution of the design of POC biosensors in the first two decades of the 21st century, through gradually including nanomaterials for signal enhancing, as well as biomaterials for attaining biocompatibility and biodegradability. There was also an increase interest in portability for facilitating the use of POC devices, including strategies using smartphones, wearable or paper-based designs. Future perspectives seem infinite in sensors field, considering the scientific work developed so far.

It is not a surprise that when a mutation occurs, proteins get impaired, which leads to a decontrol of cell activity and, consequently, to malignancy. Antigens in the cellular surface can also function as biomarkers, as it is the case of CD20 expressed in non-Hodgkin lymphomas and other leukemia (Buckner, Christiansen, Bourgeois, Lazarchick, & Lazarchick, 2007; Giles et al., 2003), CD22 in hairy cell leukemia and lung cancer (Matsushita et al., 2008; Pop et al., 2014), and CD33 for acute myeloid leukemia (Khan et al., 2017), among many other examples found in the literature. Other protein biomarkers for cancer detection include the human epidermal growth factor receptor 2 (HER2), which is found in breast tissue (Walsh, Smith, & Stearns, 2020), or protein biomarkers found in blood, where their levels are unusually high. That is the case of increased levels of epithelial cell adhesion molecule (EpCAM), carcinoembryonic antigen (CEA), cancer antigen 15−3 (CA15−3), cancer antigen 19−9 (CA19−9) and α-fetoprotein (AFP) in various types of cancers (Edoo et al., 2019; Krishna & Bekaii-Saab, 2015; Li, Xu, & Zhang, 2019; Mahato et al., 2018; Mohtar, Syafruddin, Nasir, & Yew, 2020).

Thus, POC biosensors target the aforementioned elements for specific recognition and cancer detection. The transduction event is now explored, giving insight into the newest POC devices developed by researchers in the last years.

### 4.2.1 Electrochemical POC biosensors

Briefly, the basic principle of electrochemical biosensors relies on the transduction of an electrical signal due to the biomolecule recognition, through a change in current, potential, accumulated charge or impedance. For signal characterization and quantification, it is possible to employ amperometry, linear sweep voltammetry (LSV), cyclic voltammetry (CV), differential pulse voltammetry (DPV), square wave voltammetry (SWV) or electrochemical impedance spectroscopy (EIS) (Bezinge, Suea-Ngam, Demello, & Shih, 2020).

Several electrochemical biosensors have been developed with the aim of being used in the clinical setup for early cancer diagnosis (Lim, Ryu, Yun, Park, & Park, 2017; Dong et al., 2018; Gomes, Moreira, Fernandes, & Goreti Sales, 2018; Islam et al., 2017; Smith, Newbury, Drago, Bowen, & Redman, 2017; Yen, Chao, & Yeh, 2020; Kutluk, Bruch, Urban, & Dincer, 2020; Cai et al., 2013; Benvidi et al., 2015; Mathew et al., 2020; Mohd Azmi et al., 2014). Exploring some of the most recent works, there is a prevalent enthusiasm for miRs and extracellular vesicles detection, as they constitute newfound cancer biomarkers. Hong et al. (2013) used an electrochemical biosensor with a stem-loop structure as capture probe; in the presence of miR-21, the capture probe unfolds to link to miR-21, and its terminal is available for hybridizing with a DNA concatemer. The concatemer serves as carrier for the hexaammineruthenium (III) chloride redox indicator (Hong et al., 2013). Although this method showed a LOD of 100 aM, Cardoso, Moreira, Fernandes, and Sales (2016) developed a simpler electrochemical biosensor for the detection and quantification of miR-155 in breast cancer patients, through the immobilization of the thiol-modified complementary sequence of the miR (anti-miR) in a gold electrode. EIS and SWV were applied to confirm the successful detection of miR-155, showing a LOD as low as

5.7 aM (Cardoso et al., 2016). Han, Liu, Yang, and Wang (2019) showed a label-free origami nanostructure with a gold surface supporting ssDNA probes to target miR-21. Methylene blue was chosen as the hybridization redox indicator. Even though the LOD was 79.8 fM, which is a higher value than in the other two presented works, the simple design allows a cost-effective and amplification-free biosensing platform for clinical applications (Han et al., 2019). Y. G. Zhou et al. (2016) chose to target exosomes related to prostate cancer. For that, the research team developed a sensor chip with 11 gold electrodes (for multiplex readout), which were functionalized with thiolated anti-EpCAM aptamers. After exosomes capture, silver nanoparticles and copper nanoparticles modified with anti-EpCAM and anti-PSMA (prostate-specific membrane antigen) aptamers, respectively, were added as reporters, because their oxidation potentials fall in the window of gold electrodes, but are well-separated. LSV was then applied for the oxidation of the nanoparticles, and the presence of an electrochemical peak was correlated with the amount of surface markers, reaching a LOD of 50 exosomes/sensor (Y. G. Zhou et al., 2016).

Besides nucleic acids and exosomes, protein biomarkers continue to have significant importance in cancer diagnosis. For example, CD59 is an oral cancer biomarker therefore, a simple impedimetric immunosensor based on gold electrodes modified with anti-CD59 was built, capable of detecting at least 0.38 fg/mL of CD59 (Choudhary et al., 2016) Zhou et al. (2017) also constructed an impedimetric sensor, but for the detection of CEA (Zhou et al., 2017), whereas Ribeiro, Pereira, Silva, and Sales (2018) opted for sensing CA15−3, through a different electrochemical approach: they constructed a MIP onto the gold electrode, using the electrically conducting poly(Toluidine Blue) (Fig. 4.5). The incubation with increasing concentration of CA15−3 in the MIP resulted in a decrease of redox current, showing a linear response from 0.10 to 100 U/mL (Ribeiro et al., 2018).

Nanomaterials have emerged as promising tools for signal amplification in electrochemical biosensors. They can act as nanocatalysts through facilitating the production of electroactive species, as nanoreporters acting as a redox active species, and as nanocarriers of reporter molecules (Bezinge et al., 2020).

Nanoparticles based on noble metals have high conductivity, biocompatibility, and well-established fabrication protocols, which make them a powerful choice for signal amplification. More particularly, gold nanoparticles (AuNPs) show a great potential as nanocarriers for thiolated nucleic acids. For example, Ensafi, Taei, Rahmani, and Khayamian (2011) anchored a DNA probe with a thiol group onto the AuNPs modified electrode for hybridization with a DNA sequence associated with chronic lymphocytic leukemia, showing a LOD of 1 pM Enfasi et al. (2011). Carbon-based nanomaterials are widely used in electrochemical sensors since they increase the electroactive surface area, enhance electron transfer, and promote adsorption of molecules. An example is the DNA sensor by Ye et al. (2017) that combined reduced graphene oxide functionalized with hemin and AuNPs, and allowed to reach a LOD as low as 0.14 aM, (Ye et al., 2017). Labib et al. (2013) constructed an electrochemical sensor which could detect and quantify five different miRs at the same time (miR-21, miR-32, miR-122, miR-141 and miR-200, which are biomarkers of colorectal, prostate, liver, colon, and ovarian cancers, respectively), using a AuNP-modified carbon electrode with the antisense sequences of the targeted miRs. In this work, the binding of the p19 dimer protein to dsRNA originated a decrease in current density, which improved the LOD from 0.4 fM without the p19 to 5 aM with the p19 (Fig. 4.6) (Labib et al., 2013). Azimzadeh, Rahaie, Nasirizadeh, Ashtari, and Naderi-Manesh (2016) developed a sensor based on graphene oxide and gold nanorod functionalized with a thiolated probe to detected miR-155 directly from plasma samples, without the need of sample preparation or RNA extraction. They accomplished a LOD of 0.6 fM, and this sensor could be easily transferred to a clinical setup for breast cancer diagnosis (Azimzadeh et al., 2016). Also, Islam et al. (2018) developed gold-loaded nanoporous superparamagnetic iron oxide nanocubes, which allowed detecting miR-107 from cancer cell lines or directly from tissue samples of esophageal squamous cell carcinoma, with excellent reproducibility and specificity.

AuNPs were also already used for signal amplification of electrochemical sensors for cervical tumorous cells (Wang, Di, Ma, & Ma, 2012) or breast cancer cells detection (Wang, He, et al., 2017), at concentrations lower than 10 cells/mL, as well as for protein biomarkers (Nunna et al., 2019; Zhu, Chandra, & Shim, 2013). For example, Nunna et al. (2019) designed an electrode-based microfluidic biosensor for obtaining a miniaturized low-cost and sensitive POC design for the ovarian cancer biomarker cancer antigen 125 (CA-125). The biosensor was based on the immobilization of anti-CA-125 in AuNPs. This allowed a selective detection of the protein under shear flow conditions, which enhances its potential to be used with actual patient's body fluids, such as plasma or blood (Nunna et al., 2019). Other metal and metal oxide nanoparticles can also be used for signal amplification, as it is the case of palladium (Wu, Chai, Yuan, Su, & Han, 2013) and alumina (Chuang et al., 2016).

As previously noted, carbon-based nanomaterials can be used in combination with metal nanoparticles (Wei, Mao, Liu, Zhang, & Yang, 2018; Ye et al., 2017), with metal oxides or simply as the nanocatalyst or nanoreporter themselves (Soares et al., 2019). Paul, Singh, Vanjari, and Singh (2017) constructed an interface of zinc oxide with embedded carbon nanotubes for CA-125 detection. Zinc oxide is not only chemically, thermally, and mechanically stable, but also

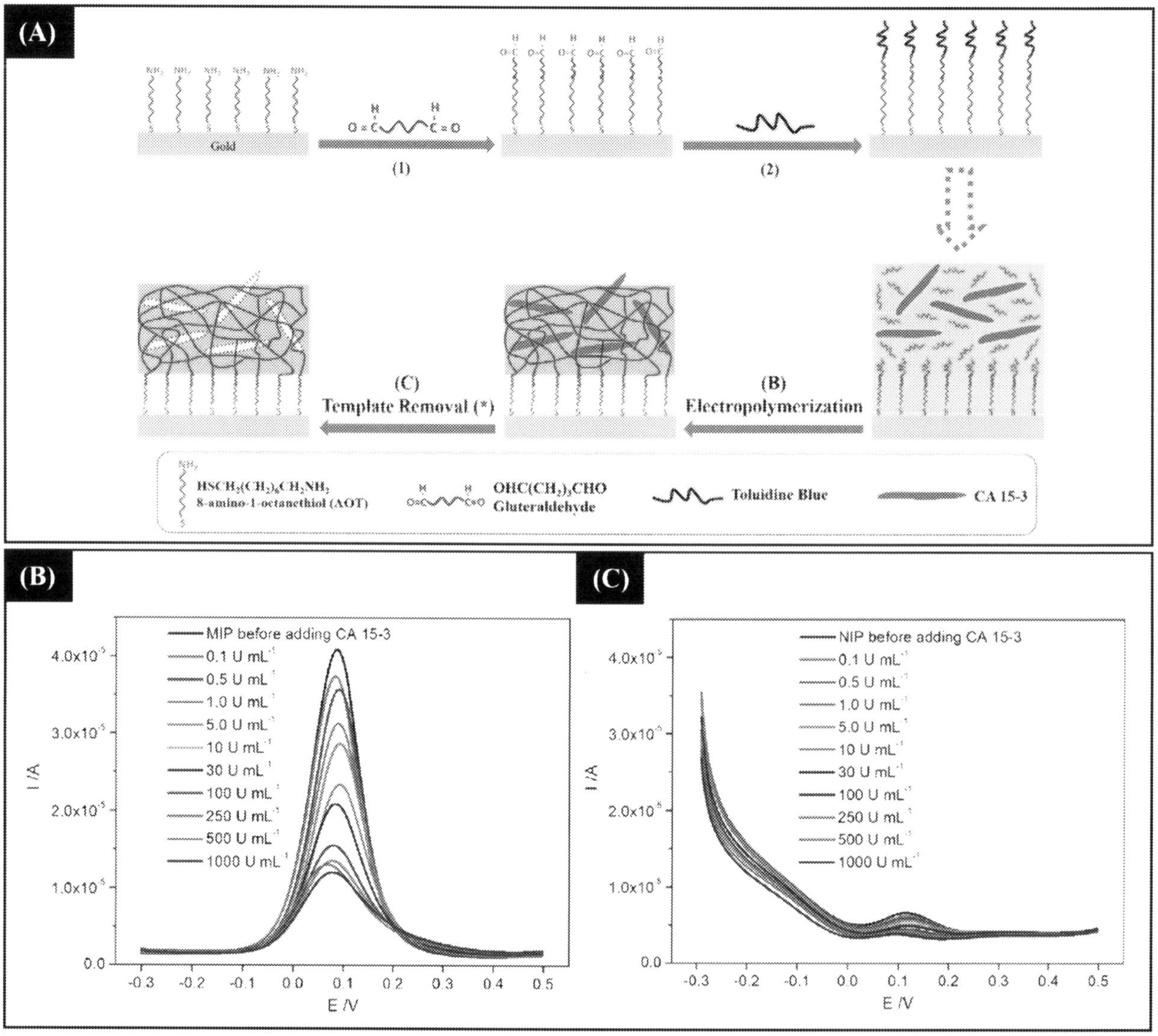


**FIGURE 4.5** (A) Preparation of a molecularly imprinted polymer (MIP) with electrically conducting poly(Toluidine Blue), through electrochemical polymerization on a gold surface, for the specific screening of breast cancer biomarker CA 15−3. Incubation of protein concentrations ranging between 0.10 and 1000 U/mL showed a decrease of the current in (B) MIP, but not in the case of (C) NIP receptor film. *Reproduced under the terms and conditions of the Creative Commons Attribution (CC BY-NC-ND 4.0), License. Ribeiro, J. A., Pereira, C. M., Silva, A. F., & Sales, M. G. F. (2018). Disposable electrochemical detection of breast cancer tumour marker CA 15−3 using poly(Toluidine Blue) as imprinted polymer receptor.* Biosensors and Bioelectronics, 109*, 246−254. https://doi.org/10.1016/j.bios.2018.03.011. Copyright Elsevier.*

presents high surface-to-volume ratio, biocompatibility, and high electron transfer ability, while carbon nanotubes show optimal electrical conductivity. The combination of both resulted into an enhanced sensing performance (Paul et al., 2017). Graphene quantum dots have also been explored for electrochemical biosensing, because they increase the sensor sensitivity by enhancing the biorecognition signal (Hu, Zhang, Wen, Zhang, & Wang, 2016).

### 4.2.2 Optical POC biosensors

Optical techniques have experienced a substantial growth for medical diagnosis, especially because they allow inexpensive miniaturized systems with reliable and fast responses. Within optical approaches, colorimetric, photonic, fluorescent, SERS, and SPR are included. Some optical biosensors have the advantage of being label-free, enable naked eye detection of the target analyte through color changes and do no require complex apparatus for reading the signal.

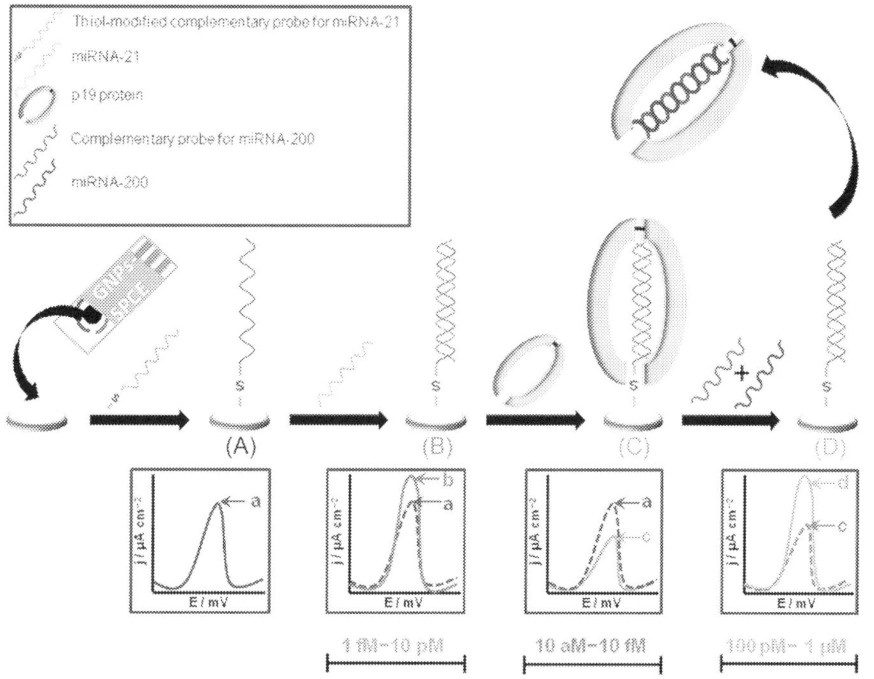


**FIGURE 4.6** Schematic representation of the 3-mode electrochemical sensor for miRs detection. (A) A thiol-modified complementary probe for miR-21 was self-assembled onto AuNPs-modified screen-printed carbon electrode. (B) The binding of the target miR-21 causes an increase in the current intensity, with a liner detection range from 1 fM to 10 pM. (C) The binding of the p19 protein causes a large decrease in current density, thus improving the detection range, from 10 aM to 10 fM. (D) The hybridization product of miR-200 and its complementary probe forced the p19 protein to dissociate from the previous immobilized hybrid, causing a shift-back in the signal. Therefore, the sensor is qualified to detect several miRs at the same time. *Reproduced with permission from Labib, M., Khan, N., Ghobadloo, S. M., Cheng, J., Pezacki, J. P., & Berezovski, M. V. (2013). Three-mode electrochemical sensing of ultralow microRNA levels.* Journal of the American Chemical Society, 135*(8), 3027–3038. https://doi.org/10.1021/ja308216z. Copyright American Chemical Society.*

Colorimetric strategies developed to detect cancer cells usually resort to metal nanoparticles (Chen, 2016; Kang et al., 2010; Yu, Dai, Xu, & Chen, 2016), or enzymatic reactions (Norouzi et al., 2018; Qu, Li, & Yang, 2011; Wang, Fan, Liu, & Wang, 2015; Wang, He, Zhai, He, & Yu, 2017; L. N. Zhang et al., 2014; Zhang et al., 2014; Zhao, Qu, Yuan, & Quan, 2016), both methods leading to color changes. To exemplify the first technique, X. Zhang et al. (2014) used a cell-triggered cyclic signal amplification method. Briefly, the hairpin aptamer probes and AuNPs functionalized with DNA linkers stably coexisted in solution, and the DNA-AuNPs would link between them, giving a purple color to the solution. When the target cells were present and bound to the hairpin probes, conformational change was triggered, resulting in linker DNA hybridization and cleavage by nicking endonuclease-strand scission cycles. The cleaved fragments of the DNA linker could not assemble into DNA-AuNPs, then the nanoparticles would no longer aggregate, resulting in a red color (X. Zhang et al., 2014). Regarding the second technique, in the work of L. N. Zhang et al. (2014), porous platinum nanoparticles were grown on graphene oxide in situ and functionalized with folic acid, which binds to the folic receptors of tumorous cells. In the presence of 3,3′,5,5′ - tetramethylbenzidin (TMB) and hydrogen peroxide, the nanocomposite catalyzed the oxidation of the previous reagent, changing the color of the reaction system (L. N. Zhang et al., 2014). The same principle was used to detect miR-21, with graphene-AuNPs hybrids functionalized with the antisense sequence of miR-21. The presence of this sequence on the graphene-AuNPs surface inhibits the peroxidase-like activity of the hybrids, due to $\pi-\pi$ stacking interactions. When miR-21 is present, it connects to the antisense sequence, releasing it from the hybrid surface, which leads to oxidation of TMB in the presence of hydrogen peroxide (Fig. 4.7) (Zhao et al., 2016).

Photonic crystals (PCs) are among one of the most interesting natural phenomena of structural colouration and have been used in colorimetric sensor approaches. PCs are composed of 1D, 2D, or 3D arrays of materials with different refractive indexes. Due to the periodic modulation of the refractive index within the crystal, there is a specific energy band where propagation of electromagnetic waves is prohibited, denominated photonic band gap (PBG). As a consequence, the crystal reflects the wavelengths located in the PBG, exhibiting vibrant colors if it is within the visible light region (Inan et al., 2017). PCs can also be used to enhance the sensitivity of other techniques, such as fluorescence

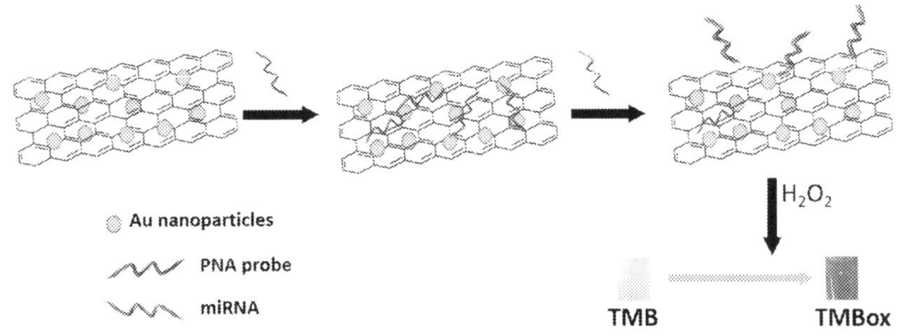


**FIGURE 4.7** Colorimetric device using the antisense sequence of miR-21 on the graphene-AuNPs surface for miR-21 detection.

(Bian et al., 2019; Chang et al., 2018; Frascella et al., 2015; Huang et al., 2011; Li et al., 2014; Liu, Sheng, Xie, Chen, & Gu, 2018; Sinibaldi et al., 2020) and optical microresonators (Graybill, Para, & Bailey, 2016; Qavi, Kindt, Gleeson, & Bailey, 2011). Microring resonators consist of an optical waveguide that is looped back on itself, and a coupling mechanism to access the loop, capable of detecting changes in the refractive index. Through the functionalization of the resonator with DNA probes complementary to the miR target, and further introduction of an anti-DNA:RNA antibody after target hybridization, Qavi et al. (2011) was able to quantify four different mouse brain tissue miRs (miR-16, miR-21, miR-24−1, and miR-26), with a time response of about 40 min and a LOD as low as 10 pM (Qavi et al., 2011). Graybill et al. (2016) also utilized a silicon photonic microring resonator to detect seven different types of human miRs without target amplification by PCR (Graybill et al., 2016). Even though optical microresonators have called attention due to their high sensitivity, their key disadvantage is the susceptibility to changes in the ambient conditions.

Coupling fluorescence with PCs is a popular technique among research studies, because the PC surface can enhance fluorescence signals if its optical resonance overlaps with the excitation or emission spectra of the fluorophore (Inan et al., 2017). In other approaches, PC beads are used as encoded carriers for reporting target biomolecules. Quantum dots (QDs) are also great coding elements, since they provide a large number of codes for multiplex assays. Li et al. (2014) embedded QDs of different sizes into polystyrene nanospheres, resulting in fluorescent and reflection codes. Antibodies specific to AFP and CEA were immobilized into the PC beads, and the detection of these two tumor markers was based on a sandwich format, where antibodies labeled with another fluorophore with a different fluorescence wavelength of the QDs were used as signal. This work achieved a LOD of 0.72 ng/mL for AFP and 0.89 ng/mL for CEA (Li et al., 2014). Bian et al. (2019) also used PC beads with QDs codes for multiplex pancreatic-cancer derived miRs screening, achieving a LOD of 4.2 nM (Bian et al., 2019). Liu, Sheng, Xie, Chen, and Gu (2018) preferred to coat silica nanoparticles with polydopamine before the construction of the barcode beads, since polydopamine provides broad light absorption, improves the mechanical strength and gives the possibility of grafting biomolecules on the surface of the beads. This way, the LOD found for AFP was 0.89 ng/mL, 0.36 ng/mL for CEA, and 1.62 ng/mL for prostate specific antigen (PSA) (P. Liu et al., 2018). Li, Zhou, Zhang, Zheng, and Tang (2020) went further and created a sensor chip with photonic nanospheres, enhancing the fluorescence signal of miR-21 recognition, which combined with cyclic enzymatic amplification enabled a detection as low as 55 fM. (Li et al., 2020).

There are also studies were fluorescence is used without PCs, showing great results in terms of strategies for future POC devices (C. Liu et al., 2018; Liu et al., 2017; Williams, Lee, & Heller, 2018; Xia et al., 2018; Yang et al., 2017). One study showed a sensitive and selective immunosensor for the detection of colon cancer cells, using fluorophore-encapsulated silica nanoparticles modified with anti-EpCAM antibodies (Tao et al., 2012). Another study demonstrated the possibility of detecting multiple breast cancer associated miRs in exosomes using miR-targeting molecular beacons. Molecular beacons are dual-labeled oligonucleotide hairpin probes with a fluorophore and a quencher at each end, whose strong fluorescence is only possible in the presence of the target, due to changes in the structure of the probe. It is possible to use them *in situ*, but molecular beacons are also capable of penetrating exosomes and hybridize with the target miR (Lee, Kim, Jeong, & Rhee, 2016). Song et al. (2018) created a 3D network of carbon nanotubes on a silicon pillar substrate, functionalized with cytokeratin-19 antibody, for oral squamous cell carcinoma (OSCC) detection. The hierarchical structure of the silicon pillars increased the density of immobilized antibody, due to increasing the surface area, and consequently the intensity of the fluorescent signal (Fig. 4.8) (Song et al., 2018).

SERS is a technique for the enhancement of Raman scattering of analyte molecules that are adsorbed on or near SERS-active surfaces, that is, rough or nanostructured metal surfaces, often relying on the classic use of gold, silver, or copper. SERS enhancing substrates range in structure and enhancement factors. It is a powerful non-destructive technique for characterization and sensing, whose enhancement factors can be high enough to enable single molecule spectroscopy. SERS has been thoroughly utilized for cancer-related miRs detection (Crawford et al., 2020; Lee et al., 2018;

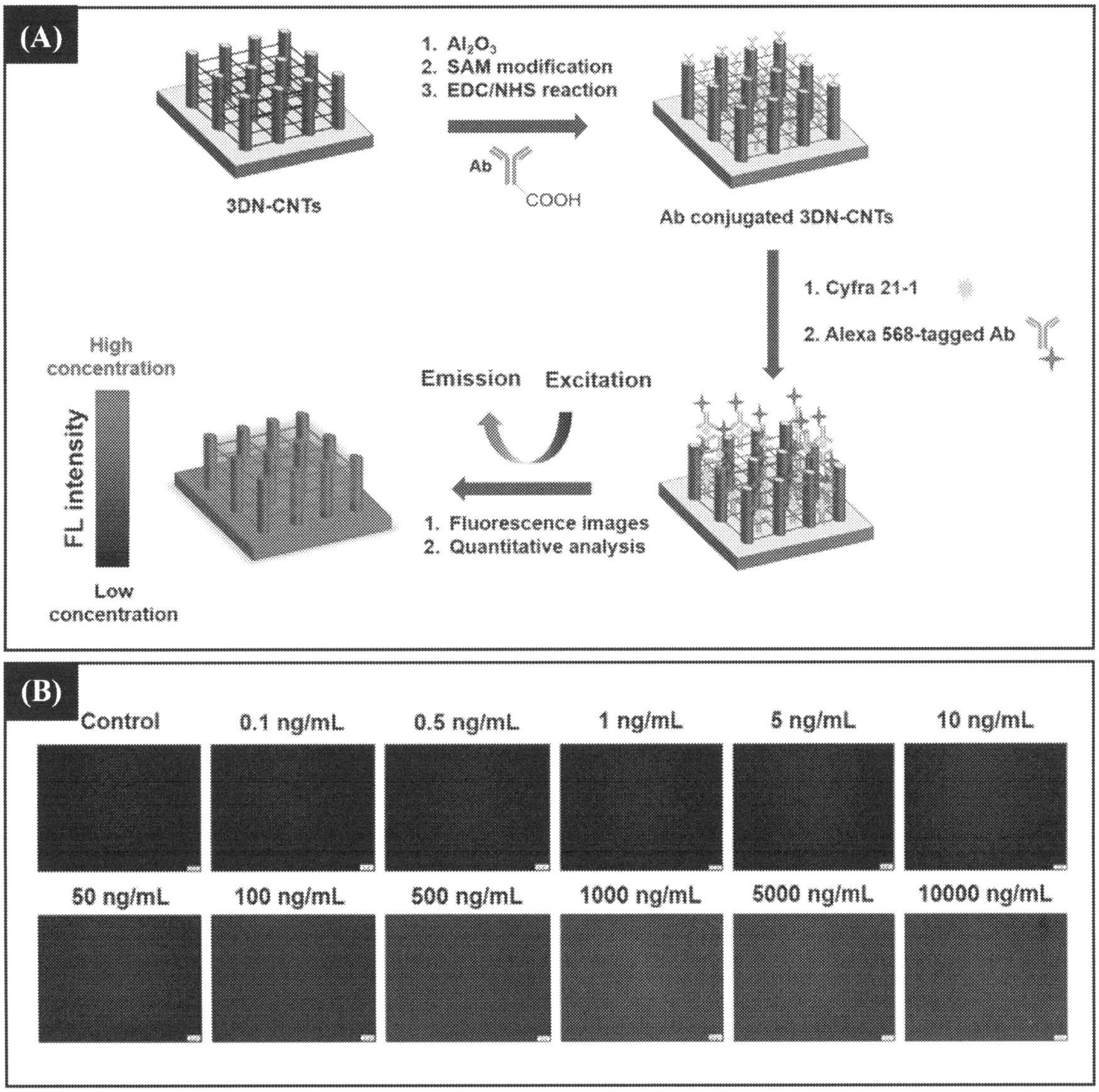


**FIGURE 4.8** (A) Schematic representation of a fluorescence-based immunosensor for the detection of OSCC in clinical saliva samples, using a 3D network of carbon nanotubes on Si pillar substrate. The nanotubes were uniformly coated with a layer of aluminum oxide to enhance structural stability during biomarker detection. Cytokeratin-19 antigen (Cyfra 21−1) was utilized as a model biomarker of OSCC. (B) Representative fluorescence images of the biosensor at different concentrations of Cyfra 21−1. *Reproduced with permission from Song, C. K., Oh, E., Kang, M. S., Shin, B. S., Han, S. Y., Jung, M., Lee, E. S., Yoon, S. Y., Sung, M. M., Ng, W. B., Cho, N. J., & Lee, H. (2018). Fluorescence-based immunosensor using three-dimensional CNT network structure for sensitive and reproducible detection of oral squamous cell carcinoma biomarker.* Analytica Chimica Acta, *1027, 101−108. https://doi.org/10.1016/j.aca.2018.04.025. Copyright Elsevier.*

Ma et al., 2018; Pang et al., 2019; Schechinger, Marks, Mabbott, Choudhury, & Cote, 2019; Zhang, Liu, Gao, & Zhen, 2015; Zhang et al., 2017), exosomes (Zong et al., 2016), cancer cells (Lee et al., 2014), as well as protein biomarkers (Chen, Guo, Liu, & Zhang, 2019; Lee et al., 2011; Wang et al., 2011). For instance, a research team combined SERS nanoprobes made of a gold core—silver shell nanorods with magnetic nanobeads. The surface of the nanoprobes were decorated with antibodies specific to the exosomes surface, while the magnetic beads were made of $Fe_3O_4$ with a shell of silica also attached to antibodies that recognized different proteins on the exosomes. Thus, in the presence of the target exosomes, a sandwich-type immunocomplex is formed, which is precipitated by a magnet followed by SERS signal measurement (Zong et al., 2016). Zhang et al. (2017) utilized a similar strategy, but for the sensing of miR-141 with a LOD of 100 fM (Zhang et al., 2017). R. Chen et al. (2019) developed a SERS-based vertical flow assay for multiplex detection of PSA, CEA, and AFP. In conventional flow assays, AuNPs conjugated with antibodies are used, due to its reddish color and easy visual determination. However, the technique suffers from poor precision, thus researchers generally combine magnetic particles, fluorescence microspheres and carbon nanoparticles with optical readers, to improve the sensitivity and precision. In this work, the advantages of SERS were combined with the ones of vertical flow assays, and the sensing was based on Raman dye-encoded core-shell SERS nanotags (Chen, Guo, Liu, & Zhang, 2019).

SPR is also included in optical transduction methods for biosensors: it originates from the oscillation of conduction-band electrons at the dielectric-metal interface, induced by incident light. As it was observed for SERS, the present technique has also been studied for the detection of cancer protein biomarkers (Aćimović et al., 2014; Chiu & Yang, 2020; Ermini, Song, Špringer, & Homola, 2019; Liang et al., 2015; Narayan et al., 2019; Tan et al., 2018; Yavas et al., 2018), as well as for miR quantification (Hu, Xu, & Chen, 2017; Šípová et al., 2010; Xue et al., 2019). For example, Liang et al. (2015) created triangular silver nanoprisms functionalized with anti-PSA to trigger oxidation of glucose, producing hydrogen peroxide, which in turn oxidizes and etches the triangular nanoprisms into smaller spherical nanoparticles. This allows a shift in the SPR peak and color change observed with at naked eye (Fig. 4.9) (Liang et al., 2015). Yavas et al. (2018) went to a more complex design of a localized surface plasmon resonance (LSPR) chip for automatic calibration and multiplexed screening of breast cancer biomarkers present in human serum. While SPR is based on flat metal films, LSPR enables the sensing volume down to the molecular scale, due to the confinement of a surface plasmon on a single metal nanoparticle. The LSPR can be tuned by using metal nanoparticles with different sizes, shapes, and compositions. In this work, a gold nanorod array was fabricated on a glass substrate with electron beam lithography, and specific antibodies were immobilized on it. Furthermore, the chip had micromechanical and electronic valves, controlled by MATLAB software, and the LSPR changes were tracked in real-time with a transmission spectroscopy setup controlled by Labview software interface (Yavas et al., 2018). For miR detection and quantification, SPR already showed to be able to detect attomolar concentrations with a 2D antimonene nanomaterial-based SPR sensor. This happens because antimonene has a stronger interaction with

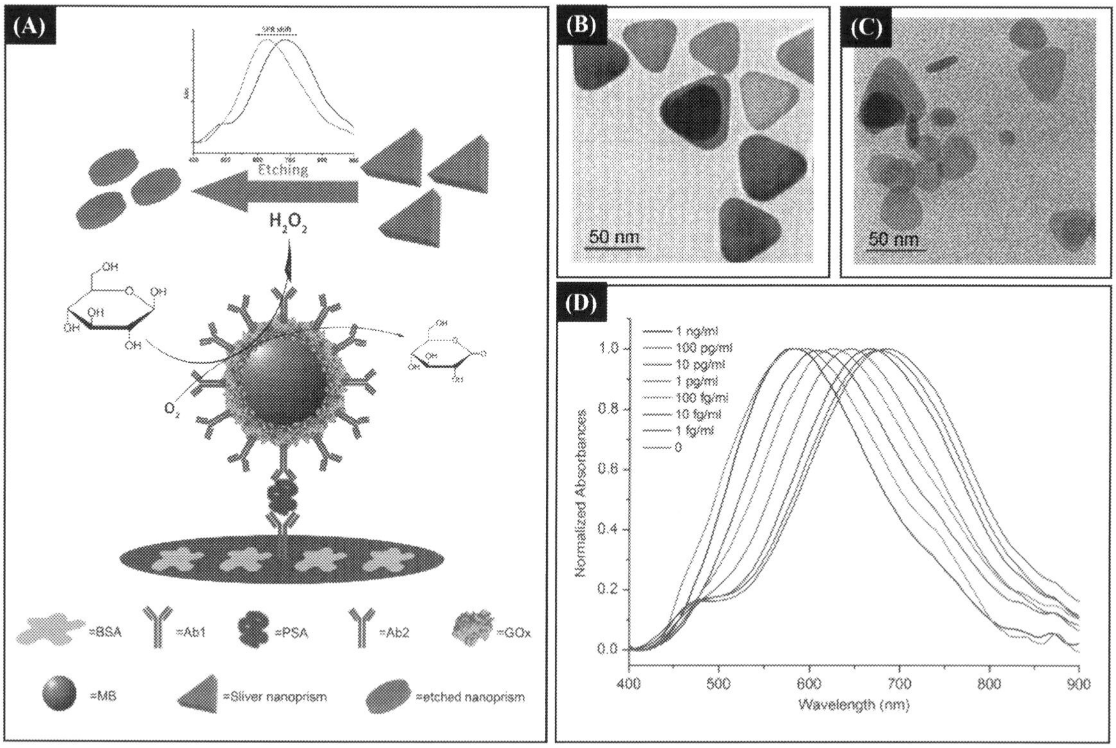


**FIGURE 4.9** (A) Schematic diagram of a SPR biosensor for the quantitative immunoassay of PSA, based on glucose oxidase (GOx)-catalyzed etching of triangular silver nanoprisms into smaller spherical silver nanoparticles. PSA is first immobilized by the capture antibody (Ab1), and then is recognized by the detection antibody (Ab2) conjugated with GOx on the surface of magnetic beads (MBs). The immobilized GOx catalyzes the oxidation of glucose (Glu) to generate hydrogen peroxide, which induces the etching of (B) triangular silver nanoprisms into (C) smaller spherical silver nanoparticles. With the etching, the solution turns blue to purple. (D) SPR peak shifts of triangular silver nanoprisms with different concentrations of PSA in fetal bovine serum. *Reproduced with permission from Liang, J., Yao, C., Li, X., Wu, Z., Huang, C., Fu, Q., Lan, C., Cao, D., & Tang, Y. (2015). Silver nanoprism etching-based plasmonic ELISA for the high sensitive detection of prostate-specific antigen.* Biosensors and Bioelectronics, *69, 128–134. https://doi.org/10.1016/j.bios.2015.02.026. Copyright Elsevier.*

ssDNA probes than the usually used graphene. Then, it was possible to detect miR-21 and miR-155 at a concentration as low as 10 aM (Xue et al., 2019).

### 4.2.3 Piezoelectric POC biosensors

Piezoelectric biosensors are an attractive approach for POC diagnosis, due to their simplicity, real-time output, high sensitivity, and specificity. A quartz crystal microbalance (QCM) is a mass sensitive piezoelectric device working on the principle that the frequency of crystal oscillation is affected by mass changes, i.e., due to a mass bound on the crystal surface reflecting a recognition event. The oscillation is also influenced by the density and viscosity of the detection media (Uludag & Tothill, 2012).

Although optical and electrochemical approaches are more popular among POC research, several authors have devoted their efforts to obtain piezoelectric biosensors for early detection of cancer (Bianco et al., 2013; Crivianu-Gaita, Aamer, Posaratnanathan, Romaschin, & Thompson, 2016; Garai-Ibabe et al., 2011; Kim et al., 2009; J. H. Lee et al., 2005). For instance, Shen, Liu, Cai, and Lu (2008) assembled a primary cysteamine monolayer onto a gold electrode associated with the piezoelectric quartz crystal, which was further modified with a hyperbranched polymer with a high density of functional groups to connect to the antibody against AFP (Shen et al., 2008). In another study, Loo et al. (2011) developed a piezoelectric microcantilever sensor functionalized with anti-HER2 antibodies, which was able to detect HER2 at clinically relevant levels found in serum of patients (Loo et al., 2011). Uludag and Tothill (2012) also analyzed serum samples, but for the detection of PSA, performing a sandwich assay. For that, they used a functionalized piezoelectric sensor with anti-PSA, which linked to the PSA, and then AuNPs of 40 nm modified with anti-PSA were also added, whose mass allowed to enhance the QCM signal. The platform was very successful, as the LOD was 0.39 ng/mL in 100% serum, (...), whereas the PSA levels in prostate cancer are above 4 ng/mL (Uludag & Tothill, 2012).

Piezoelectric biosensors were already applied in microfluidic devices for cancer cells detection. However, the passive detection mode of the QCM results in a time-consuming regeneration process of the device because it requires surface modification to bind biochemical samples and it is impossible to perform in situ clinical analysis without a manipulation component. To overcome these limitations, Zhang et al. (2010) introduced a nickel pillar array onto the surface of the QCM, to induce an active magnetic control over superparamagnetic microbeads functionalized with wheat germ agglutinin, to capture A549 cancer cells (Zhang et al., 2010). It was also possible to use piezoelectric biosensors to distinguish between normal and cancerous cells (Zhou, Marx, Dewilde, McIntosh, & Braunhut, 2012), or increase their sensitivity for tumorous cells with a $2 \times 3$ leaky surface acoustic wave aptasensor array (Chang et al. 2014). This last system could analyse 5 samples simultaneously, and without mutual interference, because each resonator crystal unit had an individual oscillator circuit (Chang et al., 2014).

### 4.2.4 Thermometric POC biosensors

Thermometric biosensors allow analyzing small quantities of samples with higher sensitivity. Microcalorimetric designs commonly used gold and nickel, whose convenience derived from their low resistance and ease of manufacturing. However, those materials are not suitable to achieve high sensitivity, because they lack high Seebeck coefficient. To overcome these limitations, the thermopile of recent designs usually incorporates materials such as titanium and bismuth, besides a low thermal conductivity polymer membrane (e.g., SU-8 and PDMS), and better vapor sealing. (Lubbers, Kazura, Dawson, Mernaugh, & Baudenbacher, 2019).

For thermometric POC biosensors for breast cancer detection, Park et al. (2007) proposed a microcalorimetric device with a sensing part composed by a thermopile consisting of 18 Cu-Ni thermocouples in a radial structure, and a PDMS microfluidic system. Then, breast cancer cells showing the receptor HER2 and several concentrations of the drug transtuzumab (which is a monoclonal antibody against HER2) were injected in the fabricated microcalorimeter. When the receptor HER2 of breast cancer cells bonded to the antibody, a temperature difference was observed between junctions of the thermopile, leading to an output voltage nearly linear according to the concentration of HER2-transtuzumab complex. The response was confirmed by fluorescence imaging (Park et al., 2007). In another study, Lubbers et al. (2019) reported a microcalorimeter for a sandwich based thermal ELISA, applied to quantify the trastuzumab drug in an optimized device requiring only nL reaction volumes. The detection step was based on the heat from the reaction of hydrogen peroxide and o-phenylenediamine dihydrochloride with horseradish peroxidase (HRP) linked to a detecting antibody (Lubbers et al., 2019). Ahmad, Towe, Wolf, Mertens, and Lerchner (2010) evolved the design of a flow-through chip calorimeter by adding magnetic beads coupled to the binding agent of interest and disposed atop the sensing junctions of the thermopile. This approach intended to achieve high sensitivity and was successfully applied to detect DNA hybridization (Ahmad et al., 2010).

More recently, other designs have been developed in coordination with optical techniques. For example, Polo, Del Pino, Pelaz, Grazu, and De la Fuente (2013) built a plasmonic-driven thermal sensor, using gold nanoprisms modified with antibodies against CEA, placed onto fax paper, which is a thermosensitive and cheap support. When the antibody-antigen reaction occurred, a thermal signal was imprinted upon illumination, which enabled detecting CEA in the attomolar range (Polo et al., 2013). Another interesting design is of a paper-based sensor for photo-thermal detection of breast cancer cells. Upon optical stimulation, AuNPs generate heat and temperature variation. This photo-thermal energy conversion is enhanced by the presence of graphene oxide therefore, the researchers developed graphene oxide-Au nanocomposites functionalized with anti-EpCAM antibody specific for surface antigen of MCF-7 cells. The number of cells could be detected by immobilizing anti-EpCAM in a nitrocellulose paper to capture the cells, which were subsequently labeled with the functionalized nanocomposite and irradiated by a laser to record the temperature contrast (J. Zhou et al., 2016).

## 4.3 Application of biomaterials in biosensors

As it has been happening in other research fields, such as therapy, diagnostics and imaging, the development of POC biosensors has focused on using biocompatible and environmentally friendly materials, considering their availability in the ecosystem and optical, mechanical, chemical and biological properties. In particular, polysaccharides, such as cellulose and chitosan, are promising biopolymers, due to their versatility in fitting different structural designs, biocompatibility and amenability to functionalization (Shan et al., 2018).

Chitosan is the deacetylated form of chitin, and one of the most studied biopolymers, since it is biocompatible, biodegradable and edible, besides having one amine group and two hydroxyl groups in the repeating hexosamine moiety, which enables it to be modified. Electrochemical immunobiosensors that incorporate chitosan for the detection of cancer biomarkers have been studied, showing the ability of chitosan to stabilize metal nanoparticles, and to immobilize enzymes or antibodies (joint with carbon nanotubes), whose importance relates to the increased sensitivity of the biosensors (Choudhary, Singh, Kaur, & Arora, 2014; Li, Yuan, Chai, & Chen, 2010; Suresh, Brahman, Reddy, & Bondili, 2018). One example is the work of Suresh et al. (2018), who constructed a screen-printed electrode with AuNPs and chitosan nanocomposite film for the immobilization of a primary anti-PSA antibody. After PSA binding, a secondary antibody tagged with HRP was added yielding a very sensitive sensor with LOD of 1 pg/mL (Suresh et al., 2018). Chitosan can also be used to encapsulate QDs (Sharma et al., 2012) for improving their interaction with biological molecules, as well as magnetic nanoparticles for cell capture (Chen et al., 2011). Utilizing this technique, Sharma et al. (2012) demonstrated the ability of QDs coated with chitosan to be functionalized with a DNA probe and used to detect leukemia. The sensor had a response time of one minute, LOD of 2.56 pM, and was reusable five to six times (Sharma et al., 2012).

Cellulose is the most abundant polymer in nature, which is mainly found in plants in the form of fibers, even though it is possible to obtain nanofibrils or nanocrystals after treatment. Kim et al. (2017) introduced a cellulose strip with silver core-gold shell nanoparticles placed on it for SERS, achieving an enhancement factor above $10^8$, which indicates that it can detect single molecules (Kim et al., 2017). In another study, a colorimetric immunosensor using a porous cellulose hydrogel with embedded AuNPs functionalized with an antibody for the tumor marker AFP, enabled a LOD of 0.46 ng/mL (Fig. 4.10) (Ma et al., 2019).

Paper-based biosensors have been a hot topic in terms of POC diagnosis, being constituted by cellulose paper with a large number of pores for storage of immobilized reagents. Usually, the detection of the bioanalyte is performed by color change, its intensity, emitted fluorescence, or others. The main advantages of paper-based biosensors are the possibility of a wide selection of paper porosity, being highly modifiable, show low cost and high sensitivity and selectivity, as well as fast speed of analysis. However, they are not reusable, absorption of moisture and interferents may occur, and the stability of biomolecules stored may be compromised, which require storage in sealed polybags (Ratajczak & Stobiecka, 2020). Some works regarding paper-based approaches in cancer diagnosis were performed with the aim of obtaining immunosensors with a simple design (Chen, Guo, Liu, & Zhang, 2019; Ge, Zhang, Yan, Huang, & Yu, 2017; Ji, Lee, & Kim, 2018; Kumar et al., 2019; Sun et al., 2018). For instance, Y. Chen et al. (2019) developed a paper-based fluorometric immunodevice for the simultaneous detection of CEA and PSA. For that, they used two kinds of CdTe QDs, with different emission peaks under the same excitation wavelength, to label the two respective antibodies against the bioanalytes in a sandwich immunoassay format, enabling simultaneous detection (Y. Chen et al., 2019). Sun et al. (2018) developed an electrochemical and optical paper microfluidic device for miR-21 detection. Briefly, they used a cellulose paper modified with gold nanorods to immobilize the hairpin probe DNA. For electrochemical signal amplification, a probe made of cerium dioxide ($CeO_2$)−AuNPs surface-modified with the enzyme GOx was used. In

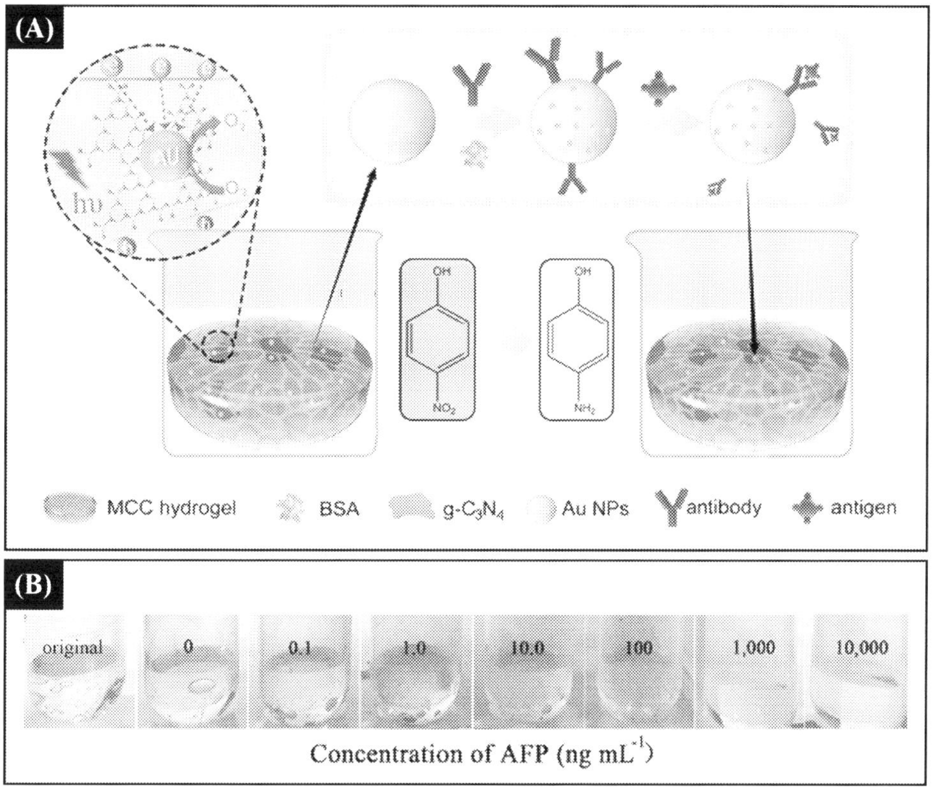


**FIGURE 4.10** (A) Colorimetric immunosensor obtained using a porous cellulose hydrogel with embedded AuNPs, which were functionalized with the antibody against the AFP. After being incubated with the antibody, the turned-off catalytic sites on AuNPs in the cellulose hydrogel would not be "turned on" until the corresponding antigen was added. When AFP was added to the system, (B) 4-nitrophenol was reduced to 4-aminophenol, catalyzed by the turned-on AuNPs, and accompanied by a color fading. The number of the recovered Au active sites was related to the amount of the antigen added. *Reproduced with permission from Ma, F., Yuan, C.W., Liu, J.N., Cao, J.H., & Wu, D.Y. (2019). Colorimetric immunosensor based on Au@g-C3N4-doped spongelike 3D network cellulose hydrogels for detecting α-fetoprotein.* ACS Applied Materials and Interfaces, 11*(22), 19902–19912. https://doi.org/10.1021/acsami.9b06769. Copyright American Chemical Society.*

the presence of the miR-21 target, the hairpin structure opened to hybridize with the miR-21 and retained a single strand fragment to hybridize with a captured DNA on the electrochemical probe of $CeO_2$-Au@GOx. The glucose oxidase catalyzed glucose to produce hydrogen peroxide, which was further electrocatalysed by $CeO_2$. The semi-quantitative chromogenic detection relied on the oxidation of TMB in the presence of hydrogen peroxide, with the help of exonuclease I (Sun et al., 2018).

Although chitosan and cellulose are the trendiest biological materials, other works studied the use of polysaccharides such as agarose (Guo, Hao, Duan, Wang, & Wei, 2012). Also, polydopamine can be considered a biopolymer, even though it is synthetic, since it derives from dopamine, a melanin-related polymer found in organisms such as bacteria and fungi, the ink of cephalopods, and in the hair, skin, and eyes of humans. Polydopamine is generally used for signal amplification, due to its ability to improve adhesion of biomolecules to the sensor by directly using polydopamine nanoparticles or using other core nanoparticles with a shell of polydopamine (N. Jiang et al., 2019; Jiang, Tang, & Miao, 2019; Liu, Xu, He, Deng, & Zhu, 2013; Xu et al., 2018).

## 4.4 New trends in POC biosensors design

As can be observed throughout the present chapter, POC biosensors have been booming in a research effort to overcome the limitations of the time-consuming and expensive techniques found in clinical setups. Beyond the transduction methods and signal amplification attempts, the integration of these methods with technologies, such as microfluidics, paper strips, and other test strips have paved the way to develop desired automated, portable, and easy-to-use POC screening devices (Qin et al., 2017; Rong et al., 2019; Wu et al., 2016; Yang et al., 2011). Moreover, the adaptation of already existent equipment such as the glucose meter (Gao et al., 2020; Sun et al., 2019; Zhou et al., 2019), or the

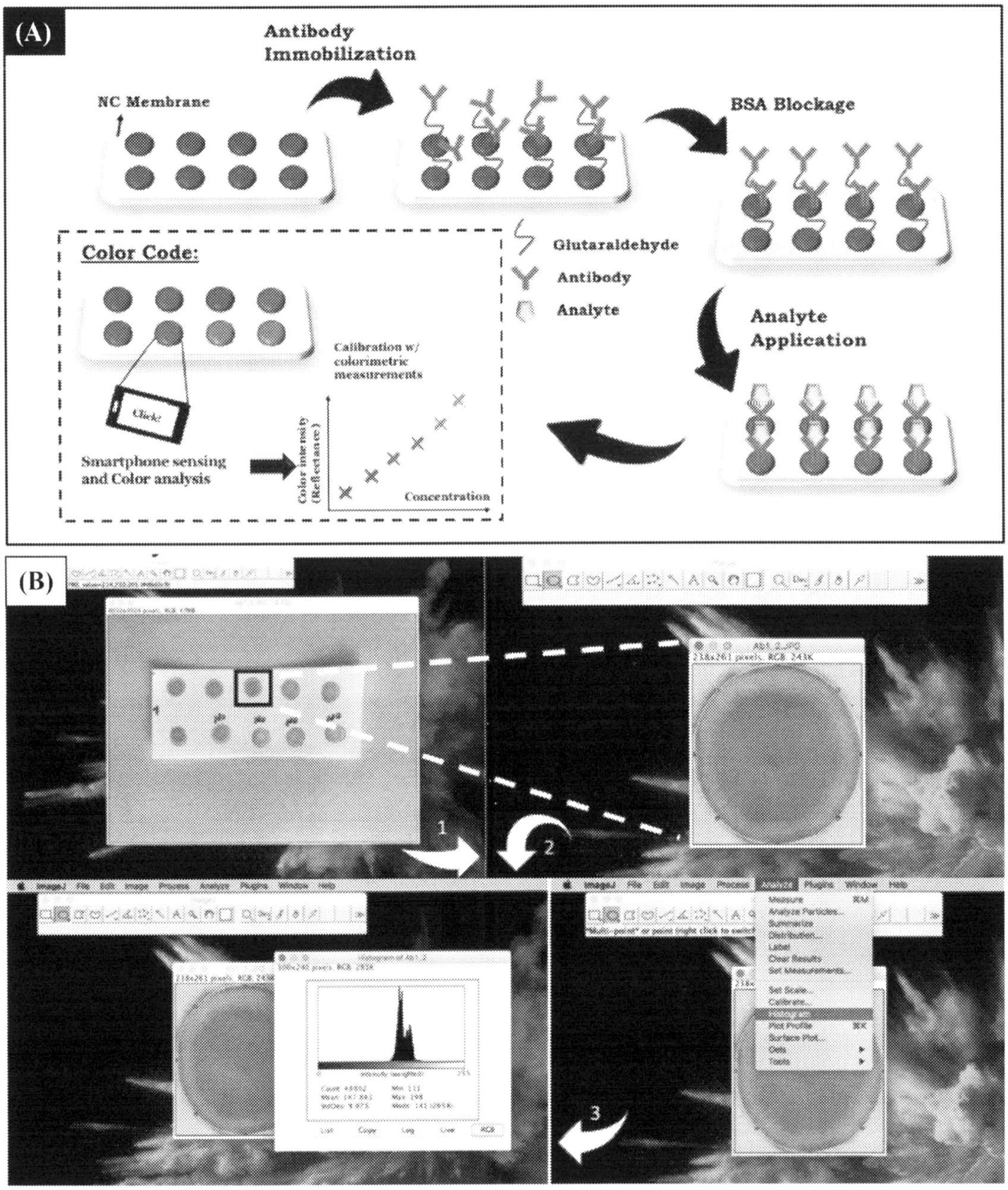


FIGURE 4.11 (A) Biosensor framework formed by a paper-based colorimetric test combined with a smartphone. AuNPs functionalized with antibodies against AFP and mucin-16 were immobilized on a nitrocellulose membrane, whose color changed in the presence of the analytes. (B) The smartphone was used for image and color analysis. *Reproduced with permission from Aydindogan, E., Ceylan, A. E., & Timur, S. (2020). Paper-based colorimetric spot test utilizing smartphone sensing for detection of biomarkers.* Talanta, *208, 120446. https://doi.org/10.1016/j.talanta.2019.120446. Copyright Elsevier.*

construction of other electrochemical portable devices for detection of cancer biomarkers present in body fluids (Bianchi et al., 2020; Sawhney & Conlan, 2019) have already been assessed to surpass the need of sample analysis in a clinical laboratory.

Smartphone-based biosensors are also one of the newest trends in portable POC design. They can be used as standalone platforms without costly reagents and with low consumption of energy (Liu, Geng, Fan, Liu, & Chen, 2019) (Junjie, Zhaoxin, Zhiyuan, Jian, & Hongda, 2019). Multiple approaches have been successfully attempted (Alizadeh, Salimi, & Hallaj, 2018; Aydindogan, Ceylan, & Timur, 2020; Barbosa, Gehlot, Sidapra, Edwards, & Reis, 2015; Fan et al., 2020; Wang, Chang, Sun, & Li, 2017; Zhang, Li, & Wei, 2020), as is the case of Barbosa et al. (2015)

portable smartphone for colorimetric and fluorescence sandwich immunoassay for the detection of PSA. The smartphone was coupled to fluoropolymer strips precoated with anti-PSA, a magnifying lens, a light source for chromogenic detection, or a UV blacklight and a dichroic additive green filter for fluorescence detection (Barbosa et al., 2015). Aydindogan et al. (2020) opted by a simpler framework, joining a paper-based colorimetric test to the smartphone. The AuNPs, functionalized with antibodies against AFP and mucin-16, were immobilized on a nitrocellulose membrane, whose color changed in the presence of the analytes. The smartphone was used for image and color analysis, and found a LOD of 1.054 and 0.413 ng/mL for AFP and mucin-16, respectively (Fig. 4.11) (Aydindogan et al., 2020).

## 4.5 Commercially available POC biosensors for cancer diagnosis

Biosensors should constitute bioanalytical tools suited for cost-effective, rapid, and sensitive analyses in low volume complex matrices, such as blood, urine, sweat, or saliva. Although some of the systems studied in laboratories cannot compete with the technologies already present in the clinical setup, due to lower accuracy, they can be used in routine testing.

For example, EPIC CTC Platform (©2019 Epic Sciences), which was developed to detect unbiased CTCs in patient blood samples, such as CTC clusters, small CTCs, and apoptotic CTCs, has been confirming its appropriate use for clinical tests. Even though CTCs exist as one in 1 billion blood cells, monitoring them has advantages over tissue biopsies, since it is a less invasive real-time biopsy, causing less hazardous to patients who may be debilitated, can be performed more often than tissue biopsies, and is cheaper than radiographic imaging (Werner et al., 2015). This platform was already used to study metastatic prostate cancer (Salami et al., 2019; Scher et al., 2017) and non-small cell lung cancer (Pirie-Shepherd et al., 2017). Despite these promising examples, there are few commercially available POC biosensors for cancer diagnostics. CellSearch Platform (© Menarini Silicon Biosystems, Inc 2021) is the only clinically and analytically FDA-approved system for identification, isolation, and enumeration of CTCs.

## 4.6 Future perspectives

Biosensors are continuously evolving and demonstrate impressive improvements in terms of reduced production costs, enhanced sensitivity and selectivity, and multiplexed miniaturized assays with user-friendly configurations. Nonetheless, the use of biosensors in POC diagnostics needs to expand its current capabilities to be commercialized and translated into the clinics. In terms of POC cancer assessment, as well as for other diseases with comorbidities, it is relevant that devices are able to differentiate conditions with similar symptoms, to allow detection at early stages, and to help the stratification of patients into different groups to improve treatment options. The expansion beyond standard and ubiquitous lateral flow format requires the use of new materials, nanocomposites, and reporting interfaces, since the cost is not the only limiting factor in the scale-up and implementation of innovative sensing technologies. The integration of responsive materials may amplify the development of reusable tests for conditions where frequent measurements are required. Moreover, this feature is also aligned with the need to invest in environmentally friendly devices. Further investigations into unexplored sensing mechanisms are also expected to continue merging with the current advances in smart, mobile technologies (e.g., smartphones, handheld readers), but also to be ground-breaking in equipment-free assays with the same accuracy of highly sensitive and fully quantitative assays. This enhanced connectivity also demands less time-consuming tests, using biological fluids from minimally invasive procedures, with associated challenges of using unprocessed samples. In conclusion, as for the past decades, the future of biosensing in POC will follow the paradigms of society needs, and great advances are expected including in niche areas of wearable, implantable, and autonomous devices.

## Acknowledgments

The authors acknowledge the financial support from the European Commission/H2020, through MindGAP/FET-Open/GA829040 project. The author RV also acknowledges Fundação para a Ciência e a Tecnologia her PhD grant (2020.09673.BD).

## References

Aćimović, S. S., Ortega, M. A., Sanz, V., Berthelot, J., Garcia-Cordero, J. L., Renger, J., … Quidant, R. (2014). LSPR chip for parallel, rapid, and sensitive detection of cancer markers in serum. *Nano Letters*, *14*(5), 2636–2641. Available from https://doi.org/10.1021/nl500574n.

Ahmad, L. M., Towe, B., Wolf, A., Mertens, F., & Lerchner, J. (2010). Binding event measurement using a chip calorimeter coupled to magnetic beads. *Sensors and Actuators, B: Chemical, 145*(1), 239–245. Available from https://doi.org/10.1016/j.snb.2009.12.012.

Alhadrami, H. A. (2018). Biosensors: Classifications, medical applications, and future prospective. *Biotechnology and Applied Biochemistry, 65*(3), 497–508. Available from https://doi.org/10.1002/bab.1621.

Alizadeh, N., Salimi, A., & Hallaj, R. (2018). Mimicking peroxidase activity of Co2(OH)2CO3-CeO2 nanocomposite for smartphone based detection of tumor marker using paper-based microfluidic immunodevice. *Talanta, 189*, 100–110. Available from https://doi.org/10.1016/j.talanta.2018.06.034.

Ankeny, J. S., Court, C. M., Hou, S., Li, Q., Song, M., Wu, D., ... Tomlinson, J. S. (2016). Circulating tumour cells as a biomarker for diagnosis and staging in pancreatic cancer. *British Journal of Cancer, 114*(12), 1367–1375. Available from https://doi.org/10.1038/bjc.2016.121.

Anwarul, H., Md, N., Mahboob, M., Arghya, P., Alessandro, P., Tapas, K., ... Jaffa, A. A. (2014). Recent advances in application of biosensors in tissue engineering. *BioMed Research International*, 1–18. Available from https://doi.org/10.1155/2014/307519.

Asal, M., Özen, Ö., Şahinler, M., Baysal, H. T., & Polatoğlu, İ. (2019). An overview of biomolecules, immobilization methods and support materials of biosensors. *Sensor Review, 39*(3), 377–386. Available from https://doi.org/10.1108/SR-04-2018-0084.

Aydindogan, E., Ceylan, A. E., & Timur, S. (2020). Paper-based colorimetric spot test utilizing smartphone sensing for detection of biomarkers. *Talanta, 208*, 120446. Available from https://doi.org/10.1016/j.talanta.2019.120446.

Azimzadeh, M., Rahaie, M., Nasirizadeh, N., Ashtari, K., & Naderi-Manesh, H. (2016). An electrochemical nanobiosensor for plasma miRNA-155, based on graphene oxide and gold nanorod, for early detection of breast cancer. *Biosensors and Bioelectronics, 77*, 99–106. Available from https://doi.org/10.1016/j.bios.2015.09.020.

Barbosa, A. I., Gehlot, P., Sidapra, K., Edwards, A. D., & Reis, N. M. (2015). Portable smartphone quantitation of prostate specific antigen (PSA) in a fluoropolymer microfluidic device. *Biosensors and Bioelectronics, 70*, 5–14. Available from https://doi.org/10.1016/j.bios.2015.03.006.

Benvidi, A., Dehghani Firouzabadi, A., Dehghan Tezerjani, M., Moshtaghiun, S. M., Mazloum-Ardakani, M., & Ansarin, A. (2015). A highly sensitive and selective electrochemical DNA biosensor to diagnose breast cancer. *Journal of Electroanalytical Chemistry, 750*, 57–64. Available from https://doi.org/10.1016/j.jelechem.2015.05.002.

Bergveld, P. (1970). Development of an Ion-sensitive solid-state device for neurophysiological measurements. *IEEE Transactions on Biomedical Engineering*, 70–71. Available from https://doi.org/10.1109/TBME.1970.4502688.

Bettazzi, F., Marrazza, G., Minunni, M., Palchetti, I., & Scarano, S. (2017). Biosensors and related bioanalytical tools. *Comprehensive Analytical Chemistry, 77*, 1–33. Available from https://doi.org/10.1016/bs.coac.2017.05.003.

Bezinge, L., Suea-Ngam, A., Demello, A. J., & Shih, C. J. (2020). Nanomaterials for molecular signal amplification in electrochemical nucleic acid biosensing: Recent advances and future prospects for point-of-care diagnostics. *Molecular Systems Design and Engineering, 5*(1), 49–66. Available from https://doi.org/10.1039/c9me00135b.

Bhalla, N., Jolly, P., Formisano, N., & Estrela, P. (2016). Introduction to biosensors. *Essays in Biochemistry, 60*(1), 1–8. Available from https://doi.org/10.1042/EBC20150001.

Bian, F., Sun, L., Cai, L., Wang, Y., Zhao, Y., Wang, S., & Zhou, M. (2019). Molybdenum disulfide-integrated photonic barcodes for tumor markers screening. *Biosensors and Bioelectronics, 133*, 199–204. Available from https://doi.org/10.1016/j.bios.2019.02.066.

Bianchi, V., Mattarozzi, M., Giannetto, M., Boni, A., Munari, I. D., & Careri, M. (2020). A self-calibrating IoT portable electrochemical immunosensor for serum human epididymis protein 4 as a tumor biomarker for ovarian cancer. *Sensors (Basel), 20*(7), 1–12. Available from https://doi.org/10.3390/s20072016.

Bianco, M., Aloisi, A., Arima, V., Capello, M., Ferri-Borgogno, S., Novelli, F., ... Rinaldi, R. (2013). Quartz crystal microbalance with dissipation (QCM-D) as tool to exploit antigen-antibody interactions in pancreatic ductal adenocarcinoma detection. *Biosensors and Bioelectronics, 42*(1), 646–652. Available from https://doi.org/10.1016/j.bios.2012.10.012.

Brown, M., Hahn, W., Bailey, B., Hall, A., Rodriguez, G., Zahn, H., ... Erickson, S. (2020). Development and evaluation of a sensitive bacteriophage-based MRSA diagnostic screen. *Viruses, 12*(6), 1–15. Available from https://doi.org/10.3390/v12060631.

Buckner, C. L., Christiansen, L. R., Bourgeois, D., Lazarchick, J. J., & Lazarchick, J. (2007). Case reports: CD20 positive T-cell lymphoma/leukemia: A rare entity with potential diagnostic pitfalls. *Annals of Clinical and Laboratory Science, 37*(3), 263–267.

Cai, L., Wang, W., Li, X., Dong, T., Zhang, Q., Zhu, B., ... Wu, S. (2018). MicroRNA-21-5p induces the metastatic phenotype of human cervical carcinoma cells in vitro by targeting the von Hippel-Lindau tumor suppressor. *Oncology Letters, 15*(4), 5213–5219. Available from https://doi.org/10.3892/ol.2018.7937.

Cai, Z., Song, Y., Wu, Y., Zhu, Z., James Yang, C., & Chen, X. (2013). An electrochemical sensor based on label-free functional allosteric molecular beacons for detection target DNA/miRNA. *Biosensors and Bioelectronics, 41*(1), 783–788. Available from https://doi.org/10.1016/j.bios.2012.10.002.

Cardoso, A. R., Moreira, F. T. C., Fernandes, R., & Sales, M. G. F. (2016). Novel and simple electrochemical biosensor monitoring attomolar levels of miRNA-155 in breast cancer. *Biosensors and Bioelectronics, 80*, 621–630. Available from https://doi.org/10.1016/j.bios.2016.02.035.

Chang, K., Pi, Y., Lu, W., Wang, F., Pan, F., Li, F., ... Chen, M. (2014). Label-free and high-sensitive detection of human breast cancer cells by aptamer-based leaky surface acoustic wave biosensor array. *Biosensors and Bioelectronics, 60*, 318–324. Available from https://doi.org/10.1016/j.bios.2014.04.027.

Chang, N., Zhai, J., Liu, B., Zhou, J., Zeng, Z., & Zhao, X. (2018). Low cost 3D microfluidic chips for multiplex protein detection based on photonic crystal beads. *Lab on a Chip, 18*(23), 3638–3644. Available from https://doi.org/10.1039/c8lc00784e.

Chen, L., Bao, C. C., Yang, H., Li, D., Lei, C., Wang, T., ... Cui, D. X. (2011). A prototype of giant magnetoimpedance-based biosensing system for targeted detection of gastric cancer cells. *Biosensors and Bioelectronics, 26*(7), 3246–3253. Available from https://doi.org/10.1016/j.bios.2010.12.034.

Chen, R., Liu, B., Ni, H., Chang, N., Luan, C., Ge, Q., . . . Zhao, X. (2019). Vertical flow assays based on core-shell SERS nanotags for multiplex prostate cancer biomarker detection. *Analyst*, *144*(13), 4051–4059. Available from https://doi.org/10.1039/c9an00733d.

Chen, Y., Guo, X., Liu, W., & Zhang, L. (2019). Paper-based fluorometric immunodevice with quantum-dot labeled antibodies for simultaneous detection of carcinoembryonic antigen and prostate specific antigen. *Microchimica Acta*, *186*(2), 112–121. Available from https://doi.org/10.1007/s00604-019-3232-0.

Chiu, N. F., & Yang, H. T. (2020). High-sensitivity detection of the lung cancer biomarker CYFRA21–1 in serum samples using a carboxyl-MoS2 functional film for SPR-based immunosensors. *Frontiers in Bioengineering and Biotechnology*, *8*, 234. Available from https://doi.org/10.3389/fbioe.2020.00234.

Chong, A. S. M., & Zhao, X. S. (2004). Design of large-pore mesoporous materials for immobilization of penicillin G acylase biocatalyst. *In Catalysis today*, *93–95*, 293–299. Available from https://doi.org/10.1016/j.cattod.2004.06.064.

Choudhary, M., Singh, A., Kaur, S., & Arora, K. (2014). Enhancing lung cancer diagnosis: Electrochemical simultaneous bianalyte immunosensing using carbon nanotubes–chitosan nanocomposite. *Applied Biochemistry and Biotechnology*, *174*(3), 1188–1200. Available from https://doi.org/10.1007/s12010-014-1020-1.

Choudhary, M., Yadav, P., Singh, A., Kaur, S., Ramirez-Vick, J., Chandra, P., . . . Singh, S. P. (2016). CD 59 targeted ultrasensitive electrochemical immunosensor for fast and noninvasive diagnosis of oral cancer. *Electroanalysis*, *28*(10), 2565–2574. Available from https://doi.org/10.1002/elan.201600238.

Chuang, C. H., Du, Y. C., Wu, T. F., Chen, C. H., Lee, D. H., Chen, S. M., . . . Shaikh, M. O. (2016). Immunosensor for the ultrasensitive and quantitative detection of bladder cancer in point of care testing. *Biosensors and Bioelectronics*, *84*, 126–132. Available from https://doi.org/10.1016/j.bios.2015.12.103.

Cooney, C. L., Weaver, J. C., Tannenbaum, S. R., Faller, D. V., Shields, A., & Jahnke, M. (1974). *Thermal enzyme probe-a novel approach to chemical analysis* (2, pp. 411–417). Plenum Press. Available from https://doi.org/10.1007/978-1-4615-8897-9_58.

Crawford, B. M., Wang, H. N., Stolarchuk, C., Von Furstenberg, R. J., Strobbia, P., Zhang, D., . . . Vo-Dinh, T. (2020). Plasmonic nanobiosensors for detection of microRNA cancer biomarkers in clinical samples. *Analyst*, *145*(13), 4587–4594. Available from https://doi.org/10.1039/d0an00193g.

Crivianu-Gaita, V., Aamer, M., Posaratnanathan, R. T., Romaschin, A., & Thompson, M. (2016). Acoustic wave biosensor for the detection of the breast and prostate cancer metastasis biomarker protein PTHrP. *Biosensors and Bioelectronics*, *78*, 92–99. Available from https://doi.org/10.1016/j.bios.2015.11.031.

Danarto, R., Astuti, I., Umbas, R., & Haryana, S. M. (2020). Urine miR-21–5p and miR-200c-3p as potential non-invasive biomarkers in patients with prostate cancer. *Turkish Journal of Urology*, *46*(1), 26–30. Available from https://doi.org/10.5152/tud.2019.19163.

Diviés, C. (1975). Remarks on ethanol oxidation by an \Acetobacter xylinum\ microbial electrode. *Annals of Microbiology*, *126*, 175–186.

Dong, H., Chen, H., Jiang, J., Zhang, H., Cai, C., & Shen, Q. (2018). Highly sensitive electrochemical detection of tumor exosomes based on aptamer recognition-induced multi-DNA release and cyclic enzymatic amplification. *Analytical Chemistry*, *90*(7), 4507–4513. Available from https://doi.org/10.1021/acs.analchem.7b04863.

Du, D., Wang, L., Shao, Y., Wang, J., Engelhard, M. H., & Lin, Y. (2011). Functionalized graphene oxide as a nanocarrier in a multienzyme labeling amplification strategy for ultrasensitive electrochemical immunoassay of phosphorylated p53 (s392). *Analytical Chemistry*, *83*(3), 746–752. Available from https://doi.org/10.1021/ac101715s.

Edoo, M. I. A., Chutturghoon, V. K., Wusu-Ansah, G. K., Zhu, H., Zhen, T. Y., Xie, H. Y., & Zheng, S. S. (2019). Serum biomarkers AFP, CEA and CA19–9 combined detection for early diagnosis of hepatocellular carcinoma. *Iranian Journal of Public Health*, *48*(2), 314–322. Available from http://ijph.tums.ac.ir/index.php/ijph/article/view/16153/6289.

Elnashar, M. (2010). Review Article: Immobilized molecules using biomaterials and nanobiotechnology. *Journal of Biomaterials and Nanobiotechnology*, *1*, 61–76. Available from https://doi.org/10.4236/jbnb.2010.11008.

Ensafi, A. A., Taei, M., Rahmani, H. R., & Khayamian, T. (2011). Sensitive DNA impedance biosensor for detection of cancer, chronic lymphocytic leukemia, based on gold nanoparticles/gold modified electrode. *Electrochimica Acta*, *56*(24), 8176–8183. Available from https://doi.org/10.1016/j.electacta.2011.05.124.

Erickson, K. A., & Wilding, P. (1993). Evaluation of a novel point-of-care system, the i-STAT portable clinical analyzer. *Clinical Chemistry*, *39*(2), 283–287. Available from https://doi.org/10.1093/clinchem/39.2.283.

Ermini, M. L., Song, X. C., Špringer, T., & Homola, J. (2019). Peptide functionalization of gold nanoparticles for the detection of carcinoembryonic antigen in blood plasma via SPR-based biosensor. *Frontiers in Chemistry*, *7*(40), 1–11. Available from https://doi.org/10.3389/fchem.2019.00040.

Fan, Z., Geng, Z., Fang, W., Lv, X., Su, Y., Wang, S., & Chen, H. (2020). Smartphone biosensor system with multi-testing unit based on localized surface plasmon resonance integrated with microfluidics chip. *Sensors (Switzerland)*, *20*(2), 446–459. Available from https://doi.org/10.3390/s20020446.

Farzin, L., Shamsipur, M., Samandari, L., & Sheibani, S. (2018). Advances in the design of nanomaterial-based electrochemical affinity and enzymatic biosensors for metabolic biomarkers: A review. *Microchimica Acta*, *185*(5), 276. Available from https://doi.org/10.1007/s00604-018-2820-8.

Fei, X., Ting-rui, D., Shu-yue, H., Qiu-hong, W., Min, X., & Zi-hui, M. (2013). A covalently imprinted photonic crystal for glucose sensing. *Journal of Nanomaterials*, 1–6. Available from https://doi.org/10.1155/2013/530701.

Frascella, F., Ricciardi, S., Pasquardini, L., Potrich, C., Angelini, A., Chiadò, A., . . . Descrovi, E. (2015). Enhanced fluorescence detection of miRNA-16 on a photonic crystal. *Analyst*, *140*(16), 5459–5463. Available from https://doi.org/10.1039/c5an00889a.

Garai-Ibabe, G., Grinyte, R., Golub, E. I., Canaan, A., de la Chapelle, M. L., Marks, R. S., & Pavlov, V. (2011). Label free and amplified detection of cancer marker EBNA-1 by DNA probe based biosensors. *Biosensors and Bioelectronics*, *30*(1), 272–275. Available from https://doi.org/10.1016/j.bios.2011.09.025.

Garcia-Cordero, J. L., & Maerkl, S. J. (2020). Microfluidic systems for cancer diagnostics. *Current Opinion in Biotechnology*, *65*, 37–44. Available from https://doi.org/10.1016/j.copbio.2019.11.022.

Gao, X., Li, X., Sun, X., Zhang, J., Zhao, Y., Liu, X., & Li, F. (2020). DNA tetrahedra-cross-linked hydrogel functionalized paper for onsite analysis of DNA methyltransferase activity using a personal glucose Meter. *Analytical Chemistry*, *92*(6), 4592–4599. Available from https://doi.org/10.1021/acs.analchem.0c00018.

Ge, S., Zhang, Y., Yan, M., Huang, J., & Yu, J. (2017). *Fabrication of lab-on-paper using porous Au-paper electrode: Application to tumor marker electrochemical immunoassays, Methods in molecular biology* (1572, pp. 125–134). Humana Press Inc. Available from https://doi.org/10.1007/978-1-4939-6911-1_9.

Ge, Y., Mu, W., Ba, Q., Li, J., Jiang, Y., Xia, Q., & Wang, H. (2020). Hepatocellular carcinoma-derived exosomes in organotropic metastasis, recurrence and early diagnosis application. *Cancer Letters*, *477*, 41–48. Available from https://doi.org/10.1016/j.canlet.2020.02.003.

Gener Lahav, T., Adler, O., Zait, Y., Shani, O., Amer, M., Doron, H., … Erez, N. (2019). Melanoma-derived extracellular vesicles instigate proinflammatory signaling in the metastatic microenvironment. *International Journal of Cancer*, *145*(9), 2521–2534. Available from https://doi.org/10.1002/ijc.32521.

Giles, F. J., Vose, J. M., Do, K. A., Johnson, M. M., Manshouri, T., Bociek, G., … Albitar, M. (2003). Circulating CD20 and CD52 in patients with non-Hodgkin's lymphoma or Hodgkin's disease. *British Journal of Haematology*, *123*(5), 850–857. Available from https://doi.org/10.1046/j.1365-2141.2003.04683.x.

Gomes, R. S., Moreira, F. T. C., Fernandes, R., & Goreti Sales, M. F. (2018). Sensing CA 15–3 in point-of-care by electropolymerizing O-phenylenediamine (oPDA) on Au-screen printed electrodes. *PLoS ONE*, *13*(5), e0196656. Available from https://doi.org/10.1371/journal.pone.0196656.

Gowda, R., Robertson, B., Iyer, S., Barry, J., Dinavahi, S., & Robertson, G. P. (2020). The role of exosomes in metastasis and progression of melanoma. *Cancer Treat Rev*, *85*, 101975. Available from https://doi.org/10.1016/j.ctrv.2020.101975.

Graybill, R. M., Para, C. S., & Bailey, R. C. (2016). PCR-Free, multiplexed expression profiling of microRNAs using silicon photonic microring resonators. *Analytical Chemistry*, *88*(21), 10347–10351. Available from https://doi.org/10.1021/acs.analchem.6b03350.

Guilbault, G. G., & Montalvo, J. G. (1969). A Urea-specific enzyme electrode. *Journal of the American Chemical Society*, *91*(8), 2164–2165. Available from https://doi.org/10.1021/ja01036a083.

Guo, Z., Hao, T., Duan, J., Wang, S., & Wei, D. (2012). Electrochemiluminescence immunosensor based on graphene-CdS quantum dots-agarose composite for the ultrasensitive detection of alpha fetoprotein. *Talanta*, *89*, 27–32. Available from https://doi.org/10.1016/j.talanta.2011.11.017.

Han, S., Liu, W., Yang, S., & Wang, R. (2019). Facile and label-free electrochemical biosensors for microRNA detection based on DNA origami nanostructures. *ACS Omega*, *4*(6), 11025–11031. Available from https://doi.org/10.1021/acsomega.9b01166.

Hassan, S. H. A., Van Ginkel, S. W., Hussein, M. A. M., Abskharon, R., & Oh, S. E. (2016). Toxicity assessment using different bioassays and microbial biosensors. *Environment International*, *92–93*, 106–118. Available from https://doi.org/10.1016/j.envint.2016.03.003.

He, W., Yuan, S., Zhong, W. H., Siddikee, M. A., & Dai, C. C. (2016). Application of genetically engineered microbial whole-cell biosensors for combined chemosensing. *Applied Microbiology and Biotechnology*, *100*(3), 1109–1119. Available from https://doi.org/10.1007/s00253-015-7160-6.

Hong, C. Y., Chen, X., Liu, T., Li, J., Yang, H. H., Chen, J. H., & Chen, G. N. (2013). Ultrasensitive electrochemical detection of cancer-associated circulating microRNA in serum samples based on DNA concatamers. *Biosensors and Bioelectronics*, *50*, 132–136. Available from https://doi.org/10.1016/j.bios.2013.06.040.

Hu, T., Zhang, L., Wen, W., Zhang, X., & Wang, S. (2016). Enzyme catalytic amplification of miRNA-155 detection with graphene quantum dot-based electrochemical biosensor. *Biosensors and Bioelectronics*, *77*, 451–456. Available from https://doi.org/10.1016/j.bios.2015.09.068.

Hu, F., Xu, J., & Chen, Y. (2017). Surface plasmon resonance imaging detection of sub-femtomolar microRNA. *Analytical Chemistry*, *89*(18), 10071–10077. Available from https://doi.org/10.1021/acs.analchem.7b02838.

Huang, C. S., George, S., Lu, M., Chaudhery, V., Tan, R., Zangar, R. C., & Cunningham, B. T. (2011). Application of photonic crystal enhanced fluorescence to cancer biomarker microarrays. *Analytical Chemistry*, *83*(4), 1425–1430. Available from https://doi.org/10.1021/ac102989n.

Inan, H., Poyraz, M., Inci, F., Lifson, M. A., Baday, M., Cunningham, B. T., & Demirci, U. (2017). Photonic crystals: Emerging biosensors and their promise for point-of-care applications. *Chemical Society Reviews*, *46*(2), 366–388. Available from https://doi.org/10.1039/c6cs00206d.

Islam, F., Haque, M. H., Yadav, S., Islam, M. N., Gopalan, V., Nguyen, N. T., … Shiddiky, M. J. A. (2017). An electrochemical method for sensitive and rapid detection of FAM134B protein in colon cancer samples. *Scientific Reports*, *7*(1), 133–141. Available from https://doi.org/10.1038/s41598-017-00206-8.

Islam, M. N., Masud, M. K., Nguyen, N. T., Gopalan, V., Alamri, H. R., Alothman, Z. A., … Shiddiky, M. J. A. (2018). Gold-loaded nanoporous ferric oxide nanocubes for electrocatalytic detection of microRNA at attomolar level. *Biosensors and Bioelectronics*, *101*, 275–281. Available from https://doi.org/10.1016/j.bios.2017.09.027.

Ji, S., Lee, M., & Kim, D. (2018). Detection of early stage prostate cancer by using a simple carbon nanotube@paper biosensor. *Biosensors and Bioelectronics*, *102*, 345–350. Available from https://doi.org/10.1016/j.bios.2017.11.035.

Jiang, N., Hu, Y., Wei, W., Zhu, T., Yang, K., Zhu, G., & Yu, M. (2019). Detection of microRNA using a polydopamine mediated bimetallic SERS substrate and a re-circulated enzymatic amplification system. *Microchimica Acta*, *186*(2), 65–74. Available from https://doi.org/10.1007/s00604-018-3174-y.

Jiang, Y., Tang, Y., & Miao, P. (2019). Polydopamine nanosphere@silver nanoclusters for fluorescence detection of multiplex tumor markers. *Nanoscale*, *11*(17), 8119–8123. Available from https://doi.org/10.1039/c9nr01307e.

Jönsson, U., Fägerstam, L., Ivarsson, B., Johnsson, B., Karlsson, R., Lundh, K., ... Rönnberg, I. (1991). Real-time biospecific interaction analysis using surface plasmon resonance and sensor chip technology. *BioTechniques*, *11*, 620–627.

Junjie, L., Zhaoxin, G., Zhiyuan, F., Jian, L., & Hongda, C. (2019). Point-of-care testing based on smartphone: The current state-of-the-art (2017–2018). *Biosensors and Bioelectronics*, 17–37. Available from https://doi.org/10.1016/j.bios.2019.01.068.

Justino, C. I. L., Freitas, A. C., Pereira, R., Duarte, A. C., & Rocha Santos, T. A. P. (2015). Recent developments in recognition elements for chemical sensors and biosensors. *TrAC – Trends in Analytical Chemistry*, *68*, 2–17. Available from https://doi.org/10.1016/j.trac.2015.03.006.

Kadhem, A. J., Xiang, S., Nagel, S., Lin, C. H., & de Cortalezzi, M. F. (2018). Photonic molecularly imprinted polymer film for the detection of testosterone in aqueous samples. *Polymers*, *10*(4), 1–13. Available from https://doi.org/10.3390/polym10040349.

Kalluri, R., & LeBleu, V. S. (2020). The biology, function, and biomedical applications of exosomes. *Science*, *367*(6478). Available from https://doi.org/10.1126/science.aau6977.

Kang, J.-H., Asami, Y., Murata, M., Kitazaki, H., Sadanaga, N., Tokunaga, E., ... Katayama, Y. (2010). Gold nanoparticle-based colorimetric assay for cancer diagnosis. *Biosensors and Bioelectronics*, *25*(8), 1869–1874. Available from https://doi.org/10.1016/j.bios.2009.12.022.

Kawamura, A., & Miyata, T. (2016). *Biosensors. In Biomaterials nanoarchitectonics* (pp. 157–176). Elsevier Inc. Available from https://doi.org/10.1016/B978-0-323-37127-8.00010-8.

Khan, N., Hills, R. K., Virgo, P., Couzens, S., Clark, N., Gilkes, A., ... Freeman, S. D. (2017). Expression of CD33 is a predictive factor for effect of gemtuzumab ozogamicin at different doses in adult acute myeloid leukaemia. *Leukemia*, *31*(5), 1059–1068. Available from https://doi.org/10.1038/leu.2016.309.

Kim, D. M., Noh, H. B., Park, D. S., Ryu, S. H., Koo, J. S., & Shim, Y. B. (2009). Immunosensors for detection of Annexin II and MUC5AC for early diagnosis of lung cancer. *Biosensors and Bioelectronics*, *25*(2), 456–462. Available from https://doi.org/10.1016/j.bios.2009.08.007.

Kim, W., Lee, J. C., Lee, G. J., Park, H. K., Lee, A., & Choi, S. (2017). Low-cost label-free biosensing bimetallic cellulose strip with SILAR-synthesized silver core-gold shell nanoparticle structures. *Analytical Chemistry*, *89*(12), 6448–6454. Available from https://doi.org/10.1021/acs.analchem.7b00300.

Krishna, K., & Bekaii-Saab, T. (2015). *CA 19–9 as a serum biomarker in cancer. Biomarkers in disease: Methods, discoveries and applications: Biomarkers in cancer* (pp. 179–201). Netherlands: Springer. Available from https://doi.org/10.1007/978-94-007-7681-417.

Kumar, S., Umar, M., Saifi, A., Kumar, S., Augustine, S., Srivastava, S., & Malhotra, B. D. (2019). Electrochemical paper based cancer biosensor using iron oxide nanoparticles decorated PEDOT:PSS. *Analytica Chimica Acta*, *1056*, 135–145. Available from https://doi.org/10.1016/j.ac.2018.12.053.

Kutluk, H., Bruch, R., Urban, G. A., & Dincer, C. (2020). Impact of assay format on miRNA sensing: Electrochemical microfluidic biosensor for miRNA-197 detection. *Biosensors and Bioelectronics*, *148*, 111824. Available from https://doi.org/10.1016/j.bios.2019.111824.

Labib, M., Khan, N., Ghobadloo, S. M., Cheng, J., Pezacki, J. P., & Berezovski, M. V. (2013). Three-mode electrochemical sensing of ultralow microRNA levels. *Journal of the American Chemical Society*, *135*(8), 3027–3038. Available from https://doi.org/10.1021/ja308216z.

Labuda, J., Brett, A. M. O., Evtugyn, G., Fojta, M., Mascini, M., Ozsoz, M., ... Wang, J. (2010). Electrochemical nucleic acid-based biosensors: Concepts, terms, and methodology (IUPAC Technical Report). *Pure and Applied Chemistry*, *82*(5), 1161–1187. Available from https://doi.org/10.1351/PAC-REP-09-08-16.

Larrea, E., Sole, C., Manterola, L., Goicoechea, I., Armesto, M., Arestin, M., ... Lawrie, C. H. (2016). New concepts in cancer biomarkers: Circulating miRNAs in liquid biopsies. *International Journal of Molecular Sciences*, *17*(5). Available from https://doi.org/10.3390/ijms17050627.

Lee, J. H., Hwang, K. S., Park, J., Yoon, K. H., Yoon, D. S., & Kim, T. S. (2005). *Immunoassay of prostate-specific antigen (PSA) using resonant frequency shift of piezoelectric nanomechanical microcantilever. In, Biosensors and Bioelectronics* (20, pp. 2157–2162). Elsevier Ltd. 10. Available from https://doi.org/10.1016/j.bios.2004.09.024.

Lee, J. H., Kim, J. A., Jeong, S., & Rhee, W. J. (2016). Simultaneous and multiplexed detection of exosome microRNAs using molecular beacons. *Biosensors and Bioelectronics*, *86*, 202–210. Available from https://doi.org/10.1016/j.bios.2016.06.058.

Lee, M., Lee, S., Lee, J.-h., Lim, H.-W., Seong, G. H., Lee, E. K., ... Choo, J. (2011). Highly reproducible immunoassay of cancer markers on a gold-patterned microarray chip using surface-enhanced Raman scattering imaging. *Biosensors and Bioelectronics*, *26*(5), 2135–2141. Available from https://doi.org/10.1016/j.bios.2010.09.021.

Lee, S., Chon, H., Lee, J., Ko, J., Chung, B. H., Lim, D. W., & Choo, J. (2014). Rapid and sensitive phenotypic marker detection on breast cancer cells using surface-enhanced Raman scattering (SERS) imaging. *Biosensors and Bioelectronics*, *51*, 238–243. Available from https://doi.org/10.1016/j.bios.2013.07.063.

Lee, T., Wi, J.-S., Oh, A., Na, H.-K., Lee, J., Lee, K., ... Haam, S. (2018). Highly robust, uniform and ultra-sensitive surface-enhanced Raman scattering substrates for microRNA detection fabricated by using silver nanostructures grown in gold nanobowls. *Nanoscale*, *10*(8), 3680–3687. Available from https://doi.org/10.1039/c7nr08066b.

Li, J., Wang, H., Dong, S., Zhu, P., Diao, G., & Yang, Z. (2014). Quantum-dot-tagged photonic crystal beads for multiplex detection of tumor markers. *Chemical Communications*, *50*(93), 14589–14592. Available from https://doi.org/10.1039/c4cc07019d.

Li, Q., Zhou, S., Zhang, T., Zheng, B., & Tang, H. (2020). Bioinspired sensor chip for detection of miRNA-21 based on photonic crystals assisted cyclic enzymatic amplification method. *Biosensors and Bioelectronics*, *150*, 111866. Available from https://doi.org/10.1016/j.bios.2019.111866.

Li, W., Yuan, R., Chai, Y., & Chen, S. (2010). Reagentless amperometric cancer antigen 15–3 immunosensor based on enzyme-mediated direct electrochemistry. *Biosensors and Bioelectronics*, *25*(11), 2548–2552. Available from https://doi.org/10.1016/j.bios.2010.04.011.

Li, X., Xu, Y., & Zhang, L. (2019). *Serum CA153 as biomarker for cancer and noncancer diseases, Progress in molecular biology and translational science* (162, pp. 265–276). Elsevier B.V. Available from https://doi.org/10.1016/bs.pmbts.2019.01.005.

Liang, J., Yao, C., Li, X., Wu, Z., Huang, C., Fu, Q., ... Tang, Y. (2015). Silver nanoprism etching-based plasmonic ELISA for the high sensitive detection of prostate-specific antigen. *Biosensors and Bioelectronics*, *69*, 128–134. Available from https://doi.org/10.1016/j.bios.2015.02.026.

Liedberg, B., Nylander, C., & Lunström, I. (1983). Surface plasmon resonance for gas detection and biosensing. *Sensors and Actuators, 4(C)*, 299–304. Available from https://doi.org/10.1016/0250-6874(83)85036-7.

Lim, J. M., Ryu, M. Y., Yun, J. W., Park, T. J., & Park, J. P. (2017). Electrochemical peptide sensor for diagnosing adenoma-carcinoma transition in colon cancer. *Biosensors and Bioelectronics*, *98*, 330–337. Available from https://doi.org/10.1016/j.bios.2017.07.013.

Liu, H., Xu, S., He, Z., Deng, A., & Zhu, J. J. (2013). Supersandwich cytosensor for selective and ultrasensitive detection of cancer cells using aptamer-DNA concatamer-quantum dots probes. *Analytical Chemistry*, *85*(6), 3385–3392. Available from https://doi.org/10.1021/ac303789x.

Liu, P., Sheng, T., Xie, Z., Chen, J., & Gu, Z. (2018). Robust, highly visible, and facile bioconjugation colloidal crystal beads for bioassay. *ACS Applied Materials and Interfaces*, *10*(35), 29378–29384. Available from https://doi.org/10.1021/acsami.8b11472.

Liu, R., Shi, R., Zou, W., Chen, W., Yin, X., Zhao, F., & Yang, Z. (2020). Highly sensitive phage-magnetic-chemiluminescent enzyme immunoassay for determination of zearalenone. *Food Chemistry*, *325*, 126905. Available from https://doi.org/10.1016/j.foodchem.2020.126905.

Liu, C., Xu, X., Li, B., Situ, B., Pan, W., Hu, Y., ... Zheng, L. (2018). Single-exosome-counting immunoassays for cancer diagnostics. *Nano Letters*, *18*(7), 4226–4232. Available from https://doi.org/10.1021/acs.nanolett.8b01184.

Liu, H., Tian, T., Zhang, Y., Ding, L., Yu, J., & Yan, M. (2017). Sensitive and rapid detection of microRNAs using hairpin probes-mediated exponential isothermal amplification. *Biosensors and Bioelectronics*, *89*, 710–714. Available from https://doi.org/10.1016/j.bios.2016.10.099.

Liu, J, Geng, Z, Fan, Z, Liu, J, & Chen, H (2019). Point-of-care testing based on smartphone: The current state-of-the-art (2017-2018). *Biosens Bioelectron*, *132*, 17–37. Available from https://doi.org/10.1016/j.bios.2019.01.068.

Loo, L., Capobianco, J. A., Wu, W., Gao, X., Shih, W. Y., Shih, W.-H., ... Adams, G. P. (2011). Highly sensitive detection of HER2 extracellular domain in the serum of breast cancer patients by piezoelectric microcantilevers. *Analytical Chemistry*, *83*(9), 3392–3397. Available from https://doi.org/10.1021/ac103301r.

Lubbers, B., Kazura, E., Dawson, E., Mernaugh, R., & Baudenbacher, F. (2019). Microfabricated calorimeters for thermometric enzyme linked immunosorbent assay in one-nanoliter droplets. *Biomedical Microdevices*, *21*(4), 1–7. Available from https://doi.org/10.1007/s10544-019-0429-2.

Lübbers, D., & Opitz, N. (1975). [The pCO2-/pO2-optode: a new probe for measurement of pCO2 or pO in fluids and gases (authors transl)]. Zeitschrift fur Naturforschung. Section C. *Biosciences*, *30*(4), 532–533. Available from http://europepmc.org/abstract/MED/126595.

Ma, D., Huang, C., Zheng, J., Tang, J., Li, J., Yang, J., & Yang, R. (2018). Quantitative detection of exosomal microRNA extracted from human blood based on surface-enhanced Raman scattering. *Biosensors and Bioelectronics*, *101*, 167–173. Available from https://doi.org/10.1016/j.bios.2017.08.062.

Ma, F., Yuan, C. W., Liu, J. N., Cao, J. H., & Wu, D. Y. (2019). Colorimetric immunosensor based on Au@g-C3N4-doped spongelike 3D network cellulose hydrogels for detecting α-fetoprotein. *ACS Applied Materials and Interfaces*, *11*(22), 19902–19912. Available from https://doi.org/10.1021/acsami.9b06769.

Mahato, K., Kumar, A., Maurya, P. K., & Chandra, P. (2018). Shifting paradigm of cancer diagnoses in clinically relevant samples based on miniaturized electrochemical nanobiosensors and microfluidic devices. *Biosensors and Bioelectronics*, *100*, 411–428. Available from https://doi.org/10.1016/j.bios.2017.09.003.

Martins, G. V., Marques, A. C., Fortunato, E., & Sales, M. G. F. (2020). Paper-based (bio)sensor for label-free detection of 3-nitrotyrosine in human urine samples using molecular imprinted polymer. *Sensing and Bio-Sensing Research*, *28*, 100333. Available from https://doi.org/10.1016/j.sbsr.2020.100333.

Mathew, D. G., Beekman, P., Lemay, S. G., Zuilhof, H., Le Gac, S., & Van Der Wiel, W. G. (2020). Electrochemical detection of tumor-derived extracellular vesicles on nanointerdigitated electrodes. *Nano Letters*, *20*(2), 820–828. Available from https://doi.org/10.1021/acs.nanolett.9b02741.

Matsushita, K., Margulies, I., Onda, M., Nagata, S., Stetler-Stevenson, M., & Kreitman, R. J. (2008). Soluble CD22 as a tumor marker for hairy cell leuker. *Blood*, *112*(6), 2272–2277. Available from https://doi.org/10.1182/blood-2008-01-131987.

Mohammadniaei, M., Nguyen, H. V., Tieu, M. V., & Lee, M. H. (2019). 2D materials in development of electrochemical point-of-care cancer screening devices. *Micromachines*, *10*(10), 662–702. Available from https://doi.org/10.3390/mi10100662.

Mohd Azmi, M. A., Tehrani, Z., Lewis, R. P., Walker, K.-A. D., Jones, D. R., Daniels, D. R., ... Guy, O. J. (2014). Highly sensitive covalently functionalised integrated silicon nanowire biosensor devices for detection of cancer risk biomarker. *Biosensors and Bioelectronics*, *52*, 216–224. Available from https://doi.org/10.1016/j.bios.2013.08.030.

Mohtar, M. A., Syafruddin, S. E., Nasir, S. N., & Yew, L. T. (2020). Revisiting the roles of pro-metastatic epcam in cancer. *Biomolecules*, *10*(2), 255–275. Available from https://doi.org/10.3390/biom10020255.

Narayan, T., Kumar, S., Kumar, S., Augustine, S., Yadav, B. K., & Malhotra, B. D. (2019). Protein functionalised self assembled monolayer based biosensor for colon cancer detection. *Talanta*, *201*, 465–473. Available from https://doi.org/10.1016/j.talanta.2019.04.039.

Newman, J.D., & Turner, A.P.F. (2008). Historical Perspective of Biosensor and Biochip Development. In Handbook of Biosensors and Biochips (eds R.S. Marks, D.C. Cullen, I. Karube, C.R. Lowe and H.H. Weetall). Available from https://doi.org/10.1002/9780470061565.hbb002

Norouzi, A., Ravan, H., Mohammadi, A., Hosseinzadeh, E., Norouzi, M., & Fozooni, T. (2018). Aptamer–integrated DNA nanoassembly: A simple and sensitive DNA framework to detect cancer cells. *Analytica Chimica Acta*, *1017*, 26–33. Available from https://doi.org/10.1016/j.ac.2018.02.037.

Nunna, B. B., Mandal, D., Lee, J. U., Singh, H., Zhuang, S., Misra, D., ... Lee, E. S. (2019). Detection of cancer antigens (CA-125) using gold nano particles on interdigitated electrode-based microfluidic biosensor. *Nano Convergence*, *6*(1), 3–15. Available from https://doi.org/10.1186/s40580-019-0173-6.

Pang, Y., Wang, C., Lu, L. C., Wang, C., Sun, Z., & Xiao, R. (2019). Dual-SERS biosensor for one-step detection of microRNAs in exosome and residual plasma of blood samples for diagnosing pancreatic cancer. *Biosensors and Bioelectronics*, *130*, 204–213. Available from https://doi.org/10.1016/j.bios.2019.01.039.

Park, S. C., Cho, E. J., Moon, S. Y., Yoon, S. I., Kim, Y. J., Kim, D. H., & Suh, J. S. (2007). *A calorimetric biosensor and its application for detecting a cancer cell with optical imaging*, *IFMBE proceedings* (14, pp. 637–640). Springer Verlag 1. Available from https://doi.org/10.1007/978-3-540-36841-0_147.

Paul, K. B., Singh, V., Vanjari, S. R. K., & Singh, S. G. (2017). One step biofunctionalized electrospun multiwalled carbon nanotubes embedded zinc oxide nanowire interface for highly sensitive detection of carcinoma antigen-125. *Biosensors and Bioelectronics*, *88*, 144–152. Available from https://doi.org/10.1016/j.bios.2016.07.114.

Piloto, A. M. L., Ribeiro, D. S. M., Rodrigues, S. S. M., Santos, J. L. M., & Ferreira Sales, M. G. (2020). Label-free quantum dot conjugates for human protein IL-2 based on molecularly imprinted polymers. *Sensors and Actuators, B: Chemical*, *304*, 127343. Available from https://doi.org/10.1016/j.snb.2019.127343.

Piñeiro, R., Martínez-Pena, I., & López-López, R. (2020). *Relevance of CTC clusters in breast cancer metastasis*, *Advances in experimental medicine and biology* (1220, pp. 93–115). Springer. Available from https://doi.org/10.1007/978-3-030-35805-17.

Pirie-Shepherd, S. R., Painter, C., Whalen, P., Vizcarra, P., Roy, M., Qian, J., ... Powell, E. L. (2017). Detecting expression of 5T4 in CTCs and tumor samples from NSCLC patients. *PLoS ONE*, *12*(7), e0179561–e0179587. Available from https://doi.org/10.1371/journal.pone.0179561.

Pirzada, M., & Altintas, Z. (2019). Nanomaterials for healthcare biosensing applications. *Sensors (Switzerland)*, *19*(23). Available from https://doi.org/10.3390/s19235311.

Polo, E., Del Pino, P., Pelaz, B., Grazu, V., & De la Fuente, J. M. (2013). Plasmonic-driven thermal sensing: Ultralow detection of cancer markers. *Chemical Communications*, *49*(35), 3676–3678. Available from https://doi.org/10.1039/c3cc39112d.

Pop, L. M., Barman, S., Shao, C., Poe, J. C., Venturi, G. M., Shelton, J. M., ... Vitetta, E. S. (2014). A reevaluation of CD22 expression in human lung cancer. *Cancer Research*, *74*(1), 263–271. Available from https://doi.org/10.1158/0008-5472.CAN-13-1436.

Qavi, A. J., Kindt, J. T., Gleeson, M. A., & Bailey, R. C. (2011). Anti-DNA:RNA antibodies and silicon photonic microring resonators: Increased sensitivity for multiplexed microRNA detection. *Analytical Chemistry*, *83*(15), 5949–5956. Available from https://doi.org/10.1021/ac201340s.

Rastislav, M., Miroslav, S., & Ernest, Š. (2012). Biosensors – Classification, characterization and new trends. *Acta Chimica Slovaca*, 109–120. Available from https://doi.org/10.2478/v10188-012-0017-z.

Qin, W., Wang, K., Xiao, K., Hou, Y., Lu, W., Xu, H., ... Cui, D. (2017). Carcinoembryonic antigen detection with "Handing"-controlled fluorescence spectroscopy using a color matrix for point-of-care applications. *Biosensors and Bioelectronics*, *90*, 508–515. Available from https://doi.org/10.1016/j.bios.2016.10.052.

Qu, F., Li, T., & Yang, M. (2011). Colorimetric platform for visual detection of cancer biomarker based on intrinsic peroxidase activity of graphene oxide. *Biosensors and Bioelectronics*, *26*(9), 3927–3931. Available from https://doi.org/10.1016/j.bios.2011.03.013.

Ratajczak, K., & Stobiecka, M. (2020). High-performance modified cellulose paper-based biosensors for medical diagnostics and early cancer screening: A concise review. *Carbohydrate Polymers*, *229*, 115463. Available from https://doi.org/10.1016/j.carbpol.2019.115463.

Rebelo, T. S. C. R., Costa, R., Brandão, A. T. S. C., Silva, A. F., Sales, M. G. F., & Pereira, C. M. (2019). Molecularly imprinted polymer SPE sensor for analysis of CA-125 on serum. *Analytica Chimica Acta*, *1082*, 126–135. Available from https://doi.org/10.1016/j.aca.2019.07.050.

Rechnitz, G. A. (1978). Biochemical electrode uses tissue slices. *Chemical & Engineering News*, *56*(41), 16. Available from https://doi.org/10.1021/cen-v056n041.p016.

Ribeiro, J. A., Pereira, C. M., Silva, A. F., & Sales, M. G. F. (2018). Disposable electrochemical detection of breast cancer tumour marker CA 15–3 using poly(Toluidine Blue) as imprinted polymer receptor. *Biosensors and Bioelectronics*, *109*, 246–254. Available from https://doi.org/10.1016/j.bios.2018.03.011.

Rong, Z., Bai, Z., Li, J., Tang, H., Shen, T., Wang, Q., ... Wang, S. (2019). Dual-color magnetic-quantum dot nanobeads as versatile fluorescent probes in test strip for simultaneous point-of-care detection of free and complexed prostate-specific antigen. *Biosensors and Bioelectronics*, *145*, 111719. Available from https://doi.org/10.1016/j.bios.2019.111719.

Sadana, A., & Sadana, N. (2015). *Biosensor economics and manufacturing*, 653–680. Available from https://doi.org/10.1016/B978-0-444-53794-2.00014-8.

Sagredo, A. I., Sepulveda, S. A., Roa, J. C., & Oróstica, L. (2017). Exosomes in bile as potential pancreatobiliary tumor biomarkers. *Translational Cancer Research*, *6*, S1371–S1383. Available from https://doi.org/10.21037/tcr.2017.10.37.

Salami, S. S., Singhai, U., Spratt, D. E., Palapattu, G. S., Hollenbeck, B. K., Schonhoft, J. D., ... Morgan, T. M. (2019). Circulating tumor cells as a predictor of treatment response in clinically localized prostate cancer. *JCO Precision Oncology*, *3*, 1–9. Available from https://doi.org/10.1200/PO.18.00352.

Sawhney, M. A., & Conlan, R. S. (2019). POISED-5, a portable on-board electrochemical impedance spectroscopy biomarker analysis device. *Biomedical Microdevices*, *21*(3), 70–83. Available from https://doi.org/10.1007/s10544-019-0406-9.

Schechinger, M., Marks, H., Mabbott, S., Choudhury, M., & Cote, G. (2019). A SERS approach for rapid detection of microRNA-17 in the picomolar range. *Analyst*, *144*(13), 4033–4044. Available from https://doi.org/10.1039/c9an00653b.

Scher, H. I., Graf, R. P., Schreiber, N. A., McLaughlin, B., Jendrisak, A., Wang, Y., ... Dittamore, R. (2017). Phenotypic heterogeneity of circulating tumor cells informs clinical decisions between AR signaling inhibitors and taxanes in metastatic prostate cancer. *Cancer Research*, *77*(20), 5687–5698. Available from https://doi.org/10.1158/0008-5472.CAN-17-1353.

Schultz, J. S. (1982). Optical sensor of plasma constituents. *United States Pat 4*, *344*. Vol https://patents.google.com/patent/US4344438A/en.

Shan, D., Gerhard, E., Zhang, C., Tierney, J. W., Xie, D., Liu, Z., & Yang, J. (2018). Polymeric biomaterials for biophotonic applications. *Bioactive Materials*, *3*(4), 434–445. Available from https://doi.org/10.1016/j.bioactmat.2018.07.001.

Shandilya, R., Bhargava, A., Bunkar, N., Tiwari, R., Goryacheva, I. Y., & Mishra, P. K. (2019). Nanobiosensors: Point-of-care approaches for cancer diagnostics. *Biosensors and Bioelectronics*, *130*, 147–165. Available from https://doi.org/10.1016/j.bios.2019.01.034.

Sharma, A., Pandey, C. M., Sumana, G., Soni, U., Sapra, S., Srivastava, A. K., . . . Malhotra, B. D. (2012). Chitosan encapsulated quantum dots platform for leukemia detection. *Biosensors and Bioelectronics*, *38*(1), 107–113. Available from https://doi.org/10.1016/j.bios.2012.05.010.

Shen, G., Liu, M., Cai, X., & Lu, J. (2008). A novel piezoelectric quartz crystal immnuosensor based on hyperbranched polymer films for the detection of α-fetoprotein. *Analytica Chimica Acta*, *630*(1), 75–81. Available from https://doi.org/10.1016/j.aca.2008.09.053.

Shons, A., Dorman, F., & Najarian, J. (1972). An immunospecific microbalance. *Journal of Biomedical Materials Research*, *6*(6), 565–570. Available from https://doi.org/10.1002/jbm.820060608.

Šípová, H., Zhang, S., Dudley, A. M., Galas, D., Wang, K., & Homola, J. (2010). Surface plasmon resonance biosensor for rapid label-free detection of microribonucleic acid at subfemtomole level. *Analytical Chemistry*, *82*(24), 10110–10115. Available from https://doi.org/10.1021/ac102131s.

Sinibaldi, A., Doricchi, A., Pileri, T., Allegretti, M., Danz, N., Munzert, P., . . . Michelotti, F. (2020). Bioassay engineering: A combined label-free and fluorescence approach to optimize HER2 detection in complex biological media. *Analytical and Bioanalytical Chemistry*, *412*(14), 3509–3517. Available from https://doi.org/10.1007/s00216-020-02643-3.

Sirisha, V. L., Jain, A., & Jain, A. (2016). *Enzyme immobilization: An overview on methods, support material, and applications of immobilized enzymes, Advances in food and nutrition research* (79, pp. 179–211). Academic Press Inc. Available from https://doi.org/10.1016/bs.afnr.2016.07.004.

Smith, D. A., Newbury, L. J., Drago, G., Bowen, T., & Redman, J. E. (2017). Electrochemical detection of urinary microRNAs via sulfonamide-bound antisense hybridisation. *Sensors and Actuators, B: Chemical*, *253*, 335–341. Available from https://doi.org/10.1016/j.snb.2017.06.069.

Soares, J. C., Soares, A. C., Rodrigues, V. C., Melendez, M. E., Santos, A. C., Faria, E. F., . . . Oliveira, O. N. (2019). Detection of the prostate cancer biomarker PCA3 with electrochemical and impedance-based biosensors. *ACS Applied Materials and Interfaces*, *11*(50), 46645–46650. Available from https://doi.org/10.1021/acsami.9b19180.

Song, C. K., Oh, E., Kang, M. S., Shin, B. S., Han, S. Y., Jung, M., . . . Lee, H. (2018). Fluorescence-based immunosensor using three-dimensional CNT network structure for sensitive and reproducible detection of oral squamous cell carcinoma biomarker. *Analytica Chimica Acta*, *1027*, 101–108. Available from https://doi.org/10.1016/j.aca.2018.04.025.

Sun, F., Sun, X., Jia, Y., Hu, Z., Xu, S., Li, L., . . . Ouyang, J. (2019). Ultrasensitive detection of prostate specific antigen using a personal glucose meter based on DNA-mediated immunoreaction. *Analyst*, *144*(20), 6019–6024. Available from https://doi.org/10.1039/c9an01558b.

Sun, X., Wang, H., Jian, Y., Lan, F., Zhang, L., Liu, H., . . . Yu, J. (2018). Ultrasensitive microfluidic paper-based electrochemical/visual biosensor based on spherical-like cerium dioxide catalyst for miR-21 detection. *Biosensors and Bioelectronics*, *105*, 218–225. Available from https://doi.org/10.1016/j.bios.2018.01.025.

Suresh, L., Brahman, P. K., Reddy, K. R., & Bondili, J. S. (2018). Development of an electrochemical immunosensor based on gold nanoparticles incorporated chitosan biopolymer nanocomposite film for the detection of prostate cancer using PSA as biomarker. *Enzyme and Microbial Technology*, *112*, 43–51. Available from https://doi.org/10.1016/j.enzmictec.2017.10.009.

Tan, F., Yang, Y., Xie, X., Wang, L., Deng, K., Xia, X., . . . Huang, H. (2018). Prompting peroxidase-like activity of gold nanorod composites by localized surface plasmon resonance for fast colorimetric detection of prostate specific antigen. *Analyst*, *143*(20), 5038–5045. Available from https://doi.org/10.1039/c8an00664d.

Tao, L., Zhang, K., Sun, Y., Jin, B., Zhang, Z., & Yang, K. (2012). Anti-epithelial cell adhesion molecule monoclonal antibody conjugated fluorescent nanoparticle biosensor for sensitive detection of colon cancer cells. *Biosensors and Bioelectronics*, *35*(1), 186–192. Available from https://doi.org/10.1016/j.bios.2012.02.044.

Uludag, Y., & Tothill, I. E. (2012). Cancer biomarker detection in serum samples using surface plasmon resonance and quartz crystal microbalance sensors with nanoparticle signal amplification. *Analytical Chemistry*, *84*(14), 5898–5904. Available from https://doi.org/10.1021/ac300278p.

Walsh, E. M., Smith, K. L., & Stearns, V. (2020). Management of hormone receptor-positive, HER2-negative early breast cancer. *Seminars in Oncology*, *47*(4), 187–200. Available from https://doi.org/10.1053/j.seminoncol.2020.05.010.

Wang, G., Lipert, R. J., Jain, M., Kaur, S., Chakraboty, S., Torres, M. P., . . . Porter, M. D. (2011). Detection of the potential pancreatic cancer marker MUC4 in serum using surface-enhanced Raman scattering. *Analytical Chemistry*, *83*(7), 2554–2561. Available from https://doi.org/10.1021/ac102829b.

Wang, L. J., Chang, Y. C., Sun, R., & Li, L. (2017). A multichannel smartphone optical biosensor for high-throughput point-of-care diagnostics. *Biosensors and Bioelectronics*, *87*, 686–692. Available from https://doi.org/10.1016/j.bios.2016.09.021.

Wang, R., Di, J., Ma, J., & Ma, Z. (2012). Highly sensitive detection of cancer cells by electrochemical impedance spectroscopy. *Electrochimica Acta*, *61*, 179–184. Available from https://doi.org/10.1016/j.electacta.2011.11.112.

Wang, K., Fan, D., Liu, Y., & Wang, E. (2015). Highly sensitive and specific colorimetric detection of cancer cells via dual-aptamer target binding strategy. *Biosensors and Bioelectronics*, *73*, 1–6. Available from https://doi.org/10.1016/j.bios.2015.05.044.

Wang, K., He, M. Q., Zhai, F. H., He, R. H., & Yu, Y. L. (2017). A novel electrochemical biosensor based on polyadenine modified aptamer for label-free and ultrasensitive detection of human breast cancer cells. *Talanta*, *166*, 87–92. Available from https://doi.org/10.1016/j.talanta.2017.01.052.

Wei, B., Mao, K., Liu, N., Zhang, M., & Yang, Z. (2018). Graphene nanocomposites modified electrochemical aptamer sensor for rapid and highly sensitive detection of prostate specific antigen. *Biosensors and Bioelectronics*, *121*, 41–46. Available from https://doi.org/10.1016/j.bios.2018.08.067.

Werner, S. L., Graf, R. P., Landers, M., Valenta, D. T., Schroeder, M., Greene, S. B., . . . Marrinucci, D. (2015). Analytical validation and capabilities of the Epic CTC platform: Enrichment-free circulating tumour cell detection and characterization. *Journal of Circulating Biomarkers*, *4*, 3–16. Available from https://doi.org/10.5772/60725.

Williams, R. M., Lee, C., & Heller, D. A. (2018). A fluorescent carbon nanotube sensor detects the metastatic prostate cancer biomarker uPA. *ACS Sensors*, *3*(9), 1838–1845. Available from https://doi.org/10.1021/acssensors.8b00063.

World Health Organization. (2020). Cancer. https://www.who.int/news-room/fact-sheets/detail/cancer.

Wu, X., Chai, Y., Yuan, R., Su, H., & Han, J. (2013). A novel label-free electrochemical microRNA biosensor using Pd nanoparticles as enhancer and linker. *Analyst*, *138*(4), 1060–1066. Available from https://doi.org/10.1039/c2an36506e.

Wu, Z., Fu, Q., Yu, S., Sheng, L., Xu, M., Yao, C., . . . Tang, Y. (2016). Pt@AuNPs integrated quantitative capillary-based biosensors for point-of-care testing application. *Biosensors and Bioelectronics*, *85*, 657–663. Available from https://doi.org/10.1016/j.bios.2016.05.074.

Xia, Y., Wang, L., Li, J., Chen, X., Lan, J., Yan, A., . . . Chen, J. (2018). A ratiometric fluorescent bioprobe based on carbon dots and acridone derivate for signal amplification detection exosomal microRNA. *Analytical Chemistry*, *90*(15), 8969–8976. Available from https://doi.org/10.1021/acs.analchem.8b01143.

Xu, S., Nie, Y., Jiang, L., Wang, J., Xu, G., Wang, W., & Luo, X. (2018). Polydopamine nanosphere/gold nanocluster (Au NC)-based nanoplatform for dual color simultaneous detection of multiple tumor-related microRNAs with DNase-I-assisted target recycling amplification. *Analytical Chemistry*, *90*(6), 4039–4045. Available from https://doi.org/10.1021/acs.analchem.7b05253.

Xue, T., Liang, W., Li, Y., Sun, Y., Xiang, Y., Zhang, Y., . . . Bao, Q. (2019). Ultrasensitive detection of miRNA with an antimonene-based surface plasmon resonance sensor. *Nature Communications*, *10*(28), 1–9. Available from https://doi.org/10.1038/s41467-018-07947-8.

Yang, Q., Gong, X., Song, T., Yang, J., Zhu, S., Li, Y., . . . Chang, J. (2011). Quantum dot-based immunochromatography test strip for rapid, quantitative and sensitive detection of alpha fetoprotein. *Biosensors and Bioelectronics*, *30*(1), 145–150. Available from https://doi.org/10.1016/j.bios.2011.09.002.

Yang, T., Hou, P., Zheng, L. L., Zhan, L., Gao, P. F., Li, Y. F., & Huang, C. Z. (2017). Surface-engineered quantum dots/electrospun nanofibers as a networked fluorescence aptasensing platform toward biomarkers. *Nanoscale*, *9*(43), 17020–17028. Available from https://doi.org/10.1039/c7nr04817c.

Yavas, O., Aćimović, S. S., Garcia-Guirado, J., Berthelot, J., Dobosz, P., Sanz, V., & Quidant, R. (2018). Self-calibrating on-chip localized surface plasmon resonance sensing for quantitative and multiplexed detection of cancer markers in human serum. *ACS Sensors*, *3*(7), 1376–1384. Available from https://doi.org/10.1021/acssensors.8b00305.

Yen, Y., Chao, C., & Yeh, Y. S. (2020). A Graphene-PEDOT: PSS modified paper-based aptasensor for electrochemical impedance spectroscopy detection of tumor marker. *Sensors (Basel)*, *20*(5), 1372–1383. Available from https://doi.org/10.3390/s20051372.

Ye, Y., Gao, J., Zhuang, H., Zheng, H., Sun, H., Ye, Y., . . . Cao, X. (2017). Electrochemical gene sensor based on a glassy carbon electrode modified with hemin-functionalized reduced graphene oxide and gold nanoparticle-immobilized probe DNA. *Microchimica Acta*, *184*(1), 245–252. Available from https://doi.org/10.1007/s00604-016-1999-9.

Yu, T., Dai, P. P., Xu, J. J., & Chen, H. Y. (2016). Highly sensitive colorimetric cancer cell detection based on dual signal amplification. *ACS Applied Materials and Interfaces*, *8*(7), 4434–4441. Available from https://doi.org/10.1021/acsami.5b12117.

Yu, X., Liang, J., Xu, J., Li, X., Xing, S., Li, H., . . . Du, H. (2018). Identification and validation of circulating microRNA signatures for breast cancer early detection based on large scale tissue-derived data. *Journal of Breast Cancer*, *21*(4), 363–370. Available from https://doi.org/10.4048/jbc.2018.21.e56.

Zdarta, J., Meyer, A., Jesionowski, T., & Pinelo, M. (2018). A general overview of support materials for enzyme immobilization: Characteristics, properties, practical utility. *Catalysts*, *8*(2), 1–27. Available from https://doi.org/10.3390/catal8020092.

Zhang, H., Liu, Y., Gao, J., & Zhen, J. (2015). A sensitive SERS detection of miRNA using a label-free multifunctional probe. *Chemical Communications*, *51*(94), 16836–16839. Available from https://doi.org/10.1039/c5cc06225j.

Zhang, H., Yi, Y., Zhou, C., Ying, G., Zhou, X., Fu, C., . . . Shen, Y. (2017). SERS detection of microRNA biomarkers for cancer diagnosis using gold-coated paramagnetic nanoparticles to capture SERS-active gold nanoparticles. *RSC Advances*, *7*(83), 52782–52793. Available from https://doi.org/10.1039/c7ra10918k.

Zhang, K., Zhao, L.-B., Guo, S.-S., Shi, B.-X., Lam, T.-L., Leung, Y.-C., . . . Wang, Y. (2010). A microfluidic system with surface modified piezoelectric sensor for trapping and detection of cancer cells. *Biosensors and Bioelectronics*, *26*(2), 935–939. Available from https://doi.org/10.1016/j.bios.2010.06.039.

Zhang, L.-N., Deng, H.-H., Lin, F.-L., Xu, X.-W., Weng, S.-H., Liu, A.-L., . . . Chen, W. (2014). In situ growth of porous platinum nanoparticles on graphene oxide for colorimetric detection of cancer cells. *Analytical Chemistry*, *86*(5), 2711–2718. Available from https://doi.org/10.1021/ac404104j.

Zhang, S., Li, Z., & Wei, Q. (2020). Smartphone-based cytometric biosensors for point-of-care cellular diagnostics. *Nami Jishu Yu Jingmi Gongcheng/Nanotechnology and Precision Engineering*, *3*(1), 32–42. Available from https://doi.org/10.1016/j.npe.2019.12.004.

Zhang, X., Xiao, K., Cheng, L., Chen, H., Liu, B., Zhang, S., & Kong, J. (2014). Visual and highly sensitive detection of cancer cells by a colorimetric aptasensor based on cell-triggered cyclic enzymatic signal amplification. *Analytical Chemistry*, *86*(11), 5567–5572. Available from https://doi.org/10.1021/ac501068k.

Zhao, H., Qu, Y., Yuan, F., & Quan, X. (2016). A visible and label-free colorimetric sensor for miRNA-21 detection based on peroxidase-like activity of graphene/gold-nanoparticle hybrids. *Analytical Methods*, *8*(9), 2005–2012. Available from https://doi.org/10.1039/c5ay03296b.

Zhou, T., Marx, K. A., Dewilde, A. H., McIntosh, D., & Braunhut, S. J. (2012). Dynamic cell adhesion and viscoelastic signatures distinguish normal from malignant human mammary cells using quartz crystal microbalance. *Analytical Biochemistry*, *421*(1), 164–171. Available from https://doi.org/10.1016/j.ab.2011.10.052.

Zhou, J., Zheng, Y., Liu, J., Bing, X., Hua, J., & Zhang, H. (2016). A paper-based detection method of cancer cells using the photo-thermal effect of nanocomposite. *Journal of Pharmaceutical and Biomedical Analysis*, *117*, 333–337. Available from https://doi.org/10.1016/j.jpba.2015.09.017.

Zhou, L., Mao, H., Wu, C., Tang, L., Wu, Z., Sun, H., ... Zhao, J. (2017). Label-free graphene biosensor targeting cancer molecules based on non-covalent modification. *Biosensors and Bioelectronics*, *87*, 701–707. Available from https://doi.org/10.1016/j.bios.2016.09.025.

Zhou, Y.-G., Mohamadi, R. M., Poudineh, M., Kermanshah, L., Ahmed, S., Safaei, T. S., ... Kelley, S. O. (2016). Interrogating circulating microsomes and exosomes using metal nanoparticles. *Small*, *12*(6), 727–732. Available from https://doi.org/10.1002/smll.201502365.

Zhou, J., Duan, L., Huang, J., Zuo, Z., Tang, T., Cao, D., ... Zhang, L. (2019). Portable detection of colorectal cancer SW620 cells by using a personal glucose meter. *Analytical Biochemistry*, *577*, 110–116. Available from https://doi.org/10.1016/j.ab.2019.04.018.

Zhu, Y., Chandra, P., & Shim, Y. B. (2013). Ultrasensitive and selective electrochemical diagnosis of breast cancer based on a hydrazine-Au nanoparticle-aptamer bioconjugate. *Analytical Chemistry*, *85*(2), 1058–1064. Available from https://doi.org/10.1021/ac302923k.

Zong, S., Wang, L., Chen, C., Lu, J., Zhu, D., Zhang, Y., ... Cui, Y. (2016). Facile detection of tumor-derived exosomes using magnetic nanobeads and SERS nanoprobes. *Analytical Methods*, *8*(25), 5001–5008. Available from https://doi.org/10.1039/c6ay00406g.

Chapter 5

# Early detection of lung cancer biomarkers through biosensor

Mehdi Dadmehr[1], Pouria Jafari[2] and Morteza Hosseini[2]
[1]*Department of Biology, Payame Noor University, Tehran, Iran*
[2]*Department of Life Science Engineering, Faculty of New Sciences & Technologies, University of Tehran, Tehran, Iran*

## 5.1 Introduction

### 5.1.1 Lung cancer

So far, more than 200 distinct types of cancer have been diagnosed in 60 human organs. Most of these cancers led to death if their tumors reach the metastasis stage. Lung cancer as one of the most threating cancers had a great impact on worldwide human health (Arya & Bhansali, 2011). It had been estimated that more than 1.76 million people worldwide are affected by lung cancer disease of which 25% of them resulted in death. Several abnormalities had been observed in lung cancer tumor cells which include genetic and epigenetic mutations. These mutations could alter the mechanism of regular cell division control which finally results in uncontrolled cell proliferation and forms the cell mass known as tumors.

Proliferated tumors are classified in two types: benign or malignant. While the malignant tumors show the tendency to invade and spread to other organs, the benign tumors will be treated by routine surgery without metastasis process. Tumor metastasis involve spreading of cancer cells to the other parts of body through bloodstream or lymphatic systems. Normally this step will occur in the late stages of cancers, but exceptionally, the metastasis in lung cancer tends to happen in early stages which make the treatment strategies difficult. Spreading cancer cells can move through the body and tend to spread to other organs, specifically adrenal glands, liver, brain, and bones. According to patient condition and preference, and also type and stage of the diagnosed lung cancer, the most appropriate treatment strategy will be performed. At the early stages of lung cancer, the effective recommendation will be surgery due to confinement of cancer in the lung tissue. Administration of chemotherapy or radiotherapy is recommended before and after surgery to obtain the maximum tumor shrinkage and inhibit the reoccurrence of tumors before and after surgery, respectively (Khanmohammadi et al., 2020).

In the histological view, the tumor cells are classified into two groups including small cell lung carcinomas (SCLC) and nonsmall cell lung carcinomas (NSCLC). About 15%–20% of lung cancers are small cell lung carcinomas (SCLC) which form malignant and aggressive subtype of lung tumors (Fig. 5.1).

The patients with SCLC diagnosis respond effectively to chemotherapy and radiotherapy, but early diagnosis of this type of cancers is very important due to its frequent reoccurrence in all treated patients. SCLCs are classified into three categories including small cell carcinoma, mixed small cell/large cell carcinoma, and combined small cell carcinoma. NSCLC type is the common form of pulmonary carcinoma involved in about 80% of cases of the total lung malignancies (Roointan et al., 2019). NSCLC types are classified according to their stages from I to IV. In stage I, only the lung tissue cells are cancerous while during the next two stages (II and III), the cancer progresses; the formed tumor is larger but noninvasive and limited to the chest. At stage IV cancer, the tumor spreads from the chest to the other parts of the body (Khanmohammadi et al., 2020).

### 5.1.2 Epidemiology of lung cancer

As it was discussed before, the lung cancer has the highest mortality rate of worldwide cancer deaths among other types of cancers in the past three decades. The GLOBOCAN 2018 has reported that the most diagnosed malignancy cases

Biosensor Based Advanced Cancer Diagnostics. DOI: https://doi.org/10.1016/B978-0-12-823424-2.00008-9

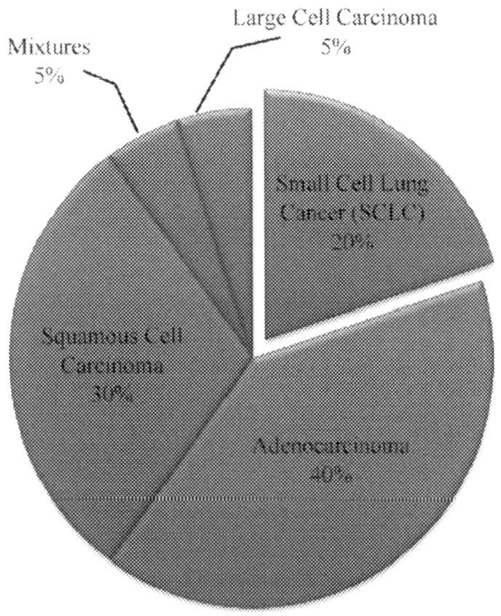


**FIGURE 5.1** The lung cancer tumor cell types percentage according to their histological view. *From Khanmohammadi, A., Aghaie, A., Vahedi, E., Qazvini, A., Ghanei, M., Afkhami, A., ... Bagheri, H. (2020). Electrochemical biosensors for the detection of lung cancer biomarkers: A review.* Talanta, 206. *https://doi.org/10.1016/j.talanta.2019.120251.*

were for lung cancer samples (Lin et al., 2019). But lung cancer still is the main cause of cancer death. There is a close relationship between the increased number of tobacco smokers and lung cancer, climbing since 2011 in developing countries. Recently, the GLOBOCAN reported that 11.6% of total diagnosed cancer cases include 2.09 million cases and 18.4% of total cancer deaths include 1.76 million deaths as belonging to lung cancer, which was higher than 2012 reported cases. These statistical data represent the frequent rate of lung cancer among other cancer cases.

### 5.1.3 Causes, genetic changes, and traditional screening of lung cancer

Among several factors that led to lung cancer, smoking (e.g., tobacco smoking) and also second hand smoke are the main factors. The environmental pollution and airborne chemical and industrial pollutants such as arsenic, asbestos, and radon are considered as other important risk factors. Also, it had been demonstrated that the genetic profile of a person has a significant role in lung cancer development (Khanmohammadi et al., 2020). There are no specific symptoms at early stages of cancer but later, the signs such as coughing, shortness of breath, and sudden weight loss may occur. In most cases, the symptoms will appear after tumor formation at late stages when treatment strategies are not so effective. Despite the many attempts made for early diagnosis of lung cancer in past decade, it is very limited (Arya & Bhansali, 2011). Early stages of lung cancer without specific symptoms is the main challenge. Sometimes using conventional techniques such as x-ray, optical coherence, spiral and virtual bronchoscopy led to misdiagnoses that reflected their dependency on morphological interpretation and illustrated the drawbacks of these methods. These wrong diagnosis may also lead to significant delay in the initiation of the right medical treatment and expose the patient to inappropriate and even harmful medication to treat lung cancer sickness. Due to the complicated nature and incurability of lung cancer, there is a real need for design and fabrication of biosensor devices with highly sensitive detection ranges for screening the patients which are at the premalignant and premetastatic stages of lung cancer.

Typically, a biosensor is portable device composed of two main parts including a biological recognition section and a signal transducer section for detection of target molecules through signal conversion into detectable and easily readable output signal. Most of developed biosensing platforms induce and generate measurable signals after subjecting of target biomolecules with the detector part of biosensor. Biological analyte of interest includes proteins, DNA, enzymes, antibodies, polysaccharides, some drugs, toxins, tissue, cell and cell receptors. Most of the applied biosensors are classified according to their signal transducer types including optical, thermal, piezoelectric, magnetic, and DNA-based. The lung cancer biosensors should be set in the manner to approach toward a highly sensitive, fast, and cost-effective way for efficient detection at the early stages (Roointan et al., 2019).

## 5.2 Lung cancer biomarkers

The biomarkers are the specific biomolecules which their presence at defined level maybe illustrate the cancer disease. Any abnormality in body function originated from cancer presence is associated with the secretion of these biomarkers.

Also, they also can use to determine the body response to employed specific treatment. Analysis of soluble biomarkers in body fluids such as serum, plasma, urine, saliva, sputum, and tears could provide a safe, inexpensive and noninvasive method towards efficient early cancer detection. These biomarkers include protein, nucleic acid, intermediate metabolites, and hormones which their level expression may be regarded as early cancer symptoms. Among them, the proteins and nucleic acids are more informative for cancer diagnosis and its course of reappearance. So, in this chapter, we will discuss these two remarkable and frequent nucleic acid-based biomarkers and protein-based biomarkers (Khanmohammadi et al., 2020).

### 5.2.1 Nucleic acid-based biomarkers

The nucleic acid-based biomarkers include genetic or epigenetic alteration which are the most specified character of cancer cells. The modification in genomic DNA includes insertion, deletions, amplifications, rearrangements or other epigenetic modification such as hypermethylation. These modification are recognized genetic variations in lung cancer which can be considered as a nucleic acid-based biomarker (Roointan et al., 2019). Cancer occurrence is a multistep process where some genes modify during tumor formation. Successive occurence of molecular, genetic, and epigenetic abnormalities results in cancer development. So genetic and epigenetic alteration, such as mutation or hypermethylation which are permanent and irreversible, results to gain or lose gene function (Arya & Bhansali, 2011). Detection of epigenetic modification such as methylation is an emerging field to investigate normal and pathological conditions. DNA methylation is the main epigenetic biomarker, which can inhibit the gene expression of some specific genes in cancer cells. Hypermethylation usually results in loss of function or gene silencing of specific CpG DNA sites known as CpG islands in promotion of tumor suppressor genes. So this process was highly investigated in early cancer diagnosis research and modified methylated genes studied as cancer markers. Micro RNAs, which are small noncoding RNA (miRNA), are another type of nucleic acid marker that interrupts proper function of gene expression (Roointan et al., 2019).

### 5.2.2 Protein-based biomarkers

The proteomics science provides important information about critical proteins involved in cancer development. Some of them are use as protein-based markers in early lung cancer diagnosis. The level of above mentioned expressed proteins in lung cancer is a determining factor for choosing them as the biomarker. Some well-known and common lung cancer protein biomarkers in serum are carcinoembryonic antigen (CEA), CYFRA 21−1 (cytokeratine 19 fragment), neuron-specific enolase (NSE), progastrin-releasing peptide (ProGRP), and squamous cell carcinoma antigen (SCCA). Among them, CEA and NSE are more important and used more frequently in conducted researches. Haptoglobin (Hp), annexin II, and enolase 1 (ENO1) are other highly expressing proteins known as biomarkers. Detection of biomarkers in blood samples could be a noninvasive and cost-effective tool to screen the people who they are at high risk of lung cancer and guide for subsequent therapy. Despite the enhanced level of these proteins in serum of lung cancer patients, they are at the minimum level that only biosensors with high detection limit could diagnose (Broodman, Lindemans, Van Sten, Bischoff, & Luider, 2017). Table 5.1 shows a list of nucleic acid and protein-based biomarkers that have been used in lung cancer detection.

**TABLE 5.1** Nucleic acid and protein-based biomarkers used in lung cancer detection.

| Category | Biomarker |
|---|---|
| Genetic biomarkers | COX2, RASSFIA, IL-8 mRNA, FHIT, K-ras mutant, p53 mutant, epidermal growth factor receptor (EGFR) such as c-ErbB-1 and c-ErbB-2 |
| Protein biomarkers | CEA, CYFRA 21−1, TPA, tumor M2-pyruvate kinase, Haptoglobin-R 2, APOA1, KLKB1, ProGRP, R-enolase (ENO1), R-1-acid glycoprotein, chromogranin A, bombesin-like gastrin-releasing peptide, cytokeratin-7, carbohydrate antigens 19−9, carbohydrate antigen 125 (CA 125), vascular endothelial growth factor (VEGF), nitrated ceruloplasmin, Annexin II, CD59 glycoproteins, transthyretin (TTR), GM2 activator protein (GM2AP) |

## 5.3 Biosensors for lung cancer biomarker detection

Due to trace levels of existing biomarkers in early stages of cancer, sensitivity and selectivity of applied methods for screening are very crucial. Several biosensors based on biochemical and immunological methodologies have been developed for early detection of cancer biomarkers. However, some of these conventional approaches lack sufficient sensitivity in detection of biomarkers. Using alternative methods, such as optical-based techniques, are dominant in biomarker detection and despite their time-consuming complexity and expensive properties they can provide a reasonable selectivity and sensitivity (Roointan et al., 2019). The unique properties of efficient biosensors are their rapidness, sensitivity, specificity, stability, and cost effectiveness while they are noninvasive detections for early lung cancer diagnosis. Biorecognition molecules in biosensor consist of antibodies, complementary nucleic acid probes, or another immobilized biomolecule on a transducer surface. The interaction of biorecognition molecules with the biomarkers (targets) generated biological responses that are convertible to measurable analytical signals by transducer (Fig. 5.2).

Regarding to different type of biological response, various transducers had been utilized in the construction of biosensors based on distinctive analytical approaches such as electrochemical, optical, and mass-based methods (Khanmohammadi et al., 2020).

### 5.3.1 Electrochemical-based approaches

Basically, in electrochemical-based assays an electrochemical signal (current, potential, conductance, impedance) is generated during bio-interaction of recognition molecules and biomarkers. There is a proportional relation between generated electrochemical signals with analyte concentration. The significant interaction should exist between biorecognition molecules and biomarker so that they are able to change the observable electrical properties of the biosensor surface. The electrical signals that their changes are detectable in electrochemical platforms are surface conductivity, electron transfer rate, and potential of redox reaction. After completion of redox reaction, the generated current will be affected by any changes in the electron transfer rate. Also, alterations in the surface impedance or redox potential are induced through changes in surface conductivity. The extent of these alterations is proportional to the presence of analyte concentration.

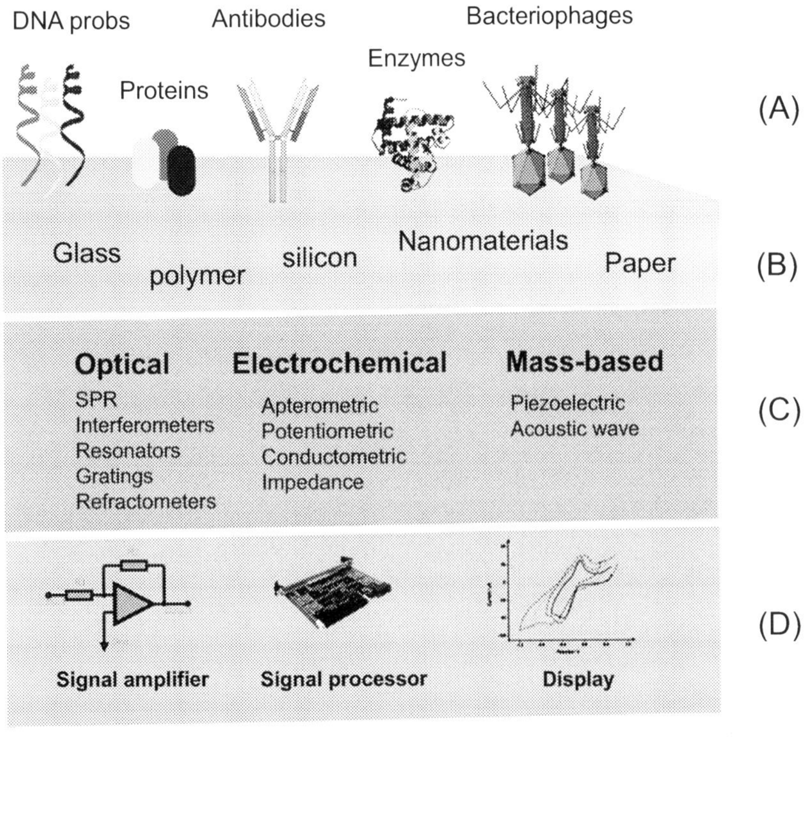


**FIGURE 5.2** Schematic structure of biosensor platform includes (A) biorecognition element, (B) matrixes, (C) transducer type, and (D) electronic part. *From Roointan, A., Ahmad Mir, T., Ibrahim Wani, S., Mati-ur-Rehman, Hussain, K. K., Ahmed, B., Abrahim, S., Savardashtaki, A., Gandomani, G., Gandomani, M., Chinnappan, R., & Akhtar, M. H. (2019). Early detection of lung cancer biomarkers through biosensor technology: A review.* Journal of Pharmaceutical and Biomedical Analysis, 164, *93–103. https://doi.org/10.1016/j.jpba.2018.10.017.*

The various advantages of electrochemical biosensors are their distinctive properties such as portability, operation simplicity, easy to use, cost effectiveness, and, in most cases, disposable character. The principle of electrochemical biosensors function is based on electrochemical processes which perform at the surface of an electrode and the changes in electrical signals on electrode surface are monitored. The above mentioned changes include the changes in applied potential, current, or frequency through an electrode. An electrochemical biosensor typically is composed of three sections including a recognition element for specific interaction with the biomarker, a signal transducer which generate a measurable signal from the analyte-recognition element interaction, and an electronic system for the data analysis and processing. The first part is the most important section that has a great impact on sensitivity and specificity of the biomolecules.

Among different recognition elements that have been used for detection of biomarkers, the antibodies, enzymes and synthetic elements such as nucleic acids, proteins, and aptamers are the most widely used recognition molecules. The transducers of electrochemical biosensors can be classified as amperometry, potentiometric, conductometric, and impedimetric. If an electrochemical biosensor was constructed in a small size as a portable device, it would be applicable for home use. The potentiometric and amperometric transducers are the common transducers, but in order to achieve the best sensitivity, voltammetric transducers are preferred (Khanmohammadi et al., 2020). In this chapter, we survey the electrochemical biosensors according to the type of their biorecognition elements toward biomarker detection.

#### 5.3.1.1 Protein analyte-based electrochemical approaches

Electrochemical immunosensors are the biosensors that are working based on electrochemical antibody−antigen affinity interaction utilizing the immunological molecule, such as an antibody, to perform its role as a recognition element. The higher sensitivity and specificity of antibody−antigen reaction is the main advantage that made this platform so attractive (Roointan et al., 2019). Li et al. reported the fabrication of the antibody-based biosensor using an electrode with AuNPs and a graphene functionalized surface (Li et al., 2017). The specific antibody for detection of CEA (a well-known biomarker in lung cancer) was used in the recognition section of the biosensor. The AuNPs have the unique property of electron transfer enhancement and could also increase the amounts of immobilized antibodies on the surface. In order to increase the AuNPs immobilization and also the induction of hydrophilicity of graphene, the polyethylene glycol (PEG) molecules are functionalized on the graphene surface. The superior electrical property was obtained by this modified GR sheet. The proposed biosensor showed a linear relationship for detection of a CEA biomarker in the range of 0.1−1000 ng $mL^{-1}$, while the detection limit was determined at 0.06 ng $mL^{-1}$. The results of the applied method were meaningful while the immunobiosensor was tested for clinical samples.

In another attempt, a labeled electrochemical-based biosensor was used for detection of CYFRA21−1. Zeng et al. (2018), fabricated a novel platform composed of 3D-GO, chitosan (CS), and glutaraldehyde (GA). Also, amino-functionalized carbon nanotubes were used for development of Ab-modified nanocomposites. The Au NPs, and horseradish peroxidase-labeled antibodies (HPR-Ab) were used as a label for detection of CYFRA21−1. The mentioned modification provide excellent conductivity and good biocompatibility for applied nanomaterials in biosensor which had a linear range of 0.1 − 150 ng $mL^{-1}$ and the LOD of 43 pg $mL^{-1}$ was estimated for detection of the CYFRA21−1 biomarker. The CYFRA21−1 biomarker was also used as a marker by Gao, Lu, Wang, and Li (2016) in label-free electrochemical biosensors. They reported an immunosensor which was designed based on a bottom-up method that was constructed according to a silicon nanowire tunneling field-effect transistor (SiNW-TFET). As shown in Fig. 5.3, the proposed biosensor could detect the negative-charged biomolecules through increasing the tunneling current of SiNW-TFET and this "bottom-up" approach was able to detect CYFRA21−1 concentrations as low as 0.65 fg $mL^{-1}$.

Various label-free electrochemical biosensors have been developed for detection of CEA biomarker detection. For instance, Su et al. (2016) introduced a label-free immunoassay based on Prussian blue nanocube-loaded molybdenum disulfide nanocomposites ($MoS_2$-PBNCs) (Fig. 5.4). The proposed biosensor presented superior electrocatalytic feature, which enhanced the sensitivity of label-free biosensor for detection of CEA in experimental and also in human serum sample with satisfied results.

So far, electrochemiluminescence (ECL) biosensors have shown high sensitivity and wide dynamic signal response range. Based on this strategy, Pang et al. (2015) constructed a label-free ECL immunosensor composed of nanocomposites with graphene oxide/carboxylated multiwall carbon nanotubes/gold/cerium oxide nanoparticles (GO/MWCNTs-COOH/Au@$CeO_2$) that introduced specific features including high electron conduction, convenient stability, and higher specific surface as a sensing matrix for CEA detection (Fig. 5.5).

In another alternative strategy a controlled release system-based labeled immunosensor was developed by Ma et al. (2016) to detect SCCA. They fabricated a sandwich-type biosensor using β-cyclodextrin functionalized gold decorated

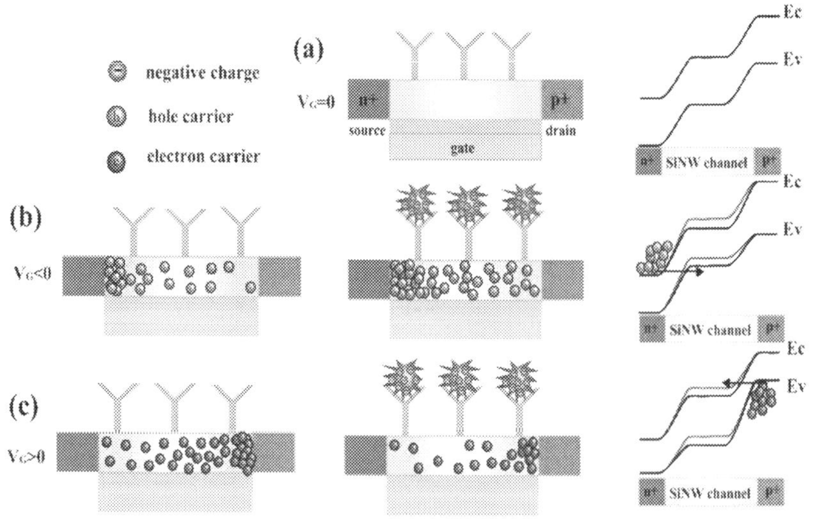


FIGURE 5.3 Schematic illustration of applied strategy in SiNW-TFET-based biosensor for detection of charged biomolecules. *From Gao, A., Lu, N., Wang, Y., & Li, T. (2016). Robust ultra-sensitive tunneling-FET biosensor for point-of-care diagnostics.* Scientific Reports, *6. https://doi.org/10.1038/srep22554.*

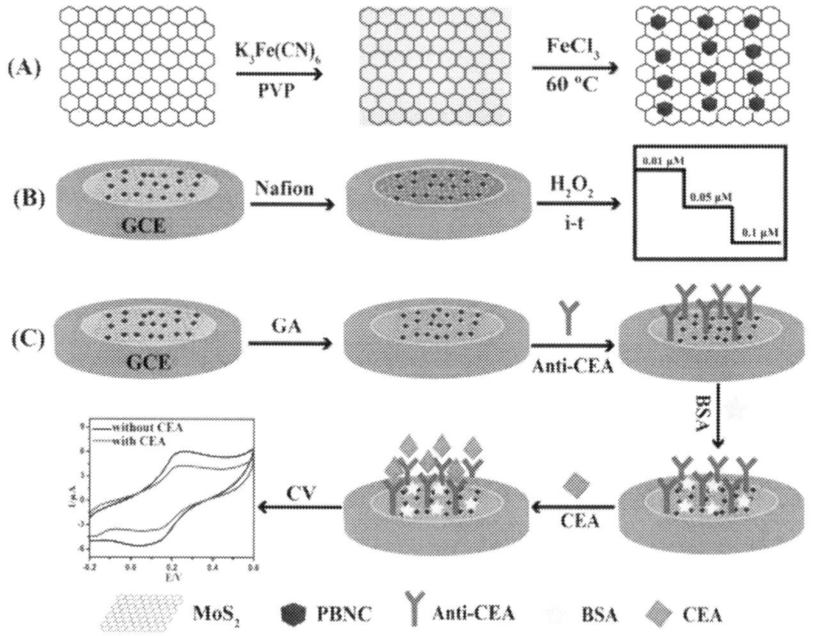


FIGURE 5.4 Synthesis of $MoS_2$-PBNCs nanocomposite and its detection strategy toward H2O2 and CEA detection. *From Su, S., Han, X., Lu, Z., Liu, W., Zhu, D., Chao, J., Fan, C., Wang, L., Song, S., Weng, L., & Wang, L. (2017). Facile synthesis of a $MoS_2$-Prussian Blue nanocube nanohybrid-based electrochemical sensing platform for hydrogen peroxide and carcinoembryonic antigen detection.* ACS Applied Materials and Interfaces, *9(14), 12773–12781. https://doi.org/10.1021/acsami.7b01141.*

SiO2 (CD-Au@$SiO_2$) to perform as a tag and Ab1 immobilized gold electrode (AuE) as a sensing platform for ultrasensitive detection of SCCA (Fig. 5.6). 1-methyl-1H-benzimidazole functionalized mesoporous $SiO_2$ (MBI-Ms) was also used for encapsulation of methylene blue (MB) with CD-Au@$SiO_2$ which entrapped the adamantly modified Ab2 (ADA-Ab2). The acidic environment resulted to release of MB from MBI-Ms while the SCCA and functional Ab2 subjected to an immune reaction. This novel biosensor displayed good linear range from 0.001 to 20 ng · $mL^{-1}$ with a low LOD of 0.25 pg · $mL^{-1}$.

### *5.3.1.2 Nucleic acid analyte-based electrochemical approaches*

Although antibodies had been recognized as the determining agent in the bioreceptor section of biosensor, it also shows the drawbacks in some aspects and it is suggested to find alternative methods to be fast, more robust, highly sensitive and low cost. One of the alternative ways is the application of nucleic acid-based structures as recognition elements in

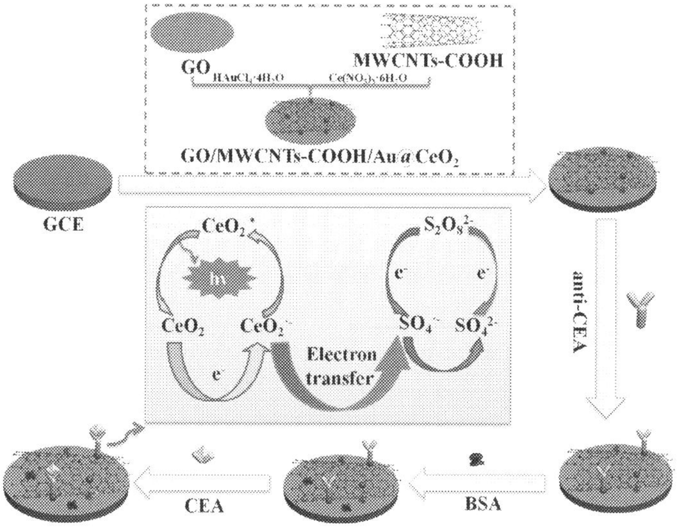


FIGURE 5.5 Schematic representation of graphene based label-free ECL based biosensor. *From Pang, X., Li, J., Zhao, Y., Wu, D., Zhang, Y., Du, B., Ma, H., & Wei, Q. (2015). Label-free electrochemiluminescent immunosensor for detection of carcinoembryonic antigen based on nanocomposites of GO/MWCNTs-COOH/Au@CeO₂.* ACS Applied Materials and Interfaces, 7*(34), 19260–19267. https://doi.org/10.1021/acsami.5b05185.*

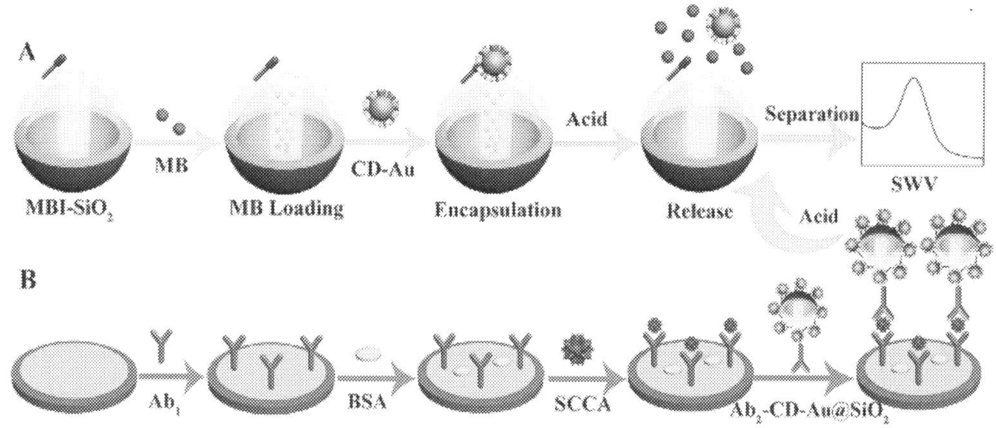


FIGURE 5.6 Graphical demonstration of Au@SiO2 based on pH responsive biosensor. *From Ma, H., Wang, Y., Wu, D., Zhang, Y., Gao, J., Ren, X., Du, B., & Wei, Q. (2016). A novel controlled release immunosensor based on benzimidazole functionalized $SiO_2$ and cyclodextrin functionalized gold.* Scientific Reports, 6*(1), 1–8. https://doi.org/10.1038/srep19797.*

biosensor platforms that predict to overcome the immunoassay-based failures. The nucleic acids can perform their recognition role as a substitute for antibodies and change the design of biosensing strategies.

Among different kinds of applied nucleic acids, the specific sequence of single-stranded DNA-named aptamers provide the binding sites with a specific target. The main advantage of aptamers compared to antibodies are their stability and also reproducible usage by recovering their active sites for several usages (Khanmohammadi et al., 2020). The biomolecular target of a functionalized aptamer is the determining factor for applied strategy for the construction of the electrochemical aptasensor. Some aptamers are designed so that switch their structure in specific form after interaction with biomolecule targets and subsequently result to detectable signals. Using a AuNPs−$MoS_2$ nanocomposites, Su et al. (2016), developed the electrochemical aptasensor for detection of thrombin and adenosine triphosphate (ATP). In this research, they designed two different labeled aptamers and subsequently immobilized them on an AuNPs−$MoS_2$ film-modified electrode (Fig. 5.7). After applying of target molecules includes thrombin and ATP the distance between methylene blue and ferrocence redox tag was occurred and attached to the aptamers due to the structural switches in the aptamers. So, a dual signal detection strategy was developed in the detection system to simultaneously detection of thrombin and ATP.

Sandwich or sandwich-like mode are other types of strategies used in aptasensors. Almost a protein target such as thrombin with dual binding sites will use in these platforms. The above mentioned proteins are able to interact with the recognition molecules such as aptamers and antibodies. The sandwich-type aptasensors can form three structures including aptamer–protein–antibody, aptamer–protein–aptamer, and antibody–protein–aptamer. The aptamer-based biosensors are constructed based on two types of structures.

In a study conducted by Chen et al. (2015), a sandwich-type electrochemical aptasensor were used with dual signal amplification strategy for the determination of MUC 1 in serum. In this strategy a carrier layer composed of hybrid film of poly(o-phenylenediamine)–Au nanoparticles (PoPD–AuNPs) was integrated with a tag of AuNPs functionalized silica/multiwalled carbon nanotubes core–shell nanocomposites (AuNPs/$SiO_2$@MWCNTs). First aptamer was immobilized on the surface of PoPD–AuNPs hybrid film. Then, the second aptamer and thionin as electrochemical probes were immobilized on AuNPs/$SiO_2$@MWCNTs film which has provided an enhanced surface area for immobilization. After the addition of MUC 1 target, Thi-AuNPs/$SiO_2$@MWCNTs nanopores and surface of PoPD–AuNPs modified electrode formed sandwich-type reaction on aptasensor surface to form a biocomplex. Immobilized AuNPs and MWCNTs facilitated the electron transfer from Thi to the electrode which resulted in amplified detection response (Fig. 5.8).

### 5.3.2 Optical-based approaches

The optical-based lung biosensors are fabricated by using of biorecognition element and an optical transducer. Most of strategies which use in optical biosensor are based on surface plasmon resonance (SPR) phenomenon which interaction

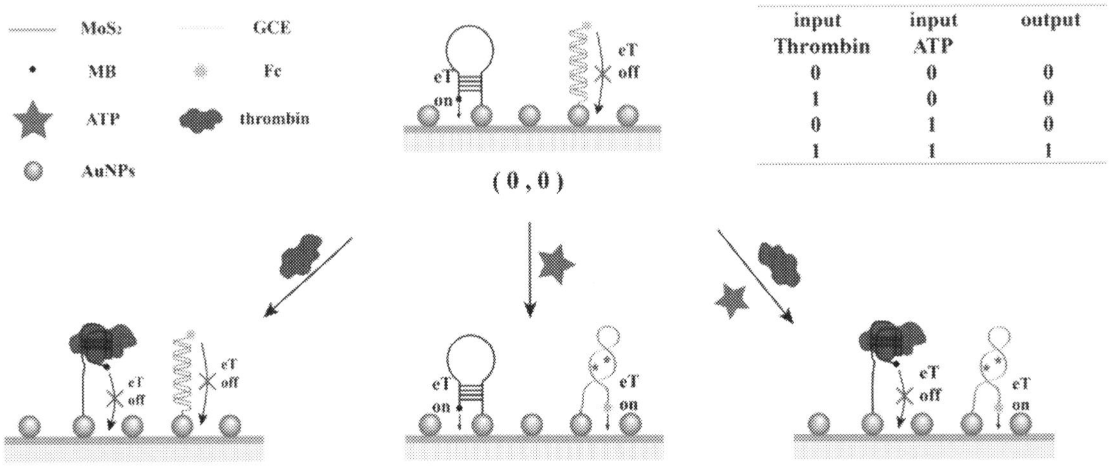


**FIGURE 5.7** Schematic representation of target-induced structure switching mode electrochemical aptasensor based on gold nanoparticles-decorated $MoS_2$ nanosheets. *From Su, S., Sun, H., Cao, W., Chao, J., Peng, H., Zuo, X., Yuwen, L., Fan, C., & Wang, L. (2016). Dual-target electrochemical biosensing based on DNA structural switching on gold nanoparticle-decorated $MoS_2$ nanosheets.* ACS Applied Materials and Interfaces, *8(11), 6826–6833. https://doi.org/10.1021/acsami.5b12833.*

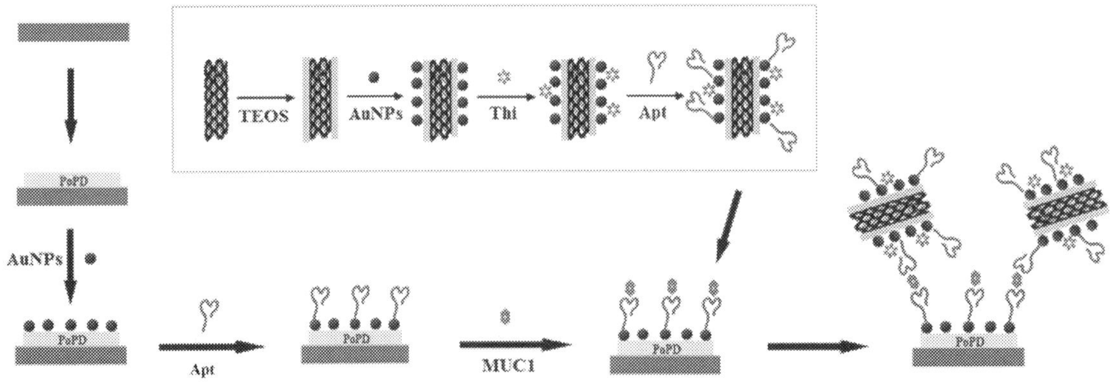


**FIGURE 5.8** Schematic representation of the sandwich-based electrochemical aptasensor strategy. *From Chen, X., Zhang, Q., Qian, C., Hao, N., Xu, L., & Yao, C. (2015). Electrochemical aptasensor for mucin 1 based on dual signal amplification of poly(o-phenylenediamine) carrier and functionalized carbon nanotubes tracing tag.* Biosensors and Bioelectronics, 64, *485–492. https://doi.org/10.1016/j.bios.2014.09.052.*

of biorecognition element and target element cause some changes in refractive index at the surface of sensor. Other strategies such as fluorescence resonance energy transfer (FRET), photoelectrochemical (PEC), chemiluminescence (CL) and surface-enhanced Raman scattering (SERS) are also alternative methods which had been used in construction of optical biosensors. Several studies implied fabrication of optical biosensor for detection of lung cancer biomarkers. Optical biosensors despite electrochemical biosensors have the minimum interference with detection system and then it suggest for detection of biological target elements (Yang et al., 2019).

### 5.3.2.1 *Protein analyte-based optical approaches*

The advantages of optical-based immunosensors such as low interference, low background nature, and high sensitivity can considerably enhance the sensitivity of detection system. Lin et al. (2019), used same strategy to construct an enzyme-free multicolor immunosensor for sensitive detection of NSE. In this research $Cu2^{+}$-modified carbon nitride nanosheet ($Cu2^{+}-C3N4$) was used to perform its catalytic function and gold nanobipyramid (Au NBP) applied as the chromogenic substrate to display multicolor properties (Fig. 5.9). The multicolor variation results when $TMB2^{+}$ etches Au NPB.

The optical immunoassay also had been used for detection of CEA due to their ability in enhancing their stability, sensitivity, and antifouling. Danesh et al. (2018) fabricated an optical aptasensor based on fluorescence strategy which 5,6,7-trimethyl-1,8-naphthyridin-2-amine (ATMND) compound exploited as a fluorescent dye and three-way junction pocket as fluorescence quenching probe. They pointed out that CEA aptamer of three-way junction pocket was stripped in the presence of CEA. The obtained results showed that aptasensor could detect CEA concentration effectively and efficient recovery was obtained from clinical application in human serum. Upconversion nanoparticles (UCNPs) presenting advantages such as high sensitivity and simple operation. The surface-enhanced Raman scattering (SERS) is another applicable approach toward immunosensing. In this strategy, the surface modified for passivation or the conjugation of biomolecules. Based on this strategy Gao et al. (2016) fabricated SERS based biosensor composed of Au nanostar@malachite green isothiocyanate@silica nanoparticles (Au@MGITC@Si) as SERS probes. The biosensor platform prepared by applying disposable paper-based lateral flow strip (PLFS) and subsequently Au@MGITC@Si used as probes for the detection of NSE marker.

In another SPR-based strategy, a novel PEC biosensor fabricated which showed low interference and high sensitivity advantages. In this biosensor (Yu et al., 2019), biofunctional polydopamine/tungsten oxide nanocomposites (pDA/WO3 NCs) synthesized to perform as a sensing platform for detection of CYFRA21−1. The light-stimulated oscillation of electrons at the modifiable metal films, can be directly used to detect the CYFRA21−1. Applicability of proposed SPR-based biosensors was confirmed in plasma samples and exhibit excellent stability and high sensitivity for CYFRA21−1. In same way, Chiu, Lin, and Kuo (2018), constructed another SPR based immunosensor for detection of CYFRA21−1 biomarker. Cystamine (Cys) was immobilized on chip surface as a linker on the GO sheet. They also proved that biosensor have convenient stability and sensitivity in human plasma. Photoelectrochemical (PEC) based methods are also used to detect lung cancer cells and related biomarkers. A renewable CuFeSe2/Au heterostructured and porous nanospheres was synthesized by Wen et al. (2019) for specific and sensitive detection of lung cancer

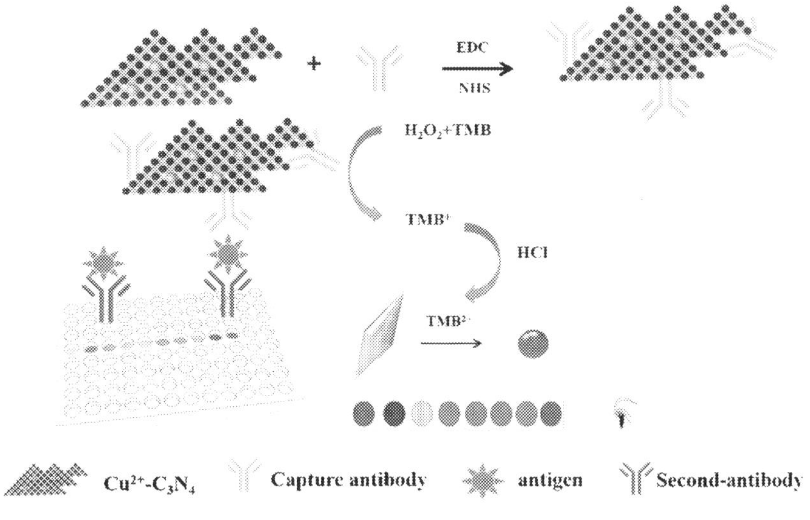


**FIGURE 5.9** Schematic illustration of the principle of the multicolor immunosensor. *From Lin, Y., Kannan, P., Zeng, Y., Qiu, B., Guo, L., & Lin, Z. (2019). Enzyme-free multicolor biosensor based on Cu2 + -modified carbon nitride nanosheets and gold nanobipyramids for sensitive detection of neuron specific enolase.* Sensors and Actuators, B: Chemical, 283, *138–145. https://doi.org/10.1016/j.snb.2018.12.007.*

biomarkers of aldehydes and lung cancer cells. In this process, Au shell deposited on heterostructured nanospheres by photoreduction on the CuFeSe2 framework. Then CuFeSe2/Au nanospheres were functionalized with P-amino thiophenol (4-ATP) as a Raman-active probe molecule, and finally aldehyde gaseous target captured on the nanosphere surface through C = N bond with a detection limit of 1.0 ppb (Fig. 5.10). Also, resulting folic acid (FA)-bonded nanospheres showed a high SERS activity toward rhodamine B isothiocyanate (RBITC), and this characteristic enable it to detect the A549 cells with high sensitivity.

### 5.3.3 DNA analyte based optical approaches

One of the most promising noninvasive biomarkers for early cancer diagnosis are circulating tumor DNA (ctDNA). Detection of ctDNA is challenging issue, because the applied biosensor should introduce an accurate and sensitive method with pico-to-femtomolar detection limit for serum ctDNA, even in the presence of its related analogs that produce strong background noise. So far, many attempts had been made to presents the sensitive detection of cancer nucleic acid biomarkers using colorimetric based approaches (Dadmehr, Hosseini, Hosseinkhani, Reza Ganjali, & Sheikhnejad, 2015; Fakhri et al., 2020; Karimi, Dadmehr, Hosseini, Korouzhdehi, & Oroojalian, 2019; Rafiei, 2019). Detection of cancer nucleic acid biomarkers gain increasing attention due to its strong potential in lung cancer early diagnosis. For example, Zhang et al. (2019) introduced a DNA-rN1-DNA-mediated surface-enhanced Raman scattering frequency shift assay which was able to discriminate mutated KRAS G12D with one single base pair mutation from the normal KARS G12D in lung cancer detection. In this strategy a hairpin DNA-rN1-DNA probe was designed for detection of specific ctDNA. They employed specific RNase HII enzyme that hydrolyzes the DNA-rN1-DNA/ctDNA hybrid and led to ctDNA recycling and subsequent signal amplification. The applied strategy showed sensitivity at subfemtomolar concentration in the phosphate-buffered saline (PBS) solution and is efficiency and applicability were well demonstrated in real sample test in both fetal bovine serum and human physiological media. In another approach, Dadmehr et al. (2014) developed a very sensitive fluorescence-based nanobiosensor to detect the abnormal DNA methylation pattern in specific sites of adenomatous polyposis coli(APC), a well-studied tumor suppressor gene. They used $Fe_3O_4$ core/shell nanoparticles functionalized by single Stranded DNA (ssDNA) probe (Fig. 5.11). So, designed unmethylated and methylated DNA counterparts which used as a target were hybridized with the functionalized ssDNA surface probe. The interaction of the applied fluorescent marker (dipyridamole) with methylated and unmethylated DNA targets were

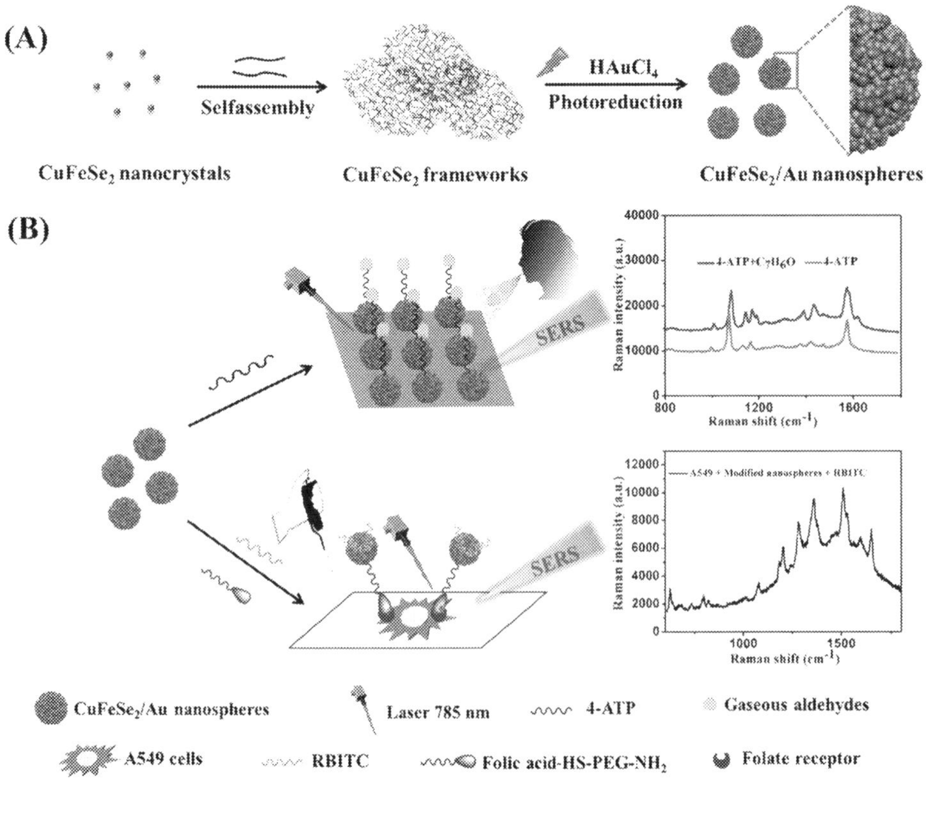


**FIGURE 5.10** Schematic illustration of synthesis of CuFeSe2/Au Nanospheres (A) and the used strategy to SERS based biosensor. *From Wen, H., Wang, H., Hai, J., He, S., Chen, F., & Wang, B. (2019). Photochemical synthesis of porous CuFeSe₂/Au heterostructured nanospheres as SERS sensor for ultrasensitive detection of lung cancer cells and their biomarkers.* ACS Sustainable Chemistry and Engineering, *7(5), 5200–5208. https://doi.org/10.1021/acssuschemeng.8b06116.*

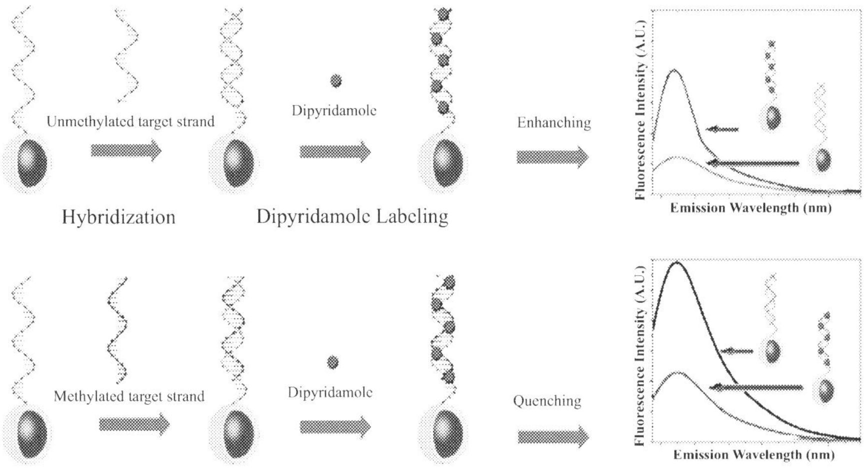


**FIGURE 5.11** Schematic illustration of the fluorescence-based biosensor for DNA methylation detection. *From Dadmehr, M., Hosseini, M., Hosseinkhani, S., Ganjali, M. R., Khoobi, M., Behzadi, H., Hamedani, M., & Sheikhnejad, R. (2014). DNA methylation detection by a novel fluorimetric nanobiosensor for early cancer diagnosis.* Biosensors and Bioelectronics, 60, *35–44. https://doi.org/10.1016/j.bios.2014.03.033.*

surveyed by fluorescence spectra emission and it was demonstrated that hybridized dsDNA results to detectable signal changes in the fluorescence intensity and provide a platform to distinguish methylated from unmethylated DNA. Reported detection limit in this approach was $3.1 \times 10^{-16}$ M and also real sample assay confirmed that the nanobiosensor have practical application for methylation detection in human plasma sample.

## 5.4 Conclusion and future perspectives

Detection of lung cancer at early stages via biomarkers provide the opportunity to perform more efficient therapeutic strategies in order to prevent the lung cancer spreading. These approaches also help us to have better understanding of pulmonary malignancies and to introduce efficient medical care for recuperating and recovery of patients. So, various researches conducted which was based on several analysis methods for constructing of biosensors for the early diagnosis of lung cancer. These attempts resulted in great achievements that presented applicable and promising results for efficient and early detection of lung cancer biomarkers. In this chapter, the important and distinguished lung cancer biomarkers were introduced and described briefly. According to different type of applied analysis methods and recognition strategies, many biosensors have been developed. Most of surveyed biosensors showed a wide detection range with the acceptable detection limits toward different types of lung cancer biomarkers. It was clearly demonstrated that applying nanomaterials in designing strategies of biosensors provided signal amplification with several orders of magnitudes. The obtained results showed that fabricated biosensors could have great effects in clinical application, which may have great effects in the near future in this field. Novel molecular markers including blood circulating cell surface proteins and nucleic based markers such as microRNAs would be regarded as novel targets. Novel methods should be fabricated as the future perspective toward early lung cancer diagnosis. Further achievements in nano-based biosensors leading to miniaturization approaches will significantly enhance diagnostic efficiency for cancer biomarkers detection with high sensitivity, selectivity, robustness, and cost effectiveness.

## References

Arya, S. K., & Bhansali, S. (2011). Lung cancer and its early detection using biomarker-based biosensors. *Chemical Reviews*, *111*(11), 6783–6809. Available from https://doi.org/10.1021/cr100420s.

Broodman, I., Lindemans, J., Van Sten, J., Bischoff, R., & Luider, T. (2017). Serum protein markers for the early detection of lung cancer: A focus on autoantibodies. *Journal of Proteome Research*, *16*(1), 3–13. Available from https://doi.org/10.1021/acs.jproteome.6b00559.

Chen, X., Zhang, Q., Qian, C., Hao, N., Xu, L., & Yao, C. (2015). Electrochemical aptasensor for mucin 1 based on dual signal amplification of poly (o-phenylenediamine) carrier and functionalized carbon nanotubes tracing tag. *Biosensors and Bioelectronics*, *64*, 485–492. Available from https://doi.org/10.1016/j.bios.2014.09.052.

Chiu, N. F., Lin, T. L., & Kuo, C. T. (2018). Highly sensitive carboxyl-graphene oxide-based surface plasmon resonance immunosensor for the detection of lung cancer for cytokeratin 19 biomarker in human plasma. *Sensors and Actuators, B: Chemical*, *265*, 264–272. Available from https://doi.org/10.1016/j.snb.2018.03.070.

Dadmehr, M., Hosseini, M., Hosseinkhani, S., Ganjali, M. R., Khoobi, M., Behzadi, H., ... Sheikhnejad, R. (2014). DNA methylation detection by a novel fluorimetric nanobiosensor for early cancer diagnosis. *Biosensors and Bioelectronics*, *60*, 35–44. Available from https://doi.org/10.1016/j.bios.2014.03.033.

Dadmehr, M., Hosseini, M., Hosseinkhani, S., Reza Ganjali, M., & Sheikhnejad, R. (2015). Label free colorimetric and fluorimetric direct detection of methylated DNA based on silver nanoclusters for cancer early diagnosis. *Biosensors and Bioelectronics*, *73*, 108–113. Available from https://doi.org/10.1016/j.bios.2015.05.062.

Danesh, N. M., Yazdian-Robati, R., Ramezani, M., Alibolandi, M., Abnous, K., & Taghdisi, S. M. (2018). A label-free aptasensor for carcinoembryonic antigen detection using three-way junction structure and ATMND as a fluorescent probe. *Sensors and Actuators, B: Chemical*, *256*, 408–412. Available from https://doi.org/10.1016/j.snb.2017.10.126.

Fakhri, N., Abarghoei, S., Dadmehr, M., Hosseini, M., Sabahi, H., & Ganjali, M. R. (2020). Paper based colorimetric detection of miRNA-21 using Ag/Pt nanoclusters. *Spectrochimica Acta - Part A: Molecular and Biomolecular Spectroscopy*, *227*. Available from https://doi.org/10.1016/j.saa.2019.117529.

Gao, A., Lu, N., Wang, Y., & Li, T. (2016). Robust ultrasensitive tunneling-FET biosensor for point-of-care diagnostics. *Scientific Reports*, *6*. Available from https://doi.org/10.1038/srep22554.

Karimi, M. A., Dadmehr, M., Hosseini, M., Korouzhdehi, B., & Oroojalian, F. (2019). Sensitive detection of methylated DNA and methyltransferase activity based on the lighting up of FAM-labeled DNA quenched fluorescence by gold nanoparticles. *RSC Advances*, *9*(21), 12063–12069. Available from https://doi.org/10.1039/c9ra01564g.

Khanmohammadi, A., Aghaie, A., Vahedi, E., Qazvini, A., Ghanei, M., Afkhami, A., ... Bagheri, H. (2020). Electrochemical biosensors for the detection of lung cancer biomarkers: A review. *Talanta*, *206*. Available from https://doi.org/10.1016/j.talanta.2019.120251.

Li, Y., Chen, Y., Deng, D., Luo, L., He, H., & Wang, Z. (2017). Water-dispersible graphene/amphiphilic pyrene derivative nanocomposite: High AuNPs loading capacity for CEA electrochemical immunosensing. *Sensors and Actuators, B: Chemical*, *248*, 966–972. Available from https://doi.org/10.1016/j.snb.2017.02.138.

Lin, Y., Kannan, P., Zeng, Y., Qiu, B., Guo, L., & Lin, Z. (2019). Enzyme-free multicolor biosensor based on Cu2 + -modified carbon nitride nanosheets and gold nanobipyramids for sensitive detection of neuron specific enolase. *Sensors and Actuators, B: Chemical*, *283*, 138–145. Available from https://doi.org/10.1016/j.snb.2018.12.007.

Ma, H., Wang, Y., Wu, D., Zhang, Y., Gao, J., Ren, X., ... Wei, Q. (2016). A novel controlled release immunosensor based on benzimidazole functionalized SiO 2 and cyclodextrin functionalized gold. *Scientific Reports*, *6*(1), 1–8. Available from https://doi.org/10.1038/srep19797.

Pang, X., Li, J., Zhao, Y., Wu, D., Zhang, Y., Du, B., ... Wei, Q. (2015). Label-free electrochemiluminescent immunosensor for detection of carcinoembryonic antigen based on nanocomposites of GO/MWCNTs-COOH/Au@$CeO_2$. *ACS Applied Materials and Interfaces*, *7*(34), 19260–19267. Available from https://doi.org/10.1021/acsami.5b05185.

Rafiei, S. (2019). A fluorometric study on the effect of DNA methylation on DNA interaction with graphene quantum dots. *Methods and Applications in Fluorescence*, *7*, Vol.

Roointan, A., Ahmad Mir, T., Ibrahim Wani, S., Mati-ur-Rehman., Hussain, K. K., Ahmed, B., ... Akhtar, M. H. (2019). Early detection of lung cancer biomarkers through biosensor technology: A review. *Journal of Pharmaceutical and Biomedical Analysis*, *164*, 93–103. Available from https://doi.org/10.1016/j.jpba.2018.10.017.

Su, S., Sun, H., Cao, W., Chao, J., Peng, H., Zuo, X., ... Wang, L. (2016). Dual-target electrochemical biosensing based on DNA structural switching on gold nanoparticle-decorated $MoS_2$ nanosheets. *ACS Applied Materials and Interfaces*, *8*(11), 6826–6833. Available from https://doi.org/10.1021/acsami.5b12833.

Wen, H., Wang, H., Hai, J., He, S., Chen, F., & Wang, B. (2019). Photochemical synthesis of porous CuFeSe 2/Au heterostructured nanospheres as SERS sensor for ultrasensitive detection of lung cancer cells and their biomarkers. *ACS Sustainable Chemistry and Engineering*, *7*(5), 5200–5208. Available from https://doi.org/10.1021/acssuschemeng.8b06116.

Yang, G., Xiao, Z., Tang, C., Deng, Y., Huang, H., & He, Z. (2019). Recent advances in biosensor for detection of lung cancer biomarkers. *Biosensors and Bioelectronics*, *141*. Available from https://doi.org/10.1016/j.bios.2019.111416.

Yu, Y., Huang, Z., Zhou, Y., Zhang, L., Liu, A., Chen, W., ... Peng, H. (2019). Facile and highly sensitive photoelectrochemical biosensing platform based on hierarchical architectured polydopamine/tungsten oxide nanocomposite film. *Biosensors and Bioelectronics*, *126*, 1–6. Available from https://doi.org/10.1016/j.bios.2018.10.026.

Zeng, Y., Bao, J., Zhao, Y., Huo, D., Chen, M., Qi, Y., ... Hou, C. (2018). A sandwich-type electrochemical immunoassay for ultrasensitive detection of non-small cell lung cancer biomarker CYFRA21–1. *Bioelectrochemistry (Amsterdam, Netherlands)*, *120*, 183–189. Available from https://doi.org/10.1016/j.bioelechem.2017.11.003.

Zhang, J., Dong, Y., Zhu, W., Xie, D., Zhao, Y., Yang, D., & Li, M. (2019). Ultrasensitive detection of circulating tumor DNA of lung cancer via an enzymatically amplified SERS-based frequency shift assay. *ACS Applied Materials and Interfaces*, *11*(20), 18145–18152. Available from https://doi.org/10.1021/acsami.9b02953.

Chapter 6

# Biosensor-based early diagnosis of hepatic cancer

**Nikita Sehgal, Ruchi Jakhmola Mani, Nitu Dogra and Deepshikha Pande Katare**
*Proteomic & Translational Research Lab, Centre for Medical Biotechnology, Amity Institute of Biotechnology, Amity University Noida, Noida, India*

## 6.1 Introduction

Liver cancer accounts for the second leading cause of cancer-related deaths in men, and the sixth leading cause in women, globally, with a significantly lower survival rate, i.e., a 5-year survival rate $< 20\%$ (Allemani et al., 2018). 80% of the primary liver cancers are hepatocellular carcinoma (HCC) (McGlynn et al., 2015). The symptoms of HCC are noticeable mainly when the advanced stages are reached; by that time, the only possible solution is tumor resection. However, most patients do not match the criteria for safe resection. Therefore, it necessary to screen or diagnose this condition at the earliest to reduce the chances of mortalities. Current screening techniques include radiological tests, i.e., ultrasonography (US), multiphase computerized tomography (CT), and magnetic resonance imaging (MRI) with contrast; the use of serological markers and liver biopsy (Attwa & El-Etreby, 2015). A lot of research in the HCC domain is currently revolving around the development of new diagnostic strategies for early diagnosis and laying down specific criteria for accurate identification of the size and number of lesions. Biosensors are a new wave in the diagnostic era. The versatility, ultra-sensitivity, high specificity, quick readout, reproducibility, accuracy, wide linear range, portability, etc., are some of the features which make biosensors a reliable diagnostic tool for varying underlying medical conditions (Patel et al., 2016). They can convert a biological signal into a quantifiable signal with the help of its bioelement and transducer, which aid in various clinical usages. The feasibility of biosensors in cancer detection is not new. A wide range of tumor-specific biomarkers in small concentrations has been successfully detected and estimated using biosensing devices that have helped in monitoring the condition, further providing cues on the particular treatment approaches to be employed for the improvement of the patient's conditions. Biosensors for HCC-specific serum biomarkers like alpha-fetoprotein (AFP), circulating tumor cells (CTCs), miRNAs, etc., have been fabricated and have emerged successful. Based on the transducing element of the sensor, HCC biomarkers can be identified and measured in various ways, including electrochemically, optically, piezo-electrically, etc. Since HCC occurs at the nanoscale, nanotechnology has a high potential in accurately and timely detection of HCC so that measures to prevent its progression into a deadly condition can be taken at the earliest (Mohammadinejad et al., 2020). This chapter focuses on the currently employed medical technologies used for HCC detection, various serum-based biomarkers specific to HCC, and the preexisting applications of biosensors in HCC.

## 6.2 Hepatocellular carcinoma

HCC is a type of liver malignancy and one of the leading causes of cancer-related mortalities. Notable progress has been made to throw light upon molecular mechanisms and the pathogenesis of HCC, which aids the clinicians develop cost-effective strategies for timely surveillance of HCC. Males are more likely to be affected than females, predominantly, because of their higher body mass index (BMI), higher alcohol consumption, higher testosterone levels, etc. HCC is mainly asymptomatic, with symptoms visible only in the chronic stages. The symptoms vary depending upon the size of the tumor, vascular invasion, presence of cirrhosis, and metastases. Commonly observed symptoms include Fever, weakness, pale urine, weight loss, yellowish eyes, etc.

Biosensor Based Advanced Cancer Diagnostics. DOI: https://doi.org/10.1016/B978-0-12-823424-2.00009-0

## 6.2.1 Leading causes of HCC

### *6.2.1.1 Cirrhosis*

Cirrhotic tissues have damaged cells that are replaced with scar tissue. Patients with cirrhosis have 20 times higher chances of developing HCC. Hepatitis B virus (HBV) and hepatitis C virus (HCV) are notable causes of cirrhosis. Nonalcoholic fatty liver disease (NAFLD) and nonalcoholic steatohepatitis (NASH) associated cirrhosis eventually leads to the onset of HCC (Desai et al., 2019).

### *6.2.1.2 Hepatitis virus*

Hepatitis B virus (HBV) and hepatitis C virus (HCV) are the most common causative viruses as they transform normal liver cells into tumor cells without causing cirrhosis. HBV is a ds-DNA virus that penetrates liver cells, acting as a mutagen causing chromosomal rearrangements (Szabó et al., 2004). HCV, on the other hand, is an RNA virus that cannot integrate itself with the host genome, hence it opts for indirect ways of forming a tumor, like modifying the apoptotic process (Sheikh et al., 2008). For patients with hepatitis B infection, the chances to develop a tumor increase 100-fold. The ones with HBV and cirrhosis both are even more vulnerable (Arbuthnot & Kew, 2001).

### *6.2.1.3 Alcohol abuse and smoking*

Consumption of 40–60 g alcohol daily is associated with higher chances of HCC. A link between smoking and HCC is attributed to the formation of tumors because of the DNA adducts of 4-aminobiphenyl and polycyclic aromatic hydrocarbons from the smoke of cigarettes as these are potential carcinogens (Lee et al., 2009).

### *6.2.1.4 Aflatoxin-contaminated food*

Aflatoxins are a group of fungal metabolites secreted by *Aspergillus flavus* on improper stored food such as rice, corn, soybeans, and peanuts. Aflatoxin is known to damage DNA of hepatic cells and mutate the tumor suppressor gene, p53 (Bressac et al., 1991).

Other underlying conditions such as hereditary hemochromatosis, Wilson's disease, biliary cirrhosis, autoimmune hepatitis, diabetes and obesity, consumption of anabolic steroids, schistosomiasis causing viral infection, are also causative agents (Sanyal et al., 2010). The risk of HCC and its associated mortality rates can be reduced by proper sensitization and raising awareness about the potential toxins and suggesting minor lifestyle changes.

## 6.2.2 Currently used HCC diagnosis techniques

### *6.2.2.1 Imaging*

Ultrasound (US), contrast-enhanced computed tomography (CT), or magnetic resonance imaging (MRI) are the most reliable imaging techniques employed for HCC screening with very few false positive diagnoses registered. The choice of imaging technique is influenced by certain factors like availability, expertise, patient preferences, accuracy, and size and type of lesion to be detected (Tanaka, 2020).

#### 6.2.2.1.1 Ultrasound

Ultrasound is preferably the first imaging modality to assess the presence of a lesion, hence it is more of a screening technique than a diagnostic one. For lesions detected by US, those with a size $<1$ cm should repeat the ultrasound every 3–6 months; whereas, for lesions $>1$ cm, either a three-phase/four-phase CT scan or an MRI is strongly recommended. Contrast enhanced US (CEUS) is more sensitive and adds a diagnostic value to the test. The imaging criteria for HCC diagnosis is kept the same as arterial phase hyper vascularity, but only half of the total cases show portal or delayed phase washout (Jang et al., 2007). CEUS is performed under two phases: venous and arterial. Recently, the diagnostic power of CEUS has been enhanced by using Sonazoid, a gaseous perflubutane microbubble, as it allowed the imaging of Kupffer cells (Kudo, 2016). Ultrasound is not the imaging choice for patients with highly cirrhotic heterogeneous livers and, overall, it is less accurate in diagnosing HCC. AFP detection coupled with unenhanced ultrasound (US) is most widely used and most recommended of all mainly because of their easy accessibility and affordability (Yu et al., 2011).

#### 6.2.2.1.2 Computed tomography

Generally, hyper-enhancement in the hepatic arterial phase and washout appearance in the portal venous is observed as the output of HCC imaging. After a tumor nodule is found through US, they are validated using MRI or CT. In CT, usually the mass increases during late arterial ($\sim$35 s) and then washes out rapidly, hypo-attenuating itself in the portal venous phase, compared to the rest of the liver. Because of the arterio-portal shunts (APS), they may even be associated with a wedge-shaped perfusion defect resulting in a focal fatty change in the normal liver or focal fatty sparing in the diffusely fatty liver. A loss of increased intracellular liver fat may also be seen around an HCC in an otherwise fatty liver (Kim et al., 2008).

#### 6.2.2.1.3 Magnetic resonance imaging

The outstanding sensitivity and selectivity of MRI with respect to HCC diagnosis has been proven by many studies, particularly with small lesions. An MRI of an HCC tumor reveals many features including tumor architecture, tumor stage and grade, the amount of fat the tissue has, etc. It is necessary to distinguish non-neoplastic nodules, i.e., regenerative nodules, and hepatocyte nodules with diameter $>$1 mm, i.e., dysplastic nodules. Conventionally, MRI T1 signal is iso- or hypointense in cystic fibrosis or liver disease (Cho & Choi, 2015) and T2 signal is variable, typically moderately hyperintense (Willatt et al., 2008).

### *6.2.2.2 Liver biopsy*

Multiphase imaging of HCC via CT and MRI for high-risk patients shows the following features: phase CT and MRI, key imaging features include size $\geq$ 1 cm, arterial phase hyper-enhancement, threshold growth, and capsule appearance. Even if the mentioned features are absent, the possibility of HCC cannot be ruled out, therefore, liver biopsy could be used as a confirmation tool. In the primary cases of HCC, which is curable and radiologically diagnosed, biopsies are not required. Mainly, in patients with lesions $>$1 cm who are prone to malignancy a biopsy is recommended. Positive results point towards the need for resection, transplantation, or surgical removal of the affected tissue. However, there are chances of false-negative results in monitoring small lesions. Its sensitivity and specificity are superior to those of noninvasive diagnostic techniques. Under the microscope, HCC lesions have trabecular architecture, high nuclear to cytoplasmic ratio, atypical naked nuclei, and peripheral endothelial wrapping, etc. Histologically, in well-differentiated tumors, cells appear normal whereas, in poorly differentiated tumors, cells appear giant, anaplastic, and multinucleate. A few risks associated with liver biopsies like bleeding, tumor seeding, sampling errors, etc. limits its use (Navin & Venkatesh, 2019).

### *6.2.2.3 Angiography*

It has been used for HCC diagnosis because it possesses high vascularity but is mainly used before resectioning as a tool to analyze liver anatomy. It is less used due to its ineffectiveness in detecting tumors $<$2 cm in diameter (Rasool et al., 2014).

## 6.3 Biosensors in cancer

There has been a considerable drop in the number of deaths related to cancer since 1991 which is mainly attributed to the treatment breakthroughs and growing cancer-related awareness around the globe. Despite that, cancer remains the leading cause of death worldwide. The first and foremost key for halting the cancer progression is early and accurate screening and diagnosis. Various biomarkers found in different body fluids including proteins, nucleic acids, sugars, whole cells, cytokines, cytogenetic parameters, and small metabolites that point towards any abnormality should be detected and measured accordingly to assist clinicians in any treatment-related decision-making.

### 6.3.1 Conventional techniques for cancer diagnosis and their limitations

Traditionally, cancer diagnosis requires a biopsy sample which is subjected to staining and microscopy, but these invasive techniques come with the limitations of patient discomfort, longer healing times, and the extreme care of the patient. Immunoassays like ELISA (Enzyme-linked immunosorbent assay) came into a diagnostic domain which is non-invasive, more specific, but are not pocket friendly and require larger marker concentration. Genomics and proteomics tools like Ms, gene profiling, etc. are also employed in diagnostic labs, but they produce a lot of complex data. A cancer diagnosis is not straightforward as multiple factors contribute to the onset and progression of cancer. The causes and effects of cancer are multidimensional, both at genetic and epigenetic levels and not a single, but many genes are affected in the process. Hence, identifying one biomarker for a particular condition would be misleading and an array of biomarkers needs to be simultaneously detected. So rapid, sensitive, and accurate analysis is the need of the hour.

### 6.3.2 Biosensors as a new wave in cancer prognosis

Biosensors are analytical devices that determine the presence and concentration of a specific substance in a biological analyte. The three main components of a biosensor are: a biological recognition element (that binds to the analyte; could be a tissue, microorganisms, organelles, cell receptors, enzymes, antibodies or nucleic acids); a transducer (a physico-chemical element that converts the signal obtained from the interaction between bioelement and the analyte into a readable signal); and the signal conditioning circuit consisting of an amplifier, a processor, and a display unit. Biosensors possess several ideal characteristics, making them suitable for disease diagnosis in humans. Based on the type of transducer, biosensors can be classified into:

1. Electrochemical biosensor: In this, the electrochemical species generated or consumed because of the interaction between analyte and bioelement are measured. The current or the potential difference generated as a result of redox reaction is measured for the quantification of analyte in the biosample. It can be further subdivided into:
   a. Potentiometric biosensor: Ions are detected by measuring the potential when there is no supply of external current, where potential is directly proportional to the analyte concentration.
   b. Amperometric biosensor: A constantly increasing/decreasing potential is applied to oxidize or reduce the analyte until a peak current is reached, where the peak is directly proportional to the analyte concentration.
   c. Conductimetric biosensor: A change in the solution's electrical conductivity occurred due to the change in solution's composition is measured.
   d. FET-based biosensor: Silicon-chip-based field-effect transistor can be utilized to construct a miniature version of the electrochemical sensors.
2. Optical biosensor: Based on the property of the molecules to show fluorescence, absorbance, total internal reflection, emittance, surface plasma resonance, scattering, etc., but these sensors are only applicable to the analytes which are optically active and can show a marked change in the phase, amplitude, polarization, or frequency of the input light as a result of the interaction. It has four main components including a source of light, a light transmission medium like optical fiber, a bioelement, and an optical detection system. Some of its advantages are no electrical interference, multi-parametric detection, remote sensing, elimination of reference electrode, minimally invasive, fast, and real-time measurements.
3. Piezoelectric biosensor: Based on the property of crystals to deform mechanically when an external current is applied and vice-versa, i.e., when mechanically stressed, they produce electrical signals. Thus, with an oscillating electrical potential the crystal will oscillate mechanically. This resonance frequency is characteristic for each piezoelectric crystal and is called natural resonance frequency. It is dependent on the mass of the crystal. If any such crystal is attached to any other molecule/compound, etc. it will no longer resonate on its natural resonance frequency. This property can be exploited. Detection of gaseous, liquid analytes, ultra-high sensitivity of this biosensor is advantageous.
4. Acoustic biosensor: Uses a piezoelectric crystal (PC) coated with a biomaterial. When the analyte binds to the coated crystal and is exposed to external current, some waves are produced whose resonating frequency is different from those generated alone by the crystal. This change can be noted and quantified.
5. Thermal/calorimetric biosensor: It works on the principle of measuring the change in heat after an endothermal or exothermal reaction has taken place. The change in heat measured by a thermostat is proportional to the number of reactants consumed or products formed.

## 6.4 Clinical studies on HCC serum biomarkers and their sensor-based detection

HCC diagnosis is not straight-forward as the detection and quantification of only a single marker is not enough to reveal the presence and stage of the tumor. Multiple parameters and/or serological markers should be simultaneously tested to enhance the accuracy rates of surveillance. Early HCC diagnosis is of prime importance to prevent the tumor from becoming aggressive and increase the survival rates. Biochemical parameters including serum biomarkers, when evaluated provide a reliable diagnosis. Therefore, more and more biomarkers are being recognized and their ideal quantity in the blood is being identified, too (Table 6.1 and Fig. 6.1).

**TABLE 6.1** HCC based clinical studies.

| S. no | Biomarker detected | Type of biosensor | Principle/Property exploited | Reference |
|---|---|---|---|---|
| 1 | AFP-L3 | Electrochemical (Electroactive NPs based) | Binding affinity to LCA (Lens culinaris agglutinin) | |
| 2 | AFP | Electrochemical | Conductive gold nanoparticles as immobilization matrix of anti-AFP | Wu et al. (2013) |
| 3 | AFP | Electrochemical | HRP-label conjugation onto magnetic glassy carbon electrode (MGCE) surface | Yuan et al. (2017) |
| 4 | AFP | Electrochemical (G-FET) | AFP—anti-AFP reaction | Kim et al. (2018) |
| 5 | AFP | Optical | Localized surface plasmon resonance (LSPR) | Kim et al. (2016) |
| 6 | AFP | Piezoelectric | SAM (self-assembly monolayer) coated QCM attached to AFP antibody | |
| 7 | CD133 | Acoustic | CD133-CD133 antibody reaction | Damiati et al. (2017) |
| 8 | miRNA-122 | | Double-strand displacement biosensor with 2-aminopurine (2-AP) quenching in ds-DNA | |
| 9 | miRNA-221 | Electrochemical (isothermal sensor) | Dual signal amplification strategy | |
| 10 | CTCs | | Anti-CD45 antibody modified immune-magnetic nanosphere | Chen et al. (2016) |
| 11 | Glypican-3 (GPC3) | Piezoelectric (QCM) | Reaction between GPC3 and GPC3 antibody | Ogi et al. (2008) |
| 12 | GGT | Electrochemical | Potential for voltammetric response by formation of Cu2 + -glutathione complex | Chen et al. (2012) |
| 13 | Lipocalin | Aptamer based sensor | Single-stranded, DNA aptamer-based with 9 main aptamers (LCN2_apta1 to LCN2_apta9) | Lee et al. (2015) |

Abbreviations: AFP—Alpha fetoprotein, G-FET—graphene field-effect transistor, CTCs—circulating tumor cells, GGT—gamma-glutamyl transpeptidase, QCM—quartz crystal microbalance, NP-Nanoparticles.

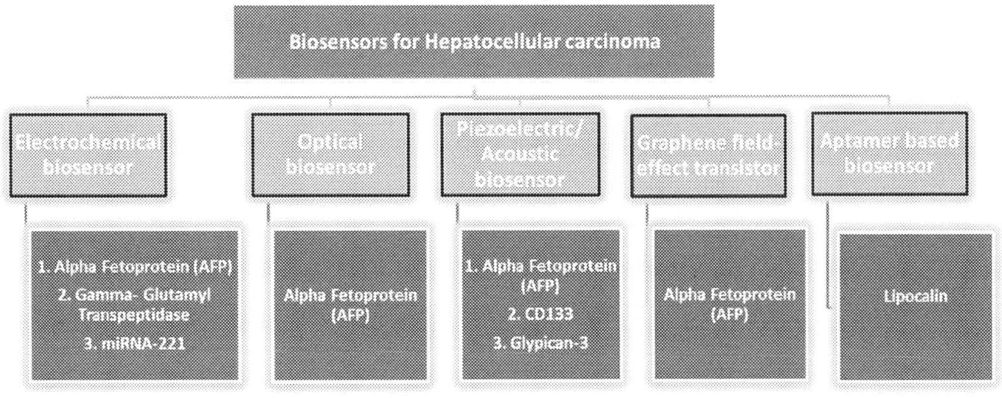


**FIGURE 6.1** Types of HCC biosensors and serum markers reported in case studies.

### 6.4.1 Alpha fetoprotein

AFP was the first glycoprotein to be recognized as a potent preclinical HCC biomarker and is till date the most widely detected biochemical element. Monitoring of AFP levels (most common serum marker for HCC diagnosis) is done mainly in the high-risk population in combination with other markers or ultrasonography. AFP is a 70 kD serum glycoprotein generated in maximum amounts by the fetal yolk sac and fetal liver. Its levels keep on declining with days, postbirth

(Kashyap et al., 2001). Conveniently, a serum level >400–500 ng/mL is suggestive of HCC malignancy. However, AFP elevation is not specific and is common in patients with tumors of organs derived from the same endodermal lining as the AFP producing tissues, i.e., stomach, biliary tract, and pancreas, etc. Hence, AFP detection is a nonspecific marker of HCC and has a potential for yielding false-positive results (Talerman et al., 1980). There are three main glycoforms of AFP based on their affinity to bind to Lens culinaris agglutinin (LCA)—AFP-L1, AFP-L2, and AFP-L3—out of which AFP-L3 is HCC-specific having high LCA binding affinity (Li et al., 2001). However, AFP was found to be of lower diagnostic value in HCC patients with total blood AFP below 20 ng/mL and a small-sized tumor. Therefore, it could not be used in early-stage HCC diagnosis (Naraki et al., 2002). AFP has a higher sensitivity (because of AFP-L3 relatively lesser quantity in the blood) than AFP-L3 for early HCC detection (Marrero et al., 2009).

Most of the HCC-centered biosensors have been made for AFP detection. Attempts are constantly being made to increase the accuracy and sensitivity of the sensors. Various electrochemical biosensors with the aim of sensitive and selective analysis of AFP have been constructed. An electrochemically active nanoparticle platform was utilized for sensitive detection of AFP-L3, an HCC biomarker. As mentioned earlier, AFP-L3 binds intimately to LCA, this property was exploited to make a highly reliable and sensitive sensor. So a biotinylated LCA-AgNP conjugate was synthesized as a recognition nanoprobe. Avidin, a protein naturally found in eggs was also added to the system because it is known to have a strong interaction affinity towards biotin, which could attract more AgNPs aggregates onto the working electrode (GCE, in this study). This amplifies the signals reaching the transducer. The system was characterized via EIS and SWV (square wave voltammetry). The data supported that this electrochemical sensor has very low detection limits (12 pg/mL) and a linear correlation of 25 –15,000 pg/mL.

Kashefi, et al., selected an aptamer, TLS11a, known to specifically bind membrane surface of HCC cells. The sensor had a sandwich architecture with the HepG2 cells captured between a TLS11a aptamer immobilized to an MPA (3-mercaptopropionic acid)-modified Au surface and a secondary TLS11a aptamer. TLS11a was used as a recognition layer, which was highly selective for HepG2 cells. This aptamer-based, label-free biosensing platform was characterized by EIS (electrochemical impedance spectroscopy) and CV (cyclic voltammetry) and had an excellent linear dynamic range between $10-10^6$ cells/mL and could detect as low as 2 cells/mL (Kashefi-Kheyrabadi et al., 2014). A silver–graphene oxide (Ag–GO) based label-free electrochemical immune-sensor was fabricated by Wu et al. for the detection of AFP. With glucose as a reducing agent, silver was allowed to reduce onto graphene oxide (GO), in situ. The large surface area of GO allowed the attachment of many Ag–GO nanocomposites, enhancing the sensitivity. Highly conductive AuNPs were used as immobilization matrix of antibody (anti-AFP), which could amplify the electrochemical signal obtained and stabilizing the system. Lastly, Nafion was used as a dispersant of Ag–GO nanocomposite because of its excellent film-forming property and chemical stability. This immune-sensor had very low detection limits of 3 pg/mL and a wide linear response (Wu et al., 2013). An electrochemical sandwich immune-complex for the detection of AFP was constructed with components like Ab1 immobilized core (Fe3O4) @ shell (Au) NPs (prepared via hydrothermal method using H2O2 as an oxidizer) functioning as capture probe, HRP, and Ab2 functionalized GNPs acting as a detection probe. AFP was sandwiched between these two components, and this whole setup was on magnetic glassy carbon electrode (MGCE) via magnetic effect. HRP was attached to the surface of MGCE and was used as a label which catalyzed H2O2 breakdown, releasing current. This current was quantified to reveal the quantity of AFP in the sample. The final sensor showed excellent anti-interference ability, high reproducibility, and stability (Yuan et al., 2017). AFP detection in the patient plasma was also tried using a G-FET (graphene field-effect transistor) functionalized with 1-pyrenebutyric acid N-hydroxysuccinimide ester (PBASE). Anti-AFP antibody was immobilized over the functionalized surface and AFP was detected by monitoring the shift in the voltage of the Dirac point ($\Delta V_{Dirac}$) after antigen-antibody interaction. The limit of detection of the G-FET biosensor was 0.1 ng mL − 1 of AFP in PBS and sensitivity was 16.91 mV. In plasma of HCC patients, detection limit and sensitivity were 12.9 ng mL − 1 and 5.68 mV, respectively (Kim et al., 2008). AFP was also identified using an optical sensor based on the principle of localized surface plasmon resonance (LSPR). A transparent plasmonic active 4 substrate was surface modified with amine group and allowed to react with Au colloid, causing self-assembly of AuNPs on the amine-containing surface. This Nanopatterned, Au-coated chip was then reacted with various concentrations of AFP antibody (AFP-Ab) (from 1 ng/mL to 1 mg/mL). However, 10 μg/mL of anti-AFP came out to be the ideal Ab concentration for coupling the antibody and AuNPs coated biosensor chip. LSPR signals were recorded via UV-vis spectroscopy. The detection limit of the biosensor was 1 ng/mL to 1 μg/mL and the detection took only 20 min as opposed to 2 h taken by ELISA (Kim et al., 2016).

Li et al. attempted to fabricate a piezoelectric sensor for the detection of AFP in very low amounts in human serum for early HCC detection (Li et al., 2001). A quartz crystal microbalance (QCM)-based AFP sensor was coated with cystamine self-assembly monolayer (SAM). Some cystamine-coated QCM chips were also dipped into a cross-linking agent, glutaraldehyde, to strengthen the covalent binding between SAM and AFP antibody. Onto this, the AFP antibody was attached and blocked via BSA to avoid non-specific antigen absorption. This system had high mass sensitivity (0.35 Hz mL/mg) and yielded a good linear response (98.6%). The addition of glutaraldehyde improved the sensitivity to 0.128 Hz mL $mg^{-1}$ with the AFP Ab concentration as low as 0.1 mg/mL (Liu et al., 2013). AFP was sensitively quantified by Sheta, et al. by using a novel copper-nano-magnetic-metal organic framework (Cu-MOF-NPs). This was the first study investigating AFP in due presence of its competitor biomarkers. The intensity of emission of photoluminescence (PL) was quenched post-AFP addition. The samples from healthy patients and those suffering from hepatitis A, B, or C having AFP in the range of 1–520 ng/mL could be accurately determined with this Cu-MOF-NP system. With no considerable competitor, wide linear dynamic range, and accurate results even at room temperature, this could be a promising approach for early HCC detection (Sheta et al., 2019).

### 6.4.2 Glypican-3 (GPC3)

GPC3 is a heparin sulfate proteoglycan (Filmus, 2002) playing a key role in cell proliferation and tumor suppression (Jiatao et al., 2017) via interaction with growth factors. GPC3 is particularly expressed in HCC, facilitating diagnostic differentiation among normal cells, cirrhosis and HCC cases (Capurro et al., 2003; Yamauchi et al., 2005). It binds to the cell membrane with the help of a glycosylphosphatidylinositol anchor. It is also involved in inhibiting the growth of HCC tumors as they are capable of removing HGF and VEGF, the GFs actively participating in cell proliferation. GPC3 is more sensitive and specific than AFP and HCCR (human cervical cancer oncogene). Moreover, their combination has a much higher diagnostic value than any of them alone (Qiao et al., 2011). GPC3 exerts different roles in different tissues. Its activity is decreased in breast and ovarian cancer (Capurro et al., 2003; Xiang et al., 2001; Yamauchi et al., 2005). A pathway called "Wnt signaling," frequently observed in HCC, is activated by GPC3, promoting tumorigenesis (Capurro et al., 2005; Feitelson et al., 2002). Another pathway, Hedgehog signaling is downregulated by GPC3, inhibiting ovarian and breast cancer progression. An upregulation in GPC3 level promotes FGF signaling (Fibroblast growth factor), thereby reducing HCC survival rate (Lai, et al., 2008). GPC3 was sensed using a wireless-electrodeless QCM (quartz-crystal microbalance) by Ogi, et al. by monitoring the reaction between GPC3 and antibody of GPC3. The QCM method has an advantage over conventional bio sensing methods like ELISA and the surface-plasmon-resonance as it directly measures the mass absorbed onto it to determine the affinity constant; hence, it does not require detailed calibration. The affinity constant as determined using a Langmuir plot was $K_A = 7.1*10^7\ M^{-1}$. The detection limit was approx. 100 pg/mL (Ogi et al., 2008).

### 6.4.3 miRNA

Various researchers around the globe have realized the diagnostic and prognostic capabilities of multiple miRNAs. miRNA known to be upregulated in the HCC tissues are: miR-222, -135a, -155, -182, -10b and -17–5p, miR-21, miR -221 (Berretta et al., 2017; Karakatsanis et al., 2013), miR-10a, -18b, -143, -210, -216a, -224, -301a, -550a, -590–5p (Chuang et al., 2015; D'Anzeo et al., 2014), miR-122–5p, miR-125b-5p, miR-885–5p, miR100–5p and miR-148a-3p (Jin et al., 2019; Liu et al., 2013), whereas the ones downregulated were: miR- 122, miR-16, miR-92a, miR-26 (Ji et al., 2009), miR-99a, -124, -139, -145 and -199b, miR-1, -7a, -195, -200a, -203, -214, -219–5p, -376a, -449, -450a and -520e, miR-26a/b, -125a/125b, -223, miR-34a, -101, -122, -139 (D'Anzeo et al., 2014; De Stefano et al., 2018). The role of miRNA has also been investigated in HCC and its related pathologies. Many miRNAs act as signature markers of HCC, being present in stable concentrations and can easily help a clinician differentiate between normal samples and cancerous samples. MiRNA-122, is reported to be downregulated in HCC patients. Liao et al., attempted to develop a simple, novel, double-strand displacement biosensor for miRNA-122 detection in a quencher-free manner. 2-aminopurine (2-AP), a fluorescent analog of adenine was used as a fluorophore. As it is known that the fluorescence of (2-AP) gets highly quenched in ds-DNA due to base stacking. A DNA strand complementary to the template miRNA-122 and partially complementary to paring with the 2-AP probe was designed forming a cDNA\2-AP duplex. The duplex was then competed off to form cDNA\RNA heteroduplex, meanwhile releasing 2-AP, giving off a good fluorescence. As a result, detection limit of 5 nM and a linear range from 5 nM to 1000 nM was obtained within 2 hours. Another miRNA was electrochemically quantitatively detected using an isothermal biosensor. MiRNA-221, known to be upregulated in HCC cases was detected using a dual signal amplification strategy, i.e., target catalyzed hairpin assembly (CHA) and super sandwich amplification strategies (Zhang, Shao, et al., 2015).

### 6.4.4 Cancer stem/tumor cells

CSCs, just like stem cells, are capable of self-renewal and differentiation into multiple cancer cell types. They are important for processes like cancer initiation, progression, metastasis, recurrence, and resistance to a potential drug.

These are rare biomarkers and are now considered targets for cancer therapy (Chen et al., 2013). Efficient isolation of circulating tumor cells (CTCs) from the clinical HCC blood samples can aid in early HCC detection as they point towards distant metastases or early recurrence (Colombo et al., 2011). But, because of CTC's heterogeneity and sparseness in the serum, their isolation and detection becomes challenging. Chen, et al., devised a cell sorting method based on anti-CD45 antibody modified magnetic nanospheres for isolation, detection, and subtype analyses of HCC CTCs. They made five-layer magnetic polystyrene acrylamide copolymer nanospheres, i.e., immunomagnetic nanospheres (IMNs) and modified their surface with anti-CD45 antibodies. Cells were sorted based on the recognition of IMNs (CD45). The applied got depleted from the mixed cell sample, while the non-labeled ones were separated out. This way, HCC cells magnetic field, many magnetically labeled leukocytes, and other CTC subtypes could be separately collected within 30 minutes and remained viable. Coupling this technique with ICC identification via specific HCC biomarkers like AFP and GPC3 allowed HCC-specific CTC isolation in peripheral blood samples (Chen et al., 2013). CD44, a hyaluronic acid receptor and a co-receptor for growth factors and cytokines, is a poor marker for many cancer types. Some of the CD44 isoforms are expressed specifically in higher grade cancer (Chen et al., 2013).

A study by Henry, et al., indicated that AFP levels increased with the increase in CD44 expression (Henry et al., 2010). CD133, a human prominin-1, is a transmembrane surface glycoprotein whose expression was observed in the normal human liver or HCC tissues by northern blot analysis. However, its expression was confirmed via immunohistochemistry and was found as tumorigenic in immune-deficient mice and highly clonogenic in vitro (Yin et al., 2007). Another marker CD90 (also c/s-Thy-1) regulates multiple cancer-related processes in the body. Yang et al., reported the increase in CD90 + cells with the tumorigenicity of HCC cell lines (Yang et al., 2008). CD90 + are known to be related to multi-drug resistance in HCC (Sukowati et al., 2013). An acoustic, hybrid three-dimensional (3D)-printed, label-free electrochemical biosensor was fabricated for the detection of CD133, an HCC surface marker. For the detection, the gold sensor's surface was coated with recombinant S-layer fusion protein (rSbpA/ZZ) which allowed effortless immobilization of CD-133 antibody which acted as a sensing layer. This recombinant protein could minimize any nonspecific protein adsorption and ensured the correct orientation of Ab over it. This allowed the efficient capture of HepG2 cells. EpCAM (epithelial cell adhesion/activating molecule) (Nio et al., 2015), CD13 (Yang et al., 2008), OV-6 (Yang et al., 2008), IB50−1 (Zhao et al., 2013), ALD4 (aldehyde dehydrogenase) (Qin et al., 2012; Shmelkov et al., 2008), SALL4 (Sal-like protein 4) (Oikawa et al., 2013), ICAM-1 (intercellular adhesion molecule 1) (Van De Stolpe & Van Der Saag, 1996), are some other commonly studied biosensors in correlation with HCC and associated drug resistance. (Fig. 6.1).

## 6.5 Other clinically relevant biomarkers for HCC

### 6.5.1 Des-ϒ-carboxyprothrombin (DCP) or PIVKA II (prothrombin induced by vitamin K deficiency)

It is an abnormal form of plasma protein, the prothrombin produced by a fault in the vitamin K-dependent post-translational carboxylation reaction system (Tara & Sitki, 2012). A DCP level of more than 100 ng/mL is indicative of HCC. DCP is a better HCC biomarker than AFP and AFP-L3 combined as it allows clear differentiation between HCC and nonmalignant liver cirrhosis (Jiatao et al., 2017). Besides, a case-control study suggested that DCP-AFP coupled immunoassay is highly sensitive, especially for the early stage HCC diagnosis (Lok et al., 2010) (Fig. 6.2).

### 6.5.2 Alpha L fucosidase

A mammalian cell lysosomal enzyme involved in the hydrolysis of fucose glycosidic linkages of glycoproteins and glycolipids. The activity of Serum Alpha L fucosidase (AFU) was reported to be increased in HCC or patients suffering with chronic hepatitis when compared to healthy adults (Deugnier et al., 1984). The reason for its elevation could be the production of proteins by the tumors with high fucose turnover. Moreover, AFU levels are not dependent on tumor size and have mostly been observed in early HCC stages (Zhang, Lin, & Li, 2015). Ultrasonography in 85% of patients showed an elevated AFU at least 6 months before HCC detection (Tangkijvanich et al., 1999).

### 6.5.3 Human carbonyl reductase 2 (HCR2)

Involved in the detoxification of reactive alpha-dicarbonyl compounds and ROS, produced as a result of oxidative stress during HCC. It is mainly expressed in the human liver and kidney, and its levels are reduced after the incidence of HCC (Liu et al., 2006).

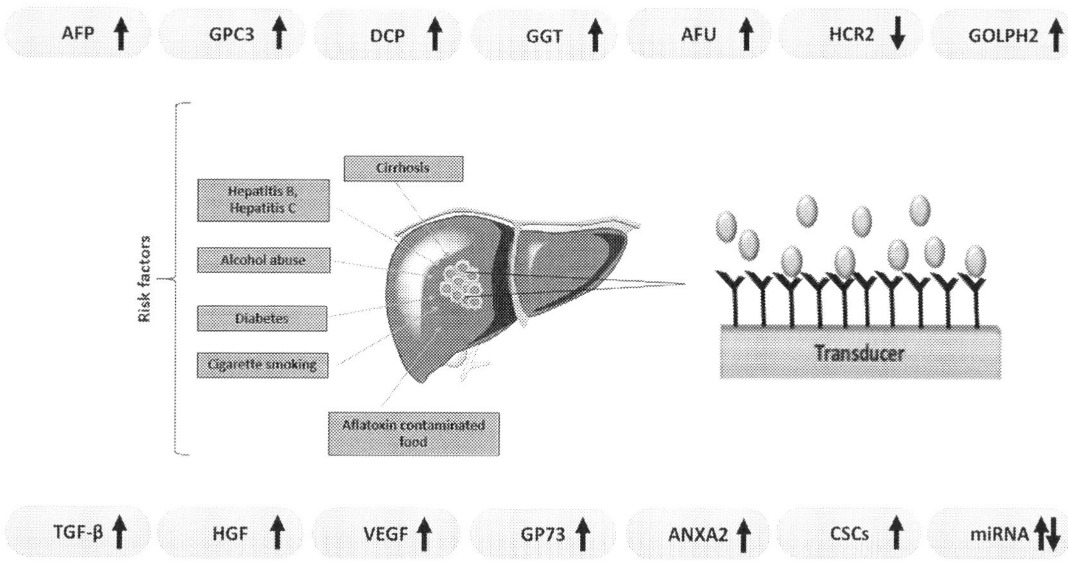


**FIGURE 6.2** Diagrammatic representation of HCC detection using a biosensor. Various biomarkers and their respective up-regulation/down-regulation status along with the main risk factors contributing to the development of HCC.

### 6.5.4 Golgi phosphoprotein 2 (GOLPH2)

A protein associated with golgi apparatus whose serum levels have been quantified using ELISA. Its levels are found to be elevated in the cases of HBV, cirrhosis, and bile duct carcinoma, but they are maximum in the sera of HCC affected patients. Its levels considerably fall after HCC tumor resection (Mao et al., 2010; Marrero et al., 2005).

### 6.5.5 Transforming growth factor-beta

A chief human growth factor and cytokine which is critical for cell growth, differentiation, angiogenesis, immune suppression, fibrogenesis, and cancer cell proliferation, etc. Transforming growth factor-beta (TGF-β) may be an initiator of HCC when it gets transformed from a tumor suppressor to an oncogenic factor. Increased TGF-β exposure may lead to a higher expression of miR-181b in the liver cells which results in the growth, invasion, and migration of HCC (Wang et al., 2018). Hence, TGF-β can be exploited as a target for HCC therapy (Yang et al., 2013).

### 6.5.6 Hepatocyte growth factor/scatter factor (HGF/SF)

A cytokine primarily involved in liver regeneration and the development of the embryo. In association with the c-Met receptor (a receptor tyrosine kinase which regulates processes like cell scattering, growth, stimulation, and branching morphogenesis of cells in tissues) (Wade & Vande Pol, 2006), it aids in the development of HCC. High c-met levels point towards invasive HCC, metastasis, and reduce chances of survival (Korn, 2001; Osada et al., 2008).

### 6.5.7 Fibroblast growth factor (FGF)

A soluble, heparin-binding polypeptide cytokines having mitogenic, morphological, endocrine, and regulatory effects of multiple cell types. In a study by Yang, et al., co- silencing of TGFBR1 and FGF-9 could have inhibitory effects on HCC tumor growth (Yang et al., 2013).

### 6.5.8 Vascular endothelial growth factors

A growth factor required for neo-angiogenesis. It is also known to increase vascular permeability which aids in the process of metastasis. Vascular endothelial growth factors (VEGFs) elevation has recently been recognized as a risk factor for HCC development as its levels in the body are proportional to tumor size, venous infiltration, and metastasis

(Mao et al., 2010). Also, the latest drug for treating HCC, Sorafenib, targets VEGF, making it an all the more essential biomarker to be explored.

### 6.5.9 Golgi protein 73

Golgi protein 73 (GP73) type II Golgi-specific transmembrane glycoprotein which is normally absent in healthy hepatocytes, but is found in high levels in serum of patients with liver disorder, mainly HCC (Kladney et al., 2002). In a comparative study between AFP in healthy subjects, HBV carriers, cirrhotic patients, and cancer patients, GPC3 was found to be superior in terms of sensitivity and specificity for the diagnosis of small, early-stage, and low AFP HCC (Tian et al., 2011) and other types of cancers.

### 6.5.10 Osteopontin (OPN)

Osteopontin is a glycosylated, phosphorylated, non-collagenous, chemokine-like extracellular matrix protein which mediates processes like chemotaxis, degradation of ECM (Extracellular matrix), angiogenesis, cell—cell adhesion, negative regulation of apoptosis, calcification of bone, tissue remodeling, pro-inflammatory processes, etc. (Denhardt et al., 2001; Estko et al., 2015; McAllister et al., 2008; Zhu et al., 2005). An overexpressed OPN is indicative of metastatic tumor phenotype (Chen et al., 1997; Lunxiu, 2014; Sharp et al., 1999; Urquidi et al., 2002). OPN is a novel, but attractive HCC marker and a therapeutic target against HCC metastasis, however, it is less specific as it is raised in almost more than 30 cancer types (Weber, 2011).

### 6.5.11 Annexin A2

Annexin A2 is a calcium-dependent membrane-binding protein that regulates cell growth, proliferation, adhesion, apoptosis, and cell surface fibrinolysis (Chiang et al., 1999; Fritz et al., 2010; Hajjar & Acharya, 2000; Jacovina et al., 2009; Mai et al., 2000). It is a known marker for immune liver fibrosis in a rat liver fibrosis, i.e., for liver cirrhosis and fibrosis (Yin et al., 2007). ELISA experiments performed on HCC patients, chronic disease patients, and healthy subjects revealed that the serum ANXA2 was elevated the most in HCC subjects, however, the sample studies were too few to conclude (El-Abd et al., 2016). Further study by Liu, et al., lent credence to the observation that ANXA2 is a poor HCC prognostic biomarker as it was upregulated in a correlated amount in both HCC and cirrhosis, i.e., HCC is not only produced as a result of cancer (Liu et al., 2013).

### 6.5.12 Squamous cell carcinoma antigen

Squamous cell carcinoma antigen (SCCA) is a serine protease inhibitor physiologically present in squamous epithelium, mainly expressed by epithelial neoplastic cells, and SCCA-IgM is an antigen-antibody complex (Pontisso et al., 2006). SCCA levels are raised on the cases of epithelial cancers of the neck, cervix, and lungs (Catanzaro et al., 2011; Kim et al., 2005; Stenman et al., 2001). For HCC, it is considered to be of moderate value (Zhang, Yao et al., 2015). A combination of SCCA with AFP, however, yielded 90.83% correct analysis in HCC subjects highlighting its potential as an HCC marker (Giannelli et al., 2005). Raising SCCA-IgM in cirrhotic patients pointed towards a higher probability of HCC development (Pontisso et al., 2004).

### 6.5.13 Midkine

Midkine (MDK) is a heparin-binding growth factor which is produced by the fetal liver and is minimally expressed during adulthood (Mashaly et al., 2018). It elevates drastically as the inflammation worsens and is a factor contributing to tumor development, hence can be used for HCC diagnosis (Vongsuvanh et al., 2016). While differentiating between Cirrhosis and HCC patients, MDK was found 50% more sensitive than AFP (Shaheen et al., 2015).

### 6.5.14 mRNAs

mRNAs of various well known HCC biomarkers like GPC3, TGF-Beta, AFP, gamma-glutamyl transferase (GGT), IGF-II, etc. are present in higher amount in the serum and liver biopsies of HCC patients than in healthy subjects. A few mRNAs like that of AFP and GGT indicates the chances of recurrence of HCC. AFP mRNA in serum 1 week after the tumor removal points towards a recurrence of HCC (Ding et al., 2005). Similarly, Type-II isoform of GGT in the blood of patients is correlated with more recurrence related deaths (Zhou et al., 2006). mRNA of the most abundantly present protein in the blood, i.e., albumin is also a sensitive pathological indicator for liver-related morbidities. Albumin upsurge in patients after liver transplant is characteristic of posttransplant HCC recurrence (Cheung et al., 2008) (Fig. 6.2).

## 6.6 Conclusion

The complexity and lack of specific knowledge regarding HCC poses a significant challenge in the medical domain, which calls for a rapid, robust, sensitive, and specific technology while maintaining cost-effectiveness and easy availability. Biosensors are a new wave in cancer diagnostics and monitoring, which allows overcoming multiple challenges like multiplexing, accuracy, etc. Identifying a specific biomarker or an array of biomarkers involved in this cancer's pathophysiology is indispensable for developing and modifying HCC-specific sensing platforms. Various nanomaterials are being tested to fabricate the most reliable and efficient biosensor to obtain an improved signal after the biomarker-bioelement interaction. A higher degree of investment and commercialization of the promising bioanalytical devices is the need of an hour to prevent this malady from progressing towards a life-threatening condition.

## 6.7 Future Prospects

In recent years, many efforts have been made to find out the biomarkers specific to HCC. Enormous success has been achieved in developing novel materials ideal for biosensing applications. But the progress in generating clinically applicable biosensors for HCC detection is made at a relatively sedate pace. The practical applicability of the relevant scientific information for solving real-life diagnostic problems needs to be addressed on a prompt basis. More studies are warranted on the stability, safety, and toxicity aspects of the materials used to fabricate the biosensor. Aptamer-based biosensors could be a solution as aptamers have higher thermal stability, less immunogenicity, and less toxicity as these oligonucleotide moieties are already present in the human body. Further, miniaturization of the biosensors can be done to enhance the user experience. Detection of multiple biomarkers simultaneously would improve the sensitivity and diagnostic reliability. The inclusion of biomarker pattern software and microfluidics can offer superior predictive values. The synergistic combination of biology and microelectronics will open doors for "smart biosensing devices" with improved sensitivity, specificity, and reliability.

## References

Allemani, C., Matsuda, T., Di Carlo, V., Harewood, R., Matz, M., Nikšić, M., ... Visser, O. (2018). Global surveillance of trends in cancer survival 2000–14 (CONCORD-3): Analysis of individual records for 37 513 025 patients diagnosed with one of 18 cancers from 322 population-based registries in 71 countries. *The Lancet*, *391*(10125), 1023–1075. Available from https://doi.org/10.1016/S0140-6736(17)33326-3.

Arbuthnot, P., & Kew, M. (2001). Hepatitis B virus and hepatocellular carcinoma. *International Journal of Experimental Pathology*, *82*(2), 77–100. Available from https://doi.org/10.1046/j.1365-2613.2001.00178.x.

Attwa, M. H., & El-Etreby, S. A. (2015). Guide for diagnosis and treatment of hepatocellular carcinoma. *World Journal of Hepatology*, *7*(12), 1632–1651. Available from https://doi.org/10.4254/wjh.v7.i12.1632.

Berretta, M., Cavaliere, C., Alessandrini, L., Stanzione, B., Facchini, G., Balestreri, L., ... Canzonieri, V. (2017). Serum and tissue markers in hepatocellular carcinoma and cholangiocarcinoma: Clinical and prognostic implications. *Oncotarget*, *8*(8), 14192–14220. Available from https://doi.org/10.18632/oncotarget.13929.

Bressac, B., Kew, M., Wands, J., & Ozturk, M. (1991). Selective G to T mutations of p53 gene in hepatocellular carcinoma from southern Africa. *Nature*, *350*(6317), 429–431. Available from https://doi.org/10.1038/350429a0.

Capurro, M., Wanless, I. R., Sherman, M., Deboer, G., Shi, W., Miyoshi, E., & Filmus, J. (2003). Glypican-3: A novel serum and histochemical marker for hepatocellular carcinoma. *Gastroenterology*, *125*(1), 89–97. Available from https://doi.org/10.1016/S0016-5085(03)00689-9.

Capurro, M. I., Xiang, Y. Y., Lobe, C., & Filmus, J. (2005). Glypican-3 promotes the growth of hepatocellular carcinoma by stimulating canonical Wnt signaling. *Cancer Research*, *65*(14), 6245–6254. Available from https://doi.org/10.1158/0008-5472.CAN-04-4244.

Catanzaro, J. M., Guerriero, J. L., Liu, J., Ullman, E., Sheshadri, N., Chen, J. J., & Zong, W. X. (2011). Elevated expression of squamous cell carcinoma antigen (SCCA) is associated with human breast carcinoma. *PLoS One*, *6*(4), e19096. Available from https://doi.org/10.1371/journal.pone.0019096.

Chen, G., Ni, S., Zhu, S., Yang, J., & Yin, Y. (2012). An electrochemical method to detect gamma glutamyl transpeptidase. *International Journal of Molecular Sciences*, *13*(3), 2801–2809.

Chen, H., Ke, Y., Oates, A. J., Barraclough, R., & Rudland, P. S. (1997). Isolation of and effector for metastasis-inducing DNAs from a human metastatic carcinoma cell line. *Oncogene*, *14*(13), 1581–1588. Available from https://doi.org/10.1038/sj.onc.1200993.

Chen, K., Huang, Y. H., & Chen, J. L. (2013). Understanding and targeting cancer stem cells: Therapeutic implications and challenges. *Acta Pharmacologica Sinica*, *34*(6), 732–740. Available from https://doi.org/10.1038/aps.2013.27.

Chen, L., Wu, L. L., Zhang, Z. L., Hu, J., Tang, M., Qi, C. B., … Pang, D. W. (2016). Biofunctionalized magnetic nanospheres-based cell sorting strategy for efficient isolation, detection and subtype analyses of heterogeneous circulating hepatocellular carcinoma cells. *Biosensors and Bioelectronics*, *85*, 633–640. Available from https://doi.org/10.1016/j.bios.2016.05.071.

Cheung, S. T., Fan, S. T., Lee, Y. T., Chow, J. P., Ng, I. O., Fong, D. Y., & Lo, C. M. (2008). Albumin mRNA in plasma predicts post-transplant recurrence of patients with hepatocellular carcinoma. *Transplantation*, *85*(1), 81–87. Available from https://doi.org/10.1097/01.tp.0000298003.88530.11.

Chiang, Y., Rizzino, A., Sibenaller, Z. A., Wold, M. S., & Vishwanatha, J. K. (1999). Specific down-regulation of annexin II expression in human cells interferes with cell proliferation. *Molecular and Cellular Biochemistry*, *199*(1–2), 139–147. Available from https://doi.org/10.1023/A:1006942128672.

Cho, E. S., & Choi, J. Y. (2015). MRI features of hepatocellular carcinoma related to biologic behavior. *Korean Journal of Radiology*, *16*(3), 449–464. Available from https://doi.org/10.3348/kjr.2015.16.3.449.

Chuang, K. H., Whitney-Miller, C. L., Chu, C. Y., Zhou, Z., Dokus, M. K., Schmit, S., & Barry, C. T. (2015). ). MicroRNA-494 is a master epigenetic regulator of multiple invasion-suppressor microRNAs by targeting ten eleven translocation 1 in invasive human hepatocellular carcinoma tumors. *Hepatology (Baltimore, Md.)*, *62*(2), 466–480.

Colombo, F., Baldan, F., Mazzucchelli, S., Martin-Padura, I., Marighetti, P., Cattaneo, A., & Porretti, L. (2011). Evidence of distinct tumor-propagating cell populations with different properties in primary human hepatocellular carcinoma. *PLoS One*, *6*(6), e21369.

D'Anzeo, M., Faloppi, L., Scartozzi, M., Giampieri, R., Bianconi, M., Del Prete, M., … Cascinu, S. (2014). The role of micro-RNAs in hepatocellular carcinoma: From molecular biology to treatment. *Molecules*, *19*(5), 6393–6406, Basel, Switzerland.

Damiati, S., Küpcü, S., Peacock, M., Eilenberger, C., Zamzami, M., Qadri, I., … Schuster, B. (2017). Acoustic and hybrid 3D-printed electrochemical biosensors for the real-time immunodetection of liver cancer cells (HepG2). *Biosensors and Bioelectronics*, *94*, 500–506. Available from https://doi.org/10.1016/j.bios.2017.03.045.

De Stefano, F., Chacon, E., Turcios, L., Marti, F., & Gedaly, R. (2018). Novel biomarkers in hepatocellular carcinoma. *Digestive and Liver Disease*, *50*(11), 1115–1123. Available from https://doi.org/10.1016/j.dld.2018.08.019.

Denhardt, D. T., Giachelli, C. M., & Rittling, S. R. (2001). Role of osteopontin in cellular signaling and toxicant injury. *Annual Review of Pharmacology and Toxicology*, *41*, 723–749. Available from https://doi.org/10.1146/annurev.pharmtox.41.1.723.

Desai, A., Sandhu, S., Lai, J. P., & Sandhu, D. S. (2019). Hepatocellular carcinoma in non-cirrhotic liver: A comprehensive review. *World Journal of Hepatology*, *11*(1), 1–18. Available from https://doi.org/10.4254/wjh.v11.i1.1.

Deugnier, Y., David, V., Brissot, P., Mabo, P., Delamaire, D., Messner, M., … Legall, J. (1984). Serum α-L-fucosidase: A new marker for the diagnosis of primary hepatic carcinoma? *Hepatology (Baltimore, Md.)*, *4*(5), 889–892. Available from https://doi.org/10.1002/hep.1840040516.

Ding, X., Yang, L. Y., Huang, G. W., Yang, J. Q., Liu, H. L., Wang, W., … Ling, X. S. (2005). Role of AFP mRNA expression in peripheral blood as a predictor for postsurgical recurrence of hepatocellular carcinoma: A systematic review and *meta*-analysis. *World Journal of Gastroenterology*, *11*(17), 2656–2661. Available from https://doi.org/10.3748/wjg.v11.i17.2656.

El-Abd, N., Fawzy, A., Elbaz, T., & Hamdy, S. (2016). Evaluation of annexin A2 and as potential biomarkers for hepatocellular carcinoma. *Tumor Biology*, *37*(1), 211–216. Available from https://doi.org/10.1007/s13277-015-3524-x.

Estko, M., Baumgartner, S., Urech, K., Kunz, M., Regueiro, U., Heusser, P., & Weissenstein, U. (2015). Tumour cell derived effects on monocyte/macrophage polarization and function and modulatory potential of Viscum album lipophilic extract in vitro. *BMC Complementary and Alternative Medicine*, *15*(1). Available from https://doi.org/10.1186/s12906-015-0650-3.

Feitelson, M. A., Sun, B., Satiroglu Tufan, N. L., Liu, J., Pan, J., & Lian, Z. (2002). Genetic mechanisms of hepatocarcinogenesis. *Oncogene*, *21*(16), 2593–2604. Available from https://doi.org/10.1038/sj.onc.1205434.

Filmus, J. (2002). The contribution of in vivo manipulation of gene expression to the understanding of the function of glypicans. *Glycoconjugate Journal*, *19*(4–5), 319–323. Available from https://doi.org/10.1023/A:1025312819804.

Fritz, K., Fritz, G., Windschiegl, B., Steinem, C., & Nickel, B. (2010). Arrangement of Annexin A2 tetramer and its impact on the structure and diffusivity of supported lipid bilayers. *Soft Matter*, *6*(17), 4084–4094. Available from https://doi.org/10.1039/c0sm00047g.

Giannelli, G., Marinosci, F., Trerotoli, P., Volpe, A., Quaranta, M., Dentico, P., & Antonaci, S. (2005). SCCA antigen combined with alpha-fetoprotein as serologic markers of HCC. *International Journal of Cancer*, *117*(3), 506–509. Available from https://doi.org/10.1002/ijc.21189.

Hajjar, K. A., & Acharya, S. S. (2000). *Annexin II and regulation of cell surface fibrinolysis*, . *Annals of the New York Academy of Sciences* (Vol. 902, pp. 265–271). New York Academy of Sciences. Available from https://doi.org/10.1111/j.1749-6632.2000.tb06321.x.

Henry, J. C., Park, J. K., Jiang, J., Kim, J. H., Nagorney, D. M., Roberts, L. R., … Schmittgen, T. D. (2010). miR-199a-3p targets CD44 and reduces proliferation of CD44 positive hepatocellular carcinoma cell lines. *Biochemical and Biophysical Research Communications*, *403*(1), 120–125. Available from https://doi.org/10.1016/j.bbrc.2010.10.130.

Jacovina, A. T., Deora, A. B., Ling, Q., Broekman, M. J., Almeida, D., Greenberg, C. B., … Hajjar, K. A. (2009). Homocysteine inhibits neoangiogenesis in mice through blockade of annexin A2-dependent fibrinolysis. *Journal of Clinical Investigation*, *119*(11), 3384–3394. Available from https://doi.org/10.1172/JCI39591.

Jang, H. J., Tae, K. K., Burns, P. N., & Wilson, S. R. (2007). Enhancement patterns of hepatocellular carcinoma at contrast-enhanced United States: Comparison with histologic differentiation. *Radiology*, *244*(3), 898–906. Available from https://doi.org/10.1148/radiol.2443061520.

Ji, J., Shi, J., Budhu, A., Yu, Z., Forgues, M., Roessler, S., … Wang, X. W. (2009). MicroRNA expression, survival, and response to interferon in liver cancer. *New England Journal of Medicine*, *361*(15), 1437–1447. Available from https://doi.org/10.1056/NEJMoa0901282.

Jiatao, L., LingFei, Z., Shaogang, L., Chenzi, Z., & Shuai, J. (2017). Biomarkers for hepatocellular carcinoma. *Biomarkers in Cancer, 1179299*, X1668464. Available from https://doi.org/10.1177/1179299X16684640.

Jin, Y., Wong, Y. S., Goh, B., Chan, C. Y., Cheow, P. C., Chow, P., ... Lee, C. (2019). Circulating microRNAs as potential diagnostic and prognostic biomarkers in hepatocellular carcinoma. *Scientific reports*, *9*(1), 10464. Available from https://doi.org/10.1038/s41598-019-46872-8.

Karakatsanis, A., Papaconstantinou, I., Gazouli, M., Lyberopoulou, A., Polymeneas, G., & Voros, D. (2013). Expression of microRNAs, miR-21, miR-31, miR-122, miR-145, miR-146a, miR-200c, miR-221, miR-222, and miR-223 in patients with d or intrahepatic cholangiocarcinoma and its prognostic significance. *Molecular Carcinogenesis*, *52*(4), 297–303. Available from https://doi.org/10.1002/mc.21864.

Kashefi-Kheyrabadi, L., Mehrgardi, M. A., Wiechec, E., Turner, A. P. F., & Tiwari, A. (2014). Ultrasensitive detection of human liver hepatocellular carcinoma cells using a label-free aptasensor. *Analytical Chemistry*, *86*(10), 4956–4960. Available from https://doi.org/10.1021/ac500375p.

Kashyap, R., Jain, A., Nalesnik, M., Carr, B., Barnes, J., Vargas, H. E., ... Fung, J. (2001). Clinical significance of elevated α-fetoprotein in adults and children. *Digestive Diseases and Sciences*, *46*(8), 1709–1713. Available from https://doi.org/10.1023/A:1010605621406.

Kim, D., Kim, J., Kwak, C. H., Heo, N. S., Oh, S. Y., Lee, H., & Huh, Y. S. (2016). Rapid and label-free bioanalytical method of alpha fetoprotein detection using LSPR chip. *Journal of Crystal Growth*.

Kim, D. H., Oh, H. G., Park, W. H., Jeon, D. C., Lim, K. M., Kim, H. J., ... Song, K. S. (2018). Detection of alpha-fetoprotein in hepatocellular carcinoma patient plasma with graphene field-effect transistor. *Sensors (Switzerland)*, *18*(11), 4032. Available from https://doi.org/10.3390/s18114032.

Kim, K. W., Min, J. K., Seung, S. L., Hyoung, J. K., Yong, M. S., Kim, P. N., & Lee, M. G. (2008). Sparing of fatty infiltration around focal hepatic lesions in patients with hepatic steatosis: Sonographic appearance with CT and MRI correlation. *American Journal of Roentgenology*, *190*(4), 1018–1027. Available from https://doi.org/10.2214/AJR.07.2863.

Kim, Y. T., Yoon, B. S., Kim, J. W., Kim, S. H., Kwon, J. Y., & Kim, J. H. (2005). Pretreatment levels of serum squamous cell carcinoma antigen and urine polyamines in women with squamous cell carcinoma of the cervix. *International Journal of Gynecology and Obstetrics*, *91*(1), 47–52. Available from https://doi.org/10.1016/j.ijgo.2005.06.010.

Kladney, R. D., Cui, X., Bulla, G. A., Brunt, E. M., & Fimmel, C. J. (2002). Expression of GP73, a resident golgi membrane protein, in viral and nonviral liver disease. *Hepatology (Baltimore, Md.)*, *35*(6), 1431–1440. Available from https://doi.org/10.1053/jhep.2002.32525.

Korn, W. M. (2001). Moving toward an understanding of the metastatic process in hepatocellular carcinoma. *World Journal of Gastroenterology*, *7*(6), 777–778. Available from https://doi.org/10.3748/wjg.v7.i6.777.

Kudo, M. (2016). Defect reperfusion imaging with sonazoid(r): A breakthrough in hepatocellular carcinoma. *Liver Cancer*, *5*(1), 1–7. Available from https://doi.org/10.1159/000367760.

Lai, J. P., Sandhu, D. S., Yu, C., Han, T., Moser, C. D., Jackson, K. K., ... Roberts, L. R. (2008). ). Sulfatase 2 upregulates glypican 3, promotes fibroblast growth factor signaling, and decreases survival in hepatocellular carcinoma. *Hepatology (Baltimore, Md.)*, *47*(4), 1211–1222.

Lee, K.-A., Ahn, J.-Y., Lee, S.-H., Singh Sekhon, S., Kim, D.-G., Min, J., & Kim, Y.-H. (2015). Aptamer-based sandwich assay and its clinical outlooks for detecting lipocalin-2 in hepatocellular carcinoma (HCC). *Scientific Reports*, *5*(1).

Lee, Y. C. A., Cohet, C., Yang, Y. C., Stayner, L., Hashibe, M., & Straif, K. (2009). *Meta*-analysis of epidemiologic studies on cigarette smoking and liver cancer. *International Journal of Epidemiology*, *38*(6), 1497–1511. Available from https://doi.org/10.1093/ije/dyp280.

Li, D., Mallory, T., & Satomura, S. (2001). AFP-L3: A new generation of tumor marker for hepatocellular carcinoma. *Clinica Chimica Acta*, *313* (1–2), 15–19.

Liu, S., Ma, L., Huang, W., Shai, Y., Ji, X., Ding, L., ... Zhao, S. (2006). Decreased expression of the human carbonyl reductase 2 gene HCR2 in hepatocellular carcinoma. *Cellular and Molecular Biology Letters*, *11*(2), 230–241. Available from https://doi.org/10.2478/s11658-006-0022-6.

Liu, Z., Ling, Q., Wang, J., Xie, H., Xu, X., & Zheng, S. (2013). Annexin A2 is not a good biomarker for hepatocellular carcinoma in cirrhosis. *Oncology Letters*, *6*(1), 125–129. Available from https://doi.org/10.3892/ol.2013.1337.

Lok, A. S., Sterling, R. K., Everhart, J. E., Wright, E. C., Hoefs, J. C., Di Bisceglie, A. M., ... Dienstag, J. L. (2010). Des-γ-carboxy prothrombin and α-fetoprotein as biomarkers for the early detection of hepatocellular carcinoma. *Gastroenterology*, *138*(2), 493–502. Available from https://doi.org/10.1053/j.gastro.2009.10.031.

Lunxiu, Q. (2014). Osteopontin is a promoter for hepatocellular carcinoma metastasis: A summary of 10 years of studies. *Frontiers of Medicine*, 24–32. Available from https://doi.org/10.1007/s11684-014-0312-8.

Mai, J., Waisman, D. M., & Sloane, B. F. (2000). Cell surface complex of cathepsin B/annexin II tetramer in malignant progression. *Biochimica et Biophysica Acta - Protein Structure and Molecular Enzymology*, *1477*(1–2), 215–230. Available from https://doi.org/10.1016/S0167-4838(99)00274-5.

Mao, Y., Yang, H., Xu, H., Lu, X., Sang, X., Du, S., ... Zhang, H. (2010). Golgi protein 73 (GOLPH2) is a valuable serum marker for hepatocellular carcinoma. *Gut*, *59*(12), 1687–1693. Available from https://doi.org/10.1136/gut.2010.214916.

Marrero, J. A., Feng, Z., Wang, Y., Nguyen, M. H., Befeler, A. S., Roberts, L. R., ... Schwartz, M. (2009). α-Fetoprotein, Des-γ carboxyprothrombin, and lectin-bound α-fetoprotein in early hepatocellular carcinoma. *Gastroenterology*, *137*(1), 110–118. Available from https://doi.org/10.1053/j.gastro.2009.04.005.

Marrero, J. A., Romano, P. R., Nikolaeva, O., Steel, L., Mehta, A., Fimmel, C. J., ... Block, T. M. (2005). GP73, a resident Golgi glycoprotein, is a novel serum marker for hepatocellular carcinoma. *Journal of Hepatology*, *43*(6), 1007–1012. Available from https://doi.org/10.1016/j.jhep.2005.05.028.

Mashaly, A. H., Anwar, R., Ebrahim, M. A., Eissa, L. A., & El Shishtawy, M. M. (2018). Diagnostic and prognostic value of Talin-1 and midkine as tumor markers in hepatocellular carcinoma in Egyptian patients. *Asian Pacific Journal of Cancer Prevention*, *19*(6), 1503–1508. Available from https://doi.org/10.22034/APJCP.2018.19.6.1503.

McAllister, S. S., Gifford, A. M., Greiner, A. L., Kelleher, S. P., Saelzler, M. P., Ince, T. A., ... Weinberg, R. A. (2008). Systemic endocrine instigation of indolent tumor growth requires osteopontin. *Cell, 133*(6), 994–1005. Available from https://doi.org/10.1016/j.cell.2008.04.045.

McGlynn, K. A., Petrick, J. L., & London, W. T. (2015). Global epidemiology of hepatocellular carcinoma: An emphasis on demographic and regional variability. *Clinics in Liver Disease, 19*(2), 223–238. Available from https://doi.org/10.1016/j.cld.2015.01.001.

Mohammadinejad, A., Kazemi Oskuee, R., Eivazzadeh-Keihan, R., Rezayi, M., Baradaran, B., Maleki, A., ... de la Guardia, M. (2020). Development of biosensors for detection of alpha-fetoprotein: As a major biomarker for hepatocellular carcinoma. *TrAC - Trends in Analytical Chemistry*, 130. Available from https://doi.org/10.1016/j.trac.2020.115961.

Naraki, T., Kohno, N., Saito, H., Fujimoto, Y., Ohhira, M., Morita, T., & Kohgo, Y. (2002). $\gamma$-Carboxyglutamic acid content of hepatocellular carcinoma-associated des-$\gamma$-carboxy prothrombin. *Biochimica et Biophysica Acta - Molecular Basis of Disease, 1586*(3), 287–298. Available from https://doi.org/10.1016/S0925-4439(01)00107-7.

Navin, P. J., & Venkatesh, S. K. (2019). Hepatocellular carcinoma: State of the art imaging and recent advances. *Journal of Clinical and Translational Hepatology, 7*(1), 72–85. Available from https://doi.org/10.14218/JCTH.2018.00032.

Nio, K., Yamashita, T., Okada, H., Kondo, M., Hayashi, T., Hara, Y., ... Kaneko, S. (2015). Defeating EpCAM + liver cancer stem cells by targeting chromatin remodeling enzyme CHD4 in human hepatocellular carcinoma. *Journal of Hepatology, 63*(5), 1164–1172. Available from https://doi.org/10.1016/j.jhep.2015.06.009.

Ogi, H., Omori, T., Hatanaka, K., Hirao, M., & Nishiyama, M. (2008). Detection of glypican-3 proteins for hepatocellular carcinoma marker using wireless-electrodeless quartz-crystal microbalance. *Japanese Journal of Applied Physics, 47*(5), 4021–4023. Available from https://doi.org/10.1143/JJAP.47.4021.

Oikawa, T., Kamiya, A., Zeniya, M., Chikada, H., Hyuck, A. D., Yamazaki, Y., ... Nakauchi, H. (2013). Sal-like protein 4 (SALL4), a stem cell biomarker in liver cancers. *Hepatology (Baltimore, MD), 57*(4), 1469–1483. Available from https://doi.org/10.1002/hep.26159.

Osada, S., Kanematsu, M., Imai, H., & Goshima, S. (2008). Clinical significance of serum HGF and c-Met expression in tumor tissue for evaluation of properties and treatment of hepatocellular carcinoma. *Hepato-gastroenterology, 55*(82–83), 544–549.

Patel, S., Nanda, R., Sahoo, S., & Mohapatra, E. (2016). Biosensors in health care: The milestones achieved in their development towards lab-on-chip-analysis. *Biochemistry Research International, 2016*. Available from https://doi.org/10.1155/2016/3130469.

Pontisso, P., Calabrese, F., Benvegnù, L., Lise, M., Belluco, C., Ruvoletto, M. G., ... Fassina, G. (2004). Overexpression of squamous cell carcinoma antigen variants in hepatocellular carcinoma. *British Journal of Cancer, 90*(4), 833–837. Available from https://doi.org/10.1038/sj.bjc.6601543.

Pontisso, P., Quarta, S., Caberlotto, C., Beneduce, L., Marino, M., Bernardinello, E., ... Chemello, L. (2006). Progressive increase of SCCA-IgM immune complexes in cirrhotic patients is associated with development of hepatocellular carcinoma. *International Journal of Cancer, 119*(4), 735–740. Available from https://doi.org/10.1002/ijc.21908.

Qiao, S. S., Cui, Z. Q. Q., Gong, L., Han, H., Chen, P. G., Guo, L. M., ... Leng, X. S. (2011). Simultaneous measurements of serum AFP, GPC-3 and HCCR for diagnosing hepatocellular carcinoma. *Hepato-Gastroenterology, 58*(110–111), 1718–1724. Available from https://doi.org/10.5754/hge11124.

Qin, Q., Sun, Y., Fei, M., Zhang, J., Jia, Y., Gu, M., ... Deng, A. (2012). Expression of putative stem marker nestin and CD133 in advanced serous ovarian cancer. *Neoplasma, 59*(3), 310–315. Available from https://doi.org/10.4149/neo_2012_040.

Rasool, M., Rashid, S., Arooj, M., Ansari, S., Khan, K., Malik, A., ... Razzaq, Z. (2014). New possibilities in hepatocellular carcinoma treatment. *Anticancer Research, 34*(4), 1563–1571.

Sanyal, A. J., Yoon, S. K., & Lencioni, R. (2010). The etiology of hepatocellular carcinoma and consequences for treatment. *The Oncologist, 15*, 14–22. Available from https://doi.org/10.1634/theoncologist.2010-S4-14.

Shaheen, K. Y., Abdel-Mageed, A. I., Safwat, E., & AlBreedy, A. M. (2015). The value of serum midkine level in diagnosis of hepatocellular carcinoma. *International Journal of Hepatology, 2015*, 146389.

Sharp, J. A., Sung, V., Slavin, J., Thompson, E. W., & Henderson, M. A. (1999). Tumor cells are the source of osteopontin and bone sialoprotein expression in human breast cancer. *Laboratory Investigation; a Journal of Technical Methods and Pathology, 79*(7), 869–877.

Sheikh, M. Y., Choi, J., Qadri, I., Friedman, J. E., & Sanyal, A. J. (2008). Hepatitis C virus infection: Molecular pathways to metabolic syndrome. *Hepatology (Baltimore, MD), 47*(6), 2127–2133. Available from https://doi.org/10.1002/hep.22269.

Sheta, S. M., El-Sheikh, S. M., Abd-Elzaher, M. M., Salem, S. R., Moussa, H. A., Mohamed, R. M., & Mkhalid, I. A. (2019). A novel biosensor for early diagnosis of liver cancer cases using smart nano-magnetic metal–organic framework. *Applied Organometallic Chemistry, 33*(12), e5249.

Shmelkov, S. V., Butler, J. M., Hooper, A. T., Hormigo, A., Kushner, J., Milde, T., St, ... Rafii, S. (2008). CD133 expression is not restricted to stem cells, and both CD133 + and CD133- metastatic colon cancer cells initiate tumors. *Journal of Clinical Investigation, 118*(6), 2111–2120. Available from https://doi.org/10.1172/JCI34401.

Stenman, J., Hedström, J., Grénman, R., Leivo, I., Finne, P., Palotie, A., & Orpana, A. (2001). Relative levels of SCCA2 and SCCA1 mRNA in primary tumors predicts recurrent disease in squamous cell cancer of the head and neck. *International Journal of Cancer, 95*(1), 39–43, https://doi.org/10.1002/1097-0215(20010120)95:1 < 39::AID-IJC1007 > 3.0.CO;2-N.

Sukowati, C. H., Anfuso, B., Torre, G., Francalanci, P., Crocè, L. S., & Tiribelli, C. (2013). The expression of CD90/Thy-1 in hepatocellular carcinoma: an in vivo and in vitro study. *PLoS One, 8*(10), e76830.

Szabó, E., Páska, C., Kaposi Novák, P., Schaff, Z., & Kiss, A. (2004). Similarities and differences in Hepatitis B and C Virus induced hepatocarcinogenesis. *Pathology and Oncology Research, 10*(1), 5–11. Available from https://doi.org/10.1007/BF02893401.

Talerman, A., Haije, W. G., & Baggerman, L. (1980). Serum alphafetoprotein (AFP) in patients with germ cell tumors of the gonads and extragonadal sites: Correlation between endodermal sinus (yolk sac) tumor and raised serum AFP. *Cancer, 46*(2), 380–385, https://doi.org/10.1002/1097-0142(19800715)46:2 < 380::AID-CNCR2820460228 > 3.0.CO;2-U.

Tanaka, H. (2020). Current role of ultrasound in the diagnosis of hepatocellular carcinoma. *Journal of Medical Ultrasonics*, *47*(2), 239–255. Available from https://doi.org/10.1007/s10396-020-01012-y.

Tangkijvanich, P., Tosukhowong, P., Bunyongyod, P., Lertmaharit, S., Hanvivatvong, O., Kullavanijaya, P., & Poovorawan, Y. (1999). Alpha-L-fucosidase as a serum marker of hepatocellular carcinoma in Thailand. *Southeast Asian Journal of Tropical Medicine and Public Health*, *30*(1), 110–114.

Tara, B., & Sitki, C. M. (2012). Biomarkers for hepatocellular carcinoma. *International Journal of Hepatology*, 1–7. Available from https://doi.org/10.1155/2012/859076.

Tian, L., Wang, Y., Xu, D., Gui, J., Jia, X., Tong, H., ... Tian, Y. (2011). Serological AFP/Golgi protein 73 could be a new diagnostic parameter of hepatic diseases. *International Journal of Cancer*, *129*(8), 1923–1931. Available from https://doi.org/10.1002/ijc.25838.

Urquidi, V., Sloan, D., Kawai, K., Agarwal, D., Woodman, A. C., Tarin, D., & Goodison, S. (2002). Contrasting expression of thrombospondin-1 and osteopontin correlates with absence or presence of metastatic phenotype in an isogenic model of spontaneous human breast cancer metastasis. *Clinical cancer research: an official journal of the American Association for Cancer Research*, *8*(1), 61–74.

Van De Stolpe, A., & Van Der Saag, P. T. (1996). Intercellular adhesion molecule-1. *Journal of Molecular Medicine*, *74*(1), 13–33. Available from https://doi.org/10.1007/BF00202069.

Vongsuvanh, R., Van Poorten, D. D., Iseli, T., Strasser, S. I., McCaughan, G. W., & George, J. (2016). Midkine increases diagnostic yield in AFP negative and NASH-related hepatocellular carcinoma. *PLoS One*, *11*(5). Available from https://doi.org/10.1371/journal.pone.0155800.

Wade, R., & Vande Pol, S. (2006). Minimal features of paxillin that are required for the tyrosine phosphorylation of focal adhesion kinase. *Biochemical Journal*, *393*(2), 565–573. Available from https://doi.org/10.1042/BJ20051241.

Wang, X., Hassan, W., Jabeen, Q., Khan, G. J., & Iqbal, F. (2018). Interdependent and independent multidimensional role of tumor microenvironment on hepatocellular carcinoma. *Cytokine*, *103*, 150–159. Available from https://doi.org/10.1016/j.cyto.2017.09.026.

Weber, G. F. (2011). The cancer biomarker osteopontin: Combination with other markers. *Cancer Genomics and Proteomics*, *8*(6), 263–288. Available from http://cgp.iiarjournals.org.

Willatt, J. M., Hussain, H. K., Adusumilli, S., & Marrero, J. A. (2008). MR imaging of hepatocellular carcinoma in the cirrhotic liver: Challenges and controversies. *Radiology*, *247*(2), 311–330. Available from https://doi.org/10.1148/radiol.2472061331.

Wu, Y., Xu, W., Wang, Y., Yuan, Y., & Yuan, R. (2013). Silver-graphene oxide nanocomposites as redox probes for electrochemical determination of α-1-fetoprotein. *Electrochimica Acta*, *88*, 135–140. Available from https://doi.org/10.1016/j.electacta.2012.10.081.

Xiang, Y. Y., Ladeda, V., & Filmus, J. (2001). Glypican-3 expression is silenced in human breast cancer. *Oncogene*, *20*(50), 7408–7412. Available from https://doi.org/10.1038/sj.onc.1204925.

Yamauchi, N., Watanabe, A., Hishinuma, M., Ohashi, K. I., Midorikawa, Y., Morishita, Y., ... Fukayama, M. (2005). The glypican 3 oncofetal protein is a promising diagnostic marker for hepatocellular carcinoma. *Modern Pathology*, *18*(12), 1591–1598. Available from https://doi.org/10.1038/modpathol.3800436.

Yang, H., Fang, F., Chang, R., & Yang, L. (2013). MicroRNA-140–5p suppresses tumor growth and metastasis by targeting transforming growth factor β receptor 1 and fibroblast growth factor 9 in hepatocellular carcinoma. *Hepatology (Baltimore, Md.)*, *58*(1), 205–217. Available from https://doi.org/10.1002/hep.26315.

Yang, W., Yan, H. X., Chen, L., Liu, Q., He, Y. Q., Yu, L. X., ... Wang, H. Y. (2008). Wnt/β-catenin signaling contributes to activation of normal and tumorigenic liver progenitor cells. *Cancer Research*, *68*(11), 4287–4295. Available from https://doi.org/10.1158/0008-5472.CAN-07-6691.

Yin, S., Li, J., Hu, C., Chen, X., Yao, M., Yan, M., ... Gu, J. (2007). CD133 positive hepatocellular carcinoma cells possess high capacity for tumorigenicity. *International Journal of Cancer*, *120*(7), 1444–1450. Available from https://doi.org/10.1002/ijc.22476.

Yu, N. C., Chaudhari, V., Raman, S. S., Lassman, C., Tong, M. J., Busuttil, R. W., & Lu, D. S. K. (2011). CT and MRI improve detection of hepatocellular carcinoma, compared with ultrasound alone, in patients with cirrhosis. *Clinical Gastroenterology and Hepatology*, *9*(2), 161–167. Available from https://doi.org/10.1016/j.cgh.2010.09.017.

Yuan, Y., Li, S., Xue, Y., Liang, J., Cui, L., Li, Q., ... Zhao, Y. (2017). A Fe3O4@Au-basedpseudo-homogeneous electrochemical immunosensor for AFP measurement using AFP antibody-GNPs-HRP as detection probe. *Analytical Biochemistry*, *534*, 56–63. Available from https://doi.org/10.1016/j.ab.2017.07.015.

Zhang, H., Yao, M., Wu, W., Qiu, L., Sai, W., Yang, J., ... Yao, D. (2015). Up-regulation of annexin A2 expression predicates advanced clinicopathological features and poor prognosis in hepatocellular carcinoma. *Tumor Biology*, *36*(12), 9373–9383. Available from https://doi.org/10.1007/s13277-015-3678-6.

Zhang, J., Shao, C., Zhou, Q., Zhu, Y., Zhu, J., & Tu, C. (2015). Diagnostic accuracy of serum squamous cell carcinoma antigen and squamous cell carcinoma antigen-immunoglobulin M for hepatocellular carcinoma: A *meta*-analysis. *Molecular and Clinical Oncology*, *3*(5), 1165–1171.

Zhang, S. Y., Lin, B. D., & Li, B. R. (2015). Evaluation of the diagnostic value of alpha-l-fucosidase, alpha-fetoprotein and thymidine kinase 1 with ROC and logistic regression for hepatocellular carcinoma. *FEBS Open Bio*, *5*, 240–244. Available from https://doi.org/10.1016/j.fob.2015.03.010.

Zhao, W., Wang, L., Han, H., Jin, K., Lin, N., Guo, T., ... Zhang, Z. (2013). 1B50–1, a mAb raised against recurrent tumor cells, targets liver tumor-initiating cells by binding to the calcium channel α2δ1 subunit. *Cancer Cell*, *23*(4), 541–556. Available from https://doi.org/10.1016/j.ccr.2013.02.025.

Zhou, L., Liu, J., & Luo, F. (2006). Serum tumor markers for detection of hepatocellular carcinoma. *World Journal of Gastroenterology*, *12*(8), 1175–1181. Available from https://doi.org/10.3748/wjg.v12.i8.1175.

Zhu, Y., Denhardt, D. T., Cao, H., Sutphin, P. D., Koong, A. C., Giaccia, A. J., & Le, Q. T. (2005). Hypoxia upregulates osteopontin expression in NIH-3T3 cells via a Ras-activated enhancer. *Oncogene*, *24*(43), 6555–6563. Available from https://doi.org/10.1038/sj.onc.1208800.

Chapter 7

# Scope and applications of biosensors in early detection of oropharyngeal cancers

Shubhangi Mhaske[1] and Monal Yuwanati[2]

[1]*Oral and Maxillofacial Pathology and Microbiology, People's College of Dental Sciences & Research Centre, People's University, Bhopal, India*

[2]*Department Of Oral Pathology and Microbiology, Saveetha Dental College and Hospitals, Saveetha Institute of Medical and Technical Sciences, Saveetha University, Chennai, India*

## 7.1 Introduction

Oropharyngeal cancer (OPC) is the seventh malignancy that affects humans and is responsible annually for more than 400,000 deaths worldwide (Bosetti et al., 2020). One of the factors leading to poor prognosis and death is the delayed detection of OPC, although other determinants play important roles in its diagnosis (Van Harten et al., 2015). General practitioners and dentists are dependant on adjunct diagnostic tools from early detection of cancer apart from the routine clinical examination. However, it is extremely challenging in early asymptomatic stages where sensitive and specific tools are must for accurate diagnosis.

Numerous techniques have been established over the decades to help diagnose OPC, however the standard prognosis to be atleast for a 5-year survival rate has been deteriorating even with targeted molecular drug therapy. However, there is a steep rise in futuristic scope for precision based early diagnosis. Traditional diagnostic approaches include the confirmation of tissue biopsy, which is now changing to minimal or noninvasive, highly meticulous and sensitive methods such as the identification of biological biomarkers in body tissue or fluid suggesting the existence of disease or diabetes.

Although it is possible to identify malignant changes in lesions on histopathology, it is mostly detected at an advanced stage. Hence, prompt and rapid diagnosis of oral cancer remains a challenge in clinical practice. Further, tissue biopsy is an invasive technique which is mostly unacceptable to patients especially in the absence of any visible cancerous change in tissue. Early diagnosis and institution of treatment can decrease the risk of oral cancer. Hence, identification of novel noninvasive diagnostic methods for early tumor detection in oral mucosa is required.

Noninvasive cancer diagnosis with cancer biomarkers is gaining popularity with minimal risk for patients due to its predictive nature. The National Cancer Institute has described a biomarker as a biological molecule found in blood, other body fluids, or tissues that is a sign of the normal or abnormal process of a condition or disease such as cancer.

Biomolecules, i.e., antigen/antibody, enzyme, nuclear acid, hormone receptor, and live cells and tissue, specifically recognize other biological entities via catalysis and affinity binding. Biomarkers may be proteins, nucleic acids, or peptides though most of the biomarkers are limited to proteins only. Biosensors are bioanalytical devices utilized in the detection of biological (cancer) markers in tissue or body fluids which are usually proteins, nucleic acids, or peptides through light or electrochemical interaction. There are different biosensors utilized for detection of cancer biomarkers (Table 7.1). These are basically based on either analysis of enzymes or signal transduction methods. A transducer is a device that converts the molecular recognition signal to an electrical signal. The transducer may be electrochemical, optical, calorimetric, or based on mass changes.

The presence of biomarkers in blood or any other body fluids confirms the presence of cancer cells in the body. There are different biomarkers for different types of cancers. The biomarkers interleukin-6, interleukin-8, p53, TNF-alpha (TNF-$\alpha$), exosomes, epidermal growth factor response (EGFR) and cytokeratin 19 fragments (CYFRA), etc., are used for detection of OPCs.

**Biosensor Based Advanced Cancer Diagnostics. DOI: https://doi.org/10.1016/B978-0-12-823424-2.00003-X**

**TABLE 7.1 Biosensors in Oropharyngeal Cancers.**

| Biomarker | Sample | Electrochemical method | Detection limits | The levels of the biomarker in normal case and cancer patient | References |
|---|---|---|---|---|---|
| Amylase | Samples spiked in potassium ferrocyanide | Cyclic voltammetry | 1.57 pg $mL^{-1}$ | | Malhotra, Patel, Vaqué, Gutkind, and Rusling (2010) |
| IL-8 protein, IL-1β protein and IL-8 mRNA | Samples spiked in buffer | Cyclic square-wave form | Protein:100–200fg $mL^{-1}$ mRNA IL-8:10 a.m. | IL-8 protein, OSCC patient: 720 pg $mL^{-1}$; Normal: 250 pg $mL^{-1}$ IL-8 mRNA, patient: 16 fM; Normal:2 fM | Malhotra et al. (2016) |
| Interleukin-6 (IL-6) | HNSCC cell lines | Amperometry | 2.5 × 10 14 M | IL-6 protein, HNSCC patient: more than 20 pg $mL^{-1}$; Normal: less than 6 pg $mL^{-1}$ | Morteza, Shiva, Yasaman-Sadat, and Reza (2017) |
| microRNA | Artificial saliva | Cyclic voltammetry chronoamperometry | 2.2 × 10 19 M | | Riedel et al. (2005) |
| IL-6 protein, IL-8 protein | Serum | Amperometry | IL-6:5 fg $mL^{-1}$, IL-8:7 fg $mL^{-1}$ | IL-6 protein, HNSCC patient: more than 20 pg$mL^{-1}$; Normal: less than 6 pg $mL^{-1}$ IL-8 protein, patient: 720 pg $mL^{1}$; Normal: 250 pg $mL^{-1}$ | Rivlin, Brosh, Oren, and Rotter (2011) |
| Oral Cancer Overexpressed 1 | Human saliva | Differential pulse voltammetry | 0.35p.m. | | Li, Fukumoto, and Liu (2013) |
| IL-8 protein, TNF-α | Artificial saliva | I-V Curve | 100 fg $mL^{-1}$ | IL-8 protein, patient: 720 pg $mL^{-1}$; Normal: 250 pg $mL^{-1}$ | Singh, Barpande, Bhavthankar, Mandale, and Bhagwat (2017) |
| CYFRA-21-1 | Samples spiked in PBS buffer | Cyclic voltammetry Electrochemical impedance spectroscopy | 0.21 ng $mL^{-1}$ (calculated) | CYFRA-21–1 protein, normal: 3.8 ng $mL^{-1}$; patient:17.46 ± 1.46 ng $mL^{-1}$ | Tan et al. (2008) |
| CYFRA-21-1 | Samples spiked in PBS bu_er | Cyclic voltammetry Differential pulse voltammetry | 0.122 ng $mL^{-1}$ (calculated) | CYFRA-21-1 protein, normal: 3.8 ng $mL^{-1}$; Patient:17.46 ± 1.46 ng $mL^{-1}$ | St. John et al. (2004) |
| CYFRA-21-1 | Artificial Saliva | Differential pulse voltammetry | 0.001 ng $mL^{-1}$ | CYFRA-21-1 protein, normal: 3.8 ng $mL^{-1}$; Patient:17.46 ± 1.46 ng $mL^{-1}$ | (Tiwari, Gupta, Bagbi, Sarkar, & Solanki, 2017) |
| CD59 | Human saliva | Cyclic voltammetry Electrochemical impedance spectroscopy | Treated Saliva: 0.84 ± 0.04 fg $mL^{-1}$ Raw saliva: 1.46 ± 0.05 fg $mL^{-1}$ | | (Choudhary et al., 2016) |

*(Continued)*

**TABLE 7.1 (Continued)**

| Biomarker | Sample | Electrochemical method | Detection limits | The levels of the biomarker in normal case and cancer patient | References |
|---|---|---|---|---|---|
| IL-8 protein | Human saliva | Cyclic voltammetry Differential pulse voltammetry | 72.73 ± 0.18 $pg\ mL^{-1}$ (calculated) | IL-8 protein, patient: 720 $pg\ mL^{-1}$; normal: 250 $pg\ mL^{-1}$ | (Verma et al., 2017) |
| Oral Cancer Overexpressed 1 | Artificial saliva | Alternating current voltammetric Electrochemical impedance spectroscopy | 12.8 fM | | Liao et al. (2018) |
| CIP2A | Human saliva | Cyclic voltammetry Electrochemical impedance spectroscopy | 0.24 $pg\ mL^{-1}$ | | (Ding et al., 2018) |
| CYFRA-21-1 | Human saliva | Differential pulse voltammetry Electrochemical impedance spectroscopy | 0.625 $pg\ mL^{-1}$ | CYFRA-21-1 protein, normal: 3.8 $ng\ mL^{-1}$; Patients: 17.46 ± 1.46 $ng\ mL^{-1}$ | (Pachauri, Dave, Dinda, & Solanki, 2018) |
| IL-8 protein | Human serum and saliva | Cyclic voltammetry Electrochemical impedance spectroscopy Single Frequency Impedance | 3.3 $fg\ mL^{-1}$ | IL-8 protein, patient: 720 $pg\ mL^{-1}$; Normal: 250 $pg\ mL^{-1}$ | (Aydın, Aydın, & Sezgintürk, 2018) |
| IL-8 protein | Human saliva | Cyclic voltammetry Differential pulse voltammetry | 51.53 ± 0.43 $pg\ mL^{-1}$ (calculated) | IL-8 protein, patient: 720 $pg\ mL^{-1}$; Normal: 250 $pg\ mL^{-1}$ | (Verma & Singh, 2019) |

## 7.1.1 Biosensors in detection of oropharyngeal cancer

Tissue biopsy and histopathology examination is a traditional and gold standard method for establishment of OPC diagnosis. Over the decades, clinicians and oncologists heavily rely on these techniques for planning of treatment and follow-up. However, it is impractical to perform them periodically as it is invasive and may not only cause discomfort to patients but may aid in spreading cancer by metastasis or local invasion and field cancerization. Biosensors have recently become popular among clinicians for identification of carcinogenesis in OPC due its sensitivity and specificity. These biosensors are designed to detect the biomarkers, usually chemical or protein molecules, secreted or formed during carcinogenesis. Detecting them may give hints at the presence of cancer. Several cancer biomarkers related to oral and pharyngeal cancer have been reported in literature, and current researchers are busy in identifying the more sensitive and specific biomarker for early detection of OPC (Table 7.1). There are several biosensors mentioned in literature promising advantages of one biosensor over another in early detection of cancer biomarkers of OPC. These include but are not limited to vizilight, Raman spectroscopy, elastic scattering spectroscopy, diffuse reflectance spectroscopy, narrow-band imaging, and confocal reflectance microscopy. However, limited cancer biomarkers are detectable on biosensor.

### *7.1.1.1 Role of biosensors in oropharyngeal cancer*

Most cancers in the body are result of carcinogenic changes in tissue. It involves multiple step modulating the biophysiology of cells in acquiring six hallmarks of cancer (Fig. 7.1; Lin et al., 2020). During these early events of processes, the body produces several changes that biosensors can detect. Biophysiological changes involve the various biomolecules, enzymes, genes, DNA, RNA machinery, etc.; identifying changes in these elements can give early hints to

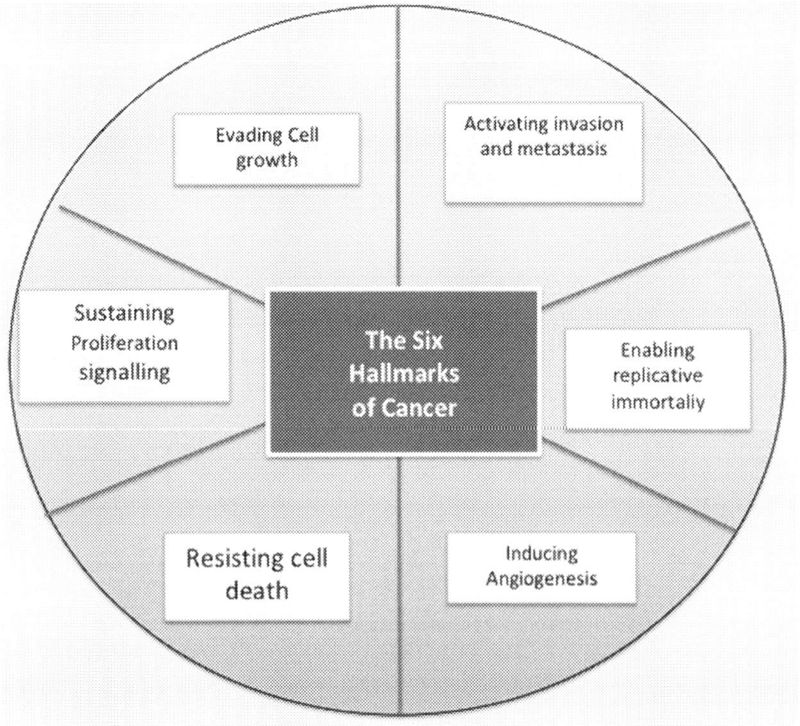


**FIGURE 7.1** Electrochemical immunosensors based on the detection of biomarkers IL-6 and IL-8 for targeting oral cancer. *From Lin, n T., Darvishi, S., Preet, A., Huang, T., Lin, S., Girault, H., & Wang, L. (2020). A review: electrochemical biosensors for oral cancer.* Chemosensors, *8(3). Electrochemical immunosensors.*

carcinogenic changes in the person. There are various technologies such as MRIs and PET scans that can assist oncologists during diagnosis and treatment. However, these are expensive and complex equipments with certain limitations. There is a need for tests as simple as the Papanicolaou test (PAP). The PAP smear is a method of cervical screening used to detect potentially precancerous and cancerous processes in the cervix or colon. This simple low-cost, noninvasive test has saved thousands of women's lives from cervical cancer 8. However, there is absence of similar types of tests for other body cancers. Over decades with advances in technology, the medical field saw a surge in tools for early detection of cancer. Optical property-based biosensors are gaining traction due to its noninvasive, sensitivity, specificity, low costs and fast turn-around time features.

### *7.1.1.2 Biosensor types and techniques*

#### 7.1.1.2.1 Optical biosensors

Optical biosensors measure the variations in wavelengths of light. These transducers can be colorimetric, fluorescent, interferometric, or luminescent. Changes of the wavelengths in response to their cognition of the analyte are converted by optical transducers and electrical readings/provide digital (Tan et al., 2008). A new type of sensor with the use of an optical transducer is a biosensor which has photonic crystals. Such sensors are designed for capturing the very tiny volumes or light areas, which allows measurements at a high susceptibility, and to display results the light is transmitted to a higher electromagnetic field. This technique detects where and when the molecules or cells dissociate or bind to the surface made of crystal through the measurement of the light reflected in the crystal. Some of the examples of Optical biosensors are Fluorescence spectroscopy, surface plasmon resonance, interferometry and spectroscopy, photoionic crystals, optical ring resonators etc. (Liao et al., 2018). However many emerging optical sensing technologies are currently under res0earch.

#### 7.1.1.2.2 Electrochemical biosensors

These biosensors are one of oldest biochemical analyzers utilized for diagnosis and evaluation of treatment outcome. However, they are most widely chosen for their low cost, portability, compact scale, and ease of use. Latest electrochemical biosensors are based on a screen-printed amperometric disposable electrode. A range of electrochemical transducers are used which include potentiometric, impedimetric/conductivity and amperometric devices. Amperometric and potentiometers biosensors are the two most common electrochemical biosensors. Potentiometric biosensors utilize ion selective electrodes that detect an electrical response when there is a molecular recognition of a specific element. Such biosensors have great potential for use in the detection of cancer. SA test (Fig. 7.1).

#### 7.1.1.2.3 Mass sensitive biosensors

Acoustic and piezoelectric biosensors are two types of biosensors that rely on mass changes. Piezoelectric sensors are based on changes in the mass of a quartz crystal when the potential energy is applied to them. The frequencies generated by mass changes can be converted to signals. Micro cantilever and immunosensors biosensors which are based upon piezoelectric technologies are useful for detecting tumor biomarkers. Overexpression of human p53 gene's point mutations in several types of tumors has been reported with piezoelectric biosensors coupled with PCR amplification

#### 7.1.1.2.4 Nanobiochip cellular analysis (bio)sensor

These point-of-care (POC) biosensors have evolved in the past few years due to innovation and advancement in nanotechnology. A novel bio-nanochip (BNC) sensor was developed to estimate the surface biomarker from rapid oral-cytology assay simultaneously allowing the analysis of cytomorphology using intensity within 45 min. The University of California, Los Angeles, Collaborative Oral Fluid Diagnostic Research Laboratory, developed a POC device used to detect oral cancer in saliva. Oral fluid nanosensor test (OFNASET) enables simultaneous and rapid detection of multiple salivary proteins and nucleic acid targets. The OFNASET technology platform combines leading technologies such as self-assembled monolayers, microfluidics, and cyclic enzymatic amplification.

#### 7.1.1.2.5 Calorimetric biosensors

Calorimetric biosensors measure exothermic reactions and are not used widely in the detection of cancer. Temperature changes due to heat produced during many enzymatic reactions are used in measurement of the desired molecules. By measuring enthalpy the reactions are monitored, which indirectly gives the information required to measure the amount of the molecules. They are less widely used in the detection of cancer, even though there have been few useful characteristics of these biosensors for diagnosis of cancer examples.

### *7.1.1.3 Most common biomarkers associated with OPC*

#### 7.1.1.3.1 Interleukins

Interleukins (ILs) are groups of cytokines which play a role in various disease processes including oral cancer. IL-8 and IL-6 are inflammatory cytokines involved in immune defense. Increases in IL-8 and IL-6 levels in tissues and body fluids especially in serum and saliva were reported in OPC patients as compared to levels in premalignant lesions/conditions.

IL-6 and IL-8 are cellular proteins that are particularly relevant to oral squamous cell carcinoma (OSCC) (Babiuch et al., n.d.). The uncontrolled expression of these cytokines could be a sign of the development of tumor growth and metastasis (Almeida et al., 2019). Usually, the normal expression levels of these cytokines in keratinocytes are low. IL-6 is an interleukin in that behaves as both a pro-inflammatory cytokine and an antiinflammatory cytokine, that is involved in acute-phase reaction, growth regulation and differentiation of cells (Osman, Costea, & Johannessen, 2012).

Latest studies have demonstrated the diagnostic value of IL-6 and IL-8 for oral cancer. The detection of IL-8 allows an accurate diagnosis for oral cancer detection (Wei et al., 2010). Although they may act on the antitumor immune response pathways, IL-6 and IL-8 tend to present a response associated with the intensification of carcinogenesis and poor prognosis in patients with OSCC (Almeida et al., 2019). IL-6 and IL-8 detection in tissue and body fluids can be helpful in early diagnosis and follow-up in OSCC (Arshad et al., 2017). These can be detected in saliva using various techniques (Osman et al., 2012). However, sensitivity to detection is concerned in those techniques. Optical protein sensors can detect IL at very low concentration (Tan et al., 2008). In another study, human saliva is used for the detection of protein IL-8 and IL-8 mRNA by the development of electrochemical magnetobiosensor (St. John et al., 2004; Tonello et al., 2019; Torrente-Rodríguez, Campuzano, Montiel, Gamella, & Pingarrón, 2016; Wei et al., 2010)

Oral cancer patients have 20–1000 pg $mL^{-1}$ of IL-6 in the serum while it is <6 pg $mL^{-1}$ in healthy people. (Riedel et al., 2005). The serum levels in healthy individuals are lower than 13 pg $mL^{-1}$ compared with 20–1000 pg $mL^{-1}$ or more in patients (Gokhale et al., 2005).

Ultrasensitive electrochemical biomarker IL-6 with carbon nanotube forest electrodes and multilabel amplification immunosensors were used which accurately measured secreted IL-6 in a wide range of oral cancer cells. This was correlated in a study with standard enzyme-linked immunosorbent assays (ELISAs), suggesting that immunosensors combined with multilabel detection have excellent promise for detecting IL-6 in research and clinical applications (Malhotra et al., 2010).

#### 7.1.1.3.2 p53

Alteration of the p53 tumor suppressor gene (TP53) is probably one of the most common genetic features of oral cancer (Li et al., 2013).

p53 gene, the tumor suppressor gene is widely mutated in oral cancer and has been correlated with poor prognosis.

Level of mutation in p53 gene could predict the locoregional recurrence, lymph node metastasis, and distant spread in oral cancers. In addition, p53 expression and levels can predict the response to the oncotherapy. p53 antibodies can be detected in saliva and tissue using the biosensors (Li et al., 2013).

High expression of the P53 gene has been reported in oral cancer precursors which have a high malignant transformation rate (Jaber & Elameen, 2020).

Serum and saliva in oral precancer and cancer patients have shown the presence of P53 protein in early stage (Rivlin et al., 2011).

Usually p53 is investigated in serum or saliva by various conventional procedures and techniques including traditional methods like ELISA, electrophoretic immunoassay, mass spectrometry based proteomics and radioimmunoassay (RIA). However, sensitivity and specificity is widely varied due impact of various factors (Tavassoli, Brunel, Maher, Johnson, & Soussi, 1998).

A variety of electrochemical, optical, and mass-sensitive biosensors is being used to detect p53 in body tissues and fluids. Tonello et al. developed an electrochemical biosensor that can detect the p53 protein mutation along with p53 conformation using nanostructured surfaces (Tonello et al., 2019).

A point mutation in the p53 gene can be identified using a DNA biotinylated capture probe in electrochemical genosensor (Ezat et al., 2015). Another potentiometric biosensor, discovered by Al-Ogaidi et al., can be used in measurement of p53 in blood with $3.37 \times 10-10\,\mu g/mL$ level of detection (LOD) (Al-Ogaidi et al., 2017). Similarly, electrochemical impedance spectra (EIS), a label free immunosensor, can measure p53 with a LOD of 3 fg/mL in blood (Burcu, Muhammet, & Kemal, 2018).

Various optical biosensors have been developed recently due to their advantages. A highly sensitive detection system for tumor suppressor protein p53 based on fiberlight-coupled optofluidic waveguide (FLOW) immunosensor was developed by Liang et al. it can detect the p53 in range of 10fg/mL to 10 ng/mL (Lili, Long, Yang, Li-Peng, & Bai-Ou, 2018). A fluorescent silver nanocluster-based biosensor can detect single base mismatch in p53 gene as low as 1.3 nM levels (Morteza et al., 2017). Considering the p53 gene role in oral carcinogenesis, it is important to develop point-of-care, novel, and cost-effective biosensors targeting the p53 gene.

## 7.2 DNA (ct DNA)

Circulating tumor DNA (ctDNA) has emerged as a potential body fluid biomarker to track the OSCC especially during treatment. ctDNA is shed into body fluids from OPC. OPC at advanced stages shows higher levels of ctDNA, whereas those at early-stages of the disease have lower levels. ctDNA levels rise steadily with disease development and decline after successful clinical intervention.

ctDNA can be detected with 80% and 100% in serum and saliva respectively in oral cancer patients.

Unique DNA sequence or mutation detection is important in the fields of oncology. The rapid identification of genetic mutations, with the prospect of accurate early diagnosis. DNA biosensors are of significant importance due to their immense potential to acquire sequence-specific information in a quicker, easier and cheaper way than conventional hybridization techniques.

DNA biosensors are made of single stranded DNA probes on separate electrodes which are immobilized by electroactive signals to quantify the DNA strand.

Droplet digital PCR (ddPCR) is a recently introduced technology that may facilitate miRNA measurement, especially in liquid biopsy, since it has proved to be more sensitive, to offer highly reproducible results, and to be less susceptible to inhibitors than conventional RT-qPCR. cDNA can be detected on label-free electrochemical biosensors using a dual enzyme assisted multiple amplification strategy (Alevizos et al., 2001) It has a detection spectrum from 0.01 fM to 1 p.m. and a detection limit of as little as 2.4 a.m. Moreover, ctDNA can be detected by the platform in biological fluids such as plasma.

## 7.3 Tumor necrosis factor

TNF-α protein is a cytokine serving as potent protein biomarker for electrochemical ELISA, which uses the selectivity of an antigen–antibody interaction on an electrochemical platform, is seen as one of the most promising methods. Electrochemical ELISA has the added advantage of not being affected by particles, chromophores, and fluorophores that might be present in a sample and which might cause interference with optical detection.

TNF are protein molecules released by inflammatory cells and involved in various physiological and pathological processes (Chakraborty, Sharma, Sharma, & Lee, 2020). TNF has been found to be diagnostically significant in oral precancer and cancer. Further, they have correlation with different grades or stages of cancer (Ameena & Rathy, 2019). It can influence the survival of oral cancer patients. It was suggested that the serum concentration of TNF-α can be an indicator of response to chemotherapy and could be used in clinical decision-making for patients (Berberoglu, Yildirim, & Celen, 2004). Therefore, the critical need for advanced TNF sensors has now grown rapidly and will continue to grow to support clinical testing, and disease studies. TNF levels are traditionally measured in saliva and serum with biosensors, these body fluids can interfere with the detection of target molecules and may give rise to false positive signals. Hence, it is a critical issue for biosensor development. Barhoumi et al. showed that an applied chronoamperometric immunosensor can detect tumor necrosis factor-α (TNF-α) within the clinically relevant concentration range with a precision of 8% and a LOD of 0.3 pg $mL^{-1}$ (Barhoumi et al., 2019) Similarly, Kongsuphol et al. developed a highly sensitive and selective TNF-α biosensor to detect TNF-α from nondiluted human serum using magnetic bead coupled antibody and electrochemical impedance spectroscopy (EIS) techniques. It has the limit of detection (LOD) at 1 pg/mL (57 fM) (Kongsuphol et al., 2014). Silicon nanowire (SiNW) field-effect transistor (FET) biosensors was proposed by Zhang et al. (2015). It is a label-free and multiplexed detection, noninvasive analysis, highly sensitive and specific determination biosensor, developed for a multiplexed detection methodology for IL-8 and TNF-α detection in saliva. Electrochemical impedance spectroscopy (EIS) has also been used for TNF-α cytokines detection. it can detected the TNF-α in saliva within the range 1–100 pg/mL (Bellagambi et al., 2017) TNF-α can be measured by another ultrasensitive impedimetric immunosensor was constructed by using poly(3-thiophene acetic acid) (P3), a conjugated polymer as an immobilization matrix (Berberoglu et al., 2004).

## 7.4 Epidermal growth factor receptor

Epidermal growth factor receptor (EGFR) is a type I tyrosine kinase receptor and active in the modulation of cell process during normal and neoplastic tissue growth in humans. A number of tumors, especially oral cancer, are known to exhibit increased EGFR activity leading to poor prognosis and resistance to anticancer therapy. EGFR has a dual role: it can help in diagnosis as well as treatment planning.

Various biosensors, e.g., saliva-based EGFR mutation detection (SABER), have been developed (Wei et al., 2010). These biosensors use different platforms for accurate measurement of EGFR in tissue body fluids. Saliva-based EGFR mutation detection (SABER) is an electrochemical biosensor based on electric field-induced release and measurement (EFIRM) technology. It can determine epidermal growth factor receptor (EGFR) in the saliva and serum of oral cancer patients (Kallempudi, Altintas, Niazi, & Gurbuz, 2012). A microfluidic paper-based electrochemical DNA biosensor can be constructed for sensitive detection of EGFR mutations in saliva with a detection limit as low as 0.167 nM (Tian et al., 2017)

## 7.5 Exosomes

Exosomes play critical roles in carcinogenesis and modulation of tumor microvironment (O'Loughlin, Woffindale, & Wood, 2012). Exosomes is a bi-layered lipid cell surface structure containing main transmembrane proteins such as CD 9. These are considered as ideal biomarkers for noninvasive early detection of oral cancer. Early identification of cancer by exosomes can be achieved by detecting them in body fluids. Cancer cells secreted them into bodily fluids in great quantity. Although detection methods based on exosomes are important, they require extensive sample purification, have high false-positive rates, and encounter labeling difficulties due to the small size of exosomes. Electrochemical biosensors, optical biosensors, and electro-chemiluminescence biosensors are three major types of biosensors which can be used for the detection of exosomes. Yap et al.'s systematic review comprehensively presented the current state of knowledge of exosomes in OSCC and OPMDs (Yap et al., 2020).

Over the years, many basic biosensors have been modified with different materials to enhance exosome detection efficiency. Xia et al. demonstrated an aptasensor based on DNA-capped-SWCNTs for detection of exosomes (Xia et al., 2017). Double imprinting-based methods with high potential in the separation and detection of complex biosamples to detect mimetic exosomes in small body fluid is another example of a biosensor that can be used (Zhu et al., 2020)

A rapid and multiple-targeted lateral flow immunoassay (LFIA) system can assist in the detection of EVs isolated from human plasma reported (Oliveira-Rodríguez et al., 2016). Another novel lateral flow immunoassay (LFIA) based on the use of tetraspanins as targets was developed by Myriam Oliveira-Rodríguez et al. It can be completed within 15 min with a detection limit of $8.54 \times 105$ exosomes/μL when used with gold nanoparticles labeled antibodies.

## 7.6 Cyfra 21-1

Cyfra 21-1 is a soluble fragment of cytokeratin-19, released during cell apoptosis, is a potential biomarker in oral cancer (Malhotra et al., 2016). They were found to be highly sensitive and specific in their detection potential and prognostic value and can be used as an adjunctive serological marker for the OSCC disease staging (Singh et al., 2017). Salivary CyfFa 21-1 was reported as a better diagnostic marker than serum Cyfra 21-1. Cyfra 21-1 can be measured in OSCC using electro-chemiluminescent immunoassay (ECLIA) similar to other biomarkers in saliva and serum of oral cancer patients (Liu et al., 2019). LC-MRM-Ms was reported as an effective method of analysis of Cyfra 21-1 in serum.

Immunosensor (Cys-GA-anti-Cyfra 21.1 antibody-BSA-Cyfra 21-1 antigen/AuE) can detect and determine the Cyfra 21-1biomarker in unprocessed human saliva samples at low concentrations and thereby can assist in diagnosis of oral cancer at an early stage. Zeng et al. developed an immunosensor based on a 3D-G@Au which has the capacity to detect Cyfra 21-1 with a wide linear range of 0.25–800 ng $mL^{-1}$ and low detection limit of 100 pg $mL^{-1}$ (Jafari & Hasanzadeh, 2020).

## 7.7 Conclusion

The survival rate of oral and foreign gene cancers patients can be enhanced by using novel methods that can provide faster, non-invasive, and ultra-precise modes of detection in an early stage. The different kinds of biosensors and biomolecules—DNA biosensors, RNA biosensor, and protein biosensors have proven diagnostic and prognostic applications. More insights into this upcoming research area are required for early diagnosis of OPCs.

## References

Alevizos, I., Mahadevappa, M., Zhang, X., Ohyama, H., Kohno, Y., Posner, M., . . . Wong, D. T. W. (2001). Oral cancer in vivo gene expression profiling assisted by laser capture microdissection and microarray analysis. *Oncogene*, *20*(43), 6196–6204. Available from https://doi.org/10.1038/sj.onc.1204685.

Almeida, V. L., Santana, I. T. S., Santos, J. N. A., Fontes, G. S., Lima, I. F. P., Matos, A. L. P., . . . Paranhos, L. R. (2019). Influence of interleukins on prognosis of patients with oral squamous cells carcinoma. *Jornal Brasileiro de Patologia e Medicina Laboratorial*. Available from https://doi.org/10.5935/1676-2444.20190051.

Al-Ogaidi, A. J. M., Stefan-Van Staden, R. I., Gugoasa, L. A., Van Staden, J. F., Yank, H., Göksel, M., & Durmuş, M. (2017). A new potentiometric sensor for the assay of P53 in blood samples. UPB scientific bulletin. *Series B: Chemistry and Materials Science*, *79*(2), 113–120. Available from http://www.scientificbulletin.upb.ro/?page = revistaonline&a = 2&cat = B.

Ameena, M., & Rathy, R. (2019). Evaluation of tumor necrosis factor: Alpha in the saliva of oral cancer, leukoplakia, and healthy controls - A comparative study. *Journal of International Oral Health*, *11*(2), 92–99. Available from https://doi.org/10.4103/jioh.jioh_202_18.

Arshad, S. H., Zohaib, K., Sannam, K. R., Mustafa, N., Mahmood, S. K., Maria, M., & Sohail, Z. M. (2017). Salivary IL-8, IL-6 and TNF-α as potential diagnostic biomarkers for oral cancer. *Diagnostics*, 21. Available from https://doi.org/10.3390/diagnostics7020021.

Aydın, M., Aydın, E. B., & Sezgintürk, M. K. (2018). A highly selective electrochemical immunosensor based on conductive carbon black and star PGMA polymer composite material for IL-8 biomarker detection in human serum and saliva. *Biosens. Bioelectron.*, *117*, 720–728.

Babiuch, K., Kuśnierz-Cabala, B., Kęsek, B., Okoń, K., Darczuk, D., & Chomyszyn-Gajewska, M. (2020). Evaluation of proinflammatory, NF-KappaB dependent cytokines: IL-1α, IL-6, IL-8, and TNF-α in tissue specimens and saliva of patients with oral squamous cell carcinoma and oral potentially malignant disorders. (n.d.) *J. Clin. Med.*

Barhoumi, L., Bellagambi, F. G., Vivaldi, F. M., Baraket, A., Clément, Y., Zine, N., . . . Errachid, A. (2019). Ultrasensitive immunosensor array for TNF-α detection in artificial saliva using polymer-coated magnetic microparticles onto screen-printed gold electrode. *Sensors (Switzerland)*, *19* (3). Available from https://doi.org/10.3390/s19030692.

Bellagambi, F. G., Baraket, A., Longo, A., Vatteroni, M., Zine, N., Bausells, J., . . . Errachid, A. (2017). Electrochemical biosensor platform for TNF-α cytokines detection in both artificial and human saliva: Heart failure. *Sensors and Actuators, B: Chemical*, *251*, 1026–1033. Available from https://doi.org/10.1016/j.snb.2017.05.169.

Berberoglu, U., Yildirim, E., & Celen, O. (2004). Serum levels of tumor necrosis factor alpha correlate with response to neoadjuvant chemotherapy in locally advanced breast cancer. *International Journal of Biological Markers*, *19*(2), 130–134. Available from https://doi.org/10.1177/172460080401900207.

Bosetti, C., Carioli, G., Santucci, C., Bertuccio, P., Gallus, S., Garavello, W., ... La Vecchia, C. (2020). Global trends in oral and pharyngeal cancer incidence and mortality. *International Journal of Cancer*, *147*(4), 1040–1049. Available from https://doi.org/10.1002/ijc.32871.

Burcu, A. E., Muhammet, A., & Kemal, S. M. (2018). Electrochemical immunosensor based on chitosan/conductive carbon black composite modified disposable ITO electrode: An analytical platform for p53 detection. *Biosensors and Bioelectronics*, 80–89. Available from https://doi.org/10.1016/j.bios.2018.09.008.

Chakraborty, C., Sharma., Sharma, G., & Lee, S. S. (2020). The interplay among miRNAs, major cytokines, and cancer-related inflammation. *Molecular Therapy-Nucleic Acids*.

Choudhary, M., Yadav, P., Singh, A., Kaur, S., Ramirez-Vick, J., Chandra, P., Arora, K., Singh, S.P. (2016). CD 59 targeted ultrasensitive electrochemical immunosensor for fast and noninvasive diagnosis of oral cancer. Electroanalysis 2016, 28, 2565–2574.

Ding, S., Das, S. R., Brownlee, B. J., Parate, K., Davis, T. M., Stromberg, L. R., ... Claussen, J. C. (2018). CIP2A immunosensor comprised of vertically-aligned carbon nanotube interdigitated electrodes towards point-of-care oral cancer screening. *Biosens. Bioelectron.*, *117*, 68–74.

Ezat, H.-A., Bakhsh, R. J., Saeid, H. M., Simin, S., Mahdi, G. S., Ilaria, P., & Marco, M. (2015). A Genosensor for point mutation detection of P53 Gene PCR product using magnetic particles. *Electroanalysis*, 1378–1386. Available from https://doi.org/10.1002/elan.201400660.

Gokhale, A. S., Haddad, R. I., Cavacini, L. A., Wirth, L., Weeks, L., Hallar, M., ... Posner, M. R. (2005). Serum concentrations of interleukin-8, vascular endothelial growth factor, and epidermal growth factor receptor in patients with squamous cell cancer of the head and neck. *Oral Oncology*, *41*(1), 70–76. Available from https://doi.org/10.1016/j.oraloncology.2004.06.005.

Jaber, M. A., & Elameen, E. M. (2020). Long-term follow-up of oral epithelial dysplasia: A hospital based cross-sectional study. *Journal of Dental Sciences*. Available from https://doi.org/10.1016/j.jds.2020.04.003.

Jafari, M., & Hasanzadeh, M. (2020). Non-invasive bioassay of Cytokeratin Fragment 21.1 (Cyfra 21.1) protein in human saliva samples using immunoreaction method: An efficient platform for early-stage diagnosis of oral cancer based on biomedicine. *Biomedicine and Pharmacotherapy*, 131. Available from https://doi.org/10.1016/j.biopha.2020.110671.

Kallempudi, S. S., Altintas, Z., Niazi, J. H., & Gurbuz, Y. (2012). A new microfluidics system with a hand-operated, on-chip actuator for immunosensor applications. *Sensors and Actuators, B: Chemical*, *163*(1), 194–201. Available from https://doi.org/10.1016/j.snb.2012.01.034.

Kongsuphol, P., Ng, H. H., Pursey, J. P., Arya, S. K., Wong, C. C., Stulz, E., & Park, M. K. (2014). EIS-based biosensor for ultra-sensitive detection of TNF-α from non-diluted human serum. *Biosensors and Bioelectronics*, *61*, 274–279. Available from https://doi.org/10.1016/j.bios.2014.05.017.

Li, L., Fukumoto, M., & Liu, D. (2013). Prognostic significance of p53 immunoexpression in the survival of oral squamous cell carcinoma patients treated with surgery and neoadjuvant chemotherapy. *Oncology Letters*, *6*(6), 1611–1615. Available from https://doi.org/10.3892/ol.2013.1627.

Liao, Z., Zhang, Y., Li, Y., Miao, Y., Gao, S., Lin., ... Geng, L. (2018). Microfluidic chip coupled with optical biosensors for simultaneous detection of multiple analytes: A review. *Biosensors & Bioelectronics*, 126. Available from https://doi.org/10.1016/j.bios.2018.11.032, Epub.

Lili, L., Long, J., Yang, R., Li-Peng, S., & Bai-Ou, G. (2018). Fiber light-coupled optofluidic waveguide (FLOW) immunosensor for highly sensitive detection of p53 protein. *Analytical Chemistry*, 10851–10857. Available from https://doi.org/10.1021/acs.analchem.8b02123.

Lin, Y., Darvishi, S., Preet, A., Huang, T., Lin, S., Girault, H., ... Lin, T. (2020). A review: Electrochemical biosensors for oral cancer. Chemosensors, 8(3), 54.

Liu, L., Xie, W., Xue, P., Wei, Z., Liang, X., & Chen, N. (2019). Diagnostic accuracy and prognostic applications of CYFRA 21-1 in head and neck cancer: A systematic review and *meta*-analysis. *PLoS One*, *14*(5). Available from https://doi.org/10.1371/journal.pone.0216561.

Malhotra, R., Patel, V., Vaqué, J. P., Gutkind, J. S., & Rusling, J. F. (2010). Ultrasensitive electrochemical immunosensor for oral cancer biomarker IL-6 using carbon nanotube forest electrodes and multilabel amplification. *Analytical Chemistry*, *82*(8), 3118–3123. Available from https://doi.org/10.1021/ac902802b.

Malhotra, R., Urs, A. B., Chakravarti, A., Kumar, S., Gupta, V. K., & Mahajan, B. (2016). Correlation of Cyfra 21–1 levels in saliva and serum with CK19 mRNA expression in oral squamous cell carcinoma. *Tumor Biology*, *37*(7), 9263–9271. Available from https://doi.org/10.1007/s13277-016-4809-4.

Morteza, H., Shiva, M., Yasaman-Sadat, B., & Reza, G. M. (2017). Detection of p53 gene mutation (single-base mismatch) using a fluorescent silver nanoclusters. *Journal of Fluorescence*, 1443–1448. Available from https://doi.org/10.1007/s10895-017-2083-5.

O'Loughlin, A. J., Woffindale, C. A., & Wood, M. J. A. (2012). Exosomes and the emerging field of exosome-based gene therapy. *Current Gene Therapy*, *12*(4), 262–274. Available from https://doi.org/10.2174/156652312802083594.

Oliveira-Rodríguez, M., López-Cobo, S., Reyburn, H. T., Costa-García, A., López-Martín, S., Yáñez-Mó, M., ... Blanco-López, M. C. (2016). Development of a rapid lateral flow immunoassay test for detection of exosomes previously enriched from cell culture medium and body fluids. *Journal of Extracellular Vesicles*, *5*(1). Available from https://doi.org/10.3402/jev.v5.31803.

Osman, T. A., Costea, D. E., & Johannessen, A. C. (2012). The use of salivary cytokines as a screening tool for oral squamous cell carcinoma: A review of the literature. *Journal of Oral and Maxillofacial Pathology*, *16*(2), 256–261. Available from https://doi.org/10.4103/0973-029X.99083.

Pachauri, N., Dave, K., Dinda, A., & Solanki, P. R. (2018). Cubic CeO2 implanted reduced graphene oxide-based highly sensitive biosensor for non-invasive oral cancer biomarker detection. *J. Mater. Chem. B*, *6*, 3000–3012.

Riedel, F., Zaiss, I., Herzog, D., Götte, K., Naim, R., & Hörmann, K. (2005). Serum levels of interleukin-6 in patients with primary head and neck squamous cell carcinoma. *Anticancer Research*, *25*(4), 2761–2766.

Rivlin, N., Brosh, R., Oren, M., & Rotter, V. (2011). Mutations in the p53 tumor suppressor gene: Important milestones at the various steps of tumorigenesis. *Genes and Cancer*, *2*(4), 466–474. Available from https://doi.org/10.1177/1947601911408889.

Singh, P., Barpande, S. R., Bhavthankar, J. D., Mandale, M. S., & Bhagwat, A. U. (2017). Serum Cyfra 21-1 levels in oral squamous cell carcinoma patients and its clinicopathologic correlation. *Indian Journal of Dental Research*, *28*(2), 162–168. Available from https://doi.org/10.4103/0970-9290.207789.

St. John, M. A. R., Li, Y., Zhou, X., Denny, P., Ho, C. M., Montemagno, C., ... Wong, D. T. W. (2004). Interleukin 6 and interleukin 8 as potential biomarkers for oral cavity and oropharyngeal squamous cell carcinoma. *Archives of Otolaryngology - Head and Neck Surgery*, *130*(8), 929–935. Available from https://doi.org/10.1001/archotol.130.8.929.

Tan, W., Sabet, L., Li, Y., Yu, T., Klokkevold, P. R., Wong, D. T., & Ho, C. M. (2008). Optical protein sensor for detecting cancer markers in saliva. *Biosensors and Bioelectronics*, *24*(2), 266–271. Available from https://doi.org/10.1016/j.bios.2008.03.037.

Tavassoli, M., Brunel, N., Maher, R., Johnson, N. W., & Soussi, T. (1998). p53 Antibodies in the saliva of patients with squamous cell carcinoma of the oral cavity [3]. *International Journal of Cancer*, *78*(3), 390–391. Available from https://doi.org/10.1002/(SICI)1097–0215(19981029)78:3 < 390::AID-IJC23 > 3.0.CO;2–9.

Tian, T., Liu, H., Li, L., Yu, J., Ge, S., Song, X., & Yan, M. (2017). Paper-based biosensor for noninvasive detection of epidermal growth factor receptor mutations in non-small cell lung cancer patients. *Sensors and Actuators, B: Chemical*, *251*, 440–445. Available from https://doi.org/10.1016/j.snb.2017.05.082.

Tiwari, S., Gupta, P. K., Bagbi, Y., Sarkar, T., & Solanki, P. R. (2017). L-Cysteine capped lanthanum hydroxide nanostructures for non-invasive detection of oral cancer biomarker. *Biosens. Bioelectron.*, *89*, 1042–1052.

Tonello, S., Stradolini, F., Abate, G., Uberti, D., Serpelloni, M., Carrara, S., & Sardini, E. (2019). Electrochemical detection of different p53 conformations by using nanostructured surfaces. *Scientific Reports*, *9*(1). Available from https://doi.org/10.1038/s41598-019-53994-6.

Torrente-Rodríguez, R. M., Campuzano, S., Montiel, V. R.-V., Gamella, M., & Pingarrón, J. M. (2016). Electrochemical bioplatforms for the simultaneous determination of interleukin (IL)-8 mRNA and IL-8 protein oral cancer biomarkers in raw saliva. *Biosensors and Bioelectronics*, 543–548. Available from https://doi.org/10.1016/j.bios.2015.10.016.

Van Harten, M. C., Hoebers, F. J. P., Kross, K. W., Van Werkhoven, E. D., Van Den Brekel, M. W. M., & Van Dijk, B. A. C. (2015). Determinants of treatment waiting times for head and neck cancer in the Netherlands and their relation to survival. *Oral Oncology*, *51*(3), 272–278. Available from https://doi.org/10.1016/j.oraloncology.2014.12.003.

Verma, S., & Singh, S. P. (2019). Non-invasive oral cancer detection from saliva using Zinc oxide-reduced graphene oxide nanocomposite based bioelectrode. *MRS Commun*, *9*, 1227–1234.

Verma, S., Singh, A., Shukla, A., Kaswan, J., Arora, K., Ramirez-Vick, J., ... Singh, S. P. (2017). Anti-IL8/AuNPs-RGO/ITO as an immunosensing platform for noninvasive electrochemical detection of oral Cancer. *ACS Appl. Mater. Interfaces*, *9*, 27462–27474.

Wei, Q., Mao, K., Wu, D., Dai, Y., Yang, J., Du, B., ... Li, H. (2010). A novel label-free electrochemical immunosensor based on graphene and thionine nanocomposite. *Sensors and Actuators, B: Chemical*, *149*(1), 314–318. Available from https://doi.org/10.1016/j.snb.2010.06.008.

Xia, Y., Liu, M., Wang, L., Yan, A., He, W., Chen, M., ... Chen, J. (2017). A visible and colorimetric aptasensor based on DNA-capped single-walled carbon nanotubes for detection of exosomes. *Biosensors and Bioelectronics*, *92*, 8–15. Available from https://doi.org/10.1016/j.bios.2017.01.063.

Yap, T., Pruthi, N., Seers, C., Belobrov, S., McCullough, M., & Celentano, A. (2020). Extracellular vesicles in oral squamous cell carcinoma and oral potentially malignant disorders: A systematic review. *International Journal of Molecular Sciences*, *21*(4). Available from https://doi.org/10.3390/ijms21041197.

Zhang, Y., Chen, R., Xu, L., Ning, Y., Xie, S., & Zhang, G. J. (2015). Silicon nanowire biosensor for highly sensitive and multiplexed detection of oral squamous cell carcinoma biomarkers in saliva. *Analytical Sciences*, *31*(2), 73–78. Available from https://doi.org/10.2116/analsci.31.73.

Zhu, Y., An, Y., Li, R., Zhang, F., Wang, Q., & He, P. (2020). Double imprinting-based electrochemical detection of mimetic exosomes. *Journal of Electroanalytical Chemistry*, 862. Available from https://doi.org/10.1016/j.jelechem.2020.113969.

Chapter 8

# Electrochemical biosensors for early detection of cancer

Meenakshi Choudhary[1] and Kavita Arora[2]

[1]*Centre for Biomedical Engineering, Indian Institute of Technology Delhi, New Delhi, India*

[2]*Advanced Instrumentation & Research Facility (AIRF) and School of Computational & Integrative Sciences (SCIS), Jawaharlal Nehru University, New Delhi, India*

## 8.1 Introduction

Cancer is the leading cause of death in developed countries and the second leading cause of death in developing countries after heart diseases as per American Cancer Society, 2011 (Harding et al., 2018). Cancer deaths are projected to rise continuously worldwide, with an estimate of 12 million deaths in 2030, where poor prognosis of the disease has contributed significantly to its monstrous nature (Globocan, 2018). Knowledge about the causes of cancer, availability of interventions for its prevention, appropriate diagnosis, and management of disease still remain a challenge to be resolved throughout the world. A database for cancer treatment can be created by understanding the relationships between measurable biological processes and clinical outcomes for a better understanding of normal and healthy physiology.

As per the International Agency for Research on Cancer (IARC) Global Cancer Observatory at Cancer Today [which is a specialized agency of the World Health Organization (WHO)] (Globocan, 2018) each year worldwide 18.1 million new cancer cases are reported, out of which 9.6 million cancer patients die (about 53%) and in India about 785,000 cancer patients die out of 1.2 million newly reported cases (about 67.8%). In other words it can also be stated that one-in-five men and one-in-six women worldwide will develop cancer over the course of their lifetime, and that one-in-eight men and one-in-eleven women will die from this disease. It is evident that cancer mortality rate is alarmingly high and requires immediate diagnostic and therapeutic interventions as shown in Fig. 8.1. Sadighbayan et al. (2019) also indicated that only $\sim$40% of people with cancer survive, i.e., a life expectancy is a little more than one-in-three, therefore, it is essential to detect this malignancy in early stages and predict therapeutic responses in patients in order to enhance the efficiency of treatment, decrease side-effects, and optimize the cure process.

Cancers can be controlled and cured by implementing evidence-based strategies for cancer prevention, appropriately effective and timely treatment, and management of cancer patients. About one-third of the global burden of cancer can be lowered if cancer cases are detected and treated early. Detecting cancer in its early stages (i.e., premalignant state) means that current or future treatment modalities might have a higher likelihood of a true cure. Existing cancer diagnostics include invasive serial random biopsies which must be performed in suspected individuals to confirm the presence of malignancy by cytologic or histologic evaluation conducted after imaging methods (like CT, MRI, etc.) (Itoh & Ichihara, 2001; Toyoda et al., 2008). Therefore, noninvasive early-stage or precancer stage detection via new screening methodologies is the need of the hour. Cancer diagnosis can be improved by converging biomolecular electronics with biomedical research and exploring new diagnostic methods targeting specific disease biomarker(s). The applications of biomarkers have become a necessity as surrogate outcomes in large trials of major diseases, like cancer (Ellenberg & Hamilton, 1989), heart diseases (Wittes et al., 1989), etc.

Cancer detected through expression of a clinically significant molecule, which exhibits abnormal levels of its expression is referred to as "a cancer biomarker." As per WHO, a true definition of a biomarker includes "Almost any measurement reflecting an interaction between a biological system and a potential hazard, which may be chemical, physical, or biological. The measured response may be functional and physiological, biochemical at the cellular level,

Biosensor Based Advanced Cancer Diagnostics. DOI: https://doi.org/10.1016/B978-0-12-823424-2.00024-7

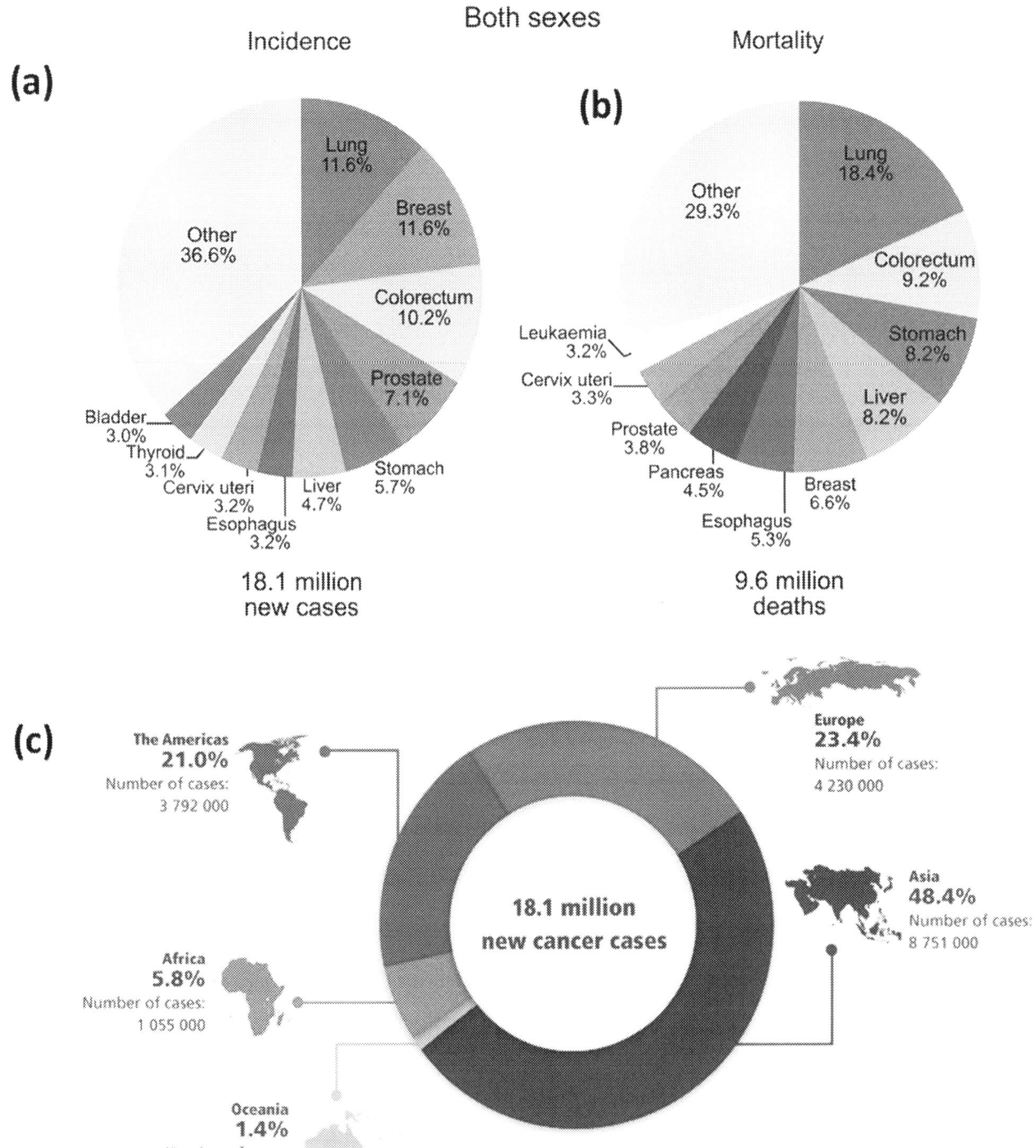


**FIGURE 8.1** (A) New cases and (B) Number of deaths and (C) Global cancer incidences in 2018. *From Globocan. (2018). Graph production. In* Global cancer observatory. *International Agency for Research on Cancer (IARC), World Health Organisation (WHO). https://gco.iarc.fr/today/online-analysis-pieAU.*

or a molecular interaction" (Strimbu & Tavel, 2010). Biomarkers may be present in malignant tissues or released in biological fluids (serum, plasma, sputum, saliva, urine, sweat, etc.). Generally, complexity of biological samples creates a big challenge for quantitative analysis and is a subject of struggle for a rapid diagnosis and clinical applications. Targeting early-stage biomarkers may facilitate early detection and subsequent appropriate treatments. Therefore, biomarkers for cancer diagnosis are majorly categorized through change in cell constituents which can be a protein,

**TABLE 8.1 Potential biomarkers for cancer diagnosis.**

| Type of cancer disease | Common biomarkers |
|---|---|
| Breast | BRCA1, BRCA2, CA 15-3, CA125, CA 27.29, CEA, NY-BR-1, Ing-1, HER2/NEU, ER/PR |
| Colon | CEA, EGF, p53 |
| Liver | AFP, CEA |
| Esophageal | SCC |
| Lung | CEA, CA 19-9, SCC, NSE, NY-ESO-1 |
| Ovarian | CA125, HCG, p53, CEA, CA 549, CASA, CA 19-9, CA 15-3, MCA, MOV-1, TAG72 |
| Melanoma tyrosinase | NY-ESO-1 |
| Prostate | PSA |

*Source*: Adapted from Sohrabi, N., Valizadeh, A., Farkhani, S. M., & Akbarzadeh, A. (2016). Basics of DNA biosensors and cancer diagnosis. *Nanomedicine, and Biotechnology, 44*(2), 654–663. https://doi.org/10.3109/21691401.2014.976707.

metabolite, nucleic acid, a hormone, or a transcription/translation factor. Genetic biomarkers can be at two levels, i.e., chromosomal and gene through deletions, additions, alterations, methylation, etc. These are epigenetic markers which lead to changes in genomes and gene expression through intensive methylations, histone modifications, expression of micro-RNA (miRNA), etc. which can lead to early-stage cancer diagnosis. The easiest examples include DNA point mutations of the p53 gene, point mutation analysis of K-ras, which is related to colorectal cancers. Proteins can represent a state of cancer through its expression/absence of antigens/antibodies in biological fluids/cancer cells. In this row, a very interesting review has summarized this concept through examples of lung cancer diagnosis by describing various biomarkers and types of biosensors demonstrated for the same (Roointan et al., 2019). Some examples of important biomarkers are included in Table 8.1 (Sohrabi et al., 2016). Carbohydrates are mainly represented through changes in cell receptors, i.e., peptidoglycan layer onto cancerous cells and changes in hormones by change in transcription/translation factors such as interleukins (Dominika et al., 2016). Changes in both glycan types (*N*- and *O*-glycans) often influence tumor cells and their interactions with stromal cell types, including leukocytes, platelets, fibroblasts, and endothelial cells and there are further changes with tumor progression and metastasis. This also explained that disease biomarkers play a crucial role in prognosis, diagnosis and in the understanding of cancer etiology.

New cancer biomarkers expressed in conveniently drawn biological fluids are potentially the most valuable tool for development of simple, reliable, and low-cost diagnostics. In recent times, biosensors provide a simple and miniaturized system for detection of the desired analyte by using biomolecules as capturing agents. A biosensor uses a biomolecule (e.g., tissue, microorganisms, organelles, cell receptors, enzymes, antibodies, nucleic acids, etc.) interfaced to a desired transducer component (physicochemical/optical/piezoelectric or electrochemical) and signals generated due to biomolecular interaction are converted into another signal via transducer to be quantified and processed to associated electronics and displayed as output in a user friendly form (Cavalli, 2006). Biosensors indeed provide a rapid and convenient alternative to conventional analytical methods for measuring an analyte in a complex medium (Arora, 2018, 2019; Kumar & Arora, 2020), A great deal of work in the development and design of biosensors has been explored and explained during past years, making use of different transducer systems while exploring benefits of miniaturization and large scale fabrication methodologies to ensure their commercial success/viability. In this context, frequently used transducer mechanisms for biosensing (ruling out feasibility of application of thermal transducer) are: electrochemical, optical, and piezoelectric (Tothill & Turner, 2003). Discussion of optical, piezoelectric and thermal biosensors would be beyond the scope of this chapter; therefore, further sections will throw a spotlight on electrochemical biosensors. In a nutshell, this chapter is aimed to focus on electrochemical biosensors for diagnosis of cancer using various novel and unique materials/biomarkers, describing their basic principal, applications, research trends, and future prospects.

## 8.2 Biosensors

A biosensor is a self-contained analytical device that incorporates a biologically active material in intimate contact with an appropriate transduction element for the purpose of detecting (reversible and selective) the concentration or activity of chemical/biological species in any of sample. It converts a biological response into an electrical signal (Arora et al.,

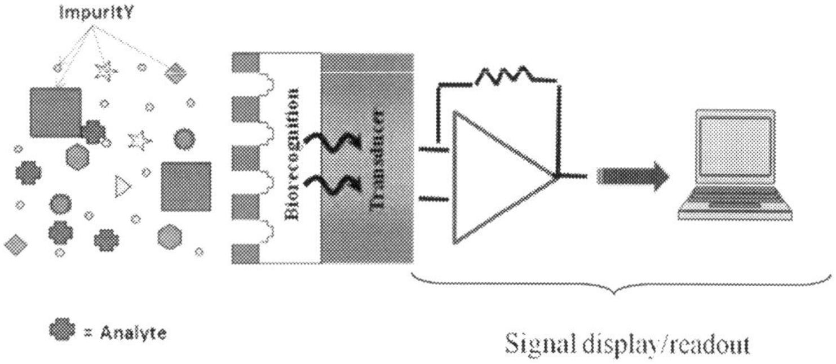


FIGURE 8.2 Schematic showing arrangement of a simple biosensor.

2006). The schematic arrangement of a typical biosensor has been shown in Fig. 8.2. According to IUPAC, "A biosensor is a self-contained integrated device which is capable of providing specific quantitative or semi-quantitative analytical information using a biological recognition element (biochemical receptor) which is in direct spatial contact with a transducer element" (Thevenot et al., 1999). On the basis of different biological recognition elements, biosensors can be categorized into two categories viz. catalytic biosensors and affinity biosensors. Catalytic biosensors use enzyme and microbiological cells as recognition element and affinity biosensors use antigen, antibodies, nucleic acids, lectins and hormones etc. as biorecognition element.

A biosensor device using antibodies or antigens as biorecognition element is called an "immunosensor" (Ghindilis et al., 1998; Itoh & Ichihara, 2001) and a biosensor device using nucleic acid as biorecognition element is called "genosensor" (Sassolas et al., 2008). Ever since the first ever biosensor was invented in 1962 (Clark & Lyons, 1962), new advancements are being made using various biomolecules and materials for qualitative and quantitative assays; this is well established and can be traced back to late 1950s and early 1960s (Ekins, 1960; Yalow & Berson, 1996). The first immunosensor was developed in 1975 by Janata to detect an ovalbumin antibody through immobilized ovalbumin protein on platinum wire (Janata, 1975). At present, an extensive range of analytes can be detected and measured by biosensors, e.g., medical diagnostic markers such as hormones (steroids and pituitary hormones), drugs (therapeutic and abused), bacteria, and environmental pollutants such as pesticides, aerosols, etc. This includes changes in physicochemical properties subsequent to biomolecular interaction which can be detected via direct and indirect methods of transduction. Direct methods include electrochemical impedance spectroscopy (EIS), cyclic voltammetry (CV), electrochemical differential pulse voltammetry (DPV), square wave voltammetry (SWV), coulometry, amperometry, conductometry, capacitometry, microgravimetry and surface plasmon resonance (SPR). These do not require any labeling and also provide information about the kinetics of biomolecular interaction occurring at the transducer (North, 1985). Indirect detection approaches may include labeling of biomolecular secondary molecules to detect the biorecognition event depending on the assay format (sandwich type, competition, capture). Most often, optical detection is achieved through the use of properties like fluorescence, chemiluminescence, electro-chemiluminescence, absorbance, etc. Electrochemical methods make use of electroactive molecule, e.g., dye, nanomaterial, conducting polymer, etc. Another way is to use competitive immunosensor for detection of desired analyte with labeled-competitor for binding onto the biosensor electrode surface. Competitive immunoassays can be carried out in two different ways: homogenous or heterogeneous approach. In the homogenous assay, the amount of labeled unbound antigen and no separation steps are followed, but in the heterogeneous assay, the labeled bound antigen is measured after removing the labeled unbound antigen by using a washing step. This adds to better sensitivity but an additional step. Further sections describe types of electrochemical biosensors that have been developed for detection of cancer using various biomolecular recognition elements.

## 8.3 Electrochemical biosensors

Electrochemical biosensors are diagnostic devices, which make use of electrochemical transducers for biorecognition event occurring on the biosensor surface upon exposure to the analyte of interest. Electrochemical transducers convert receptor–analyte complex formation events into a measurable electrical signal directly proportional to the magnitude of the biorecognition event (e.g., enzymatic reaction or surface masking of electron flow or changes in resistance/impedance/current/charge/capacitance/potential of the surface, etc. Nakazato, 2013). Electrochemical system provides a great deal of flexibility and ease of use for a wide range of materials (nanoparticles, nanocomposites, conducting polymers, sol–gel materials, etc.) for detection of biomolecular interaction. Additionally, this system provides provisions to decrease the cost while eliminating the need of an additional labeling step. Electrochemical biosensors also offer advantages of specificity, selectivity, sensitivity, reusability, miniaturization, less power requirements, compatibility with microfabrication technology, and the potential for point-of-care testing compared to existing traditional biochemical assays. In addition, detection performances have been largely enhanced by use of nanomaterials as electrode materials, label or signal carriers, tracers, separators and collectors, mediators, and catalysts owing to their large surface area, excellent conductivity, and catalytic activity.

As mentioned above, in an electrochemical transducer an electrical signal is measured, representing magnitude directly proportional to receptor−analyte complex formation event; these are widely used in sensing technology and in point-of-care devices providing additional features like portability, simplicity, ease, low cost, and disposability. Electrochemical devices are also reported to be miniaturized pocket-size devices for applications in home or the doctor's table/surgery (Tothill, 2009). These biosensors are inexpensive, robust, relatively simple to operate, have the possibility to work in a reagent-less mode, and/or in complex matrices, like serum, urine, blood, and milk (Caygill et al., 2012; Conzuelo et al., 2012). Generally, electrochemical transducers can be clubbed into amperometric (measuring the current), potentiometric (measuring the electrode potential or voltage differences), and impedimetric [measuring the conductivity or resistance or impedance (Tothill & Turner, 2003)]. Advances in various cancer-based electrochemical biosensors reported in literature are explained in the following sections. These sections explain in detail the principle and use of various nanomaterials that have been customized to achieve desired performance characteristics as well as provide the suitable conducive environment for biomolecular stability and electroactivity. Various classes of electrochemical biosensors for cancer diagnosis are described here under.

## 8.3.1 Amperometric biosensors for cancer diagnosis

Amperometric biosensors are self-contained integrated devices based on the measurement of the current resulting from biorecognition event or oxidation/reduction of an electroactive biological element providing specific quantitative analytical information (Sadeghi, 2013). Generally, electroactive species are used as redox analytes or redox labels and the resulting current change is directly proportional to the analyte concentration (Chaubey & Malhotra, 2002). These biosensors exhibit high sensitivity, direct relation between concentration of analyte, and current signal that offers the possibility to select different working potentials (Ramírez et al., 2009). Depending on the biorecognition element or biomolecule on transducer, surface amperometric biosensors for cancers developed so far can be generally categorized into immunosensors or genosensors.

### *8.3.1.1 Amperometric immunosensors*

The most common types of amperometric immunosensors can be described as ELISA integrated to electrochemical detection, where redox species (generated by a redox enzyme) are converted into a measurable current. A glucose oxidase enzyme-based biosensor fabricated by Clark and Lyon to monitor glucose was the first reported amperometric biosensor (Clark & Lyons, 1962). The Aizawa group investigated a breakthrough in the biosensor field by fabricating the first "amperometric enzyme immunosensor," The idea behind this biosensor was to immobilize the antibody on a membrane that can detect $2 \times 10^{-2}$−102 IU $mL^{-1}$ of HCG (Aizawa et al., 1979). Using a similar concept, hepatitis B surface antigen was detected using immobilized glucose oxidase labeled antibodies through a gelatin membrane and the specific detection was performed in standard glucose solutions (Boitieux et al., 1984).

Two amperometric immunosensors to detect lung cancer biomarkers r-Enolase (ENO1) and p48 molecule were developed where Biomarker ENO1 (correlated to small cell lung cancer, nonsmall cell lung cancer, and head and neck cancer) was used as a potential diagnostic marker for lung cancer (Ho et al., 2010). They fabricated a simple and sensitive, electrochemical sandwich immunosensor by immobilizing anti-ENO1 monoclonal antibodies onto polyethylene glycol-modified disposable screen-printed electrodes as the detection platform, with polyclonal secondary anti-ENO1-tagged, gold nanoparticle (AuNP) congregates as electrochemical signal probes. The resulting sigmoidal shaped dose-response curve possessed a linear dynamic working range from $10^{-8}$ to $10^{-12}$ g $mL^{-1}$. This AuNP congregate-based assay provides an amplification approach for detecting ENO1 at trace levels, leading to a detection limit as low as 11.9 fg $mL^{-1}$.

Shi and Ma (2011), reported a label-free immunosensor for the detection of carcinoembryonic antigen (CEA) targeting anti-CEA antibodies. Glassy carbon (GC) electrodes were modified using a mixture of ferricyanide and chitosan, and used as the transducer element, covered with nafion membrane-contained gold nanoparticles (Shi & Ma, 2011). Glutaraldehyde (GA) was used to crosslink anti-CEA to the surface and binding of CEA to the surface-bound epitope resulted in attenuation of the ferricyanide electrochemistry. Under optimal conditions, the response of the label-free immunosensor had a linear range of 0.01–150 ng $mL^{-1}$ with a detection limit of 3 pg $mL^{-1}$ ($S/N = 3$).

Huang et al. (2010) and Song et al. (2010) developed a novel strategy for fabrication of sensitive reagent-less amperometric immunosensor for detection of carcinoembrogenic antigens (CEAs). Prussian blue nanoparticles (PBNPs) were immobilized onto three-dimensional structured membrane of the gold colloidal nanoparticles (AuNPs) doped chitosan-multiwall carbon nanotubes (CS-MWNTs) to produce homogeneous composites (CS-MWNTs-AuNPs). Subsequently, the gold nanoparticles (GNPs) were electro-deposited on the surface of the composite by electrochemical reduction of gold chloride tetrahydrate ($HAuCl_4$) to immobilize antibody biomolecules (anti-CEA) and avoid the leakage of PBNPs. The developed immunoelectrode showed the determination of CEA in the concentration ranges from 0.3 to 120 ng $mL^{-1}$, with a detection limit of 0.1 ng $mL^{-1}$ at a signal-to-noise ratio of 3.

Kumar et al. (2016) reported the fabrication of nanostructured zirconia on reduced graphene oxide ($ZrO_2$–RGO) based noninvasive and label-free biosensing platform for detection of the oral cancer biomarker (CYFRA-21-1). These $ZrO_2$–RGO were functionalized using 3-aminopropyl triethoxy saline (APTES) and deposited on the indium tin oxide (ITO) coated glass substrate to further conjugate with anti-CYFRA-21-1 to detect CYFRA-21-1 in a wide linear range of 2–22 ng $mL^{-1}$ with a detection limit of 0.122 ng $mL^{-1}$ (Kumar & Arora, 2020).

One of the first electrochemical biosensors reported to monitor the humoral immune response of cancer patients through the determination of p53 autoantibodies with HaloTag technology using magnetic capture onto screen-printed carbon electrodes and amperometric measurements on patient sera from 24 persons of colorectal/ovarian cancer (Garranzo-Asensio et al., 2016). This method was meant to monitor patient-specific levels of p53 antibodies during the course of the disease or therapeutic manipulations.

Breast cancer is one of the leading cancers among females and many immunosensors have been developed for this (Mouffouk et al., 2017). As an example, breast cancer is detected using CV measurements by monitoring the release of ferrocene molecules in the presence of target tumor/cancer cells (MUC1 biomarker) from antibody labeled–ferrocene encapsulated nanomicelle (poly[ethylene glycol-b-trimethylsilyl methacrylate] [PEG-b-PTMSMA]) and antibody labeled magnetic beads to detect target MUC1, within 10 cells/mL in less than 1 min (Mouffouk et al., 2017). This entire experiment was demonstrated using breast cancer cell line MCF-7 and nonspecific S17 cell line. The process is shown in Fig. 8.3, which is facilitated by first releasing the encapsulated ferrocene molecules into the solution via

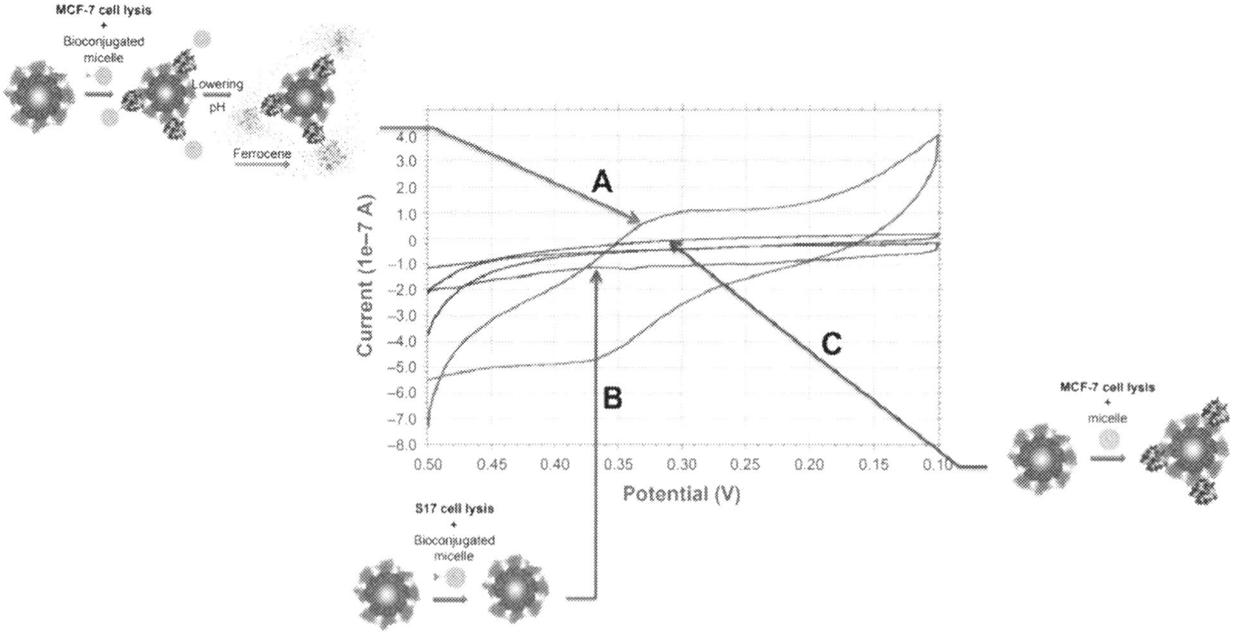


**FIGURE 8.3** Cyclic voltammetry of ferrocene indicating binding between loaded polymeric micelle and MUC1 biomarker for: (A) Detection of breast cancer biomarker MUC1 using MCF-7 breast cancer cells and ferrocene-loaded micelles; (B) detection of breast cancer biomarker MUC1 using S17 cell line instead of MCF-7 breast cancer cells (Control experiment 1); and (C) plain loaded polymeric micelles (no antibody is present on the surface) as control experiment 2. *From under creatives common (CC) by Mouffouk F., Aouabdi S., Al-Hetlani E., Serrai H., Alrefae T, & Chen LL. (2017). New generation of electrochemical immunoassay based on polymeric nanoparticles for early detection of breast cancer.* International Journal of Nanomedicine, 12, *3037–3048.*

decreasing the pH and then measuring the electrochemical signal (current intensity). This method can also be translated to facilitate therapy along with diagnostics as target drug would also be released as soon as the immunosensor will be exposed to target cancer cells.

One of the biosensors was developed for the determination of fibroblast growth factor receptor 4 (FGFR4) biomarker using a sandwich format (Torrente-Rodriguez et al., 2017). In this, magnetic immunocarriers [activated carboxylic-modified magnetic microcarriers (HOOC-MBs)] assembled onto disposable carbon screen-printed electrodes (SPCEs) where amperometric transduction using the HRP/$H_2O_2$/HQ system showed highly sensitive (Limit of detection, LOD of 48.2 pg $mL^{-1}$) and selective determination of the target protein receptor in just 15 min using a single step protocol using 2.5 μg of the raw cancer cell lysate. This method showed simplicity, analysis time, portability, and cost-affordability of the required instrumentation for the accurate determination of FGFR4 in cell lysates.

A novel strategy for electrochemical cancer cell (VCaP cells or human prostate cancer) identification using multi-marker analysis demonstrated a family of metal nanoparticle (MNP) labels specific to desired antigens via a unique linear sweep voltammetric signal. Based on reported oxidation potentials of Cu, Ag, and Pd as reporters and their redox chemistry, different target cells could be resolved on a novel microfabricated chip where different MNPs modified with antibodies or aptamers for the specific recognition of different biomarkers on the cancer cells are introduced to allow electrochemical detection. The electrochemical immunosensor using anti-hTERT immobilized onto graphene oxide was fabricated to detect hTERT a lung cancer biomarker using DPV exhibiting low detection up to 10 ag $mL^{-1}$ ($10 \times 10^{-18}$ g $mL^{-1}$) in wide detection range (10 ag $mL^{-1}$ to 50 ng $mL^{-1}$). This immunosensor showed ability to detect hTERT in spiked sputum samples up to 100 fg $mL^{-1}$ in dynamic detection range of 100 fg $mL^{-1}$ to 10 ng $mL^{-1}$ (Choudhary et al., 2013). Later, a label-free electrochemical bianalyte immunosensor was designed for simultaneous detection of lung cancer biomarkers (anti-MAGE A2 and anti-MAGE A11) using carbon nanotubes–chitosan (CNTs-CHI) composite detecting both anti-MAGE A2 and anti-MAGE A11 in the range from 5 fg $mL^{-1}$ to 50 ng $mL^{-1}$ using DPV (Choudhary et al., 2014)

Going through these exemplary amperometric immunosensors it is evident that lot more has been developed and newer efforts are being made to achieve multiple analytes using novel materials/methods at unprecedented levels of performance. Various immunosensors do exist for detection of different analytes in various applications and these novel methods have immense potential of application in the field of clinical diagnosis of cancer too.

### 8.3.1.2 Amperometric genosensors

Genosensors are also known to be as nucleic acid biosensors which make use of nucleic acids as biorecognition element. Specific DNA sequences are generally used as biomarker sequences which provide highly specific information about the disease. This interaction may be DNA–DNA, DNA–RNA or DNA–protein or synthetic molecules like PNA (peptide nucleic acids), while nucleic acids being electroactive molecules, the biorecognition event for detection of target analyte may be direct or mediator enhanced for electrochemical biosensing (Arora & Malhotra, 2008).

Nonsmall cell lung cancer (NSCLC) was diagnosed using the DPV-based DNA hybridization genosensor made up of carboxyl-functionalized graphene oxide (GO-COOH) and copper oxide nanowires (CuO NWs) using methylene blue as an electrochemical indicator with a low detection limit of $1.18 \times 10^{-13}$ M (at $S/N = 3$) of CYFRA-21-1 in linear range ($R^2 = 0.9750$) from $1.0 \times 10^{-12}$ to $1.0 \times 10^{-6}$ M (Chen et al., 2015). The sensor had good stability and selectivity, and discriminates between ssDNA sequences with one- or three-base mismatches. PCR-amplified CYFRA-21-1 from a clinical sample was successfully detected, indicating potential application of the biosensor in clinical research and practice.

Self-assembled ferrocene-cored poly(amidoamine) dendrimers were prepared on a gold electrode surface for the detection of a gene relevant to breast cancer using DPV (Senel et al., 2019). This first electrochemical DNA biosensor demonstrated excellent analytical performance, with a linear range of 10–150 ng $mL^{-1}$ (1.3–20 nM), a low detection limit of 2.9 ng $mL^{-1}$ (0.38 nM), and high sensitivity of 0.13 μA/(ng $mL^{-1}$) and offered promising results for early and easy diagnosis of breast cancer in BRCA1 gene mutation.

Raji et al. (2015) also described an aptasensor for colon cancer cells using cyclic voltammetric measurements having limit of detection up to seven cancer cells having aptamer immobilized onto a gold electrode (Raji et al., 2015).

Among nucleic acid biosensors or genosensors microRNAs (miRNAs) have emerged as new target candidates as diagnostic and prognostic biomarkers for the detection of a wide range of cancers. These offer sensitive and specific detection of cancer at early stage. microRNAs (miRNAs) are a class of noncoding RNAs of 19–25 nucleotides which play important role in post-transcriptional regulation of gene expression. Therefore, miRNA sensing technologies can allow quantitative, ultrasensitive and highly specific method with challenges of being small size, low natural abundance, and the high degree of sequence similarity among family members (Philip et al., 2019; Zeng et al., 2017). An electrochemical miRNA biosensor using DNA-1 probe gold nanoparticles modified reduced graphene oxide electrode, biotin-labeled reporter probe DNA-2 and avidin labeled $Cd^{2+}$ modified TiP nanoparticles was fabricated. This biosensor provided amplification and enhanced electron transfer using $Ru(NH_3)_6^{3+}$ as electronic wires through SWV. A proportional relationship was observed between the SWV peak currents and the logarithm of target miRNA21 concentration in a linear range from 1.0 aM to 10.0 pM with a detection limit of 0.76 aM (Cheng et al., 2015). An ultrasensitive electrochemical biosensor for multiple pancreatic carcinoma (PC)-related microRNA biomarkers (miRNA21, miRNA155, miRNA196a, and miRNA210) was fabricated using thiolated 3D DNA tetrahedral nanostructures modified gold electrodes, biotin-tagged signal probes to capture of target miRNA and generated catalytic amperometric signal with detection sensitivity as low as 10 fM.

Recently, a review article compiled electrochemical biosensors using different materials for detection of miRNAs for various purposes. Table 8.2 summarizes the trends in miRNA detection as specific miRNAs have been found to receive altered expression in a variety of cancers, too; therefore, such genosensors are frontiers of diagnostic medicine translating it to applications in point-of-care testing that have implications to real world samples.

#### *8.3.1.3 Amperometric biosensors for hormones*

Hormones are chemicals synthesized and produced by the specialized glands known as endocrine glands to control and regulate the activity of cells and organs for metabolism. There are two types of hormones in the body: peptide and steroid. Examples include interleukins, cytokines, estrogen, progesterone, testosterone, and thyroid hormones. Hence, any irregularity in the hormones, being exposed for a long time and/or to high levels of these hormones, has been linked to an increased risk of cancer (Key, 1995). Hormones play a major role in the etiology of several of the commonest cancers worldwide, including cancers of the endometrium, breast, and ovary in women and cancer of the prostate in men. Hormones can affect both early and late stages of carcinogenesis. Various biosensors have been reported for detection of hormone levels in biological matrices. In this context, (Lin et al., 2015) a reusable magnetic graphene oxide (MGO)-modified Au electrode based biosensor was developed to detect vascular endothelial growth factor (VEGF) in human plasma for cancer diagnosis. The fabricated biosensor provided appropriate sensitivity for clinical diagnostics and had wide linear detection range of 31.25–2000 pg $mL^{-1}$.

Malhotra et al. (2010) developed an electrochemical immunosensor for human IL-6 via immobilizing anti-IL-6 onto single wall carbon nanotubes (SWNT) and using horseradish peroxidase (HRP) enzyme labeling to measure very low and increased levels of IL-6 with a detection limit of 0.5 pg $m^{L-1}$ (25 fM) for IL-6 in serum.

IFN-gamma or interferon gamma (Ying et al., 2010) a cytokine (which has antiviral, antiproliferative, differentiation inducing, and immunoregulatory properties) was detected using SWV with a nucleic acid-based reusable aptamer biosensor in the 0.06 nM to 10 nM linear range. Binding of IFN-$\gamma$ caused the aptamer hairpin to unfold, pushing MB redox molecules away from the gold electrode and decreasing electron-transfer efficiency (Ying et al., 2010). IFN-gamma has also been detected in a microchip-based immunosensor having multiple channels of detection (Sato et al., 2002).

Collectively, many review publications in recent years describe use of electrochemical immunosensors for biomarkers of clinical interest. Table 8.3 indicates one such compilation of electrochemical immunosensors for cancer diagnosis at one place using hormones as a measurable indicator (Mollarasouli et al., 2019). It is suggested that hormones are critical biomarkers for cancer diagnosis and prognosis, where very minute changes in their levels plays a crucial role. Here, biosensors are promising tools that can serve the desired levels of performance parameters such as ease of operation, quick, extremely sensitive, and very low level detection limits.

### 8.3.2 Potentiometric biosensors for cancer diagnosis

Potentiometry is one of the oldest (early 1960s) instrumental methods that has well-established position in the sensoric analysis. Potentiometric biosensors are devices that use potentiometric transducers that convert biorecognition process into potential signal. In this process, the transducer measures the potential difference between a reference and a working electrode (Younus et al., 2013). In short, electrical signal generated by a potentiometric transducer is electrical potential generated due to biochemical reaction occurring at biosensor electrodes surface. This happens in biochemical processes leading to a simpler chemical species usually proton and its subsequent electrochemical detection ($NH_4OH$, $CO_2$, pH, $H_2O_2$, etc.). The best examples of potentiometric biosensors are ion-selective electrodes (ISEs) where use of new materials and new breakthroughs has encouraged innovations in ion sensing and biosensing applications. Their operation principle detects variation of concentration of a specific ion. This change can be correlated with specific analyte

**TABLE 8.2** miRNA detection using electrochemical biosensors.

| Detection method | LOD (M) | Dynamic range | Biological sample | Techniques used |
|---|---|---|---|---|
| **Backbone binding** | | | | |
| DPV | $1 \times 10^{-16}$ | 100 aM to 100 pM | Serum | Stem loop opening, DNA concatemers |
| Coulometric | $1 \times 10^{-16}$ | 100 aM to 1 nM | RNA extracts | Magnetic beads, nanoparticles |
| **Blocking electrode surface** | | | | |
| SWV | $2 \times 10^{-18}$ | 10 aM to 1 fM | serum | Four-way junction |
| SWV | $5 \times 10^{-18}$ | 10 aM to 1 μM | Serum | Four-way junction |
| EIS | $6 \times 10^{-17}$ | 0.5 fM to 40 fM | serum | Magnetic beads, nuclease |
| DPV | $1.7 \times 10^{-16}$ | 0.5 fM to 100 fM | RNA extract | Nulcease |
| EIS | $4.5 \times 10^{-16}$ | 10 fM to 1 nM | Spiked serum | Nanostructure formation, Two methods of detection |
| EIS | $1 \times 10^{-15}$ | 2 fM to 2 pM | RNA extracts and serum | Nuclease |
| SWV | $8 \times 10^{-15}$ | 1 fM to 1 nM | – | Antibody based removal |
| DPV | $1 \times 10^{-14}$ | 5 fM to 5 pM | RNA extracts | Magnetic beads, poly(A)polymerase |
| Coulometric | $2 \times 10^{-14}$ | 10 fM to 10 nM | Urine | – |
| EIS | $7 \times 10^{-13}$ | 1 pM to 100 pM | RNA extracts | – |
| DPV | $1 \times 10^{-12}$ | 1 pM to 10 nM | RNA extracted from exosomes | Magnetic beads |
| EIS | $1 \times 10^{-11}$ | 80 nM to 150 μM | Serum | – |
| **DNA enzyme** | | | | |
| CA | $5 \times 10^{-19}$ | 1 aM to 100 pM | None | Polymerase and endonuclease cascade reaction |
| DPV | $1.2 \times 10^{-15}$ | 10 fM to 1 nM | Cell lysates | Strand displacement reaction, Junction probes |
| DPV | $2.0 \times 10^{-15}$ | 1 nM to 10 fM | Spiked serum | Nanostructure formation, Nanoparticles, DNA tetrahedra |
| DPV | $5.4 \times 10^{-15}$ | 20 fM to 50 pM | RNA extracts | Polymerase |
| DPV | $6 \times 10^{-15}$ | 0.01 to 500 pM | RNA extract | Nanoparticles, stem loop opening, nanostructure formation |
| SWV | $5.2 \times 10^{-12}$ | 0.01 nM to 1 μM | Spiked serum | Strand displacement reaction |
| **Enzyme** | | | | |
| CA | $2.2 \times 10^{-19}$ | 1 aM to 10 fM | Spiked saliva | Magnetic beads, Junction probes |
| CA | $5 \times 10^{-18}$ | Approximately 5 orders of magnitude | RNA extracts | Junction probes |
| CA | $1 \times 10^{-17}$ | 10 aM to 1 pM | None | DNA tetrahedra, hybridization chain reaction |
| CA | $1.4 \times 10^{-16}$ | 1 fM to 100 pM | Spiked serum | Nanoparticles |
| DPV | $1.6 \times 10^{-16}$ | 0.1 fM to 0.1 nM | Diluted serum | Nanoparticles, Stem loop opening, Strand displacement reaction |
| DPV | $2 \times 10^{-16}$ | 3 fM to 1 nM | None | Nanoparticles, Stem loop opening, Junction probes, Rolling circle amplification |

(Continued)

**TABLE 8.2 (Continued)**

| Detection method | LOD (M) | Dynamic range | Biological sample | Techniques used |
|---|---|---|---|---|
| CA | $2 \times 10^{-16}$ | 0.5 fM to 1 pM | Serum | Stem loop opening |
| DPV | $3.5 \times 10^{-16}$ | 1 fM to 10 nM | Spiked serum | Stem loop opening, Strand displacement reaction, Nanoparticles |
| CV | $1 \times 10^{-15}$ | "6 orders of magnitude" | None | DNA tetrahedra, Stem loop opening |
| DPV | $1.7 \times 10^{-15}$ | 10 fM to 1000 fM | RNA extract from rice seedlings | Poly(U) polymerase |
| CA | $3 \times 10^{-15}$ | 10 fM to 5 pM | None | Nanoparticles, Labeling microRNA |
| DPV | $3.6 \times 10^{-15}$ | 10 fM to 0.1 nM | Cell lysates | Nanoparticles, Stem loop opening |
| CA | $4 \times 10^{-15}$ | 8 fM to 10 pM | RNA extracts | Stem loop opening |
| CA | $6 \times 10^{-15}$ | 0.01 pM to 7 pM | RNA extracts | Stem loop opening, Nanoparticles, Nanostructure formation |
| DPV | $9 \times 10^{-15}$ | 10 fM to 10 nM | RNA extracts | Polymerase recycling, Stem loop opening, Magnetic beads |
| DPV | $9 \times 10^{-15}$ | 10 fM to 10 nM | RNA extracts | Polymerase recycling, Stem loop opening, Magnetic beads |
| CA | $1 \times 10^{-14}$ | 10 fM to 1 nM | Serum | DNA tetrahedra |
| DPV | $2.5 \times 10^{-13}$ | 0.5 pM to 12.5 nM | RNA extract | Strand displacement, Rolling circle amplification |
| EIS | $4 \times 10^{-13}$ | 1.7 pM to 900 pM | Diluted spiked serum | Pre-labeled miRNA, Liposomes |
| CA | $6 \times 10^{-13}$ | 0.6 pM to 20 nM | RNA extract, serum | Strand displacement reaction |
| DPV | $6 \times 10^{-13}$ | 1 pM to 25 nM | Diluted RNA extracts | Strand displacement reaction |
| CA | $1 \times 10^{-12}$ | 1 pM to 1 nM | RNA extracts | – |
| Pulse voltammetry | $2 \times 10^{-12}$ | 2 pM to 200 nM | RNA extracts | – |
| CA | $7 \times 10^{-12}$ | 7 pM to 2.5 nM | RNA extracts | Magnetic beads |
| CA | $1.1 \times 10^{-11}$ | 100 pM to 100 nM | 10% serum | Stem loop opening, Secondary structure change |
| CA | $4 \times 10^{-11}$ | 0.14 to 100 nM | RNA extracts | Magnetic beads |
| DPV | $1 \times 10^{-7}$ | – | RNA extracts | Attaching RNA to electrode |
| DPV | $1.6 \times 10^{-7}$ | – | RNA extracts | RNA specific protein |
| CV integrating peaks | – | – | Serum | Stem loop opening, Four-way junction |
| DPV | $6 \times 10^{-17}$ | 0.1 fM to 100 pM | Serum | Nanoparticles |
| **Inorganic catalyst labelled** | | | | |
| CA | $2 \times 10^{-13}$ | 0.5 pM to 400 pM | RNA extracts | Pre-labeled microRNA, Guanine oxidation |
| DPV | $1 \times 10^{-12}$ | 1 pM to 1 nM; 1 nM to 10 nM | None | Carbon nanotubes (Two dynamic ranges reported) |
| DPV | $5 \times 10^{-9}$ | – | None | – |

(Continued)

**TABLE 8.2** (Continued)

| Detection method | LOD (M) | Dynamic range | Biological sample | Techniques used |
|---|---|---|---|---|
| DPV | $7 \times 10^{-7}$ | 1.4 μM to 5.6 μM | None | – |
| **Intercalators** | | | | |
| DPV | $1.5 \times 10^{-17}$ | 50 aM to 50 pM | spiked plasma | Nuclease, Polymerase |
| DPV | $3.3 \times 10^{-17}$ | 100 aM to 100 nM | Serum | Nanoparticles |
| DPV | $6 \times 10^{-16}$ | 2 fM to 8 pM | Plasma | Nanoparticles |
| DPV | $8.4 \times 10^{-14}$ | 0.1 to 500 pM | – | Carbon nanotubes |
| DPV | $5 \times 10^{-13}$ | 10 fM to 1 nM and 1 nM to 100 nM | None | Carbon nanotubes |
| DPV | $1 \times 10^{-12}$ | 1 pM to 800 pM | None | Hybridization chain reaction |
| **Nanoparticle (others)** | | | | |
| CV | $1.6 \times 10^{-14}$ | 0.05 to 0.9 pM | RNA extracts | – |
| **Nanofoam acting as a redox couple** | | | | |
| CV | $2 \times 10^{-16}$ | 0.2 fM to 1 nM | Plasma with and without dilution and spiking | – |
| **Nanoparticles as a direct redox couple** | | | | |
| DPV | $6.7 \times 10^{-14}$ | 10 nM to 100 pM | None | Stem loop opening |
| LSV | $2 \times 10^{-17}$ | 0.1 to 50 fM | Serum | Stem loop opening, Labeled miRNA |
| LSV | $5 \times 10^{-17}$ | 1 fM to 10 pM | RNA extracts and serum samples | DNA tetrahedra, Rolling circle amplification |
| LSV | $4 \times 10^{-16}$ | 1 fM to 1 nM | Spiked serum | DNA tetrahedra, Stem loop opening, Polymerase |
| DPV | $6.4 \times 10^{-16}$ | 1 fM to 0.1 nM | Spiked serum | Polymerase, Hybridization chain reaction |
| DPV stripping voltammetry | $1 \times 10^{-17}$ | 0.1 fM to 10 pM | Spiked blood | Strand displacement reaction, Nuclease, Hybridization chain reaction |
| SWV | $1.2 \times 10^{-14}$ | 50 fM to 30 pM | Spiked serum | Magnetic beads, Polymerase |
| **Mesoporous silica** | | | | |
| Potentiometric | $2.7 \times 10^{-18}$ | 10 aM to 1 pM | Serum | Nanoparticles |
| DPV | $3.8 \times 10^{-17}$ | 0.05 fM to 100 fM | Tumor cells | – |
| Potentiometric | $1.9 \times 10^{-11}$ | 50 pM to 5 nM | Cell lysates | – |
| DPV | $3.3 \times 10^{-17}$ | 0.1 fM to 1500 fM | Cell lysates | – |
| **Nanoparticles labeled with redox couple** | | | | |
| SWV | $7.6 \times 10^{-19}$ | 1 aM to 10 pM | Serum | Nanoparticles |
| DPV | $1.1 \times 10^{-17}$ | 0.1 fM to 100 pM | Spiked serum | Stem loop opening, Nuclease, Hybridization chain reaction |
| CV | $1.4 \times 10^{-16}$ | 5 fM to 100 fM | Serum | Magnetic beads, Stem loop opening, Nanoparticles |
| DPV | $4.4 \times 10^{-16}$ and $4.6 \times 10^{-16}$ | 1 fM to 1 nM | Cell lysates | Stem loop opening, Hybridization chain reaction |

*(Continued)*

**TABLE 8.2 (Continued)**

| Detection method | LOD (M) | Dynamic range | Biological sample | Techniques used |
|---|---|---|---|---|
| SWV | $8 \times 10^{-15}$ | $10^{-15}$ to $10^{-10}$ M | Diluted serum | – |
| CV | $1 \times 10^{-14}$ | 10 fM to 2 pM | Serum | Pre-labeled microRNA |
| DPV | $4.5 \times 10^{-14}$ | 0.1 to 10 pM | None | Pre-labeled microRNA |
| **Nanoparticle (catalytic)** | | | | |
| SWV | $3 \times 10^{-16}$ | 1 fM to 2 nM | RNA extracts | Magnetic beads, Rolling circle amplification |
| DPV | $1.92 \times 10^{-15}$ | 0.5 fM to 5 pM | Cell lysates | Stem loop opening |
| CA | $8 \times 10^{-14}$ | 0.3 pM to 200 pM | RNA extracts | – |
| CV | $1.87 \times 10^{-12}$ | 5.6 pM to 560 nM | Serum | – |
| CA frequency analysis | $1 \times 10^{-10}$ | 0.1 nM to 10 nM | None | Nuclease |
| **Polymer deposition** | | | | |
| EIS | $5 \times 10^{-16}$ | 1 fM to 5 pM | RNA extracts | DNA enzyme |
| DPV | $5 \times 10^{-16}$ | 1 fM to 100 pM | Serum | Stem loop opening, Hybridization chain reaction |
| EIS | $2 \times 10^{-15}$ | 5 fM to 2 pM | RNA extracts and serum | – |
| EIS | $3 \times 10^{-15}$ | 6 fM to 2 pM | RNA extracts | pre-labeled microRNA |
| EIS and DPV | $1.7 \times 10^{-10}$ | – | RNA extracts | – |
| **Redox labeled** | | | | |
| SWV | $1 \times 10^{-17}$ | 10 aM to 1 nM | 50% diluted blood | Magnetic nanoparticles, Conformational changes |
| SWV | $3 \times 10^{-17}$ | 100 aM to 1 nM | Spiked serum | Nuclease Junction probes |
| DPV | $6.7 \times 10^{-17}$ | 0.1 to 100 fM | RNA extract from exosomes | Magnetic beads, Strand displacement reaction, Stem loop opening |
| CV | $5 \times 10^{-16}$ | 1 to 50 fM | – | DNA flexibility |
| SWV | $1.4 \times 10^{-15}$ | 5 fM to 500 pM | Cell lysates | Strand displacement reaction |
| SWV | $4.2 \times 10^{-15}$ and $3 \times 10^{-15}$ | 5 fM to 50 pM | Cell lysates | Steam loop opening, Nuclease (two sequences detected) |
| SWV | $3 \times 10^{-14}$ | 100 fM to 2 nM | Tumor cells | Strand displacement reaction |
| DPV | $1 \times 10^{-13}$ | 0.1 pM to 10 nM | None | Hybridization chain reaction, Dendrimer formation |
| DPV | $1 \times 10^{-11}$ | 100 pM to 1 μM | Spiked cell lysis | DNA tetrahedra, nanoparticles, Stem loop opening |
| DPV | $1 \times 10^{-8}$ | 10 nM to 200 nM | RNA extracts | Magnetic beads |

*CA*, Chronoamperometry; *CV*, cyclic voltammetry; *DPV*, differential pulse voltammetry; *SWV*, square wave voltammetry; *EIS*, electrochemical impedance spectroscopy; *LSV*, linear sweep voltammetry.
*Source*: From Philip, G., Sylvain, L., & Danny, O. (2019). Molecular methods in electrochemical microRNA detection. *The Analyst*, 114–129. https://doi.org/10.1039/c8an01572d.

**TABLE 8.3** Electrochemical biosensors for hormones used for cancer diagnosis.

| Biomarker | Type of illness | Biosensor | Electrochemical technique | Linear range | Limit of Detection (LOD) |
|---|---|---|---|---|---|
| AFP (alpha feto protein) | Liver, ovarian, testicular cancers | Nanogold/TH-f GRNS-AuE | Amperometry | 0.1–200 ng $mL^{-1}$ | 0.05 ng $mL^{-1}$ |
| AFP | Liver, ovarian, testicular cancers | AuNPs/graphene-doped CS/TH-GCE | Amperometry | 1–10 ng $mL^{-1}$ | 0.7 ng $mL^{-1}$ |
| AFP | Liver, ovarian, testicular cancers | GRS/CS-CE | Square wave voltammetry | 0.05–6 ng $mL^{-1}$ | 0.02 ng $mL^{-1}$ |
| AXL (a tyrosine kinase receptor) | Prostate | Anti-AXL/GQDs/SPCE | Differential pulse voltammetry | 1.7–1000 pg $mL^{-1}$ | 0.5 pg $mL^{-1}$ |
| CEA (carcino embryonic antigen) | Breast, colorectal and pancreatic, liver, lung, ovarian, colon, bladder cancers | GOx/HRP MWCNT anti/CEA TH/AuNPs-decorated dendrimer-cysteamine/AuE | Square wave voltammetry | 10 pg $mL^{-1}$–50 ng $mL^{-1}$ | 4.4 ± 0.1 pg $mL^{-1}$ |
| CEA | Breast, colorectal, and pancreatic, liver, lung, Ovarian, colon, bladder cancers | Nitrogen-doped GQDs/Pt-PdBiMNP AuNPs/GCE | Amperometry | 5 fg $mL^{-1}$–50 ng $mL^{-1}$ | 2 fg $mL^{-1}$ |
| CEA | Breast, colorectal and pancreatic, liver, lung, ovarian, colon, bladder cancers | Trimetallic NiAuPt NPs (GRNS/–cyclodextrin/GONS/GCE) | Amperometry | 0.001–100 ng $mL^{-1}$ | 0.27 pg $mL^{-1}$ |
| CEA | Breast, colorectal and pancreatic, liver, lung, ovarian, colon, bladder cancers | AgNPs/MWCNTs/$MnO_2$ labeled anti-CEA antibodies-cyclodextrin/ MWCNT/GCE | Amperometry | NS | 0.03 pg $mL^{-1}$ |
| CEA | Breast, colorectal and pancreatic, liver, lung, ovarian, colon, bladder cancers | Bismuth film/GCE | Square wave voltammetry | 0.05–25 ng $mL^{-1}$ | 5 pg $mL^{-1}$ |
| CEA | Breast, colorectal and pancreatic, liver, lung, ovarian, colon, bladder cancers | AuNPs/poly(styrene-*co*-acrylic acid) microbead labeled anti-CEA antibodies CS/graphene oxide film/GCE | Linear sweep voltammetry | 0.5 pg $mL^{-1}$–0.5 ng $mL^{-1}$ | 0.12 pg $mL^{-1}$ |
| CEA, AFP | Breast, colorectal and pancreatic, liver, lung, ovarian, colon, bladder cancers, testicular cancers | Cd (II)/Au NPs@MWCNTs labeled anti-CEA antibodies Pb (II)/Au NPs@MWCNTs labeled anti-AFP antibodies AuNPs/AuE | Square wave voltammetry | 0.01–60 ng $mL^{-1}$ | 3 pg $mL^{-1}$ CEA<br>4.5 pg $mL^{-1}$ AFP |

*(Continued)*

**TABLE 8.3** (Continued)

| Biomarker | Type of illness | Biosensor | Electrochemical technique | Linear range | Limit of Detection (LOD) |
|---|---|---|---|---|---|
| CEA, AFP | Breast, colorectal and pancreatic, liver, lung, ovarian, colon, bladder cancers testicular cancers | PtPNPs/Cd(II) labeled anti-CEA antibodies PtPNPs/Cu(II) labeled anti-AFP antibodies Graphene oxide/GCE | Amperometry | 0.05–200 ng $mL^{-1}$0.05–200 ng $mL^{-1}$ | 0.002 ng $mL^{-1}$ CEA<br>0.05 ng $mL^{-1}$ AFP |
| CAS | Bladder Cancer | Bismuth film-modified nylon membrane-foldable SPCE | Square wave voltammetry | 0–5 g $mL^{-1}$ CAS | 0.04 g $mL^{-1}$ CAS |
| CEA, CA125, CA 153, CA 199 | Breast, colorectal and pancreatic, liver, lung, ovarian, colon, bladder cancers | 4-electrode SPCE array | Differential pulse voltammetry | 0.16–9.2 ng $mL^{-1}$ CEA<br>0.084–16 U $mL^{-1}$ CA 153<br>0.11–13 U $mL^{-1}$CA 125<br>0.16–15 U $mL^{-1}$CA 199 | 0.04 ng $mL^{-1}$ CEA<br>0.06 U $mL^{-1}$ CA 153<br>0.03 U $mL^{-1}$ CA125<br>0.1 U $mL^{-1}$ CA 199 |
| HER2-ECD, CA 15-3 | Breast cancer | SPCE-AuNPs | Square wave voltammetry | 0–50 ng $mL^{-1}$(HER2-ECD)<br>0–70 U $mL^{-1}$ (CA 15-3) | 2.9 ng $mL^{-1}$ (HER2-ECD) 5.0 U $mL^{-1}$; (CA 15-3) |
| IL-6 | Rheumatoid arthritis, Systemic lupus Erythematosus | Poly-HRP labeled anti-IL-6 antibodies Gold compact disk 8-electrode array | Amperometry | 10–1300 fg $mL^{-1}$ | 10 fg $mL^{-1}$ |
| IL-6 | Rheumatoid arthritis, Systemic lupus erythematosus | HRP labeled anti-IL-6 antibodies AuNPs-modified 8-electrode array | Amperometry | 20–400 pg $mL^{-1}$ | 20 pg $mL^{-1}$ |
| IL-6 | Rheumatoid arthritis, Systemic lupus erythematosus | HRP/MWCNT labeled anti-IL-6 antibodies Single-wall-PGDE | Amperometry | NS | 0.5 pg $mL^{-1}$ |
| PSA | Prostate | Silver hybridized mesoporous silica nanoparticles Signal amplifier-modified GCE | Amperometry | 0.15–20 ng $mL^{-1}$ | 0.06 ng $mL^{-1}$ |
| PSA | Prostate | Multi-HRP/MWCNT labeled anti-PSA antibodies SWCN-CE | Amperometry | NS | – |
| PSA | Prostate | AuNPs/PAMAM/HRP labeled PSA aptamer Graphene oxide/ CS/TH film-modified GCE | Electrochemical impedance spectroscopy | 5 pg $mL^{-1}$–35 ng $mL^{-1}$ | 5 pg $mL^{-1}$ |
| PSA | Prostate | Au/Ag-graphene oxide/GQDs labeled anti-PSA antibodies Signal amplifier-modified GCE | Electrochemiluminescent | 1 pg $mL^{-1}$–10 ng $mL^{-1}$ | 0.29 pg $mL^{-1}$ |

| | | | | | |
|---|---|---|---|---|---|
| PSA | Prostate | HRP-modified magnetic particles labeled anti-PSA antibodies AuNPs-modified pyrolytic graphite disk electrode | Amperometry | 4–10 ng $mL^{-1}$ | 0.5 pg $mL^{-1}$ |
| PSA | Prostate | Anti-PSA/MWCNTs/IL/GCE | Differential pulse voltammetry | 0.2–1.0 ng $mL^{-1}$–40 ng $mL^{-1}$ | 20 pg $mL^{-1}$ |
| PSA, IL-6 | Prostate, Rheumatoid arthritisSystemic lupus Erythematosus | AuNPs-microfluidic 8-electrode SPCE array | Amperometry | | 0.23 pg $mL^{-1}$ PSA, 0.30 pg $mL^{-1}$ IL-6 |
| PSA, hCG | Prostate cancer, ovarian, testicular, trophoblastic Cancers | porous membrane-coated 2-electrode gold array | Amperometry | NS | 0.4 g $L^{-1}$ PSA, 2.5 U $L^{-1}$ hCG |
| PSA, IL-8 | Prostate cancer, rheumatoid arthritis, inflammatory bowel disease, psoriasis, Acute respiratory distress syndrome | 16-electrode SPCE array | Amperometry | – | 5 pg $mL^{-1}$ PSA, 8 pg $mL^{-1}$ IL-8 |
| PthA | Citrus bacterial Cancer disease | AuNP/PB/CILE/GCE | Square wave voltammetry | 0.03–100 nM | 0.01 nM |
| TNF-alpha | Rheumatoid arthritis | PA + PAA/GCE | Amperometry | 0.02–200.00 ng $mL^{-1}$ | 0.01 ngm $L^{-1}$ |
| TNF-alpha | Rheumatoid arthritis | $K_3[Fe(CN)_6]^-$CHT/GA/NA/mouse anti-human TNF-alpha | Cyclic voltammetry | 0.02–34 ng $mL^{-1}$ | 10 pg $mL^{-1}$ |
| TNF-alpha | Rheumatoid arthritis | C60-fMWCNT-IL | Differential pulse voltammetry | 5.0–75 pg $mL^{-1}$ | 2.0 pg $mL^{-1}$ |
| TNF-alpha | Rheumatoid arthritis | Microfluidic | Differential pulse voltammetry | 3.25–50 ng $mL^{-1}$ | 4.1 ng $mL^{-1}$ |
| TNF-alpha | Rheumatoid arthritis | Dibutyl phthalate/polyvinyl chloride matrix | Potentiometry | 0.1–1.0 mg $L^{-1}$ | 0.015 mg $L^{-1}$ |

*TNF-α*, tumor necrosis factor-alpha; *IL*, interleukin; *n.s.*, not stated.; *CA*, cancer/carbohydrate antigen; *CEA*, carcinoembryonic antigen; *HER2/NEU*, human epidermal growth factor receptor 2; *PSA*, prostate specific antigen; *CAS*, cellular apoptosis susceptibility; *MWCNT*, multiwalled carbon nanotubes; *PA*, Polyaniline; *PAA*, polyacrylic acid; *CHT*, Chitosan; *GA*, glutaraldehyde; *NA*, nafion; *GRNS*, graphene nanosheets; *Pt–PdBiMNP*, platinum–palladium bimetallic nanoparticles; *AuNPs*, gold nanoparticles; *AgNPs*, silver nanoparticles; *CE*, Carbon electrode; *GONS*, Graphene oxide nanosheets; *PGDE*, pyrolytic graphite disk electrode; $C_{60}$, fullerene; *PB*, prussian blue; *SPCE*, screen printed carbon electrode; *HRP*, horseradish peroxidase; *Cd*, cadmium; $MnO_2$, manganese dioxide; *GCE*, glassy carbon electrode.

*Source*: Under creatives commons (CC) from Mollarasouli, F., Kurbanoglu, S. A., & Ozkan, S. A. (2019). The role of electrochemical immunosensors in clinical analysis biosensors. *Biosensors (Basel)*, *9*(3), 1–19.

concentrations present in a sample and the change is logarithmically proportional to the specific ion activity according to the Nernst equation with working conditions always near zero of current flow (Luppa et al., 2001). Mostly, potentiometric biosensors are enzyme based and rely on formation of ions as mentioned above for detection of desired analyte. However, contrary to biocatalytic systems, specific bioaffinity-based recognition (immunosensing) electroactive products enabling potentiometric detection are not formed. Therefore, enzymatic labeling is often needed to detect such events, e.g., HRP labeled secondary antibodies (Koncki, 2007). Common approach to overcome this problem is the use of enzyme-labeled immuno-reagents resulting in the development of enzyme immunoassays in the biosensor format. Field-effect transistors (FETs) are also being produced, where the reference and working electrodes are integrated by semiconductor technology, comprising a "source" and a "drain." Electrical current flows along a semiconductor path (the channel) connected to two electrodes (the source and drain), that are in direct contact with the sample solution. Hence, keeping in view these advancements of ISEs and FETs, the following sections describe some examples of reported potentiometric immunosensors and genosensors for the detection of cancer.

#### 8.3.2.1 Potentiometric immunosensors

Potentiometric immunosensors convert the immuno-recognition process into potential signal and as a result, measure the potential difference generated by the recognition event between a reference and a working electrode. A potentiometric bioaffinity immunoassay of proteins was performed via CdSe quantum dots (Thurer et al., 2007) or gold nanoparticle labeling (Chumbimuni-Torres et al., 2006) in a sandwich immunoassay in microtiter plate format. In this process, nanoparticle-based quantum dot labels were oxidatively dissolved with hydrogen peroxide and released cations are detected with cadmium ion-selective microelectrode. Signals were further enhanced using silver enlargement, i.e., catalytic deposition of silver on gold nanoparticle labels followed by oxidative dissolving with $H_2O_2$ and release of cations detected via silver ion-selective membrane microelectrode.

In the field of potentiometric immunosensors (Wang et al., 2010), a molecular imprinting-based potentiometric sensor system for the determination of cancer biomarker was developed. In the imprinting method, a pattern on a layer of polymer molecule creates a cavity with geometric patterns similar to properties of the molecule carrying the groove. A self-assembled monolayer (SAM) was used as a surface molecular imprinting where, under optimal conditions, this method determined CEA with linear determination between range of CEA 2.5 and 75 ng $mL^{-1}$. In another work, CEA was detected from CEA-producing LoVo human colon cancer cells using surface molecular-imprinted SAMs of hydroxyl-terminated alkanethiol and template biomolecules on gold-coated silicon chip as the biorecognition element for a potentiometric biosensor with linear range of 2.5–250 ng $mL^{-1}$ (Wang et al., 2010).

A light addressable potentiometric sensor (LAPS) was reported for the detection of human phosphatase of regenerating liver-3 (hPRL-3), a prognostic biomarker of liver cancer. In this work, a linear detection range of hPRL-3 was found to be 0.04–400 nM and of mammary adenocarcinoma cells was found to be $0-10^5$ cells $mL^{-1}$ (Jia et al., 2007). In malignant pleural mesotheliomas (MPM), the over expressed proteinaceous biomarker hyaluronan-linked protein 1 (HAPLN1) has also been targeted for using label-free potentiometric detection and achieved the LOD in pM range with a response time of 2–5 min in a real sample (Mathur et al., 2013).

Among all the potentiometric immunosensors, a major drawback lies in the requirement of either enzyme labeling or nanoparticle-based labeling which can provide the release of ions for ISEs to detect the presence of the desired analyte. This labeling adds to the multiple steps of the analytical procedures consisting of the consecutive incubations of sensor with sample, conjugate, substrate, washing and regeneration steps, etc. Additionally, labels impose additional costs and sometimes unstable material with a low shelf life.

#### 8.3.2.2 Potentiometric genosensors

The development of potentiometric genosensors has remain retarded and could not reach real world applications. There were various initial reports during early 2000 demonstrating field-effect devices (Poghossian et al., 2005; Schoning & Poghossian, 2006; Uslu et al., 2004) as well as membrane ion-selective electrodes (Shishkanova et al., 2007), modified with single-stranded oligonucleotides for the detection of hybridization events. These potentiometric biosensors for the detection of complementary DNA sequences couldn't transform the laboratory demonstration to a demonstration in the biological matrices such as serum, urine, blood, etc. due to interference caused by undesired molecules and superficial analysis. Various other methods such as CV, DPV, SWV, and chrono-potentiometry offered detection of intrinsic changes of electroactive species. Some of the potentiometric genosensors reported are described next.

A nucleic acid hybridization-based potentiometric microarray was developed to detect the exosomal miRNA (Goda et al., 2012) where in cancerous cells they were targeted in proximal micro-environments due to release of lactate induced pH fluctuations. Using this concept, cancer cells (MDA-MB-231) were detected with a limit of detection $10^3$ cells $mL^{-1}$. Interesting changes in pH flux surrounding the neoplasm consisting of cancer cells was shown to have the correlation to the metabolism of altered cells (Shaibani et al., 2017). Similarly, the LAPS technique was used for anti-EpCAM functionalized graphene oxide potentiometric biosensor for the selective detection of CTCs of prostate cancer (Gu et al., 2015).

Potentiometric measurements of biosensors where the relationship between the concentration and the potential is measured, is governed by the Nernst equation. Generally speaking, this offers a low limit of detection ranging from $10^{-8}$ and $10^{-11}$ M. This is important for cancer detection as the concentration of biomarkers are very low in the early stages; potentiometric measurements have a good potential for future applications (Cui et al., 2020). Despite this fact, relatively few publications exist for potentiometric biosensors and researchers favor amperometric techniques due to availability of diverse amperometric techniques such as CV, SWV, DPV, etc., that are known to offer much better insight in the electrochemical/biochemical reactions occurring on the biosensor surface.

An additional disadvantage of potentiometric monitoring is the need of labeling for affinity interaction, like antigen—antibody or DNA hybridization, that add to cost, procedural complexity, and instability. Above all this, sensitivity of whole potentiometric systems towards electrolyte composition present in real world samples due to nonspecific adsorption events significantly changes the ion exchangeable properties of bio-layers. Also, the mechanism of the generation of the potentiometric signal is unclear and limits the development of such biosensors. The interference problem is also suspected to be connected with both these response mechanisms and seems to be analogous to pH-based enzyme biosensor and electrode poisoning due to faradic reactions (Clark & Lyons, 1962; Focus, 2012). Nowadays, the selectivity and sensitivity of label-free bioaffinity-based biosensing with potentiometric signal generation and transduction is rather not acceptable for real biomedical applications (Koncki, 2007).

## 8.3.3 Impedimetric biosensors for cancer diagnosis

Impedimetric analysis measures the alterations of electrical resistance in a solution under AC voltage excitation over a range of frequencies. This technique is also known as electrochemical impedance spectroscopy (EIS) which is label free and can detect analytes down to a single molecule level. This method describes the response of an electrochemical cell to small amplitude sinusoidal voltage signal as a function of frequency. The impedance is then calculated as the ratio between voltage and current with complex impedance being a sum of the real ($Z'$) and imaginary ($Z''$) impedance, where biosensor electrode surface is modeled through equivalent circuits (Randles and Ershler circuit model) to calculate charge transfer resistance (Rct) and double-layer capacitance of electrode surface-electrolyte solution interface. Rct levels are dependent on biorecognition events or redox reactions on the electrode surface, which is influenced by the electrostatic and/or steric hindrance by adsorbed/desorbed species. Generally, changes in Rct as a function of a biorecognition event is used to quantify the analyte concentration. This technique had initially attracted more attention as a rapid technique in the characterization of biosensors, and has now succeeded towards a reagent-less biosensor transduction alternative (Farace et al., 2002). EIS permits the evaluation of the dielectric properties at the surface in a sensitive, nondestructive, and label-free manner that attracts more attention every day in the biosensors field (Liu et al., 2011; Tsekenis et al., 2008). Two EIS modes can be used: faradaic or nonfaradaic measurements. In the first, the redox species needs to be oxidized and reduced at the surface of the electrode; in the nonfaradaic mode, no additional reagent is needed (Daniels & Pourmand, 2007). Although, amperometric and potentiometric transducers are the most commonly used among all the electrochemical transducers, in recent years, more attention has been devoted to impedance transducers with the capacity of label-free detection (Zhang et al., 2013). Some examples of trending impedimetric immunosensors and genosensors are mentioned hereafter.

### *8.3.3.1 EIS-based immunosensors for cancers*

Cancer diagnosis using EIS biosensors has been extensively used and explored for advancements. Following are some of the important examples of immunosensors. Altintas et al. (2012) reported magnetic particle-modified gold electrodes as a capacitive sensor for detection of cancer markers: CEA, CA15-3, and hEGFR (Altintas et al., 2012). The transducer was modified using magnetic beads (MB) for signal enhancement and the optimal frequency range and the magnetic bead amount was determined. CEA and hEGFR could successfully be detected in the concentration range of 5 pg $mL^{-1}$ to 1 ng $mL^{-1}$ while CA15-3 was detected in the range of 1—200 U $mL^{-1}$.

Kavosi et al. (2014) demonstrated the testing of AFP using corresponding antibody, where an increased ESI response was recorded because of the hindered electron-transfer reaction on the AuNP-modified electrode surface from the binding of the target AFP (Kavosi et al., 2014). To enhance the ESI signal, the author also used AuNP/

polyamidoamine dendrimer nanocompounds to increase the electrode surface area and conductivity for a highly efficient Au electrode sensing surface. Oral cancer is one of the important cancers which go undetected despite the fact that its site is available for visible examination. For this purpose, CD59 was qualitatively detected using EIS, having wide dynamic range between 1 and 1000 fg $mL^{-1}$ with a detection limit of $0.38 \pm 0.03$ fg $mL^{-1}$ using anti-CD59 immobilized onto gold electrode with cysteine as immobilization linker (Choudhary et al., 2016). The gold nanoparticle-reduced graphene oxide (AuNPs-rGO) composite-based immunosensor for IL-8 was fabricated to show very fast detection (9 min) having experimental linear dynamic range of 500 fg $mL^{-1}$ to 4 ng $mL^{-1}$, and a detection limit of $72.73 \pm 0.18$ pg $mL^{-1}$ was reported. This immunosensor worked in a label-free and noninvasive manner in salivary samples for oral cancer biomarker interleukin-8 (IL-8) (Verma et al., 2017).

Graphene as an immunosensor substrate has limited applications because the influence of grain boundary and the scattering from substrate drastically degrades the properties of graphene and conceals the performance of intrinsic graphene as a sensor. Therefore, Li et al. (2015) reported a multimarker (ANXA2, ENO1, and VEGF) targeted label-free lung cancer biosensor based on suspended single crystalline graphene (SCG). In this work, they could get rid of grain boundary and substrate scattering, revealing the biosensing mechanism of intrinsic graphene for the first time. Monolayer SCG flakes were derived from low pressure chemical vapor deposition (LPCVD) method and the suspended structure improved sensitivity and the detection limit (0.1 pg $mL^{-1}$) of the sensor. The single crystalline nature of SCG enabled the biosensor to have superior uniformity compared to polycrystalline ones while facilitating multimarker detection. The SCG sensor exhibited good specificity and large linear detection range from 1 pg $mL^{-1}$ to 1 μg $mL^{-1}$, showing the prominent advantages of graphene as a sensing material.

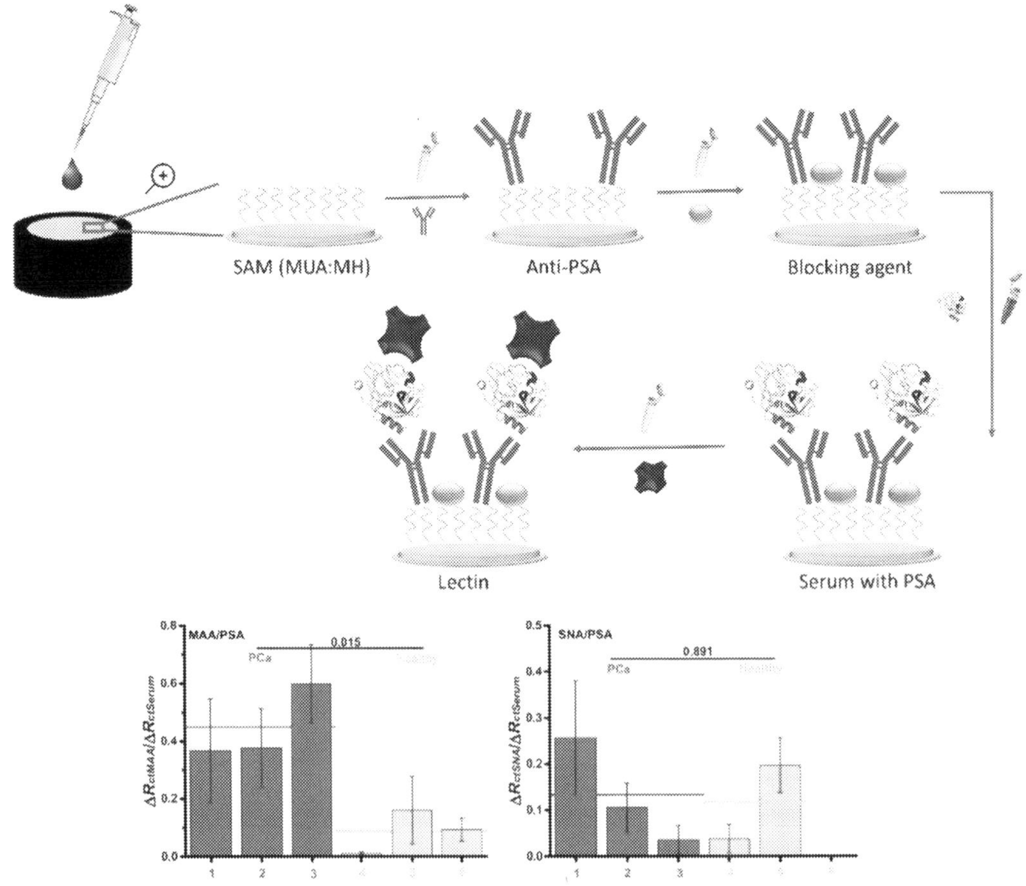


**FIGURE 8.4** EIS biosensor device for prostate-specific antigen (PSA) and glycoprofiling of PSA by application of lectin. *SAM*: self-assembled monolayer, *MUA*: 11-mercaptoundecanoic acid, *MH*: 6-mercapto-1-hexanol, Lectins: Sambucus nigra agglutinin type I (SNA, specific for α-2,6-sialic acid), Maackia amurensis agglutinin II (MAA, recognizing α-2,3-sialic acid), PCa (prostate cancer). *From Dominika, P., Peter, K., Petra, K., Roman, S., & Jan, T. (2016). Aberrant sialylation of a prostate-specific antigen: Electrochemical label-free glycoprofiling in prostate cancer serum samples.* Analytica Chimica Acta, *72–79. https://doi.org/10.1016/j.ac.2016.06.043.*

As shown in Fig. 8.4 below an immunosensor (Dominika et al., 2016) was reported using EIS to overcome the shortcomings of existing ELISA method. Prostate cancer is detected through ELISA measuring prostate-specific antigen (PSA) in the range 7.1–77.0 ng $mL^{-1}$ for patients and in the range 1.1–2.5 ng $mL^{-1}$ for healthy individuals. This immunosensor was prepared using SAM onto gold substrate for immobilization of anti-PSA for further analyte detection using EIS spectroscopy with an intent to specifically distinguish sub-glycoproteome of a tumor/normal state which is not possible through conventional methods like ELISA. The biosensor could detect PSA down to 100 ag $mL^{-1}$ with a linear concentration working range from 100 ag $mL^{-1}$ up to 1 mg $mL^{-1}$ and the sensitivity of (5.5 ± 0.2)%/decade. In addition of the presence of various glycans and even the indication of various linkages between sialic acids and the rest of glycan was successfully distinguished using EIS in a label-free way through binding of lectins (Sambucus nigra agglutinin type I [SNA, specific for α-2,6-sialic acid], Maackia amurensis agglutinin II [MAA, recognizing α-2,3-sialic acid]). Through, EIS Maackia amurensis agglutinin (MAA) recognizing α-2,3-terminal sialic acid and distinguish between these two sets of samples, since the MAA/PSA response obtained from the analysis of the PCa samples was significantly higher (5.3 fold) compared to the MAA/PSA response obtained by the analysis of samples from healthy individuals. Generally, such distinctions are conventionally analyzed using mass spectrometric techniques and analysis of glycosidic linkages (i.e., the differentiation between α-2,3- and α-2,6-terminated sialic acids). These methods are quite challenging, time consuming, requiring extensive sample pretreatment, chemical/enzymatic release of glycans, glycan derivatization, and subsequent manual data interpretation by skilled operators.

#### 8.3.3.2 EIS-based genosensors for cancers

In the series of available genosensors for cancers, some important ones are cited here as examples. Wang et al. (2015) presented an EIS-based electrochemical aptasensor for quantification of CEA (Wang et al., 2015). In this method, binding ability of GO toward ssDNA and dsDNA on the electrochemical impedance transducer at a GO-modified glassy carbon electrode was measured from the amplification system of an aptamer-switched bidirectional DNA polymerization reaction. Wang et al. (2016) reported thrombin sensor operated upon the formation of a sandwich complex among a sulfhydryl-conjugated capture probe, a biotin-labeled reporter probe, and target thrombin coupled with a dual signal amplifier generated from the nanocomposites of AuNP-decorated GO and CoPd binary nanoparticles, where the AuNP-decorated GO exhibits excellent electron-transfer capacity and large specific surface area, and CoPd shows fine catalytic activity toward $H_2O_2$ (Wang et al., 2016).

Lung cancer patient samples were distinguished from normal persons via measuring the resistance on the surface of a gold electrode modified with the specific aptamer (Zamay et al., 2016). The electrochemical index (cell index) was measured every 5 min in the RTCA Station xCELLigence system after 100 uL blood plasma sample was added in the well and incubated for 30 min.

In this category, a review by Huang et al. (2010) and Song et al. (2010) has included various amperometric, voltammetric, and impedimetric biosensors for diagnosis of cancer as part of no wash biosensors that offer advantage in terms of simplified steps for cancer patient samples using novel nanomaterials and innovative ways.

### 8.3.4 Capacitive biosensors for cancer

Capacitive biosensors belong to the subcategory of impedance biosensors where in dielectric properties and/or thickness of the dielectric layer at the electrolyte-electrode interface with the interaction of analyte with receptor immobilized on the insulating dielectric layer (Ertürk & Mattiasson, 2017). The basic principle is that when a target molecule binds to the receptor, displacement of the counter ions around the capacitive electrode results in a decrease in the capacitance since when the distance between the plates increases, the total capacitance decreases as per following Formula (8.1):

$$C = (\varepsilon_o \varepsilon A)/d \tag{8.1}$$

where $\varepsilon$ is the dielectric constant of the medium between plates, $\varepsilon_0$ is the permittivity of the free space ($8.85 \times 10^{-12}$ F/m), A is the surface area of the plates ($m^2$) and $d$ is the thickness of the insulating layer (m). For a biosensor surface Eq. (8.1) can be represented by two capacitors in series with inner part made of immobilization support with dielectric layer ($C_{dl}$) and the outer made up of biomolecule layer ($C_{bm}$) so that total capacitance ($C_t$) can be described as Formula (8.2).

$$\frac{1}{C_t} = \frac{1}{C_{dl}} + \frac{1}{C_{bm}} \tag{8.2}$$

The electrochemical capacitors are frequency dependent which are known as constant phase element (CPE) and described as Formula (8.3).

$$Z = \frac{1}{\omega C} \tag{8.3}$$

These capacitive systems when compared to Faradaic methods provide advantages as there is no need of nonpolarized reference electrodes leading to a simpler instrumental setup and avoiding hurdles related to the cell potential stability (Shimizu et al., 2017).

Low-cost microfluidic chips containing electrical double-layer capillary capacitors (μ EDLC) were fabricated using antibody-anchored magnetic beads and used for the successful quantification of CA 15-3, a biomarker protein for breast cancer, in serum samples from cancer patients with limit of detection as low as 92.0 μU $mL^{-1}$ in low sample volume (5 μL). High sensitivity using bare capillaries in a new design for double-layer capacitors with in $<2$ min was made without using redox probes, antibody on electrode (sandwich immunoassay), or signal amplification strategies (Ricardo et al., 2018).

A review by Ertürk & Mattiasson (2017) explains the applications of capacitive biosensors for measure of protein, nucleic acids, nucleotides, heavy metals, saccharides, small organic molecules and microbial cells, and various other biomolecular interactions in a simplified way. These biosensors are yet to be explored for cancer diagnosis.

### 8.3.5 Futuristic trends

Cancer diagnosis has grown from culturing cell/tissue biopsy to detection of biomarkers, which are expressed by cancerous or tumor cells in different biological fluids termed as liquid biopsy, which can be conveniently withdrawn from patients (Chang et al., 2018), i.e., analysis and quantification of cancer-related biomarkers from various body fluids, including blood, urine, saliva and cerebrospinal fluid. In addition to this, researchers have also drifted from circulating biomarkers to cell-free and circulating tumor DNA, microRNAs, and circulating tumor cells and exosomes. Further, multiplexed detection of carefully selected biomarker candidates is gaining attention as these are expected to lead to an earlier, more reliable, and personalized cancer diagnosis, expected to be free of false positives. Development of electrochemical platforms which will sense biomarkers at various levels/arenas such as the genetic (DNAs), regulatory (RNAs), functional (proteins), metabolic small regulatory (RNAs), functional (proteins) and metabolic (small molecules) levels, etc., are much needed (Lin et al., 2015). Further, in the pursuance of performance parameters of an electrochemical biosensor, in addition to choice of biomarkers and their source, there are also unique opportunities through use of novel nanomaterials and their combinations using metals, polymers (natural/synthetic), etc. These nanomaterials offer size/shape dependent physicochemical properties that offer very sensitive transduction and/or support medium for the detection of biorecognition events. This biorecognition event can be translated to a wide range of transduction mechanisms ranging from, color change, to fluorescence modulation, change in magnetic properties, spectrophotometric, electrochemical, etc. In addition to this, these interesting materials offer favorable biocompatibility, flexibility to "nanotune" various physic-chemical properties, ease of immobilization, aid in catalytic properties, and electroactivity for electrochemical biosensors. A combination of unique desired nanomaterials with biomarkers available in a liquid biopsy offer unique clinical opportunities to be tapped for diagnosis and prognosis of cancer, where biomarkers stably exist in high stability, i.e., in body fluids, and their disease dependent concentration can almost offer real-time information from the patient. In this next section, some novel combinations of biomarkers, nanomaterials and methods that have been used to further enhance the performance parameters of electrochemical biosensors for cancer diagnosis are discussed.

#### *8.3.5.1 Nanomaterial-based electrochemical biosensors for cancer*

Electrochemical biosensors for cancer diagnosis are the method of choice due to the ease of fine tuning a detection event to label-free and reagent-free, and the possibility of developing a biosensor which can sense extremely low levels of analyte directly in biological fluids. In this perspective, nanomaterials are the most promising tools to pave the way towards highly sensitive biosensors for the precise detection of cancer biomarkers. These include using metal and carbon nanomaterials as components of biosensors either in the form of providing suitable constructional support, signaling transducer, electrochemical labeling and/or sometimes the biorecognition receptor. A review by Rusling et al. (2009) and Wang et al. (2017) describes the use of various nanomaterials for improving the electrochemical biosensors for cancer diagnosis. It has included almost all kinds of nanomaterials (mainly metal and carbon nanomaterials) that had been utilized for cancer diagnosis.

MNP-based biosensors may include gold nanoparticles, silver nanoparticles, copper nanoparticles, Zn nanoparticles, multiple MNPs (Wan et al., 2014), etc., both as modified electrodes and signaling probes for detection of different biomarkers such as epidermal growth factor receptor (EGFR), carcino embryonic antigen (CEA), prostate-specific antigen (PSA), antiheat shock protein 70 (anti-HSP 70), the epithelial cancer biomarker (EpCAM), CA125, alpha feto protein (AFP), BRAC1, IL-6, etc. Carbon-based nanomaterials include the use of carbon nanotubes and graphene-based electrochemical biosensors, both as modified electrodes and signaling probes. CNT was reported to be used for detection of Concanavalin A (Con A), human chorionic gonadotropin (hCG), breast cancer biomarker mucein (MUC1), CEA, α-2,3-sialylated glycans, metalloproteinase-3 (MMP-3), osteopontine, C reactive protein (CRP), PSA, etc. for various kinds of cancers. Graphene in its various forms (reduced, quantum dots, screen-printed electrodes, N-doped composite with metals or other materials, etc.) was reported to be used for detection of hCG, AFP, CEA, CA125, CA153, oral cancer biomarker CYFRA-21-1, CD59, hTERT, MAGE A2, MAGE A 11, etc. using various electrochemical methods (Wang et al., 2017). The biosensor showed promising detection range and limit of detection touching subpico grams per mL levels. It can be concluded that use of nanomaterials substantially enhanced the electrochemical signals of biosensors via efficient transduction and generating signaling labels through high surface area-to-weight ratio and facile surface modifications limiting the biosensing process to multistep fabrication and measurement protocols (Arora, 2018, 2019). The performance characteristics of a electrochemical biosensor, including reusability, stability, sensitivity, repeatability, and ability to measure within any given biological matrix needs further investigation and developments to warrant it realization to real world low-cost diagnosis.

#### 8.3.5.2 Electrochemical exosome biosensors

Exosomes or exosomal proteins have emerged as promising biomarkers for cancer diagnosis as these are small extracellular vesicles released by cells that contains specific protein/receptors/motifs on their surface that can be detected using antibodies/aptamers. Exosomes can be generally found in blood, urine, breast milk, and saliva. Exosomes are also known to act as carriers which transport; proteins, lipids, nucleic acids, etc. from one cell to another during cancer development to regulate the immune system and metastasis. Recently, most exosomal biosensors have been reviewed that have cited electrochemical biosensors for detection of cancer (Chang et al., 2018). A few examples include the use of antibodies mostly for CD63 (using electrical field induced release and measurement: EFIRM biosensor for human lung cancer cell line H460) (Wei et al. 2013), CD24/EpCAM (using portable and multiplex integrated magnetic electrochemical exosome [iMEX] biosensor) (Jeong et al., 2016) and CD9 (for MCF-7 breast cancer having LOD of 200 exosomes $\mu L^{-1}$ and very low samples size 1.5 μL) using HRP labeling for detection of antibody-exosomal complex (Doldán et al., 2016). Later, an electrochemical aptasensor (aptamer-based biosensor) was also developed for detection of CD63 for liver cancer cells, where aptamer-binding to the target protein was measured inversely proportional to released methylene blue linked to apatmers (Zhou et al., 2016).

Researches are still working to establish the understanding of molecular mechanism of cancer in context to exosomes which are present in body fluids such as serum, saliva, urine, and biopsy withdrawn from the suspected patient. Despite the fact that exosomes are considered cellular trash bags, apoptotic bodies and microvesicles interfere as these have similar biomarkers expressed on their surfaces. In addition to this, much research in terms of new transducer materials/multi biomarker detection is needed to improve the performance of existing biosensing system.

#### 8.3.5.3 Electrohydrodynamic fluid flow-based biosensors

In the quest to achieve improved performance parameters, i.e., more sensitivity and specificity, different innovative ways are being adopted to increase the possibilities of a desired analyte or target to get attached to the biosensor transducer surface. In this context, Khondakar et al. (2019) described in a review, the use of alternating current electrohydrodynamic (ac-EHD) fluid flow as an efficient strategy to decrease nonspecific nontarget binding rendering faster assay times to be developed. ac-EHD provided fluid motion induced by an electric field with the ability to generate surface shear forces in nanometer distance to the biosensing surface (known as nanoshearing phenomenon). This was carried out to increase the collision frequency of cancer biomarkers with the biosensing surface and minimizing nonspecific binding. This group described and emphasized fundamentals and applications of ac-EHD-enhanced miniaturized systems. They described promising detection concepts for comprehensive cancer biomarker profiling based on the type of cancer biomarkers and circulating tumor cells, proteins, extracellular vesicles, and nucleic acids. A proof of concept was demonstrated to simultaneously quantify multiple biomarkers (e.g., protein and DNA) of the same target and,

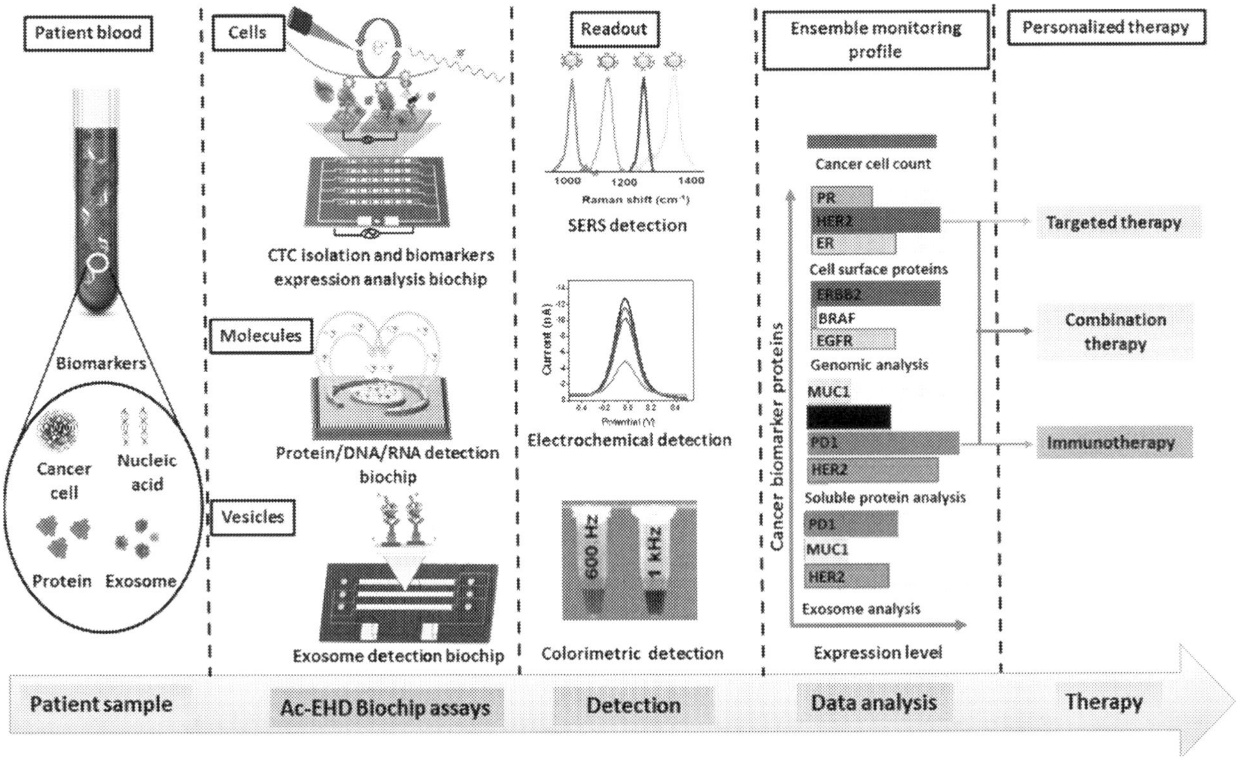


**FIGURE 8.5** ac-EHD miniaturized systems for personalized cancer treatment. Cancer biomarkers (cells, proteins, DNA/RNA, exosomes, etc.) are obtained from a minimally invasive liquid biopsy sample and are analyzed using a suite of multiplexed platforms with informative detection read-outs that facilitate the creation of a patient-specific cancer profile and personalized treatment plan. *From Khondakar, K. R., Dey, S., Wuethrich, A., & Sina, A. A. I. (2019). Toward personalized cancer treatment: From diagnostics to therapy monitoring in miniaturized electrohydrodynamic systems.* Accounts of Chemical Research, 52, *2113–2123. https://doi.org/10.1021/acs.accounts.9b00192.*

hence, improve diagnostic performance, thereby decreasing false positive or false negative results. Fig. 8.5 shows a proof of concept wherein integrated multimolecular sensors perform an entire sample-to-answer workflow of capturing melanoma cells, followed by on-chip cell lysis, and quantification of analyte using lab on a chip electrochemical biosensor system (Dey et al., 2019). However, such concepts need to be affirmed for larger sets of patients to assure real world applications. They also described improved differentiation between methylated/nonmethylated DNA and phosphorylated/nonphosphorylated protein with multiple samples simultaneously carried out under controlled nanomixing and a potential to facilitate maximum collisions of desired biomolecules with the sensing surface time to 3 min and providing detection sensitivity down to pg/μL level (Wuethrich et al., 2018). Similarly, an integrated biochip for on-chip RNA target isolation from complex biological samples using ac-EHD followed by subsequent addition of streptavidin–magnetic beads to bind biotinylated probe–target molecules for magnetic target purification via subsequent polyA extensions of purified targets and their subsequent electrochemical detection of mRNA onto Au electrodes after removal of magnetic beads bound probe via thermal release (Koo et al., 2018) as shown in Fig. 8.6. Later, extra cellular vesicles (EV) and exosomes were also detected along with ac-EHD facilitating analyte binding for detection of breast cancer and prostate cancer via HRP conjugated capture antibodies leading to visual color change (Vaidyanathan et al., 2014). The ac-EHD-enhanced EV capture and improved the assay sensitivity three times (compared to hydrodynamic flow assays) and provided a visible color change as low as 2,760 EVs per milliliter. It is shown that alternating current electrohydrodynamic fluid flow can facilitate and aid to faster and selective analyte binding to the biosensor surface. This is an important discovery which will decrease sample preprocessing time. It is a promising technique compatible with wide range of analytes and transducer systems (electrochemical optical and mass based) and therefore will have applications in a wide range of sensors for different arenas.

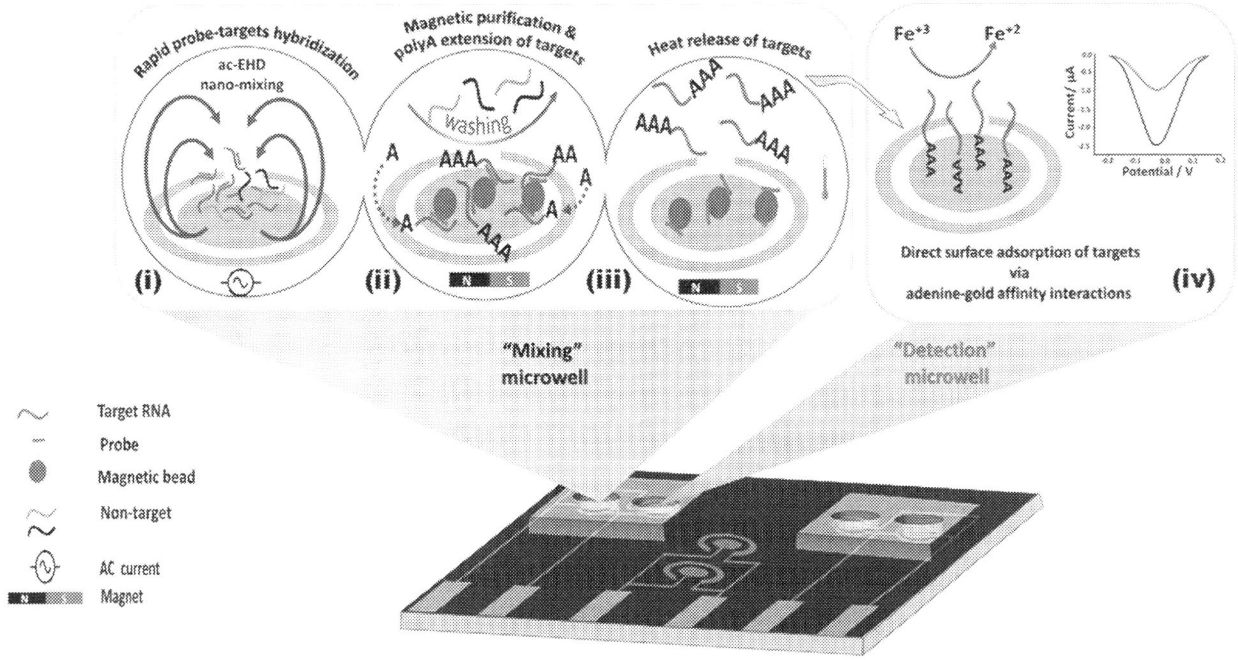


**FIGURE 8.6** Biochip for genomic target purification and detection. (i) ac-EHD nanomixing to hasten the probe–target hybridization; (ii) subsequent addition of streptavidin–magnetic beads to bind biotinylated probe–target molecules for magnetic target purification and subsequent polyA extensions of purified targets; (iii) heat release of targets from capture probes; (iv) polyA sequences facilitated rapid target adsorption onto bare gold microelectrode surface for electrochemical detection. *From Koo, K. M., Dey, S., & Trau, M. (2018). Amplification-free multi-RNA-type profiling for cancer risk stratification via alternating current electrohydrodynamic nanomixing.* Small, *14*.

### *8.3.5.4 Wearable contact-less based electrochemical biosensors for cancer*

Tears being a relatively simple biological fluid have been described to contain a wide range of proteins suitable for diagnosis of various diseases including cancers. Some of the biomarkers for cancer diagnosis which are expressed in tears include: Lacryglobin, sulf-1, cystatin SA, 5-AMP-activated protein kinase subunit-3, triosephosphate isomerase, microtubule-associated tumor suppressor 1, keratin (type I) putative LCN-1 like protein, malate dehydrogenase, Ig-alpha-2 chain c region, Ig heavy chain VIII region, protein S100-A4, keratin (type II), pericentrin, and complement C1q subcomponent subunit C (Tseng et al., 2018).

Till today only enzyme-based contact lenses as wearable biosensors for glucose, lactate, and ascorbate have been demonstrated. As an example, Chu et al. (2011) developed a contact lens biosensor for the in situ monitoring of tear glucose in rabbits, using PDMS as the contact lens material (with Pt and Ag/AgCl as the working and reference/counter electrodes, respectively). Glucose concentration in the range of 0.03– 5.0 mM (covering normal tear glucose levels in humans −0.14 mM) was also able to sense change in glucose in tears with 10 min delay and peaked value after 55 min compared to blood glucose after oral intake of glucose to rabbits. Efforts are ongoing to fabricate power sources for functioning sensor circuitry on electrochemical sensor-based contact lenses via use of wireless circuitry, induction, solar power, or biofuel cells (using the tear's glucose/lactate/ascorbate to be used to generate energy). The advances in fabricating stable power sources and low-powered electronics can aid in facilitating the realization of self-powered, minimally invasive, and continuous monitoring of diseases. There are no immediate solutions to these constraints; however, due to availability of nanofabrications and microfluidics (Rossier et al., 2002; Shimizu et al., 2017), discoveries of novel biomarkers in tears and availability of novel biocompatible materials, like graphene (optically transparent material with exciting physicochemial properties) nanomaterial, and/or composites of different materials, has a strong potential to be utilized for this purpose. This way, contact lens-based biosensors possess capability to occupy consumer market niche for the development of new gadgets and need clinical trials. In fact, integrating this with mobile applications warrants its realization of point-of-care devices to facilitate real-time data acquisition and transfer to physicians for efficient diagnosis. Not only contact lenses but wearable devices like smart watches, tattoos, rings, etc., also have immense potential for wearable biosensor devices.

Keeping in view all the futuristic trends discussed above, in a nutshell, cancer diagnosis is better and successfully achieved by employing electrochemical transduction systems integrated with advanced nanomaterials or composites and specific novel biomarker(s) to realize development of an efficient sensing platform. Firstly, compared to other biomolecules such as DNA, protein biomarkers are easy to detect in biological matrices with less interference of other undesired moieties (e.g., coagulation factors, ionic species, proteins, lipids, and various other substances including nutrients, hormones, metabolic waste, and external substances such as drug, viruses, bacteria, etc.). Owing to highly specific antigen–antibody interaction and availability of aptamers phenomenon, a biomarker-targeted nanomolecular biosensing system offers potential advantages for early cancer detection. This is also expected to provide high physicochemical stability and the possibility of detecting the desired analyte in biological fluids directly.

Secondly, response characteristics of the desired biosensor can be improved by the use of advanced nanomaterials as matrices that can largely increase surface areas of the electrode, enhanced biomolecule immobilization, and interaction efficiency. Integrating the fabrication process to microfluidics and nanofabrication technologies provides an additional edge towards improving the biosensor performance. Optimizing the amount of the immobilized biomarker and controlling the molecular orientation of binding sites can markedly upgrade the detection limit of biomarker-targeted sensing platforms which may also be controlled via appropriate immobilization matrix. Also, response time is an important parameter for immunosensors, as it is necessary to avoid mishandling or degradation of samples so that reliable detection can be achieved within time to start early treatment. Fine tuning and modulating interlinked parameters of all integrated units and careful selection of specific biomarker can facilitate desired level of performance.

Thirdly and lastly, the opportunity to customize the performance and detection of the desired analyte in a simple label-free process provides ease of operation, low processing time, and cost effectiveness. Nondestructive measurement sometimes works as the cherry on the top of the cake as the same biosensor can be regenerated and reused for following samples.

## 8.4 Conclusion

This chapter is an attempt to describe developments made among available types of electrochemical biosensors for clinical diagnosis of cancer. It is important to note that nanoinspired electrochemical biosensors using novel biomarkers (nucleic acids, miRNAs, antibodies-antigens, hormones, exosomes, etc.) and novel methods such as alternating current electrohydrodynamic fluid flow, microarrays, microfluidics, wearable biosensor devices etc., have been implemented and demonstrated for monitoring biomolecular interactions occurring at the nanoscale for cancer diagnosis. These features helped early and easy diagnosis/prognosis of cancers to provide better implementation of treatment modalities and, hence, true cure.

With the use of electrochemical transduction, various other advantages such as realizing reagent-free, label-free, noninvasive, *on site*, in situ and *online* measurements of parameters of interest in a variety of matrices or mediums was made possible. Electrochemical biosensors not only provided ease of implementation on conveniently drawn samples termed as "liquid biopsies," but also gave the opportunity to customize fabrication of a biosensor and the measurement of the desired analyte to tap unprecedented levels of performance (sensing ultra-trace amounts). Further, it has also been demonstrated that with the use of novel nanomaterials, biomolecules can be stabilized for a longer shelf life. Although many researchers are still at the level of laboratory demonstration, or at its conceptual stage, further implementation to pilot-level studies and clinical trials would warrant direct implementation of electrochemical biosensors for early-stage cancer detection, its prognosis, and relapse studies in the area of clinical diagnosis. Further, biosensors have enormous applications not only restricted to clinical diagnosis of other diseases but also environmental monitoring, food industry, and quality control.

## Acknowledgments

Thanks to Prof. M. Jagdesh Kumar, Vice Chancellor, Jawaharlal Nehru University, New Delhi, India. Financial support received under the DST PURSE, DRDO and UPoE-II, UGC is sincerely acknowledged.

## References

Aizawa, M., Morioka, A., Suzuki, S., & Nagamura, Y. (1979). Enzyme immunosenser. III. Amperometric determination of human cherienic gonadotropin by membrane-bound antibody. *Analytical Biochemistry*, *94*(1), 22–28. Available from https://doi.org/10.1016/0003-2697(79)90784-X.

Altintas, Z., Kallempudi, S. S., Sezerman, U., & Gurbuz, Y. (2012). A novel magnetic particle-modified electrochemical sensor for immunosensor applications. *Sensors and Actuators, B: Chemical, 174*, 187–194. Available from https://doi.org/10.1016/j.snb.2012.08.052.

Arora, K. (2018). Advances in Nano Based Biosensors for Food and Agriculture. In K. Gothandam, S. Ranjan, N. Dasgupta, N. Ramalingam, & E. Lichtfouse (Eds.), *Food security and water treatment. environmental chemistry for a sustainable world* (pp. 1–52). Springer Science and Business Media LLC. Available from https://doi.org/10.1007/978-3-319-70166-0_1.

Arora, K. (2019). Chapter 10: Recent biosensing applications of Graphene based nanomaterials. In B. Palys (Ed.), *Handbook of graphene: Biosensors and advanced sensors* (Vol. 6, p. 52). Wiley and Scrivener Publishing.

Arora, K., & Malhotra, B. D. (2008). *Chapter 17, Application of conducting polymer based nucleic acid biosensors. A book on applied physics in the 21st century* (pp. 473–498). Research Signpost.

Arora, K., Chand, S., & Malhotra, B. D. (2006). Recent developments in bio-molecular electronics detection techniques for food pathogens. *Analytica Chimica Acta, 568*, 259–272.

Boitieux, J. L., Romette, J. L., Aubry, N., & Thomas, D. (1984). A computerised enzyme immunosensor: application for the determination of antigens. *Clinica Chimica Acta, 136*(1), 19–28. Available from https://doi.org/10.1016/0009-8981(84)90243-2.

Cavalli, F. (2006). Cancer in the developing world: Can we avoid the disaster? *Nature Clinical Practice. Oncology, 3*(11), 582–583. Available from https://doi.org/10.1038/ncponc0611.

Caygill, J. S., Davis, F., & Higson, S. P. J. (2012). Current trends in explosive detection techniques. *Talanta, 88*, 14–29. Available from https://doi.org/10.1016/j.talanta.2011.11.043.

Chang, L., Yunchen, Y., & Yun, W. (2018). Recent advances in exosomal protein detection via liquid biopsy biosensors for cancer screening, diagnosis, and prognosis. *The AAPS Journal, 20*, 41. Available from https://doi.org/10.1208/s12248-018-0201-1.

Chaubey, A., & Malhotra, B. D. (2002). Mediated biosensors. *Biosensors and Bioelectronics, 17*(6–7), 441–456. Available from https://doi.org/10.1016/S0956-5663(01)00313-X.

Chen, M., Hou, C., Huo, D., Yang, M., & Fa, H. (2015). Huanbao Fab A highly sensitive electrochemical DNA biosensor for rapid detection of CYFRA21–1, a marker of non-small cell lung cancer. *Anal. Methods, 7*, 9466–9473.

Cheng, F. F., Ting-Ting, H., Hai-Tiao, M., Jian-Jun, S., Li-Ping, J., & Jun-Jie, Z. (2015). Electron transfer mediated electrochemical biosensor for microRNAs detection based on metal ion functionalized titanium phosphate nanospheres at attomole level. *ACS Applied Materials & Interfaces*, 2979–2985. Available from https://doi.org/10.1021/am508690x.

Choudhary, M., Kumar, V., Singh, A., Singh, M. P., Kaur, S., Reddy, G. B., & Pasricha, R. (2013). Graphene oxide based label free ultrasensitive immunosensor for lung cancer biomarker, hTERT. *Journal of Biosensors and Bioelectronics, 4*, 1–9.

Choudhary, M., Singh, A., Kaur, S., & Arora, K. (2014). Enhancing lung cancer diagnosis: Electrochemical simultaneous bianalyte immunosensing using carbon nanotubes–chitosan nanocomposite. *Applied Biochemistry and Biotechnology, 174*, 1188–1200. Available from https://doi.org/10.1007/s12010-014-1020-1.

Choudhary, M., Yadav, P., Singh, A., Kaur, S., Ramirez-Vick, J., Chandra, P., ... Singh, S. P. (2016). CD 59 targeted ultrasensitive electrochemical immunosensor for fast and noninvasive diagnosis of oral cancer. *Electroanalysis, 28*(10), 2565–2574. Available from https://doi.org/10.1002/elan.201600238.

Chu, M. X., Miyajima, K., Takahashi, D., Arakawa, T., Sano, K., Sawada, S. I., ... Mitsubayashi, K. (2011). Soft contact lens biosensor for in situ monitoring of tear glucose as non-invasive blood sugar assessment. *Talanta, 83*(3), 960–965. Available from https://doi.org/10.1016/j.talanta.2010.10.055.

Chumbimuni-Torres, K. Y., Dai, Z., Rubinova, N., Xiang, Y., Pretsch, E., Wang, J., & Bakker, E. (2006). Potentiometric biosensing of proteins with ultrasensitive ion-selective microelectrodes and nanoparticle labels. *Journal of the American Chemical Society, 128*(42), 13676–13677. Available from https://doi.org/10.1021/ja065899k.

Clark, L. C., & Lyons, C. (1962). Electrode systems for continuous monitoring in cardiovascular surgery. *Annals of the New York Academy of Sciences, 102*(1), 29–45. Available from https://doi.org/10.1111/j.1749-6632.1962.tb13623.x.

Conzuelo, F., Gamella, M., Campuzano, S., Pinacho, D. G., Reviejo, A. J., Marco, M. P., & Pingarrón, J. M. (2012). Disposable and integrated amperometric immunosensor for direct determination of sulfonamide antibiotics in milk. *Biosensors and Bioelectronics, 36*(1), 81–88. Available from https://doi.org/10.1016/j.bios.2012.03.044.

Cui, F., Zhou, Z., & Zhou, H. S. (2020). Review—Measurement and analysis of cancer biomarkers based on electrochemical biosensors. *Journal of the Electrochemical Society, 167*, 37525–37543.

Daniels, J. S., & Pourmand, N. (2007). Label-free impedance biosensors: Opportunities and challenges. *Electroanalysis, 19*(12), 1239–1257. Available from https://doi.org/10.1002/elan.200603855.

Dey, S., Koo, K. M., Wang, Z., Sina, A. A. I., Wuethrich, A., & Trau, M. (2019). An integrated multi-molecular sensor for simultaneous BRAFV600E protein and DNA single point mutation detection in circulating tumour cells. *Lab on a Chip, 19*(5), 738–748. Available from https://doi.org/10.1039/c8lc00991k.

Doldán, X., Fagúndez, P., Cayota, A., Laíz, J., & Tosar, J. P. (2016). Electrochemical sandwich immunosensor for determination of exosomes based on surface marker-mediated signal amplification. *Analytical Chemistry, 88*(21), 10466–10473. Available from https://doi.org/10.1021/acs.analchem.6b02421.

Dominika, P., Peter, K., Petra, K., Roman, S., & Jan, T. (2016). Aberrant sialylation of a prostate-specific antigen: Electrochemical label-free glycoprofiling in prostate cancer serum samples. *Analytica Chimica Acta*, 72–79. Available from https://doi.org/10.1016/j.ac.2016.06.043.

Ekins, R. P. (1960). The estimation of thyroxine in human plasma by an electrophoretic technique. *Clinica Chimica Acta, 5*(4), 453–459. Available from https://doi.org/10.1016/0009-8981(60)90051-6.

Ellenberg, S., & Hamilton, J. M. (1989). Surrogate endpoints in clinical trials: cancer. *Statistics in Medicine*, *8*(4), 405–413. Available from https://doi.org/10.1002/sim.4780080404.

Ertürk, G., & Mattiasson, B. (2017). Capacitive biosensors and molecularly imprinted electrodes. *Sensors*, *17*, 390. Available from https://doi.org/10.3390/s17020390.

Farace, G., Lillie, G., Hianik, T., Payne, P., & Vadgama, P. (2002). Reagentless biosensing using electrochemical impedance spectroscopy. *Bioelectrochemistry (Amsterdam, Netherlands)*, *55*(1–2), 1–3. Available from https://doi.org/10.1016/S1567-5394(01)00166-9.

Focus. (2012). Biosensors: Potentiometric and amperometric. *Analytical Chemistry*, 1091A–1098A. Available from https://doi.org/10.1021/ac00145a727.

Garranzo-Asensio, M., Guzman-Aranguez, A., Poves, C., Fernandez-Acenero, M. J., Torrente-Rodríguez, R. M., Montiel, V. R. V., ... Barderas, R. (2016). Toward liquid biopsy: Determination of the humoral immune response in cancer patients using halotag fusion protein-modified electrochemical bioplatforms. *Analytical Chemistry*, *88*(24), 12339–12345. Available from https://doi.org/10.1021/acs.analchem.6b03526.

Ghindilis, A. L., Atanasov, P., Wilkins, M., & Wilkins, E. (1998). Immunosensors: Electrochemical sensing and other engineering approaches. *Biosensors and Bioelectronics*, *13*(1), 113–131. Available from https://doi.org/10.1016/S0956-5663(97)00031-6.

Globocan. (2018). *Graph production. Global cancer observatory*. International Agency for Research on Cancer (IARC), World Health Organisation (WHO). Available from https://gco.iarc.fr/today/online-analysis-pie.

Goda, T., Masuno, K., Nishida, J., Kosaka, N., Ochiya, T., Matsumoto, A., & Miyahara, Y. (2012). A label-free electrical detection of exosomal microRNAs using microelectrode array. *Chemical Communications*, 48.

Gu, Y., Ju, C., Li, Y., Shang, Z., Wu, Y., Jia, Y., & Niu, Y. (2015). Detection of circulating tumor cells in prostate cancer based on carboxylated graphene oxide modified light addressable potentiometric sensor. *Biosensors and Bioelectronics*, *66*, 24–31. Available from https://doi.org/10.1016/j.bios.2014.10.070.

Harding, M. C., Sloan, C. D., Merrill, R. M., Harding, T. M., Thacker, B. J., & Thacker, E. L. (2018). Transitions from heart disease to cancer as the leading cause of death in United States, 1999–2016. *Preventing Chronic Disease Public Health Research, Practice, And Policy*, *15 E158*(12), 1–11. Available from https://doi.org/10.5888/pcd15.180151.

Ho, J. A. A., Chang, H. C., Shih, N. Y., Wu, L. C., Chang, Y. F., Chen, C. C., & Chou, C. (2010). Diagnostic detection of human lung cancer-associated antigen using a gold nanoparticle-based electrochemical immunosensor. *Analytical Chemistry*, *82*(14), 5944–5950. Available from https://doi.org/10.1021/ac1001959.

Huang, K. J., Niu, D. J., Xie, W. Z., & Wang, W. (2010). A disposable electrochemical immunosensor for carcinoembryonic antigen based on nano-Au/multi-walled carbon nanotubes-chitosans nanocomposite film modified glassy carbon electrode. *Analytica Chimica Acta*, *659*(1–2), 102–108. Available from https://doi.org/10.1016/j.ac.2009.11.023.

Itoh, Y., & Ichihara, K. (2001). Standardization of immunoassay for CRM-related proteins in Japan: From evaluating CRM 470 to setting reference intervals. *Clinical Chemistry and Laboratory Medicine*, *39*(11), 1154–1161. Available from https://doi.org/10.1515/CCLM.2001.182.

Janata, J. (1975). An Immunoelectrode. *Journal of the American Chemical Society*, *97*(10), 2914–2916. Available from https://doi.org/10.1021/ja00843a058.

Jeong, S., Park, J., Pathania, D., Castro, C. M., Weissleder, R., & Lee, H. (2016). Integrated magneto-electrochemical sensor for exosome analysis. *ACS Nano*, *10*(2), 1802–1809. Available from https://doi.org/10.1021/acsnano.5b07584.

Jia, Y., Qin, M., Zhang, H., Niu, W., Li, X., Wang, L., ... Feng, X. (2007). Label-free biosensor: A novel phage-modified Light Addressable Potentiometric Sensor system for cancer cell monitoring. *Biosensors & Bioelectronics*, 22.

Kavosi, B., Hallaj, R., Teymourian, H., & Salimi, A. (2014). Au nanoparticles/PAMAM dendrimer functionalized wired ethyleneamine-viologen as highly efficient interface for ultra-sensitive α-fetoprotein electrochemical immunosensor. *Biosensors and Bioelectronics*, *59*, 389–396. Available from https://doi.org/10.1016/j.bios.2014.03.049.

Key, T. J. A. (1995). Hormones and cancer in humans. *Mutation Research – Fundamental and Molecular Mechanisms of Mutagenesis*, *333*(1–2), 59–67. Available from https://doi.org/10.1016/0027-5107(95)00132-8.

Khondakar, K. R., Dey, S., Wuethrich, A., & Sina, A. A. I. (2019). Toward personalized cancer treatment: From diagnostics to therapy monitoring in miniaturized electrohydrodynamic systems. *Accounts of Chemical Research*, *52*, 2113–2123. Available from https://doi.org/10.1021/acs.accounts.9b00192.

Koncki, R. (2007). Recent developments in potentiometric biosensors for biomedical analysis. *Analytica Chimica Acta*, *599*(1), 7–15. Available from https://doi.org/10.1016/j.ac.2007.08.003.

Koo, K. M., Dey, S., & Trau, M. (2018). Amplification-free multi-RNA-type profiling for cancer risk stratification via alternating current electrohydrodynamic nanomixing. *Small (Weinheim an der Bergstrasse, Germany)*, 14.

Kumar, S., Sharma, J. G., Maji, S., & Malhotra, B. D. (2016). Nanostructured zirconia decorated reduced graphene oxide based efficient biosensing platform for non-invasive oral cancer detection. *Biosensors and Bioelectronics*, *78*, 497–504. Available from https://doi.org/10.1016/j.bios.2015.11.084.

Kumar, V., & Arora, K. (2020). Trends in nano-inspired biosensors for plants. *Materials Science for Energy Technologies*, *3*, 255–273. Available from https://doi.org/10.1016/j.mset.2019.10.004.

Li, P., Zhang, B., & Cui, T. (2015). Towards intrinsic graphene biosensor: A label-free, suspended single crystalline graphene sensor for multiplex lung cancer tumor markers detection. *Biosensors and Bioelectronics*, *72*, 168–174. Available from https://doi.org/10.1016/j.bios.2015.05.007.

Lin, C. W., Wei, K. C., Liao, S. S., Huang, C. Y., Sun, C. L., Wu, P. J., ... Ma, C. C. M. (2015). A reusable magnetic graphene oxide-modified biosensor for vascular endothelial growth factor detection in cancer diagnosis. *Biosensors and Bioelectronics*, *67*, 431–437. Available from https://doi.org/10.1016/j.bios.2014.08.080.

Liu, G., Liu, J., Davis, T. P., & Gooding, J. J. (2011). Electrochemical impedance immunosensor based on gold nanoparticles and aryl diazonium salt functionalized gold electrodes for the detection of antibody. *Biosensors and Bioelectronics*, *26*(8), 3660–3665. Available from https://doi.org/10.1016/j.bios.2011.02.026.

Luppa, P. B., Sokoll, L. J., & Chan, D. W. (2001). Immunosensors – Principles and applications to clinical chemistry. *Clinica Chimica Acta*, *314*(1–2), 1–26. Available from https://doi.org/10.1016/S0009-8981(01)00629-5.

Malhotra, R., Patel, V., Vaqué, J. P., Gutkind, J. S., & Rusling, J. F. (2010). Ultrasensitive electrochemical immunosensor for oral cancer biomarker IL-6 using carbon nanotube forest electrodes and multilabel amplification. *Analytical Chemistry*, *82*(8), 3118–3123. Available from https://doi.org/10.1021/ac902802b.

Mathur, A., Blais, S., Goparaju, C. M. V., Neubert, T., Pass, H., & Levon, K. (2013). Development of a biosensor for detection of pleural mesothelioma cancer biomarker using surface imprinting. *PLoS One*, *8*(3). Available from https://doi.org/10.1371/journal.pone.0057681.

Mollarasouli, F., Kurbanoglu, S. A., & Ozkan, S. A. (2019). The role of electrochemical immunosensors in clinical analysis biosensors. *Biosensors (Basel)*, *9*(3), 1–19.

Mouffouk, F., Aouabdi, S., Al-Hetlani, E., Serrai, H., Alrefae, T., & Chen, L. L. (2017). New generation of electrochemical immunoassay based on polymeric nanoparticles for early detection of breast cancer. *International Journal of Nanomedicine*, *12*, 3037–3048.

Nakazato, K. (2013). Potentiometric, amperometric, and impedimetric CMOS biosensor array. In T. Rinken (Ed.), *State of the art in biosensors – General aspects*. Intech Open. Available from https://doi.org/10.5772/53319.

North, J. R. (1985). Immunosensors: Antibody-based biosensors. *Trends in Biotechnology*, *3*(7), 180–186. Available from https://doi.org/10.1016/0167-7799(85)90119-2.

Philip, G., Sylvain, L., & Danny, O. (2019). Molecular methods in electrochemical microRNA detection. *The Analyst*, 114–129. Available from https://doi.org/10.1039/c8an01572d.

Poghossian, A., Cherstvy, A., Ingebrandt, S., Offenhäusser, A., & Schöning, M. J. (2005). Possibilities and limitations of label-free detection of DNA hybridization with field-effect-based devices. *Sensors and Actuators, B: Chemical*, *111–112*, 470–480. Available from https://doi.org/10.1016/j.snb.2005.03.083.

Raji, M. A., Amoabediny, G., Tajik, P., Hosseini, M., & Ghafar-Zadeh, E. (2015). An apta-biosensor for colon cancer diagnostics. *Sensors (Switzerland)*, *15*(9), 22291–22303. Available from https://doi.org/10.3390/s150922291.

Ramírez, N. B., Salgado, A. M., & Valdman, B. (2009). The evolution and developments of immunosensors for health and environmental monitoring: Problems and perspectives. *Brazilian Journal of Chemical Engineering*, *26*(2), 227–249. Available from https://doi.org/10.1590/s0104-66322009000200001.

Ricardo, A. G., Oliveira., Nicoliche, A. M., Pasqualeti, F. M., Shimizu, I. R., Ribeiro, M. E., … Faria, R. S. (2018). Lima low-cost and rapid-production microfluidic electrochemical double-layer capacitors for fast and sensitive breast cancer diagnosis. *Analytical Chemistry*, *90*, 12377–12384.

Roointan, A., Ibrahim Wani, S., Mati-ur-Rehman., Hussain, K. K., Ahmed, B., Abrahim, S., … Akhtar, M. H. (2019). Early detection of lung cancer biomarkers through biosensor technology: A review. *Journal of Pharmaceutical and Biomedical Analysis*, *164*, 93–103. Available from https://doi.org/10.1016/j.jpba.2018.10.017.

Rossier, J., Reymond, F., & Michel, P. E. (2002). Polymer microfluidic chips for electrochemical and biochemical analyses. *Electrophoresis*, *23*(6), 858–867, https://doi.org/10.1002/1522-2683(200203)23:6 < 858::AID-ELPS858 > 3.0.CO;2-3.

Rusling, J. F., Sotzing, G., & Papadimitrakopoulosa, F. (2009). Designing nanomaterial-enhanced electrochemical immunosensors for cancer biomarker proteins. *Bioelectrochemistry (Amsterdam, Netherlands)*, *76*(1–2), 189–194. Available from https://doi.org/10.1016/j.bioelechem.2009.03.011.

Sadeghi, S. J. (2013). Amperometric Biosensors. In A. E. Cass (Ed.), *Encyclopedia of biophysics* (pp. 61–67). Berlin: Springer. Available from https://doi.org/10.1007/978-3-642-16712-6_713.

Sadighbayan, D., Sadighbayan, K., Tohid-kia, M. R., & Hasanzadeh, M. (2019). Development of electrochemical biosensors for tumor marker determination towards cancer diagnosis: Recent progress. *Trends in Analytical Chemistry*, *118*, 73–88.

Sassolas, A., Leca-Bouvier, B. D., & Blum, L. J. (2008). DNA biosensors and microarrays. *Chemical Reviews*, *108*, 109–139.

Sato, K., Yamanaka, M., Takahashi, H., Tokeshi, M., Kimura, H., & Kitamori, T. (2002). Microchip-based immunoassay system with branching multichannels for simultaneous determination of interferon-γ. *Electrophoresis*, *23*(5), 734–739, https://doi.org/10.1002/1522-2683(200203)23:5 < 734::AID-ELPS734 > 3.0.CO;2-W.

Schoning, M. J., & Poghossian, A. (2006). Bio FEDs (field-effect devices): State-of-the-art and new directions. *Electroanalysis*, *18*(19–20), 1893–1900.

Senel, M., Dervisevic, M., & Kokkokoğlu, F. (2019). Electrochemical DNA biosensors for label-free breast cancer gene marker detection. *Analytical and Bioanalytical Chemistry*, *411*(13), 2925–2935. Available from https://doi.org/10.1007/s00216-019-01739-9.

Shaibani, P. M., Etayash, H., Naicker, S., Kaur, K., & Thundat, T. (2017). Metabolic study of cancer cells using a pH sensitive hydrogel nanofiber light addressable potentiometric sensor. *ACS Sensors*, *2*(1), 151–156. Available from https://doi.org/10.1021/acssensors.6b00632.

Shi, W., & Ma, Z. (2011). A novel label-free amperometric immunosensor for carcinoembryonic antigen based on redox membrane. *Biosensors and Bioelectronics*, *26*(6), 3068–3071. Available from https://doi.org/10.1016/j.bios.2010.11.048.

Shimizu, F. M., Todão, F. R., Gobbi, A. L., Oliveira, O. N., Garcia, C. D., & Lima, R. S. (2017). Functionalization-free microfluidic electronic tongue based on a single response. *ACS Sensors*, *2*(7), 1027–1034. Available from https://doi.org/10.1021/acssensors.7b00302.

Shishkanova, T. V., Volf, R., Krondak, M., & Král, V. (2007). Functionalization of PVC membrane with ss oligonucleotides for a potentiometric biosensor. *Biosensors and Bioelectronics*, *22*(11), 2712–2717. Available from https://doi.org/10.1016/j.bios.2006.11.014.

Sohrabi, N., Valizadeh, A., Farkhani, S. M., & Akbarzadeh, A. (2016). Basics of DNA biosensors and cancer diagnosis. *Nanomedicine, and Biotechnology*, *44*(2), 654–663. Available from https://doi.org/10.3109/21691401.2014.976707.

Song, Z., Yuan, R., Chai, Y., Yin, B., Fu, P., & Wang, J. (2010). Multilayer structured amperometric immunosensor based on gold nanoparticles and Prussian blue nanoparticles/nanocomposite functionalized interface. *Electrochimica Acta*, *55*(5), 1778–1784. Available from https://doi.org/10.1016/j.electacta.2009.10.067.

Strimbu, K., & Tavel, J. A. (2010). What are biomarkers? *Current Opinion in HIV and AIDS*, *5*(6), 463–466.

Thevenot, D. R., Toth, K., Durst, R. A., & Wilson, G. S. (1999). *Recommended definitions and classification: A technical report* (Vol. 71, pp. 2333–2348).

Thurer, R., Vigassy, T., Hirayama, M., Wang, J., Bakker, E., & Pretsch, E. (2007). Potentiometric immunoassay with quantum dot labels. *Analytical Chemistry*, *79*, 5107–5110.

Torrente-Rodriguez, R. M., Montiel, V. R. V., Campuzano, S., Pedrero, M., Farchado, M., Vargas, E., ... Pingarron, J. M. (2017). Electrochemical sensor for rapid determination of fibroblast growth factor receptor 4 in raw cancer cell lysates. *PLoS One*, *12*(4). Available from https://doi.org/10.1371/journal.pone.0175056.

Tothill, I. E. (2009). Biosensors for cancer markers diagnosis. *Seminars in Cell and Developmental Biology*, *20*(1), 55–62. Available from https://doi.org/10.1016/j.semcdb.2009.01.015.

Tothill, I. E., & Turner, A. P. F. (2003). In B. Caballero, L. Trugo, & P. Finglas (Eds.), *Biosensors* (pp. 489–499). Academic Press. Elsevier BV. Available from https://doi.org/10.1016/b0-12-227055-x/01374-2.

Toyoda, Y., Nakayama, T., Kusunoki, Y., Iso, H., & Suzuki, T. (2008). Sensitivity and specificity of lung cancer screening using chest low-dose computed tomography. *British Journal of Cancer*, *98*(10), 1602–1607. Available from https://doi.org/10.1038/sj.bjc.6604351.

Tsekenis, G., Garifallou, G. Z., Davis, F., Millner, P. A., Pinacho, D. G., Sanchez-Baeza, F., ... Higson, S. P. J. (2008). Detection of fluoroquinolone antibiotics in milk via a labeless immunoassay based upon an alternating current impedance protocol. *Analytical Chemistry*, *80*(23), 9233–9239. Available from https://doi.org/10.1021/ac8014752.

Tseng, R. C., Chen, C.-C., Hsu, S.-M., & Chuang, H.-S. (2018). Contact-lens biosensors. *Sensors*, *18*, 2651. Available from https://doi.org/10.3390/s18082651.

Uslu, F., Ingebrandt, S., Mayer, D., Böcker-Meffert, S., Odenthal, M., & Offenhäusser, A. (2004). Labelfree fully electronic nucleic acid detection system based on a field-effect transistor device. *Biosensors and Bioelectronics*, *19*(12), 1723–1731. Available from https://doi.org/10.1016/j.bios.2004.01.019.

Vaidyanathan, R., Naghibosadat, M., Rauf, S., Korbie, D., Carrascosa, L. G., Shiddiky, M. J. A., & Trau, M. (2014). Detecting exosomes specifically: A multiplexed device based on alternating current electrohydrodynamic induced nanoshearing. *Analytical Chemistry*, *86*(22), 11125–11132. Available from https://doi.org/10.1021/ac502082b.

Verma, S., Singh, A., Shukla, A., Kaswan, J., Arora, K., Ramirez-Vick, J., ... Singh, S. P. (2017). Anti-IL8/AuNPs-rGO/ITO as an immunosensing platform for noninvasive electrochemical detection of oral cancer. *ACS Applied Materials & Interfaces*, *9*(33), 27462–27474. Available from https://doi.org/10.1021/acsami.7b06839.

Wan, Y., Zhou, Y. G., Poudineh, M., Safaei, T. S., Mohamadi, R. M., Sargent, E. H., & Kelley, S. O. (2014). Highly specific electrochemical analysis of cancer cells using multi-nanoparticle labeling. *Angewandte Chemie – International Edition*, *53*(48), 13145–13149. Available from https://doi.org/10.1002/anie.201407982.

Wang, B., Akiba, U., & Anzai, J. (2017). A review on recent progress in nanomaterial-based electrochemical biosensors for cancer biomarkers. *Molecules (Basel, Switzerland)*, *22*, 1048.

Wang, W., Ge, L., Sun, X., Hou, T., & Li, F. (2015). Graphene-assisted label-free homogeneous electrochemical biosensing strategy based on aptamer-switched bidirectional DNA polymerization. *ACS Applied Materials and Interfaces*, *7*(51), 28566–28575. Available from https://doi.org/10.1021/acsami.5b09932.

Wang, Y., Zhang, Y., Yan, T., Fan, D., Du, B., Ma, H., & Wei, Q. (2016). Ultrasensitive electrochemical aptasensor for the detection of thrombin based on dual signal amplification strategy of Au@GS and DNA-CoPd NPs conjugates. *Biosensors and Bioelectronics*, *80*, 640–646. Available from https://doi.org/10.1016/j.bios.2016.02.042.

Wang, Y., Zhang, Z., Jain, V., Yi, J., Mueller, S., Sokolov, J., ... Rafailovich, M. H. (2010). Potentiometric sensors based on surface molecular imprinting: Detection of cancer biomarkers and viruses. *Sensors and Actuators, B: Chemical*, *146*(1), 381–387. Available from https://doi.org/10.1016/j.snb.2010.02.032.

Wei, F., Yang, J., & Wong, D. T. W. (2013). Detection of exosomal biomarker by electric field-induced release and measurement (EFIRM). *Biosensors and Bioelectronics*, *44*(1), 115–121. Available from https://doi.org/10.1016/j.bios.2012.12.046.

Wittes, J., Lakatos, E., & Probstfield, J. (1989). Surrogate endpoints in clinical trials: Cardiovascular diseases. *Statistics in Medicine*, *8*(4), 415–425. Available from https://doi.org/10.1002/sim.4780080405.

Wuethrich, A., Sina, A. A. I., Ahmed, M., Lin, T. Y., Carrascosa, L. G., & Trau, M. (2018). Interfacial nano-mixing in a miniaturised platform enables signal enhancement and: In situ detection of cancer biomarkers. *Nanoscale*, *10*(23), 10884–10890. Available from https://doi.org/10.1039/c7nr09496e.

Yalow, R. S., & Berson, S. A. (1996). Immunoassay of endogenous plasma insulin in man. *Obesity Research*, *4*(6), 583–600. Available from https://doi.org/10.1002/j.1550-8528.1996.tb00274.x.

Ying, L., Nazgul, T., Erlan, R., & Alexander, R. (2010). Aptamer-based electrochemical biosensor for interferon gamma detection. *Analytical Chemistry*, *82*(19), 8131–8136. Available from https://doi.org/10.1021/ac101409t.

Younus, S., Jonas, A. M., & Lakard, B. (2013). Potentiometric biosensors. In G. C. K. Roberts (Ed.), *Encyclopedia of biophysics*. Springer. Available from https://doi.org/10.1007/978-3-642-16712-6_714.

Zamay, G. S., Zamay, T. N., Kolovskaya, O. S., Krat, A. V., Glazyrin, Y. E., Dubinina, A. V., & Zamay, A. S. (2016). Development of a biosensor for electrochemical detection of tumor-associated proteins in blood plasma of cancer patients by aptamers. *Doklady. Biochemistry and Biophysics*, *466*(1), 70–73. Available from https://doi.org/10.1134/S1607672916010208.

Zeng, D., Wang, Z., Meng, Z., Wang, P., San, L., Wang, W., . . . Mi, X. (2017). DNA tetrahedral nanostructure-based electrochemical miRNA biosensor for simultaneous detection of multiple miRNAs in pancreatic carcinoma. *ACS Applied Materials and Interfaces*, 9.

Zhang, Y., Yang, D., Weng, L., & Wang, L. (2013). Early lung cancer diagnosis by biosensors. *International Journal of Molecular Sciences*, *14*(8), 15479–15509. Available from https://doi.org/10.3390/ijms140815479.

Zhou, Q., Rahimian, A., Son, K., Shin, D. S., Patel, T., & Revzin, A. (2016). Development of an aptasensor for electrochemical detection of exosomes. *Methods (San Diego, Calif.)*, *97*, 88–93. Available from https://doi.org/10.1016/j.ymeth.2015.10.012.

## Further reading

Campuzano, S., Pedrero, M., & Pingarrón, J. M. (2017). Non-invasive breast cancer diagnosis through electrochemical biosensing at different molecular levels. *Sensors (Switzerland)*, *17*(9), 1993. Available from https://doi.org/10.3390/s17091993.

Huang, X., Liu, Y., Yung, B., Xiong, Y., & Chen, X. (2017). Nanotechnology-enhanced no-wash biosensors for in vitro diagnostics of cancer. *ACS Nano*, *11*(6), 5238–5292. Available from https://doi.org/10.1021/acsnano.7b02618.

Lin, M., Song, P., Zhou, G., Zuo, X., Aldalbahi, A., Lou, X., . . . Fan, C. (2016). Electrochemical detection of nucleic acids, proteins, small molecules and cells using a DNA-nanostructure-based universal biosensing platform. *Nature Protocols*, *11*(7), 1244–1263. Available from https://doi.org/10.1038/nprot.2016.071.

Chapter 9

# Colorimetric technique-based biosensors for early detection of cancer

Kosar Shahsavar, Aida Alaei and Morteza Hosseini
*Department of Life Science Engineering, Faculty of New Sciences & Technologies, University of Tehran, Tehran, Iran*

## 9.1 Introduction

Cancer refers to a large number of diseases characterized by the abnormal and uncontrollable growth of cells caused by multiple changes in gene expression. Cancer cells can be spread by the bloodstream or lymphatic system and invade other organs or tissues (Maimaitiyiming, Hong, Yang, & Naranmandura 2019). The hallmarks of cancer include resistance to cell death and apoptosis, continuous proliferation even in the presence of growth suppressors, angiogenesis and the construction of new vessels, metastasis, self-renewal and immortality (Cosphiadi et al., 2018). It was reported that cancer remains the second leading cause of death around the world (responsible for one in six deaths in 2018) (Bray et al., 2018). Common types of cancer in men are lung, prostate, colorectal, stomach and liver cancer while breast, colorectal, lung, cervical, and thyroid cancer are the most frequent among women. Cancer can be classified in different ways based on the primary site, tissue type, stage and grade. It is vital to detect the disease in early stages to improve patient survival.

Biomarkers are defined as a measurable characteristic and evaluated as an indicator of normal biological processes, pathogenic processes, or pharmacologic responses to a therapeutic intervention. So many tumor markers have been identified as a clinical cancer biomarker. These molecular biomarkers include three main categories (Fig. 9.1) DNA-based

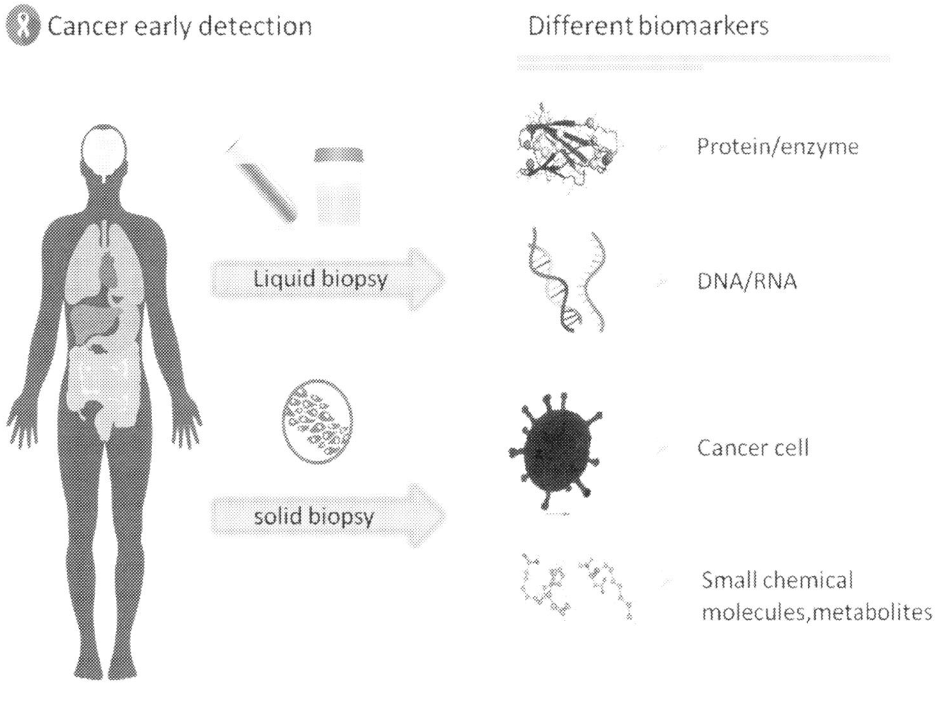


FIGURE 9.1 Biomarkers in cancer early detection.

Biosensor Based Advanced Cancer Diagnostics. DOI: https://doi.org/10.1016/B978-0-12-823424-2.00012-0

biomarkers like sequence variations, epigenetic variations and genome rearrangements; RNA-based biomarkers such as mRNAs and miRNAs signature, protein; and peptide-based biomarkers, which are other biomarkers such as circulating tumor cells and metabolites. There are two types of biopsy approaches based on biomarker origin: liquid biopsy (in the body fluids circulation) and solid biopsy (tumor tissue).

Cancer diagnosis in the early stage led to an increase in the chances for successful treatment that resulted in research for cancer biomarkers via tissue screening or blood samples. Progress in biomarker discovery and development is directly dependent on the capacities of the technologies available. Several detection strategies including the enzyme-linked immunosorbent assay (ELISA) (De La Rica & Stevens, 2012), electrophoresis (Lee, Park, & Nam 2014), polymerase chain reaction (PCR) (Kazane et al., 2012), electrochemical assays (Hasanzadeh, Shadjou, & de la Guardia, 2017; Karimi Pur, Hosseini, Faridbod, Ganjali, & Hosseinkhani 2016; Karimi Pur, Hosseini, Faridbod, Dezfuli, & Ganjali, 2016), fluorescence (Salehnia, Hosseini, & Ganjali, 2017; Shahsavar, Shokri, & Hossein et al., 2020), and colorimetric assays (Abarghoei, Fakhri, Borghei, Hosseini, &Ganjali, 2019; Borghei, Hosseini, & Ganjali, 2018; Borghei, Hosseini, Ganjali, & Ju, 2018; Kermani, Hosseini, Dadmehr, & Ganjali, 2016; Wang, Wu, Ren, & Qu, 2012) have been developed for cancer biomarker detection. Various new bioanalytical technologies are being developed, and a great effort is being done in the area of sensitivity enhancement for reliable and accurate detection of cancer biomarkers.

## 9.2 Colorimetric-based strategy

Many optical-based sensors have been developed to detect different cancer biomarkers based on fluorescence, colorimetric, and chemiluminescence which have been discussed in detail elsewhere in the book. The colorimetric method has advantages compared to other optical methods, such as low cost, no need for complex tools, and specialized operators. The color changes can also be detected with the naked eye. In addition, it does not require advanced and specialized tools for data analysis (Abarghoei et al., 2019; Borghei, Hosseini, & Ganjali, 2018; Borghei, Hosseini, Ganjali, & Hu, 2018; Kermani et al., 2016; Wang et al., 2012). The colorimetric method can be divided in many different ways. Herein, colorimetric biosensors will be categorized and discussed in three groups: nanomaterial-based approach, DNA-based approach, and other approaches.

## 9.3 Nanomaterial-based approach

Nanomaterials with small biological sensing elements such as nucleic acids (DNA, aptamers, and PNA), proteins (enzymes and antibodies), and even whole cells (microorganisms, neurons, and tissue slices) are used to make various biosensors that are accurate devices for detecting a wide range of analytes. The integration of these nanomaterials, which have different structures and properties, with electronic devices has created advanced tools for detecting different materials, which is much more efficient than previous methods (Lee et al., 2014). Until now, a wide range of different nanomaterials, including gold nanoparticles (Assah et al., 2018), nanoclusters (Li et al., 2019), carbon-based nanomaterials (Peng et al., 2019), and nanocomposites (Amanulla et al., 2017), have been reported in colorimetric biosensors to detect cancer biomarkers leading to the early detection of cancers.

### 9.3.1 AuNPs-based colorimetric biosensor

A variety of bio-functionalized metal nanoparticles has been produced and investigated for their potential applications in biological fields. From those investigations, gold nanoparticles (AuNPs) have emerged as the favorite nanomaterials in biomedical, biosensing, and imaging. AuNPs are used in biological and clinical diagnosis and therapy applications due to their higher extinction coefficient compared to organic dye molecules and their unique optical properties that endorse surface plasmon resonance (SPR) and localized SPR (LSPR). Furthermore, AuNPs are utilized in the design of biosensors owing to their biocompatibility, ease of characterization, and high chemical reactivity at the nanometer scale, allowing surface modification and reactions with a wide variety of chemical and biochemical vectors. The SPR of AuNPs has a vigorous absorption band at a wavelength of about 520 nm (Scaffardi, Lester, Skigin, & Tocho, 2007). The colorimetric behavior of AuNPs is affiliated with their SPR, which strongly depends on particle size, shape, geometry, the refractive index of the medium, interparticle distance, and aggregation state in solution. Any shift in these parameters changes the plasmon-resonance frequency, influencing the color and the strength of the plasmonic increment that affects severity (Lin, Huang, & Chang, 2011). Since the size of the nanoparticles affects their optical properties, gold nanoparticles with diameters of 13 and 56 nm show the highest SPR extinction at 520 and 530 nm, respectively, and the color of the corresponding solution is light red to dark red (Huang & Chang, 2007). AuNPs are synthesized via two methods including chemical reduction and ligand passivation. In both methods, gold (III) salt is converted to gold (0), which is done in the presence of a reducing agent. This reducing agent not only reduces but also acts as a stabilizing agent to prevent the accumulation of gold nanoparticles (Kumar, Gandhi, & Kumar, 2007). Glucose, gallic acid, and citrate are used as reducing and stabilizing agents, but these can be replaced by other factors such as DNA, peptides, and thiols that bind more strongly to AuNPs. In the second approach is the synthesis of AuNPs is biphasic, which is mostly used for thiol-stabilized AuNPs in organic solvents (Palazzo, Facchini, & Mallardi, 2012). In colorimetric biosensors based on AuNPs, depending on the system designed, the nanoparticles can be functionalized by materials that can detect a specific analyte. These materials can bond to the surface of gold nanoparticles in a variety of methods: (1) electrostatic interactions that occur between the negative charge of AuNPs and the positive charge of the domain of the biomolecule; (2) the covalent bond formed by the thiol biomolecule group and the surface of AuNPs; (3) amide bond formation between carboxyl groups on the AuNPs surfaces and amine groups present on the biomolecule; and (4) affinity interactions between streptavidin labeled AuNPs and biotin-modified biomolecules. There are two main approaches to controlled aggregation of colloidal AuNPs: (1) crosslinking aggregation mechanism, and (2) noncrosslinking aggregation mechanism (Yang, Luo, Tian, Qian, & Duan, 2019).

#### 9.3.1.1 Cross-linked aggregation

According to the assays, it seems that crosslinking compared to noncrosslinking is a more common approach to bring gold nanoparticles closer to each other. This approach may be induced either by (1) target molecules possessing multiple binding sites for the receptor molecules previously attached to the nanoparticles; (2) by the direct interaction between receptor-modified nanoparticles and antireceptor-modified nanoparticles. Subsequently, a color change from red to purple or dark blue associated with a shift of the SPR's peak to higher wavelengths can be seen. The crosslinking aggregation-based assays include strategies which use DNA hybridization (Cai, 2019), aptamer-target interactions (Borghei et al., 2016), antibody-antigen interactions (Karami et al., 2019), streptavidin-biotin interactions (Abnous, Danesh, Ramezani, Emrani, & Taghdisi, 2016), lectin-sugar interactions (Hu, Li, & Guo, 2015) and metal-ligand coordination (Jin, 2015) to induce the aggregation of the colloidal solutions (or the redispersion of aggregated colloids). This platform for the detection of targets relies on the presence of crosslinkers that bind to the sites on the AuNP surface. This method has also been utilized for the detection of a variety of biological molecules (Li et al., 2019).

#### 9.3.1.2 Salt-induced aggregation

Noncrosslinking aggregation or salt-induced aggregation is another method for controlling the aggregation of AuNPs. Van der Waals attraction forces between gold nanoparticles lead to their accumulation without forming bonds between particles. Electrostatic, steric, or electrosteric forces prevent gold nanoparticles aggregation by overcoming the van der Waals attractive forces. Decreasing the stability of colloidal NPs is possible in three ways; (1) induced loss of electrostatic stabilization; (2) induced loss of (electro) steric stabilization; and (3) induced loss of colloidal stability upon (charged) polymer conformational transitions (Li & Rothberg, 2004; Li, Schluesener, & Xu, 2010).

### 9.3.2 Nanoclusters

Nanoclusters are monodispersed particles that are less than 10 nm in diameter. Nanoclusters have physical properties varying substantially from their bulk and single-particle species. For this reason, they have unique and versatile applications including quantum computers, devices, chemical sensors, light-emitting diodes, ferrofluids for cell separations, industrial lithography, and photochemical pattern applications such as flat-panel displays. Nanoclusters also have significant potential as new types of catalysts with higher activity and selectivity. There are five methods for synthesizing nanoclusters: transition metal salt reduction, thermal decomposition and photochemical methods, ligand reduction and displacement from organometallics, metal vapor synthesis, and electrochemical synthesis. Nanoclusters must be stabilized against aggregation and the formation of larger particles and eventually, bulk material. Stabilization can be accomplished in two precedented ways: electrostatic charge, or "inorganic" stabilization and steric "organic" stabilization. Electrostatic stabilization occurs by the adsorption of ions to the often electrophilic metal surface. Steric stabilization is achieved by surrounding the metal center by layers of material that are sterically bulky, such as polymers or surfactants. Metal clusters which have a complete regular outer geometry are designated full-shell or "magic number" clusters. Full-shell clusters are constructed by successively packing layers—or shells—of metal atoms around a single metal atom (Wilcoxon & Abrams, 2006). In general, the detection of cancer biomarkers by nanoclusters has been widely studied. A novel colorimetric method based on Ag/Pt nanoclusters was developed to assay DNA methyl transferase

(MTase) activity. DNA MTase unbalanced levels are associated with cancer and bacterial diseases. According to the inhibition of the peroxidase reaction that occurred in the $TMB-H_2O_2$ system, in the presence of MTase, a highly sensitive and selective colorimetric biosensor was designed with a detection limit (LOD) of 0.05 U/mL and a linear range from 0.5 to 10 U/mL (Kermani, n.d.). Furthermore, a sensitive and simple colorimetric method was designed for the detection of miRNA-21 with a limit of detection of 0.6pM. The assay mechanism was based on the inhibitory effect of miRNA-21 on the peroxidase-like activity of DNA−Ag/Pt NCs (Fakhri et al., 2020).

### 9.3.3 Carbon nanomaterial-based biosensor

Carbon-based nanomaterials (CBNs) have received much attention in various fields due to their unique physical and chemical properties. CBNs, including carbon nanotubes (CNTs), graphene oxide (GO), and graphene quantum dots (GQDs), have been extensively investigated in biomedical applications (Xie, 2019). One of the most widely used is graphene which consists of one atom-thick planar sheets of sp2-bound carbon atoms densely packed into a honeycomb crystal lattice. The essential physical properties of defect-free graphene of particularly excellent mechanical resilience with a Young's modulus of 1100 GPa, high planar surface area, high thermal conductivity, elevated courier motility and capability, make it a perfect aspirant for a variety of uses containing energy, electronics, molecular sensing, and catalysis. Apart from its superior mechanical and electronic structure, lately, a study that concentrated on graphene toxicity on human cells proved that graphene-based nanomaterials have lower toxicities compared with carbon nanotubes. This finding endorses the terrific guarantee for the performance of graphene into the biomedical applications such as biomarker sensors, delivery of chemotherapeutic for the treatment of the cancer cells, and hyperthermia treatment. The production of graphene-based nanofluids is highly efficient due to their stability and high concentration. Unlike graphene oxide, which has good hydrophilicity due to functional oxygen groups at the edges and basal planes, graphene and reduced graphene oxide sheets do not disperse, owing to hydrophobicity in water and organic solvents. Therefore, they must be functionalized by covalent or noncovalent interaction (Ossonon & Bélanger, 2017). By virtue of their brilliant charge mobility, high surface area-to-volume ratio, geometric structure, and optical and electrical properties, graphene-based nanomaterials are highly multipurpose and provide the perfect nanoplatforms for nanobiosensors (nanoscale biosensors). Consequently, they are broadly utilized in diverse views for cancer diagnosis. Graphene-based nanobiosensors are highly sensitive and specific, expand the limit of detection, and are able to detect various cancer biomarkers simultaneously, thereby enhancing the early cancer diagnosis and screening (Eskiizmir, Baskin, & Yapici, 2018). In a study published in 2016, a colorimetric method was performed to diagnose cancer. They used the peroxidase activity of graphene/gold-nanoparticle (Au-NP) hybrids which was controlled by using single-stranded PNA-21 (ssPNA-21) for the detection of microRNA21. In the absence of microRNA21, the spontaneous absorption of ssPNA-21 on graphene/Au-NP hybrid surfaces caused the peroxidase-like catalytic activity of hybrids to be almost completely deactivated via $\pi-\pi$ stacking interactions between ssPNA-21 and graphene, thus TMB is not converted to the TMBox, resulting in no visible blue discoloration. On the other hand, in the presence of target microRNA due to the interaction between microRNA and its complementary ssPNA-21, graphene/Au-NP hybrids can convert TMB to TMBox by peroxidase-like activity which is accompanied by a change in color to blue. This sensor emitted a low background signal and responded linearly to miRNA-21 from 10 nM to 0.98 mM with a detection limit of 3.2 nM under optimal conditions (Zhao, Qu, Yuan, & Quan, 2016).

### 9.3.4 Nanocomposite-based biosensor

Nanocomposites are multiphase nanomaterials that involve components, one of which shows dimensions in the nanometer range. These complex materials can be considered in three main subgroups based on their matrix: polymeric, ceramic, and metallic-based nanocomposite. Different nanoparticles can be used as fillers in the nanocomposite matrix to improve the optical, electrical, mechanical, catalytic, and thermal properties, and carbon nanotube-reinforced composites are one of the most interesting ones which have been widely studied. Generally, human beings have made efforts to improve the quality and efficiency of materials for different applications while they did not know the exact structure of materials. Nowadays, with the advancement of science and technology, the improvement of properties in nanocomposites compared to traditional microscale composites can be justified. There are several methods for synthesizing nanocomposites, including mechanical alloying, sol-gel synthesis, thermal spraying, and microwave-induced synthesis. Nanocomposites demonstrate numerous applications in various fields such as environmental application (water purification), electronics industry (supercapacitor development), biotechnology and biomedical application. The most important nanocomposite applications in the biomedical field are comprised of medical implants, drug delivery, antimicrobial properties, tissue engineering, wound healing, cancer therapy, biosensors, and bioimaging (Ajayan, Schadler, & Braun 2006). For the sake of designing a sensitive colorimetric diagnostic system, Alizadeh et al. have developed a paper-based immune-device based on $Co_2(OH)_2CO_3{-}CeO_2$ nanocomposite. Carcinoembryonic antigen (CEA), an important tumor marker, was detected with a detection limit of 0.51 pg $mL^{-1}$. $Co_2(OH)_2CO_3{-}CeO_2$ nanocomposites were synthesized through the hydrothermal method and showed excellent catalytic activity (Alizadeh, Salimi, Hallaj, Fathi, & Soleimani, 2018). As mentioned before, the increase in catalytic activities is one of the nanocomposite properties which has been frequently used in biosensor design. Teng et al. have proposed a specific colorimetric sensor based on folate-conjugated gold-iron oxide composite nanoparticles ($Au{-}Fe_2O_3$ CNPs) for cancer cell detection (Teng, Shi, & Pong, 2019).

## 9.4 DNA-based approach

DNA molecules have been utilized as a biomaterial in diagnostic fields due to their specific properties like self-assembly, programmability, biocompatibility, and thermodynamic stability (Chen et al., 2018; Seeman & Sleiman, 2017). There are two different approaches for nanomaterial fabrication, top-down and bottom-up. DNA as biomaterials can be used to manufacture nanostructure through the bottom-up approach. The bottom-up approach uses internal information like chemical properties of DNA molecules to lead their autonomous self-assembly into nanostructures.

The nucleic acid sequence can act as a hybridization probe or DNA aptamer to recognize target molecules. DNA oligonucleotides also show catalytic activity, including ribonucleases (RNA-cleavage DNAzymes) and peroxidase-like activity (G-quadruplex DNAzymes). RNA-cleavage and G-quadruplex DNAzymes are capable of playing a role in target recognition and signal transduction in the biosensing process, respectively. It is worth mentioning that different strategies have been used to improve the sensitivity of sensor platforms like conjugation of DNA and nanoparticles, utilization of DNA nanostructure, and DNA-based amplification technique (HCR, PCR, RCA, etc.). The application of DNA-based sensing platforms in cancer diagnosis will be discussed briefly in this section.

### 9.4.1 DNA aptamer platform

Aptamers are single-stranded DNA/RNA molecules that can detect specific target molecules through their unique tertiary structures. Due to their molecular recognition properties, aptamers are used in diagnostics, imaging, and therapeutic applications. These molecules are referred to as chemical alternatives for antibodies and have advantages over protein antibodies such as easy production and modification, small size, lower immunogenicity, and physical and thermal stability. Aptamers are generated by an in vitro process called systematic evolution of ligands by exponential enrichment (SELEX). Numerous specific aptamers against cancer biomarkers have been developed through this iterative process and used in sensor strategies for early cancer detection. Many colorimetric aptamer-based biosensors (aptasensors) have been widely developed for the early detection of cancer biomarkers. The major part of colorimetric aptasensors includes Au nanoparticle aptasensors and G-quadruplex DNAzyme aptasensors.

#### *9.4.1.1 AuNp-based aptasensors*

Gold nanoparticles (AuNPs) have shown tremendous potential in colorimetric aptasensors, because of their interesting photophysical properties. The surface plasmon resonance (SPR) of AuNps is influenced by particle size, morphology, and interparticle distance. The AuNp solution will be changed from wine red to blue due to nanoparticle aggregation. AuNP aptasensors are divided into two categories: DNA-functioned and label-free. AuNp-based aptasensors possess several advantages such as high sensitivity, simplicity, and need for only inexpensive equipment. There are numerous gold nanoparticle-based aptasensors for different cancer biomarker detections (Hu et al., 2015). Cancer cell detection could be useful for analyzing tissue samples and capturing circulating tumor cells in order to detect cancer at early stages. The tumor cells spread from the primary tumor site into body fluids like blood, urine, ascites, and pleural fluid during early stages in cancer progression. Different studies are developed for cancer cell detection via AuNp nanoparticles (Medley et al., 2008; Medley, Bamrungsap, Tan, & Smith, 2011). Borghei et al. have proposed a novel colorimetric biosensor for visual detection of cancer cells based on gold nanoparticles (AuNPs) aggregation via DNA hybridization (Fig. 9.2). The specific interaction between nucleolin receptors on the cancer cell surface and its specific aptamer (AS1411) leads to the removal of AS1411 from the solution. Therefore, there is no linker to assemble AuNps aggregation and the red color of separated AuNPs was observed (Borghei et al., 2016).

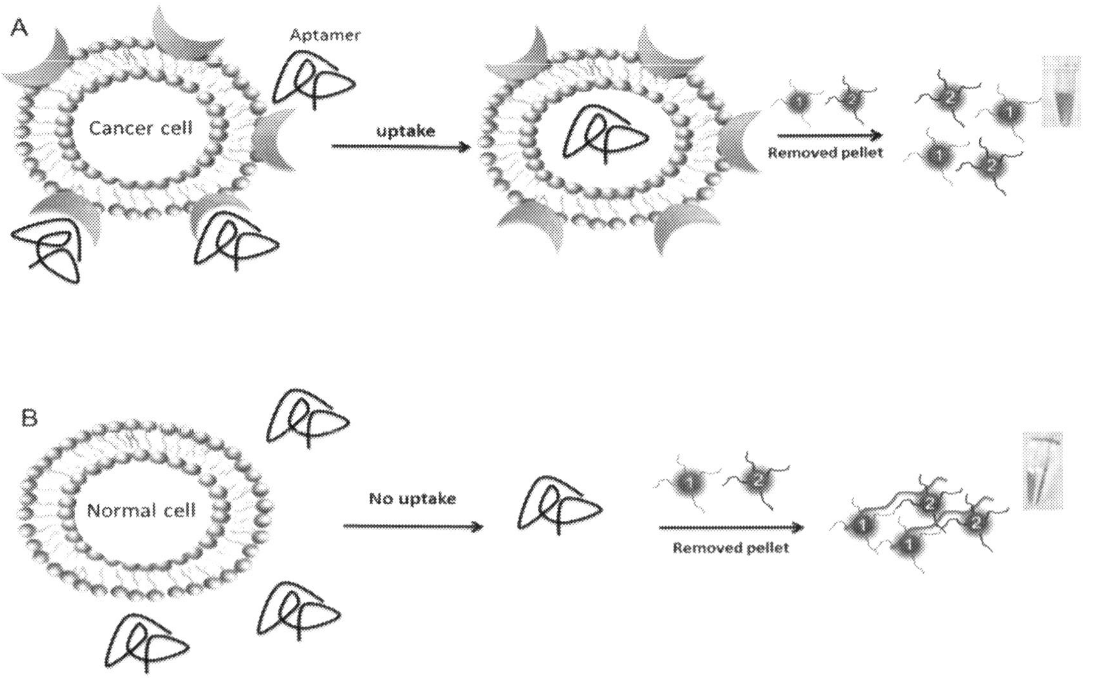


**FIGURE 9.2** Schematic representation of the selective colorimetric method for detection of cancer cells by employing DNA probe 1,2 -functionalized gold nanoparticles and AS1411 aptamer. *From Borghei, Y. S., Hosseini, M., Dadmehr, M., Hosseinkhani, S., Ganjali, M. R., & Sheikhnejad, R. (2016). Visual detection of cancer cells by colorimetric aptasensor based on aggregation of gold nanoparticles induced by DNA hybridization.* Analytica Chimica Acta, 904, *92–97. https://doi.org/10.1016/j.ac.2015.11.026.*

### *9.4.1.2 G-quadruplex DNAzyme-based aptasensors*

G-quadruplex DNAzyme is a single-stranded guanine rich DNA oligonucleotide which folds into the quadruplex structure and shows peroxidase-like activity. The presence of monovalent cations such as $K^+$ and $Na^+$ leads to the formation of stable quadruplex structures. G-quadruplex DNAzyme with peroxidase activity can be used with different chromogenic substrates like 2,2-azino-bis (3-ethylbenzothiazo-line-6-sulfonicacid) diammonium salt (ABTS) or 3,3′,5,5′-tetramethylbenzidine sulfate (TMB) to produce a colorimetric signal in the presence of $H_2O_2$. Compared to native horseradish peroxidase, DNAzymes show several advantages like high chemical and thermal stability, low cost, simple preparation, and easy modification. G-quadruplex DNAzyme aptasensors can be classified into different subclasses including colorimetric, chemiluminescence and fluorescence sensors. The typical design of DNAzyme-based aptasensors is such that the presence of the target molecule and subsequent interaction with the aptamer causes the switch structure from OFF to ON mode and a robust colorimetric signal is generated (Zhou, Xu, Wang, & Ye, 2020). As a type of on-off switch biosensor, Zhou et al. have designed a sensitive and low-cost colorimetric aptasensor based on hairpin-like structures for breast cancer-derived exosome detection (Fig. 9.3). This hairpin-like structure is composed of MUC1 aptamer tethered with mimicking the DNAzyme sequence. Interactions between the MUC1 aptamer and target lead to the opening of the hairpin structure and the formation of a quadruplex structure. Quadruplex DNAzyme generated a strong colorimetric signal in the presence of $H_2O_2$ and ABTS. The LOD for exosomes was $3.94 \times 10^5$ particles per mL.

### *9.4.1.3 Other colorimetric aptasensors*

There is a vast range of nanomaterials, including carbon nanomaterials, metal-based nanoparticles, and polymer-based nanoparticles which have been reported with intrinsic enzyme activity (nanozymes). Colorimetric aptasensors in cancer biomarker detection are not limited to the AuNps-based or DNAzyme-based approach; there are plenty of colorimetric aptasensors conjugated to other nanomaterials (nanozymes). Briefly, nanomaterials like graphitic carbon nitride nanosheets (Wang et al., 2017), porous platinum nanoparticles on graphene oxide (Zhang et al., 2014), single-walled carbon nanotubes (Xia et al., 2017), and cobalt hydroxide decorated mesoporous carbon (Mesgari et al., 2020) have been designed and used for the colorimetric detection of cancer in early stages.

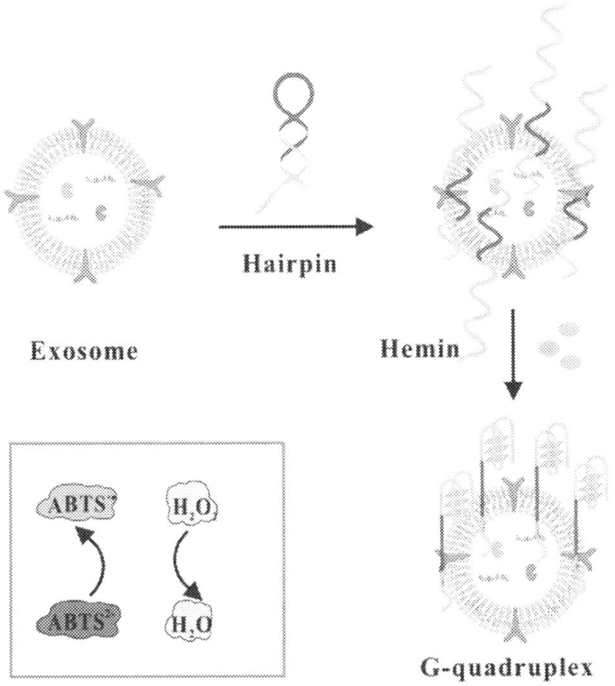


**FIGURE 9.3** Schematic illustration of the colorimetric assay for the detection of exosomes by binding to hairpin structures based on the combination of the MUC1 aptamer and G-quadruplex-mimetic enzyme. *From Zhou, Y., Xu, H., Wang, H., & Ye, B. C. (2020). Detection of breast cancer-derived exosomes using the horseradish peroxidase-mimicking DNAzyme as an aptasensor.* Analyst, *145(1), 107–114. https://doi.org/10.1039/c9an01653h.*

### 9.4.2 DNA probe platform

DNA probe refers to single-stranded DNA which is used to detect complementary DNA sequences through hybridization. Chemical methods usually produce these hybridization probes. Recently, DNA-based detection tests have received much attention due to their applications in different fields, including DNA-related biomarker diagnosis, gene analysis, forensics, and biological warfare agent detection. The hybridization and duplex formation between the DNA probe and DNA target can be detected by optical and electrochemical devices as a signal transducer. DNA probes are usually labeled with radioisotopes and fluorophores and these labels enable DNA detection via fluorometric methods. However, in the case of colorimetric methods, probes are usually coupled with nanomaterials, nanostructures, and nanozymes which can produce the colored product. Cancer-related genes and microRNAs (miRNA) are two important DNA-related biomarkers in the early detection of cancer. miRNAs are small (18–23 nt), single-stranded, noncoding RNA molecules that are involved in many biological processes like regulation of gene expression at the posttranscriptional level, organ development, and differentiation. These small molecules have been introduced as a novel generation of biomarkers in many diseases such as cancer, cardiovascular disorders, viral infections, neurological disorders, hematologic diseases, and other diseases. The limitations of conventional miRNA detection methods and a growing interest in miRNA detection have led to the development of numerous diagnostic strategies. DNA-based colorimetric strategies in the miRNA diagnostic field can be classified into two subclasses: (1) Molecular beacons (Li et al., 2016), using G-quadruplex DNAzymes for transduction (Lan, Wang, Liu, & Cheng, 2019), spherical nucleic acids, or DNA incorporates with other materials such as nanoparticles, nanoclusters, nanotubes, and graphene; and (2) DNA nanostructures like the DNA Ferris wheel (Zhou et al., 2020).

### 9.4.3 Nucleic acid amplification techniques

Fast and sensitive diagnosis of disease-associated nucleic acids has great importance in early disease detection and improvement of patient survival. Nucleic acid amplification strategies have been widely used in diagnosis, gene cloning, and research for accurate DNA detection. These methods can be used for target amplification and signal amplification. Among amplification techniques, the PCR was the first and most popular one. Limitations of PCR methods like high cost and their dependence on thermocycling gave birth to new alternative methods. Numerous enzyme-based and enzyme-free amplification strategies have been employed to exponentially increase minuscule amounts of DNA and to improve biosensors' sensitivity. The former includes enzyme-assisted nucleic acid amplification technologies, such as nuclease-assisted hairpin assembly, enzyme-catalyzed reaction, and the latter methods include catalyzed hairpin

**TABLE 9.1** Some of the colorimetric methods for cancer biomarkers detection based on nucleic acid amplification techniques.

| Signal amplification method | Target DNA | Limit of detection | Reference |
|---|---|---|---|
| Universal hybridization chain reaction (uHCR) and hemin/G-quadruplex | miR-21,miR-125b, KRAS-Q61K, and BRAF-V600E | 5.67 nM | (Park, 2019) |
| Target-triggering rolling circle amplification (RCA) strategy and DNAzyme | microRNA-378 | 80 fM | (Liu, 2019) |
| Versatile palindromic molecule beacon and triple cyclical strand displacement amplification (SDA) with DNAzyme as transducer | P53 gene | Down to 10 pM | (Shen, 2018) |
| Strand displacement amplification (SDA) with G-quadruplex/hemin DNAzyme | PIK3CA gene | As low as 0.2% PIK3CA mutation | (Shen, 2018) |
| Two-step amplification of DNA machine | K-ras gene | 10 pM | (Xu, 2017) |

assembly (CHA), hybridization chain reaction (HCR), and toehold-triggered strand displacement reaction (TSDR). In addition to the above classification, isothermal amplification techniques are an enormous subgroup of DNA amplification strategies. These methods are comprised of processes that amplify and accumulate DNA sequences in constant temperatures. Isothermal amplification can be enzyme-associated or enzyme-free. There are three types of isothermal techniques based on their kinetic reaction; exponential, linear, and cascade amplification. Among the isothermal amplification methods, nucleic acid sequence-based amplification (NASBA), strand displacement amplification (SDA), rolling circle amplification (RCA), and loop-mediated isothermal amplification (LAMP) have been applied in the colorimetric platform for cancer biomarker detection. DNA amplification strategies are extremely diverse and complex and, in this chapter, we have tried to present an introduction of this vast area. A list of colorimetric sensing strategies which use the amplification techniques is provided in Table 9.1.

## 9.5 Other approaches

Besides nanomaterials and DNA-nanostructure, various compounds have been used for the colorimetric detection of cancer such as metal-organic framework (MOF), dyes (Zhao et al., 2016), and zwitterionic compounds. Metal-organic frameworks (MOFs) formed by self-assembly of metal ions and organic linkers have recently emerged as new versatile materials. A new Ni-hemin MOF nanostructure via an one-step hydrothermal method was developed for sensitive colorimetric detection of cancer cells (Alizadeh, Salimi, & Hallaj, 2018). Zwitterionic compounds are a kind of charge-neutral molecules with positive/negative charge balanced in the same segment. Zwitterionic sulfopropy-betaines was used for the first time as naked-eye visible colorimetric pH sensors for cancer cell discriminator (He, Xin, & Chen, 2017).

## 9.6 Conclusion

Improvement of patient life quality and survival can be achieved by the detection of cancer at an early stage. Because of that, many researchers are interested in introducing new, reliable, high-throughput methods for cancer detection despite the conventional standard methods. Colorimetric-based sensing strategies compared to other optical-based strategies show advantages like rapidness, simplicity, and high sensitivity. Simple step detection and naked-eye detection in the colorimetric sensors make them ideal candidates to develop a novel generation of biosensors for cancer detection. Although ELISA and PCR remain as the gold standard methods for protein and nucleic acid assays in clinical diagnosis, each of them still has deficiencies for advanced diagnostic applications. Nowadays, nanomaterial and nanostructure with their specific physical and chemical properties improve biosensor sensitivity and efficiency. Furthermore, colorimetric biosensors utilize nucleic acid amplification techniques to detect a very low level of the biomarker. In this chapter, we aimed to review the current colorimetric methods for the sensitive detection of cancer biomarkers including proteins, enzymes, nucleic acids, cancer cells, and small molecules. With respect to the number of investigations on cancer biomarkers profile and cancer detection methods, comprehensive work still needs to be carried out.

## References


Abarghoei, S., Fakhri, N., Borghei, Y. S., Hosseini, M., & Ganjali, M. R. (2019). A colorimetric paper sensor for citrate as biomarker for early stage detection of prostate cancer based on peroxidase-like activity of cysteine-capped gold nanoclusters. *Spectrochimica Acta – Part A: Molecular and Biomolecular Spectroscopy, 210*, 251–259. Available from https://doi.org/10.1016/j.saa.2018.11.026.

Abnous, K., Danesh, N. M., Ramezani, M., Emrani, A. S., & Taghdisi, S. M. (2016). A novel colorimetric sandwich aptasensor based on an indirect competitive enzyme-free method for ultrasensitive detection of chloramphenicol. *Biosensors and Bioelectronics, 78*, 80–86. Available from https://doi.org/10.1016/j.bios.2015.11.028.

Ajayan, P.M., Schadler, L.S., & Braun, P.V. (2006). Nanocomposite science and technology.

Alizadeh, N., Salimi, A., & Hallaj, R. (2018). Mimicking peroxidase activity of Co2(OH)2CO3-CeO2 nanocomposite for smartphone based detection of tumor marker using paper-based microfluidic immunodevice. *Talanta, 189*, 100–110. Available from https://doi.org/10.1016/j.talanta.2018.06.034.

Alizadeh, N., Salimi, A., Hallaj, R., Fathi, F., & Soleimani, F. (2018). Ni-hemin metal-organic framework with highly efficient peroxidase catalytic activity: Toward colorimetric cancer cell detection and targeted therapeutics. *Journal of Nanobiotechnology, 16*(1). Available from https://doi.org/10.1186/s12951-018-0421-7.

Amanulla, B., Palanisamy, S., Chen, S. M., Chiu, T. W., Velusamy, V., Hall, J. M., ... Ramaraj, S. K. (2017). Selective colorimetric detection of nitrite in water using chitosan stabilized gold nanoparticles decorated reduced graphene oxide. *Scientific Reports, 7*(1). Available from https://doi.org/10.1038/s41598-017-14584-6.

Assah, E., Goh, W., Zheng, X. T., Lim, T. X., Li, J., Lane, D., ... Tan, Y. N. (2018). Rapid colorimetric detection of p53 protein function using DNA-gold nanoconjugates with applications for drug discovery and cancer diagnostics. *Colloids and Surfaces B: Biointerfaces, 169*, 214–221. Available from https://doi.org/10.1016/j.colsurfb.2018.05.007.

Borghei, Y. S., Hosseini, M., Dadmehr, M., Hosseinkhani, S., Ganjali, M. R., & Sheikhnejad, R. (2016). Visual detection of cancer cells by colorimetric aptasensor based on aggregation of gold nanoparticles induced by DNA hybridization. *Analytica Chimica Acta, 904*, 92–97. Available from https://doi.org/10.1016/j.ac.2015.11.026.

Borghei, Y. S., Hosseini, M., & Ganjali, M. R. (2018). Oxidase-like catalytic activity of Cys-AuNCs upon visible light irradiation and its application for visual miRNA detection. *Sensors and Actuators, B: Chemical, 273*, 1618–1626. Available from https://doi.org/10.1016/j.snb.2018.07.061.

Borghei, Y. S., Hosseini, M., Ganjali, M. R., & Ju, H. (2018). Colorimetric and energy transfer based fluorometric turn-on method for determination of microRNA using silver nanoclusters and gold nanoparticles. *Microchimica Acta, 185*(6). Available from https://doi.org/10.1007/s00604-018-2825-3.

Bray, F., Ferlay, J., Soerjomataram, I., Siegel, R. L., Torre, L. A., & Jemal, A. (2018). Global cancer statistics 2018: GLOBOCAN estimates of incidence and mortality worldwide for 36 cancers in 185 countries. *CA Cancer Journal for Clinicians, 68*(6), 394–424. Available from https://doi.org/10.3322/caac.21492.

Cai, J. (2019). A colorimetric detection of microRNA-148a in gastric cancer by gold nanoparticle–RNA conjugates. *Nanotechnology, 31*, 9.

Chen, T., Ren, L., Liu, X., Zhou, M., Li, L., Xu, J., & Zhu, X. (2018). DNA nanotechnology for cancer diagnosis and therapy. *International Journal of Molecular Sciences, 19*(6). Available from https://doi.org/10.3390/ijms19061671.

Cosphiadi, I., Atmakusumah, T. D., Siregar, N. C., Muthalib, A., Harahap, A., & Mansyur, M. (2018). Bone metastasis in advanced breast cancer: Analysis of gene expression microarray. *Clinical Breast Cancer, 18*(5), e1117–e1122. Available from https://doi.org/10.1016/j.clbc.2018.03.001.

De La Rica, R., & Stevens, M. M. (2012). Plasmonic ELISA for the ultrasensitive detection of disease biomarkers with the naked eye. *Nature Nanotechnology, 7*(12), 821–824. Available from https://doi.org/10.1038/nnano.2012.186.

Eskiizmir, G., Baskin, Y., & Yapici, K. (2018). *Graphene-based nanomaterials in cancer treatment and diagnosis. Fullerenes, graphenes and nanotubes: A pharmaceutical approach* (pp. 331–374). Elsevier. Available from https://doi.org/10.1016/B978-0-12-813691-1.00009-9.

Fakhri, N., Abarghoei, S., Dadmehr, M., Hosseini, M., Sabahi, H., & Ganjali, M. R. (2020). Paper based colorimetric detection of miRNA-21 using Ag/Pt nanoclusters. *Spectrochimica Acta– Part A: Molecular and Biomolecular Spectroscopy*, 227. Available from https://doi.org/10.1016/j.saa.2019.117529.

Hasanzadeh, M., Shadjou, N., & de la Guardia, M. (2017). Early stage screening of breast cancer using electrochemical biomarker detection. *TrAC - Trends in Analytical Chemistry, 91*, 67–76. Available from https://doi.org/10.1016/j.trac.2017.04.006.

He, L., Xin, J. H., & Chen, S. (2017). Robust and low cytotoxic betaine-based colorimetric pH sensors suitable for cancer cell discrimination. *Sensors and Actuators, B: Chemical, 252*, 277–283. Available from https://doi.org/10.1016/j.snb.2017.05.180.

Hu, X. L., Jin, H. Y., He, X. P., James, T. D., Chen, G. R., & Long, Y. T. (2015). Colorimetric and plasmonic detection of lectins using core-shell gold glyconanoparticles prepared by copper-free click chemistry. *ACS Applied Materials and Interfaces, 7*(3), 1874–1878. Available from https://doi.org/10.1021/am5076293.

Hu, Y., Li, L., & Guo, L. (2015). The sandwich-type aptasensor based on gold nanoparticles/DNA/magnetic beads for detection of cancer biomarker protein AGR2. *Sensors and Actuators, B: Chemical, 209*, 846–852. Available from https://doi.org/10.1016/j.snb.2014.12.068.

Huang, C. C., & Chang, H. T. (2007). Parameters for selective colorimetric sensing of mercury(II) in aqueous solutions using mercaptopropionic acid-modified gold nanoparticles. *Chemical Communications, 12*, 1215–1217. Available from https://doi.org/10.1039/b615383f.

Jin, W. (2015). Colorimetric detection of Cr 3+ using gold nanoparticles functionalized with 4-amino hippuric acid. *Journal of Nanoparticle Research, 17*, 9.

Karami, P., Khoshsafar, H., Johari-Ahar, M., Arduini, F., Afkhami, A., & Bagheri, H. (2019). Colorimetric immunosensor for determination of prostate specific antigen using surface plasmon resonance band of colloidal triangular shape gold nanoparticles. *Spectrochimica Acta – Part A: Molecular and Biomolecular Spectroscopy*, 222. Available from https://doi.org/10.1016/j.saa.2019.117218.

Karimi Pur, M. R., Hosseini, M., Faridbod, F., Ganjali, M. R., & Hosseinkhani, S. (2016). Early detection of cell apoptosis by a cytochrome C label-free electrochemiluminescence aptasensor. *Sensors and Actuators B: Chemical*.

Kazane, S. A., Sok, D., Cho, E. H., Uson, M. L., Kuhn, P., Schultz, P. G., & Smider, V. V. (2012). Site-specific DNA-antibody conjugates for specific and sensitive immuno-PCR. *Proceedings of the National Academy of Sciences*, 3731–3736. Available from https://doi.org/10.1073/pnas.1120682109.

Kermani, H.A. (n.d.). A colorimetric assay of DNA methyltransferase activity based on peroxidase mimicking of DNA template Ag/Pt bimetallic nanoclusters. Analytical and Bioanalytical Chemistry, 410, 4943–4952.

Kermani, H. A., Hosseini, M., Dadmehr, M., & Ganjali, M. R. (2016). Rapid restriction enzyme free detection of DNA methyltransferase activity based on DNA-templated silver nanoclusters. *Analytical and Bioanalytical Chemistry*, *408*(16), 4311–4318. Available from https://doi.org/10.1007/s00216-016-9522-z.

Kumar, S., Gandhi, K. S., & Kumar, R. (2007). Modeling of formation of gold nanoparticles by citrate method. *Industrial and Engineering Chemistry Research*, *46*(10), 3128–3136. Available from https://doi.org/10.1021/ie060672j.

Lan, L., Wang, R. L., Liu, L., & Cheng, L. (2019). A label-free colorimetric detection of microRNA via G-quadruplex-based signal quenching strategy. *Analytica Chimica Acta*, *1079*, 207–211. Available from https://doi.org/10.1016/j.ac.2019.06.063.

Lee, H., Park, J. E., & Nam, J. M. (2014). Bio-barcode gel assay for microRNA. *Nature Communications*, *5*, 3367. Available from https://doi.org/10.1038/ncomms4367.

Li, D., Cheng, W., Yan, Y., Zhang, Y., Yin, Y., Ju, H., & Ding, S. (2016). A colorimetric biosensor for detection of attomolar microRNA with a functional nucleic acid-based amplification machine. *Talanta*, *146*, 470–476. Available from https://doi.org/10.1016/j.talanta.2015.09.010.

Li, M., Lao, Y. H., Mintz, R. L., Chen, Z., Shao, D., Hu, H., ... Leong, K. W. (2019). A multifunctional mesoporous silica-gold nanocluster hybrid platform for selective breast cancer cell detection using a catalytic amplification-based colorimetric assay. *Nanoscale*, *11*(6), 2631–2636. Available from https://doi.org/10.1039/c8nr08337a.

Li, H., & Rothberg, L. (2004). Colorimetric detection of DNA sequences based on electrostatic interactions with unmodified gold nanoparticles. *Proceedings of the National Academy of Sciences*, 14036–14039. Available from https://doi.org/10.1073/pnas.0406115101.

Li, Y., Schluesener, H. J., & Xu, S. (2010). Gold nanoparticle-based biosensors. *Gold Bulletin*, *43*(1), 29–41. Available from https://doi.org/10.1007/BF03214964.

Lin, Y. W., Huang, C. C., & Chang, H. T. (2011). Gold nanoparticle probes for the detection of mercury, lead and copper ions. *Analyst*, *136*(5), 863–871. Available from https://doi.org/10.1039/c0an00652a.

Liu., et al. (2019). *Analytical Methods*.

Maimaitiyiming, Y., Hong, D. F., Yang, C., & Naranmandura, H. (2019). Novel insights into the role of aptamers in the fight against cancer. *Journal of Cancer Research and Clinical Oncology*, *145*(4), 797–810. Available from https://doi.org/10.1007/s00432-019-02882-7.

Medley, C. D., Bamrungsap, S., Tan, W., & Smith, J. E. (2011). Aptamer-conjugated nanoparticles for cancer cell detection. *Analytical Chemistry*, *83* (3), 727–734. Available from https://doi.org/10.1021/ac102263v.

Medley, C. D., Smith, J. E., Tang, Z., Wu, Y., Bamrungsap, S., & Tan, W. (2008). Gold nanoparticle-based colorimetric assay for the direct detection of cancerous cells. *Analytical Chemistry*, *80*(4), 1067–1072. Available from https://doi.org/10.1021/ac702037y.

Mesgari, F., Beigi, S. M., Fakhri, N., Hosseini, M., Aghazadeh, M., & Ganjali, M. R. (2020). Paper-based chemiluminescence and colorimetric detection of cytochrome c by cobalt hydroxide decorated mesoporous carbon. *Microchemical Journal*. Available from https://doi.org/10.1016/j.microc.2020.104991.

Ossonon, B. D., & Bélanger, D. (2017). Synthesis and characterization of sulfophenyl-functionalized reduced graphene oxide sheets. *RSC Advances*, *7* (44), 27224–27234.

Palazzo, G., Facchini, L., & Mallardi, A. (2012). Colorimetric detection of sugars based on gold nanoparticle formation. *Sensors and Actuators, B: Chemical*, *161*(1), 366–371. Available from https://doi.org/10.1016/j.snb.2011.10.046.

Park., et al. (2019). *Biotechnology and Bioengineering*.

Peng, C., Hua, M. Y., Li, N. S., Hsu, Y. P., Chen, Y. T., Chuang, C. K., ... Yang, H. W. (2019). A colorimetric immunosensor based on self-linkable dual-nanozyme for ultrasensitive bladder cancer diagnosis and prognosis monitoring. *Biosensors and Bioelectronics*, *126*, 581–589. Available from https://doi.org/10.1016/j.bios.2018.11.022.

Pur, M. R. K., Hosseini, M., Faridbod, F., Dezfuli, A. S., & Ganjali, M. R. (2016). A novel solid-state electrochemiluminescence sensor for detection of cytochrome c based on ceria nanoparticles decorated with reduced graphene oxide nanocomposite. *Analytical and Bioanalytical Chemistry*, *408* (25), 7193–7202. Available from https://doi.org/10.1007/s00216-016-9856-6.

Salehnia, F., Hosseini, M., & Ganjali, M. R. (2017). A fluorometric aptamer based assay for cytochrome C using fluorescent graphitic carbon nitride nanosheets. *Microchimica Acta*, *184*(7), 2157–2163. Available from https://doi.org/10.1007/s00604-017-2130-6.

Scaffardi, L. B., Lester, M., Skigin, D., & Tocho, J. O. (2007). Optical extinction spectroscopy used to characterize metallic nanowires. *Nanotechnology*, *18*(31). Available from https://doi.org/10.1088/0957-4484/18/31/315402.

Seeman, N.C., & Sleiman, H.F. (n.d.). Nature Reviews Materials, 2017. 3(1), 1–23.

Shahsavar, K., Shokri, E., & Hosseini, M. (2020). A fluorescence-readout method for miRNA-155 detection with double-hairpin molecular beacon based on quadruplex DNA structure. *Microchemical Journal*. Available from https://doi.org/10.1016/j.microc.2020.105277.

Shen., et al. (2018). *Analytical Chemistry*.

Shen., et al. (2018). *Sensors and Actuators B: Chemical*.

Teng, Y., Shi, J., & Pong, P. W. T. (2019). Sensitive and specific colorimetric detection of cancer cells based on folate-conjugated gold-iron-oxide composite nanoparticles. *ACS Applied Nano Materials*, *2*(11), 7421–7431. Available from https://doi.org/10.1021/acsanm.9b01947.

Wang, J., Wu, L., Ren, J., & Qu, X. (2012). Visualizing human telomerase activity with primer-modified Au nanoparticles. *Small (Weinheim an der Bergstrasse, Germany)*, *8*(2), 259–264. Available from https://doi.org/10.1002/smll.201101938.

Wang, Y. M., Liu, J. W., Adkins, G. B., Shen, W., Trinh, M. P., Duan, L. Y., … Zhong, W. (2017). Enhancement of the intrinsic peroxidase-like activity of graphitic carbon nitride nanosheets by ssDNAs and its application for detection of exosomes. *Analytical Chemistry*, *89*(22), 12327–12333. Available from https://doi.org/10.1021/acs.analchem.7b03335.

Wilcoxon, J. P., & Abrams, B. L. (2006). Synthesis, structure and properties of metal nanoclusters. *Chemical Society Reviews*, *35*(11), 1162–1194. Available from https://doi.org/10.1039/b517312b.

Xia, Y., Liu, M., Wang, L., Yan, A., He, W., Chen, M., … Chen, J. (2017). A visible and colorimetric aptasensor based on DNA-capped single-walled carbon nanotubes for detection of exosomes. *Biosensors and Bioelectronics*, *92*, 8–15. Available from https://doi.org/10.1016/j.bios.2017.01.063.

Xie, F., et al. (2019). Carbon-based nanomaterials – A promising electrochemical sensor toward persistent toxic substance. *TrAC Trends in Analytical Chemistry*, 119. Available from https://doi.org/10.1016/j.trac.2019.115624.

Xu., et al. (2017). *Biosensors and Bioelectronics*.

Yang, T., Luo, Z., Tian, Y., Qian, C., & Duan, Y. (2019). Design strategies of AuNPs-based nucleic acid colorimetric biosensors. *TrAC Trends in Analytical Chemistry*, *124*, 115795.

Zhang, L. N., Deng, H. H., Lin, F. L., Xu, X. W., Weng, S. H., Liu, A. L., … Chen, W. (2014). In situ growth of porous platinum nanoparticles on graphene oxide for colorimetric detection of cancer cells. *Analytical Chemistry*, *86*(5), 2711–2718. Available from https://doi.org/10.1021/ac404104j.

Zhao, H., Qu, Y., Yuan, F., & Quan, X. (2016). A visible and label-free colorimetric sensor for miRNA-21 detection based on peroxidase-like activity of graphene/gold-nanoparticle hybrids. *Analytical Methods*, *8*(9), 2005–2012. Available from https://doi.org/10.1039/c5ay03296b.

Zhou, Y., Xu, H., Wang, H., & Ye, B. C. (2020). Detection of breast cancer-derived exosomes using the horseradish peroxidase-mimicking DNAzyme as an aptasensor. *Analyst*, *145*(1), 107–114. Available from https://doi.org/10.1039/c9an01653h.

## Further reading

Ahirwar, R., & Nahar, P. (2016). Development of a label-free gold nanoparticle-based colorimetric aptasensor for detection of human estrogen receptor alpha. *Analytical and Bioanalytical Chemistry*, *408*(1), 327–332. Available from https://doi.org/10.1007/s00216-015-9090-7.

Aldewachi, H., Chalati, T., Woodroofe, M. N., Bricklebank, N., Sharrack, B., & Gardiner, P. (2018). Gold nanoparticle-based colorimetric biosensors. *Nanoscale*, *10*(1), 18–33. Available from https://doi.org/10.1039/c7nr06367a.

Lee, S. H., Sung, J. H., & Park, T. H. (2012). Nanomaterial-based biosensor as an emerging tool for biomedical applications. *Annals of Biomedical Engineering*, *40*(6), 1384–1397. Available from https://doi.org/10.1007/s10439-011-0457-4.

Li, N., et al. (2020). A SERS-colorimetric dual-mode aptasensor for the detection of cancer biomarker MUC1. *Analytical and Bioanalytical Chemistry*, 5707–5718.

Wang, D., Guo, R., Wei, Y., Zhang, Y., Zhao, X., & Xu, Z. (2018). Sensitive multicolor visual detection of telomerase activity based on catalytic hairpin assembly and etching of Au nanorods. *Biosensors and Bioelectronics*, *122*, 247–253. Available from https://doi.org/10.1016/j.bios.2018.09.064.

Xiong, L. H., He, X., Xia, J., Ma, H., Yang, F., Zhang, Q., … Cheng, J. (2017). Highly sensitive naked-eye assay for enterovirus 71 detection based on catalytic nanoparticle aggregation and immunomagnetic amplification. *ACS Applied Materials and Interfaces*, *9*(17), 14691–14699. Available from https://doi.org/10.1021/acsami.7b02237.

Zhao, S., Lei, J., Huo, D., Hou, C., Luo, X., Wu, H., … Yang, M. (2018). A colorimetric detector for lung cancer related volatile organic compounds based on cross-response mechanism. *Sensors and Actuators, B: Chemical*, *256*, 543–552. Available from https://doi.org/10.1016/j.snb.2017.10.091.

Zhou, W., Liang, W., Li, X., Chai, Y., Yuan, R., & Xiang, Y. (2015). MicroRNA-triggered, cascaded and catalytic self-assembly of functional \dNAzyme ferris wheel\ nanostructures for highly sensitive colorimetric detection of cancer cells. *Nanoscale*, *7*(19), 9055–9061. Available from https://doi.org/10.1039/c5nr01405k.

Chapter 10

# Magnetic properties-based biosensors for early detection of cancer

Sagar Narlawar, Samraggi Coudhury and Sonu Gandhi
*DBT-National Institute of Animal Biotechnology (DBT-NIAB), Hyderabad, Telangana, India*

## 10.1 Introduction

Cancer is one of the leading causes of death in the world, around 9.8 million deaths in 2018 according to the World Health Organization (WHO). Cancers can take various forms, some of them are lung, breast, ovarian, skin, and colon cancer, and leukemia. It can be caused by either environmental factors such as chemicals, radiation, alcohol, smoke, tobacco, or due to genetic mutations; bacterial and viral infections can also lead to stomach and cervical cancer in certain people. One of the major cancers in men and women is prostate and breast cancer, respectively; it is one of the major causes of death due to detection in later stages which decreases the rate of survival. A cancer diagnosis at an early stage is very important as it can prevent metastasis. At present, patients visit doctors only when they feel discomfort in their body; the tumor at that point might have metastasized. Frequent diagnostic procedures are expensive and time consuming; hence, efficient and cost-effective methods are required to improve the current dynamics of cancer diagnosis around the globe. This can reduce the death rate around the world due to cancer (Lin et al., 2015; Roberts, Tripathi, & Gandhi, 2019). Biosensors are devices designed to detect the presence of bioanalyte in samples which can be of environmental or biological sources. Biosensors are being employed currently as detection tools because they can be used for the development of point-of-care (POC) diagnostics, and they produce sensitive and rapid responses compared to the traditional detection mechanism which has low sensitivity, is time-consuming (Banga, Tyagi, Shahdeo, & Gandhi, 2019). In this chapter, magnetic particle-based biosensors that have been designed to detect protein biomarkers are highlighted as these biomarkers are found at an elevated level after the onset of cancer; their presence in the body fluid are in very low concentrations (Tripathi, Arami, Banga, Gupta, & Gandhi, 2018) at the start so an effective sensor is required which can detect those minute concentrations. These biosensors are also used to detect the effectiveness of the drugs at tumor sites after administration into the body and samples can be checked after frequent interval as biosensors are economical and noninvasive in nature in comparison to the current detection mechanisms. Magnetic particles are gaining attention due to their varied properties. They are being used in various detection mechanisms, such as MRIs, due to their magnetic properties, where the previously used gadolinium is being replaced with iron oxide nanoparticles which act as a better contrast agent and is nontoxic to the body. Another property which is valuable for the development of magnetic particle-based biosensors is magnetic separation, which help in effective separation, enrichment, and quantification of biomarkers in the patient samples. These properties along with signal amplification helps magnetic particle-based biosensors achieve ultra-sensitivity. Over the passage of time many magnetic particle-based biosensors have been reported for early diagnosis of protein biomarkers specific for cancer. In the section below, various types of biosensors and their applications in the detection of cancer are described briefly.

## 10.2 Biosensors and their types

### 10.2.1 Bioreceptor based biosensors

The recognition element is one of the major elements in the biosensor; they can be purified or modified from biological systems or artificially synthesized in laboratories. According to the IUPAC classifications, biosensors are analytical devices that detect the presence of analytes in a biochemical reaction that takes place in a medium of biological origin

Biosensor Based Advanced Cancer Diagnostics. DOI: https://doi.org/10.1016/B978-0-12-823424-2.00010-7

and are majorly mediated by enzymes, tissues, organelles via electrical, thermal, or optical signals (McNaught & Wilkinson, 1997). In medicine, they are developing to be a potential alternative for the previously present alternative such as glucose level detection in diabetics, harmful pathogen detection. Biosensors are being used internationally for monitoring of various diseases such as cancer. In the case of cancer detection, the tumor biomarkers act as analytes such as BRCA1, BRCA2, EGF, etc. Biosensors are designed to detect the levels of these biomarkers that are expressed or secreted by the tumor cells which can be used for determining whether there is a tumor present, whether the cancer is malignant or benign, and if the treatment given has been effective against the cancerous cells. Biosensors can be modified in the future to detect multiple biomarkers in one go, which can be a great improvement in the field of diagnostics (Tothill, 2009). Biosensors consist of three regions: the bioreceptor region, transducer region, and signal processor region which converts the signals to a readable format. The bioreceptor region detects the signal in the form of analytes, the transducer then switches the biological signal into an electrical signal (Kissinger, 2005). Biosensors can be classified based on the bioreceptor region and/or the transducer region. Analytes mostly detected by the recognition elements are enzymes, nucleic acids, antigens, antibodies, and proteins. Based on the transducers, they can be organized into four main classes: electrochemical, optical, piezoelectric, and thermal detection.

#### 10.2.1.1 Enzymes

Enzymes are one of the most widely used recognition elements, within which allosteric enzymes have shown greater potential as recognition elements. The regulatory subunit region present acts as the recognition elements and the catalytic site act as the transducer (O'Connell & Guilbault, 2001). One of the major examples of an enzyme-based sensor is the glucose sensor where the enzyme glucose oxidase is being used as the recognition element. The catalysis of glucose takes place in the presence of oxygen by the glucose oxidase enzyme leading to the production of gluconolactone and hydrogen peroxide as products. An amperometric transducer can measure the rate of removal of oxygen and the rate of production of hydrogen peroxide and convert them into readable signals (Chambers, Arulanandam, Matta, Weis, & Valdes, 2008). There are many factors responsible for optimum performances of enzyme-based biosensors such as suitable temperature and pH for the enzyme reactivity, the thickness of the enzyme layer, immobilization of enzymes on the sensor some of these factors affect the electrode performances (Rastislav, Miroslav, & Ernest, 2012).

#### 10.2.1.2 Nucleic acids

DNA, RNA, and nucleic acids are used for highly sensitive complementary base-pair binding in biosensors (Alkire, Kolb, Lipkowski, & Ross, 2013). Nucleic acid-based biosensors use oligonucleotides or polynucleotides as recognition elements which can be of natural or biomimetic forms. Aptamers are artificially synthesized single-stranded oligonucleotides of DNA or RNA, with sizes up to 100 bp. They are selected from a pool of DNA and RNA oligonucleotide sequences that have a high affinity for the target sequences by a process called SELEX (systematic evolution of ligands by exponential enrichment). Aptamer-based biosensors are capable of detecting new biomarkers in cases of early cancer diagnosis (Chen et al., 1997). Some of the previously identified cancer biomarkers using SELEX are NY-BR-1 and Ing-1, aptamers have also been used for performing sensing arrays for certain cancer proteins (Jäger et al., 2001). Electrochemical DNA-based biosensors produce electrical signals when a base-pair recognition event takes place, which can be a rapid and inexpensive detection of genetic diseases and pathological samples (Lucarelli et al., 2003).

#### 10.2.1.3 Antibodies/antigen complex

Antibodies are biomolecules made up of amino acids arranged in sequential order (Talan et al., 2018). Antigen–antibody based recognition elements are found to be more efficient for rapid detection assays due to its specificity in nature; the target molecules don't require purification to be identified by the recognition element (Hirsch, Jackson, Lee, Halas, & West, 2003). Immunosensors can be used for cancer cells/marker detection (Ehrhart et al., 2008). Antigen–antibody based biosensors are widely used for the detection of prostate cancer. Prostate specific antigen (PSA) biosensors are developed where anti-PSA antibodies are recognized by anti-PSA recognition elements. SPR based as well as microcantilever based biosensors are used where there is a change in vibrational frequency when an antigen-antibody reaction takes place (Wu et al., 2001).

### 10.2.2 Transducer-based biosensors

The transducer is the component in biosensors which provides an output signal to the biological input signal received from the reactions in the recognition elements (McNaught & Wilkinson, 1997). The transducer converts the biological

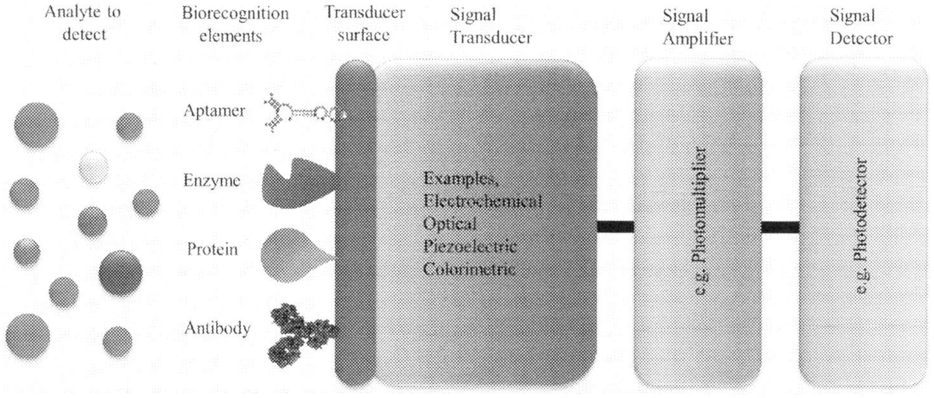


**FIGURE 10.1** Schematic representation of the biosensor construct, components, and working principle.

signal into an electrical or digital signal which can be measured as shown in Fig. 10.1. Biosensors are also categorized based on the means of transduction of signals.

#### *10.2.2.1 Electrochemical biosensors*

Electrochemical biosensors are the most common and widely available biosensors in the market currently. They are very portable, user-friendly, and more cost-effective than the other biosensors hence they are used for developing POC devices. As mentioned previously, the glucose biosensor is an example of electrochemical biosensors which has paved the path for biosensors in medical diagnostics (Tothill, 2009). Potentiometric and amperometric biosensors are the most common types of electrochemical biosensors. In potentiometric biosensors, the transducer measures the change in potential across an ion-selective membrane (Koncki, 2007). They are not yet in clinical use for cancer detection but shows promising results the light addressable potentiometric biosensor (LAPS) uses phage as a recognition element which has the potential for detecting a cancer biomarker hPRL-3 and breast cancer cell lines MDA/MB231and has been proposed for cancer detection (Jia et al., 2007). In amperometric biosensors, the transducer measures the current which is produced when there is an electric potential between the two electrodes (Heller, 1996). Amperometric biosensors measure the change in current when there is an oxidation/reduction reaction, specific DNA sequences are used for cancer detection using amperometric biosensors. These sensors can detect genetic mutations specifically associated with cancer. Wang and Kawde (2001) developed chrono potentiometric biosensors which helped in determining mutations in BRCA1 and BRCA2 genes that are associated with breast cancer (Wang & Kawde, 2001). Asphahani and Zhang (2007) demonstrated cell-based electrochemical sensors also known as cytosensors which monitor changes in a cell-induced by various stimuli. Cytosensors can be used to monitoring the effect of anticancer agents on targeted cells (Asphahani & Zhang, 2007).

#### *10.2.2.2 Optical biosensors*

They measure the change in a specific wavelength of light signals produced by the transducer; the transducer can be fluorescence, luminescence, or colorimetric-based (Gandhi, Caplash, Sharma, & Raman Suri, 2009). Transducers detect the shift in wavelength or SPR due to a reaction with an analyte and converts them into electrical or digital output signals (Tothill, 2009). New kinds of optical sensors called photonic crystal biosensors are capable of capturing light from smaller areas which makes it highly sensitive. The light reflected by the crystal is measured when a cell or molecule binds/moves from the crystal surface using a photonic crystal biosensor. Chan, Gosangari, Watkin, and Cunningham (2007) monitored the proliferation and apoptosis of breast cancer cells when treated with doxorubicin to determine the $IC_{50}$ of doxorubicin. Cancer detection in the throat using an esophageal laser fluorescence-based optical sensor; the device is swallowed by the patient and when it reaches the esophagus emits light at a specific wavelength. The device emits different lights of specific wavelengths for normal and cancerous cells. This device has been tested on about 200 patients and is found to be accurate for 98% of the time. This kind of biosensor reduces the pain of biopsies or surgeries.

#### *10.2.2.3 Calorimetric biosensors*

Calorimetric biosensors have temperature sensors. When the analyte reacts with the recognition element there is heat production that is proportional to the concentration of an analyte. The enthalpy change of mostly exothermic reactions is monitored which provides an estimated concentration of the substrate present in the biological samples (Mohanty & Kougianos, 2006). Calorimetric biosensors are not that prominent for cancer detection yet, but the use of nanotechnology enhances the applications of these sensors. Medley et al. (2008) illustrated the use of a gold nanoparticle-based aptamer biosensor which is capable of detecting cancer. They used this device for detecting two kinds of cells: Burkitt's lymphoma cells and leukemia cells. Their work showed that aptamer-based calorimetric sensors can potentially differentiate between normal and cancer cells (Medley et al., 2008).

#### *10.2.2.4 Mass-based biosensors*

Acoustic-based and piezoelectric-based biosensors fall under this category of biosensors. Piezoelectric-based biosensors have their biological component coupled with a piezoelectric component which is usually a quartz crystal coated with gold electrodes (Cooper & Singleton, 2007). When the analyte is present, potential energy is produced which leads to the change in mass of the quartz crystal. The change in mass of the crystal causes a change in frequency which is converted into a signal. Immunosensors and microcantilever sensors mentioned previously use piezoelectric technology for the detection of cancer biomarkers in biological samples (Makower, Halámek, Skládal, Kernchen, & Scheller, 2003). Dell'Atti et al. developed a piezoelectric-based sensor coupled with PCR amplification which helped in detecting the various point mutations which can take place in the p53 gene in human. p53 gene mutation plays a critical role in cancer development in humans as well as to check for the effectiveness of cancer treatments (Dell'Atti, Tombelli, Minunni, & Mascini, 2006).

## 10.3 Cancer detection and diagnostics

Cancer triggers uncontrolled growth of cells which leads to the formation of a mass of cells called tumors. After a while, tumor cells become resistant to apoptosis (Rasooly & Jacobson, 2006). With the progression of cancer the tumor starts spreading from the site of its origin; this phenomenon is called metastasis. After the cancer cells have metastasized, it is considered incurable. Various cancers are identified in different ways, mostly after clinical examination by the clinician. Most patients come with symptoms and some visit for regular examinations. Some of the major symptoms observed are a lump that doesn't heal or the presence of blood in stool urine, or cough. After diagnosis by a clinician, the patient undergoes some imaging procedures and laboratory testing of patient samples.

### 10.3.1 Computed tomography

Computed tomography (CT) is one of the conventional methods to detect cancer and monitor its progression. It helps in determining the size and shape of the tumor, CT scans are a noninvasive procedure, free of any painful steps and it takes 10–30 min. CT is majorly done using x-rays that provides more detailed image as CT scans are cross-sectional images of the body such as the bones, organs and soft tissues produced by the computer. The CT scan results helps in obtaining clear images to identifying the blood vessels which act as the source of nutrients for the tumor cells. CT-guided biopsy uses CT scans to direct the needles to individual tumor locations to remove a specific pair of tissue. CT scans can also be used to inject potential cancer therapeutic drugs on to the tumor, combined with a process called radiofrequency ablation (RFA), where heat is used to destroy the tumor. Continuous CT scans are done on the patients to monitor the effect of the cancer treatments on the tumor cells. Modern CT procedures use special contrast materials for clearer images; these contrast agents can be injected into the veins or introduced into the rectum via enema (CT Scan for Cancer. American Cancer Society, 2015).

### 10.3.2 Positron emission tomography

The positron emission tomography (PET) scan is also used for cancer screening and monitoring, and it can also be combined with an CT scan for clinal uses. PET is a semiinvasive technique which is important for determining the stages of cancer and whether it has metastasized or not. A PET scan is different from a CT scan as it can detect the anomalies in the body. In the case of PET imaging, the patient is given a radioactive tracer (fluorodoxyglucose-18), which is injected into the bloodstream. After injection it takes about 30–90 min to reach the desired organs; cells in the body absorb the sugar. Cells which need more energy such as the tumor cells pick up larger quantities of sugar. It is observed that the cancer cells have a higher metabolic rate than normal cells, which leads to a larger sugar uptake and identifying the presence or absence of cancer. The PET scan has proven to significantly be more accurate than CT scans as it can clearly determine between a benign or malignant tumor so that invasive surgery can be omitted (Schrevens, Lorent, Dooms, & Vansteenkiste, 2004)

### 10.3.3 Isotopic diagnostics

Radioisotopes possess radioactivity at an atomic level and they are used for cancer diagnosis and treatment. Radioisotopes emitting gamma rays are chosen for diagnosis due to its penetrating nature and isotope-emitting alpha and beta rays are chosen for treatment due to its cytotoxic nature. Some of the radioisotopes used are iodine-123, fluorine-18, gallium-67, rubidium-82. A radioactive tracer is introduced into the patient's body and the marker reaches up to the tumorous organ. It can be examined using various imaging methods and also provides information regarding metastasis of cancer. After the imaging procedure, the radioactive components are washed out of the system by injecting saline solution into the veins (Ogawa et al., 2019).

### 10.3.4 Magnetic resonance imaging

MRI uses an external magnetic field for imaging and is mostly used for examining cancer in the head or neck region (Tomitaka, Arami, Gandhi, & Krishnan, 2015). MRI is mostly used to screen for the metastasized cancer and it helps doctors plan out treatment procedures, radiations, or surgeries. MRI is a pain-free procedure and not much preparation is required in the patients; sometimes a contrast agent is injected in the veins which speeds up the tissue response to magnetic or radio waves and clearer images are obtained. The clinician should be informed about the presence of any metal in the body such as metal implants or pacemakers. MRI provides cross sectional images of the targeted site and uses a magnet rather than a harmful radioisotope. MRI helps in identifying certain cancers specifically, and the clinician diagnoses whether the tumor is cancerous of benign (MRI for Cancer. American Cancer Society, 2019). Magnetic nanoparticles are presently being considered as MRI contrast agents and some of them are commercially available in the market. Gadolinium was majorly used as a contrast agent and is being replaced by Resovist (carboxydextran-coated ultra-superparamagnetic iron oxide nanoparticles), Ferumoxsil (silica-coated super paramagnetic iron oxide nanoparticles), and others (Li et al., 2013).

### 10.3.5 Mammography

It is an x-ray imaging which is mostly done for examining breast tumors and for breast cancer screening. Mammograms can show tumors at the initial stage even before it can be felt physically. If there is a lump formation in the breast, various clinical examinations can be conducted to determine its type. The cancerous tumors have different physical appearances (Joensuu et al., 2014).

#### *10.3.5.1 Cancer biomarkers*

Biomarkers originate from varying sources; they can be DNA, RNA, or proteins (antibodies, enzymes, hormones, tumor suppressor or oncogenes). The National Cancer Institute defined biomarkers as "a biological molecule found in blood, other body fluids, or tissues that is a sign of a normal or abnormal process or a condition or disease. A biomarker may be used to see how well the body responds to a treatment for a disease or condition" (Mishra & Verma, 2010). For early detection of cancer, these biomarkers play an important role if identified at an early stage of cancer; chemotherapeutic medications can be provided to the patients and their progression can be monitored (Basil et al., 2006). Biomarkers are mainly found in human fluids such as blood, serum, cerebrospinal fluid, which originate mostly from the tumor cells. A list of cancer biomarkers is present in Table 10.1.

### 10.3.6 Prostate-specific antigen

It was one of the first identified biomarkers for prostate cancer and used for screening and detection. It has been studied that the normal level of prostate-specific antigen (PSA) is 4 ng/mL and an increase in the PSA level directly correlates to prostate cancer (Smith, Humphrey, & Catalona, 1997).

**TABLE 10.1** Common biomarkers which are presently being used for detection of various cancers.

| Serial number | Biomarkers | Type of cancer |
|---|---|---|
| 1 | PSA | Prostate |
| 2 | BRCA1, BRCA2, CA 15-3, CA 125, CA 27.29, CEA, NY-BR-1, Ing-1, HER2/NEU, ER/PR | Breast |
| 3 | AFP, CEA | Liver |
| 4 | CEA, EGF, p53 | Colon |
| 5 | CEA, CA 19-9, SCC, NSE, NY-ESO-1 | Lung |
| 6 | CA 125, HCG, p53, CEA, CA 549, CASA CA 19–9, CA 15–3, MCA, MOV-1, TAG72 | Ovarian |
| 7 | Tyrosinase, NY-ESO-1 | Skin |
| 8 | SCC | Esophageal |

### 10.3.7 CA 15-3

It is one of the important biomarkers for breast cancer. Some of the other biomarkers related to breast cancer are carcinoembryonic antigen (CEA), BRCA1, BRCA2, and CA 27.29. It is seen that the CA15-3 level increases with cancer progression and directly correlates to metastasis (Duffy, 2006).

### 10.3.8 Cancer antigen 125

An increase in levels of cancer antigen 125 is associated with ovarian cancer mostly, but also found in the case of uterus, cervix, lung, breast, liver, and colon cancer. In 90% of the cases, women with ovarian cancer have elevated CA 125 levels (Meyer & Rustin, 2000).

### 10.3.9 RCAS1 (EBAG-9)

RCAS1 stands for receptor-binding cancer antigen expressed on SiSo cells. It is overexpressed in the case of gastric carcinoma and gastric tumor progression. It also occurs during esophageal, gall bladder, and endometrial cancer (Kubokawa et al., 2001). Nanotechnology is having a major impact on the biosensors and helping in diagnosis, prognosis, and monitoring of metastasis. Major cancers are identified only after they have metastasized which makes them deadly and difficult for treatment. The application of nanotechnology in biosensor development has a high chance of improvement in the early detection of cancer, thus increasing the survival rates around the globe (Grodzinski, Silver, & Molnar, 2006).

## 10.4 Applications of a magnetic properties-based biosensor for cancer detection

The recent decade has seen a rise in the use of biomagnetism and magnetic biosensors for medical applications. Magnetic nanoparticles are being used for biomedicine production and synthesis of magnetic biosensors which helps in highly sensitive and accurate detection of biomarkers that are essential for disease diagnosis (Gandhi, Arami, & Krishnan, 2016). These properties of magnetic nanoparticles help in proficient targeting paired with signal amplification that leads to the development of ultrasensitive biosensors (Zhang & Zhou, 2012). The sooner cancer can be detected is there a higher chance of cure, so sensitive and accurate detection mechanisms are in demand. Biosensors have the potential for providing the faster and effective prognosis of the disease, whether the cancer cells have metastasized or not, and the efficacy of the anticancer treatments (Bohunicky & Mousa, 2011). Table 10.2 discusses the various biosensor techniques for cancer biomarker detection.

### 10.4.1 Magnetic barcode assay

This particular assay was developed by the Nam group, where they conjugated the antibody with magnetic microparticle (Ab1-MMP) as well as gold nanoparticles (Ab2-GNP), and both of them were tagged with hundreds of barcoded

**TABLE 10.2 Comparison of different magnetic particle-based biosensors and their detection limits.**

| Serial number | Detection method | Surface coating of particles | Analyte | Limit of detection (LOD) | References |
|---|---|---|---|---|---|
| 1 | Magnetic field sensor | Streptavidin | S100ßß protein | 27 pg/mL | De Palma et al. (2007) |
| 2 | Magnetic field sensor | Streptavidin | Carcinoembryonic antigen (CEA), Tumor necrosis factor (TNF)-α, lactoferrin | 1 pg/mL of each biomarker | Osterfeld et al. (2008) |
| 3 | Magnetic barcode assay | Polyamine | Prostate specific antigen (PSA) | 3:00 a.m. | Nam, Thaxton, and Mirkin (2003) |
| 4 | Magnetic barcode assay | Tosyl particles | PSA, Human chorionic gonadotropin (HCG), α-Fetoprotein (AFP) | 170 fM of each biomarker in the serum sample | Stoeva, Lee, Smith, Rosen, and Mirkin (2006) |
| 5 | SPR-based assay | Streptavidin | Brain natriuretic peptide (BNP) | 25 pg/mL | Teramura, Arima, and Iwata (2006) |
| 6 | SPR-based assay | Amphiphilic polymer | Adenosine triphosphate (ATP) | 10 nM | Wang, Munir, Zhu, and Zhou (2010) |
| 7 | ECL-based sensor | Gold | Cancer cells | 56 cell/ mL | Bi, Zhou, and Zhang (2010) |
| 8 | ECL-based sensor | Streptavidin | CEA | 1.6 pg/mL | Li et al. (2010) |

DNAs (Nam et al., 2003). A sandwich ELISA was prepared for the target prostate-specific antigen (PSA) which is a protein-based cancer biomarker. The sandwich had Ab1-MMP-PSA-Ab2-GNP, the unbound species were removed by washing and the sandwiched component was magnetically separated. The barcoded DNAs which were attached to the antibodies were separated which helped in converting PSA into hundreds of barcoded DNAs (shown in Fig. 10.2C) These DNAs were detected via a scanometric assay followed by silver amplification. Scanometric assay helps in the efficient capturing of the target molecules, target conversion (protein detection to DNA detection), and amplification of the product. This assay helps in label-free detection of PSA as low as 30 a.m. which is very precise compared to normal ELISA (Rosi & Mirkin, 2005; Thaxton et al., 2009).

## 10.4.2 Nanostructured immunosensor

Munge et al., developed an electrochemical biosensor where the primary antibodies were labeled with glutathione modified gold nanoparticles and the secondary antibodies were labeled with magnetic nanoparticles and tagged with HRP enzymes (shown in Fig. 10.2B). A sandwich was performed to detect the target cancer marker IL-8 via a highly sensitive electrochemical redox reaction. The HRP tags present on the magnetic nanoparticle-coated antibodies amplifies the signal received from the redox reaction which helps in detecting IL-8 to a low level of 100 a.m. in the blood serum of the patients, which is fourfold lower than the detection limit of an ELISA (Munge et al., 2011). Many other strategies are used for magnetic particle-based immunoassays; some of them are thermoresponsive polymers along with magnetic particle loaded antibodies (Gandhi, Sharma, Capalash, Verma, & Raman Suri, 2008; Gandhi et al., 2015; Nagaoka et al., 2011).

## 10.4.3 Giant magnetoresistive sensors

Wang et al. designed a giant magnetoresistive sensor which (shown in Fig. 10.3), other than using magnetic particles for target molecule capturing and enrichment, will also help in detecting the magnetic field of the magnetic particles by applying an external magnetic field. This sensor has the capability of quantitatively detecting various cancer biomarkers in clinical samples, with a lower limit of detection (LOD) of approximately 50 a.m. (Wang et al., 2010). It helps in

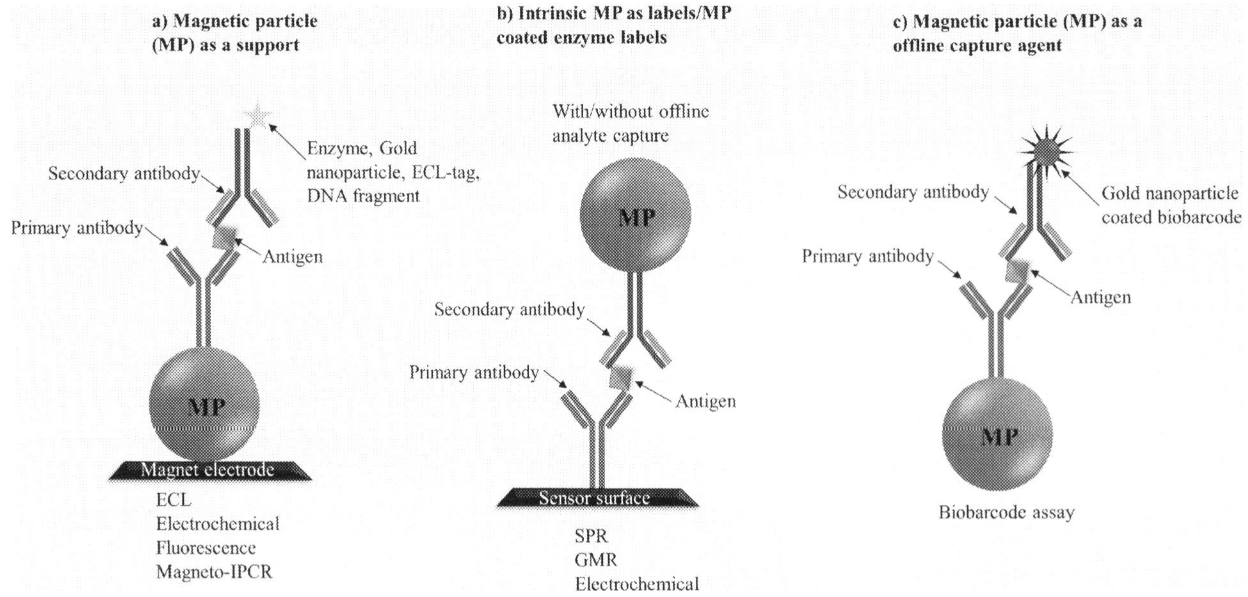


**FIGURE 10.2** Magnetic barcode assay: Schematic illustration of the use of magnetic particles as supports (A & C) and labels (B) for protein biomarker measurements using various types of detectors. *MP*: Magnetic particle, *IPCR*: immuno PCR, *ECL*: electrochemiluminescence, *GMR*: giant magnetoresistance, *SPR*: surface plasmon resonance.

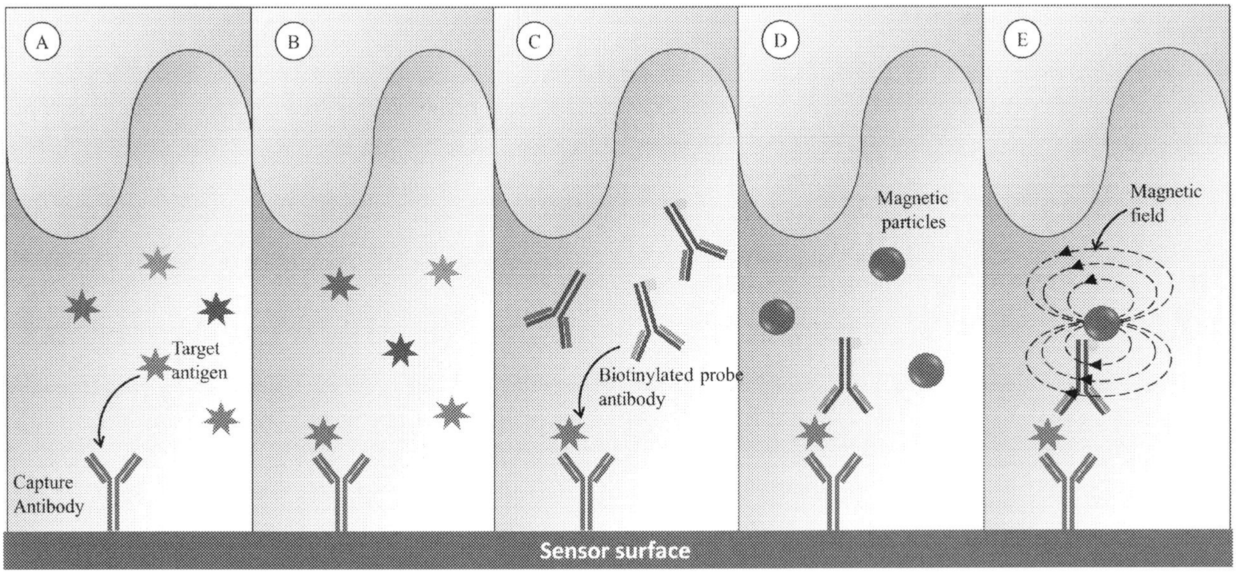


**FIGURE 10.3** Working principle of giant magneto-resistive sensor (GMR sensor). A schematic representation of GMR sensor for the detection of cancer biomarker (A) Capture antibodies immobilized on the sensor surface are used to capture the target antigen [red]; (B) after washing away free unbound species; (C) biotinylated probe antibodies are added into system, where only antibodies specific for target antigens will bind and any unbound free antibodies are washed away; (D) streptavidin-modified magnetic particles are bound to sandwiches via streptavidin-biotin interactions; (E) as the magnetic tag diffuses to GMR sensor surface and bind to detection antibodies (target specific biotinylated antibodies), the magnetic fields from magnetic particles are detected by underlying GMR sensor in real time in the presence of small external modulation of magnetic field.

real-time monitoring of multiple protein biomarkers. Giant magnetoresistive (GMR) sensors are considered to be matrix insensitive which means that their sensing ability is not influenced by any environmental changes such as a change in the pH, temperature, or the various proteins present in the serum, making them an ultrasensitive detection method for early detection of cancer biomarkers (Gaster et al., 2009). A GMR sensor chip array was developed by The Naval Research Laboratory which has 64 sensors called the bead array counter (BARC) which detects biomolecules using superparamagnetic labels (Baselt et al., 1998). Mulvaney et al. reported a magneto electronic detection of proteins in complex matrices using magnetic bead labels as well as fluid flow discrimination (FFD) (Mulvaney et al., 2007). The magnetic beads which are captured on the sensor surface area were calculated via microscopy and BARC sensors. FFD reduced the nonspecific binding and facilitated the selective detection of various proteins. Eight protein biomarkers related to different cancers were identified by this sensor and it showed matrix insensitivity, they can be used for the development of POC devices (Kricka & Park, 2009).

### 10.4.4 Electrochemiluminescence detection

Whole cancer cell detection of Burkitt's lymphoma or Ramos was performed using this electrochemiluminescence detection method. An aptamer sensor was developed using magnetic beads for selective separation and DNA aptamers for signal recognition. A magnetic particle-gold-cadmium sulfide (MP-Au-CdS) complex or CuS/DNA/Au/DNA/MNP nanoprobe was used to cause a chemiluminescent or electro chemiluminescent signal amplification which is detected by the magnetic particle-based aptamer sensors. The detection limit was 67 cells/mL which was highly sensitive. This setup could differentiate between the target cells and the regular control cells due to the attachment of specific aptamer (Ding, Ge, & Zhang, 2010) (shown in Fig. 10.4). These sensors have a very low LOD; in this case, the specific target DNA sequences were sensitively detected as low as in the 6.8 a.m. range (Bi et al., 2010). Wang et al. used the ECL-based peptide biosensor for the detection of cyclin $A_2$ ($CA_2$) which is a predictive indicator for early detection of multiple cancers. Hollow magnetic nanosheets made up of manganese oxide nanocrystals with ECL properties were immobilized on to platinum nanoparticles in presence of cyclin $A_2$ the luminescent signal increases. This sensor can detect $CA_2$ from a range of 0.001 to 100 ng/L and the detection limit of 0.3 pg/L (Wang & Kawde, 2001).

### 10.4.5 Magnetic bead-based biosensors

Paramagnetic beads are now commercially available which helps in magnetic separation in the presence of an external magnetic field, but possesses zero magnetization in the absence of a magnetic field. Magnetic bead-based assays are capable of protein marker identification. Willner and Katz worked on magnetic beads that controlled the magneto switchable electrocatalytic process which is useful for immunosensors; for example, the oxidation of glucose by the enzyme glucose oxidase was enhanced when a ferrocene coated magnetic particle was used on a magnetic electrode

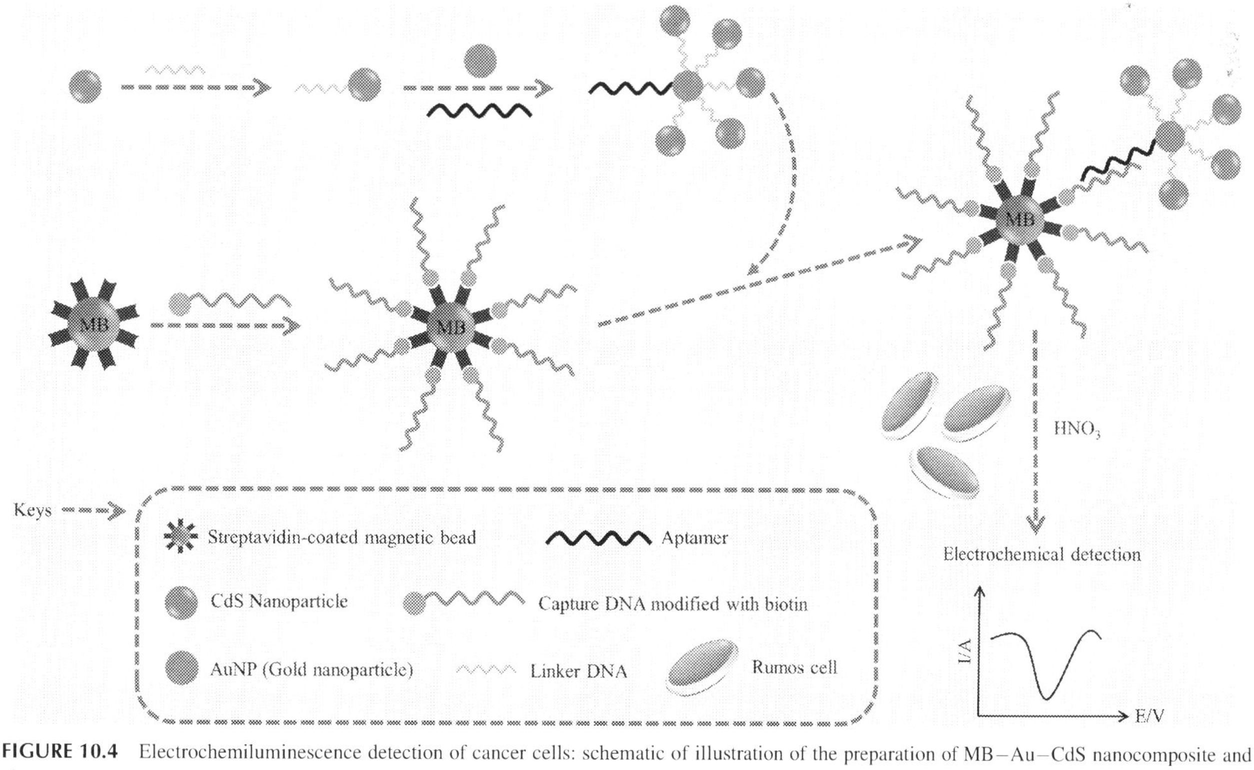


**FIGURE 10.4** Electrochemiluminescence detection of cancer cells: schematic of illustration of the preparation of MB–Au–CdS nanocomposite and electrochemical detection of cancer cells.

where ferrocene acted as a mediator (Katz & Willner, 2005; Willner & Katz, 2003). Sarkar et al. used electrochemical detection to detect the free PSA in the biological sample using a magnetic bead on screen-printed electrodes (SPCEs). In a cuvette, magnetic beads were used to capture the protein, then the beads were transferred on the SPCEs. The magnetic beads were coated with HRP enzyme and in the presence of hydrogen peroxide caused amperometric changes that were detected up to a lower limit of 0.1 ng/mL of free PSA (Sarkar, Ghosh, Bhattacharyay, Setford, & Turner, 2008).

### 10.4.6 Magnetic PCR-based assay

Wacker et al. first reported about magnetic immune PCR (IPCR) where the immunogen is sandwiched onto magnetic particles. They contain antibodies conjugated biogenic magneto, some of which are advantageous over synthetic nanoparticles. The basic principle for magnetic immune PCR is similar to that of the double-sided immunoassay. These magnetic particles conjugated with specific antibodies and were labeled with DNA fragments in the presence of an external magnetic field. The DNA labels were detected and amplified using real-time PCR to find the protein concentration (shown in Fig. 10.5). In this study, the hepatitis B surface antigen was to be detected with a detection limit of 320 pg/mL; it was observed to be 100-fold more sensitive than analogous magneto ELISA (Wacker et al., 2007). Csordas et al. developed a micromagnetic aptamer-based PCR which is capable of detecting protein biomarkers in serum samples. In this method, the antibodies are coated with magnetic nanoparticles which help in the identification of the protein biomarker (platelet-derived growth factor) by binding to aptamer specific for the target biomarker. The immunocomplex is separated magnetically and the aptamer is amplified using a PCR which is highly sensitive in detection. The growth factor can be detected in the serum sample as low as the range 62 fM to 1 nM using this technology (Csordas et al., 2010).

### 10.4.7 Surface plasmon resonance-based assay

Surface plasmon resonance (SPR) is a resonant oscillation of electrons in presence of an incident light. SPR helps in measuring the absorption of material on planar metals surfaces such as gold and silver or onto the surface of metal nanoparticles. SPR is highly sensitive to changes in the refractive index (RI) on the nanoparticle surfaces. In SPR based immune assays, when there is a protein binding onto a surface of the antibody it causes a change in the refractive index

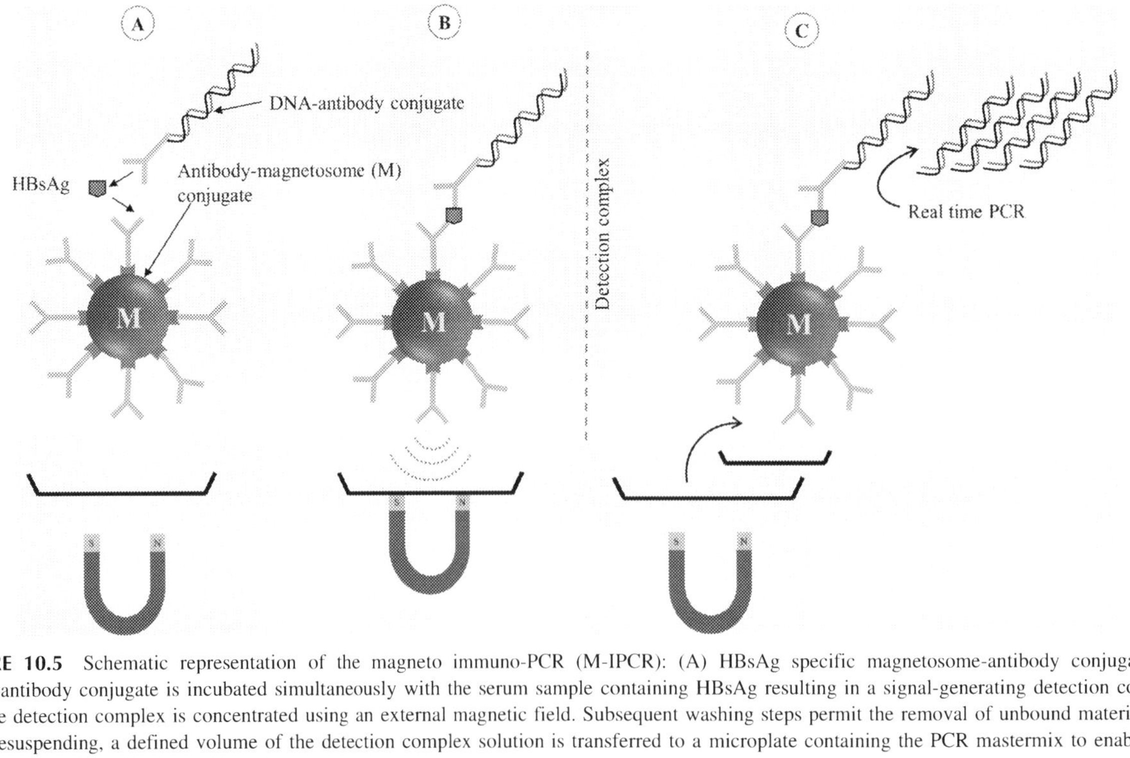


**FIGURE 10.5** Schematic representation of the magneto immuno-PCR (M-IPCR): (A) HBsAg specific magnetosome-antibody conjugate and DNA–antibody conjugate is incubated simultaneously with the serum sample containing HBsAg resulting in a signal-generating detection complex. (B) The detection complex is concentrated using an external magnetic field. Subsequent washing steps permit the removal of unbound materials. (C) After resuspending, a defined volume of the detection complex solution is transferred to a microplate containing the PCR mastermix to enable real-time PCR detection of the immobilized antigen.

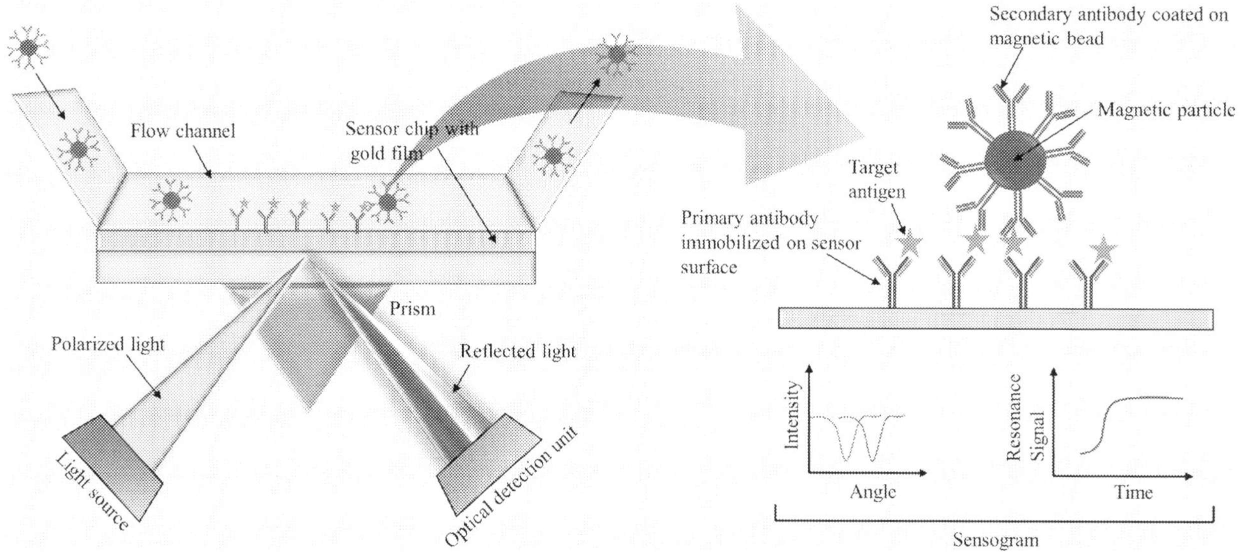


**FIGURE 10.6** Surface plasmon resonance biosensor: A schematic illustration showing the SPR-based detection of a cancer biomarkers (target antigen in this illustration). Specific antibody against target antigen immobilized on sensor surface detect the antigen and in turn, the secondary antibody conjugated to magnetic bead detects the bound antigen. This binding changes the refractive index which induces a shift of the SPR angle (Li et al., 2013). *FromLi, L., Jiang, W., Luo, K., Song, H., Lan, F., Wu, Y., & Gu, Z. (2013). Superparamagnetic iron oxide nanoparticles as MRI contrast agents for non-invasive stem cell labeling and tracking.* Theranostics, *3(8), 595–615. https://doi.org/10.7150/thno.5366.*

due to the increase in the thickness of the film (shown in Fig. 10.6). In a classic sandwich immunoassay, the sensor is labeled with primary antibody and it captures the target protein, if a secondary antibody is coated with a magnetic particle binds to the immunocomplex it leads to the change in RI. Due to the increase in the surface thickness, the RI increases and a bigger signal is produced due to the higher RI of the magnetic particles (Ishii & Kitamoto, 2016). This approach has been used for highly sensitive protein detection, and magnetic beads can reduce the nonspecific binding effect in the samples. Teramura et al. reported a SPR-based sensor for the detection of proteins such as brain natriuretic peptide by using biotin-labeled secondary antibodies that are captured using streptavidin-coated magnetic nanobeads; the limit of detection was 25 pg/mL in buffer solution (Teramura et al., 2006). Krishnan et al. used 1 μm superparamagnetic beads conjugated to antibodies for determining the presence of prostate cancer biomarkers in blood serum samples. An SPR-based sensor was used to capture the protein biomarker with an ultra-low detection limit of 10 fg/mL. The detection limit was greatly amplified using magnetic particles conjugated to secondary antibodies as they aggregate on the sensor surface that leads to an increase in the refractive index which is specific for detecting changes per bound protein. The accuracy of the device was validated using a cancer patient serum sample (Krishnan, Mani, Wasalathanthri, Kumar, & Rusling, 2011).

## 10.5 Conclusion

Magnetic particles possess very unique properties which make them one of the essential choices for efficient target capture and magnetic separation from unwanted substances; hence, magnetic particles are chosen for ultrasensitive assays. Magnetic particles combined with the immunocomplex provides sensitive detection and signal amplification. The limit of detection was found to be near femtomolar to the attomolar range (Singh et al., 2018). These assays help in the early detection of biomarkers present in the very low levels, which makes it impossible to be detected by current clinical assays. Some of the significant examples mentioned in this chapter are magnetic nanoparticle barcode assays, magnetic immunosensors, and target amplification coupled with chemiluminescence (Rosi & Mirkin, 2005). The magnetoresistive sensor which explores other qualities of magnetic particles, such as magnetic relaxation and magnetic resistance which makes it an ultrasensitive biosensor combined with matrix insensitive detection, makes it a suitable choice for early diagnosis of cancer biomarkers (Gaster et al., 2009). Cancer biomarkers are vital indicators for tumor growth but also the diagnosis and monitoring of the disease. They are widely used in POC testing or diagnostics. Incorporating biomarker detection by biosensors is going to provide rapid and easy results. Currently, biosensors detect one marker at a time; however, to use biosensors for POC devices they should be capable of detecting multiple biomarkers simultaneously to facilitate the detection process. Besides, POC biosensor technology should bring POC devices to the patient's bedside. It is evident that within the next 5–10 years nanotechnology is going to revolutionize cancer detection and diagnosis by integrating nanomaterials/magnetic particles with biosensor technology which will help in quicker detection, improved imaging, and prognosis of disease after treatment with low contrary reactions. Cancer has put forward many challenges to the medical field. The biosensor technology has the potential for overcoming them to provide cost-effective and faster results.

## References


Alkire, R. C., Kolb, D. M., Lipkowski, J., & Ross, P. N. (2013). *Advances in electrochemical science and engineering. Advances in Electrochemical Science and Engineering* (13, pp. 1–359). wiley. Available from https://doi.org/10.1002/9783527633227.

Asphahani, F., & Zhang, M. (2007). Cellular impedance biosensors for drug screening and toxin detection. *Analyst*, *132*(9), 835–841. Available from https://doi.org/10.1039/b704513a.

Banga, I., Tyagi, R., Shahdeo, D., & Gandhi, S. (2019). *Biosensors and their application for the detection of avian influenza virus. Nanotechnology in modern animal biotechnology: Concepts and applications* (pp. 1–16). Elsevier. Available from https://doi.org/10.1016/B978–0–12–818823-1.00001–6.

Baselt, D. R., Lee, G. U., Natesan, M., Metzger, S. W., Sheehan, P. E., & Colton, R. J. (1998). A biosensor based on magnetoresistance technology. *Biosensors and Bioelectronics*. Available from https://doi.org/10.1016/S0956-5663(98, 37–2.

Basil, C. F., Zhao, Y., Zavaglia, K., Jin, P., Panelli, M. C., Voiculescu, S., ... Wang, E. (2006). Common cancer biomarkers. *Cancer Research*, *66*(6), 2953–2961. Available from https://doi.org/10.1158/0008-5472.CAN-05-3433.

Bi, S., Zhou, H., & Zhang, S. (2010). A novel synergistic enhanced chemiluminescence achieved by a multiplex nanoprobe for biological applications combined with dual-amplification of magnetic nanoparticles. *Chemical Science*, *1*(6), 681–687. Available from https://doi.org/10.1039/c0sc00341g.

Bohunicky, B., & Mousa, S. A. (2011). Biosensors: The new wave in cancer diagnosis. *Nanotechnology, Science and Applications*, *4*(1), 1–10. Available from https://doi.org/10.2147/NSA.S13465.

Chambers, J. P., Arulanandam, B. P., Matta, L. L., Weis, A., & Valdes, J. J. (2008). Biosensor recognition elements. *Current Issues in Molecular Biology*, *10*(1), 1–12. Available from http://www.horizonpress.com/cimb/v/v10/01.pdf.

Chan, L. L., Gosangari, S. L., Watkin, K. L., & Cunningham, B. T. (2007). A label-free photonic crystal biosensor imaging method for detection of cancer cell cytotoxicity and proliferation. *Apoptosis: An International Journal on Programmed Cell Death*, *12*(6), 1061–1068. Available from https://doi.org/10.1007/s10495-006-0031-y.

Chen, Y. T., Scanlan, M. J., Sahin, U., Türeci, O., Gure, A. O., Tsang, S., ... Old, L. J. (1997). A testicular antigen aberrantly expressed in human cancers detected by autologous antibody screening. *Proceedings of the National Academy of Sciences of the United States of America*, *94*(5), 1914–1918. Available from https://doi.org/10.1073/pnas.94.5.1914.

Cooper, M. A., & Singleton, V. T. (2007). A survey of the 2001 to 2005 quartz crystal microbalance biosensor literature: Applications of acoustic physics to the analysis of biomolecular interactions. *Journal of Molecular Recognition*, *20*(3), 154–184. Available from https://doi.org/10.1002/jmr.826.

Csordas, A., Gerdon, A. E., Adams, J. D., Qian, J., Oh, S. S., Xiao, Y., & Soh, H. T. (2010). Detection of proteins in serum by micromagnetic aptamer PCR (MAP) technology. *Angewandte Chemie – International Edition*, *49*(2), 355–358. Available from https://doi.org/10.1002/anie.200904846.

CT Scan for Cancer. American Cancer Society. (2015). Available from https://www.cancer.org/treatment/understanding-your-diagnosis/tests/ct-scan-for-cancer.html

De Palma, R., Reekmans, G., Liu, C., Wirix-Speetjens, R., Laureyn, W., Nilsson, O., & Lagae, L. (2007). Magnetic bead sensing platform for the detection of proteins. *Analytical Chemistry*, *79*(22), 8669–8677. Available from https://doi.org/10.1021/ac070821n.

Dell'Atti, D., Tombelli, S., Minunni, M., & Mascini, M. (2006). Detection of clinically relevant point mutations by a novel piezoelectric biosensor. *In Biosensors and bioelectronics*, *21*(10), 1876–1879. Available from https://doi.org/10.1016/j.bios.2005.11.023.

Ding, C., Ge, Y., & Zhang, S. (2010). Electrochemical and electrochemiluminescence determination of cancer cells based on aptamers and magnetic beads. *Chemistry – A European Journal*, *16*(35), 10707–10714. Available from https://doi.org/10.1002/chem.201001173.

Duffy, M. J. (2006). Serum tumor markers in breast cancer: Are they of clinical value? *Clinical Chemistry*, *52*(3), 345–351. Available from https://doi.org/10.1373/clinchem.2005.059832.

Ehrhart, J. C., Bennetau, B., Renaud, L., Madrange, J. P., Thomas, L., Morisot, J., ... Tran, P. L. (2008). A new immunosensor for breast cancer cell detection using antibody-coated long alkylsilane self-assembled monolayers in a parallel plate flow chamber. *Biosensors and Bioelectronics*, *24*(3), 467–474. Available from https://doi.org/10.1016/j.bios.2008.04.027.

Gandhi, S., Arami, H., & Krishnan, K. M. (2016). Detection of Cancer-Specific Proteases Using Magnetic Relaxation of Peptide-Conjugated Nanoparticles in Biological Environment. *Nano Letters*, *16*(6), 3668–3674. Available from https://doi.org/10.1021/acs.nanolett.6b00867.

Gandhi, S., Caplash, N., Sharma, P., & Raman Suri, C. (2009). Strip-based immunochromatographic assay using specific egg yolk antibodies for rapid detection of morphine in urine samples. *Biosensors and Bioelectronics*, *25*(2), 502–505. Available from https://doi.org/10.1016/j.bios.2009.07.018.

Gandhi, S., Sharma, P., Capalash, N., Verma, R. S., & Raman Suri, C. (2008). Group-selective antibodies based fluorescence immunoassay for monitoring opiate drugs. *Analytical and Bioanalytical Chemistry*, *392*(1–2), 215–222. Available from https://doi.org/10.1007/s00216-008-2256-9.

Gandhi, S., Suman, P., Kumar, A., Sharma, P., Capalash, N., & Raman Suri, C. (2015). Recent advances in immunosensor for narcotic drug detection. *BioImpacts*, *5*(4), 207–213. Available from https://doi.org/10.15171/bi.2015.30.

Gaster, R. S., Hall, D. A., Nielsen, C. H., Osterfeld, S. J., Yu, H., MacH, K. E., … Wang, S. X. (2009). Matrix-insensitive protein assays push the limits of biosensors in medicine. *Nature Medicine*, *15*(11), 1327–1332. Available from https://doi.org/10.1038/nm.2032.

Grodzinski, P., Silver, M., & Molnar, L. K. (2006). Nanotechnology for cancer diagnostics: Promises and challenges. *Expert Review of Molecular Diagnostics*, *6*(3), 307–318. Available from https://doi.org/10.1586/14737159.6.3.307.

Heller, A. (1996). Amperometric biosensors. *Current Opinion in Biotechnology*, *7*(1), 50. Available from http://www.elsevier.com/locate/copbio.

Hirsch, L. R., Jackson, J. B., Lee, A., Halas, N. J., & West, J. L. (2003). A whole blood immunoassay using gold nanoshells. *Analytical Chemistry*, *75* (10), 2377–2381. Available from https://doi.org/10.1021/ac0262210.

Ishii, Y., & Kitamoto, Y. (2016). Fabrication of Bi-YIG/Au composite particles for magneto-optical devices. *Funtai Oyobi Fummatsu Yakin/Journal of the Japan Society of Powder and Powder Metallurgy*, *63*(10), 882–886. Available from https://doi.org/10.2497/jjspm.63.882.

Jäger, D., Stockert, E., Güre, A. O., Scanlan, M. J., Karbach, J., Jäger, E., … Chen, Y. T. (2001). Identification of a tissue-specific putative transcription factor in breast tissue by serological screening of a breast cancer library. *Cancer Research*, *61*(5), 2055–2061.

Jia, Y., Qin, M., Zhang, H., Niu, W., Li, X., Wang, L., … Feng, X. (2007). Label-free biosensor: A novel phage-modified Light Addressable Potentiometric Sensor system for cancer cell monitoring. *Biosensors and Bioelectronics*, *22*(12), 3261–3266. Available from https://doi.org/10.1016/j.bios.2007.01.018.

Joensuu, H., Jyrkkiö, S., Kellokumpu-Lehtinen, P.-L., Kouri, M., Roberts, P.J., & Teppo, L. (2014). Cytotoxic drugs or cytostatics | All about cancer.

Katz, E., & Willner, I. (2005). Enhancement of bioelectrocatalytic processes by the rotation of mediator-functionalized magnetic particles on electrode surfaces: Comparison with a rotating disk electrode. *Electroanalysis*, *17*(18), 1616–1626. Available from https://doi.org/10.1002/elan.200503266.

Kissinger, P. T. (2005). Biosensors – A perspective. *Biosensors and Bioelectronics*, *20*(12), 2512–2516. Available from https://doi.org/10.1016/j.bios.2004.10.004.

Koncki, R. (2007). Recent developments in potentiometric biosensors for biomedical analysis. *Analytica Chimica Acta*, *599*(1), 7–15. Available from https://doi.org/10.1016/j.ac.2007.08.003.

Kricka, L. J., & Park, J. Y. (2009). Magnetism and magnetoresistance: Attractive prospects for point-of-care testing? *Clinical Chemistry*, *55*(6), 1058–1060. Available from https://doi.org/10.1373/clinchem.2009.123927.

Krishnan, S., Mani, V., Wasalathanthri, D., Kumar, C. V., & Rusling, J. F. (2011). Attomolar detection of a cancer biomarker protein in serum by surface plasmon resonance using superparamagnetic particle labels. *Angewandte Chemie – International Edition*, *50*(5), 1175–1178. Available from https://doi.org/10.1002/anie.201005607.

Kubokawa, M., Nakashima, M., Yao, T., Ito, K. I., Harada, N., Nawata, H., & Watanabe, T. (2001). Aberrant intracellular localization of RCAS1 is associated with tumor progression of gastric cancer. *International Journal of Oncology*, *19*(4), 695–700. Available from https://doi.org/10.3892/ijo.19.4.695.

Li, L., Jiang, W., Luo, K., Song, H., Lan, F., Wu, Y., & Gu, Z. (2013). Superparamagnetic iron oxide nanoparticles as MRI contrast agents for non-invasive stem cell labeling and tracking. *Theranostics*, *3*(8), 595–615. Available from https://doi.org/10.7150/thno.5366.

Li, M., Sun, Y., Chen, L., Li, L., Zou, G., Zhang, X., & Jin, W. (2010). Ultrasensitive eletrogenerated chemiluminescence immunoassay by magnetic nanobead amplification. *Electroanalysis*, *22*(3), 333–337. Available from https://doi.org/10.1002/elan.200900351.

Lin, C. W., Wei, K. C., Liao, S. S., Huang, C. Y., Sun, C. L., Wu, P. J., … Ma, C. C. M. (2015). A reusable magnetic graphene oxide-modified biosensor for vascular endothelial growth factor detection in cancer diagnosis. *Biosensors and Bioelectronics*, *67*, 431–437. Available from https://doi.org/10.1016/j.bios.2014.08.080.

Lucarelli, F., Authier, L., Bagni, G., Marrazza, G., Baussant, T., Aas, E., & Mascini, M. (2003). DNA biosensor investigations in fish bile for use as a biomonitoring tool. *Analytical Letters*, *36*(9), 1887–1901. Available from https://doi.org/10.1081/AL-120023620.

Makower, A., Halámek, J., Skládal, P., Kernchen, F., & Scheller, F. W. (2003). New principle of direct real-time monitoring of the interaction of cholinesterase and its inhibitors by piezolectric biosensor. *Biosensors and Bioelectronics*. Available from https://doi.org/10.1016/S0956-5663(03, 89–7.

McNaught, A. D., & Wilkinson, A. (1997). *Gold Book. Compendium of chemical terminology*. IUPAC. Available from https://doi.org/10.1351/goldbook.

Medley, C. D., Smith, J. E., Tang, Z., Wu, Y., Bamrungsap, S., & Tan, W. (2008). Gold nanoparticle-based colorimetric assay for the direct detection of cancerous cells. *Analytical Chemistry*, *80*(4), 1067–1072. Available from https://doi.org/10.1021/ac702037y.

Meyer, T., & Rustin, G. J. S. (2000). Role of tumour markers in monitoring epithelial ovarian cancer. *British Journal of Cancer*, *82*(9), 1535–1538.

Mishra, A., & Verma, M. (2010). Cancer biomarkers: Are we ready for the prime time? *Cancers*, *2*(1), 190–208. Available from https://doi.org/10.3390/cancers2010190.

Mohanty, S.P., & Kougianos, E. (2006). Steady and transient state analysis of gate leakage current in nanoscale CMOS logic gates. In *IEEE International Conference on Computer Design, ICCD 2006* (pp. 210–215). Available from https://doi.org/10.1109/ICCD.2006.4380819

MRI for Cancer. American Cancer Society. (2019). Available from https://www.cancer.org/treatment/understanding-your-diagnosis/tests/mri-for-cancer.html

Mulvaney, S. P., Cole, C. L., Kniller, M. D., Malito, M., Tamanaha, C. R., Rife, J. C., … Whitman, L. J. (2007). Rapid, femtomolar bioassays in complex matrices combining microfluidics and magnetoelectronics. *Biosensors and Bioelectronics*, *23*(2), 191–200. Available from https://doi.org/10.1016/j.bios.2007.03.029.

Munge, B. S., Coffey, A. L., Doucette, J. M., Somba, B. K., Malhotra, R., Patel, V., … Rusling, J. F. (2011). Nanostructured immunosensor for attomolar detection of cancer biomarker interleukin-8 using massively labeled superparamagnetic particles. *Angewandte Chemie – International Edition*, *50*(34), 7915–7918. Available from https://doi.org/10.1002/anie.201102941.

Nagaoka, H., Sato, Y., Xie, X., Hata, H., Eguchi, M., Sakurai, N., ... Ohnishi, N. (2011). Coupling stimuli-responsive magnetic nanoparticles with antibody-antigen detection in immunoassays. *Analytical Chemistry*, *83*(24), 9197–9200. Available from https://doi.org/10.1021/ac201814n.

Nam, J. M., Thaxton, C. S., & Mirkin, C. A. (2003). Nanoparticle-based bio-bar codes for the ultrasensitive detection of proteins. *Science (New York, N.Y.)*, *301*(5641), 1884–1886. Available from https://doi.org/10.1126/science.1088755.

O'Connell, P. J., & Guilbault, G. G. (2001). Future trends in biosensor research. *Analytical Letters*, *34*(7), 1063–1078. Available from https://doi.org/10.1081/AL-100104953.

Ogawa, K., Takeda, T., Mishiro, K., Toyoshima, A., Shiba, K., Yoshimura, T., ... Odani, A. (2019). Radiotheranostics coupled between an At-211-labeled RGD peptide and the corresponding radioiodine-labeled RGD peptide. *ACS Omega*, *4*(3), 4584–4591. Available from https://doi.org/10.1021/acsomega.8b03679.

Osterfeld, S. J., Yu, H., Gaster, R. S., Caramuta, S., Xu, L., Han, S. J., ... Wang, S. X. (2008). Multiplex protein assays based on real-time magnetic nanotag sensing. *Proceedings of the National Academy of Sciences of the United States of America*, *105*(52), 20637–20640. Available from https://doi.org/10.1073/pnas.0810822105.

Rasooly, A., & Jacobson, J. (2006). Development of biosensors for cancer clinical testing. *Biosensors and Bioelectronics*, *21*(10), 1851–1858. Available from https://doi.org/10.1016/j.bios.2006.01.003.

Rastislav, M., Miroslav, S., & Ernest, Š. (2012). Biosensors – Classification, characterization and new trends. *Acta Chimica Slovaca*, 109–120. Available from https://doi.org/10.2478/v10188-012-0017-z.

Roberts, A., Tripathi, P. P., & Gandhi, S. (2019). Graphene nanosheets as an electric mediator for ultrafast sensing of urokinase plasminogen activator receptor – A biomarker of cancer. *Biosensors and Bioelectronics*, 141. Available from https://doi.org/10.1016/j.bios.2019.111398.

Rosi, N. L., & Mirkin, C. A. (2005). Nanostructures in biodiagnostics. *Chemical Reviews*, *105*(4), 1547–1562. Available from https://doi.org/10.1021/cr030067f.

Sarkar, P., Ghosh, D., Bhattacharyay, D., Setford, S. J., & Turner, A. P. F. (2008). Electrochemical immunoassay for free prostate specific antigen (f-PSA) using magnetic beads. *Electroanalysis*, *20*(13), 1414–1420. Available from https://doi.org/10.1002/elan.200804194.

Schrevens, L., Lorent, N., Dooms, C., & Vansteenkiste, J. (2004). The role of PET scan in diagnosis, staging, and management of non-small cell lung cancer. *The Oncologist*, *9*(6), 633–643. Available from https://doi.org/10.1634/theoncologist.9-6-633.

Singh, S., Mishra, P., Banga, I., Parmar, A. S., Tripathi, P. P., & Gandhi, S. (2018). Chemiluminescence-based immunoassay for the detection of heroin and its metabolites. *BioImpacts*, *8*(1), 53–58. Available from https://doi.org/10.15171/bi.2018.07.

Smith, D. S., Humphrey, P. A., & Catalona, W. J. (1997). The early detection of prostate carcinoma with prostate specific antigen. *The Washington University Experience*, *80*. Available from https://doi.org/10.1002/(SICI)1097-0142.

Stoeva, S. I., Lee, J. S., Smith, J. E., Rosen, S. T., & Mirkin, C. A. (2006). Multiplexed detection of protein cancer markers with biobarcoded nanoparticle probes. *Journal of the American Chemical Society*, *128*(26), 8378–8379. Available from https://doi.org/10.1021/ja0613106.

Talan, A., Mishra, A., Eremin, S. A., Narang, J., Kumar, A., & Gandhi, S. (2018). Ultrasensitive electrochemical immuno-sensing platform based on gold nanoparticles triggering chlorpyrifos detection in fruits and vegetables. *Biosensors and Bioelectronics*, *105*, 14–21. Available from https://doi.org/10.1016/j.bios.2018.01.013.

Teramura, Y., Arima, Y., & Iwata, H. (2006). Surface plasmon resonance-based highly sensitive immunosensing for brain natriuretic peptide using nanobeads for signal amplification. *Analytical Biochemistry*, *357*(2), 208–215. Available from https://doi.org/10.1016/j.ab.2006.07.032.

Thaxton, C. S., Elghanian, R., Thomas, A. D., Stoeva, S. I., Lee, J. S., Smith, N. D., ... Mirkin, C. A. (2009). Nanoparticle-based bio-barcode assay redefines\undetectable\PSA and biochemical recurrence after radical prostatectomy. *Proceedings of the National Academy of Sciences of the United States of America*, *106*(44), 18437–18442. Available from https://doi.org/10.1073/pnas.0904719106.

Tomitaka, A., Arami, H., Gandhi, S., & Krishnan, K. M. (2015). Lactoferrin conjugated iron oxide nanoparticles for targeting brain glioma cells in magnetic particle imaging. *Nanoscale*, *7*(40), 16890–16898. Available from https://doi.org/10.1039/c5nr02831k.

Tothill, I. E. (2009). Biosensors for cancer markers diagnosis. *Seminars in Cell and Developmental Biology*, *20*(1), 55–62. Available from https://doi.org/10.1016/j.semcdb.2009.01.015.

Tripathi, P. P., Arami, H., Banga, I., Gupta, J., & Gandhi, S. (2018). Cell penetrating peptides in preclinical and clinical cancer diagnosis and therapy. *Oncotarget*, *9*(98), 37252–37267. Available from https://doi.org/10.18632/oncotarget.26442.

Wacker, R., Ceyhan, B., Alhorn, P., Schueler, D., Lang, C., & Niemeyer, C. M. (2007). Magneto immuno-PCR: A novel immunoassay based on biogenic magnetosome nanoparticles. *Biochemical and Biophysical Research Communications*, *357*(2), 391–396. Available from https://doi.org/10.1016/j.bbrc.2007.03.156.

Wang, J., & Kawde, A. N. (2001). Pencil-based renewable biosensor for label-free electrochemical detection of DNA hybridization. *Analytica Chimica Acta*, *431*(2), 219–224. Available from https://doi.org/10.1016/S0003-2670(00)01318-0.

Wang, J., Munir, A., Zhu, Z., & Zhou, H. S. (2010). Magnetic nanoparticle enhanced surface plasmon resonance sensing and its application for the ultrasensitive detection of magnetic nanoparticle-enriched small molecules. *Analytical Chemistry*, *82*(16), 6782–6789. Available from https://doi.org/10.1021/ac100812c.

Willner, I., & Katz, E. (2003). Magnetic control of electrocatalytic and bioelectrocatalytic processes. *Angewandte Chemie – International Edition*, *42*(38), 4576–4588. Available from https://doi.org/10.1002/anie.200201602.

Wu, G., Ji, H., Hansen, K., Thundat, T., Datar, R., Cote, R., ... Majumdar, A. (2001). Origin of nanomechanical cantilever motion generated from biomolecular interactions. *Proceedings of the National Academy of Sciences of the United States of America*, *98*(4), 1560–1564. Available from https://doi.org/10.1073/pnas.98.4.1560.

Zhang, Y., & Zhou, D. (2012). Magnetic particle-based ultrasensitive biosensors for diagnostics. *Expert Review of Molecular Diagnostics*, *12*(6), 565–571. Available from https://doi.org/10.1586/erm.12.54.

Chapter 11

# Next generation biosensors as a cancer diagnostic tool

**Deepshikha Shahdeo and Sonu Gandhi**
*DBT-National Institute of Animal Biotechnology (DBT-NIAB), Hyderabad, Telangana, India*

## 11.1 Introduction

Cancer is one of the major causes of death all over the world. It is a genetic disorder caused due to alteration in the epigenetic of somatic cells. Factors, including exposure to UV radiation, a cancer-causing chemical, infection, and in some cases genetic modification causes abnormal proliferation and generation of cancer in different parts of the body (Arya & Estrela, 2018; Zhao et al., 2015). Recently, cancer has overtaken heart disease as the common cause of death in the United Kingdom (Kmietowicz, 2015). More than 200 types of cancer have been identified on the basis of cells, tissues, organs, among them lung, breast, colorectal, stomach, head, and neck cancer are the most common (Nagai & Kim, 2017). The study of GBD 2015 reported that most common type of cancer leading to death in females is breast, lung, and colorectal cancer. According to one survey of the World Health Organization (WHO), breast cancer is one of the leading causes of cancer-related death in the world after cervical cancer (Azamjah, Soltan-Zadeh, & Zayeri, 2019; Freitas, Nouws, Keating, & Delerue-Matos, 2020; Topkaya, Azimzadeh, & Ozsoz, 2016). On the other hand lung, prostate, and colorectal are the most common cancer in males. In young adults, the most common cancer are breast, cervical, leukemia, and liver cancer (Densmore et al., 2008; Wang et al., 2015). Hepatitis B virus (HBV) and Hepatitis C virus are the main causes of liver cancer, which exhibit a high mortality rate in the world. Immunization against HBV and HCV would lead to a reduction in the incidence of liver cancer (Bruno et al., 2007; Qu et al., 2014). Current cancer diagnostic techniques and treatment are primarily focused on the determination and identification of the pattern of genetic abnormalities result from chromosomal aberration. Numerous invasive and noninvasive methods are available for the diagnosis of cancer. Imaging and screening tests are the most commonly used technique for cancer diagnosis. Although as mentioned in Fig. 11.1 there are several traditional diagnostic methods for cancer, primarily based on endoscopy (René, 2012), computed tomography, positron emission tomography (Hu et al., 2016), magnetic resonance imaging (Haris et al., 2015), x-rays (Frangioni, 2008) and invasive tissue biopsy (Cowling & Loshak, 2016). These methods are neither accessible to a large population nor practical for the repeated screening. These techniques are limited as they are dependent on the phenotypic properties of the tumor, highly expensive, and not accessible to the large population of developing countries such as India.

Cancer biomarkers are emerging as one of the promising strategies for cancer diagnostic and therapeutics. Biomarkers are biomolecules produced by the tumor cells or other cells in response to the tumor. These biological molecules can also be detected in blood, urine, and other body fluids which undergo several alterations and overexpressed due to abnormalities and commencement of the disease (Goossens, Nakagawa, Sun, & Hoshida, 2015). They act as a "molecular signature" that provides accurate information about the different stages and mechanisms of action of different types of cancer (Sethi, Ali, Philip, & Sarkar, 2013). Several novel biomarkers have been identified. Some of the important cancer biomarkers and their normal range are mentioned in Table 11.1. However only very few of them have seen the face of clinical trials for diagnostic and prognostic purposes. Some of the important cancer biomarkers used for treatment and early diagnostic are EGF, FGF, VEGF, TGF-$\beta$, bax, bid caspases, interferons, TNF, COX-2, prostaglandins, MMPs, pRB, BRCA1, BRCA2, and p53, Ras, Raf, and Src (Maruvada, Wang, Wagner, & Srivastava, 2005).

Cancer biomarkers could be used for the early and rapid detection of disease progression (Mor et al., 2005; Shapira et al., 2014). ELISA and PCR are highly sensitive techniques for detection of cancer biomarker; however, these

Biosensor Based Advanced Cancer Diagnostics. DOI: https://doi.org/10.1016/B978-0-12-823424-2.00016-8

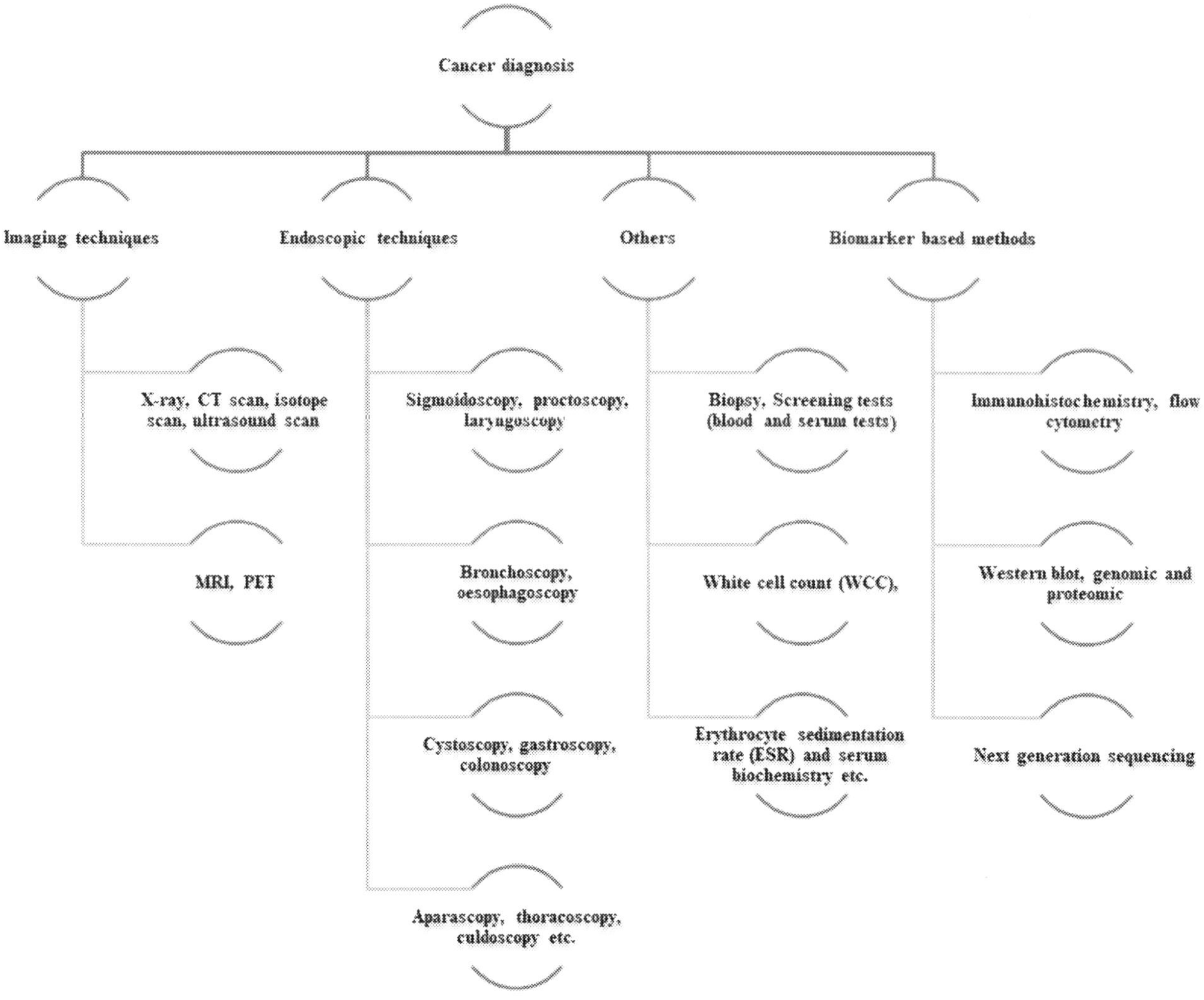


**FIGURE 11.1** Different methods and tools available for the diagnosis of cancer.

**TABLE 11.1 Some of the important cancer biomarkers for the treatment and diagnosis of the cancer progression.**

| Biomarker | Cancer | Sample | Normal level | References |
|---|---|---|---|---|
| Prostate-Specific Antigen (PSA) | Prostate cancer | Serum | <20 ng/mL | Nogueira, Corradi, and Eastham (2009) |
| Alpha-fetoprotein (AFP) | Hepatocellular carcinoma (HCC) | Serum | 5–10 µg/L | Debruyne and Delanghe (2008) |
| Calcitonin | Medullary thyroid cancer (MTC) | Blood | <10 pg/mL | Bae, Schaab, and Kratzsch (2015) |
| Carcinoembryonic antigen (CEA) | CRC and lung cancer | Serum | < 5 ng/mL | Świderska et al. (2014) |
| Cancer antigen 125 (CA 125) | Ovarian cancer | Serum | <35 U/mL | Bottoni and Scatena (2015) |
| Hormone receptor (ER, PR, HER2) | Thyroid cancer | Serum | <2 ng/mL | Grebe (2009) |
| Thyroglobulin (Tg) | Breast cancer | Breast tissue | <1% positive strain | Yip and Rhodes (2014) |
| Human epididymis protein 4 (HE4) | Ovarian cancer | Serum | <2 ng/mL | Bottoni and Scatena (2015) |

methods are expensive and suffer from the technological limitation of slow detection and are not practical enough for continuous monitoring of patients during treatment (Eftimie & Hassanein, 2018; Zhao et al., 2015). Also, these techniques are not convenient for the detection of cancer with multiple biomarkers (Damiati, Haslam, Peacock, & Awan, 2018; G, 2014). Thus, it is required to detect multiple biomarkers simultaneously for efficient diagnosis (Chatterjee & Zetter, 2005; Florea, Ravalli, Cristea, Săndulescu, & Marrazza, 2015). With the increase in the expertise and technological advancement in medical science and cancer biology, several theranostic agents and biosensors have emerged (Zhao et al., 2015). An appropriate small-sized theranostic agent with favorable surface chemistry for targeting and cellular imaging can be used as a leading tool for cancer treatment. Different theranostic agents are available such as CT imaging, photoluminescence, photo-thermal therapy, SERS-based, magnetic hyperthermia, and several combined therapies (Choi, Kwak, & Park, 2010). Mesoporous silica, quantum dots, AuNPs, graphene are some of the nanoparticles that provide platforms for the development of nanocomposite theranostic. Point-of-care (POC) diagnosis is vital as it reduces the overall cost and provides a portable method for "on-site" tests for patients (Roberts, Tripathi, & Gandhi, 2019; Zhang, Li, Gao, Chen, & Liu, 2019). Usually, the development of a POC screen device requires sophisticated equipment and expertise. Electrochemical devices, lateral flow assays, and colorimetric paper-based are some POC screening methods that can be fabricated efficiently (Mohammadniaei, Nguyen, Tieu, & Lee, 2019; Xiao et al., 2017b). Among them, electrochemical devices are the most promising type of method (Gandhi, Arami, & Krishnan, 2016; Tripathi, Arami, Banga, Gupta, & Gandhi, 2018). However, the fabrication of POC-based devices for cancer diagnostics is still a challenging field (Chinen et al., 2015; Song et al., 2010). In this chapter, the development of next-generation biosensors for the detection and quantification of cancer biomarkers will be discussed. However, the main focus of this chapter is to explain different sensor transducers on the performance, particularly involvement of 2D nanomaterial in the improvement of the sensitivity and efficiency of the biosensor will be compared and summarized.

PSA prostate cancer serum $<20$ ng/mL (Damborska et al., 2017; Nogueira et al., 2009) AFP hepatocellular carcinoma (HCC) serum $5-10$ μg/L (Debruyne & Delanghe, 2008) BRCA1 and BRCA2 breast cancer calcitonin MTC Blood $<10$ pg/mL (Bae et al., 2015) carcinoembryonic antigen (CEA) CRC and lung cancer serum $<5$ ng/mL (Świderska et al., 2014) CA 125 ovarian cancer serum $<35$ U/mL (Bottoni & Scatena, 2015) Tg thyroid cancer serum $<2$ ng/mL (Grebe, 2009) ER/PR/HER2 breast cancer breast tissue $<1\%$ positive strain (Wu et al., 2003; Yip & Rhodes, 2014) HE4 ovarian cancer serum $<2$ ng/mL (Bottoni & Scatena, 2015) cancer biomarkers could be used for the early and rapid detection of disease progression (Mor et al., 2005; Shapira et al., 2014).

## 11.2 Biosensor transducers

The transducer is the element that converts the biological signal into the digital or electrical signal, which can be further quantified and analyzed. There are mainly four types of transducers: electrochemical, optical, mass-based, and calorimetric (Bhalla, Jolly, Formisano, & Estrela, 2016; Bohunicky & Mousa, 2011).

### 11.2.1 Electrochemical sensor

Electrochemical biosensors are the systems that converts the biological signals into electrochemical signals (Banga, Tyagi, Shahdeo, & Gandhi, 2019; Freitas, Nouws, & Delerue-Matos, 2018; Thévenot, Toth, Durst, & Wilson, 2001).

### 11.2.2 Optical biosensor

Optical sensors are light-based biosensors, which are based on the change in measurement of wavelength after interaction of the analyte with the bio-recognition element (Gandhi et al., 2015). Optical biosensors transducer can be interferometric, fluorescence, luminescence, or colorimetric (Bohunicky & Mousa, 2011). Recently, photonic crystal-based biosensors are evolving as a new class of biosensor that uses an optical transducer. Example of optical biosensors are surface-enhanced Raman scattering (SERS) biosensors (Dinish et al., 2012), reflectometric interference spectroscopy (RIfS) biosensors (Rau, Hilbig, & Gauglitz, 2014), ellipsometric biosensors (Fei et al., 2015), and SPR-based sensors (Safina, 2012).

### 11.2.3 Mass-based biosensors

Surface acoustic wave (SAW) biosensor, microcantilever (MCL)-based biosensor, and quartz crystal microbalance (QCM) are some of the common mass-based biosensors (Bohunicky & Mousa, 2011). These biosensors are commonly used sensors for the detection of cancer biomarkers. Piezoelectric biosensors work on the principle of change in the oscillations due to mass bound on the piezoelectric crystal when potential energy is applied (Pohanka, 2018; Su et al., 2013).

### 11.2.4 Calorimetric biosensor

Calorimetric biosensors are not mainly used for cancer diagnostics. It is based on the measurement of exothermic reactions. Generally, biological processes are more or less exothermic; the addition of biocatalyst results in the generation of heat, and change in the heat intensity can be used for detection of the interaction between the analyte and biorecognition element. The reaction is measured by the change in the enthalpy, which provides information about the concentration of the substrate (Danielsson, 1990). Calorimetric-based sensors are not generally used for the detection of cancer, although Medley et al. reported the aptamer-based colorimetric sensor for detection of cancer. They were able to differentiate between acute leukemia and Burkitt's lymphoma cell type using gold nanoparticles (Medley et al., 2008).

## 11.3 Biosensors for cancer biomarker detection

The sensor surface plays an important role in immobilizing the biomolecules; it can be used to enhance the signal intensity. To elevate the efficiency of biosensors by increasing the loading of antibodies, researchers have tried to increase the surface area by introducing various nanomaterials in the field of biosensor development. Currently, 2D materials such as graphene oxide (Gao et al., 2014), reduced graphene oxide (rGO) (Verma et al., 2017; Zhao et al., 2015; Zhou, Liu, Bai, & Shi, 2013), graphene sheets (Wei, Mao, Liu, Zhang, & Yang, 2018; Xu, Wu, Dai, Gao, & Xiang, 2016), carbon nanotubes (Gandhi et al., 2015; Huang, Dong, Liu, Li, & Chen, 2011; Istamboulie et al., 2010; Stephen Inbaraj & Chen, 2016), magnetic beads (MB), molybdenum disulfide ($MoS_2$), and bismuth selenide ($Bi_2Se_3$) are accelerating the performance of conventional devices toward more practical approaches. Some of the biosensor for detection of cancer biomarkers is given in the below Table 11.2.

### 11.3.1 Graphene-based biosensors

Graphene is the single layer of the carbon atom with honeycomb lattice; it possesses remarkable electrochemical, optical, and thermal properties (Freitas et al., 2018; Janire, N, F, & R, 2018; Nogueira et al., 2009). It has been utilized as a novel material for cancer diagnostic and biosensor development due to numerous advantages such as good compatibility, large surface area, chemical stability, and high electron transfer rate. It provides a biocompatible surface to load for the recognition unit. Several graphene-based electrochemical biosensors have been developed for the reliable and accurate sensing of the target molecules. Yin et al. have developed the first graphene-based electrochemical biosensor (Yin et al., 2012) where a glassy carbon electrode was modified with graphene and the thin layer of dendritic gold nanostructure (Den-Au) was deposited onto it, which provides a highly conductive electrode surface. The electrode surface was further functionalized with the capture probe (locked nucleic acid-integrated hairpin molecular beacon [MB]). When the target miRNA hybridizes with MB, that was complementary to the target results in the formation of the sandwich with gold particles (Den-Au) strep conjugate (Yin, Zhou, Zhang, Meng, & Ai, 2012); this results in electrochemical detection. Several sensitive electrochemical biosensors have been reported for the miRNA detection, such as GO/gold-nanorod (Coutinho & Somoza, 2019), magnesium oxide nanoflower (Azimzadeh, Rahaie, Nasirizadeh, Ashtari, & Naderi-Manesh, 2016) and rGO/carbon nanotube@screen printed gold electrodes (Densmore et al., 2008; Wang et al., 2015). Ruiyi et al. introduced an electrochemical biosensor for the detection of liver tumor cells (HepG2) in whole blood cells. They synthesized folic acid and octadecylamine-functionalized graphene aerogel microspheres, which has excellent electrocatalytic activity (Ruiyi, Fangchao, Haiyan, Xiulan, & Zaijun, 2018). Recently, Verma et al. fabricated electrochemical immunosensors for the noninvasive detection of oral cancer biomarker interleukin-8 (Gandhi et al., 2015; Huang et al., 2011; Istamboulie et al., 2010; Stephen Inbaraj & Chen, 2016). It was comprised of gold nanoparticles (AuNPs) conjugated with reduced graphene oxide (AuNPs-rGO) immobilized on the surface of ITO coated glass as a transducer matrix, along with anti-IL8 antibody onto the activated surface and their electrochemical activity was measured using CV as shown in Fig. 11.2. The sensor showed fast detection and high sensitivity with LOD of $72.73 \pm 0.18$ pg/mL (Verma et al., 2017). Chen et al. designed an electrochemical biosensor based on the 3D graphene functionalized with silver nanoparticles for the detection of DNA samples (Chen et al., 2018). The electrochemical biosensor was also developed cancer biomarkers such as CA-125, PSA (Khan et al., 2018), squamous cell carcinoma antigen (SCCA), folic acid protein (FP) (He et al., 2016) and human growth factor receptor 2 (HER2) (Tabasi et al., 2017).

**TABLE 11.2 Different types of biosensors for the detection of common cancer biomarker utilized for cancer diagnosis.**

| Biomarker | Detection method | Biomaterial/Biosensing element | Limit of detection | References |
|---|---|---|---|---|
| Interleukin 6 (IL6) | Electrochemical/ Amperometric sensor | Monolayer graphene oxide (GO) | ~10 pg/mL | Huang, Harvey, Derrick Fam, Nimmo, and Alfred Tok (2013) |
| Prostate specific antigen (PSA) | Piezoelectric | Titanate zirconate (PZT) | 0.25 ng/mL | Su et al. (2013) |
| Prostate specific antigen (PSA) | Paper microfluidic device | Apt/AuNPs/rGO/THI/ | 10 pg/mL | Wei et al. (2018) |
| Prostate specific antigen (PSA) | Optical | AuNPs | 0.15 ng/mL | Damborska et al. (2017) |
| Human chronic gonadotropin (hCG) | Electrochemical | 1-pyrenebutyric acid-N-hydroxysuccinimide ester (PANHS) | 1 pg/mL | Damiati et al. (2018) |
| Carcinoembryonic antigen (CEA) | Microfluidic | (Thionin (THI)- gold nanoparticles (AuNPs) | 10 pg/mL | Wang et al. (2019) |
| Carcinoembryonic antigen (CEA) | Electrochemical | Gold nanoclusters | 5.38 pg/mL | Zheng, Wang, Song, Xu, and Zhang (2020) |
| Cancer antigen 125 (CA-125) | Electrochemical | $MoS_2$ nanosheets | 0.36 pg/mL | Wang, Deng, Shen, Yan, and Yu (2016) |
| Cancer antigen 125 (CA-125) | Microfluidic paper based device | $SiO_2$ nanobiohybrid | 0.001 ng/mL | Wu, Xue, Kang, and Hui (2013) |
| α-Fetoprotein (AFP) | Electrochemical | AuNPs/glassy carbon electrode (GCE) | 2 pg/mL | Wu et al. (2019) |
| Human epidermal growth factor receptor-2 (HER-2) | Electrochemical | Reduced graphene oxide-chitosan (rGO-Chit) | 0.21 ng/mL | Tabasi, Noorbakhsh, and Sharifi (2017) |
| Ferritin | Optical | SPR (AuNPs) | 0.2 ng/mL | Chou, Hsu, Hwang, and Chen (2004) |
| Breast cancer susceptibility gene (BRAC-1) | Electrochemical | cDNA immobilized chitosan-co- polyaniline functionalized | 0.05fM | Tiwari and Gong (2009) |
| Mucin I (MUC 1) | Electrochemical | Magnetic bead | 0.07 nM | Florea et al. (2015) |

Poly (3,4-ethylenedioxythiophene): Poly (styrenesulfonate) (PEDOT: PSS) based paper device was fabricated for sensitive detection of CEA (Kumar et al., 2015).

Later, sensitive electrochemical biosensor was developed based on the microfluidics for detection multiple markers (PSA, PSMA). $Fe_3O_4$ nanoparticles were immobilized on the surface of graphene oxide nanosheets as shown in Fig. 11.3. The antibody against the PSA and PSMA were immobilized on the surface of the nanosheet and used to capture the target analyte and deliver them to 8 sensor microfluidic system (Wu et al., 2003; Yip & Rhodes, 2014). Later, reduced graphene oxide was coated onto the screen-printed carbon sensor and a second antibody (Ab2) was allowed to assembled on it. The second antibody (Ab1) selectively binds to the biomarker immobilized on the surface of Fe3O4@GO particles, which subsequently results in the reduction of hydrogen peroxide for the detection PSA and PSMA. This fabricated device showed very low limit of detection (LOD) of 4.8 fg/mL for PSMA and 15 fg/mL for PSA in serum.

## 11.3.2 Molybdenum disulfide-based biosensor

Molybdenum disulfide ($MoS_2$) has gained attention in recent times due to its multiple application and advantages. It is comprised of S−Mo−S triple-layer and has semiconductor properties (Barua, Dutta, Gogoi, Devi, & Khan, 2018). $MoS_2$

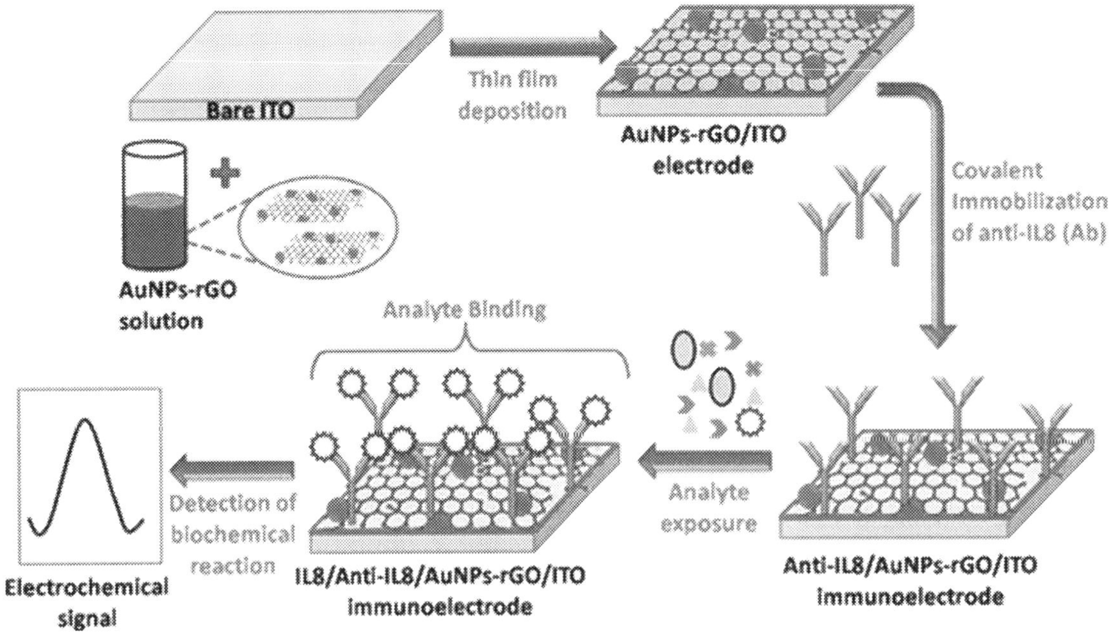


**FIGURE 11.2** Fabrication of AuNPs-rGO based immunosensor: (A) Immobilization of the AuNPs/rGO conjugate onto the surface of ITO functionalized glass; (B) introduction of anti-IL8 antibody on the surface of the AuNPs/rGO functionalized ITO via covalent bond; (C) addition of that target analyte; (D) incubation of analyte with anti-IL8 antibody for proper interaction; and (E) detection of biochemical reaction between the analyte and antibody. *Reproduced with permission from Verma, S., Singh, A., Shukla, A., Kaswan, J., Arora, K., Ramirez-Vick, J., . . . Singh, S. P. (2017). Anti-IL8/AuNPs-rGO/ITO as an immunosensing platform for noninvasive electrochemical detection of oral cancer.* ACS Applied Materials and Interfaces, *9(33), 27462–27474. Available from https://doi.org/10.1021/acsami.7b06839.*

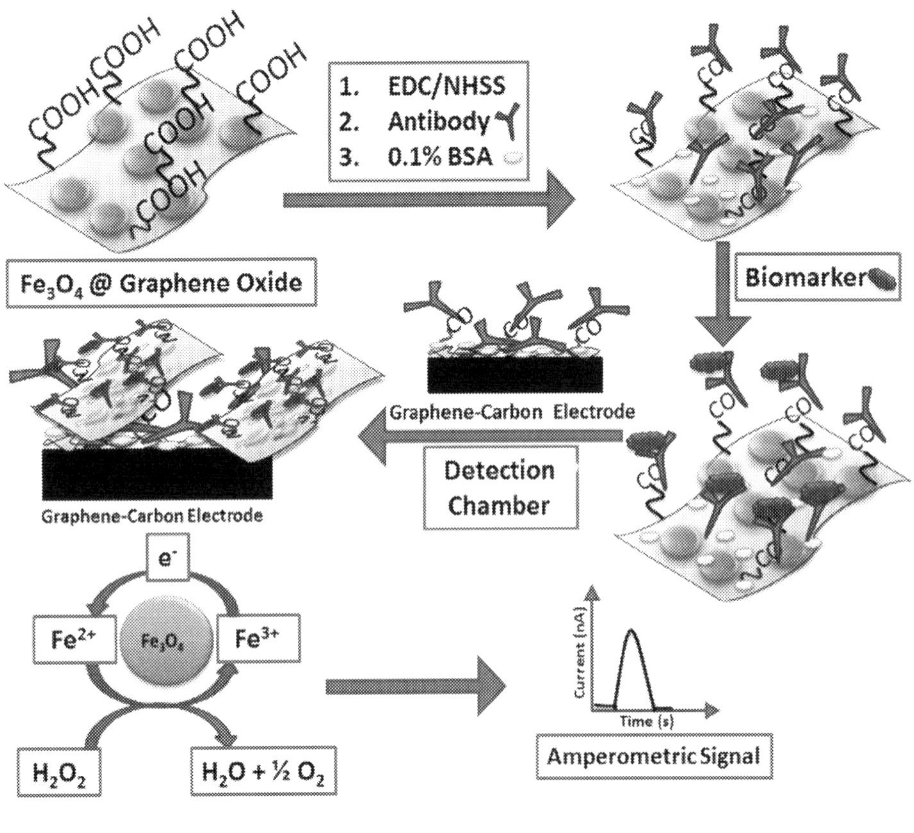


**FIGURE 11.3** Electrochemical sensor based on the principle of microfluidics for detection of PSA and PSMA; (A) $Fe_3O_4$ modified graphene oxide; (B) Conjugation of antibody to the $Fe_3O_4$@GO; (C) Addition of the biomarker; (D) Fabrication of glassy carbon electrode with second Antibody (Ab2); (E) Interaction of biomarker with Ab2; (F) Interaction of $Fe_3O_4$ @GO results in reduction of hydrogen peroxide; (G) Amperometric detection of the hydrogen peroxide. *Reproduced with permission from Sharafeldin, M., Bishop, G. W., Bhakta, S., El-Sawy, A., Suib, S. L., & Rusling, J. F. (2016). Fe3O4 nanoparticles on graphene oxide sheets for isolation and ultrasensitive amperometric detection of cancer biomarker proteins.* Biosensors and Bioelectronic, 91, *359–366. http://dx.doi.org/10.1016/j.bios.2016.12.052.*

has excellent electrochemical and luminescence properties, which makes it a noble nanomaterial as a biosensing probe (Soni, Pandey, Pandey, & Sumana, 2019). Physiochemical properties such as fast charge transfer, high conductivity, and large surface area volume ratio are some of the prime cause of the attraction. (Gan, Zhao, & Quan, 2017). Biosensors of virus, bacteria, glucose, fatty acid, and cancer biomarkers have been developed using the $MoS_2$ based nanomaterial (Roberts et al., 2019; Zhang et al., 2019). Wang et al. developed a biosensor for monitoring the DNA samples in human serum using $MoS_2$ − thionin composite (Wang et al., 2014). Huang et al. developed a biosensor for the detection of miRNA in the sample (He et al., 2016). AuNPs modified $MoS_2$ microcubes were used for quantification and detection of the sample. The mechanism of sensing was based on the hybridization of the target (miRNA) with biotin-labeled ssDNA, on the AuNP/$MoS_2$, and signal amplification was done via electrochemical−chemical−chemical redox cycle.

$MoS_2$ showed feasibility for miRNA sensing. AuNPs modified $MoS_2$ nanosheets functionalized with a captured DNA probe (Bartel, 2004). The paper-based device was also developed for the detection of miRNA (He, Huang, & Wu, 2020). Electrochemical-based biosensors are widely used for the identification of cancer biomarkers. Jiao et al. developed a biosensor based on porous anodic aluminum oxide (AAO) modified with molybdenum disulfide ($MoS_2$) for the detection of the miRNA-155 (Jiao, Liu, Wang, Ma, & Lv, 2020). Cai et al. developed a ($MoS_2$) based, highly specific, and efficient fluorescence sensor for the detection of miR21, which promotes breast cancer proliferation. $MoS_2$ was introduced to the fluorescence dye-labeled with the DNA probe. Complementary miR-21 were allowed to hybridize with the DNA probe of the sensor and a fluorescent signal was measured before and after binding and thus with change in the signal, miR-21 can be detected (Cai, Guo, & Li, 2018).

Yan et al. fabricated a cost-effective electrochemical device for cancer antigen-125 (CA-125) detection (Wang et al., 2016). The device was based on the strategy of dual signal amplification. Here, $MoS_2$ acted as solid support for the binding of the CA-125 antibody, as shown in Fig. 11.4, and gold nanoflowers (AuNF), with a high surface to volume ratio used for multiple label assays with glucose oxidase (GOx) and a second antibody to generate $H_2O_2$ for signal amplification. Firstly, μ-PADs was developed using the wax printer and Ag/AgCl and carbon ink were used to print electrodes directly on the paper sheet surface. Au-paper working electrode (AuPWE) was fabricated on the sample zone. When the analyte attached to the primary antibody, $MoS_2$ nanohybrid enhanced the catalytic activity of the GOx and increased the sensitivity.

ELISA is commonly considered as standard technique for detection and quantification of antigens in the presence of a specific antibody. Interaction of an antibody with antigen results in a change in color, which is used for measuring the analyte concentration. Although it provides sensitive and specific data with better detection range, it suffers from the drawback of the tedious procedure and centralized laboratory equipment. Recently this issue was overcome with the development of ELISA-based electrochemical sensors. Portable instrumentation, low volume, fast measurement, and easy procedure are

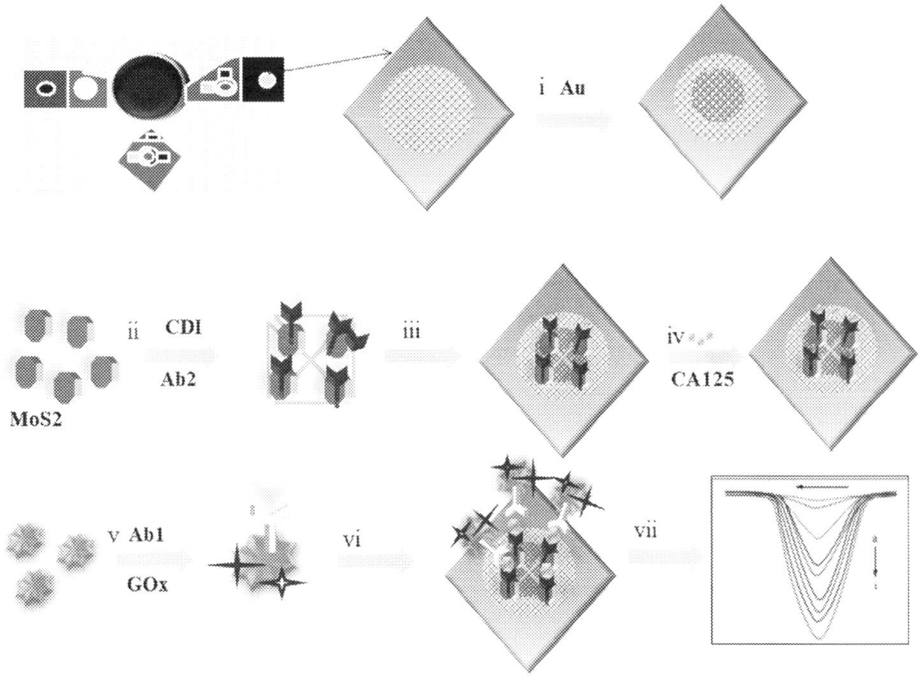


**FIGURE 11.4** Fabrication of a 3D electrochemical biosensor for detection of CA-125 cancer biomarker; (i) Au/paper working electrode was fabricated; (ii) MoS2 nanohybrid was synthesized and Ab2 was conjugated; (iii) Au/paper was functionalized on the MoS2 nanohybrid; (iv) Addition of CA-125; (v) primary antibody was attached to AuNF; (vi) captured antibody along with AuNF cofunctionalized with Ab2 after CA-125 binding; (vii) Differential pulse voltammetry response was used to measure different concentration of CA-125.

some of the advantages of using ELISA based electrochemical sensors. It has also attracted interest in cancer biomarker detection due to greater sensitivity, the ability of signal amplification, ease of handling, and low cost. Recently, silver nanoparticles (AgNPs) and molybdenum disulfide ($MoS_2$) have attracted attention due to their large surface area, accelerating electron mobility, electro-conductivity, and relatively low toxicity. Based on the above properties, Y. Wang et al. developed an $Ag/MoS_2@Fe_3O_4$ modified electrochemical immunosensor for detecting carcinoembryonic antigen (CEA) (Wang et al., 2018). Initially, primary antibody (Ab1) was immobilized on the surface of ELISA microplates as shown in Fig. 11.5A. Then BSA and target CEA were added to the surface of the plate. $Ag/MoS_2@Fe_3O_4$ nanocomposite was synthesized and secondary antibody (Ab2) was attached to it (Fig. 11.5A). Subsequently, the $Ab_2$-$Ag/MoS_2@Fe_3O_4$ were incubated with CEA antigen (Fig. 11.5A) and allowed for binding with anti CEA antibody. Further, free $Ab_2$-$Ag/MoS_2@Fe_3O_4$ were collected and attached via Ag-$NH_2$ interaction to the magnetic glassy carbon electrode (MGCE). Thus, amounts of unbound $Ab_2$-$Ag/MoS_2@Fe_3O_4$ were measured by electrochemical impedance spectroscopy (EIS) and differential pulse voltammetry (DPV). With the increase in the amount of CEA, current response decreases. LOD of this immunosensors was 0.03 pg/mL.

## 11.3.3 $Bi_2Se_3$-based electrochemical biosensor

Bismuth selenide ($Bi_2Se_3$) is a 3D topological insulator (TI), exhibiting properties such as Dirac plasmons (Di Pietro et al., 2013), thermoelectric behavior (Rui, Wang, Gu, Zhan, & Cui, 2016), photothermal conversion ability, and unique electrical conductivity (Hsieh et al., 2011) which has attracted intensive interest from the various interdisciplinary research fields (Stankiewicz et al., 2019). TIs are the new class of quantum material with great conducting surface and they have a high resistance towards crustal disorder. Several $Bi_2Se_3$ based nanoparticles and nanosheets have been used for clinical applications (Dhanjai et al., 2018), such as detection of biomarker (Chen et al., 2018), imaging, and cancer radiation therapy. The first enzymatic $Bi_2Se_3$ based electrochemical biosensor was developed by Fan et al. (2012). They developed flower-like $Bi_2Se_3$ nanostructures using the facile hydrothermal technique. Later it was also used for the development of a biosensor for the detection of IgG. Dong et al. developed an electrochemical biosensor based on $Bi_2Se_3$ nanosheets for the detection of immunoglobulin G (anti-IgG) (Dong, Li, Wei, Liu, & Huang, 2017). Carbon

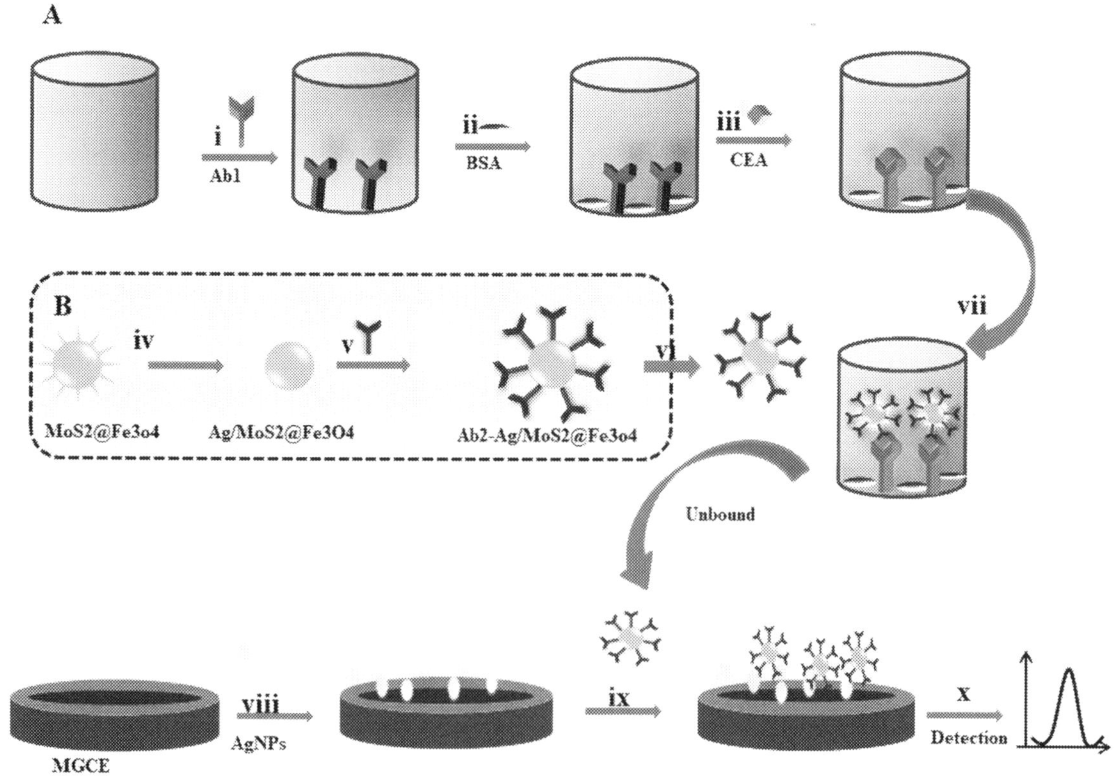


FIGURE 11.5 Schematic illustration of fabrication of immunosensor for CEA detection based on $Ag/MoS_2@Fe_3O_4$; (A) fabrication process of the immunosensors, different steps representing the attachment of Ab1 to the surface of ELSA plate and binding of CEA to the antibody; (B) it shows the preparation process of $Ab_2$-$Ag/MoS_2@Fe_3O_4$.

based electrode was modified with $Bi_2Se_3$ nanosheets and human immunoglobulin G (anti-IgG) was added via glutaraldehyde crosslinking. The fabricated sensor was highly sensitive and specific with a LOD of $0.8\ ng \cdot mL^{-1}$. Mohammadniaei et al. detected the breast cancer cells (MCF7 and MDA-MB-231). They have used a highly sensitive electrochemical biosensor based on the difference in extracellular $H_2O_2$ produced by two different cell lines. As shown in Figure 11.6, $Bi_2Se_3$ were self-assembled on an Au electrode, and later on (Fig. 11.6) AuNPs were deposited on the surface of $Bi_2Se_3$. Sandwiched $Bi_2Se_3$ provides a platform for the $Ag^+$-modified DNA's modification. The encapsulation of $Bi_2Se_3$ protects the TI surface state and also helps in improving the stability of the electrochemical signal. The developed biosensor showed a fast response and significantly with LOD of 10 nM (Mohammadniaei et al., 2018).

miRNA are the small noncoding regulatory sequence (average 22 nucleotides) of RNA. They act as a regulator of genes, responsible for the fundamental biological process of the cell cycle, cell division, and apoptosis (O'Brien, Hayder, Zayed, & Peng, 2018). miRNA suppresses the gene expression by binding to the target mRNA. These miRNAs are mainly present in plasma, serum, saliva, urine, and other body fluid (Keshavarz, Behpour, & Rafiee-Pour, 2015). They are also resistant to the extreme pH, temperature, and ribonuclease (RNase), which makes them highly stable and an easily accessible biomarker for clinical diagnosis (Yu, Li, Ding, & Ding, 2011). Strategies such as microarray (Lin et al., 2011; Santarelli et al., 2013), locked nucleic acid-based northern blot (Tripathi et al., 2018; Válóczi et al., 2004), MB-based flow-cytometry, and real-time PCR (C. Chen et al., 2005) have been developed for the detection of miRNA. However, these strategies suffer from the disadvantages of high assay cost, time-consuming, and tedious process. In recent years, miRNA-based biosensors are attracting great attention. Several miRNA-based biosensors have been developed in the recent years for rapid and reliable detection.

Magnetic bead (MBs)-based biosensors are widely used for the enhancement of the sensitivity of molecule detection. Wang et al. proposed an electrochemical biosensor for the detection of oral cancer (Wang et al., 2019). Herein, a biotin-labeled capture probe was attached on the surface of streptavidin-modified MBs via streptavidin-biotin interaction. In the presence of target miRNA (target probe), the capture probe hybridizes with the target probe and a signal probe with biotin tag at both ends to form a ternary structure on the surface of MBs. A biotin-labeled signal probe

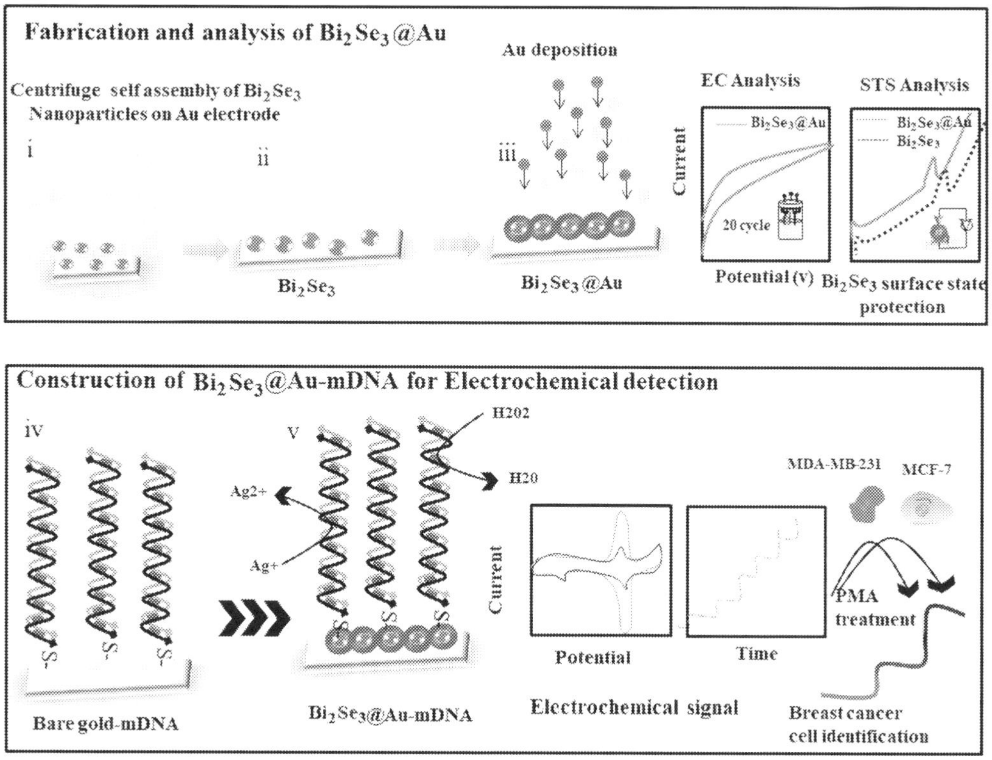


FIGURE 11.6 Schematic for the fabrication of $Bi_2Se_3$@Au biosensor for profiling breast cancer cells on the basis of their $H_2O_2$ content. *Reproduced with permission from Mohammadniaei M., Yoon, J., Lee, T., Bharate, B. G., Jo, J., Lee, D., & Choi, J. W. (2018). Electrochemical biosensor composed of silver ion-mediated dsDNA on Au-encapsulated $Bi_2Se_3$ nanoparticles for the detection of $H_2O_2$ released from breast cancer cells.* Electrochemical Sensors, *14, 16.*

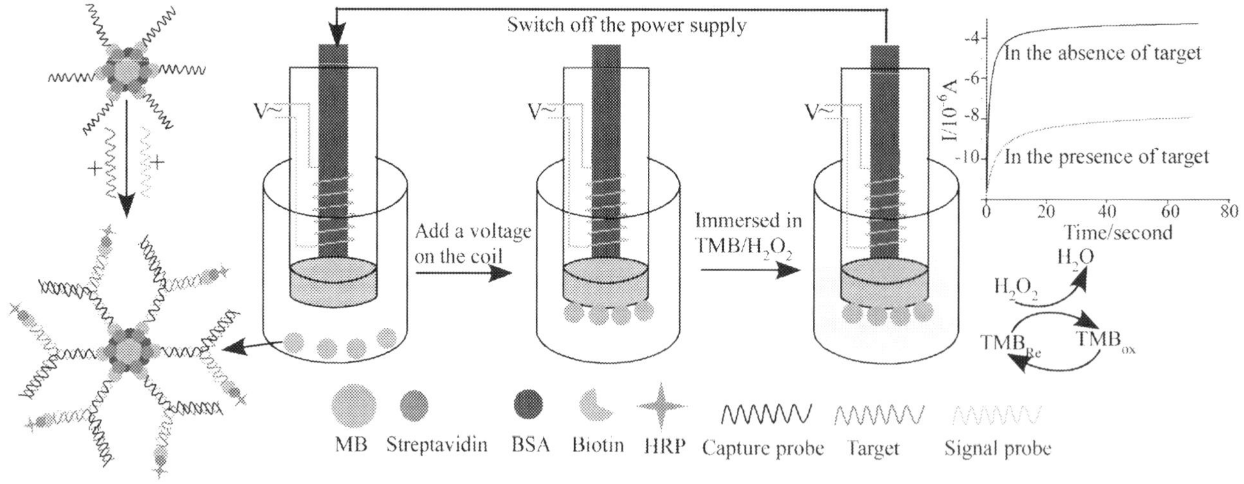


FIGURE 11.7 Schematic of the magnetic-controllable electrochemical biosensor for detection of oral cancer. *Used with permission from Wang, Z. W., Zhang, J., Guo, Y., Wu, X. Y., Yang, W. J., Xu, L. J., . . . Fu, F. F. (2013). A novel electrically magnetic-controllable electrochemical biosensor for the ultra sensitive and specific detection of attomolar level oral cancer-related microRNA.* Biosensors and Bioelectronics, 45*(1), 108–113. Available from https://doi.org/10.1016/j.bios.2013.02.007.*

results in the $H_2O_2$ mediated oxidation of 3,3′,5,5′-Tetramethylbenzidine (TMB). Later, Horseradish Peroxidase (HRP) was tagged to the MBs and were adsorbed on the surface of electrically magnetic controllable electrodes. When the working electrode dipped into the TMB-$H_2O_2$ solution, the HRP labeled MBs catalyze the $H_2O_2$, which gives rise to an increased electrochemical signal as shown in Fig. 11.7 (Wang et al., 2013).

### 11.3.4 Surface plasmon resonance-based biosensor

Surface plasmon resonance (SPR) biosensors are based on label-free optical biosensing technologies (Kooyman, Kolkman, Van Gent, & Greve, 1988; Piliarik, Vaisocherová, & Homola, 2009; Wu et al., 2008). This method is based on the refractive index after binding of the site with a biorecognition element (Ghindilis, Atanasov, Wilkins, & Wilkins, 1998; Homola, 2003). Recently, various biosensors were developed for the detection of multiple tumor markers (Chou et al., 2004). Cheng et al. developed a field-effect transistor (FET)-based biosensor for label-free detection of tumor markers of cancer (CYFRA 21−1 and neuron-specific enolase [NSE]) (Cheng, Hideshima, Kuroiwa, Nakanishi, & Osaka, 2015). It is a sensitive method for the detection label-free bio-molecular interaction (Souto, Volpe, Gonçalves, Ramos, & Kubotaet al., 2019). Although the concentration of tumor markers in the serum is very low, signal amplification for the SPR biosensor is necessary (Shpacovitch & Hergenröder, 2020). Various approaches have been used to amplify the signal in SPR assay, such as functionalization of nanoparticles and quantum dots (Ermini, Mariani, Scarano, & Minunni, 2014; Martinez-Perdiguero, Retolaza, Bujanda, & Merino, 2014). Martinez-Perdiguero et al. used biotin-labeled antibody and streptavidin−functionalized nanoparticles for enhancing the detection signal (Martinez-Perdiguero et al., 2014). Chuang et al. developed a loop-mediated isothermal amplification (LAMP) sensing system based on the SPR technique for detection of hepatitis B virus (HBV) (Chuang, Wei, Lee, & Lin, 2012). In the present scenario, quantum dots (QD) are emerging as an attractive tool for biological imaging, efficient donors in fluorescence resonance energy transfer (FRET) mechanisms, and diagnosis (Anderson et al., 2013). QDs are luminescence semiconductors with unique optical and electrical properties such as photostability, tunable optical property, strong luminescence, and broad excitation (Yong et al., 2009). Many researchers have been working on the development of biosensors based on the combination of QDs and SPR to achieve the high sensitivity and low LOD (Zhang & Johnson, 2006).

Wang et al. has developed an SPR-based method for the quantitative determination of multiple tumor markers (Qu et al., 2014; Sang et al., 2016). They designed antibody-quantum dots conjugated for the signal amplification and quantitative detection of α-fetoprotein (AFP) (Wu et al., 2019), carcinoembryonic antigen (CEA), and cytokeratin fragment 21-1 (CYFRA 21-1) in clinical samples as shown in Fig. 11.8. Initially, a self-assembled monolayer of hexanedithiol (HDT) was developed (Fig. 11.8Ai) and was used as a chemical linker for the fabrication of AuNPs monolayer on the surface of the chip (Fig. 11.8Aii) in order to improve the sensitivity of detection. Ab1 molecule (anti-$AFP_1$, anti-$CEA_1$, and anti-CYFRA 21-1) (Fig. 11.8) were attached to the modified chip via amide bond formation and further AuNP@Ab1 modified chip was incubated with BSA to block the nonspecific site (Fig. 1.8A). Target samples (AFP, CEA, and CYFRA 21-1 were piped into

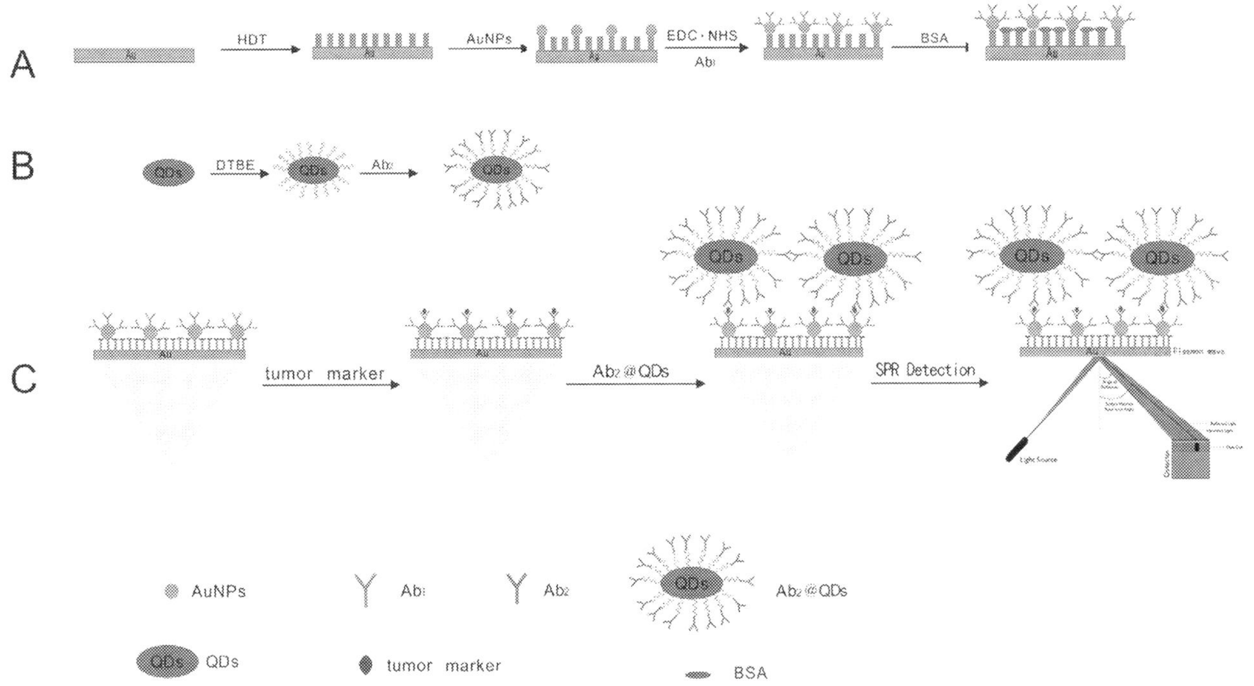


**FIGURE 11.8** Schematic representation of the fabrication of SPR based biosensor for detection of multiple tumor biomarkers (A) Coating of a SPR chip with AuNP@Ab1 conjugates. (B) Preparation of $Ab_2$@QD conjugates. (C) Sample was piped on the detection channel followed by addition of $Ab_2$@QD conjugate to amplify the signal, which could be detected by the SPR biosensor. *Reproduced with permission from Wang, H., Wang, X., Wang, J., Fu, W., & Yao, C. (2016). A SPR biosensor based on signal amplification using antibody-QD conjugates for quantitative determination of multiple tumor markers, 6, 33140.*

the detection channel (Fig. 11.8C) and allowed to react with the AuNPs@Ab1. After completion of the reaction, refractive angle change with addition of $Ab_2$@QD conjugates was measured in a real-time manner. The mass of the analyte was directly proportional to changes in the refractive angles. Enhancement in the mass of QDs, cause a change in the SPR signal, and this change in the signal can be used for the determination of the target (Fig. 11.8C).

### 11.3.5 Silicon photonic-based biosensors

Silicon photonic biosensors consist of waveguides (i.e., dielectric material used to guide high-frequency waves, such as electromagnetic waves) and are getting popular in the field of biosensor development. These optical waveguides are usually used as an optical analogy to electrical wire in integrated photonic circuits, but they can also be used as a transducer in the biosensor (Du, Chen, Song, Li, & Chen, 2008; Soler, Huertas, & Lechuga, 2019; Song et al., 2010). Techniques such as the etching process and photolithography are used for the fabrication of silicon photonics sensors. Recently, several technologies are available for the fabrication of photonic biosensors. These silicon-based sensors have stability, good optical properties, with temperature, and inertness to different chemicals (Densmore et al., 2008; Huertas, Calvo-Lozano, Mitchell, & Lechuga, 2019; Lechuga, 2005; Rastislav, Miroslav, & Ernest, 2012). It has good compatibility with a metal oxide semiconductor (CMOS), silicon photonics integrated circuit can be synthesized with the high efficiency (Soref, 2006). High refractive index contrast between the silicon and silicon dioxide provides the ability to be easily miniaturized with the additional possibility of multiple sensors on a single chip (Zinoviev et al., 2008). These silicon photonics can be an excellent transducer for biosensing the response of interaction between the analyte and biorecognition element in real-time. Mach–Zehnder interferometers (MZIs) (H. Fan et al., 2012), microring resonators (MRRs) (Bogaerts et al., 2012), Bragg grating resonators (Jugessur, Dou, Aitchison, De La Rue, & Gnan, 2009; Prabhathan, Murukeshan, Jing, & Ramana, 2009), microdisk resonators (Mandai, Serey, & Erickson, 2010) and 1D (Lee & Fauchet, 2007) or 2D photonic crystals (PhCs) (Washburn & Bailey, 2011) are some of the silicon photonics devices developed over the past decade for cancer diagnostic applications (Gavela, García, Ramirez, & Lechuga, 2016)

### 11.3.6 Colorimetric biosensors

Calorimetric biosensors are not mainly used for the diagnosis of cancer. However, the emergence of advanced nanoparticles in the field of sensor development has led to enhance range application. According to Medley et al. (2008), AuNPs functionalized aptamer can be used for cancer detection (Kasoju et al., 2020; Talan et al., 2018; Tiwari & Gong, 2009).

AuNPs are widely used for the fabrication of colorimetric based biosensors. However, enhancement of the catalytic property still remains a challenge to enhance the catalytic property of AuNPs and resolving their aggregation issue. Recently, several 2D carbon-based nanomaterials such as graphene, reduced graphene oxide are become popular to support the AuNP for the construction of colorimetric biosensors. $Bi_2Se_3$ nanosheets are the other tool which is used for the development of biosensor. Xiao et al. developed AuNPs decorated $Bi_2Se_3$ nanosheets ($Au/Bi_2Se_3$) for tumor markers detection of (Xiao et al., 2017a). An ultrathin $Bi_2Se_3$ nanosheet was synthesized according to the method described by Zang et al. (2020) and an AuNPs layer was immobilized on the surface which results in enhancement of $Au/Bi_2Se_3$ catalytic activity. Nanosheets made up of $Au/Bi_2Se_3$ (Fig. 11.9A) were used as a catalyst by reducing 4-NP (4-nitrophenol) to 4-AP (4-aminophenol) by NaBH4, which results in a change in the color from yellow to colorless as shown in the Figure 11.9B. Meanwhile, the addition of (anti-CEA) antibody results in "switching off" the catalytic activity of nanosheets ($Au/Bi_2Se_3$) (Fig. 11.9C) and the addition of antigen (CEA) (Fig. 11.9D); this results in binding of antibody (anti-CEA) present on the surface of $Au/Bi_2Se_3$. Anti-CEA antibody undergoes conformation, leads to weak affinity toward $Au/Bi_2Se_3$ nanosheet. Dissociation from the surface of $Au/Bi_2Se_3$ nanosheet results in "switching on" the catalytic reaction. Changes in the color of the solution and the absorbance value were used to determine the concentration of CEA antigen (Fig. 11.9E).

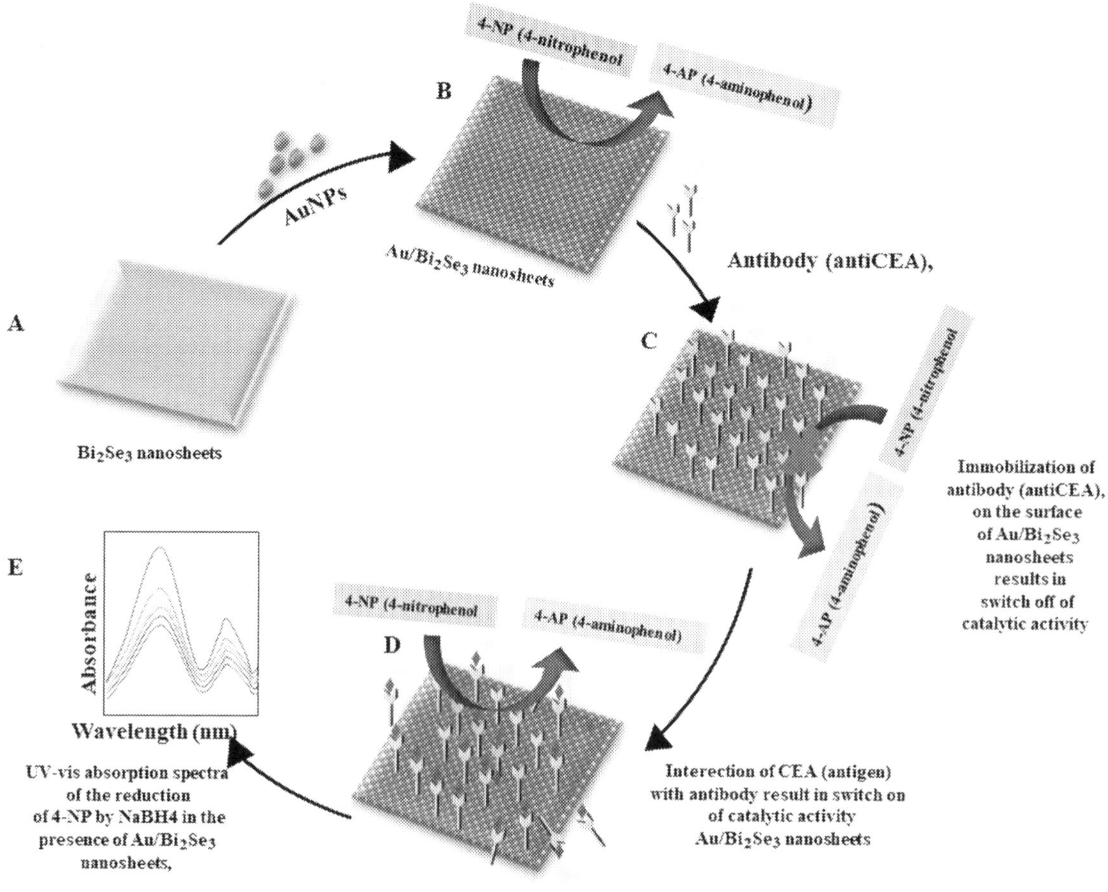


FIGURE 11.9 Fabrication of $Au/Bi_2Se_3$ nanosheet-based colorimetric biosensor for detection of CEA.

## 11.4 Conclusion and discussion

Cancer biomarkers are one of the important biomarkers for tumor growth. It not only helps in monitoring the disease, but also offers a prognostic approach for treating the disease. The emergence of sensors has revolutionized the field of detection. It allows a faster rapid detection of the analyte. Development of the advanced biosensors for the fast, accurate, and cheap detection of biomarkers creates a great opportunity in the field of diagnostics. Herein we reported the most recent efforts for the development of the advanced biosensor for the analysis of cancer biomarker. Various 2D and 3D nanomaterials such as graphene oxide, quantum dots, and hybrid nanocomposites are used for the enhancement of the sensitivity; however, weak reproducibility still remains a challenge to overcome. Properties of the nanomaterials mentioned above are of great importance in the field of nanotechnology as it has the possibility to contribute in the construction of devices besides medical diagnostics. In addition, research and development are still required in the construction of a highly reliable cancer screening devices. Most studies are focused on method development rather than the manufacturing of devices. Technology-based microelectronics, microfluidic, and device manufacturing are some of the fields that still need to be focused on for development.

## References


Anderson, G. P., Glaven, R. H., Algar, W. R., Susumu, K., Stewart, M. H., Medintz, I. L., & Goldman, E. R. (2013). Single domain antibody-quantum dot conjugates for ricin detection by both fluoroimmunoassay and surface plasmon resonance. *Analytica Chimica Acta*, *786*, 132–138. Available from https://doi.org/10.1016/j.ac.2013.05.010.

Arya, S. K., & Estrela, P. (2018). Recent advances in enhancement strategies for electrochemical ELISA-based immunoassays for cancer biomarker detection. *Sensors (Switzerland)*, *18*(7). Available from https://doi.org/10.3390/s18072010.

Azamjah, N., Soltan-Zadeh, Y., & Zayeri, F. (2019). Global trend of breast cancer mortality rate: A 25-year study. *Asian Pacific Journal of Cancer Prevention*, *20*(7), 2015–2020. Available from https://doi.org/10.31557/APJCP.2019.20.7.2015.

Azimzadeh, M., Rahaie, M., Nasirizadeh, N., Ashtari, K., & Naderi-Manesh, H. (2016). An electrochemical nanobiosensor for plasma miRNA-155, based on graphene oxide and gold nanorod, for early detection of breast cancer. *Biosensors and Bioelectronics*, *77*, 99–106. Available from https://doi.org/10.1016/j.bios.2015.09.020.

Bae, Y. J., Schaab, M., & Kratzsch, J. (2015). *Calcitonin as biomarker for the medullary thyroid carcinoma*, . *Recent results in cancer research* (Vol. 204, pp. 117–137). New York LLC: Springer. https://doi.org/10.1007/978-3-319-22542-5_5.

Banga, I., Tyagi, R., Shahdeo, D., & Gandhi, S. (2019). *Biosensors and their application for the detection of avian influenza virus. Nanotechnology in modern animal biotechnology: Concepts and applications* (pp. 1–16). Elsevier. https://doi.org/10.1016/B978-0-12-818823-1.00001-6.

Bartel, D. P. (2004). MicroRNAs: Genomics, biogenesis, mechanism, and function. *Cell*, *116*(2), 281–297. Available from https://doi.org/10.1016/S0092-8674(04)00045-5.

Barua, S., Dutta, H. S., Gogoi, S., Devi, R., & Khan, R. (2018). Nanostructured MoS2-based advanced biosensors: A review. *ACS Applied Nano Materials*, *1*(1), 2–25. Available from https://doi.org/10.1021/acsanm.7b00157.

Bhalla, N., Jolly, P., Formisano, N., & Estrela, P. (2016). Introduction to biosensors. *Essays in Biochemistry*, *60*(1), 1–8. Available from https://doi.org/10.1042/EBC20150001.

Bogaerts, W., de Heyn, P., van Vaerenbergh, T., de Vos, K., Kumar Selvaraja, S., Claes, T., … Baets, R. (2012). Silicon microring resonators. *Laser and Photonics Reviews*, *6*(1), 47–73. Available from https://doi.org/10.1002/lpor.201100017.

Bohunicky, B., & Mousa, S. A. (2011). Biosensors: The new wave in cancer diagnosis. *Nanotechnology, Science and Applications*, *4*(1), 1–10. Available from https://doi.org/10.2147/NSA.S13465.

Bottoni, P., & Scatena, R. (2015). *The role of CA 125 as tumor marker: Biochemical and clinical aspects*, . *Advances in experimental medicine and biology* (Vol. 867, pp. 229–244). Springer New York LLC, https://doi.org/10.1007/978-94-017-7215-0_14.

Bruno, S., Crosignani, A., Maisonneuve, P., Rossi, S., Silini, E., & Mondelli, M. U. (2007). Hepatitis C virus genotype 1b as a major risk factor associated with hepatocellular carcinoma in patients with cirrhosis: A seventeen-year prospective cohort study. *Hepatology (Baltimore, Md.)*, *46*(5), 1350–1356. Available from https://doi.org/10.1002/hep.21826.

Cai, B., Guo, S., & Li, Y. (2018). MoS2-based sensor for the detection of miRNA in serum samples related to breast cancer. *Analytical Methods*, *10* (2), 230–236. Available from https://doi.org/10.1039/c7ay02329d.

Chatterjee, S. K., & Zetter, B. R. (2005). Cancer biomarkers: Knowing the present and predicting the future. *Future Oncology*, *1*(1), 37–50. Available from https://doi.org/10.1517/14796694.1.1.37.

Chen, C., Ridzon, D. A., Broomer, A. J., Zhou, Z., Lee, D. H., Nguyen, J. T., … Guegler, K. J. (2005). Real-time quantification of microRNAs by stem-loop RT-PCR. *Nucleic Acids Research*, *33*(20), e179.9. Available from https://doi.org/10.1093/nar/gni178.

Chen, M., Wang, Y., Su, H., Mao, L., Jiang, X., Zhang, T., & Dai, X. (2018). Three-dimensional electrochemical DNA biosensor based on 3D graphene-Ag nanoparticles for sensitive detection of CYFRA21–1 in non-small cell lung cancer. *Sensors and Actuators, B: Chemical*, *255*, 2910–2918. Available from https://doi.org/10.1016/j.snb.2017.09.111.

Cheng, S., Hideshima, S., Kuroiwa, S., Nakanishi, T., & Osaka, T. (2015). Label-free detection of tumor markers using field effect transistor (FET)-based biosensors for lung cancer diagnosis. *Sensors and Actuators, B: Chemical, 212*, 329–334. Available from https://doi.org/10.1016/j.snb.2015.02.038.

Chinen, A. B., Guan, C. M., Ferrer, J. R., Barnaby, S. N., Merkel, T. J., & Mirkin, C. A. (2015). Nanoparticle probes for the detection of cancer biomarkers, cells, and tissues by fluorescence. *Chemical Reviews, 115*(19), 10530–10574. Available from https://doi.org/10.1021/acs.chemrev.5b00321.

Choi, Y. E., Kwak, J. W., & Park, J. W. (2010). Nanotechnology for early cancer detection. *Sensors, 10*(1), 428–455. Available from https://doi.org/10.3390/s100100428.

Chou, S. F., Hsu, W. L., Hwang, J. M., & Chen, C. Y. (2004). Development of an immunosensor for human ferritin, a nonspecific tumor marker, based on surface plasmon resonance. *Biosensors and Bioelectronics, 19*(9), 999–1005. Available from https://doi.org/10.1016/j.bios.2003.09.004.

Chuang, T. L., Wei, S. C., Lee, S. Y., & Lin, C. W. (2012). A polycarbonate based surface plasmon resonance sensing cartridge for high sensitivity HBV loop-mediated isothermal amplification. *Biosensors and Bioelectronics, 32*(1), 89–95. Available from https://doi.org/10.1016/j.bios.2011.11.037.

Coutinho, C., & Somoza, Á. (2019). MicroRNA sensors based on gold nanoparticles. *Analytical and Bioanalytical Chemistry, 411*(9), 1807–1824. Available from https://doi.org/10.1007/s00216-018-1450-7.

Cowling, T., & Loshak. (2016). An overview of liquid biopsy for screening and early detection of cancer. In *CADTH issues in emerging health technologies*.

Damborska, D., Bertok, T., Dosekova, E., Holazova, A., Lorencova, L., Kasak, P., & Tkac, J. (2017). Nanomaterial-based biosensors for detection of prostate specific antigen. *Microchimica Acta, 184*(9), 3049–3067. Available from https://doi.org/10.1007/s00604-017-2410-1.

Damiati, S., Haslam, C., Peacock, M., & Awan, S. (2018). PO-088 A modified electrochemical biosensor for sensitive detection of human chorionic gonadotropin (hCG) biomarker. In *ESMO Open* (Vol. 3). https://doi.org/10.1136/esmoopen-2018-eacr25.616.

Danielsson, B. (1990). Calorimetric biosensors. *Journal of Biotechnology, 15*(3), 187–200. Available from https://doi.org/10.1016/0168-1656(90)90026-8.

Debruyne, E. N., & Delanghe, J. R. (2008). Diagnosing and monitoring hepatocellular carcinoma with alpha-fetoprotein: New aspects and applications. *Clinica Chimica Acta, 395*(1–2), 19–26. Available from https://doi.org/10.1016/j.cc.2008.05.010.

Densmore, A., Xu, D.-X., Janz, S., Waldron, P., Mischki, T., Lopinski, G., ... Schmid, J. H. (2008). Spiral-path high-sensitivity silicon photonic wire molecular sensor with temperature-independent response. *Optics Letters*, 596. Available from https://doi.org/10.1364/OL.33.000596.

Dhanjai., Sinha, A., Wu, L., Lu, X., Chen, J., & Jain, R. (2018). Advances in sensing and biosensing of bisphenols: A review. *Analytica Chimica Acta, 998*, 1–27. Available from https://doi.org/10.1016/j.ac.2017.09.048.

Di Pietro, P., Ortolani, M., Limaj, O., Di Gaspare, A., Giliberti, V., Giorgianni, F., ... Lupi, S. (2013). Plasmonic excitations in $Bi_2Se_3$ topological insulator. In *International Conference on Infrared, Millimeter, and Terahertz Waves*, IRMMW-THz. https://doi.org/10.1109/IRMMW-THz.2013.6665572.

Dinish, U. S., Fu, C. Y., Soh, K. S., Ramaswamy, B., Kumar, A., & Olivo, M. (2012). Highly sensitive SERS detection of cancer proteins in low sample volume using hollow core photonic crystal fiber. *Biosensors and Bioelectronics, 33*(1), 293–298. Available from https://doi.org/10.1016/j.bios.2011.12.056.

Dong, S., Li, M., Wei, W., Liu, D., & Huang, T. (2017). An convenient strategy for IgG electrochemical immunosensor: the platform of topological insulator materials Bi2Se3 and ionic liquid. *Journal of Solid State Electrochemistry, 21*(3), 793–801. Available from https://doi.org/10.1007/s10008-016-3420-3.

Du, D., Chen, S., Song, D., Li, H., & Chen, X. (2008). Development of acetylcholinesterase biosensor based on CdTe quantum dots/gold nanoparticles modified chitosan microspheres interface. *Biosensors and Bioelectronics, 24*(3), 475–479. Available from https://doi.org/10.1016/j.bios.2008.05.005.

Eftimie, R., & Hassanein, E. (2018). Improving cancer detection through combinations of cancer and immune biomarkers: A modelling approach. *Journal of Translational Medicine, 16*(1). Available from https://doi.org/10.1186/s12967-018-1432-8.

Ermini, M. L., Mariani, S., Scarano, S., & Minunni, M. (2014). Bioanalytical approaches for the detection of single nucleotide polymorphisms by surface plasmon resonance biosensors. *Biosensors and Bioelectronics, 61*, 28–37. Available from https://doi.org/10.1016/j.bios.2014.04.052.

Fan, H., Zhang, S., Ju, P., Su, H., & Ai, S. (2012). Flower-like Bi2Se3 nanostructures: Synthesis and their application for the direct electrochemistry of hemoglobin and H2O2 detection. *Electrochimica Acta, 64*, 171–176. Available from https://doi.org/10.1016/j.electacta.2012.01.010.

Fei, Y., Sun, Y. S., Li, Y., Yu, H., Lau, K., Landry, J. P., ... Zhu, X. (2015). Characterization of receptor binding profiles of influenza a viruses using an ellipsometry-based label-free glycan microarray assay platform. *Biomolecules, 5*(3), 1480–1498. Available from https://doi.org/10.3390/biom5031480.

Florea, A., Ravalli, A., Cristea, C., Săndulescu, R., & Marrazza, G. (2015). An optimized bioassay for mucin1 detection in serum samples. *Electroanalysis, 27*(7), 1594–1601. Available from https://doi.org/10.1002/elan.201400689.

Frangioni, J. V. (2008). New technologies for human cancer imaging. *Journal of Clinical Oncology, 26*(24), 4012–4021. Available from https://doi.org/10.1200/JCO.2007.14.3065.

Freitas, M., Nouws, H. P. A., & Delerue-Matos, C. (2018). Electrochemical biosensing in cancer diagnostics and follow-up. *Electroanalysis, 30*(8), 1576–1595. Available from https://doi.org/10.1002/elan.201800193.

Freitas, M., Nouws, H. P. A., Keating, E., & Delerue-Matos, C. (2020). High-performance electrochemical immunomagnetic assay for breast cancer analysis. *Sensors and Actuators, B: Chemical*, 308. Available from https://doi.org/10.1016/j.snb.2020.127667.

G, C. J. (2014). In 2014, can we do better than CA125 in the early detection of ovarian cancer? *World Journal of Biological Chemistry*, 286. Available from https://doi.org/10.4331/wjbc.v5.i3.286.

Gan, X., Zhao, H., & Quan, X. (2017). Two-dimensional MoS2: A promising building block for biosensors. *Biosensors and Bioelectronics*, *89*, 56–71. Available from https://doi.org/10.1016/j.bios.2016.03.042.

Gandhi, S., Arami, H., & Krishnan, K. M. (2016). Detection of cancer-specific proteases using magnetic relaxation of peptide-conjugated nanoparticles in biological environment. *Nano Letters*, *16*(6), 3668–3674. Available from https://doi.org/10.1021/acs.nanolett.6b00867.

Gandhi, S., Suman, P., Kumar, A., Sharma, P., Capalash, N., & Raman Suri, C. (2015). Recent advances in immunosensor for narcotic drug detection. *BioImpacts*, *5*(4), 207–213. Available from https://doi.org/10.15171/bi.2015.30.

Gao, J., Du, B., Zhang, X., Guo, A., Zhang, Y., Wu, D., ... Wei, Q. (2014). Ultrasensitive enzyme-free immunoassay for squamous cell carcinoma antigen using carbon supported Pd-Au as electrocatalytic labels. *Analytica Chimica Acta*, *833*, 9–14. Available from https://doi.org/10.1016/j.ac.2014.05.004.

Gavela, A. F., García, D. G., Ramirez, J. C., & Lechuga, L. M. (2016). Last advances in silicon-based optical biosensors. *Sensors (Switzerland)*, *16*(3). Available from https://doi.org/10.3390/s16030285.

Ghindilis, A. L., Atanasov, P., Wilkins, M., & Wilkins, E. (1998). Immunosensors: Electrochemical sensing and other engineering approaches. *Biosensors and Bioelectronics*, *13*(1), 113–131. Available from https://doi.org/10.1016/S0956-5663(97)00031-6.

Goossens, N., Nakagawa, S., Sun, X., & Hoshida, Y. (2015). Cancer biomarker discovery and validation. *Translational Cancer Research*, *4*(3), 256–269. Available from https://doi.org/10.3978/j.issn.2218-676X.2015.06.04.

Grebe, S. K. G. (2009). Diagnosis and management of thyroid carcinoma: A focus on serum thyroglobulin. *Expert Review of Endocrinology and Metabolism*, *4*(1), 25–43. Available from https://doi.org/10.1586/17446651.4.1.25.

Haris, M., Yadav, S. K., Rizwan, A., Singh, A., Wang, E., Hariharan, H., ... Marincola, F. M. (2015). Molecular magnetic resonance imaging in cancer. *Journal of Translational Medicine*, *13*(1). Available from https://doi.org/10.1186/s12967-015-0659-x.

He, L., Wang, Q., Mandler, D., Li, M., Boukherroub, R., & Szunerits, S. (2016). Detection of folic acid protein in human serum using reduced graphene oxide electrodes modified by folic-acid. *Biosensors and Bioelectronics*, *75*, 389–395. Available from https://doi.org/10.1016/j.bios.2015.08.060.

He, W., Huang, Y., & Wu, J. (2020). Enzyme-Free Glucose Biosensors Based on MoS2 Nanocomposites. *Nanoscale Research Letters*, *15*(1). Available from https://doi.org/10.1186/s11671-020-3285-3.

Homola, J. (2003). Present and future of surface plasmon resonance biosensors. *Analytical and Bioanalytical Chemistry*, *377*(3), 528–539. Available from https://doi.org/10.1007/s00216-003-2101-0.

Hsieh, D., Xia, Y., Qian, D., Wray, L., Dil, J.H., Meier, F., & Samarth, N. (2011). A tunable topological insulator in the spin helical Dirac transport regime\Observation of quantum-tunneling-modulated spin texture in ultrathin topological insulator Bi2Se3 thin films. *Nature Nature Communications*.

Hu, C., Liu, C. P., Cheng, J. S., Chiu, Y. L., Chan, H. P., & Peng, N. J. (2016). Application of whole-body FDG-PET for cancer screening in a cohort of hospital employees. *Medicine*, *95*(44). Available from https://doi.org/10.1097/MD.0000000000005131.

Huang, J., Harvey, J., Derrick Fam, W. H., Nimmo, M. A., & Alfred Tok, I. Y. (2013). *Novel biosensor for interleukin-6 detection, . Procedia engineering* (Vol. 60, pp. 195–200). Elsevier Ltd., https://doi.org/10.1016/j.proeng.2013.07.042.

Huang, Y., Dong, X., Liu, Y., Li, L. J., & Chen, P. (2011). Graphene-based biosensors for detection of bacteria and their metabolic activities. *Journal of Materials Chemistry*, *21*(33), 12358–12362. Available from https://doi.org/10.1039/c1jm11436k.

Huertas, C. S., Calvo-Lozano, O., Mitchell, A., & Lechuga, L. M. (2019). Advanced evanescent-wave optical biosensors for the detection of nucleic acids: An analytic perspective. *Frontiers in Chemistry*, 7. Available from https://doi.org/10.3389/fchem.2019.00724.

Istamboulie, G., Sikora, T., Jubete, E., Ochoteco, E., Marty, J. L., & Noguer, T. (2010). Screen-printed poly (3,4-ethylenedioxythiophene) (PEDOT): A new electrochemical mediator for acetylcholinesterase-based biosensors. *Talanta*, *82*(3), 957–961. Available from https://doi.org/10.1016/j.talanta.2010.05.070.

Janire, P.-B., N, H. N., F, S. K., & R, D. F. (2018). Recent advances in graphene-based biosensor technology with applications in life sciences. *Journal of Nanobiotechnology*. Available from https://doi.org/10.1186/s12951-018-0400-z.

Jiao, S., Liu, L., Wang, J., Ma, K., & Lv, J. (2020). A novel biosensor based on molybdenum disulfide (MoS2) modified porous anodic aluminum oxide nanochannels for ultrasensitive microRNA-155 detection. *Small (Weinheim an der Bergstrasse, Germany)*, *16*(28). Available from https://doi.org/10.1002/smll.202001223.

Jugessur, A. S., Dou, J., Aitchison, J. S., De La Rue, R. M., & Gnan, M. (2009). A photonic nano-Bragg grating device integrated with microfluidic channels for bio-sensing applications. *Microelectronic Engineering*, *86*(4–6), 1488–1490. Available from https://doi.org/10.1016/j.mee.2008.12.002.

Kasoju, A., Shrikrishna, N. S., Shahdeo, D., Khan, A. A., Alanazi, A. M., & Gandhi, S. (2020). Microfluidic paper device for rapid detection of aflatoxin B1 using an aptamer based colorimetric assay. *RSC Advances*, *10*(20), 11843–11850. Available from https://doi.org/10.1039/d0ra00062k.

Keshavarz, M., Behpour, M., & Rafiee-Pour, H. A. (2015). Recent trends in electrochemical microRNA biosensors for early detection of cancer. *RSC Advances*, *5*(45), 35651–35660. Available from https://doi.org/10.1039/c5ra01726b.

Khan, M. S., Dighe, K., Wang, Z., Srivastava, I., Daza, E., Schwartz-Dual, A. S., ... Pan, D. (2018). Detection of prostate specific antigen (PSA) in human saliva using an ultra-sensitive nanocomposite of graphene nanoplatelets with diblock-: Co -polymers and Au electrodes. *Analyst*, *143*(5), 1094–1103. Available from https://doi.org/10.1039/c7an01932g.

Kmietowicz, Z. (2015). Most common cause of death in England and Wales in 2013 was heart disease in men and dementia in women. *BMJ (Clinical Research ed.)*, *350*, h1156. Available from https://doi.org/10.1136/bmj.h1156.

Kooyman, R. P. H., Kolkman, H., Van Gent, J., & Greve, J. (1988). Surface plasmon resonance immunosensors: sensitivity considerations. *Analytica Chimica Acta*, *213*(C), 35–45. Available from https://doi.org/10.1016/S0003-2670(00)81337-9.

Kumar, S., Kumar, S., Srivastava, S., Yadav, B. K., Lee, S. H., Sharma, J. G., ... Malhotra, B. D. (2015). Reduced graphene oxide modified smart conducting paper for cancer biosensor. *Biosensors and Bioelectronics*, *73*, 114–122. Available from https://doi.org/10.1016/j.bios.2015.05.040.

Lechuga, L. M. (2005). Chapter 5 Optical biosensors. *Comprehensive Analytical Chemistry*, *44*, 209–250. Available from https://doi.org/10.1016/S0166-526X(05)44005-2.

Lee, M., & Fauchet, P. M. (2007). Two-dimensional silicon photonic crystal based biosensing platform for protein detection. *Optics Express*, *15*(8), 4530–4535. Available from https://doi.org/10.1364/OE.15.004530.

Lin, L., Liu, Q., Wang, L., Liu, A., Weng, S., Lei, Y., ... Chen, Y. (2011). Enzyme-amplified electrochemical biosensor for detection of PML-RARα fusion gene based on hairpin LNA probe. *Biosensors and Bioelectronics*, *28*(1), 277–283. Available from https://doi.org/10.1016/j.bios.2011.07.032.

Mandai, S., Serey, X., & Erickson, D. (2010). Nanomanipulation using silicon photonic crystal resonators. *Nano Letters*, *10*(1), 99–104. Available from https://doi.org/10.1021/nl9029225.

Martinez-Perdiguero, J., Retolaza, A., Bujanda, L., & Merino, S. (2014). Surface plasmon resonance immunoassay for the detection of the TNFα biomarker in human serum. *Talanta*, *119*, 492–497. Available from https://doi.org/10.1016/j.talanta.2013.11.063.

Maruvada, P., Wang, W., Wagner, P. D., & Srivastava, S. (2005). Biomarkers in molecular medicine: Cancer detection and diagnosis. *Biotechniques* (Suppl), 9–15.

Medley, C. D., Smith, J. E., Tang, Z., Wu, Y., Bamrungsap, S., & Tan, W. (2008). Gold nanoparticle-based colorimetric assay for the direct detection of cancerous cells. *Analytical Chemistry*, *80*(4), 1067–1072. Available from https://doi.org/10.1021/ac702037y.

Mohammadniaei, M., Nguyen, H. V., Tieu, M. V., & Lee, M. H. (2019). 2D materials in development of electrochemical point-of-care cancer screening devices. *Micromachines*, *10*(10). Available from https://doi.org/10.3390/mi10100662.

Mohammadniaei, M., Yoon, J., Lee, T., Bharate, B. G., Jo, J., Lee, D., & Choi, J. W. (2018). Electrochemical biosensor composed of silver ion-mediated dsDNA on Au-encapsulated Bi2Se3 Nanoparticles for the detection of H2O2 released from breast cancer cells. *Small (Weinheim an der Bergstrasse, Germany)*, *14*(16). Available from https://doi.org/10.1002/smll.201703970.

Mor, G., Visintin, I., Lai, Y., Zhao, H., Schwartz, P., Rutherford, T., ... Ward, D. C. (2005). Serum protein markers for early detection of ovarian cancer. *Proceedings of the National Academy of Sciences of the United States of America*, *102*(21), 7677–7682. Available from https://doi.org/10.1073/pnas.0502178102.

Nagai, H., & Kim, Y. H. (2017). Cancer prevention from the perspective of global cancer burden patterns. *Journal of Thoracic Disease*, *9*(3), 448–451. Available from https://doi.org/10.21037/jtd.2017.02.75.

Nogueira, L., Corradi, R., & Eastham, J. A. (2009). Prostatic specific antigen for prostate cancer detection. *International Braz J Urol*, *35*(5), 521–529. Available from https://doi.org/10.1590/S1677-55382009000500003.

O'Brien, J., Hayder, H., Zayed, Y., & Peng, C. (2018). Overview of microRNA biogenesis, mechanisms of actions, and circulation. *Frontiers in Endocrinology*, 9. Available from https://doi.org/10.3389/fendo.2018.00402.

Piliarik, M., Vaisocherová, H., & Homola, J. (2009). Surface plasmon resonance biosensing. *Methods in Molecular Biology (Clifton, N.J.)*, *503*, 65–88. Available from https://doi.org/10.1007/978-1-60327-567-5_5.

Pohanka, M. (2018). Overview of piezoelectric biosensors, immunosensors and DNA sensors and their applications. *Materials*, *11*(3). Available from https://doi.org/10.3390/ma11030448.

Prabhathan, P., Murukeshan, V. M., Jing, Z., & Ramana, P. V. (2009). Compact SOI nanowire refractive index sensor using phase shifted Bragg grating. *Optics Express*, 15330. Available from https://doi.org/10.1364/OE.17.015330.

Qu, C., Chen, T., Fan, C., Zhan, Q., Wang, Y., Lu, J., ... Sun, Z. (2014). Efficacy of neonatal HBV vaccination on liver cancer and other liver diseases over 30-year follow-up of the qidong hepatitis B intervention study: A cluster randomized controlled trial. *PLoS Medicine*, *11*(12). Available from https://doi.org/10.1371/journal.pmed.1001774.

Rastislav, M., Miroslav, S., & Ernest, Š. (2012). Biosensors – classification, characterization and new trends. *Acta Chimica Slovaca*, 109–120. Available from https://doi.org/10.2478/v10188-012-0017-z.

Rau, S., Hilbig, U., & Gauglitz, G. (2014). Label-free optical biosensor for detection and quantification of the non-steroidal anti-inflammatory drug diclofenac in milk without any sample pretreatment. *Analytical and Bioanalytical Chemistry*, *406*(14), 3377–3386. Available from https://doi.org/10.1007/s00216-014-7755-2.

René, L. (2012). Endoscopy in screening for digestive cancer. *World Journal of Gastrointestinal Endoscopy*, 518. Available from https://doi.org/10.4253/wjge.v4.i12.518.

Roberts, A., Tripathi, P. P., & Gandhi, S. (2019). Graphene nanosheets as an electric mediator for ultrafast sensing of urokinase plasminogen activator receptor-A biomarker of cancer. *Biosensors and Bioelectronics*, 141. Available from https://doi.org/10.1016/j.bios.2019.111398.

Rui, G., Wang, X., Gu, B., Zhan, Q., & Cui, Y. (2016). Manipulation metallic nanoparticle at resonant wavelength using engineered azimuthally polarized optical field. *Optics Express*, *24*(7), 7212–7223. Available from https://doi.org/10.1364/OE.24.007212.

Ruiyi, L., Fangchao, C., Haiyan, Z., Xiulan, S., & Zaijun, L. (2018). Electrochemical sensor for detection of cancer cell based on folic acid and octadecylamine-functionalized graphene aerogel microspheres. *Biosensors and Bioelectronics*, *119*, 156–162. Available from https://doi.org/10.1016/j.bios.2018.07.060.

Safina, G. (2012). Application of surface plasmon resonance for the detection of carbohydrates, glycoconjugates, and measurement of the carbohydrate-specific interactions: A comparison with conventional analytical techniques. A critical review. *Analytica Chimica Acta*, *712*, 9–29. Available from https://doi.org/10.1016/j.ac.2011.11.016.

Sang, S., Wang, Y., Feng, Q., Wei, Y., Ji, J., & Zhang, W. (2016). Progress of new label-free techniques for biosensors: A review. *Critical Reviews in Biotechnology*, *36*(3), 465–481. Available from https://doi.org/10.3109/07388551.2014.991270.

Santarelli, D. M., Liu, B., Duncan, C. E., Beveridge, N. J., Tooney, P. A., Schofield, P. R., & Cairns, M. J. (2013). Gene-microRNA interactions associated with antipsychotic mechanisms and the metabolic side effects of olanzapine. *Psychopharmacology*, *227*(1), 67–78. Available from https://doi.org/10.1007/s00213-012-2939-y.

Sethi, S., Ali, S., Philip, P. A., & Sarkar, F. H. (2013). Clinical advances in molecular biomarkers for cancer diagnosis and therapy. *International Journal of Molecular Sciences*, *14*(7), 14771–14784. Available from https://doi.org/10.3390/ijms140714771.

Shapira, I., Oswald, M., Lovecchio, J., Khalili, H., Menzin, A., Whyte, J., ... Lee, A. T. (2014). Circulating biomarkers for detection of ovarian cancer and predicting cancer outcomes. *British Journal of Cancer*, *110*(4), 976–983. Available from https://doi.org/10.1038/bjc.2013.795.

Shpacovitch, V., & Hergenröder, R. (2020). Surface plasmon resonance (SPR)-based biosensors as instruments with high versatility and sensitivity. *Sensors (Switzerland)*, *20*(11). Available from https://doi.org/10.3390/s20113010.

Soler, M., Huertas, C. S., & Lechuga, L. M. (2019). Label-free plasmonic biosensors for point-of-care diagnostics: a review. *Expert Review of Molecular Diagnostics*, *19*(1), 71–81. Available from https://doi.org/10.1080/14737159.2019.1554435.

Song, S., Qin, Y., He, Y., Huang, Q., Fan, C., & Chen, H. Y. (2010). Functional nanoprobes for ultrasensitive detection of biomolecules. *Chemical Society Reviews*, *39*(11), 4234–4243. Available from https://doi.org/10.1039/c000682n.

Soni, A., Pandey, C. M., Pandey, M. K., & Sumana, G. (2019). Highly efficient Polyaniline-MoS 2 hybrid nanostructures based biosensor for cancer biomarker detection. *Analytica Chimica Acta*, *1055*, 26–35. Available from https://doi.org/10.1016/j.ac.2018.12.033.

Soref, R. (2006). The past, present, and future of silicon photonics. *IEEE Journal on Selected Topics in Quantum Electronics*, *12*(6), 1678–1687. Available from https://doi.org/10.1109/JSTQE.2006.883151.

Souto, D. E. P., Volpe, J., Gonçalves, Cd. C., Ramos, C. H. I., & Kubota, L. T. (2019). A brief review on the strategy of developing SPR-based biosensors for application to the diagnosis of neglected tropical diseases. *Talanta*, 205. Available from https://doi.org/10.1016/j.talanta.2019.120122.

Stankiewicz, J., Schlottmann, P., Arauzo, A., Martinez Perez, M. J., Rosa, P. F. S., Civale, L., & Fisk, Z. (2019). Localized magnetic moments in metallic SrB 6 single crystals. *Journal of Physics Condensed Matter*, *31*(6). Available from https://doi.org/10.1088/1361-648X/aaf40f.

Stephen Inbaraj, B., & Chen, B. H. (2016). Nanomaterial-based sensors for detection of foodborne bacterial pathogens and toxins as well as pork adulteration in meat products. *Journal of Food and Drug Analysis*, *24*(1), 15–28. Available from https://doi.org/10.1016/j.jfda.2015.05.001.

Su, L., Zou, L., Fong, C. C., Wong, W. L., Wei, F., Wong, K. Y., ... Yang, M. (2013). Detection of cancer biomarkers by piezoelectric biosensor using PZT ceramic resonator as the transducer. *Biosensors and Bioelectronics*, *46*, 155–161. Available from https://doi.org/10.1016/j.bios.2013.01.074.

Świderska, M., Choromańska, B., Dąbrowska, E., Konarzewska-Duchnowska, E., Choromańska, K., Szczurko, G., ... Zwierz, K. (2014). The diagnostics of colorectal cancer. *Wspolczesna Onkologia*, *18*(1), 1–6. Available from https://doi.org/10.5114/wo.2013.39995.

Tabasi, A., Noorbakhsh, A., & Sharifi, E. (2017). Reduced graphene oxide-chitosan-aptamer interface as new platform for ultrasensitive detection of human epidermal growth factor receptor 2. *Biosensors and Bioelectronics*, *95*, 117–123. Available from https://doi.org/10.1016/j.bios.2017.04.020.

Talan, A., Mishra, A., Eremin, S. A., Narang, J., Kumar, A., & Gandhi, S. (2018). Ultrasensitive electrochemical immuno-sensing platform based on gold nanoparticles triggering chlorpyrifos detection in fruits and vegetables. *Biosensors and Bioelectronics*, *105*, 14–21. Available from https://doi.org/10.1016/j.bios.2018.01.013.

Thévenot, D. R., Toth, K., Durst, R. A., & Wilson, G. S. (2001). Electrochemical biosensors: Recommended definitions and classification. *Biosensors and Bioelectronics*, *16*(1–2), 121–131. Available from https://doi.org/10.1016/S0956-5663(01)00115-4.

Tiwari, A., & Gong, S. (2009). Electrochemical detection of a breast cancer susceptible gene using cDNA immobilized chitosan-co-polyaniline electrode. *Talanta*, *77*(3), 1217–1222. Available from https://doi.org/10.1016/j.talanta.2008.08.029.

Topkaya, S. N., Azimzadeh, M., & Ozsoz, M. (2016). Electrochemical biosensors for cancer biomarkers detection: Recent advances and challenges. *Electroanalysis*, *28*(7), 1402–1419. Available from https://doi.org/10.1002/elan.201501174.

Tripathi, P. P., Arami, H., Banga, I., Gupta, J., & Gandhi, S. (2018). Cell penetrating peptides in preclinical and clinical cancer diagnosis and therapy. *Oncotarget*, *9*(98), 37252–37267. Available from https://doi.org/10.18632/oncotarget.26442.

Válóczi, A., Hornyik, C., Varga, N., Burgyán, J., Kauppinen, S., & Havelda, Z. (2004). Sensitive and specific detection of microRNAs by northern blot analysis using LNA-modified oligonucleotide probes. *Nucleic Acids Research*, *32*(22), e175. Available from https://doi.org/10.1093/nar/gnh171.

Verma, S., Singh, A., Shukla, A., Kaswan, J., Arora, K., Ramirez-Vick, J., ... Singh, S. P. (2017). Anti-IL8/AuNPs-rGO/ITO as an immunosensing platform for noninvasive electrochemical detection of oral cancer. *ACS Applied Materials and Interfaces*, *9*(33), 27462–27474. Available from https://doi.org/10.1021/acsami.7b06839.

Wang, H., Naghavi, M., Allen, C., Barber, R. M., Carter, A., Casey, D. C., & Zuhlke, L. J. (2015). Global, regional, and national life expectancy, all-cause mortality, and cause-specific mortality for 249 causes of death, 1980–2015: A systematic analysis for the Global Burden of Disease Study. *The Lancet*, *388*. Available from https://doi.org/10.1016/S0140-6736(16, 31012–1.

Wang, T., Zhu, R., Zhuo, J., Zhu, Z., Shao, Y., & Li, M. (2014). Direct detection of DNA below ppb level based on thionin-functionalized layered MoS2 electrochemical sensors. *Analytical Chemistry*, *86*(24), 12064–12069. Available from https://doi.org/10.1021/ac5027786.

Wang, X., Deng, W., Shen, L., Yan, M., & Yu, J. (2016). A 3D electrochemical immunodevice based on an Au paper electrode and using Au nanoflowers for amplification. *New Journal of Chemistry*, *40*(3), 2835–2842. Available from https://doi.org/10.1039/c5nj03222a.

Wang, Y., Luo, J., Liu, J., Sun, S., Xiong, Y., Ma, Y., ... Cai, X. (2019). Label-free microfluidic paper-based electrochemical aptasensor for ultrasensitive and simultaneous multiplexed detection of cancer biomarkers. *Biosensors and Bioelectronics*, *136*, 84–90. Available from https://doi.org/10.1016/j.bios.2019.04.032.

Wang, Y., Zhao, G., Zhang, Y., Pang, X., Cao, W., Du, B., & Wei, Q. (2018). Sandwich-type electrochemical immunosensor for CEA detection based on Ag/MoS2@Fe3O4 and an analogous ELISA method with total internal reflection microscopy. *Sensors and Actuators, B: Chemical, 266*, 561–569. Available from https://doi.org/10.1016/j.snb.2018.03.178.

Wang, Z. W., Zhang, J., Guo, Y., Wu, X. Y., Yang, W. J., Xu, L. J., ... Fu, F. F. (2013). A novel electrically magnetic-controllable electrochemical biosensor for the ultra sensitive and specific detection of attomolar level oral cancer-related microRNA. *Biosensors and Bioelectronics, 45*(1), 108–113. Available from https://doi.org/10.1016/j.bios.2013.02.007.

Washburn, A. L., & Bailey, R. C. (2011). Photonics-on-a-chip: Recent advances in integrated waveguides as enabling detection elements for real-world, lab-on-a-chip biosensing applications. *Analyst, 136*(2), 227–236. Available from https://doi.org/10.1039/c0an00449a.

Wei, B., Mao, K., Liu, N., Zhang, M., & Yang, Z. (2018). Graphene nanocomposites modified electrochemical aptamer sensor for rapid and highly sensitive detection of prostate specific antigen. *Biosensors and Bioelectronics, 121*, 41–46. Available from https://doi.org/10.1016/j.bios.2018.08.067.

Wu, M. C., Chuang, C. M., Lo, H. H., Cheng, K. C., Chen, Y. F., & Su, W. F. (2008). Surface plasmon resonance enhanced photoluminescence from Au coated periodic arrays of CdSe quantum dots and polymer composite thin film. *Thin Solid Films, 517*(2), 863–866. Available from https://doi.org/10.1016/j.tsf.2008.06.069.

Wu, X., Liu, H., Liu, J., Haley, K. N., Treadway, J. A., Larson, J. P., ... Bruchez, M. P. (2003). Immunofluorescent labeling of cancer marker Her2 and other cellular targets with semiconductor quantum dots. *Nature Biotechnology, 21*(1), 41–46. Available from https://doi.org/10.1038/nbt764.

Wu, Y., Wang, Y., Wang, X., Wang, C., Li, C., & Wang, Z. (2019). Electrochemical sensing of α-fetoprotein based on molecularly imprinted polymerized ionic liquid film on a gold nanoparticle modified electrode surface. *Sensors (Switzerland), 19*(14). Available from https://doi.org/10.3390/s19143218.

Wu, Y., Xue, P., Kang, Y., & Hui, K. M. (2013). Paper-based microfluidic electrochemical immunodevice integrated with nanobioprobes onto graphene film for ultrasensitive multiplexed detection of cancer biomarkers. *Analytical Chemistry, 85*(18), 8661–8668. Available from https://doi.org/10.1021/ac401445a.

Xiao, L., Zhu, A., Xu, Q., Chen, Y., Xu, J., & Weng, J. (2017a). Colorimetric Biosensor for Detection of Cancer Biomarker by Au Nanoparticle-Decorated Bi2Se3 Nanosheets. *ACS Applied Materials and Interfaces, 9*(8), 6931–6940. Available from https://doi.org/10.1021/acsami.6b15750.

Xiao, L., Zhu, A., Xu, Q., Chen, Y., Xu, J., & Weng, J. (2017b). Colorimetric Biosensor for Detection of Cancer Biomarker by Au Nanoparticle-Decorated Bi2Se3 Nanosheets. *ACS Applied Materials and Interfaces, 9*(8), 6931–6940. Available from https://doi.org/10.1021/acsami.6b15750.

Xu, H., Wu, L., Dai, X., Gao, Y., & Xiang, Y. (2016). An ultra-high sensitivity surface plasmon resonance sensor based on graphene-aluminum-graphene sandwich-like structure. *Journal of Applied Physics, 120*(5). Available from https://doi.org/10.1063/1.4959982.

Yin, H., Zhou, Y., Zhang, H., Meng, X., & Ai, S. (2012). Electrochemical determination of microRNA-21 based on graphene, LNA integrated molecular beacon, AuNPs and biotin multifunctional bio bar codes and enzymatic assay system. *Biosensors and Bioelectronics, 33*(1), 247–253. Available from https://doi.org/10.1016/j.bios.2012.01.014.

Yip, C. H., & Rhodes, A. (2014). Estrogen and progesterone receptors in breast cancer. *Future Oncology, 10*(14), 2293–2301. Available from https://doi.org/10.2217/fon.14.110.

Yong, K. T., Ding, H., Roy, I., Law, W. C., Bergey, E. J., Maitra, A., & Prasad, P. N. (2009). Imaging pancreatic cancer using bioconjugated inp quantum dots. *ACS Nano, 3*(3), 502–510. Available from https://doi.org/10.1021/nn8008933.

Yu, D. C., Li, Q. G., Ding, X. W., & Ding, Y. T. (2011). Circulating MicroRNAs: Potential biomarkers for cancer. *International Journal of Molecular Sciences, 12*(3), 2055–2063. Available from https://doi.org/10.3390/ijms12032055.

Zhang, C. Y., & Johnson, L. W. (2006). Quantum dot-based fluorescence resonance energy transfer with improved FRET efficiency in capillary flows. *Analytical Chemistry, 78*(15), 5532–5537. Available from https://doi.org/10.1021/ac0605389.

Zhang, Y., Li, M., Gao, X., Chen, Y., & Liu, T. (2019). Nanotechnology in cancer diagnosis: Progress, challenges and opportunities. *Journal of Hematology and Oncology, 12*(1). Available from https://doi.org/10.1186/s13045-019-0833-3.

Zhao, B., Gandhi, S., Yuan, C., Luo, Z., Li, R., Gårdsvoll, H., ... Ploug, M. (2015). Mapping the topographic epitope landscape on the urokinase plasminogen activator receptor (uPAR) by surface plasmon resonance and X-ray crystallography. *Data in Brief, 5*, 107–113. Available from https://doi.org/10.1016/j.dib.2015.08.027.

Zheng, J., Wang, J., Song, D., Xu, J., & Zhang, M. (2020). Electrochemical Aptasensor of Carcinoembryonic Antigen Based on Concanavalin A-Functionalized Magnetic Copper Silicate Carbon Microtubes and Gold-Nanocluster-Assisted Signal Amplification. *ACS Applied Nano Materials, 3*(4), 3449–3458. Available from https://doi.org/10.1021/acsanm.0c00194.

Zhou, X. H., Liu, L. H., Bai, X., & Shi, H. C. (2013). A reduced graphene oxide based biosensor for high-sensitive detection of phenols in water samples. *Sensors and Actuators, B: Chemical, 181*, 661–667. Available from https://doi.org/10.1016/j.snb.2013.02.021.

Zinoviev, K., Carrascosa, L. G., Del Río, J. S., Sepúlveda, B., Domínguez, C., & Lechuga, L. M. (2008). Silicon photonic biosensors for lab-on-a-chip applications. *Advances in Optical Technologies*. Available from https://doi.org/10.1155/2008/383927.

Chapter 12

# Microfluidics-based devices and their role on point-of-care testing

Avinash Kumar and Udwesh Panda
*Department of Mechanical Engineering, Indian Institute of Information Technology Design & Manufacturing Kancheepuram, Chennai, India*

## 12.1 Introduction

The evolution of microfluidics began nearly 20 years ago in the field of biomedical research. Since then, it is in a continuous development stage for its implementation to the more advanced and rapid surgical/medical techniques. The power of microfluidic devices in diagnosing various kinds of diseases paves the way for its enhancement and incorporation in almost all the vital types of treatments. So it has always been able to grab the keen interest of scientists and researchers all over the globe.

Point-of-care (POC) devices have been very efficient and helpful in personalized treatment in an environment which is safer and nimbler than before. The handling and collection of various kinds of data used for treatment and monitoring have been reduced to a great extent which eventually lowered the effective cost of health management.

The present chapter will begin with the advent of microfluidics in the world and its impact on various fields, especially on the development of POC devices for treating diseases that used to be untreatable before. Cancer is one of the diseases which leads to death and its treatment is very difficult if not detected at the early stages. The recent development and integration of biosensors in microfluidics helps to detect cancer at the initial stage and the ability to monitor it will be easier than before. The materials used in these microfluidic devices and the way they are fabricated are also discussed. The chapter will also deal with the advancement of these devices in the context of various diseases including the viral, parasitic, and bacterial pathogens. The primary focus of this chapter shall be adhering to the principles of microfluidic POC devices to detect cancer using rapid techniques. The applications discussed will be restricted to cancer detection and treatment. However, there will be some review regarding the use of these POC devices in other treatments as well. In this chapter a thorough review of the existing approaches and the emerging technologies will be discussed. Finally, the whole chapter will be outlined briefly in the summary section.

### 12.1.1 History of microfluidics

The science dealing with the study of fluids that are flowing in a very small volume is referred to as microfluidics, where the volume, which is generally studied, ranges below $10^{-5}$ L and in channels of sub-millimeter scale. The channels are about 1 − 1000 μm size and the flow which is encountered is laminar (Whitesides, 2006). Some would relate the microfluidics system in the dimensional range of human hair.

The driving mechanism of fluid in the microchannel or microfluidics system may be pressure-driven, electrokinetic, centrifugal, capillary-based, and acoustic as shown in Fig. 12.1 (Mark, Haeberle, Roth, Von Stetten, & Zengerle, 2010). In the early 1950s, the lithography technique named photolithography made a significant contribution in the semiconductor industries when the integration of transistors on wafers were made possible. In the later period around the 1980s to 1990s, the importance of microfluidics in the research community and its advantages as a biological tool and its application were realized; since then, it is applied and continually evolving in the biological research facilities all over the globe. The micropumps and microvalves were the first type of microfluidic devices that were fabricated (Esashi, Shoji, & Nakano, 1989; Oh & Ahn, 2006; Santiago & Laser, 2004).

Biosensor Based Advanced Cancer Diagnostics. DOI: https://doi.org/10.1016/B978-0-12-823424-2.00011-9

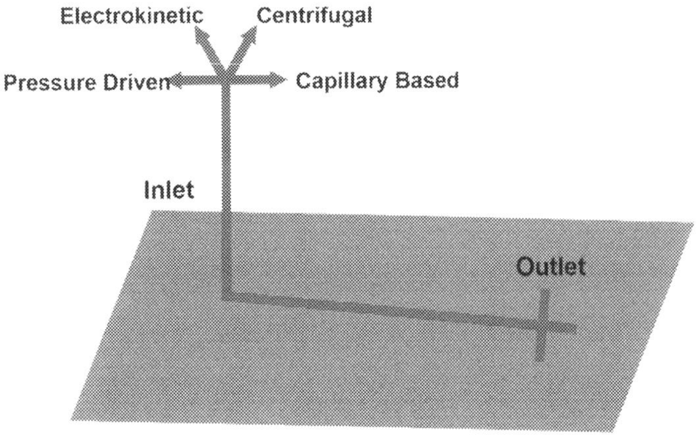


**FIGURE 12.1** Various inlets of microfluidic devices.

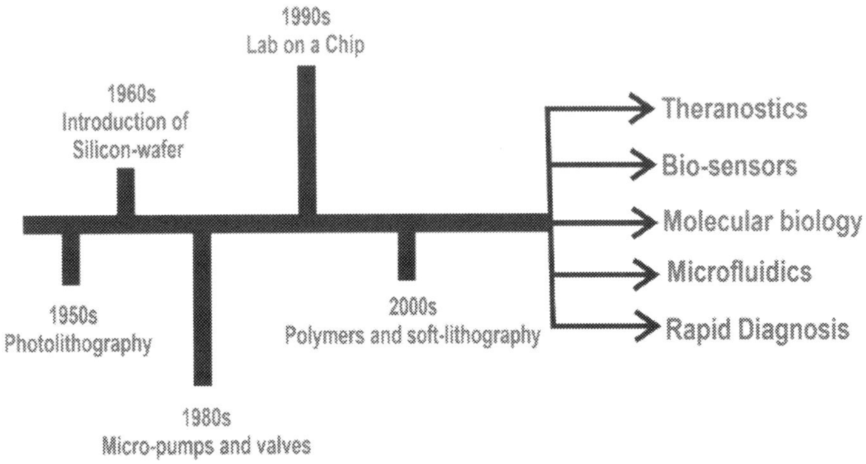


**FIGURE 12.2** Timeline of microfluidic fabrication techniques and its applications.

In the late 1990s, a new concept emerged in terms of lab on a chip (LOC) which is quite popular nowadays. This LOC technology facilitated the growth of microfluidics even more. The reactors designed for chemical synthesis proved to be very advantageous and efficient in terms of cost, safety, and scale-up. The demand for decreasing the cost and development time for the microfluidic devices favored the way for polymer-based fabrication involving a soft lithography technique, which eventually became quite popular and is still used due to its advantages over traditional techniques, as shown in Fig. 12.2 (Becker & Locascio, 2002). The materials for the devices have also been upgraded from silica and glass to bio-compatible polymers like the polydimethylsiloxane (PDMS), which is excessively used due to its optical, chemical, thermal properties, and also its ease of usability (Xia & Whitesides, 1998). This next subsection of the chapter will discuss the scientific background of these devices and how the fluid behaves when flowing inside these devices.

### 12.1.2 The behavior of fluids in microscale

There are various fluid mechanical terms which are important from the microscale perspective. The fluid dynamics concept is vast and one must know the behavior of fluid properties when it flows in different channels and\or in different types of cross-sections. Some of the basic fluid properties and the governing equations for the flow of fluids in microfluidic devices are given in the following text.

#### *12.1.2.1 Viscosity*

The intrinsic property of a fluid, which plays the most vital role in governing all kind of flows, and the distribution of shear stress in a fluid is the viscosity. The fluids are characterized generally into two types defined by Newton's law of viscosity, which describes the relationship between the rate of deformation and the shear stress. The law states that "the shear stress is directly proportional to the rate of deformation or the velocity gradient across the flow." Mathematically,

$$\tau = \mu \frac{\partial u}{\partial y} \quad (12.1)$$

where, $\tau$ is the shear stress $\mu$ is the dynamic viscosity, $u$ is the velocity of the fluids, and $y$ is the location of the fluid in the channel. Certain fluids follow this relationship and some fluids do not. The fluids which follow this law are known as Newtonian fluids and the ones who do not follow are known as non-Newtonian fluids. The viscosity that appeared in the relation is known as dynamic viscosity. There is one more type of viscosity known as kinematic viscosity and is denoted by the symbol $\nu$. It is defined as the ratio of dynamic viscosity to the density ($\rho$) of a particular fluid:

$$\nu = \frac{\mu}{\rho} \quad (12.2)$$

The unit of viscosity is pascal-second (Pa.s). Water has the lowest viscosity of about $8.94 \times 10^{-4}$ Pa.s. The viscosity can also have higher values as in the cases of polymers.

### *12.1.2.2 Reynolds number*

The Reynolds number is a dimensionless number which represents the ratio of fluid momentum to its viscosity. The number was first introduced by Sir George Stokes (1851) and was given the name by Arnold Sommerfield (1908) after Osborne Reynolds. It is a parameter to study the viscosity in a quantitative manner (Sommerfeld, 1908; Stokes, 1851; Xia & Whitesides, 1998). It is denoted by the symbol *Re*, the mathematical definition of the number is

$$\mathrm{Re} = \frac{\rho U L}{\mu} \quad (12.3)$$

where $\rho$ stands for density, $\mu$ is the dynamic viscosity, $U$ is the average velocity of the fluid, and $L$ is the characteristic length of the channel. The Re determines the type of flow in a channel; when Re is greater than ~2000 the flow is regarded as turbulent, or else the flow is laminar. The flow in microchannels is generally laminar. For example, water flowing through a passage with a diameter of 10 μm and a velocity of about 1 mm/s, will have Re given by the above formula as ~0.01. So, in microfluidics one must deal with very low-dimensional sections which are similar to the above example and so generally it is dealing with the laminar flow.

### *12.1.2.3 Navier–Stokes equation*

After discussing the viscous properties of the fluid and the dimensionless number (Reynolds number), it is important to know how these properties of fluid will affect and govern the fluid flow in a channel. So far, the discussion made it clear that the laminar flow will dominate in the microfluidic devices. Some assumptions regarding the laminar flow in the microscale level can be viewed as an incompressible flow, i.e., the density will remain constant. It can be represented in the form of an equation which is generally known as the Navier–Stokes equation:

$$\rho\left(\frac{\partial u}{\partial t} + u.\nabla u\right) = -\nabla p + \mu \nabla^2 u + f \quad (12.4)$$

where $u$ is the velocity matrix, $p$ is the pressure, $f$ is the body force matrix, and $\nabla$ is the gradient operator. The left-hand side of the equation denotes the momentum change caused by the inertia of the fluid, the right-hand side of the equation denotes the pressure drop, the viscous force, and the body forces in the respective order. The assumption to simplify the equation is:

1. Low Reynolds number flow;
2. Fully developed Newtonian fluid;
3. Continuity equation applicable to incompressible flow.

Applying the assumptions above with specific boundary conditions i.e. slip and\or no-slip. The simplified form of the Navier–Stokes equation can be solved and can obtain the velocity profile across the cross-section geometries affecting the flow can be found. There are a lot of other factors that are important and define the fluid flow in microscale, like the Peclet number, the diffusion, and mixing phenomena, etc.

### 12.1.3 Fabrication of microfluidic devices

The fabrication of devices started from the semiconductor industries where silicon was used as a prime substrate, later followed by glass and polymers. The microfluidics systems fabrication involves a list of materials that are processed using different techniques and are employed with certain fabrication methods to have the desired product.

#### *12.1.3.1 Materials*

The general category of materials is usually classified into metals, ceramics, composites, and polymers. The silicon and glass being the earliest materials that were used as a substrate. The abundance of silicon along with its resistivity towards the organic solvents made it advantageous in the early phase of microfluidic device fabrication, while the advantages of glass substrates were easy availability, bio-compatibility, and having lower nonspecific adsorption (Nge, Rogers, & Woolley, 2013; Ren, Zhou, & Wu, 2013).

The problem with the fabrication of microfluidic devices with the silicon-based substrate was the cost associated per unit area and also the hindrance in sealing silicon and glass together to provide leak-proof microchannel. The techniques which are employed for the same requires a time-consuming design and processing steps, i.e., the traditional photolithography processes. The ceramic-based material, i.e., aluminum oxide laminate sheet, was theoretically proven to fabricate microfluidic systems with good mechanical and electrical properties. This type of device became dominant in the fabrication of microfluidic devices until polymer-based materials appeared in the market. The low cost, ease of accessibility, alteration to physical and chemical properties made the polymer the most suitable materials for microfabrication (Becker & Locascio, 2002; Fakunle & Fritsch, 2010; Sommerfeld, 1909).

The materials that are used in microfluidic are categorized into five sections:

1. Elastomers (polyester, PDMS);
2. Thermoplastics for ease of fabrication and production;
3. Thermoset usually processed using photopolymerization;
4. Paper due to disposability and low cost;
5. Hydrogels are extensively used in cell molecular engineering.

The manufacturing methods are upgraded and developed constantly so that more advanced techniques are used to come up with sophisticated microfluidic devices which can be helpful for rapid operations.

#### *12.1.3.2 Common fabrication techniques*

##### 12.1.3.2.1 Photolithography

This is a commonly used lithography technique in the fabrication of microfluidic devices. The fabrication process for microfluidic devices involves the preparation of mask, generation of mold, and finally the formation of a microchannel. The silicon wafer is coated with a thin film of light-sensitive polymer known as photoresists (PRs). These PRs are of two types, positive and negative, and they dissolve in presence of light. For a specific PR composition, the thickness of the PR film may be managed by speed and spin-time. The photolithography process consists of several steps to obtain the product as shown in Fig. 12.3. Initially, the wafer is cleaned thoroughly and the barrier is formed followed by the application of PR. The soft-baking method is carried out to remove the solvents since it hinders the positive resist efficiency. Then the process of alignment is carried out which is followed by contact and projection printing. Lastly, the development phase where the solubility of the resist is checked and employed (Betancourt & Brannon-Peppas, 2006; Li, Tourovskaia, & Folch, 2003; Voldman, Gray, & Schmidt, 1999).

##### 12.1.3.2.2 Soft-lithography

This is an advanced form of the lithography technique which is employed extensively nowadays due to its low-cost and ease of accessibility. This is the process that is generally followed in medical fields because it offers bio-compatible fabrications along with better optical and mechanical characteristics. The operations like the microfluidic patterning as well as microstamping are mostly used in microfluidics fabrication. The PDMS is generally employed for soft-lithography. In Fig. 12.4, first prepolymer is poured over the substrate, then the PDMS stamp moves over it. The fabricated pattern undergoes curing for a period of time and then PDMS channels are removed by making a polymer microstructure over the substrate. New and more advanced techniques for the fabrication of microfluidics are also available; for example, the microfluidics modular using interlocking blocks which is made by the students at MIT (Owens & Hart, 2018; Xia & Whitesides, 1998).

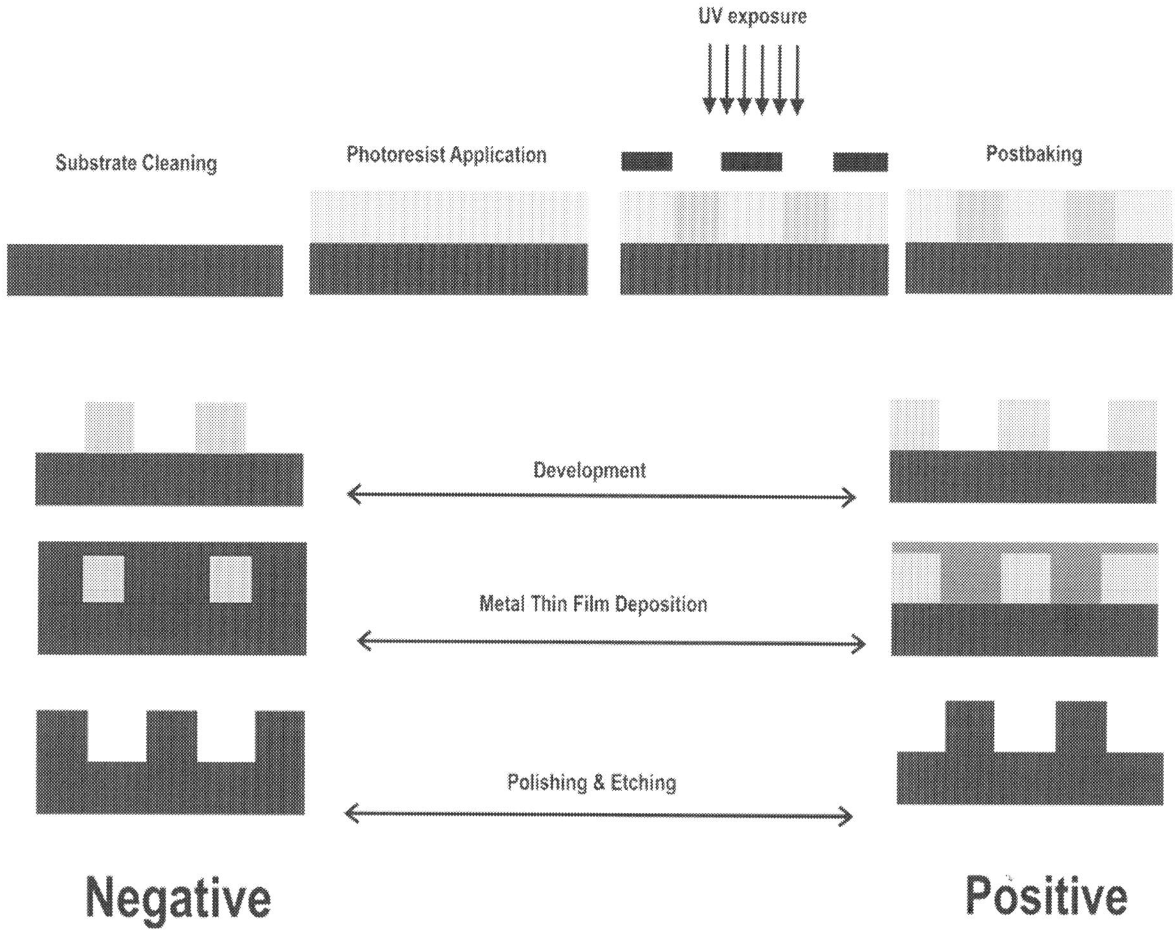


**FIGURE 12.3** Photolithography process.

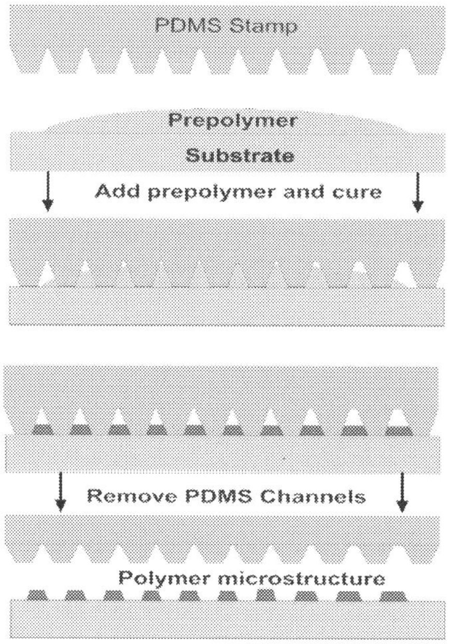


**FIGURE 12.4** Soft-lithography process.

Apart from the above techniques, there are lot more new and advanced techniques for microfluidic device fabrication, but since the discussion of the fabrication techniques is not the focus of this chapter, this just gives a brief review of techniques that are already available and still used in the market for microfluidic devices.

## 12.2 Point-of-care devices

The development of safe autonomous POC systems for diagnoses and in-field research is an important trend for biomedical engineering. The current breed of these initiatives tackles health and environmental needs more and more. It can boost the quality of life in both the developing and the resource-depleted countries and expand its interaction with daily items, such as mobile phones. New methods have been provided for a wide variety of applications, from health (diagnosis and disease management) to environmental pollution tracking, as well as for the identification of bio-warfare agents, toxins, and allergens on food and agricultural products. The innovations in the LOC have grown exponentially. The next cycle of such instruments have unimaginable opportunities. Significant social problems, for example, in human well-being and environmental protection shall be resolved by methods of fast and high-sensitiveness assessments and in-field studies. The key aim of the next technologies is to track or control many important metrics, from blood pressure to biomarkers, both in the clinical as well as at home. This is a central concept of POC diagnostics, which can be grossly categorized as (a) near-patient research, for rapid diagnosis and decision-making or long-term surveillance of pathogens, and (b) on-the-spot research for the prevention of epidemics and monitoring for water and food safety (Primiceri et al., 2018).

As per the FDA description, the "simple test" is desirable features for POC devices or on-field tests (Yager, Domingo, & Gerdes, 2008):

The test should be shorter than 1 hour with just a few steps for the process, following the fundamental principles of good laboratory. Also, the laboratory practice must be as clear as possible.

Also the test should be simple and fool proof. So that an untrained operator can do it after the minimum experience. Moreover, the outcomes should also be clear and understandable.

For public health services as well as consumers and customers, networks are meant to be accessible. The research can be less costly than expected, lowering expenses for the patient—for example, in low-income areas, where even the expense of attending health services can be reduced.

The POC device should have multiplex capabilities such as, simultaneously performing more than one test, enabling full biological sample characterization along with improvements in clinical diagnostics, which could be used to obtain a patient's molecular fingerprint that allows for precise medicine.

### 12.2.1 Point-of-care in developing countries

The POC devices can be beneficial in developing nations and in the nations where the number of resources is limited. The screening test for some common diseases and virus which are infectious and account for millions of deaths per year can be monitored easily with the help of POC devices in the countries that are still tackling these issues. Diseases like malaria, human immunodeficiency virus (HIV), an tuberculosis are spread all over the globe and can be fatal in most cases. In countries like Africa, for example, one observes that the access to medical facilities in this country is very poor compared to others, so cases are treated as a relative treatment that has a very low success rate for curing the actual disease (Chan, 2014). Most hospitals are often crowded (only one or two physicians per 100,000 people in mostly urban areas) and almost no methods are available to prevent infection because interactions are tracked but not regularly monitored (Pollock, Colby, & Rolland, 2013). Furthermore, it may be complicated and costly for people who live far away to get to hospitals. If rapid diagnostic tests (RDTs) can make a real-time diagnosis, hospitals can release patients on an effective medication earlier, prevent a second visit, and increase the care of diseases dramatically. RDT products are developed and sold for contagious diseases, but only available for a small range of individuals. Multiparameter methods for identifying microbes, DNA and RNA, protozoa, toxins, and biowarfare agents present in food and water will have a significant effect on life safety (Agarwal, Hachamovitch, & Menon, 2012; Irenge & Gala, 2012).

Heart attacks and strokes result in around 20 million deaths around the globe annually. The cardiac injury markers (myoglobin) and troponins (cTnI and cTnT) are the on-site POC tests that help in effective screening and also a reduction in overall cost. In addition to all these diseases, there is cancer which is highly responsible for most of the diagnostics market and can be estimated to about $180 billion by 2026 (Shaikh & Pajankar, 2019). It drives inventive system development, focusing on protein biomarker identification, such as the prostate-specific antigen (PSA), platelet factor 4, and the carcinoembryonic antigen.

Another factor which must be taken into account in developing countries is the rapidly increasing number of elderlies. Recent advancement in key technology enables, and in particular, in the developing field of wearable devices, will give new strategies to help people survive and smart aging, as well as the introduction of safe and tailored medication through ongoing tracking and self-management of their health (Elsherif, Hassan, Yetisen, & Butt, 2018; Salvatore et al., 2017). Several cases can be cited for portable or implantable appliances, most of which are sweat tracked with the use of noninvasive instruments and the benefits of sophisticated systems to regulate levels of glucose, electrolytes, saliva, tears, and other body fluids (Ascaso & Huerva, 2016; Dehennis, Mortellaro, & Ioacara, 2015; Rose et al., 2015; Valdés-Ramírez et al., 2014).

## 12.2.2 Personalized medicine

With the potential to develop and monitor a large amount of data, biology/medicine interacts strongly with emerging technologies to reshape conventional medicine into so-called constructive P4 medicine. Hood and Friend, were the first to recognize P4—predictive, preventive, personalized, and participatory—as the next huge step toward better health (Hood & Friend, 2011; Tian, Price, & Hood, 2012). An easier and more consistent quantity of diseases will boost follow-up sometimes not effective over time, mostly because of impractical checks, including repetitive biopsies. One essential aspect of implementation is data administration (storage, analysis, and modeling) to turn the vast quantity of details into exploitable outcomes—the so-called "data explosion" (Hood & Flores, 2012).

The treatment of cancer diseases will be one of the key areas of application. Tumor tissues are well known to have a high heterogeneity in the intratumor, changing in times and locations (differentiating from primary carcinomas to metastatic sites), which can encourage tumor development and adjustments and easily overcome treatment methods. Liquid biopsy (LB) will include tumor aggressive specifics and enhance prognostic prediction, help medical judgment and track tumor treatment results without the need for multiple biopsies (like circulating tumor cells [CTCs], and even the circulating tumor DNA [ctDNA] and exosomes [EXOs]). One clear gain is that LB only requires regular blood selection, during the disease it is quickly repeatable—an overarching consideration. Much has recently been done to introduce new methods for isolating, identifying, quantifying, and examining LB components with a LoC technology that offers multiple possibilities and major benefits. Fig. 12.5 show how the LOC undergoes cell separation, chip culture, and analysis. The understanding of all the criteria for a chronic illness is important, then illness itself can be divided into its major subtypes such that each particular patient is paired with the right medication for its subtype disorder (Gerlinger et al., 2012; Rana, Zhang, & Esfandiari, 2018; Wu et al., 2012).

A variety of factors need examination for a full diagnosis and a better understanding of the condition. Various types of biomarkers ranging from proteins and nucleic acids to whole cells must either be established or applied.

### *12.2.2.1 Cell identification*

In clinical diagnostics, the blood cell count is significant because changes in number can differentiate health status from pathology. There are also some popular POCs used for the calculation of blood cells, including HemoCue, in which

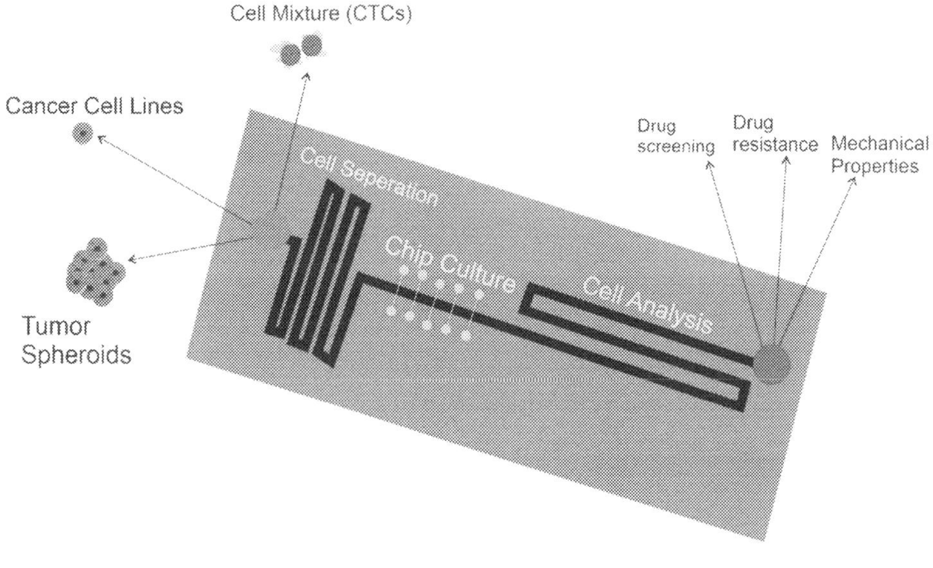


**FIGURE 12.5** Cell separation and analysis.

white cells can be measured in the blood. Photometrical measurement quantifies blood cells with an accuracy equivalent to bench-top tools in minutes. One of the significant advances in the identification at quite small quantities of CTCs, i.e., blood-shedding cells from the primary tumor site ( ~ 1 CTC per billion regular blood cells in advanced cancer cells) (Alberter, Klein, & Polzer, 2016; Allard et al., 2004; Bidard et al., 2014; Quesada-González & Merkoçi, 2015; Steeg, 2006). Concerning CTCs, their low amount of patient blood is the key limiting factor and a significant effort has been made to introduce new approaches:

1. Biochemical methods: The complexity of the CTC and the absence of commonly recognized tumor markers were restricted to these techniques. Furthermore, the positive selection generated by the capture mechanism is intrinsically biased. Besides that, binding antibodies to the surface of the CTCs can cause phenotypical changes and lead to misleading molecular studies.
2. Physical Methods: This is accessible and dependent on physical features such as scale, form, polarization, and plasticity. No unique surface biomarkers with a substantial benefit are required in this situation. However, the physical characteristics of CTCs can interfere with the physical characteristics of residential blood cells.

#### *12.2.2.2 Protein and nucleic acid analysis*

The identification of proteins is comparatively easier, quicker, and cheaper, than the identification of nucleic acid involving several steps in the preparation of the sample such as cells, cleansing of nucleic acid, and the amplification of DNA through analytical techniques focused on, for example, lateral flow or chromatography. Due to the unique association between antigen and antibody, enzyme and substrate, or the recipient and ligand, parting of analytes that pass through the pore medium takes place in lateral flow assays (LFAs) (Quesada-González & Merkoçi, 2015). Lateral flow immunoassays (LFAs) are widely distributed systems due to low cost and rapid reaction (around 15–20 min) but are also poorly responsive and unquantified. LFA could be used to tackle this problem in combination with advanced reading methods that can maximize LFA response diagnostic outcomes. Wang et al. suggested a novel method: their technique called thermal contrast amplification (TCA), was the laser stimulation of gold nanoparticles. The TCA reader will increase (8 times) sensitivity and allow LFA quantification (Wang et al., 2016).

Even for nuclear acid detection, much of the mechanisms mentioned for protein analysis were used but additional functions must be utilized to meet a real POC implementation. In addition to detecting the samples and molecular amplification, new devices for POC diagnosis are preferably required. miRNAs are especially interesting among nucleic acids since they play a key role in developing certain diseases such as cancer. They are also closely linked to drug resistance unique to the patient (Dehghanzadeh, Jadidi-Niaragh, Gharibi, & Yousefi, 2015; Fernández-Baldo et al., 2016).

## 12.3 Nanoengineered materials

Engineered micro-/nanoparticles produced by different physicochemical characteristics play important roles, in clinical applications such as biosensing, detection, and diagnosis of early illness. Significant advances have been made in recent times in the design and development of material synthesis systems. Microfluidic reactor mixing, exact reaction control, and fast chemical processes help microfluidic reactors to operate in a time-efficient and cost-effective manner, as opposed to traditional batch reactors (Hao, Nie, & Zhang, 2018). This section will be focused on microfluidics-based nanoengineered materials which are used in POC devices,

### 12.3.1 Metallic particles

The particles which are generally used for nanomaterial synthesis vary from ranges of metals, polymers, ceramics, and composites and also feature some advanced engineering material. Gold (Au) and silver (Ag), due to their appealing optoelectronic qualities and strong stability, are two of the largest categories of particles studied. For future clinical diagnostic applications, microreactors may specifically change the predictive characteristics to produce gold and silver nanoparticles with configurable scale, shape, and surface composition properties. In the past few years, gold/silver particles were synthesized through several steady flows and droplet-based microreactors. Microreactors normally improve synthesized materials sensor sensitivity and/or performance about batch synthesis. Also, microfluidic systems allow multiple stages of gold/silver particle scale-up synthesis and inline optical inspection of the synthesis method for improvement of parameters to be applied. To tackle the problem of surface accumulation of the blocking in microfluidic channels of gold/silver particles, various strategies such as salinization and pH correction have been investigated for minimal fouling. In this way, gold and silver particles are used in the synthesis of nanoparticles (Boleininger, Kurz, Reuss, & Sönnichsen, 2006;

Duraiswamy & Khan, 2009; Gomez et al., 2014; Lin, Terepka, & Yang, 2004; Parisi, Su, & Lei, 2013; Sebastian Cabeza, Kuhn, Kulkarni, & Jensen, 2012; Tsunoyama, Ichikuni, & Tsukuda, 2008; Wagner & Köhler, 2005; Wagner, Tshikhudo, & Köhler, 2008).

### 12.3.2 Quantum dots

These particles are very suitable for sensing operations because of their favorable optical and electrical properties. As the properties depend on particle size, form, and surface chemistry, an effort has been made to manufacture quantum dots (QDs) with physicochemical properties that can be tuned. Cadmium sulfide (CdS) and cadmium selenide (CdSe) are the widely studied nanocrystals that are fabricated using microreactors. The oxidation and reduction reactions of cadmium salts will result in microfluidic CdS synthesis. CdSe QDs are normally synthesized at high temperatures with a particular surfactant (Edel, Fortt, DeMello, & DeMello, 2002; Hung et al., 2006; Nakamura et al., 2002; Shestopalov, Tice, & Ismagilov, 2004)

Many surface ligands can also be bonded into QD by microreactors, including glucose, amino acids, and proteins. Biofunctionalized QDs resulting are considered to be effective fluorescent-directed labels for bioimaging and other screening procedures. The production rate will easily be more than 100 g a day at the normal flow rate on a mL/min size equipment. Microreactors with an inline detector analysis system are now allowed to simultaneously investigate the emission QDs to have an evaluation of their scale in real-time nucleation and growth concentration (Hu et al., 2014; Toyota et al., 2010).

### 12.3.3 Hydrogels

Natural and synthetic hydrogel-based materials are typically made with microreactors. These hydrogel materials are processed using different techniques such as temperature, UV light, and chemical chelators. The regulation of physicochemical properties of nanoparticles may be dominating factors such as porosity, swelling, weight deterioration, and the gelling effects of hydrogels. Fig. 12.6 explains the factors affecting the hydrogel. Those factors are pH value, temperature, redox reaction, magnetic properties, light, and humidity. The pores can be produced to synthesize porous microfibers via a range of processes, such as solvent exchanges or by using porogens. Likewise, hydrogels that are photo-cross-linkable can also be manufactured into microfibers using microfluidic applications (Chung, Lee, Khademhosseini, & Lee, 2012; Jeong et al., 2004).

The hydrogel dissolution ratio could be tuned according to a different ratio of monomers. The researchers were able to provide fluids in different flow speeds, combining this function with temperature modulation and thus dissolving drained reagents and checking the residence time of solutions in various parts of the device to achieve maximum reaction conditions (Niedl & Beta, 2015).

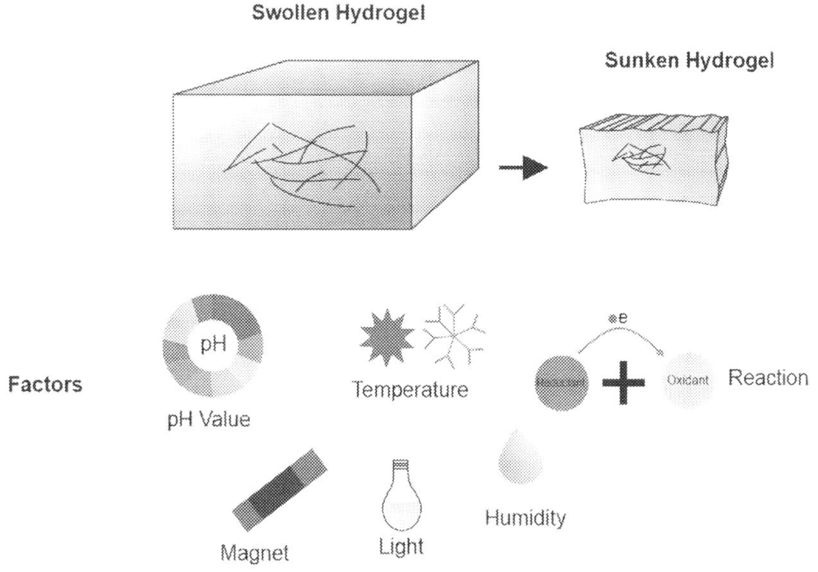


FIGURE 12.6 Factors influencing hydrogel.

Microfabricating processes have been used recently to produce microparticles in varying sizes and formats in addition to microfibers. The mono-dispersive emulsion-based hydrogel microparticles produced on microfluidic devices can be genuinely beneficial in diagnostic disciplines, like cell encapsulation and targeted delivery of medicinal products. Janus microparticles are also produced by a microfluidic system, containing two or even more biochemical functions (Prasad, Perumal, Choi, Lee, & Kim, 2009).

### 12.3.4 Nanotubes, nanopores, and nanowires

The movement of liquids via carbon nanotubes and nanopores poses a host of distinct challenges. These tubes or pores, which can detect single molecules as they move through the tube or pores are small electrically isolated. The identification of the molecule is dependent on the ionic current of the electrolyte solution which contains interesting molecules and results in a shift of the electric current (signal for events of translocation). Biochips and nanofluid with nanopores or nanotubes can introduce new clinical sequencing methods of DNA, as each DNA base has a specific molecular structure and thus a particular signal for a translocation event. Nanofluidic instruments are designed to improve the sensitivity to DNA molecules with several measurements on one molecule. Membrane integration techniques involve placing nanopores into microfluidic devices that greatly reduce noise and allow nanoporous networks to be built. At present, nanowires are considered to be special and useful for constructing a nano biosensor. Nanowires are a nanoscale channel which transmits current and can be produced from carbon nanotubes, metal oxides, or silicon, which require the syncretization of high temperatures and are normally prepared on silicon wafers. Antibodies interfere with the biological goal of interest, which leads to a shift in the current passing through the nanowire, which allows a sensitive, unique detection. The use of nanowires in an array mode, where multiple antibodies are paired with each nanowire, enables the mass identification of numerous diseases and conditions or the production of a customized molecular profile for one single disease type (Choi, Tripathi, & Singh, 2014; Choi et al., 2016).

## 12.4 Microfluidic devices based on specific substrates

POC devices, i.e., hand-held and big bench-top devices, are narrowly divided into two groups: (1) the small handheld systems are produced using futuristic micromanufacturing techniques that use automated sample processing, analysis, monitoring, and signal detection; and (2) the mainframe core workbench systems are miniaturized models with reduced complexity and scale. Moreover, various obstacles to the development of biosensors are posed by the strict demands of POC diagnostics. The identification of high accuracy and reliability target analytes, for instance, is crucial in POC diagnostics as limited sample volumes are required. The other challenge is to integrate the detection aspect on a single podium with other complex regulatory components. To overcome these challenges, microfluidics comes into the picture and the development of POC devices with the use of microfluidics technology is the primary goal (Balaji & Steve, 2015; Dixit & Kaushik, 2016; Luppa, Müller, Schlichtiger, & Schlebusch, 2011).

Microfluidics exquisiteness lies in detailed quantity management and sample and reagent rate of flow that enables separation and study of high-precision and critical analytes. Because of many advantages including ease of manufacturing, low use of reagents, low reaction time, better sensing parameters, and continuous tracking of desired analytes, microfluidics-dependent POC offer effective biochemical and chemical analysis platforms (Rackus, Shamsi, & Wheeler, 2015; Vasudev, Kaushik, Tomizawa, Norena, & Bhansali, 2013; Whitesides, 2006).

Therefore, depending on the particular application, microsystems with different properties may be produced with the option of suitable material for their chemical stability, surface properties, and thermal and electrical conductivity.

### 12.4.1 Glass-based microfluidic devices

The selection of materials for MF equipment depends on many factors including the desired purpose, the scale of incorporation, buffer quality, excellent physiochemical properties, and their possible applications. Here, the consideration is a glass substrate because of its versatility of certain characteristics. It is bio-compatible, a hydrophilic and consistent coating that allows its use in biomedical devices is considered to be accessible (Mazurczyk & Mansfield, 2013; Nge et al., 2013; Shaurya, Marie, & Bharat, 2012; Whitesides, 2006).

This glass-based polymer is fabricated using photolithography techniques as well as other methods like etching and metallization. The pyrex glass, borosilicate glass, and soda-lime glass are the commonly used substrate with borosilicate glass being the most popular because of its outstanding optical, chemical, physical, and bonding properties. There are various applications for glass-based microfluidic devices; several applications have been documented to detect enzymes, antibody, and whole cells. An antigen IgG and cTnI (a particular biomarker of myocardial infarction) were produced in 3D interdigital electrodes for immunosensors. A cell culture/biosensor system composed of an aptamer altered Au electric electrode constructed on a glass substrate using a microfluidic channel and PDMS was reported by Matharu et al. By microfluidic patterning of the CTAB coated AuNRs on a glass-based $O_2$ plasma-treated substrate, Chen et al. designed a multiarray LSPR chip assay through ionic interactions between the positively charged AuNRs and the oppositely charged glass surface. Still, the most challenging part of the use of glass substrate microfluidic devices is the brittleness, nonflexibility, and higher cost (Bange, Halsall, & Heineman, 2005; Chen et al., 2015; Grego, Gilchrist, Carlson, & Stoner, 2012; Han, Kim, Kang, & Chung, 2014; Matharu et al., 2014; Mirasoli et al., 2013).

### 12.4.2 Silicon-based microfluidic devices

Due to its high resistance to differing conditions and low bonding temperature requirements, silicon emerged as the ideal substrate for microfluidic channel manufacture. The manufacturing methods for these microfluidic devices range from the traditional techniques to a more modern and sophisticated process like the surface/bulk micromachining. The bulk micromachining is most common, where channels are formed by the removal of material on a silicon wafer that can then be sealed by chemical bonding or physical conformity with other wafers. Nanoimprint lithography and electron irradiation may be used to render still more reliable designs of a silicon substrate with a nanoscale level characteristic (Chae, Giachino, & Najafi, 2008; Iliescu, Taylor, Avram, Miao, & Franssila, 2012; James, Okandan, Mani, Galambos, & Shul, 2006; Schöning & Lüth, 2001; Stjernström & Roeraade, 1998).

The insertion of microfluidics into the silicon substrate decreases the volume of the samples, measurement time, and allows reliable controlled flow rates in biosensors. The fluid flow, however, will weaken the biomolecules attached to the substrate surface, while the surface working chemistry, the evaluation of the microfluidic system requires bioconjugation and stabilization. A suitable immobilization approach may combine various legends on the surface of MF channels, further broadening the spectrum of detection of analytes. To study the nucleotide polymorphism of human cystic transmembrane fibrosis regulator, Jenison et al. designed a silicon-based biosensor. Recently, cell-based biosensors were very common since they can specifically classify biological results articulated by a live cell. The silicon substrate's biocompatibility may be cell-binding moieties that can enhance the surface within. Besides, these results can be translated into electronic signals that establish bridge living environment and electronics. A significant range of biosensor applications has been identified in silicon nanowires and porous silicon. However, with the use of polymers, the silicon-based microfluidic devices are still expensive (Das, Dey, RoyChaudhuri, & Das, 2012; Goddard & Erickson, 2009; James et al., 2006; Jenison, La, Haeberli, Ostroff, & Polisky, 2001; Liu et al., 2014; Syshchyk, Skryshevsky, Soldatkin, & Soldatkin, 2015).

### 12.4.3 Polymer-based microfluidic devices

Owing to its low cost and simple manufacturing steps as opposed to glass and silicon, the implementation of polymers to microfluidic devices has fascinated industrial producers greatly. The most common application of microfluidic devices is PDMS and polymethacrylate (PMMA), because their electrical resistance and clarity are excellent and suggest a valuable ability to manufacture large microfluidic devices.

There are a variety of techniques and methods used to fabricate this device; polymers are well known for their formability and structure altering capabilities. Other than the soft-lithography and photolithography techniques, methods like hot embossing and imprinting, laser ablation, 3D printing, and injection molding are quite common in the fabrication of polymer microfluidic devices (Pandey et al., 2018).

At low pressure and high temperatures, hot embossing can be done while imprinting can be performed at room temperature. The exact copy of the stamp is the subsequent microchannels of the desired proportions inside the plastic substrate. Various criteria, such as time of impression, friction, and plastic characteristics affect microchannel dimensions of the room temperature. The advantage of imprinting over hot embossing is that manufacturing time is minimized and the produced system is strongly reproducible. The injection molding has shown many benefits over imprinting and hot embossing because 3D functions can be generated and the preformed components can be inserted in the plastic effectively. The PMMA and polycarbonate microfluidic channels are rendered with high accuracy with injection molding (Becker & Locascio, 2002; Xu, Locascio, Gaitan, & Lee, 2000).

In the laser ablation process, as a result of UV laser pulsation induction, the bond of the polymer backbone is broken down. Also, due to the wave pulse which forms a void indoors, the particles are ejected from the polymer substrate. Polymers with significant absorption into the wavelength of laser emission are more effective in ablation. Light passes through a mask in the laser photoablation that defines the ablated area into the polymer substrate (Sylvain & Vincent, 1989).

In the past years, 3D printing has generated tremendous interest as it can build numerous structures with high resolution and short production times. Stereolithography (SLA) and the fused deposition system (FDM) have been most popular among the various methods. UV laser is used in SLA for scanning and tracing the given area to cure the material for the liquid resin. For hardening resin materials, high-intensity lasers or UV rays are employed and the whole platform passes across a single layer in the z-direction. The 3D presses using the SLA process are most frequently used and sold (Azouz et al., 2014).

Several types of electric biosensors have been applied in polymer-based microfluidic systems, including immunoassays, DNA hybridization, and signal swapping. Actual samples are often tested until a preferred biosensor is added. Thanks to its particular binding capabilities, biosensors dependent on affinity can accommodate sample matrices and samples in complex applications. Polymer-based microfluidic devices provide benefits over traditional glass/silicon microfluidic devices as they are not complex and cost-efficient in the photolithographic process. They were found to have good electrical power/chemical and optical resistivity that is ideal for electrical/optical biosensing instruments. The limitation of polymer microfluidic systems is the durability of these systems.

### 12.4.4 Paper-based microfluidic devices

Paper is an adaptable, cheap, compact, reusable, biocompatible cellulose-based material. It is a hopeful microfluidic substrate since there are no external pumps or supplies of energy for the flowing fluid, and requires a limited reagent and sample quantities, to produce a faster rate of analysis. This paper-based microfluidic device targets to confine the flow using microchannels. The characteristic of such a channel lies in its hydrophobicity and certain specific methods are generally opted to fabricate these devices. In addition to conventional photolithography techniques, methods like plotting and wax printing are emerging advantageous for the paper-based microfluidic devices (Fu & Downs, 2017).

The plotting method makes use of a plotter in the two axes (X and Y) to print the PDMS in a paper for an expected pattern. The width between the two channels is 1 mm and is relatively less costly than photolithography. Microfluidic paper channels developed by Shangguan et al., where the paper pores with PDMS adsorption are constituted in the formation of a hydrophobic barrier prevents exposure to aqueous solutions. The biosensing method for liver function markers and serum protein was conducted to prove the functionality of this method (Bruzewicz, Reches, & Whitesides, 2008; Shangguan et al., 2017).

Wax-printing on the other hand uses the solidified ink of the printer which is melted to form the barrier to direct the fluid by capillary action. It is a simple and cost-effective methodology that can be applied to mass manufacturing. Moreover, it is more practical for use in low resource environments, conveniently accessible, and robust up to 60°C. The hemi-channel and a completely closed channel were developed by Renault et al. where the wax transport properties of paper were studied (Carrilho, Martinez, & Whitesides, 2009; Renault, Koehne, Ricco, & Crooks, 2014).

The design and immobilization of biomolecules on a specific surface are important for the manufacture of an effective paper microfluidic biosensors. Morbioli et al. used a 3D method to make the cellulose matrix of paper microfluidic glucose detectors more uniformly permeated by the fluids. Li et al. used a modified ZnO nanometer paper to produce a highly sensitive glucose detection electrode. Electrochemical DNA sensors for the identification of human papillomaviruses have been developed by Teengam et al. Despite these important advances, a major advancement is made by updating the design and alteration of the nanomaterial substrate surface to increase the limits of detection and intolerances of paper-based microfluidic systems. Humidity and variance in temperature have profoundly influenced the sensing characteristics of paper microfluidic systems. Other reasons affecting the growth of POC products, is the nature of the model, the multiple manufacturing methods, and incorporation into a single chip (Betancourt & Brannon-Peppas, 2006; Li, Tourovskaia, & Folch, 2003; Teengam et al., 2017; Voldman, Gray, & Schmidt, 1999).

## 12.5 Microfluidic-based point-of-care devices for cancer diagnosing

This topic is the primary focus of the whole chapter; how microfluidics is helping the world to diagnose fatal diseases like cancer, and the techniques involved have always been a growing field in and around the research lab. The standard workflow of cancer-diagnosed patients includes (1) symptom growth in patients, (2) first appointment to a clinic, and (3) imaging and tissue biopsy. Microfluidics could help diagnose by providing more efficient processing, decreased costs, improved cell monitoring, and reducing sample volumes needed. Several microfluidic technologies are being built to reduce existing diagnostic procedure expenses, time-to-result, and invasive characteristics. The techniques for cancer

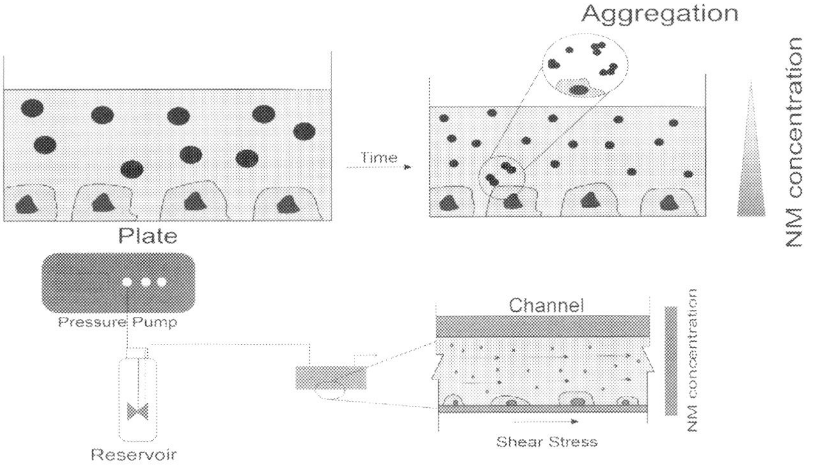


FIGURE 12.7 Static versus dynamic conditions in cell-based assays for nanotoxicology.

treatment in the clinic include liquid bioscopy, immunohistochemistry, and the POC devices. The former two are generally referred to in clinical treatment, whereas the latter is out of clinic procedure (Tadimety et al., 2017, 2018). Also, Static versus dynamic conditions in cell-based assays for nanotoxicology are discussed and shown in Fig. 12.7.

### 12.5.1 Technologies in point-of-care devices

Cancer is a chronic disease marked by irregular cell growth attributable to multiple epigenetic modifications resulting in uncontrolled proliferation, division, and invasion of surrounding tissue, contributing to severe disease and death in separate sites or organs. According to a World Cancer Report, 2014 issued by the World Health Organization's International Agency for Research on Cancer, the annual global prevalence of cancer is projected to rise to 18.3 million by 2025. Early diagnosis and the control of tumor cells that persist and invade as well as the response to treatment pose core difficulties in successful cancer care. Delay in diagnosis or advanced diagnosis of serious cancers such as breast, lung, and colorectal mean survival rates of less than 5 years. Instead, substantial survival rates were observed if the diagnosis was made at a preliminary phase (Gulland, 2014; Ruddon, 2007; Semmes et al., 2005; Zoli et al., 1996).

Gloeckler et al. reported an early diagnosis of cancer that led to a substantial rise ($>85\%$) in the rate of survival, but advanced diagnosis only indicated a small change in recoveries, such as 34% for breast cancer, 57% for colorectal, and 72% for lung cancer patients. The outcome of early diagnosis was a large increase. Early identification, accurate surveillance, and tumor staging by precise selective measurements of tumor biomarkers are also important to enhance cancer diagnosis and forecast. These approaches are, however, limited because of their difficulty, sluggish examination, and high prices. There is therefore an increasing requirement to create the cancer-diagnostic tool that is highly precise and reliable, inexpensive, and user-friendly for multiplex diagnosis in a single portable computer for use in difficult environments (Grosso, Carrara, Stagni, & Benini, 2010; Hanash, Pitteri, & Faca, 2008; Kingsmore, 2006; Manne, Srivastava, & Srivastava, 2005; Ries, Reichman, Lewis, Hankey, & Edwards, 2003; Rusling, Kumar, Gutkind, & Patel, 2010; Vashist, Luppa, Yeo, Ozcan, & Luong, 2015).

Conventional experiments, like the enzyme-linked immunosorbent assay (ELISA), also have little real use in poorly resourced conditions, such as bulky scanners or detectors which are clumsy and highly expensive. Conventional chemiluminescence methods require large-format scanners or detectors. Moreover, techniques such as surface-enhanced scattering of Raman, electrophoresis, gelatine zymography, and mass spectrometric testing are subject to intensive screening, contribute to cell death, and degradation of biomarkers throughout duration (Du et al., 2010; Hu, Zhang, Hu, Xing, & Zhang, 2007; Kricka, 2003; Lotan & Roehrborn, 2003; Lyapandra & Karpinskyy, 2004).

### 12.5.2 Chemical resistor arrays diagnostics

High specificity and sensitivity for the respiratory mixture of hazardous organs in patients that cannot be identified by typical lung cancer have been demonstrated in nanostructure-based medical POC systems just by analyzing the breath. The self-assembling or printing technology was used to construct the sensor array of golden nanoparticles (2–5 nm) as a thin film without dehumidification or breathing biomarker preconcentration during pulmonary cancer screening. A compact POC sensing nanotechnology device for early detection of lung cancer has recently been documented by

Zhao et al. A chemical resistor unit was composed of the patterned gold-copper (AuCu) electrodes on the thin film of gold nanoparticles deposited on a polyethylene terephthalate. Chuang et al. developed an optimized immunosensor chip incorporated into a gold ring microelectrode array ($3 \times 3$) for real-time impedance sensors, as a POC for supersensitive Galactin-1 (Gal-1) identification in diagnosis of bladder cancer. They have produced a mobile impedance analyzer device attached to a vaccine chip as a POC test to allow data transmission for cloud storage in distant and low infrastructure environments (Das et al., 2012; Konvalina & Haick, 2014; Luo et al., 2012; Peng et al., 2009; Zhao et al., 2016).

### 12.5.3 Near-infrared-optical diagnostics

Near-infrared (NIR) coupled with optical sensors are widely used in cancer diagnostic because of their greater penetration into body tissue and/or cells with minimal interaction with protein. Wang et al. utilized gold-nanorod surfaces with rose-bengal(4,5,6,7-tetrachloro-20,40,50,70-tetraiod derivative of fluorescein) for the detection of oral cancer at a very early stage. Considering the approach, the results indicated that this research has essential potential for conversion into accessible diagnostic POC implementation for early prostate screening and observation (Wang et al., 2016).

### 12.5.4 Biomarkers and paper microfluidics diagnostics

Fan et al. developed a POC method that detected many proteins within 10 m from tiny serum blood samples. An encoding of the DNA technique of the antibody library has been used to create the 12 tumor-related biomarkers' barcode arrays for immuno-detection. The effectiveness of the method has been studied and validated in 22 patients with breast and prostate cancer. Wu et al. has developed an electrochemical paper device for cancer biomarker detection. Target molecules were trapped on the surface utilizing immunoaffinity and electrochemical signal intensity by a controlled radical polymerization response to identifying four protein biomarkers for cancer. The limit of detection for rare biomarkers was as low as 0.01 ng/mL, which showed a lower-cost alternative. The device also had a huge spectrum of dynamics for centuries and was promising for significant clinical testing of protein in the plasma of patients. This was a step towards inexpensive and readable fast research outside the healthcare facility for small concentration biomarkers. The use of paper microfluidics is an important alternative method synthetic urinary biomarker combined with microfluidic paper complimentary trials. This idea includes a biomarker to send the patient a detectable indication in the urine whether he or she has a disease. For example, a nanoparticle unique to colorectal cancer may be inserted into the patients and creates a complex that can be read from a paper sandwich immunoassay in the urine (Gerlinger et al., 2012; Rana et al., 2018; Wu et al., 2012).

### 12.5.5 Nanowires and nanoparticle-based diagnostics

So many nanomaterials have so far been used to detect different cancer markers (proteins/peptides and DNA/RNA) in a responsive and precise way, particularly when used in the building of high-performance nanobiosensors, such as AuNPs, semi-conductive II-VI QDs, silicone-nanowire (SNWs), carbon CNTs, and graphene. To monitor and estimate the risk of premature biochemical relapse, FET-SNWs are often used to diagnose many prostate cancers, such as PSA at the fg/mL level of PSA. Ribonucleic acid (RNA), as biomarkers for active prostate cancer, has also been calculated by the use of nanowire technology (Counter Analysis System). This nanowire design allows the potential of a sensor chip, enables more than one cancer marker to be simultaneously identified, and measures a panel of biomarkers relevant to a particular form and/or patient. Tran et al., designed a portable read-out nanowire laboratory-on-a-chip to build a nanoplatform capable of detecting ALCAM serum with a detection limit of $< 30$ m at 15.5 pg/mL by adding a complementary metal-oxide-semiconductor (CMOS) in FET-SiNWs. To detect ssDNA and mil-RNAs associated with the initiation and development of different forms of cancer, FET-SiNWs are used in nanowires of zinc oxide (ZnONWs) (Gao et al., 2014; Takahashi et al., 2015; Tran et al., 2015; Warren, Kwong, Wood, Lin, & Bhatia, 2014).

### 12.5.6 Nonreusable immunosensitive diagnostics

The inexpensive design, fast accessible supply, and simple control during the manufacturing phase of cotton thread is its appeal. It was recently used for biomedical purposes including microfluidic-based quick diagnostic testing and lateral flow-based immunochromatographic assessments. Mao and his colleagues documented a fiber cotton-based POC diagnostic system to identify squamous cell carcinoma antigen (SCCA), biomarkers for lung cancer, and the tyrosinemia type-I genetic disease-related DNA sequence. Another study group has patented a POC device; an enzyme-free paper immunosensor system for multiplexed biomarker analysis using hybrid CdTe-QDS-DZ (dithiazone modified QDs). The system has 4 × 10 detection areas on Whatman Chromatography Paper (Number 1) (Lu et al., 2014; Mao, Du, Wang, & Meng, 2015; Reches et al., 2010; Safavieh, Zhou, & Juncker, 2011; Zhao et al., 2016; Zhou, Mao, & Juncker, 2012).

### 12.5.7 Micronuclear magnetic resonance diagnostics

CTCs are widely included in patients' blood during metastasis, as a clinical marker and source of tissues that are readily usable for the molecular profile of tumors. Micronuclear magnetic resonance (μ-NMR) has been reported by Ghazani et al. from biopsies of ovarian cancer for the identification of CTC by repeated multiplexed detection of four tumor markers [EpCAM, human epidermal growth factor receptor 2, epithelial growth factor receptor, and mucin-1, i.e., mucin-associated cell surface]. In whole blood samples, the μ-NMR system gives clear cellular profiling without purification of samples or separation of cells. The procedure involves implantation of a mixture of entire blood samples with the four trans-cyclooctene (TCO) named antibodies against identified biomarkers followed by the classification of antibody-positive cells using tetrazine-modified magnetic nanoparticles (TzMNPs). The POC diagnostic potential of a miniature μ-NMR system, which involves iron-oxide-based CTC tagging in combination with multi-biomarker molecular profiling, namely EGFR (epithelial growth factor receptor), EpCAM, and HER-2 (biomarker panel set, overexpressed in tumor biopsies) as well as vimentin (EMT and CTC) marker, in the whole sample of blood are also reported in various works of literature (Ge et al., 2013; Ghazani, Castro, Gorbatov, Lee, & Weissleder, 2012).

### 12.5.8 MEMS (micro-electromechanical system) based diagnostics

The ability of QDs modified ligands as selective bio-probes in cellular and molecular biology applications have been explored across numerous articles. To detect the biomarker panels of salivary and serum clinical samples (CEA, CA125, and tyrosine-protein receptor kinase proto-oncogenic and [C-Erb-2]), Jokerst et al. has engineered a standardized and quantitative microfluidic biosensor with incorporated QDs. The chip was laminated with an independent set of adhesive layers to monitor the movement of a reagent on the PMMA base. This type of MEMS-based integrated device can be used as a supersensitive sensing platform in conjunction with nano bio-probe fluorophores for POC devices of cancer (Ghazani et al., 2013). Fig. 12.8 explains the working of ciliated micropillar device. Exosomes from the cells were made to flow in silicon wafers.

### 12.5.9 Programmable bio-chip diagnostics

Brandenburg et al. developed a microfluidic POC device using polymer foil that was very cost-efficient and expendable. It was a total internal reflection-based optical read-out integrated biochip to detect provocative biomarkers. In a very short reading time (0.04–10 s), depending on the signal strength, the biochip displayed a dynamic range with a detection limit of 1 ng $mL^{-1}$, which was equivalent to commercial laser scanners. Such a read-out device has exciting implementation prospects with future medical POC diagnostics to enhance wellness and healthcare services. In ovarian cancer-diagnostics, Raamanathan and colleagues have developed a programmable bio-chip based treatment unit for

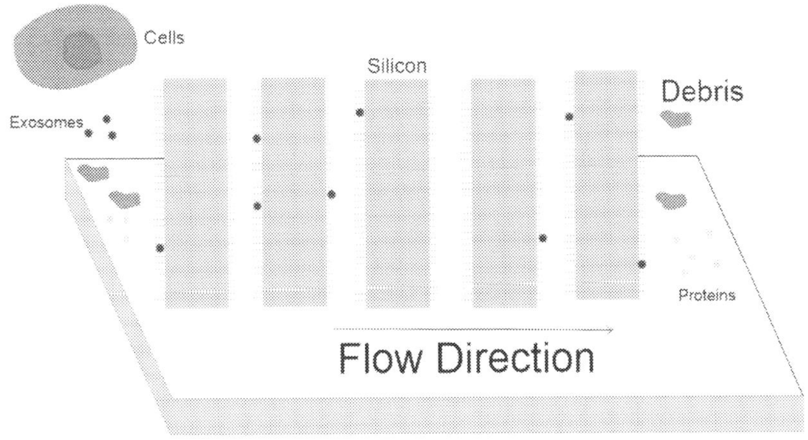


FIGURE 12.8 Ciliated micropillar device.

quantitative recognition of serum CA125. Shadfan et al., have developed a programmable bio-chip based on the identification of a novel panel of biomarkers (CA125, human epididymis protein 4, MMP7, and CA72-4) as a POC diagnostic method for ovarian cancer. Later, with the help of clinical research studies, it became evident that the programmable bio-chip helped in improving the sensitivity due to the multiplexed analysis compared to single biomarkers detections (Brandenburg et al., 2009; Jokerst et al., 2009).

## 12.6 Current trends and future prospects

A POC is a fast and instantaneous diagnostic platform that has affordable, accurate, and user-friendly diagnostic tools that can provide simple and easy results to encourage early detection and screening in real-time. Due to low sensitivity, lack of flexibility, long assay time, costly research, and difficulty in infrastructure access to restricted areas and rural areas, the regular POC devices for cancer diagnostics have restricted functional applicability. Advances in nano- and microfabrication technology through the relentless development and validation of cancer-biomarkers for accurate targets calculation opened up new horizons on POC diagnostic platforms because of their inherent benefits including miniaturization and ease of use. Microfluidics is very successful with the use of small volumes of materials and controls for miniaturization and automation. CTCs in the peripheral blood and genetic material can be measured by microfluidics for validating stage-specific markers for diagnosis and tracking patient response and relapse. The next generation of microfluidic devices could use a range of biochemically and bio-physical sensitive indicators special to cancer biomarkers, which will improve the clinical utility of microfluidic technology for the identification of cancer, high detection performance, cell viability, and high production (Raamanathan et al., 2012; Sandbhor Gaikwad & Banerjee, 2018).

Presently, QD technology is the most commonly used diagnostic nanotechnology, especially for the treatment of cancer. The in vivo toxicity is the only question about QDs. Researchers, however, recommend the use of silicon-composed QDs which is considered to be less toxic relative to cadmium in many QDs. Studies have recently reported nanoflare genetic technologies for the circulating identification of cancer cells that allow for blood-stream identification of life, CTCS. Nanoflare is engineered to hybridize with cellular variants and cancer-based oligonucleotides. The idea that nanoflare will reach the cell seems to have a great value, like any nanoparticle, as a result of its size; it allows different biomolecules to be used, not just markers embedded on the cell surface, but in the cell. Nanosensors and blood sensors sensitive enough to detect several pathogens or chemical compounds are one such example. POC diagnostics are feasible with nanosensors and also an appealing tool that will be convenient for the patient to use at home and will allow the incorporation of diagnostics with therapeutics and the growth of customized effective treatments (Choi et al., 2014). Fig. 12.9 shows the parts of a nanobiosensor. There are three parts to nanobiosensors: (1) Transducer converts one form of energy into another, (2) bio receptors are protein, antibody, and antigen, and (3) detector measures the output signal as a readable one.

Non-invasive detection is a diagnostic procedure that needs no skin or body split. Naturally produced biofluids including saliva, urine, sweat, tears, breathing condensate, etc., may be used for the manufacture of noninvasive biosensors. Disease detection by noninvasive biofluids offers various benefits: simplicity, painlessness, safer administration

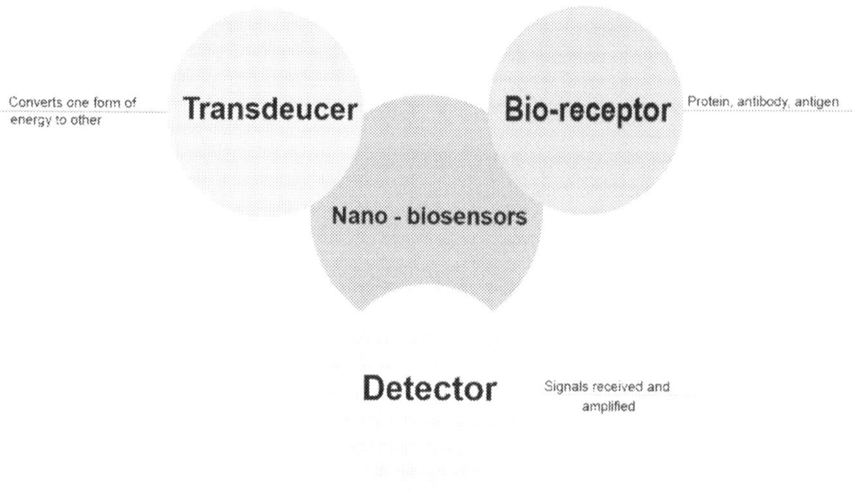


**FIGURE 12.9** Parts of a typical nanobiosensors.

than serum sampling, access to different specimens, sample collection, and home screening. A thin, wearable colorimetric microfluidic biosensor was developed using sweat as a biological analytical fluid. This multi-analytic system of four chambers could discern various glucose, lactate, chloride ions, and pH of the sweat. Efforts to differentiate diabetic from nondiabetic patients were also made by analyzing the amount of glucose in saliva, urine, and tear samples. Biosensor for multi-analyte identification of glucose and lactate in saliva biofluid was manufactured by Ji et al. This microfluidic unit was 100 nm thick and required 30 μL of biofluid for testing. The secretion of biomolecules at very low concentration is nevertheless the noninvasive biofluid usage challenge. More initiatives are required to investigate the possibilities of using these microfluidic devices further in low-resource environments (Darwazeh, MacFarlane, McCuish, & Lamey, 1991; Javaid, Ahmed, Durand, & Tran, 2016; Ji et al., 2016; Koh et al., 2016; Kumar et al., 2015; Kumar, Sharma, Maji, & Malhotra, 2016; Su et al., 2012; Zhang & Nagrath, 2013). Fig. 12.10 explains the detection of analytes using different forms of transducers. First, the analyte is converted into the form of a bio receptor (such as enzyme, antibody, microorganisms and Cdl). These bio receptors produce input signals in different forms like electrostatic substance, pH change, heat, light, and mass change. These input signals are read by the transducers, and finally it is detected as a measurable signal by the detector.

Yu et al. in their work described the detection and treatment of metastasis of lung cancer in its early stage. A portable microfluidic device was fabricated, which helped to control the temperature and the $CO_2$ environment inside a grid-based microsystem. The manipulation of microfluidics to design the device for cancer metastatic potential (MD-CaMP) was the primary concern of the research. They demonstrated that MD-CaMP would break cells to have a simple metastatic cancer index. It was a mobile device for site analysis, the characteristics were that only a small specimen, controllable temperature, and $CO_2$ were needed to sustain a reasonable growth environment; the method discriminates the cell's migration potential in operation. The unit provides a clear and reliable temperature distribution (37°C) in the cell culture market. The equipment can be used for the cell as well as the microheater in the grid shape, which is easily modified to the correct temperature. The device can also be used for counting, cell migration trajectories, or scale bar for DNA observation. A clinic can also make use of MD-CaMP as a fast screening method, such that this chip divides a biopsy specimen with infringements of uncertain, dispersed cells. Rzheversuskiy et al., in their work fabricated a spiral microfluidic chip that was capable of isolating tumor cells involving a rapid and label-free method. The results show potential in the diagnosis and projection of clustered urinary prostate cancer with fluid urinary biopsy, paving the way for an inexpensive, quick, and not invasive, diagnosis, and sampling and evaluation of prostate cancer and other urology cancers clinical outcomes. Yang et al.. also demonstrated a diagnostic test, loop-mediated isothermal amplification (LAMP), which is a super-sensitive nucleic acid test that can be used in different diagnosis including food safety, environmental monitoring, as well as public health diagnosis. One of the most effective POC diagnosis platforms is the hybrid of paper-based microfluidic instruments with LAMP. To achieve a more stable LAMP enzyme and other related reagents, biotechnologists must continue their research in recombinant microorganisms and improve nucleic acid synthesizers. With a nonenzymatic LAMP, the development of new chemical processes and techniques need to become more useful for resource-intensive regions. Smarter integration is essential for promoting LAMP-based device development (Lu et al., 2014; Mao et al., 2015; Reches et al., 2010; Safavieh et al., 2011; Zhao et al., 2016; Zhou et al., 2012).

The incorporation of ICTs (information and communication technologies) with POC systems for continued real-time health tracking, medical record collection, and retrieval (eHealth) has become a trend in biotech. Although it has all

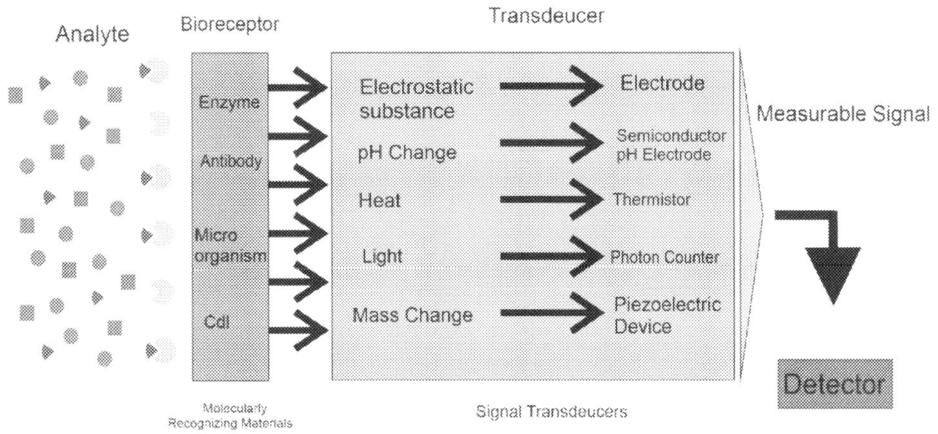


FIGURE 12.10 Detection of analytes using different form of transducers.

begun as a cost-efficient alternative to diagnostic strategies in developed countries, especially in rural, remote, and vulnerable populations, it has proliferated rapidly as a groundbreaking instrument that will encourage patients and doctors to make virtual appointments, for example, a portable electric epidermal microfluid platform with a versatile electronic board that enables the transmission of wireless real-time data. Without external fields and pumps, capillary forces and gland friction push sweat into the microfluidic canal. For continuous transmitting of the enzyme detection amperometric reaction to a desktop application, an integrated BlueTooth controller is used (Kim et al., 2011; Martín et al., 2017; Martinez et al., 2008; Mejía-Salazar, Cruz, Vásques, & de Oliveira, 2020).

The sensitivity of rare hormones can be calculated with digital microfluidic devices that integrate electrical electrodes and use electrical strengths to transfer liquids. The level of estrogen is, for example, a significant breast cancer risk factor. Mousa et al. have shown the handling of tissue samples, blood, and serum within a limited period with extremely responsive droplet estrogen studies. The instrument used electrical strengths to control samples manipulating the conductivity of most biologically important liquids. Microfluidics can imitate physiological signals in the cellular context by controlling the gradient of soluble factors and interaction between cells in the extracellular array spatially and temporally. In addition, an interconnected device may be linked with components for downstream research including imaging or molecular characterization with a cell culture module. All these aspects make good reasons to research cell biology using microfluidics. Various models to study cancer cell migration, angiogenesis and tumor microenvironment were developed for the exploration of cancer cell biology (El-Ali, Sorger, & Jensen, 2006; Mousa et al., 2009).

Another microfluidic concept for co-culture in 3D has been proposed by Huang et al. A series of juxtaposed channels separated by the regularly spaced posts serving as a partition of the cell loader channels is filled by multiple cell forms in a separate parallel channel. Macrophages were identified after 1 week of coculture toward breast cancer cells. This coculture device aims to imitate the relationships between tumor cells and their environment in the stream and to expose epigenetic modifications through various environmental indicators. The regulation of the tiny space around the cells is one of the difficulties of cultivating microfluidics cells. The architecture of the device takes into account parameters such as medium structure, shear stress, chemical gradients, and temperature. However, microfluidic instruments provide a tailor-made regulated environment for cellular studies until suitable conditions are collected (Huang et al., 2009; Sung & Shuler, 2012; Walker, Zeringue, & Beebe, 2004).

The use of electrochemical detection techniques is widely studied for reactions including antigen–antibody, inter-DNA, DNA-proteins, nucleic acid, and mRNA. Different variables, for example, can modify biosensor accuracy and precision, weak electrochemical sensor sensitivity, due to slow electron transmission, stability of immobilized molecules or transducer surface effects, and redox accessibility. Technological developments in the areas of intelligent nanomaterials and microfluid have also resulted in exponential development in the identification of molecular and cancer biomarkers and in the sensitiveness and stabilization of nonimmobilized molecules. A multiplexed electrochemical sensor (interlukin-8 mRNA and interlukin-8 protein ) was developed by Wei et al. and other companies to concurrently detect salivary biomarkers at the point for diagnosis of oral cancer treatment. The electrochemical sensor consists of an array (16) of gold electrodes on a chip that contain functioning, counter, and reference electrodes. The modified dendrimer nanocarbons (70–90 nm) of streptavidin were loaded into a conducting polymer matrix (via electrostatic interaction) which was further coated into the sensor as a supporting film. The cyclic square wave electrical field (+350 MV for 9 s and 20 cycles) subsequently performed electro polymerization, for effective and flexible regulation of hybridization and protein binding on a gold chip. The IL-8mRNA hairpin probes (dual labeled with biotin and fluorescein) and the human monoclonal antibody biotinylated IL-8 was immobilized on the chip with gold electrodes. Fifty-six (28 salivary samples and 28 control samples 100 μL each) were loaded onto an IL-8 mRNA and IL-8 enzyme detection probe-coated electrode, followed by signal read-off with an ampero-metric detection system, TMB/HRP (3,3′,5,5″-tetramethlybenzidine/horseradish peroxidase). For IL-8 and the detection limit was 7.4 pg mL − 1 and 3.9 fM for IL-8 mRNA, with almost 90% sensitivity and specificity for the detection of biomarkers of salivary cancer (Sadik, Mwilu, & Aluoch, 2010; Wei et al., 2009).

In immuno-sensing applications such as hemagglutinin type-1 identification and neuraminidase type-1, i.e., viral influenza A subtype (H1N1), carcinoembryonic antigens, etc., the ultrasensitive detection limit in previous experiments were generally implemented in laser-induced fluorescence (LIF). In addition, Fernández-Baldo and his team have used similar platform technologies to research the promise in the identification of LIF-coupling nanoparticles (compared with zinc oxide and polyvinyl alcohol) of epithelial cancer biomarkers (EpCAM) for metastatic colon cancer patients. The T-shaped microfluidic system was made of two tubes, the primary one consisting of the glass, the central one (60 mm long, 100 μm diameter), and the accessory (15 mm long, 70 μm diameter). Moreover, during salinization using 3-aminopropylo-triethoxysylane and glutaraldehyde, the microfluidic unit has been surface adapted to immuno sensor applications by adding different classes of amine and aldehyde on a central channel. The ZnONP-PVA was subsequently immobilized with aldehyde groups at 25°C by covalent binding for 12 h, while the ZnONP-PVA residual amino groups were tagged with glutaraldehyde anti-EpCAM (10% $mL^{-1}$) antibodies. The CTCs are isolated from peripheral blood samples of healthy colon cancer volunteers using an immunomagnetic CTC screen. The HRP conjugated anti-EpCAM antimicrobials were further analyzed to measure binding antibodies to anti-EpCAM modified nanoparticles from samples. The detection limit of nanoparticle-based immunosensors and ELISA was found respectively to be 1.2 pg $mL^{-1}$ and 13.9 pg $mL^{-1}$ with system durability of 1 month without severe loss of sensitivity when stored at 4°C. Thus, the overall results provide a new insight for the use of LIF-based nanointegrated microsensors to identify different epithelial cancer biomarkers as a POCdiagnostic tool (Fernández-Baldo et al., 2016; X. Li, Zhao, & Liu, 2015; Z. Li, Xu, Fang, Tong, & Zhang, 2015; Morbioli, Mazzu-Nascimento, Milan, Stockton, & Carrilho, 2017; Teengam et al., 2017).

Paper-based colorimetric immunoassays are good candidates for low-cost diagnostic instruments in resource-limited areas, but the need of compact desktop scanners and sensitive cameras hinders their use. The use of smartphone-built high-resolution cameras to easily detect analytes using limited instrumentations was encouraged by these constraints. With a paper-plastic microfluid hybrid system combined with a single channel with a micropump, the time dependent deviation of the colorimetric analysis attributable to an unregulated reaction volume can be overcome. The predefined volume of reaction chamber (urine in this case), reacts by a variety of paper dependent reagent test pads inserted in the plastic microchannels, is filled in with a finger-pressed PDMS pump. In order to dispense the light reliably on the sample and increase system precision, an imaging box with a PDMS light diffuser was used. A smartphone camera senses the colorimetric adjustment of the reagent pads after the analyses response. Exotic optics are also being used in the production of compact microfluidic PoC systems in contained plasmonic fields of nanoscale. This interest is motivated by the high sensitivity of surface plasm resonances to minor changes in the surrounding dielectrical properties in addition to CMOS (complementary plasm plane) integrability of plasmon nanostructures for chip-scale equipment. The chemical reactors close the metal-dielectric interface can then be label-free and detected in real time. In order to calculate the optical phase shift, patients' blood samples as small as 3 μL are needed to conduct a fast one-step study of the analysis method where an light emitting diode (LED) source overlaps the extraordinary transmission peak of the nanohole array. Lastly, light passes by two polarizers P1 and P2, while the SP1 and SP2 are two Savart plates (Kim et al., 2011; Martín et al., 2017; Martinez et al., 2008; Mejía-Salazar et al., 2020).

The United States Food and Drug Administration (FDA) recently approved the first microfluidic POC method to assess the PSA for prostate cancer diagnosis within less than 15 min. The PSA test kit is called Sangia (Silver Amplified NeoGold ImmunoAssay) (OPKO Health Inc.). It requires a microfluidic immunoassay assembled for a collector of samples. In the sampler and in the microfluidic region, NeoGold-labeled monoclonal anti-PSA antibodies are used to generate sandwiches of antibody-antigen-antibody, which are eventually used with silver amplifying reagents to produce a silver metal film. A microfluidic device was manufactured to diagnose papillomavirus (HPV16) which is a significant etiological factor in head and neck cancers. The unit consisted of gold inter-digited electrodes on glass BK7 made with photolithographs. Chitosan (CHT) and chondroitin sulfate (CS) were used with gold electrodes in a layer by layer (LbL) arrangement. The capability curves of the complementary ssHPV16 at 22°C were observed by heavy interactions with the probe chain (cpHPV16) (Meyer Alexa & Gorin Michael, 2019; Soares et al., 2018).

Microfluidic electronic tongues (e-tongues), i.e., an electronic system that imitates the biological recognition of human tongue papillae to compare the flavors, are another approach used in microfluidic POC systems. Such instruments are rendered as ranges of sensing devices that can be used for the identification/distinction of samples using multisensory units to create "fingerprints" from complex liquid samples. This definition could be more effective for simultaneous multiple sensory devices in comparison to the high precision of other methods in which pure or ultrapure samples are appropriate. The proper assignment of the huge volume of data to separate patterns of identification is accomplished by integrating mathematical and numerical methods. This integration is a promising solution in future computer-aided diagnostics eHealth applications, in which machine learning, big data, and IoT technology can converge on biosensing capability. In particular, e-tongues are analyzed by working nanolayers with biomolecules for biosensing applications to verify the specificity of the interactions with the analyte tested. The above will be used for substantial numbers of the biological details of the patient. Data is used to feed and continuously transform machine learning algorithms into information (through big data and machine learning methods). This method enables the user to efficiently differentiate pertinent threatening conditions in real-time with online-based (IoT) tools, and/or will be able to adapt the medications that can be achieved by gathering data from the wearable biosensing system linked to the internet (Braunger, Fier, Rodrigues, Arratia, & Riul, 2020; Martucci et al., 2018; Shimizu et al., 2018).

Other biomolecules such as circulating ctDNA, microRNAs (micro-RNA), proteins, and serum microvesicles are tracked in addition to cellular approaches. The sensitivity and precise design of microfluidics for the identification of cancer biomarkers at low concentrations is especially good. In order to achieve a regular minimally invasive clinical research, microfluidics can also be developed into POC equipment at reduced cost. In patients with non-small cell lung cancer, Yung et al. showed a wireless microfluidic polymerase chain reaction (PCR) tool to identify uncommon EGFR mutations from tumor tissue and plasma. White et al. introduced an optimized microfluidic platform for high performance and accuracy of RT-qPCR. The unit comprised 300 concurrent test chambers that handled samples of 20 μL. Single-cell mutations in metastatic breast cancer cells were found. Pekin et al. have recently been shown to identify unusual Kirsten rat sarcoma (KRAS) mutations in ctDNA encapsulated in droplets quantitatively and adaptive. By reading the fluorescence from the droplet series, mutant DNA and wild-type DNA were readily identified. This technology lowered costs and made the automated PCR process simpler. Tiny, non-coding RNA molecules which regulate gene expression and can play a vital role in cancer development are microRNAs (miRNAs). miRNAs are soluble in the blood and can be seen as a possible biomarker to diagnose early cancer. In order to perform miRNA profiling in the blood of 48 women with early-stage breast cancer, Schrauder et al. used the industrial microfluidic-based array geniom biochip. In contrast with stable controls, 46 down-regulated and 13 over-regulated miRNAs were observed in patients. They observed that miR-202 had a substantial risc in carcinogenesis and may be involved (Diehl et al., 2008; Schrauder et al., 2012; Tsujiura et al., 2010; Valenti et al., 2006; Yung et al., 2009).

Thus, the advancement of microfluidic devices can be viewed in certain perspective and research is still going on, implementing and developing the POC devices using microfluidics technology. Still due to various reasons like the present clinical technologies, regulations of various patents and experimenting procedures, the global market potential somewhat suppressed the development of microfluidic POC devices and, hence, provides ample opportunities to broaden the research field. The new technology is focusing on the use as much as fewer resources with less cost involved. The sensitivity of these devices and the efficiency and reliability will always be the scope and point of interest for future research.

## 12.7 Summary

This chapter presents an overview of the microfluidic-based POC devices. The flow of fluid which is encountered in microfluidic applications is generally laminar, which is governed by certain fluid dynamics laws. The fabrication of these devices has some special procedures involved, and the materials used can range from metals to polymers. The conventional techniques discussed in this chapter are the common lithography processes and other advanced procedures discussed explicitly in material-based microfluidic devices. The material-based fabrication of microfluidic devices presents us with new advanced devices and some articles help us to understand the prospect of the devices and how it is being extensively used in microfluidics technology.

The discussion later in the chapter dealt with the microfluidic devices related to the diagnosis of cancer technology and how the researchers can fabricate newly complex, but reliable, devices. Different types of coupled devices are discussed along with its brief principle and applications. The current trends of the microfluidic POC devices are fabricating easy to use and portable out of clinic devices. The paper-based technologies are in huge demand because of its disposability and low-cost. Some advanced techniques like the LAMP integrated microfluidic devices are also discussed.

## References

Agarwal, S., Hachamovitch, R., & Menon, V. (2012). Current trial-associated outcomes with warfarin in prevention of stroke in patients with nonvalvular atrial fibrillation: A *meta*-analysis. *Archives of Internal Medicine*, *172*(8), 623–631. Available from https://doi.org/10.1001/archinternmed.2012.121.

Alberter, B., Klein, C. A., & Polzer, B. (2016). Single-cell analysis of CTCs with diagnostic precision: Opportunities and challenges for personalized medicine. *Expert Review of Molecular Diagnostics*, *16*(1), 25–38. Available from https://doi.org/10.1586/14737159.2016.1121099.

Allard, W. J., Matera, J., Miller, M. C., Repollet, M., Connelly, M. C., Rao, C., ... Terstappen, L. W. M. M. (2004). Tumor cells circulate in the peripheral blood of all major carcinomas but not in healthy subjects or patients with nonmalignant diseases. *Clinical Cancer Research*, *10*(20), 6897–6904. Available from https://doi.org/10.1158/1078-0432.CCR-04-0378.

Ascaso, F. J., & Huerva, V. (2016). Noninvasive continuous monitoring of tear glucose using glucose-sensing contact lenses. *Optometry and Vision Science*, *93*(4), 426–434. Available from https://doi.org/10.1097/OPX.0000000000000698.

Azouz, Vazquez, M., Liu, J., Marczak, S., Slouka, Z., Chang, H. C., Diamond, D., & Brabazon, D. (2014). Advances in three-dimensional rapid prototyping of microfluidic devices for biological applications. *Biomicrofluidics*, *8*(5).

Balaji, S., & Steve, T. (2015). Development and applications of portable biosensors. *Journal of Laboratory Automation*, 365–389. Available from https://doi.org/10.1177/2211068215581349.

Bange, A., Halsall, H. B., & Heineman, W. R. (2005). Microfluidic immunosensor systems. *Biosensors and Bioelectronics*, *20*(12), 2488–2503. Available from https://doi.org/10.1016/j.bios.2004.10.016.

Becker, H., & Locascio, L. E. (2002). Polymer microfluidic devices. *Talanta*, *56*(2), 267–287. Available from https://doi.org/10.1016/S0039-9140(01)00594-X.

Betancourt, T., & Brannon-Peppas, L. (2006). Micro-and nanofabrication methods in nanotechnological medical and pharmaceutical devices. *International Journal of Nanomedicine*, *1*(4), 483–495. Available from https://doi.org/10.2147/nano.2006.1.4.483.

Bidard, F. C., Peeters, D. J., Fehm, T., Nolé, F., Gisbert-Criado, R., Mavroudis, D., ... Michiels, S. (2014). Clinical validity of circulating tumour cells in patients with metastatic breast cancer: A pooled analysis of individual patient data. *The Lancet Oncology*, *15*(4), 406–414. Available from https://doi.org/10.1016/S1470-2045(14)70069-5.

Boleininger, J., Kurz, A., Reuss, V., & Sönnichsen, C. (2006). Microfluidic continuous flow synthesis of rod-shaped gold and silver nanocrystals. *Physical Chemistry Chemical Physics*, *8*(33), 3824–3827. Available from https://doi.org/10.1039/b604666e.

Brandenburg, A., Curdt, F., Sulz, G., Ebling, F., Nestler, J., Wunderlich, K., & Michel, D. (2009). Biochip readout system for point-of-care applications. *Sensors and Actuators, B: Chemical*, *139*(1), 245–251. Available from https://doi.org/10.1016/j.snb.2009.02.052.

Braunger, M. L., Fier, I., Rodrigues, V., Arratia, P. E., & Riul, A. (2020). Microfluidic mixer with automated electrode switching for sensing applications. *Chemosensors*, *8*(1). Available from https://doi.org/10.3390/chemosensors8010013.

Bruzewicz, D. A., Reches, M., & Whitesides, G. M. (2008). Low-cost printing of poly(dimethylsiloxane) barriers to define microchannels in paper. *Analytical Chemistry*, *80*(9), 3387–3392. Available from https://doi.org/10.1021/ac702605a.

Carrilho, E., Martinez, A. W., & Whitesides, G. M. (2009). Understanding wax printing: A simple micropatterning process for paper-based microfluidics. *Analytical Chemistry*, *81*(16), 7091–7095. Available from https://doi.org/10.1021/ac901071p.

Chae, J., Giachino, J. M., & Najafi, K. (2008). Fabrication and characterization of a wafer-level MEMS vacuum package with vertical feedthroughs. *Journal of Microelectromechanical Systems*, *17*(1), 193–200. Available from https://doi.org/10.1109/JMEMS.2007.910258.

Chan, M. (2014). Ebola virus disease in West Africa - No early end to the outbreak. *New England Journal of Medicine*, *371*(13), 1183–1185. Available from https://doi.org/10.1056/NEJMp1409859.

Chen, P., Chung, M. T., McHugh, W., Nidetz, R., Li, Y., Fu, J., ... Kurabayashi, K. (2015). Multiplex serum cytokine immunoassay using nanoplasmonic biosensor microarrays. *ACS Nano*, *9*(4), 4173–4181. Available from https://doi.org/10.1021/acsnano.5b00396.

Choi, S., Tripathi, A., & Singh, D. (2014). Smart nanomaterials for biomedics. *Journal of Biomedical Nanotechnology*, *10*(10), 3162–3188. Available from https://doi.org/10.1166/jbn.2014.1933.

Chung, B. G., Lee, K. H., Khademhosseini, A., & Lee, S. H. (2012). Microfluidic fabrication of microengineered hydrogels and their application in tissue engineering. *Lab on a Chip*, *12*(1), 45–59. Available from https://doi.org/10.1039/c1lc20859d.

Darwazeh, A. M. G., MacFarlane, T. W., McCuish, A., & Lamey, P. (1991). Mixed salivary glucose levels and candidal carriage in patients with diabetes mellitus. *Journal of Oral Pathology & Medicine*, *20*(6), 280–283. Available from https://doi.org/10.1111/j.1600-0714.1991.tb00928.x.

Das, R.D., Dey, S., RoyChaudhuri, C., & Das, S. (2012). Functionalised silicon microchannel immunosensor with portable electronic readout for bacteria detection in blood. *Proceedings - ISPTS-1, 1st international symposium on physics and technology of sensors* (. 278–281). https://doi.org/10.1109/ISPTS.2012.6260946

Dehennis, A., Mortellaro, M. A., & Ioacara, S. (2015). Multisite study of an implanted continuous glucose sensor over 90 days in patients with diabetes mellitus. *Journal of Diabetes Science and Technology*, *9*(5), 951–956. Available from https://doi.org/10.1177/1932296815596760.

Dehghanzadeh, R., Jadidi-Niaragh, F., Gharibi, T., & Yousefi, M. (2015). MicroRNA-induced drug resistance in gastric cancer. *Biomedicine and Pharmacotherapy*, *74*, 191–199. Available from https://doi.org/10.1016/j.biopha.2015.08.009.

Diehl, F., Schmidt, K., Choti, M. A., Romans, K., Goodman, S., Li, M., ... Diaz, L. A. (2008). Circulating mutant DNA to assess tumor dynamics. *Nature Medicine*, *14*(9), 985–990. Available from https://doi.org/10.1038/nm.1789.

Dixit, C. K., & Kaushik, A. (2016). *Microfluidics for biologists: Fundamentals and applications. In. Microfuidics for biologists: Fundamentals and applications* (pp. 1–252). Springer International Publishing, https://doi.org/10.1007/978-3-319-40036-5.

Du, D., Zou, Z., Shin, Y., Wang, J., Wu, H., Engelhard, M. H., ... Lin, Y. (2010). Sensitive immunosensor for cancer biomarker based on dual signal amplification strategy of graphene sheets and multienzyme functionalized carbon nanospheres. *Analytical Chemistry*, *82*(7), 2989–2995. Available from https://doi.org/10.1021/ac100036p.

Duraiswamy, S., & Khan, S. A. (2009). Droplet-based microfluidic synthesis of anisotropic metal nanocrystals. *Small*, *5*(24), 2828–2834. Available from https://doi.org/10.1002/smll.200901453.

Edel, J. B., Fortt, R., DeMello, J. C., & DeMello, A. J. (2002). Microfluidic routes to the controlled production of nanoparticles. *Chemical Communications*, *2*(10), 1136–1137. Available from https://doi.org/10.1039/b202998g.

El-Ali, J., Sorger, P. K., & Jensen, K. F. (2006). Cells on chips. *Nature*, *442*(7101), 403–411. Available from https://doi.org/10.1038/nature05063.

Elsherif, M., Hassan, M. U., Yetisen, A. K., & Butt, H. (2018). Wearable contact lens biosensors for continuous glucose monitoring using smartphones. *ACS Nano*, *12*(6), 5452–5462. Available from https://doi.org/10.1021/acsnano.8b00829.

Esashi, M., Shoji, S., & Nakano, A. (1989). Normally closed microvalve and mircopump fabricated on a silicon wafer. *Sensors and Actuators*, *20*(1–2), 163–169. Available from https://doi.org/10.1016/0250-6874(89)87114-8.

Fakunle, E. S., & Fritsch, I. (2010). Low-temperature co-fired ceramic microchannels with individually addressable screen-printed gold electrodes on four walls for self-contained electrochemical immunoassays. *Analytical and Bioanalytical Chemistry*, *398*(6), 2605–2615. Available from https://doi.org/10.1007/s00216-010-4098-5.

Fernández-Baldo, M. A., Ortega, F. G., Pereira, S. V., Bertolino, F. A., Serrano, M. J., Lorente, J. A., ... Messina, G. A. (2016). Nanostructured platform integrated into a microfluidic immunosensor coupled to laser-induced fluorescence for the epithelial cancer biomarker determination. *Microchemical Journal*, *128*, 18–25. Available from https://doi.org/10.1016/j.microc.2016.03.012.

Fu, E., & Downs, C. (2017). Progress in the development and integration of fluid flow control tools in paper microfluidics. *Lab on a Chip*, *17*(4), 614–628. Available from https://doi.org/10.1039/c6lc01451h.

G., S. M., Reiner, S., Rüdiger, S.-W., L., S. P., Laura, K., R., L. C., ... D., H. J. (2012). Circulating micro-RNAs as potential blood-based markers for early stage breast cancer detection. *PLoS ONE*, e29770. Available from https://doi.org/10.1371/journal.pone.0029770.

Gao, A., Lu, N., Dai, P., Fan, C., Wang, Y., & Li, T. (2014). Direct ultrasensitive electrical detection of prostate cancer biomarkers with CMOS-compatible n- and p-type silicon nanowire sensor arrays. *Nanoscale*, *6*(21), 13036–13042. Available from https://doi.org/10.1039/c4nr03210a.

Ge, S., Ge, L., Yan, M., Song, X., Yu, J., & Liu, S. (2013). A disposable immunosensor device for point-of-care test of tumor marker based on copper-mediated amplification. *Biosensors and Bioelectronics*, *43*(1), 425–431. Available from https://doi.org/10.1016/j.bios.2012.12.047.

Gerlinger, M., Rowan, A. J., Horswell, S., Larkin, J., Endesfelder, D., Gronroos, E., ... Swanton, C. (2012). Intratumor heterogeneity and branched evolution revealed by multiregion sequencing. *New England Journal of Medicine*, *366*(10), 883–892. Available from https://doi.org/10.1056/NEJMoa1113205.

Ghazani, A. A., Castro, C. M., Gorbatov, R., Lee, H., & Weissleder, R. (2012). Sensitive and direct detection of circulating tumor cells by multimarker μ-nuclear magnetic resonance. *Neoplasia (United States)*, *14*(5), 388–395. Available from https://doi.org/10.1596/neo.12696.

Ghazani, A. A., McDermott, S., Pectasides, M., Sebas, M., Mino-Kenudson, M., Lee, H., ... Castro, C. M. (2013). Comparison of select cancer biomarkers in human circulating and bulk tumor cells using magnetic nanoparticles and a miniaturized micro-NMR system. *Nanomedicine: Nanotechnology, Biology, and Medicine*, *9*(7), 1009–1017. Available from https://doi.org/10.1016/j.nano.2013.03.011.

Goddard, J. M., & Erickson, D. (2009). Bioconjugation techniques for microfluidic biosensors. *Analytical and Bioanalytical Chemistry*, *394*(2), 469–479. Available from https://doi.org/10.1007/s00216-009-2731-y.

Gomez, L., Sebastian, V., Irusta, S., Ibarra, A., Arruebo, M., & Santamaria, J. (2014). Scaled-up production of plasmonic nanoparticles using microfluidics: From metal precursors to functionalized and sterilized nanoparticles. *Lab on a Chip*, *14*(2), 325–332. Available from https://doi.org/10.1039/c3lc50999k.

Grego, S., Gilchrist, K. H., Carlson, J. B., & Stoner, B. R. (2012). A compact and multichannel optical biosensor based on a wavelength interrogated input grating coupler. *Sensors and Actuators, B: Chemical*, *161*(1), 721–727. Available from https://doi.org/10.1016/j.snb.2011.11.020.

Grosso, P., Carrara, S., Stagni, C., & Benini, L. (2010). Cancer marker detection in human serum with a point-of-care low-cost system. *Sensors and Actuators, B: Chemical*, *147*(2), 475–480. Available from https://doi.org/10.1016/j.snb.2010.04.001.

Gulland, A. (2014). First Ebola treatment is approved by WHO. *BMJ: British Medical Journal*.

Han, D., Kim, Y. R., Kang, C. M., & Chung, T. D. (2014). Electrochemical signal amplification for immunosensor based on 3D interdigitated array electrodes. *Analytical Chemistry*, *86*(12), 5991–5998. Available from https://doi.org/10.1021/ac501120y.

Hanash, S. M., Pitteri, S. J., & Faca, V. M. (2008). Mining the plasma proteome for cancer biomarkers. *Nature*, *452*(7187), 571–579. Available from https://doi.org/10.1038/nature06916.

Hao, N., Nie, Y., & Zhang, J. X. J. (2018). Microfluidic synthesis of functional inorganic micro-/nanoparticles and applications in biomedical engineering. *International Materials Reviews*, *63*(8), 461–487. Available from https://doi.org/10.1080/09506608.2018.1434452.

Hood, L., & Flores, M. (2012). A personal view on systems medicine and the emergence of proactive P4 medicine: Predictive, preventive, personalized and participatory. *New Biotechnology*, *29*(6), 613–624. Available from https://doi.org/10.1016/j.nbt.2012.03.004.

Hood, L., & Friend, S. H. (2011). Predictive, personalized, preventive, participatory (P4) cancer medicine. *Nature Reviews Clinical Oncology*, *8*(3), 184–187. Available from https://doi.org/10.1038/nrclinonc.2010.227.

Hu, S., Zeng, S., Zhang, B., Yang, C., Song, P., Hang Danny, T. J., ... Yong, K. T. (2014). Preparation of biofunctionalized quantum dots using microfluidic chips for bioimaging. *Analyst*, *139*(18), 4681–4690. Available from https://doi.org/10.1039/c4an00773e.

Hu, S., Zhang, S., Hu, Z., Xing, Z., & Zhang, X. (2007). Detection of multiple proteins on one spot by laser ablation inductively coupled plasma mass spectrometry and application to immuno-microarray with element-tagged antibodies. *Analytical Chemistry*, *79*(3), 923–929. Available from https://doi.org/10.1021/ac061269p.

Huang, C. P., Lu, J., Seon, H., Lee, A. P., Flanagan, L. A., Kim, H. Y., ... Jeon, N. L. (2009). Engineering microscale cellular niches for three-dimensional multicellular co-cultures. *Lab on a Chip*, *9*(12), 1740–1748. Available from https://doi.org/10.1039/b818401a.

Hung, L. H., Choi, K. M., Tseng, W. Y., Tan, Y. C., Shea, K. J., & Lee, A. P. (2006). Alternating droplet generation and controlled dynamic droplet fusion in microfluidic device for CdS nanoparticle synthesis. *Lab on a Chip*, *6*(2), 174–178. Available from https://doi.org/10.1039/b513908b.

Iliescu, C., Taylor, H., Avram, M., Miao, J., & Franssila, S. (2012). A practical guide for the fabrication of microfluidic devices using glass and silicon. *Biomicrofluidics*, *6*(1).

Irenge, L. M., & Gala, J. L. (2012). Rapid detection methods for Bacillus anthracis in environmental samples: A review. *Applied Microbiology and Biotechnology*, *93*(4), 1411–1422. Available from https://doi.org/10.1007/s00253-011-3845-7.

James, C. D., Okandan, M., Mani, S. S., Galambos, P. C., & Shul, R. (2006). Monolithic surface micromachined fluidic devices for dielectrophoretic preconcentration and routing of particles. *Journal of Micromechanics and Microengineering*, *16*(10), 1909–1918. Available from https://doi.org/10.1088/0960-1317/16/10/001.

Javaid, M. A., Ahmed, A. S., Durand, R., & Tran, S. D. (2016). Saliva as a diagnostic tool for oral and systemic diseases. *Journal of Oral Biology and Craniofacial Research*, *6*(1), 67–76. Available from https://doi.org/10.1016/j.jobcr.2015.08.006.

Jenison, R., La, H., Haeberli, A., Ostroff, R., & Polisky, B. (2001). Silicon-based biosensors for rapid detection of protein or nucleic acid targets. *Clinical chemistry*, *47*(Issue 10), 1894–1900, American Association for Clinical Chemistry Inc. Available from https://doi.org/10.1093/clinchem/47.10.1894.

Jeong, W., Kim, J., Kim, S., Lee, S., Mensing, G., & Beebe, D. J. (2004). Hydrodynamic microfabrication via \on the fly\ photopolymerization of microscale fibers and tubes. *Lab on a Chip*, *4*(6), 576–580. Available from https://doi.org/10.1039/b411249k.

Ji, X., Lau, H. Y., Ren, X., Peng, B., Zhai, P., Feng, S. P., & Chan, P. K. L. (2016). Highly Sensitive Metabolite Biosensor Based on Organic Electrochemical Transistor Integrated with Microfluidic Channel and Poly(N-vinyl-2-pyrrolidone)-Capped Platinum Nanoparticles. *Advanced Materials Technologies*, *1*(5). Available from https://doi.org/10.1002/admt.201600042.

Jokerst, J. V., Raamanathan, A., Christodoulides, N., Floriano, P. N., Pollard, A. A., Simmons, G. W., ... McDevitt, J. T. (2009). Nano-bio-chips for high performance multiplexed protein detection: Determinations of cancer biomarkers in serum and saliva using quantum dot bioconjugate labels. *Biosensors and Bioelectronics*, *24*(12), 3622–3629. Available from https://doi.org/10.1016/j.bios.2009.05.026.

Kim, D. H., Lu, N., Ma, R., Kim, Y. S., Kim, R. H., Wang, S., ... Rogers, J. A. (2011). Epidermal electronics. *Science*, *333*(6044), 838–843. Available from https://doi.org/10.1126/science.1206157.

Kingsmore, S. F. (2006). Multiplexed protein measurement: Technologies and applications of protein and antibody arrays. *Nature Reviews Drug Discovery*, *5*(4), 310–320. Available from https://doi.org/10.1038/nrd2006.

Koh, A., Kang, D., Xue, Y., Lee, S., Pielak, R. M., Kim, J., ... Manco, M. C. (2016). A soft, wearable microfluidic device for the capture, storage, and colorimetric sensing of sweat. *Science Translational Medicine*, *8*, 366.

Konvalina, G., & Haick, H. (2014). Sensors for breath testing: From nanomaterials to comprehensive disease detection. *Accounts of Chemical Research*, *47*(1), 66–76. Available from https://doi.org/10.1021/ar400070m.

Kricka, L. J. (2003). Clinical applications of chemiluminescence. *Analytica chimica acta*, *500*(Issues 1–2), 279–286, Elsevier. Available from https://doi.org/10.1016/S0003-2670(03)00809-2.

Kumar, S., Kumar, S., Tiwari, S., Srivastava, S., Srivastava, M., Yadav, B. K., ... Malhotra, B. D. (2015). Biofunctionalized nanostructured zirconia for biomedical application: A smart approach for oral cancer detection. *Advanced Science*, *2*(8). Available from https://doi.org/10.1002/advs.201500048.

Kumar, S., Sharma, J. G., Maji, S., & Malhotra, B. D. (2016). A biocompatible serine functionalized nanostructured zirconia based biosensing platform for non-invasive oral cancer detection. *RSC Advances*, *6*(80), 77037–77046. Available from https://doi.org/10.1039/c6ra07392a.

Li, N., Tourovskaia, A., & Folch, A. (2003). Biology on a chip: Microfabrication for studying the behavior of cultured cells. *Critical Reviews in Biomedical Engineering*, *31*(5–6), 423–488. Available from https://doi.org/10.1615/CritRevBiomedEng.v31.i56.20.

Li, X., Zhao, C., & Liu, X. (2015). A paper-based microfluidic biosensor integrating zinc oxide nanowires for electrochemical glucose detection. *Microsystems and Nanoengineering*, *1*. Available from https://doi.org/10.1038/micronano.2015.14.

Li, Z., Xu, Y., Fang, W., Tong, L., & Zhang, L. (2015). Ultra-sensitive nanofiber fluorescence detection in a microfluidic chip. *Sensors (Switzerland)*, *15*(3), 4890–5898. Available from https://doi.org/10.3390/s150304890.

Lin, X. Z., Terepka, A. D., & Yang, H. (2004). Synthesis of silver nanoparticles in a continuous flow tubular microreactor. *Nano Letters*, *4*(11), 2227–2232. Available from https://doi.org/10.1021/nl0485859.

Liu, Q., Wu, C., Cai, H., Hu, N., Zhou, J., & Wang, P. (2014). Cell-based biosensors and their application in biomedicine. *Chemical Reviews*, *114*(12), 6423–6461. Available from https://doi.org/10.1021/cr2003129.

Lotan, Y., & Roehrborn, C. G. (2003). Sensitivity and specificity of commonly available bladder tumor markers vs cytology: Results of a comprehensive literature review and *meta*-analyses. *Urology*, *61*(1), 109–118. Available from https://doi.org/10.1016/S0090-4295(02)02136-2.

Lu, N., Gao, A., Dai, P., Song, S., Fan, C., Wang, Y., & Li, T. (2014). CMOS-compatible silicon nanowire field-effect transistors for ultrasensitive and label-free microRNAs sensing. *Small*, *10*(10), 2022–2028. Available from https://doi.org/10.1002/smll.201302990.

Luo, J., Luo, J., Wang, L., Shi, X., Yin, J., Crew, E., ... Zhong, C. J. (2012). Nanoparticle-structured thin film sensor arrays for breath sensing. *Sensors and Actuators, B: Chemical*, *161*(1), 845–854. Available from https://doi.org/10.1016/j.snb.2011.11.045.

Luppa, P. B., Müller, C., Schlichtiger, A., & Schlebusch, H. (2011). Point-of-care testing (POCT): Current techniques and future perspectives. *TrAC - Trends in Analytical Chemistry*, *30*(6), 887–898. Available from https://doi.org/10.1016/j.trac.2011.01.019.

Lyapandra, A., & Karpinskyy, M. (2004). Electronic devices for chemiluminescence assay. In modern problems of radio engineering, Telecommunications and Computer Science. In Proceedings of the international conference TCSET'2004 (p. 510).

Manne, U., Srivastava, R. G., & Srivastava, S. (2005). Keynote review: Recent advances in biomarkers for cancer diagnosis and treatment. *Drug Discovery Today*, *10*(14), 965–976. Available from https://doi.org/10.1016/S1359-6446(05)03487-2.

Mao, X., Du, T. E., Wang, Y., & Meng, L. (2015). Disposable dry-reagent cotton thread-based point-of-care diagnosis devices for protein and nucleic acid test. *Biosensors and Bioelectronics*, *65*, 390–396. Available from https://doi.org/10.1016/j.bios.2014.10.053.

Mark, D., Haeberle, S., Roth, G., Von Stetten, F., & Zengerle, R. (2010). Microfluidic lab-on-a-chip platforms: Requirements, characteristics and applications. *NATO Science for Peace and Security Series A: Chemistry and Biology*, 305–376. Available from https://doi.org/10.1007/978-90-481-9029-4_17.

Martín, A., Kim, J., Kurniawan, J. F., Sempionatto, J. R., Moreto, J. R., Tang, G., ... Wang, J. (2017). Epidermal microfluidic electrochemical detection system: Enhanced sweat sampling and metabolite detection. *ACS Sensors*, *2*(12), 1860–1868. Available from https://doi.org/10.1021/acssensors.7b00729.

Martinez, A. W., Phillips, S. T., Carrilho, E., Thomas, S. W., Sindi, H., & Whitesides, G. M. (2008). Simple telemedicine for developing regions: Camera phones and paper-based microfluidic devices for real-time, off-site diagnosis. *Analytical Chemistry*, *80*(10), 3699–3707. Available from https://doi.org/10.1021/ac800112r.

Martucci, D. H., Todão, F. R., Shimizu, F. M., Fukudome, T. M., Schwarz, S. D. F., Carrilho, E., ... Lima, R. S. (2018). Auxiliary electrode oxidation for naked-eye electrochemical determinations in microfluidics: Towards on-the-spot applications. *Electrochimica Acta*, *292*, 125–135. Available from https://doi.org/10.1016/j.electacta.2018.08.133.

Matharu, Z., Patel, D., Gao, Y., Haque, A., Zhou, Q., & Revzin, A. (2014). Detecting transforming growth factor-β release from liver cells using an aptasensor integrated with microfluidics. *Analytical Chemistry*, *86*(17), 8865–8872. Available from https://doi.org/10.1021/ac502383e.

Mazurczyk, R., & Mansfield, C. D. (2013). Introduction to glass microstructuring techniques. *Methods in Molecular Biology*, *949*, 125–140. Available from https://doi.org/10.1007/978-1-62703-134-9_9.

Mejía-Salazar, J. R., Cruz, K. R., Vásques, E. M. M., & de Oliveira, O. N. (2020). Microfluidic point-of-care devices: New trends and future prospects for ehealth diagnostics. *Sensors (Switzerland)*, *20*(7). Available from https://doi.org/10.3390/s20071951.

Meyer Alexa, R., & Gorin Michael, A. (2019). First point-of-care PSA test for prostate cancer detection. *Nature Reviews Urology*, 331–332. Available from https://doi.org/10.1038/s41585-019-0179-1.

Mirasoli, M., Bonvicini, F., Dolci, L. S., Zangheri, M., Gallinella, G., & Roda, A. (2013). Portable chemiluminescence multiplex biosensor for quantitative detection of three B19 DNA genotypes. *Analytical and Bioanalytical Chemistry*, *405*(2–3), 1139–1143. Available from https://doi.org/10.1007/s00216-012-6573-7.

Morbioli, G. G., Mazzu-Nascimento, T., Milan, L. A., Stockton, A. M., & Carrilho, E. (2017). Improving sample distribution homogeneity in three-dimensional microfluidic paper-based analytical devices by rational device design. *Analytical Chemistry*, *89*(9), 4786–4792. Available from https://doi.org/10.1021/acs.analchem.6b04953.

Mousa, N. A., Jebrail, M. J., Yang, H., Abdelgawad, M., Metalnikov, P., Chen, J., ... Casper, R. F. (2009). Droplet-scale estrogen assays in breast tissue, blood, and serum. *Science Translational Medicine*, 1ra2–1ra2. Available from https://doi.org/10.1126/scitranslmed.3000105.

Nakamura, H., Yamaguchi, Y., Miyazaki, M., Maeda, H., Uehara, M., & Mulvaney, P. (2002). Preparation of CdSe nanocrystals in a micro-flow-reactor. *Chemical Communications*, *23*, 2844–2845. Available from https://doi.org/10.1039/b208992k.

Nge, P. N., Rogers, C. I., & Woolley, A. T. (2013). Advances in microfluidic materials, functions, integration, and applications. *Chemical Reviews*, *113*(4), 2550–2583. Available from https://doi.org/10.1021/cr300337x.

Niedl, R. R., & Beta, C. (2015). Hydrogel-driven paper-based microfluidics. *Lab on a Chip*, *15*(11), 2452–2459. Available from https://doi.org/10.1039/c5lc00276a.

Oh, K. W., & Ahn, C. H. (2006). A review of microvalves. *Journal of Micromechanics and Microengineering*, *16*(5).

Owens, C. E., & Hart, A. J. (2018). High-precision modular microfluidics by micromilling of interlocking injection-molded blocks. *Lab on a Chip*, *18* (6), 890–901. Available from https://doi.org/10.1039/c7lc00951h.

Pandey, C. M., Augustine, S., Kumar, S., Kumar, S., Nara, S., Srivastava, S., & Malhotra, B. D. (2018). Microfluidics based point-of-care diagnostics. *Biotechnology Journal*, *13*(1). Available from https://doi.org/10.1002/biot.201700047.

Parisi, J., Su, L., & Lei, Y. (2013). In situ synthesis of silver nanoparticle decorated vertical nanowalls in a microfluidic device for ultrasensitive in-channel SERS sensing. *Lab on a Chip*, *13*(8), 1501–1508. Available from https://doi.org/10.1039/c3lc41249k.

Peng, G., Tisch, U., Adams, O., Hakim, M., Shehada, N., & Broza, Y. Y. (2009). *Nature Nanotechnology*, *4*.

Pollock, N. R., Colby, D., & Rolland, J. P. (2013). A point-of-care paper-based fingerstick transaminase test: Toward low-cost \lab-on-a-chip\ technology for the developing world. *Clinical Gastroenterology and Hepatology*, *11*(5), 478–482. Available from https://doi.org/10.1016/j.cgh.2013.02.022.

Prasad, N., Perumal, J., Choi, C. H., Lee, C. S., & Kim, D. P. (2009). Generation of monodisperse inorganic-organic janus microspheres in a microfluidic device. *Advanced Functional Materials*, *19*(10), 1656–1662. Available from https://doi.org/10.1002/adfm.200801181.

Primiceri, E., Chiriacò, M. S., Notarangelo, F. M., Crocamo, A., Ardissino, D., Cereda, M., ... Maruccio, G. (2018). Key enabling technologies for point-of-care diagnostics. *Sensors (Switzerland)*, *18*(11). Available from https://doi.org/10.3390/s18113607.

Quesada-González, D., & Merkoçi, A. (2015). Nanoparticle-based lateral flow biosensors. *Biosensors and Bioelectronics*, *73*, 47–63. Available from https://doi.org/10.1016/j.bios.2015.05.050.

Raamanathan, A., Simmons, G. W., Christodoulides, N., Floriano, P. N., Furmaga, W. B., Redding, S. W., ... McDevitt, J. T. (2012). Programmable bio-nano-chip systems for serum CA125 quantification: Toward ovarian cancer diagnostics at the point-of-care. *Cancer Prevention Research*, *5* (5), 706–716. Available from https://doi.org/10.1158/1940-6207.CAPR-11-0508.

Rackus, D. G., Shamsi, M. H., & Wheeler, A. R. (2015). Electrochemistry, biosensors and microfluidics: a convergence of fields. *Chemical Society Reviews*, *44*(15), 5320–5340. Available from https://doi.org/10.1039/c4cs00369a.

Rana, A., Zhang, Y., & Esfandiari, L. (2018). Advancements in microfluidic technologies for isolation and early detection of circulating cancer-related biomarkers. *Analyst*, *143*(13), 2971–2991. Available from https://doi.org/10.1039/c7an01965c.

Reches, M., Mirica, K. A., Dasgupta, R., Dickey, M. D., Butte, M. J., & Whitesides, G. M. (2010). Thread as a matrix for biomedical assays. *ACS Applied Materials and Interfaces*, *2*(6), 1722–1728. Available from https://doi.org/10.1021/am1002266.

Ren, K., Zhou, J., & Wu, H. (2013). Materials for microfluidic chip fabrication. *Accounts of Chemical Research*, *46*(11), 2396–2406. Available from https://doi.org/10.1021/ar300314s.

Renault, C., Koehne, J., Ricco, A. J., & Crooks, R. M. (2014). Three-dimensional wax patterning of paper fluidic devices. *Langmuir*, *30*(23), 7030–7036. Available from https://doi.org/10.1021/la501212b.

Ries, L. A. G., Reichman, M. E., Lewis, D. R., Hankey, B. F., & Edwards, B. K. (2003). Cancer survival and incidence from, the surveillance, epidemiology, and end results (SEER) program. *Oncologist*, *8*(6), 541–552. Available from https://doi.org/10.1634/theoncologist.8-6-541.

Rose, D. P., Ratterman, M. E., Griffin, D. K., Hou, L., Kelley-Loughnane, N., Naik, R. R., ... Heikenfeld, J. C. (2015). Adhesive RFID sensor patch for monitoring of sweat electrolytes. *IEEE Transactions on Biomedical Engineering*, *62*(6), 1457–1465. Available from https://doi.org/10.1109/TBME.2014.2369991.

Ruddon, R.W. (2007). Cancer biology.

Rusling, J. F., Kumar, C. V., Gutkind, J. S., & Patel, V. (2010). Measurement of biomarker proteins for point-of-care early detection and monitoring of cancer. *Analyst*, *135*(10), 2496–2511. Available from https://doi.org/10.1039/c0an00204f.

Sadik, O. A., Mwilu, S. K., & Aluoch, A. (2010). Smart electrochemical biosensors: From advanced materials to ultrasensitive devices. *Electrochimica Acta*, *55*(14), 4287–4295. Available from https://doi.org/10.1016/j.electacta.2009.03.008.

Safavieh, R., Zhou, G. Z., & Juncker, D. (2011). Microfluidics made of yarns and knots: From fundamental properties to simple networks and operations. *Lab on a Chip*, *11*(15), 2618–2624. Available from https://doi.org/10.1039/c1lc20336c.

Salvatore, G. A., Sülzle, J., Dalla Valle, F., Cantarella, G., Robotti, F., Jokic, P., ... Tröster, G. (2017). Biodegradable and highly deformable temperature sensors for the internet of things. *Advanced Functional Materials*, *27*(35). Available from https://doi.org/10.1002/adfm.201702390.

Sandbhor Gaikwad, P., & Banerjee, R. (2018). Advances in point-of-care diagnostic devices in cancers. *Analyst*, *143*(6), 1326–1348. Available from https://doi.org/10.1039/c7an01771e.

Santiago, J. G., & Laser, D. J. (2004). A review of micropumps. *Journal of Micromechanics and Microengineering*, *14*.

Schöning, M. J., & Lüth, H. (2001). Novel concepts for silicon-based biosensors. *Physica Status Solidi (a)*, 65–77, https://doi.org/10.1002/1521-396x(200105)185:1 < 65::aid-pssa65 > 3.0.co;2-y.

Schrauder, Michael G., Strick, Reiner, Schulz-Wendtland, Rüdiger, Strissel, Pamela L., Kahmann, Laura, Loehberg, Christian R., ... Fasching, Peter A. (2012). Circulating micro-RNAs as potential blood-based markers for early stage breast cancer detection. *PLoS One*, *7*(1), e29770. Available from https://doi.org/10.1371/journal.pone.0029770.

Sebastian Cabeza, V., Kuhn, S., Kulkarni, A. A., & Jensen, K. F. (2012). Size-controlled flow synthesis of gold nanoparticles using a segmented flow microfluidic platform. *Langmuir*, *28*(17), 7007–7013. Available from https://doi.org/10.1021/la205131e.

Semmes, O. J., Feng, Z., Adam, B. L., Banez, L. L., Bigbee, W. L., Campos, D., ... Zhu, L. (2005). Evaluation of serum protein profiling by surface-enhanced laser desorption/ionization time-of-flight mass spectrometry for the detection of prostate cancer: I. Assessment of platform reproducibility. *Clinical Chemistry*, *51*(1), 102–112. Available from https://doi.org/10.1373/clinchem.2004.038950.

Shaikh, Sohail, & Pajankar, Shreyashi (2019). *Cancer Therapeutics Market by Application (Blood Cancer, Lung Cancer, Colorectal Cancer, Prostate Cancer, Breast Cancer, Cervical Cancer, Head & Neck Cancer, Glioblastoma, Malignant Meningioma, Mesothelioma, Melanoma, and Others) and Top Selling Drugs (Revlimid, Avastin, Herceptin, Rituxan, Opdivo, Gleevec, Velcade, Imbruvica, Ibrance, Zytiga, Alimta, Xtandi, Tarceva, Perjeta, Temodar, and Others): Global Opportunity Analysis and Industry Forecast, 2019–2026*. Allied Market Research.

Shangguan, J. W., Liu, Y., Pan, J. B., Xu, B. Y., Xu, J. J., & Chen, H. Y. (2017). Microfluidic PDMS on paper (POP) devices. *Lab on a Chip*, *17*(1), 120–127. Available from https://doi.org/10.1039/c6lc01250g.

Shaurya, P., Marie, P., & Bharat, B. (2012). Theory, fabrication and applications of microfluidic and nanofluidic biosensors. *Philosophical Transactions of the Royal Society A: Mathematical, Physical and Engineering Sciences*, *370*(1967), 2269–2303. Available from https://doi.org/10.1098/rsta.2011.0498.

Shestopalov, I., Tice, J. D., & Ismagilov, R. F. (2004). Multi-step synthesis of nanoparticles performed on millisecond time scale in a microfluidic droplet-based system. *Lab on a Chip*, *4*(4), 316–321. Available from https://doi.org/10.1039/b403378g.

Shimizu, F. M., Pasqualeti, A. M., Todão, F. R., De Oliveira, J. F. A., Vieira, L. C. S., Gonçalves, S. P. C., ... Lima, R. S. (2018). Monitoring the surface chemistry of functionalized nanomaterials with a microfluidic electronic tongue. *ACS Sensors*, *3*(3), 716–726. Available from https://doi.org/10.1021/acssensors.8b00056.

Soares, Andrey Coatrini, Soares, Juliana Coatrini, Rodrigues, Valquiria Cruz, Follmann, Heveline Dal Magro, Arantes, Lidia Maria Rebolho Batista, Carvalho, Ana Carolina, ... Oliveira, Osvaldo N, Jr. (2018). Microfluidic-Based Genosensor To Detect Human Papillomavirus (HPV16) for Head and Neck Cancer. *ACS Appl Mater Interfaces*, *10*(43), 36757–36763. Available from https://doi.org/10.1021/acsami.8b14632.

Sommerfeld, A. (1909). A contribution to the hydrodynamical explanation of turbulent fluid motions. *Atti Del IV. Cong. Intern. Dei Matematici*, 116–124.

Sommerfeld, Arnold (1908). Ein Beitrag zur hydrodynamischen Erkläerung der turbulenten Flüssigkeitsbewegüngen (A Contribution to Hydrodynamic Explanation of Turbulent Fluid Motions)"(PDF). *International Congress of Mathematicians*, *3*, 116–124.

Steeg, P. S. (2006). Tumor metastasis: Mechanistic insights and clinical challenges. *Nature Medicine*, *12*(8), 895–904. Available from https://doi.org/10.1038/nm1469.

Stjernström, M., & Roeraade, J. (1998). Method for fabrication of microfluidic systems in glass. *Journal of Micromechanics and Microengineering*, *8*(1), 33–38. Available from https://doi.org/10.1088/0960-1317/8/1/006.

Stokes, George (1851). On the Effect of the Internal Friction of Fluids on the Motion of Pendulums. *Transactions of the Cambridge Philosophical Society*, *9*, 8–106.

Su, L., Feng, J., Zhou, X., Ren, C., Li, H., & Chen, X. (2012). Colorimetric detection of urine glucose based ZnFe2O4 magnetic nanoparticles. *Analytical Chemistry*, *84*(13), 5753–5758. Available from https://doi.org/10.1021/ac300939z.

Sung, J. H., & Shuler, M. L. (2012). Microtechnology for mimicking in vivo tissue environment. *Annals of Biomedical Engineering*, *40*(6), 1289–1300. Available from https://doi.org/10.1007/s10439-011-0491-2.

Sylvain, L., & Vincent, G. (1989). Ultraviolet laser photoablation of polymers: A review and recent results. *Laser Chemistry*, 25–40. Available from https://doi.org/10.1155/1989/18750.

Syshchyk, O., Skryshevsky, V. A., Soldatkin, O. O., & Soldatkin, A. P. (2015). Enzyme biosensor systems based on porous silicon photoluminescence for detection of glucose, urea and heavy metals. *Biosensors and Bioelectronics*, *66*, 89–94. Available from https://doi.org/10.1016/j.bios.2014.10.075.

Tadimety, A., Closson, A., Li, C., Yi, S., Shen, T., & Zhang, J. X. J. (2018). Advances in liquid biopsy on-chip for cancer management: Technologies, biomarkers, and clinical analysis. *Critical Reviews in Clinical Laboratory Sciences*, *55*(3), 140–162. Available from https://doi.org/10.1080/10408363.2018.1425976.

Tadimety, A., Syed, A., Nie, Y., Long, C. R., Kready, K. M., & Zhang, J. X. J. (2017). Liquid biopsy on chip: A paradigm shift towards the understanding of cancer metastasis. *Integrative Biology (United Kingdom)*, *9*(1), 22–49. Available from https://doi.org/10.1039/c6ib00202a.

Takahashi, S., Shiraishi, T., Miles, N., Trock, B. J., Kulkarni, P., & Getzenberg, R. H. (2015). Nanowire analysis of cancer-testis antigens as biomarkers of aggressive prostate cancer. *Urology*, *85*(3), 704-704.e7. Available from https://doi.org/10.1016/j.urology.2014.12.004.

Teengam, P., Siangproh, W., Tuantranont, A., Henry, C. S., Vilaivan, T., & Chailapakul, O. (2017). Electrochemical paper-based peptide nucleic acid biosensor for detecting human papillomavirus. *Analytica Chimica Acta*, *952*, 32–40. Available from https://doi.org/10.1016/j.ac.2016.11.071.

Tian, Q., Price, N. D., & Hood, L. (2012). Systems cancer medicine: Towards realization of predictive, preventive, personalized and participatory (P4) medicine. *Journal of Internal Medicine*, *271*(Issue 2), 111–121. Available from https://doi.org/10.1111/j.1365-2796.2011.02498.x.

Toyota, A., Nakamura, H., Ozono, H., Yamashita, K., Uehara, M., & Maeda, H. (2010). Combinatorial synthesis of CdSe nanoparticles using microreactors. *Journal of Physical Chemistry C*, *114*(17), 7527–7534. Available from https://doi.org/10.1021/jp911876s.

Tran, D. P., Wolfrum, B., Stockmann, R., Pai, J. H., Pourhassan-Moghaddam, M., Offenhäusser, A., & Thierry, B. (2015). Complementary metal oxide semiconductor compatible silicon nanowires-on-a-chip: Fabrication and preclinical validation for the detection of a cancer prognostic protein marker in serum. *Analytical Chemistry*, *87*(3), 1662–1668. Available from https://doi.org/10.1021/ac503374j.

Tsujiura, M., Ichikawa, D., Komatsu, S., Shiozaki, A., Takeshita, H., Kosuga, T., ... Otsuji, E. (2010). Circulating microRNAs in plasma of patients with gastric cancers. *British Journal of Cancer*, *102*(7), 1174–1179. Available from https://doi.org/10.1038/sj.bjc.6605608.

Tsunoyama, H., Ichikuni, N., & Tsukuda, T. (2008). Microfluidic synthesis and catalytic application of pvp-stabilized, ~1 nm gold clusters. *Langmuir*, *24*(20), 11327–11330. Available from https://doi.org/10.1021/la801372j.

Valdés-Ramírez, G., Bandodkar, A. J., Jia, W., Martinez, A. G., Julian, R., Mercier, P., & Wang, J. (2014). Non-invasive mouthguard biosensor for continuous salivary monitoring of metabolites. *Analyst*, *139*(7), 1632–1636. Available from https://doi.org/10.1039/c3an02359a.

Valenti, R., Huber, V., Filipazzi, P., Pilla, L., Sovena, G., Villa, A., ... Rivoltini, L. (2006). Human tumor-released microvesicles promote the differentiation of myeloid cells with transforming growth factor-β-mediated suppressive activity on T lymphocytes. *Cancer Research*, *66*(18), 9290–9298. Available from https://doi.org/10.1158/0008-5472.CAN-06-1819.

Vashist, S. K., Luppa, P. B., Yeo, L. Y., Ozcan, A., & Luong, J. H. T. (2015). Emerging technologies for next-generation point-of-care testing. *Trends in Biotechnology*, *33*(11), 692–705. Available from https://doi.org/10.1016/j.tibtech.2015.09.001.

Vasudev, A., Kaushik, A., Tomizawa, Y., Norena, N., & Bhansali, S. (2013). An LTCC-based microfluidic system for label-free, electrochemical detection of cortisol. *Sensors and Actuators, B: Chemical*, *182*, 139–146. Available from https://doi.org/10.1016/j.snb.2013.02.096.

Voldman, J., Gray, M. L., & Schmidt, M. A. (1999). Microfabrication in biology and medicine. *Annual Review of Biomedical Engineering*, *1*, 401–425. Available from https://doi.org/10.1146/annurev.bioeng.1.1.401.

Wagner, J., & Köhler, J. M. (2005). Continuous synthesis of gold nanoparticles in a microreactor. *Nano Letters*, *5*(4), 685–691. Available from https://doi.org/10.1021/nl050097t.

Wagner, J., Tshikhudo, T. R., & Köhler, J. M. (2008). Microfluidic generation of metal nanoparticles by borohydride reduction. *Chemical Engineering Journal*, *135*(1), S104–S109. Available from https://doi.org/10.1016/j.cej.2007.07.046.

Walker, G. M., Zeringue, H. C., & Beebe, D. J. (2004). Microenvironment design considerations for cellular scale studies. *Lab on a Chip*, *4*(2), 91–97. Available from https://doi.org/10.1039/b311214d.

Wang, Y., Qin, Z., Boulware, D. R., Pritt, B. S., Sloan, L. M., Gonzalez, I. J., ... Bischof, J. C. (2016). Thermal contrast amplification reader yielding 8⇑fold analytical improvement for disease detection with lateral flow assays. *Analytical Chemistry*, *88*(23), 11774–11782. Available from https://doi.org/10.1021/acs.analchem.6b03406.

Warren, A. D., Kwong, G. A., Wood, D. K., Lin, K. Y., & Bhatia, S. N. (2014). Point-of-care diagnostics for noncommunicable diseases using synthetic urinary biomarkers and paper microfluidics. *Proceedings of the National Academy of Sciences of the United States of America*, *111*(10), 3671–3676. Available from https://doi.org/10.1073/pnas.1314651111.

Wei, F., Patel, P., Liao, W., Chaudhry, K., Zhang, L., Arellano-Garcia, M., ... Wong, D. T. (2009). Electrochemical sensor for multiplex biomarkers detection. *Clinical Cancer Research*, *15*(13), 4446–4452. Available from https://doi.org/10.1158/1078-0432.CCR-09-0050.

Whitesides, G. M. (2006). The origins and the future of microfluidics. *Nature*, *442*(7101), 368–373. Available from https://doi.org/10.1038/nature05058.

Wu, L. J., Pan, Y. D., Pei, X. Y., Chen, H., Nguyen, S., Kashyap, A., ... Wu, J. (2012). Capturing circulating tumor cells of hepatocellular carcinoma. *Cancer Letters*, *326*(1), 17–22. Available from https://doi.org/10.1016/j.canlet.2012.07.024.

Xia, Y., & Whitesides, G. M. (1998). On the theories of the internal friction of fluids in motion, and of the equilibrium and motion of elastic solids. *Transactions of the Cambridge Philosophical Society*, *28*(1).

Xu, J., Locascio, L., Gaitan, M., & Lee, C. S. (2000). Room-temperature imprinting method for plastic microchannel fabrication. *Analytical Chemistry*, *72*(8), 1930–1933. Available from https://doi.org/10.1021/ac991216q.

Yager, P., Domingo, G. J., & Gerdes, J. (2008). Point-of-care diagnostics for global health. *Annual Review of Biomedical Engineering*, *10*, 107–144. Available from https://doi.org/10.1146/annurev.bioeng.10.061807.160524.

Yung, T. K. F., Chan, K. C. A., Mok, T. S. K., Tong, J., To, K. F., & Lo, Y. M. D. (2009). Single-molecule detection of epidermal growth factor receptor mutations in plasma by microfluidics digital PCR in non-small cell lung cancer patients. *Clinical Cancer Research*, *15*(6), 2076–2084. Available from https://doi.org/10.1158/1078-0432.CCR-08-2622.

Zhang, Z., & Nagrath, S. (2013). Microfluidics and cancer: Are we there yet? *Biomedical Microdevices*, *15*(4), 595–609. Available from https://doi.org/10.1007/s10544-012-9734-8.

Zhao, W., Al-Nasser, L. F., Shan, S., Li, J., Skeete, Z., Kang, N., … Harris, R. (2016). Detection of mixed volatile organic compounds and lung cancer breaths using chemiresistor arrays with crosslinked nanoparticle thin films. *Sensors and Actuators, B: Chemical*, *232*, 292–299. Available from https://doi.org/10.1016/j.snb.2016.03.121.

Zhou, G., Mao, X., & Juncker, D. (2012). Immunochromatographic assay on thread. *Analytical Chemistry*, *84*(18), 7736–7743. Available from https://doi.org/10.1021/ac301082d.

Zoli, M., Magalotti, D., Blanchi, G., Gueli, C., Marchesini, G., & Pisi, E. (1996). Efficacy of a surveillance program for early detection of hepatocellular carcinoma. *Cancer*, *78*(5), 977–985, https://doi.org/10.1002/(SICI)1097-0142(19960901)78:5 < 977:AID-CNCR6 > 3.0.CO;2-9 (2016). In Proof and Concepts in Rapid Diagnostic Tests and Technologies.

## Further reading

Bhakta, S. A., Evans, E., Benavidez, T. E., & Garcia, C. D. (2015). Protein adsorption onto nanomaterials for the development of biosensors and analytical devices: A review. *Analytica Chimica Acta*, *872*, 7–25. Available from https://doi.org/10.1016/j.ac.2014.10.031.

Coto-García, A. M., Sotelo-González, E., Fernández-Argüelles, M. T., Pereiro, R., Costa-Fernández, J. M., & Sanz-Medel, A. (2011). Nanoparticles as fluorescent labels for optical imaging and sensing in genomics and proteomics. *Analytical and Bioanalytical Chemistry*, *399*(1), 29–42. Available from https://doi.org/10.1007/s00216-010-4330-3.

Dey, P., Fabri-Faja, N., Calvo-Lozano, O., Terborg, R. A., Belushkin, A., Yesilkoy, F., … Lechuga, L. M. (2019). Label-free bacteria quantification in blood plasma by a bioprinted microarray based interferometric point-of-care device. *ACS Sensors*, *4*(1), 52–60. Available from https://doi.org/10.1021/acssensors.8b00789.

Ebbesen, T. W., Lezec, H. J., Ghaemi, H. F., Thio, T., & Wolff, P. A. (1998). Extraordinary optical transmission through sub-wavelenght hole arrays. *Nature*, *391*(6668), 667–669. Available from https://doi.org/10.1038/35570.

Fan, R., Vermesh, O., Srivastava, A., Yen, B. K. H., Qin, L., Ahmad, H., … Heath, J. R. (2008). Integrated barcode chips for rapid, multiplexed analysis of proteins in microliter quantities of blood. *Nature Biotechnology*, *26*(12), 1373–1378. Available from https://doi.org/10.1038/nbt.1507.

Jalal, U. M., Kim, S. C., & Shim, J. S. (2017). Histogram analysis for smartphone-based rapid hematocrit determination. *Biomedical Optics Express*, *8*(7), 3317–3328. Available from https://doi.org/10.1364/BOE.8.003317.

Peng, F., Zhang, Y., Wang, R., Zhou, W., Zhao, Z., Liang, H., … Gu, Y. (2016). Identification of differentially expressed miRNAs in individual breast cancer patient and application in personalized medicine. *Oncogenesis*, e194-e194. Available from https://doi.org/10.1038/oncsis.2016.4.

Tan, S. J., Lao, I. K., Ji, H. M., Agarwal, A., Balasubramanian, N., & Kwong, D. L. (2006). Microfluidic design for bio-sample delivery to silicon nanowire biosensor – A simulation study. *Journal of Physics: Conference Series*, 626–630. Available from https://doi.org/10.1088/1742-6596/34/1/103.

Jalal Uddin, M., Jin, G. J., & Shim, J. S. (2017). Paper-plastic hybrid microfluidic device for smartphone-based colorimetric analysis of urine. *Analytical Chemistry*, *89*(24), 13160–13166. Available from https://doi.org/10.1021/acs.analchem.7b02612.

Kim, H., Awofeso, O., Choi, S. M., Jung, Y., & Bae, E. (2017). Colorimetric analysis of saliva-alcohol test strips by smartphone-based instruments using machine-learning algorithms. *Applied Optics*, *56*(1), 84–92. Available from https://doi.org/10.1364/AO.56.000084.

Lee, K. G., Lee, T. J., Jeong, S. W., Choi, H. W., Heo, N. S., Park, J. Y., … Lee, S. J. (2012). Development of a plastic-based microfluidic immunosensor chip for detection of H1N1 influenza. *Sensors (Switzerland)*, *12*(8), 10810–10819. Available from https://doi.org/10.3390/s120810810.

Lee, S., Oncescu, V., Mancuso, M., Mehta, S., & Erickson, D. (2014). A smartphone platform for the quantification of vitamin D levels. *Lab on a Chip*, *14*(8), 1437–1442. Available from https://doi.org/10.1039/c3lc51375k.

Mao, K., Min, X., Zhang, H., Zhang, K., Cao, H., Guo, Y., & Yang, Z. (2020). Based microfluidics for rapid diagnostics and drug delivery. *Journal of Controlled Release*, *322*, 187–199.

Pansare, V. J., Hejazi, S., Faenza, W. J., & Prud'Homme, R. K. (2012). Review of long-wavelength optical and NIR imaging materials: Contrast agents, fluorophores, and multifunctional nano carriers. *Chemistry of Materials*, *24*(5), 812–827. Available from https://doi.org/10.1021/cm2028367.

Roda, A., Michelini, E., Cevenini, L., Calabria, D., Calabretta, M. M., & Simoni, P. (2014). Integrating biochemiluminescence detection on smartphones: Mobile chemistry platform for point-of-need analysis. *Analytical Chemistry*, *86*(15), 7299–7304. Available from https://doi.org/10.1021/ac502137s.

Rzhevskiy, A. S., Bazaz, S. R., Ding, L., Kapitannikova, A., Sayyadi, N., Campbell, D., … Zvyagin, A. V. (2020). Rapid and label-free isolation of tumour cells from the urine of patients with localised prostate cancer using inertial microfluidics. *Cancers*, *12*(1). Available from https://doi.org/10.3390/cancers12010081.

Soo, K. C., Petra, W.-S., Yeh-Chan, A., L., L. L.-H., Zhongping, C., & Jik, K. Y. (2009). Enhanced detection of early-stage oral cancer in vivo by optical coherence tomography using multimodal delivery of gold nanoparticles. *Journal of Biomedical Optics*, 034008. Available from https://doi.org/10.1117/1.3130323.

Wang, J. H., Wang, B., Liu, Q., Li, Q., Huang, H., Song, L., . . . Chu, P. K. (2013). Bimodal optical diagnostics of oral cancer based on Rose Bengal conjugated gold nanorod platform. *Biomaterials*, *34*(17), 4274–4283. Available from https://doi.org/10.1016/j.biomaterials.2013.02.012.

Wu, Y., Xue, P., Hui, K. M., & Kang, Y. (2014). A paper-based microfluidic electrochemical immunodevice integrated with amplification-by-polymerization for the ultrasensitive multiplexed detection of cancer biomarkers. *Biosensors and Bioelectronics*, *52*, 180–187. Available from https://doi.org/10.1016/j.bios.2013.08.039.

Yu, I. F., Yu, Y. H., Chen, L. Y., Fan, S. K., Chou, H. Y. E., & Yang, J. T. (2014). A portable microfluidic device for the rapid diagnosis of cancer metastatic potential which is programmable for temperature and CO2. *Lab on a Chip*, *14*(18), 3621–3628. Available from https://doi.org/10.1039/c4lc00502c.

Chapter 13

# Graphene-based devices for cancer diagnosis

**Fatemeh Nemati, Azam Bagheri Pebdeni and Morteza Hosseini**
*Department of Life Science Engineering, Faculty of New Sciences & Technologies, University of Tehran, Tehran, Iran*

## 13.1 Introduction

Cancer is the second leading cause of mortality around the globe with over 200 types of cancers identified with more than 1500 deaths occurring each day (Jayanthi, Das, & Saxena, 2017). One of the primary reasons for such a high mortality rate is the lack of effective early detection and diagnosis methods. The conventional techniques for cancer diagnosis rely on a variety of complex clinical settings, which include x-ray, computerized tomography (CT), magnetic resonance imaging (MRI), positron emission tomography, endoscopy, sonography, thermography, cytology, and biopsy (Cui, Zhou, & Zhou, 2020).

However, these approaches are expensive, invasive, laboratory-based, need time consuming step-based analysis, and skilled technicians. In recent years, there is a growing interest in developing biosensors as they show high sensitivity, specificity, portability, ease of use, and fast response. Biosensors are devices that allow analytical determinations through an analyte–receptor binding reaction to a transduction mechanism. The biological molecules which undergo prominent changes during cancer are recognized as biomarkers and have high clinical significance. The measurement of altered levels of specific biomarkers during the onset of cancers can be performed through the development of biosensors. In biosensors, the recognition event is amplified by the presence of a nanomaterial. Among various nanomaterials used in cancer applications, graphene and its derivatives are getting much more attention due to their process ability and unique properties (Al-Ani et al., 2017). This chapter focuses on the different platforms of graphene and its derivatives in order to understand which are the best nanomaterials for an aptamer-based, antibody-based, or enzyme-based biosensor.

## 13.2 Cancer biomarkers

A cancer biomarker is a biological molecule produced either by the tumor cells or by human tissues as an indicator of cancer progression within the body and can be measured objectively. The morphological changes occurring in tissues and cells that appear in the later stages of the disease are the basis of many diagnostic techniques. In contrast, changes at the molecular level occur before the clinical features. Therefore, cancer biomarkers are valuable tools for early diagnosis and therapeutic intervention. Cancer biomarkers are composed of various molecular origins, including proteins, nucleic acids, peptides, cytokeratins, hormones, and metabolites (Das, Sedighi, & Krull, 2018).

A summary of several important biomarkers associated with different cancers is listed in Table 13.1.

## 13.3 Graphene and its derivatives

Graphene is a 2D layered structure of hexagonal honeycomb arrangements of carbon atoms. Different layers in graphene are held together with Van der Waals forces, which makes it a soft material. Until now, several graphene synthesis methods have been developed. There are two main approaches with different methods that can be categorized: top-down approach (e.g., electrochemical, mechanical or chemical exfoliation of graphite) and bottom-up methods (e.g., chemical vapor deposition [CVD]) (Antiochia, Tortolini, Tasca, Gorton, & Bollella, 2018). In top-down

**Biosensor Based Advanced Cancer Diagnostics. DOI: https://doi.org/10.1016/B978-0-12-823424-2.00014-4**

**TABLE 13.1** Example of the FDA approved cancer biomarkers.

| Type of cancer disease | Common biomarkers |
|---|---|
| Colon | CEA, EGF, p53 |
| Ovarian | CA 125, HCG, p53, CEA, CA 549, CA 19–9, CA 15–3, MCA, MOV-1, TAG72 |
| Breast | BRCA1, BRCA2, CA 15–3, CA 125, CA 27.29, CEA, NY-BR-1, Ing-1, HER2/NEU, ER/PR |
| Liver | AFP, CEA |
| Lung | CEA, CA 19–9, SCC, NSE, NY-ESO-1 |
| Prostate | PSA, Pro2PSA |
| Thyroid | TG |
| pancreatic cancer | CA19.9 |
| Bladder | BTA, NMP22, MG, CK19 |
| Colorectal | KRAS, FDP |

*Abbreviations*: *CEA*, Carcinoembryonic antigen; *EGF*, epidermal growth factor; *p53*, protein suppressor gene; *CA*, cancer antigen; *HCG*, human chorionic gonadotropin; *BRCA1/BRCA2*, breast cancer antigen 1/2′; *HER2*, human epidermal growth factor receptor 2; *AFP*, alpha-fetoprotein; *SCC*, squamous cell carcinoma antigen; *NSE*, neuron-specific enolase; *PSA*, prostate-specific antigen; *TG*, thyroglobulin; *KRAS*, V-Ki-ras2 Kirsten rat sarcoma viral oncogene homolog; *FDP*, fibrin/fibrinogen degradation products.

approaches, interatomic van der Waals interactions between the graphene sheets are reduced by mechanical procedures. In bottom-up strategies, graphene can be produced by CVD from small molecular carbon precursors.

Among graphene-based nanomaterials, we can focus mainly on graphene oxide (GO), reduced graphene oxide (rGO) and graphene quantum dots (GQDs). GO is a chemically derived graphene produced from graphite by oxidative stripping, which exhibits a layered structure similar to graphene. Until now, there have been three well-known techniques for the synthesis of GO: Boride, Staudenmaier, and Hummers methods or slight modifications of these procedures. In all these processes, oxidation of graphite takes place at various levels (Ghosal & Sarkar, 2018) and GO is characterized to have high amounts of oxygen groups that are responsible for the strong reactive activity and good dispersibility (Dreyer, Park, Bielawski, & Ruoff, 2010). GO has attracted great attention for the development of optical sensors due to its unique properties for optical sensing: photoluminescence over a broad range of wavelengths and high efficiency as a fluorescence quencher. Chemical or thermal reduction of GO leads to rGO that presents high degree of defects on carbon $sp^2$ lattice and less oxygen contents (Fritea, Tertis, Sandulescu, & Cristea, 2018). Since rGO is electrochemically more active than any members of the graphene family, it is considered as a promising material for the development of electronic sensing (Pumera, 2010). As a new type of 0D material, GQDs are small graphene fragments with particle sizes around 3–20 nm, and favorable surface grafting using $\pi-\pi$ conjugation (Şenel, Demir, Büyükköroğlu, & Yıldız, 2019). Similar to graphene oxide, hydrophilic functional groups present in GQD such as hydroxyl and carboxylic are responsible for high solubility of GQDs in water and can enable modification with various organic, inorganic, and biological species according to requirement. Both top-down and bottom-up methods have also been developed to synthesize GQDs. The first method is based on size-reduction and breaking down of carbon materials through chemical or physical routes (hydrothermal/solvothermal, chemical oxidation, electrochemical exfoliation, oxygen plasma treatment, etc.). Top-down approaches give a large scale production of GQDs due to the early synthetic steps and the use of cheap carbon starting materials. Conversely, bottom-up methods rely on the pyrolysis or carbonization of organic precursors. However, these approaches suffer from several disadvantages, i.e., the involvement of toxic solvent, high temperature, and substrate concentrations.

## 13.4 Graphene-based nanomaterials in cancer diagnosis

In cancer, effective detection of several biomarkers and/or biomolecules, including cancer-specific metabolites, oligonucleotides, proteins and antigens, is an important objective for early diagnosis (Eskiizmir, Baskin, & Yapici, 2018). In addition, the mortality rate can be reduced through early detection of cancer and immediate treatment. Biosensors, a technique at the forefront of cancer diagnosis, are designed for the detection of specific targets by essentially converting a bio-recognition element into a measurable signal that can be detected and analyzed (Salek-Maghsoudi et al., 2018). Owing to their excellent properties, graphene-based nanomaterials play a crucial role in the biosensor field. It provides a natural biocompatible carrier to adsorb biomolecules due to π−π stacking interaction between its hexagonal cells and the ring structure present in the majority of biomolecules. Furthermore, higher concentrations of probe molecules achieved via either direct absorption or binding with other functional groups on graphene can enhance the sensitivity of the diagnosis system. The excellent electric and optical properties of graphene offers an amplifying signal for biosensors, which can lead to lower detection limits in biomarkers detection (Cruz, 2016; Gu, Tang, Xiong, & Zhou, 2019). Graphene-based nanobiosensors are highly sensitive and specific, enhance the limit of detection, and are able to detect multiple cancer biomarkers simultaneously, thereby improving early cancer diagnosis and screening.

## 13.5 Functionalization of graphene for sensing application

Functionalization of graphene and their subsidiaries are unequivocally significant to alter their qualities and expand the opportunities in central investigations as well as in various devices. This phenomenon can be countered through functionalization of graphene as it will make strong polar−polar interactions of hydrophilic groups to keep away from agglomeration of single layer graphene during reduction in solvent phase and assists with keeping up the inherent properties of graphene (Sengupta & Hussain, 2019). GO contains several oxygen-containing groups, such as carboxylic, hydroxyl, and epoxide functional groups on the carbon surface, making it more dispersed in water and organic solvents. The biological components (enzymes, DNA probes, cells and antibodies) are immobilized on the outside of the GO. The utilization of nano polymers, metallic nanoparticles, and the functionalities of GO and rGO themselves are some common approaches for biologic component immobilization (Tajik et al., 2020; Zhang et al., 2020). The edges of GQDs have functional groups such as amines, carboxyls, and hydroxyls that cause solubility in the aqueous solutions, and they are suitable for functionalizing with the organic, inorganic, or biological agents (Tuteja et al., 2016). There are two methodologies for surface functionalization of graphene sheets: (1) covalent and (2) noncovalent. The covalent approaches allow the engineering of the properties of graphene with respect to its bandgap and bio interfacing to a large extend. The reaction efficiency depends on parameters involving the number of graphene layers, the electrostatic charges, and the deformity thickness. Non-covalent modification employs hydrogen bonds and π−π interaction to adsorb polymers on the surface of graphene. It permits changes in graphene with the polymers bearing exceptional structures or functions that should be presynthesized and are hard to bond covalently. The existing methods of preparing graphene sheets include chemical exfoliation i.e., oxidation of graphite and subsequent reduction of the exfoliated graphite oxide sheets. However, it regularly utilizes unsafe chemicals (e.g., hydrazine) as reductants, other than indicating poor electronic properties of the synthesized material that make it incompatible for biosensing applications. The driving forces for noncovalent functionalization of carbon nanomaterials with organic molecules or polymers involve the π-π interaction, van der Waals force, electron donor-acceptor complexes, ionic interaction, and hydrogen bonding.

## 13.6 Graphene material-based sensors

Graphene exhibits great potential in biosensing, owing to its extraordinary properties, such as high electron transport rate, low cost, tunable optical properties, and mechanical strength. Biosensors usually consists of two elements: a biological recognition element (bioreceptor) and a transducer. A bioreceptor is an immobilized sensitive biological component (e.g., enzyme, DNA probe, and antibody) that can specifically recognize the target analytes (Parolo & Merkoçi, 2013). The division of graphene-based sensors into three classes: (1) aptamer based, (2) antibody based, and (3) enzyme based was requested due to the extraordinary large number of publications regarding cancer biomarker detection.

### 13.6.1 Graphene material in aptamer-based biosensors

Aptamers are synthetic oligonucleotides (DNA or RNA) that bind to a wide range of targets (proteins, enzymes, nucleotides, whole cells, drugs, toxins, and other small molecules) by folding into complex 3D structures. The SELEX (systematic evolution of ligands by exponential enrichment) approach led to the development of aptamers with increased affinity and specificity against cancer biomarkers which have been used in sensor platforms for the early detection of cancer (Tombelli, Minunni, & Mascini, 2005). The aptamer is utilized as a biorecognition element in aptamer-based biosensors (aptasensors). Upon binding to the analyte, the aptamers undergo structural rearrangements and conformational changes, which leads to signal change during biosensing. Aptamers have some advantages over antibodies including easier engineering and synthesis, smaller size, better stability (especially DNA aptamers), lower immunogenicity, longer shelf life, and easy storage. Graphene acts as a perfect platform for the combination of aptamers to construct the sensor assays for small molecules. DNA sequences can be adsorbed onto the graphene surface through the electrostatic interaction between DNA bases and the graphene surface, or the noncovalent $\pi-\pi$ stacking. Specifically, graphene can serve as the carrier for aptamers to enter cells (Dong et al., 2020). Many aptasensors have been developed to detect cancer and cancer biomarkers using a variety of transduction methods including electrochemistry, fluorescence, and chemiluminescence have been reported elsewhere (Hassan & DeRosa, 2020).

#### *13.6.1.1 Graphene-based electrochemical aptasensors*

Electrochemical biosensors operate on the principle of change in the current or voltage when the analyte is recognized by the biorecognition element. Large surface area, high electrical conductivity, good biocompatibility, and excellent electrocatalytic activity of graphene allows it to be used as a nanocarrier or modifier of electrode materials. The combination of graphene with specific aptamers has great potential for improving both the sensitivity and selectivity for electrochemical biosensors (Dong et al., 2020). To date, various electrochemical techniques, such as cyclic voltammetry (CV), differential pulse voltammetry (DPV), square wave voltammetry (SWV), and electrochemical impedance spectroscopy (EIS), have been used to prepare graphene based aptasensors for the detection of cancer biomarkers with high selectivity (Table 13.2).

Aptamer-based electrochemical biosensors are among the most common biosensors for cancer protein biomarker detection. Protein-based markers are more important than nucleic acid-based markers because proteins are the main executioner bio-molecules in cells (Mishra & Verma, 2010). Carcinoembryonic antigen (CEA) is one of the most extensively used biomarkers for the diagnosis and management of different human cancers (lung, breast, ovarian carcinoma and colorectal cancer). A DNase I enzyme aided potentiometric amplification system was developed based on GO-aptamers for the detection of CEA with a 9.4 pg/mL detection limit (Hong, Chen, Yu, Huang, & Fan, 2018). According to the study, the electrode surface was modified via graphene oxide nanosheets by physical adsorption, and then

**TABLE 13.2 Summary of the graphene-based aptasensors developed over the past 5 years.**

| Detection technique | Material | Target | Detection limit | Dynamic range | Ref. |
|---|---|---|---|---|---|
| EIS | $TiO_2$(200)-rGO | PSA | 1 pg/mL | 0.003–1000 ng/mL | Karimipour et al. (2019) |
| EIS | rGO | CEA | 0.1 fg/mL | 0.1 fg/mL–5.0 pg/mL | Wang et al. (2015) |
| DPV | Thionin/GO/Gold NPs | AFP | 0.050 μg/mL | 0.1–100.0 μg/mL | Li et al. (2018) |
| DPV | AuNPs-GO-PEDOT | mucin1 | 0.031 fM | 0.1 fM–31.25 nM | Gupta et al. (2018) |
| DPV | rGO-Chit | HER2 | 0.22 ng/mL | 0.5–2 ng/mL | Tabasi et al. (2017) |
| DPV | AuNPs/GO | Cancer exosomes | 96 particles/μL | 1.12 × 102–1.12 × 108 particles/μL | An et al. (2019) |
| DPV | Cu-MOF-RGO | Mucin 1 | 7.5 pg/mL | 25 pg/mL–2500 ng/mL | Hatami et al. (2019) |
| DPV | GQD/AuNP/NG | CEA | 3.2 fg/mL | 10 fg/mL–200 ng/mL | Shekari et al. (2019) |
| DPV | $NH_2$-GO/THI/AuNP | EGFR | 5 pg/mL | 0.05–200 ng/mL | Wang et al. (2020) |
| DPV | AuNPs/GO | MCF-7 | 8 cells/mL | 10–105 cells/mL | Wang et al. (2017) |
| DPV | AuNPs/rGO/THI | PSA | 10 pg/mL | 0.05–200 ng/mL | Wei et al. (2018) |
| CV | GQDs-IL-NF | CEA | 0.34 fg/mL | 0.5 fg/mL–0.5 ng/mL | Huang et al. (2018) |
| FET | MWCNTs-COOH/rGO | CA125 | 0.5 nU/mL | 1 nU/mL–1.0 U/mL | Mansouri et al. (2018) |
| SWV | Au/GQD | VEGF165 | 0.3 fM | 1 fM–120 p.m. | Hongxia et al. (2019) |

*Abbreviations*: *AuNPs-GO-PEDOT*, gold nanoparticles (AuNPs) and graphene oxide (GO) doped Poly (3,4-ethylenedioxythiophene) (PEDOT); *rGO-Chit*, reduced graphene oxide-chitosan; *Cu-MOF-RGO*, metal-organic framework-reduced graphene oxide nanocomposite; *(NH2-GO)/(THI)/(AuNP)*, amino-functionalized graphene/thionine/gold particle; *GQDs-IL-NF*, graphene quantum dot-ionic liquid-nafion; *MWCNTs-COOH/rGO*, carboxylated multiwalled carbon nanotubes/reduced graphene oxide.

aptamers were attached through π−stacking interaction. The aptamer could be protected from the cleavage by DNase I by GO. The change in the electrical potential is derived from the dissociation of the immobilized aptamers on the nanosheets by the introduction of target CEA. The assay can be used for the detection of CEA in human serum, but using physical adsorption of aptamers on the electrode was a disadvantage of the proposed method. Graphene materials can also be modified with other molecules to give special features to the desired platform. As an example, the hemin–graphene conjugates had the advantages of both hemin and graphene, leading to their great properties, such as different dispersibility in the high salt concentration and highly active biomimetic oxidation catalytic activity. Considering this, Zhang et al. designed an electrochemical aptasensor for the sensitive detection of the prostate specific antigen (PSA), a biomarker of prostate cancer, based on hemin-functionalized graphene-conjugated palladium nanoparticles (H-Gr/PdNPs) (Zhang et al., 2018). Hemin, an iron porphyrin derivative, can be used as electron media based on the reversible redox of Fe (III)/Fe (II). The hemin placed on graphene not only acts as a protective agent, but also as an in situ electrochemical probe through the reversible hemin ox/hemin red pair. According to the study, they immobilized the PSA aptamer using biotin-streptavidin interactions on the surface of electrodes and an LOD of 8 ng/mL was reached by DPV technique (Fig. 13.1). Nevertheless, frequent pretreatment and modification of the GCE and capability of single biomarker detection are the limitations of this platform.

Circulating tumor cells (CTCs) are metastatic cancer cells detached from the primary tumor found in the blood of patients. CTCs compared to other biomarkers such as nucleic acids and proteins provide a large surface area that presents more binding sites (Yang, Yang, Yang, & Yuan, 2018). Dou et al. reported an electrochemical sensing method for the ultrasensitive monitoring of CTCs based on the aptamer-functionalized and gold nanoparticle (AuNP) array-decorated magnetic graphene nanosheet (AuNPs-Fe3O4-GS) capture probes and the electroactive species-loaded AuNP amplification signal probes (Dou, Xu, Jiang, Yuan, & Xiang, 2019) (Fig. 13.2). The incubation of the capture probes and signal probes with the sample solutions (containing CTCs) can lead to generation of two distinct voltammetric peaks by square wave voltammetry (SWV). The advantages of the SWV method, compared to classical voltammetric methods, is the noticeable sensitivity and high speed, thus increasing the signal-to-noise ratio, having a wide dynamic range and low LOD values.

As known, high current signal is beneficial for improving the detection sensitivity. Layer by layer (LBL) assembly technique provided more effective probes to enhance amplified signals for improving the sensitivity of the detection. Wang et al. reported an electrochemical aptasensor for detecting cancer cells constructed though LBL techniques with

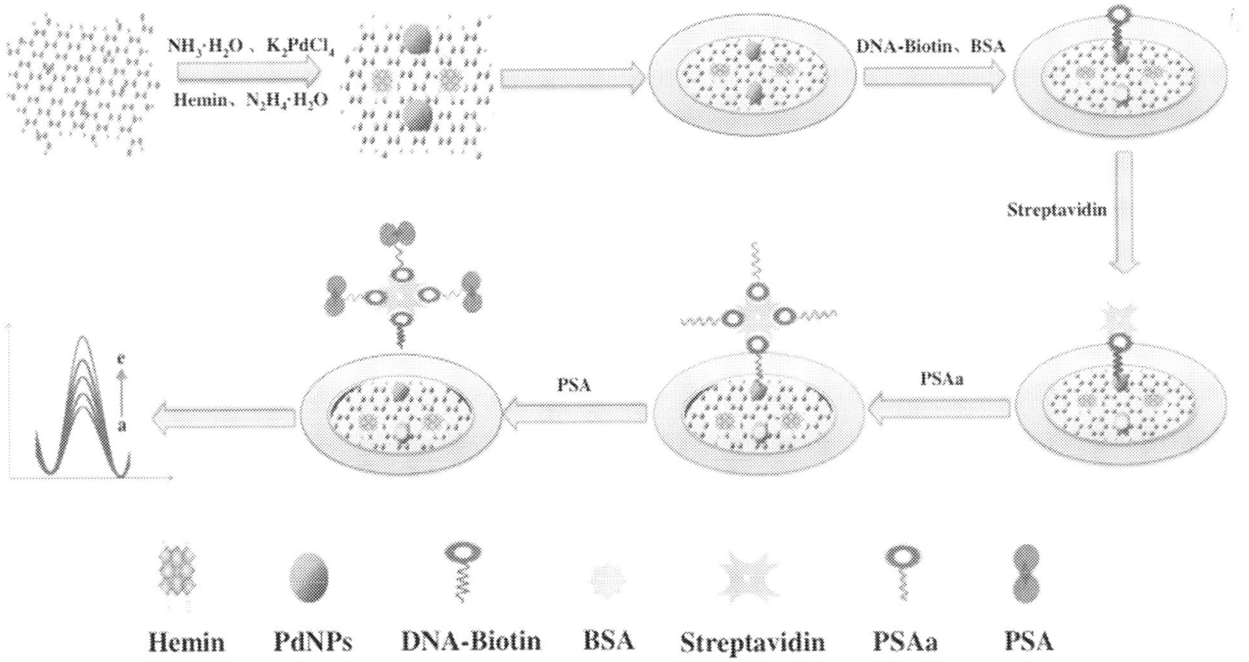


**FIGURE 13.1** Schematic illustrations of H-Gr/PdNP nanocomposites-based electrochemical PSA aptasensor. *From Zhang, Liu, Fan, and Guo (2018). Electrochemical prostate specific antigen aptasensor based on hemin functionalized graphene-conjugated palladium nanocomposites. Mikrochimica Acta, 185(3), 159. https://doi.org/10.1007/s00604-018-2686-9.*

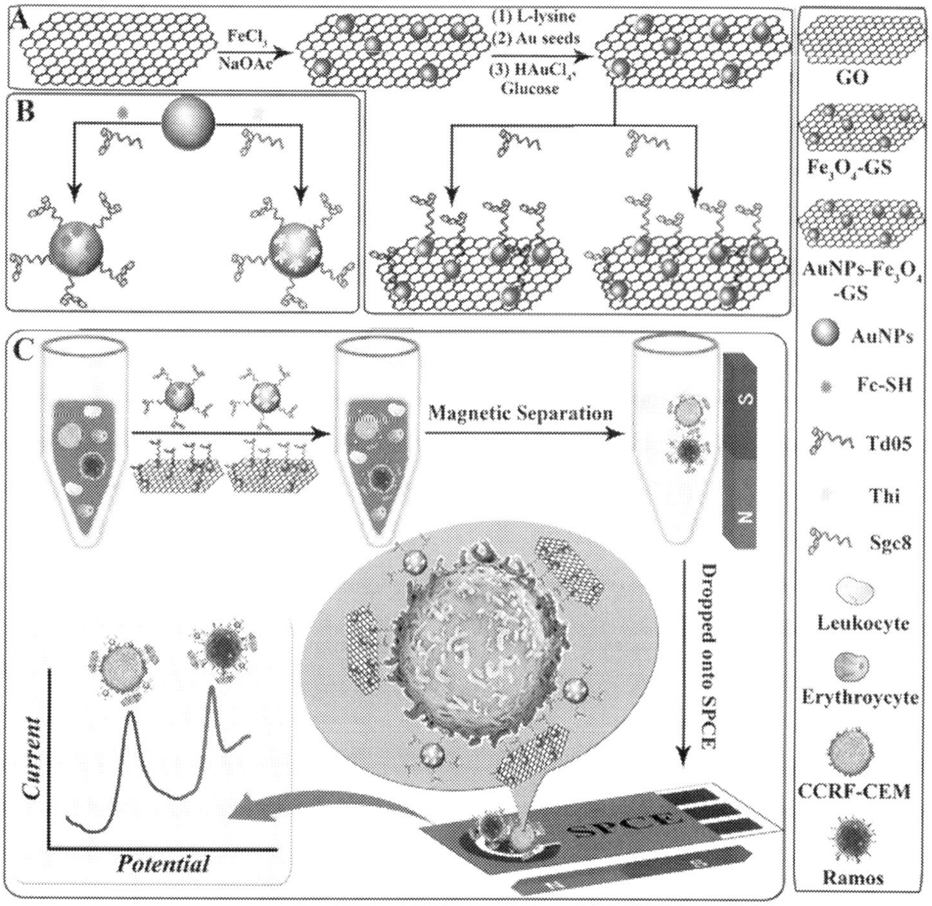


**FIGURE 13.2** Illustration of the synthesis of (A) the aptamer-functionalized AuNPs-Fe3O4-GS capture probes; (B) the aptamer/electroactive species-loaded AuNP amplification signal probes; and (C) the capture, isolation, and amplified and multiplexed detection of the target CTCs in whole blood. *From Dou, B., Xu, L., Jiang, B., Yuan, R., & Xiang, Y. (2019). Aptamer-functionalized and gold nanoparticle array-decorated magnetic graphenenanosheets enable multiplexed and sensitive electrochemical detection of rare circulating tumor cells in whole blood. Analytical Chemistry, 91(16), 10792-10799. Available from https://doi.org/10.1021/acs.analchem.9b02403.*

ferrocene-appended poly (allylamine hydrochloride) functionalized graphene (Fc-PAH-G), poly (sodium-p-styrenesulfonate) (PSS) and aptamer (AS1411). In the study, graphene was employed as the matrix for Fc-PAH to bring probes and promote electron transfer on the electrode to enhance the amplified signal. PSS, a negative polymer, was employed as the linker of Fc-PAH-G though LBL self-assembly techniques (Wang et al., 2015).

### *13.6.1.2 Graphene-based optical aptasensors*

As mentioned before, a biosensor is defined as a sensing device or a measurement system which converts a response of bioanalyte interaction and its corresponding bioreceptor into a readable form with the help of a transduction mechanism. The most common type of biosensors is the optical biosensor, which can sense phenomena related to the interaction of the optical field with their biorecognition elements. The combination of aptamer with various optical transducers developed numerous sensing methods such as surface plasmon resonance (SPR) surface-enhanced Raman spectroscopy (SERS), fluorescence and colorimetric approaches.

#### 13.6.1.2.1 Fluorescence-based platforms

Fluorescence techniques are a popular class of optical biosensors that have demonstrated a number of advantages compared to other: nonfluorescence methods such as high sensitivity, simplicity, nondestructive and rapid detection of the analyte (Sabet, Hosseini, Khabbaz, Dadmehr, & Ganjali, 2017). In diagnosis studies, the use of combined fluorescent assays and aptamers has a great potential to identify biomarkers faster and easier. In this way, through direct modification with fluorophores, the process of converting an aptamer into "structure-switching" by the binding of the target, can affect the fluorescence of a fluorescent molecule or the fluorescence resonance energy transfer (FRET) between fluorescent molecules—donor and acceptor. A change in signal: either signal generation (signal-on) or signal quenching (signal-off) reflects the extent of the binding, thereby allowing for recognition and measurement of the concentration of the target (Feagin, Maganzini, & Soh, 2018). FRET-based analytical investigations are widely reported in cancer diagnostic

(Jeong et al., 2015; Wang et al., 2018; Furukawa et al., 2016; He, Lin, Tang, & Pang, 2012). GO and GQDs have been widely used as attractive fluorophores in recent years due to their great quenching property, good photostability and biocompability, excellent conductivity, and tunable optical properties. Recently, GO-based fluorescent aptasensors were developed using GO as a quenching material, so GO can efficiently quench the fluorescence of semiconductor QDs, organic dye, metal nanoclusters, and so on. Considering this, Ding et al., 2015 proposed an aptasensor using the Mucin 1 (MUC1) aptamer conjugated to CDs (aptamer-CDs) and GO as a FRET pair for Mucin 1 protein detection (Ding et al., 2015). MUC1 is a tumor marker encoded by the MUC1 gene in humans. It is overexpressed on the surface of cancer cells like breast ovarian, lung and pancreatic cancers. In the study, the fluorescence of the aptamer-CDs was efficiently quenched by GO due to their efficient self-assembly through specific $\pi-\pi$ interactions. In the presence of target MUC1 protein, an aptamer−MUC1 complex could be formed and the interaction between aptamer-CDs and GO was weakened, leading to significant fluorescence recovery. This aptasensor exhibited a linear range from 20 to 804 nM with a detection limit of 17.1 nM.

Compared with conventional organic fluorophores, GQDs have unique optical and electronic properties: enhanced signal brightness, resistance to photo-bleaching, broad absorption spectra, and a size tunable narrow emission range (Kermani et al., 2017; Nemati, Zare-Dorabei, Hosseini, & Ganjali, 2018). For instance, Shi et al. designed a fluorescence biosensor based on GQDs and molybdenum disulfide ($MoS_2$) nanosheets for epithelial cell adhesion molecule (EpCAM) detection (Shi, Lyu, Tian, & Yang, 2017). EpCAM protein has attracted considerable attention owing to its overexpression in a variety of carcinomas such as breast, lung, and colon. In the work, the sensing platform was designed by the adsorption of GQD-PEG-aptamer on MoS2 nanosheets via van der Walls forces, which brought GQDs and $MoS_2$ into close proximity to trigger the FRET phenomena (Fig. 13.3). By monitoring the change in the fluorescence signal emitted by GQD-PEG-aptamer, the target EpCAM protein could be detected with a linear range from 3 to 54 nM and a limit of detection around 450 p.m.

In recent years, paper-based devices have been proposed as a portable, simple, inexpensive, disposable, or recyclable platforms for clinical diagnosis (Mesgari, 2020). Given the availability of high-throughput printers, such devices are also amenable to rapid and scalable manufacturing and can be set-up to perform miniaturized tests. In this regard, a paper-based heterogeneous assay for the detection of EpCAM was constructed by Das and Krull (Das & Krull, 2017). The cellulose paper was used as a substrate to immobilized aptamers linked to quantum dots (QDs-Apt) and Cy3 labeled complementary DNA (cDNA), which served as the donor and the acceptor, respectively. This bioassay allowed the sensitive quantification of EpCAM with a detection limit of 250 p.m. In another work, Liang et al. reported a fluorescence method for multiplexed monitoring of cancer cells using mesoporous silica nanoparticles (MSNs)/QDs-labeled aptamers and GO in microfluidic paper-based analytical devices (Liang et al., 2016). The fluorescence of DNA-conjugated MSNs/QDs was quenched by GO to generate a cancer cell capture probe. After the addition of target cells, the corresponding colored probe could be released, thus emitting strong fluorescence and indicating the existence of the corresponding target cell.

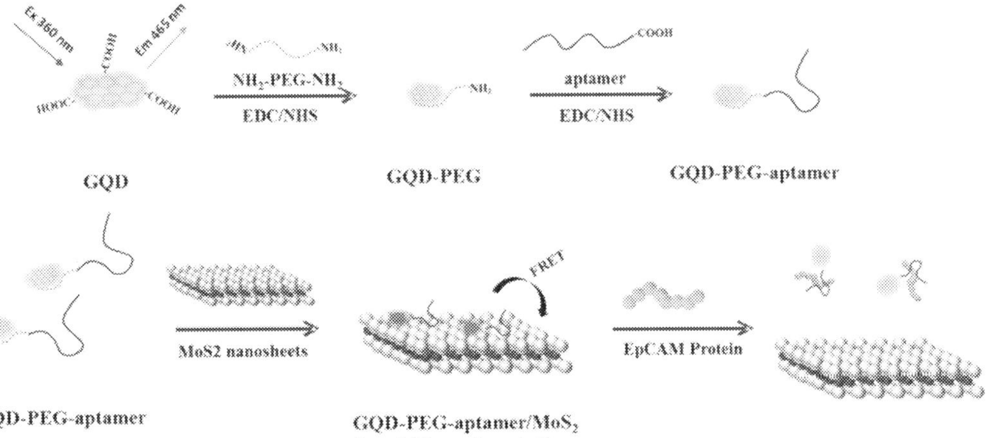


**FIGURE 13.3** The sensing mechanism of GQD-PEG-aptamer/$MoS_2$-based FRET biosensor for EpCAM protein detection. *From Shi, J., Lyu, J., Tian, F., & Yang, M. (2017). A fluorescence turn-on biosensor based on graphene quantum dots (GQDs) and molybdenum disulfide(MoS2) nanosheets for epithelial cell adhesion molecule (EpCAM) detection. Biosensors and Bioelectronics, 93, 182188. Available from https://doi.org/10.1016/j.bios.2016.09.012.*

### 13.6.1.2.2 SPR/SERS-based platforms

During the last two decades, surface plasmon resonance (SPR) sensors have been researched and developed as sensitive, specific, label-free, portable and rapid biosensing devices to detect various analytes in complex matrices. Typically, the biological macromolecules were coated on a metal layer (e.g., silver or gold). The SPR signal can be obtained from the change in the refractive index at the interface, which arises due to the binding the between target and the recognition molecule. To improve the SPR signal compared with bare metal thin films, layers of graphene have been established as an appropriate dielectric top layers for SPR sensing. In an effort to achieve more efficient sensing systems, a high sensitivity SPR biosensor was constructed based on two layers of GO-AuNPs composites (Fig. 13.4) (Li et al., 2017). In this biosensor, two layers of GO-AuNPs composites acted not only as the sensing substrate for the immobilization of captured DNA molecules (bottom layer) but also, as the signal amplification element (upper layer). By employing the dual amplification strategy, sensitive detection of miRNA and adenosine small molecules with the detection limit of 0.1 fM for miRNA and detection limit of 0.1 p.m. for adenosine were obtained.

Additionally, the feasibility of combining SPR and electrochemical techniques was also reported. For example, a graphite carbon nitride ($g\text{-}C_3N_4$) nanocomposite embedded with $MoS_2$ QDs and chitosan stabilized Au nanoparticles (CS-AuNPs) (denoted as $MoS_2QDs@g\text{-}C_3N_4@CS\text{-}AuNPs$), have been applied for designing both SPR and electrochemical aptasensors for the determination of trace levels of PSA (Duan et al., 2018). Graphitic carbon nitride is a kind of metal-free material with a 2D graphene-like structure and has broad usage in optics because of its low toxicity, high fluorescence, good stability, and biocompatibility (Nemati & Zare-Dorabei, 2019). In the aforementioned sensing

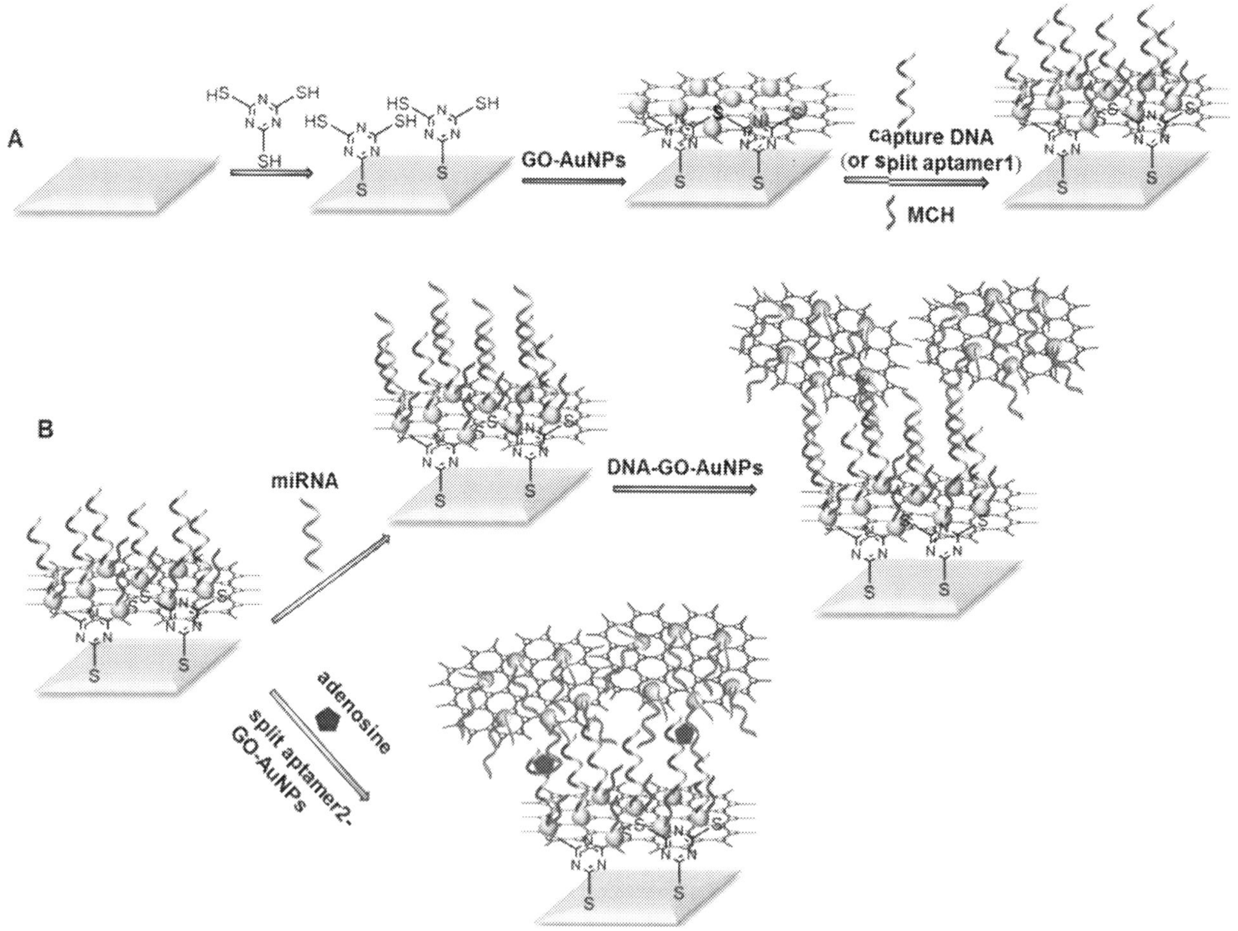


**FIGURE 13.4** Schematic illustration of the SPR biosensor based on the GO-AuNPs composites. (A) Thiolated capture DNA was covalently attached to the GO-AuNPs functionalized Au film surface (B) target miRNA and DNA functionalizedGO-AuNPs composites (DNA-GO-AuNPs) were respectively introduced, and then the sandwich structure was formed due to DNA/RNA hybridization. *From Li, Q., Wang, Q., Yang, X., Wang, K., Zhang, H., & Nie, W. (2017). High sensitivity surface plasmon resonance biosensor for detection ofmicroRNA and small molecule based on graphene oxide-gold nanoparticles composites. Talanta, 174, 521526. Available from https://doi.org/10.1016/j.talanta.2017.06.048.*

system, the SPR aptasensor provided an LOD of 0.72 ng/mL. Concurrently, the electrochemical aptasensor allowed the sensitive detection of PSA with an LOD of 0.71 pg/mL.

It has been demonstrated recently that plasmonic metal nanoparticles also have the potential to be applied to SERS for wider analysis. Based on the principle of SERS, Raman scattering intensity of target molecule could be enhanced by adsorption onto the rough metal surfaces. Until now, numerous SERS-active supports based on silver/copper/AuNPs have been fabricated by strategies like chemical reduction, self-assembly, sol–gel process, electrochemical deposition, and others. However, it is difficult to form heterogeneous adsorption of molecules at substrate surface, and the observed Raman signals are usually overwhelmed by interference from the intense fluorescence background. Graphene was employed as an amazing SERS active substrate owing to its chemical stability, biocompatibility, uniform electronic and photonic properties, and high absorptivity (Wang, Wu, Colombi Ciacchi, & Wei, 2018). Considering this, Zou et al. fabricated graphene–isolated-Au-nanocrystals (GIANs) integrated for SERS (Zou et al., 2018). An MUC1 aptamer was linked to these nanocrystals to target proteins overexpressed on the cancer cell surface. GIAN tags demonstrated high resolution multiplexed Raman imaging, both in vivo and in vitro, with low background interference due to the use of graphene and the specific targeting of cancer cells using the MUC-1aptamer.

#### 13.6.1.2.3 Colorimetric-based platforms

Colorimetric strategies for unaided eye detection in real-time are interesting because of the simplicity of the assays, the need for very little or no external supporting equipment or power, and the ability to be performed without the presence of prepared professionals. In a new study, visual and sensitive detection of miRNAs in fluorescence and colorimetric dual-sensing platform was designed based on graphene oxide and a novel hairpin probe (Shin et al., 2020). A progressed GO-based sensing system for miRNAs was applied using the newly designed hairpin-like probe embedding a G-rich DNAzyme sequence to a typical fluorescence sensing platform. G-rich Dz was folded into a G-quadruplex structure in buffer containing $K^+$ ions that in the presence of hemin showed peroxidase-like activity. Thermal stability, ease of functionalization, and cost-effectiveness of this catalytic DNA molecule has led to the utilization of this system in various colorimetric sensing processes. Changes in the fluorescence and colorimetric responses were related to the presentation of the Dz-embedded hairpin-like probe to the GO-based platform which enables estimations of miRNA concentration in a single solution without needing an additional separation procedure. They developed the color of the DNAzyme-catalyze system by adding colorimetric reagents to a mixture of FLHp6/GO in the presence of various concentrations of miRNAs. The miR-21 activated the DNAzyme in the presence of miR-21 and quickly made a green color in the high concentrations of the target, which could be observed by the naked eye. The linear range of the miRNA concentration was ( ~ 10 nM) with a detection limit of 2.60 nM.

Human telomerase is a ribonucleoprotein complex that maintains telomere lengths by adding repetitive nucleotide sequences (TTAGGG) onto the end of the human chromosomes utilizing its RNA template, reverse transcriptase and related proteins. Telomerase expression is related to cell immortalization and tumorigenesis. In this way, a simple method for the detection of telomerase activity could be significant for cancer diagnosis, screening of anticancer drugs, and assessment of cancer treatment. The hemin-graphene conjugates (H-GNs) had the benefits of both hemin and shows good properties. Thus, a simple label-free colorimetric sensor for telomerase activity dependent on the upsides of H-GNs was created (Xu et al., 2017). In the presence of TS primer, H-GNs were adjusted to coagulate to an appropriate degree via cautiously choosing the presented NaCl concentration. At that point, the supernatant of the relating solution contained little H-GNs and indicated light blue color after chromogenic reaction. Then, repeating sequences of (TTAGGG) were stretched out on the TS primer by the telomerase.

### 13.6.2 Graphene material in antibody-based sensors

Antibodies, also known as immunoglobulins, are large glycoproteins, which can specifically bind to antigens with a high binding constant exceeding $10^8$ L $mol^{-1}$. Antibodies have attracted wide research interest because of their homogeneous nature, sensitivity, affinity and specificity. They are commonly used as the recognition layers for sensing applications in many fields, particularly in the field of medical diagnostics (Farahavar, Abolmaali, Gholijani, & Nejatollahi, 2019). The conjugation of graphene-related materials with antibodies combines the properties of the graphene nanomaterials themselves with the specific and selective recognition ability of the antibodies to antigens. In this way, the graphene nanomaterials offer increased surface area for antibody binding while also providing a direct link between the bioreceptor and sensor surface (Yang, Yang, Yang, & Yuan, 2018; Yang et al., 2015).

### 13.6.2.1 Electrochemical sensors

In the antibody-based electrochemical biosensors, also called electrochemical immunosensors, immobilization of biomarker specific antibodies on the sensor surface is a critical factor (Qian et al., 2019). A suitable sensor surface should offer a set of the necessary characteristics including large surface area for high density probe immobilization, excellent electrical and thermal conductivity, low diffusion rates, and a low signal-to-noise ratio due to matrix effects. No doubt, graphene-based composite nanomaterials, which combine the physicochemical advantages of specific features of other nanomaterials with the merits of graphene, provide a superior platform for the immobilization of the immune reagents. In this sense, Verma et al. proposed an electrochemical immunosensor based on AuNPs-reduced graphene oxide (AuNPs-rGO) for the detection of salivary oral cancer biomarker interleukin-8 (IL8) as shown in Fig. 13.5 (Verma et al., 2017).

Because of the large surface area of graphene and the good biocompatibility of AuNPs, this composite provided a suitable support for the immobilization of antibodies. The synergy between rGO and AuNPs allowed the immunosensor to exhibit fast response and high sensitivity due to the improved electron transfer behavior of the composite. In the study, the attained LOD value (72.73 pg/mL) was below the clinical salivary expression level of IL8 (720 pg/mL) in an average oral cancer patient.

Antigen−antibody binding can cause a change in the detectable signal. Although the antigen−antibody binding can generate a change in the detectable signal, the change is relatively small. Thus, appropriate labels are usually used for the amplification of the detectable signal and give the highest level of sensitivity and specificity due to the use of a couple of matched antibodies. In this regard, Jian et al. designed an electrochemical immunosensor for the ultrasensitive determination of α-fetoprotein (AFP), a well-known biomarker of liver cancer (Jian et al., 2016). The immunosensor was constructed by modifying gold electrodes with electrochemical reduction of graphene oxide-carboxyl multi-walled carbon nanotube composites (ERGO-CMWCNTs) and electro-deposition of AuNPs for effective immobilization of primary antibodies ($Ab_1$). Ferroferric oxide-manganese dioxide-reduced graphene oxide nanocomposites ($Fe_3O_4$@$MnO_2$-rGO) were designed as labels for signal amplification. On one hand, the excellent electroconductivity and outstanding electron transfer capability of ERGO-CMWCNTs/AuNPs improved the sensitivity of the immunosensor. On the other hand, introduction of rGO could not only increase the specific surface area for the immobilization of secondary antibody (Ab2) but also build a synergetic effect to reinforce the electrocatalytic properties of the catalysts. The proposed electrochemical immunosensor showed a low detection limit (5.8 pg/mL), and a wide linear range (0.01−50 ng/mL) for the detection of AFP. In another work, Huang et al. demonstrated a novel approach toward the development of advanced immunosensors based on chemically functionalized Ag/Au nanoparticles coated on graphene for the detection of carcinoembryonic antigens (CEA) (Huang et al., 2015). Graphene sheets (GS) were prepared and used both for the immobilization of primary antibodies (Ab1) and as a tracer to label the secondary antibodies ($Ab_2$) to fabricate the electrochemical immunosensor. For the immobilization of $Ab_1$, 1,5-diaminonaphthalene (DN) was selected and adsorbed onto GS through stacking, which was used to coat AuNPs; then a conjugated complex between AuNPs and Ab1 was formed based on the Au−S chemistry. For the preparation of the tracer to label Ab2, DN was adsorbed onto GS, and

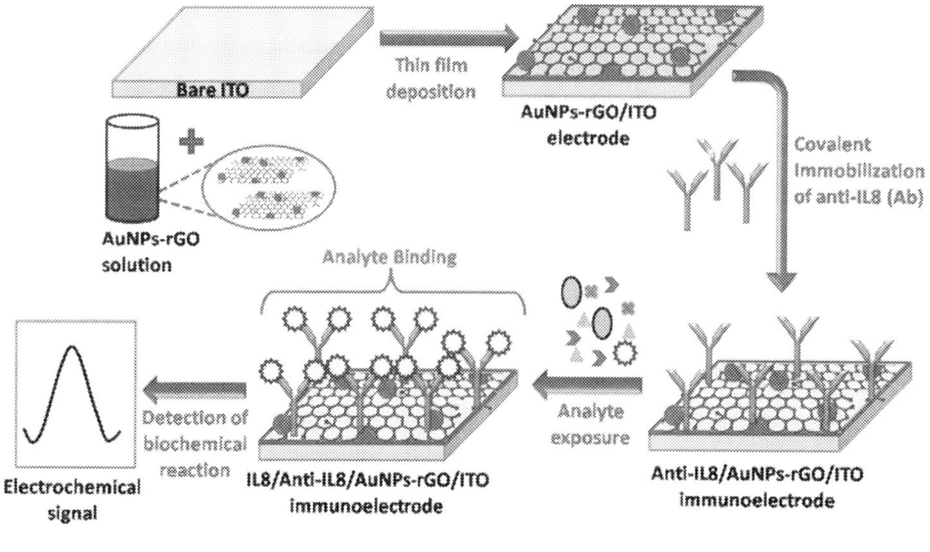


**FIGURE 13.5** Schematic of fabrication of AuNPs-rGO-based immunoelectrode for immunosensing application. *From Verma, S., Singh, A., Shukla, A., Kaswan, J., Arora, K., Ramirez-Vick, J., Singh, S. P. (2017). Anti-IL8/AuNPs-rGO/ITO as an Immunosensing Platform for Noninvasive Electrochemical Detection of Oral Cancer. ACS Applied Materials and Interfaces, 9(33), 2746227474. Available from https://doi.org/10.1021/acsami.7b06839.*

**TABLE 13.3 Summary of the graphene-based electrochemical immunosensor reported (2015 onwards) for the detection of various targets.**

| Detection technique | Material | Target | Detection limit | Dynamic range | Ref. |
|---|---|---|---|---|---|
| SWV | $Cu_2O$@GO | AFP | 0.1 fg/mL | 0.001 pg/mL – 100 ng/mL | Wang (2017) |
| SWV | TB-Au-$Fe_3O_4$-rGO | AFP | 2.7 fg/mL | 0.00001–10.0 ng/mL | Wang et al. (2017) |
| DPV | Nafion-rGO-CHO-MP | PSA | 1.6 pg/mL | 0.005–90 ng/mL | Jeong et al. (2019) |
| DPV | Cubic $CeO_2$/RGO PBSE/ Graphene/Cu | Cyfra-21–1 | 3 pg/mL 0.625 pg/mL | 0.625 pg mL–0.01 ng/mL | Pachauri et al. (2018) |
| DPV | ($NH_2$-rGO)/(COOH-AgPtPd)/ (TNPs) | PSA | 4 fg/mL | 4fg/mL–300 ng/mL | Sharifuzzaman et al. (2019) |
| DPV | Pβ-CD/GAs | CA153 | 0.03 mU/mL | 0.1 mU/mL–00 U/mL | Jia et al. (2018) |
| DPV | Au NPs/GO | PSA | 0.24 fg/mL | 0.001 fg/mL–0.02 μ g/mL | Pal et al. (2017) |
| DPV | CuS/rGO | CA 153 | 0.3 U/mL | 1.0–150 U/mL | Amani et al. (2017) |
| DPV | Porous GR/TiO2 nanofibers | ErbB2 | 0.06 ng/mL | 1fM–0.1 μM | Ali et al. (2016) |
| CV | GR/AuNPs | PSA | 0.59 ng/mL | 0–10 ng/mL | Jang et al. (2015) |
| CV | PDA-rGO | CEA | 0.23 pg/mL | 0.5 pg/mL–5 ng/mL | Miao et al. (2017) |
| CV | PtCu@rGO/g-C3N4 | PSA | 16.6 fM | 50 fM–40 nM | Feng et al. (2017) |
| CV | (M-Pd@Pt/NH2-GS) | PSA | 3.3 fM | 50 fM–40 nM | Li et al. (2017) |
| EIS | PBSE/Graphene/Cu | CEA | 0.23 ng/mL | 1.0–25.0 ng/mL | Sing et al. (2018) |
| ECL | AuNPs/g-C3N4 | SCCA | 0.4 pg/mL | 0.001–10 ng/mL | Wu et al. (2016) |
| ECL | P5FIn/rGO | CEA | 3.78 fg/mL | 0.1 pg/mL – 10 ng/mL | Nie et al. (2018) |

*Abbreviations: Nafion-rGO-CHO-MP*, nafion/reduced graphene oxide/aldehyde methyl pyridine composite; *PBSE/Graphene/Cu*, pyrenebutanoic acid succinimidyl ester/graphene/Cu; *P5FIn/rGO*, poly (5-formylindole)/reduced graphene oxide nanocomposite; *Pβ-CD/Gas*, β-cyclodextrin polymer/porous network graphene aerogel; *PDA-rGO*; polydopamine-reduced graphene oxide; *(NH2-rGO)/(COOH-AgPtPd)/(TNPs)*, amino-functionalized reduced graphene oxide/carboxylic surface modified AgPtPd/trimetallic nanoparticles; *PtCu@rGO/g-C3N4*, PtCu bimetallic loaded on reduced graphene oxide/ graphitic carbon nitride; *M-Pd@Pt/NH2-GS*, mesoporous core-shell Pd@Pt nanoparticle loaded by amino group functionalized graphene; *AuNPs/g-C3N4*, Au nanoparticles/graphitic-phase carbon nitride; *AFP*, α-fetoprotein; *PSA*, Prostate-Specific Antigen; *Cyfra-21-1*, cytokeratin fragment 21-1; *CEA*, carcinoembryonic antigen; *CA153*, Carbohydrate antigen 153; *ErbB2*, Erb-B2 Receptor Tyrosine Kinase 2; *SCCA*, Squamous cell carcinoma antigen.

the amino group of DN was then used to coat Ag/Au nanoparticles and conjugate $Ab_2$. Furthermore, Table 13.3 lists some critical features associated with several other recently reported graphene-based electrochemical immunosensors.

### *13.6.2.2 Optical sensors*

#### 13.6.2.2.1 Fluorescence-based platforms

In the case of tumorous disease diagnosis, the selection of a specific biomarker to be monitored in the body fluid is a key factor. Moreover, elevated levels of most biomarkers may be associated with more than one cancer type. Thus, monitoring a combination of several biomarkers in a single analytical run provides better predictive values for accurate cancer diagnosis, which is not only important for early stage diagnosis, but also for choosing and monitoring of effective therapy (Korecká et al., 2018). In 2017, Tsai and researchers presented a method for dual-protein detections in a thin channel by using functional magnetic GQDs and two biofunctional QDs. Magnetic GQDs enabled selective and quantitative nanoparticle deposition with blue emission. Biofunctional QDs confirmed the detection of two protein (AFP and CA125) with orange and green emission (Tsai, Lin, Chuang, Lu, & Fuh, 2017). The advantages of using functional nanoparticles in this assay were reproducible antibody labels, large reacting surfaces, magnetic selective deposition, and quantitative fluorescent detection. The "turn-off" switch mode is commonly used for a variety of fluorescence immunosensors. However, the application of the "turn-off" switch system is limited due to the low signal-to-background ratio. In this regard, Pei at al. reported a fluorescent turn-on nanoprobe for ultrasensitive detection of the

PSA based on graphene oxide quantum dots@silver (GQDs@Ag) core-shell nanocrystals (Pei et al., 2015). According to the study, magnetic beads (MBs) were employed to immobilize the capture antibody (Ab1) and for convenient separation based on the simple application of an external magnetic field (Fig. 13.6). GQDs@Ag linked to the antibody (Ab2) was utilized as the immunosensing probe. The developed immunosensor showed a good linear relationship between the fluorescence intensity and the concentration of PSA in the range of 1 pg/mL to 20 ng/mL with a detection limit of 0.3 pg/mL.

In another work, a fluorescent immunoassay for the determination of AFP was developed based on the nano graphite carbon nitride (g-$C_3N_4$) as fluorophore and immunomagnetic beads (MBs) as separation material (Li et al., 2018). In the aforementioned sensing system, capture antibodies ($Ab_1$) and detection antibodies ($Ab_2$) were conjugated to the MBs and the g-$C_3N_4$, respectively. Thus, the sandwich-type complex could be formed between magnetic part (MBs-$Ab_1$) and signal part (g-$C_3N_4$-$Ab_2$) of the immunosensor. The resulting assay had a linear response that covered the AFP concentration range from 5 to 600 ng/mL, with a 0.43 ng/mL low detection limit. Unfortunately, despite the high affinity and specificity of antibodies, their use as the recognition component of biosensors has limitations and disadvantages. They have high molecular weights and limited chemical and thermal stability and are expensive to produce.

#### 13.6.2.2.2 SPR/SERS-based platforms

In recent years, there has been evidence showing the improvement of the plasmonic coupling mechanism in GO-based surface plasmon resonance (SPR) biosensors by adding carboxyl groups. Through the functionalization of graphene with carboxyl groups, carbon can modulate its visible spectrum, and can therefore be used to improve and control the plasmonic coupling mechanism. Chiu et al. (2018) presented a carboxyl-functionalized graphene oxide (GO-COOH)-based SPR chip for the rapid and quantitative detection of cytolerayin 19 (CK19), a biomarker for lung carcinoma

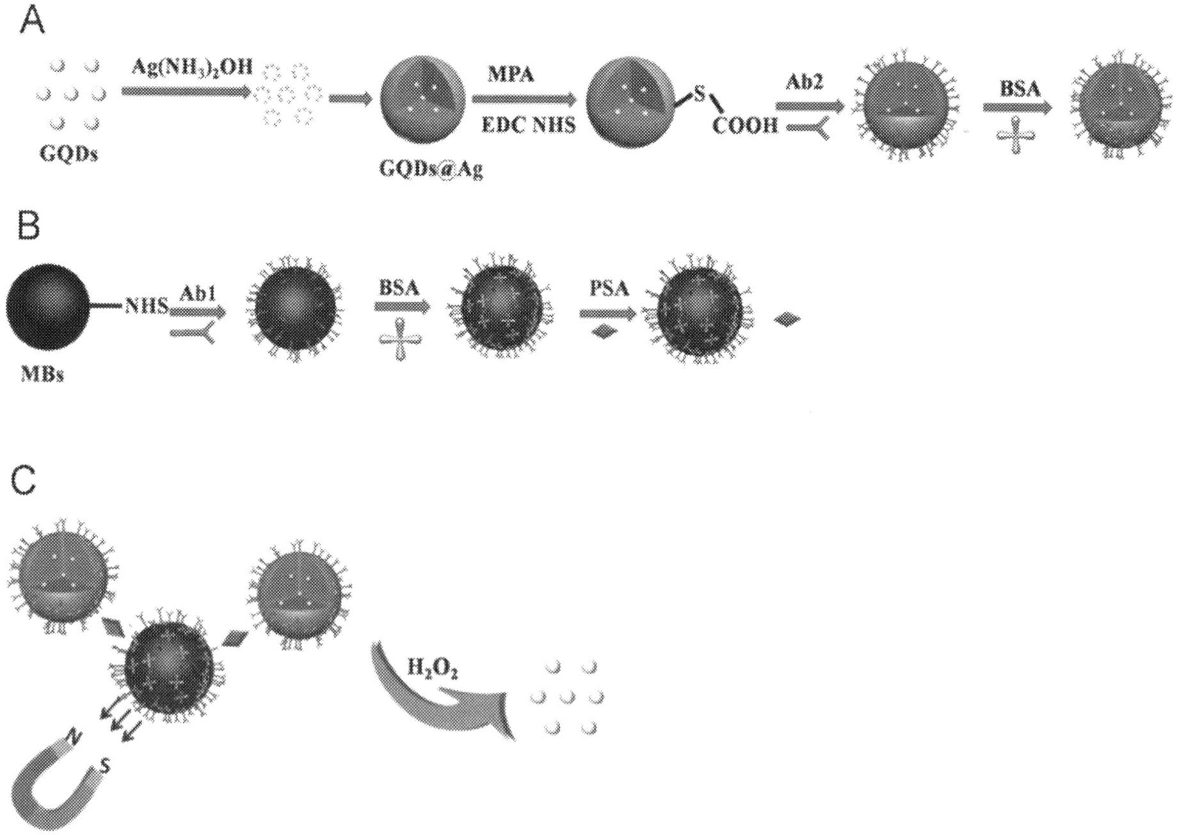


**FIGURE 13.6** Schematic illustrations of the sensing mechanism. (A) The as-prepared GQDs@Ag was utilized to immobilize detection antibodies (Ab2) after being carboxylated. (B) Capture antibodies (Ab1) were directly adsorbedonto MBs and incubated with BSA to block nonspecific bindingsites. (C) A sandwich type immunocomplex was formedafter the addition of PSA and could be separated using an externalmagnet. *From Pei, H., Zhu, S., Yang, M., Kong, R., Zheng, Y., & Qu, F. (2015). Graphene oxide quantum dots@silver core-shell nanocrystals as turn-on fluorescentnanoprobe for ultrasensitive detection of prostate specific antigen. Biosensors and Bioelectronics, 74, 909-914. Available from https://doi.org/10.1016/j.bios.2015.07.056.*

(NSCLC), in spiked human plasma (Chiu, Lin, & Kuo, 2018). The immunosensor was constructed by immobilizing a low concentration (10 μg/mL) of CK19 antibodies on an SPR chip (Fig. 13.7). In this method, the attained LOD value (0.05 pg/mL) was below the normal physiological level of the serum protein (3.3 ng/mL).

The SERS immunoassay has been widely applied in biosciences due to its advantages in sensitivity, rapid detection, and facile pre-operation. Yang at al. developed a sensitive SERS immunoassay based on the GOx assisted silver dissolution reaction for PSA detection (Yang, Zhen, et al., 2018). This strategy involved dual signal amplification of enzymes and nanocomposites to improve the detection sensitivity. With the dissolution of AgNPs, a greatly decreased SERS signal of GO was obtained. The SERS immunoassay showed good analytical performance in the range from 0.5 pg/mL to 500 pg/mL with a limit of detection of 0.23 pg/mL.

Graphene-based SERS substrates were also used for the detection of DNA, which is important for diagnosis as many diseases are strongly associated with DNA damage, such as physiological disorders and cancer cells. Methylation is an important epigenetic DNA modification that governs gene expression. It is known that the genomic level of methylated DNA is an important indicator for cancer initiation and progression. In a study conducted by Ouyang et al., SERS sensors for sensitive detection of methylated DNA was fabricated based on laser wrapped graphene-Ag array as an enhancing substrate, and modified AuNPs with the methylation-specific antibody as the target scaffold to enrich methylation related DNA (Ouyang, Hu, Zhu, Cheng, & Irudayaraj, 2017). A detection limit as low as 0.2 pg/μL was achieved.

#### 13.6.2.2.3 Colorimetric-based platforms

The specific protein biomarkers in bladder cancer (BC) are nuclear matrix protein 22 and apolipoprotein A2 and A1 proteins (ApoA1, ApoA2) that can be used for the recognition of BC. ApoA1 levels in the urine of patients with BC is very high which can enable the early detection of BC. A glassy chip was functionalized with ApoA1 antibodies, Prussian blue (PB)-incorporated magnetic graphene oxide (PMGO) with magnificent stability and self-binding associating reaction ability to build up an enzyme-free immunosensor for sensitive identification of ApoA1 concentrations in

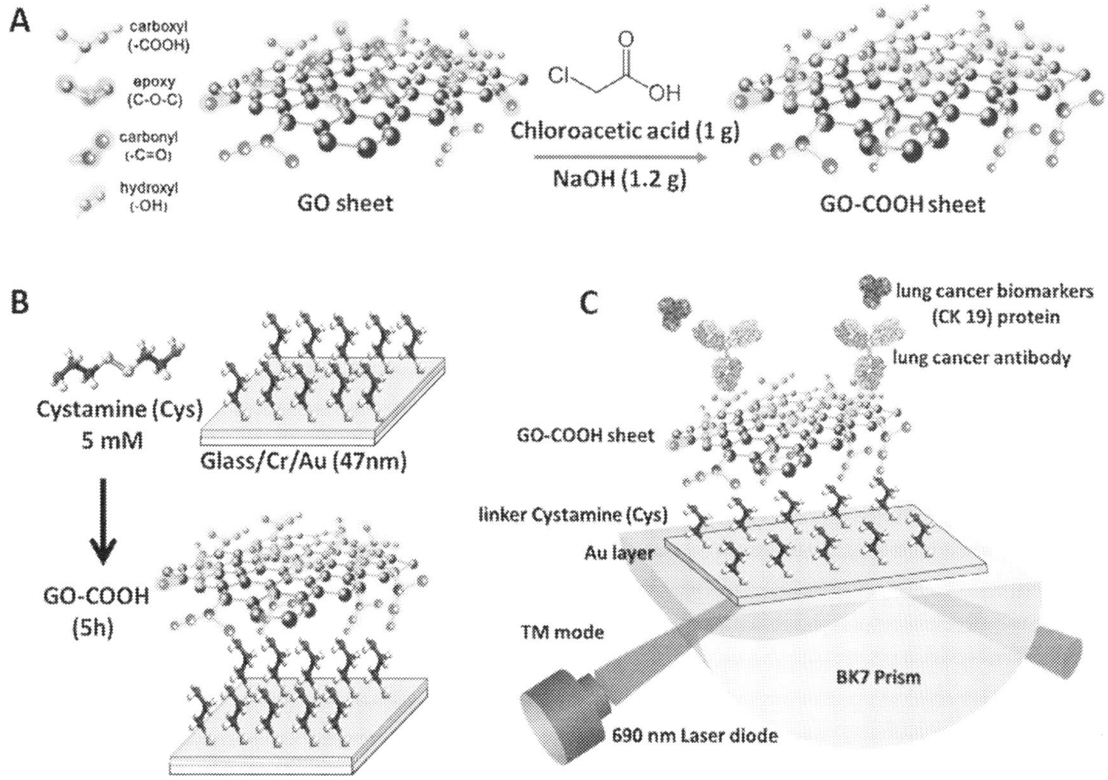


**FIGURE 13.7** (A) Surface modification of covalent links to carboxylic acid functionalized graphene oxide (GO). (B) Covalent immobilization procedure using cystamine (Cys) as a linker on the GO-COOH sheet and on the surface of gold chips. (C) Immobilization of the biomarker anti-cytolerayin-19 (anti-CK19) antibody on the GO-COOH-based SPR chip using an immunoassay to detect lung cancer antigen. *From Chiu, N. F., Lin, T. L., & Kuo, C. T. (2018). Highly sensitive carboxyl-graphene oxide-based surface plasmon resonance immunosensor for the detection of lung cancer for cytokeratin 19 biomarker in human plasma. Sensors and Actuators, B: Chemical, 265, 264272. Available from https://doi.org/10.1016/j.snb.2018.03.070.*

urine (Peng et al., 2019). This potential for easy diagnosis of BC and sensitive ApoA1 detection was attained by this colorimetric immunosensor with PMGOs structure for displaying the enhanced signal in a time of 60 min. The calibration range is clearly augmented in the presence of self-linkable PMGO structure at 0.05−100 ng/mL that compared with the tests without signal amplification (1−100 ng/mL) with a LOD of 0.02 ng/mL, corroborated the recognition of ApoA1 and BC in the clinical test of urine. In another report, a new nano hybrid was immobilized on the surface of graphene oxide that consisted of $Fe_3O_4$ magnetic nanoparticles (MNPs) and platinum nanoparticles (Pt) (Kim et al., 2014). By synergistically integrating highly catalytically active PtNPs and MNPs on GO whose frameworks possess high substrate affinity, the nanohybrid is able to achieve up to a 30-fold higher maximal reaction velocity (Vmax) compared to that of free GO for the colorimetric response of the peroxidase substrate, 3,3′,5,5′-tetramethylbenzidine (TMB), and enabled fast recognition of target cancer cells. The limit of detection for target SKBR-3 (human breast adenocarcinoma cells) was seen at around 100 cells in the linear range from 100 to 1,000 cells depending on the calibration curve.

### 13.6.3 Graphene material in enzyme-based sensors

Graphene and its derivatives with remarkable physicochemical features, as previously discussed in this chapter, are commonly used as an effective substrate for enzyme immobilization. The kind of interactions with enzymes can be determined by tuning surface chemistry of graphene derivatives. In addition, surface coverage with functional groups could influence enzyme catalytic behavior (Soozanipour & Taheri-Kafrani, 2018). In the enzyme-based biosensors, the procedure of enzyme immobilization plays an important role, due to enzyme conformational changes and loss of activity (Kuralay, 2019). Furthermore, the surface engineering for desirable grafting of functional groups can influence enzyme catalytic behavior. Electrochemical enzyme biosensors are widely used compared with other types of enzyme biosensors due to their ease of use, low cost, robust fabrication, fast response, and availability for use in raw samples. The direct electron transfer between the enzyme and the electrode is enhanced because of high conductivity of graphene. Horseradish peroxidase (HRP) is a widely used reporter enzyme which has the advantages of low cost and availability (Afsharan et al., 2016; Jing et al., 2020; Zhang, Sun, & Zhou, 2017). A facile amperometric sensor based on graphene quantum dots (GQDs) and HRP was designed by Hu et al. for miRNA-155 detection (Fig. 13.8) (Hu et al., 2016). A double-stranded DNA (dsDNA) was hybridized on the gold electrode surface via the use of thiol-tethered oligodeoxynucleotide probes (capture DNA), aminated indicator probe ($NH_2$−DNA) as well as target DNA. Subsequently, the activated carboxyl groups of GQDs were assembled on $NH_2$-DNA. They used GQDs as a new platform for HRP immobilization through noncovalent assembly. HRP-modified biosensor was able to catalyze the hydrogen peroxide ($H_2O_2$)-mediated oxidation of 3,3′,5,5′-tetramethylbenzidine (TMB) accompanied by an increased electrochemical signal. The as-prepared biosensor presented a linearity for miRNA-155 from 1fM to 100 p.m. with an LOD of 0.14 fM.

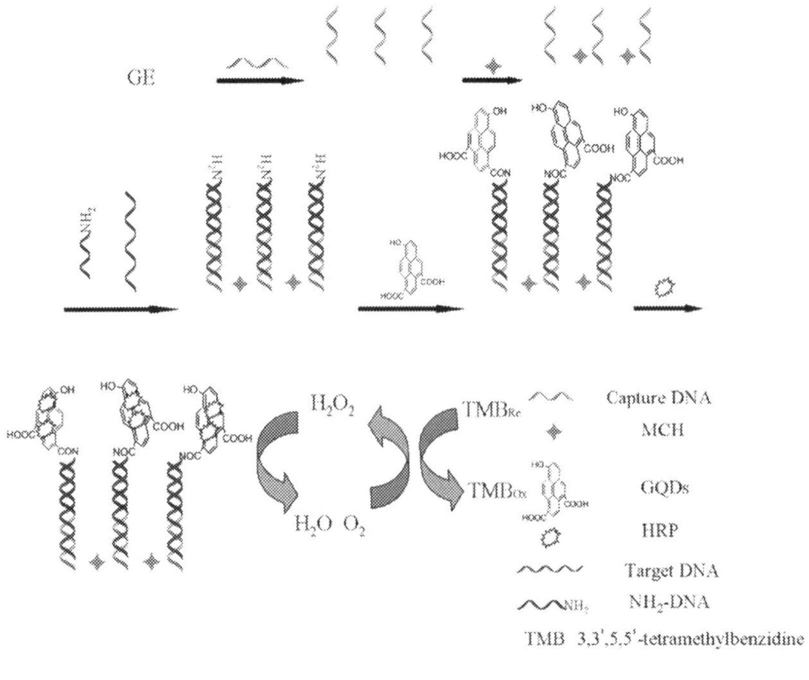


**FIGURE 13.8** Principle of the enzyme catalytic amplification of miRNA-155 detection with graphene quantum dot-based electrochemical biosensor. *From Hu, T., Zhang, L., Wen, W., Zhang, X., & Wang, S. (2016). Enzyme catalytic amplification of miRNA-155 detection with graphene quantum dot-based electrochemical biosensor. Biosensors and Bioelectronics, 77, 451-456. Available from https://doi.org/10.1016/j.bios.2015.09.068.*

In another study, Peng and co-workers designed a facile electrochemical biosensor for methyltransferase (M. SssI MTase) activity detection and inhibition screening was developed based on the amplification of GQD and enzyme-catalyzed reaction (Peng et al., 2017). The hybridization of auxiliary DNA with capture DNA formed the ds-DNA structure containing a specific recognition sequence for M. SssI MTase and HpaII. The proposed biosensor showed a good sensitivity and selectivity for M. SssI MTase detection with a detection limit of 0.3 U $mL^{-1}$.

## 13.7 Conclusion

The chapter discussed the potential of graphene-based devices for the detection of cancer biomarkers and cells. The excellent properties of graphene such as good conductivity, the rich surface, and easily modifiable functional groups give it a great advantage as a resistance sensor. In recent years, different composite materials based on graphene with polymers, metals, metal oxides, etc. have emerged in the field of cancer detection. Different biosensors based on graphene utilize aptamers, antibodies, and enzymes to target cancer cells and tumor biomarkers such as proteins, microRNAs, etc. The electrochemical and optical (fluorescence, SPR/SERS, and colorimetric) responses of these biosensors enhance their performance and indicate that graphene composites in the sensitive biosensors have excellent potentials for the development of cancer detection. The large scale production of biosensors based on graphene, along with the controllability, reproducibility, and portability methods for designing of graphene-based devices are challenges that need to be addressed and overcome. In the future, the aim is to develop multifunction graphene-based materials for high throughput and simultaneously recognition of multiple cancer biomarkers and readouts.

## References


Afsharan, H., Khalilzadeh, B., Tajalli, H., Mollabashi, M., Navaeipour, F., & Rashidi, M. R. (2016). A sandwich type immunosensor for ultrasensitive electrochemical quantification of p53 protein based on gold nanoparticles/graphene oxide. *Electrochimica Acta, 188*, 153–164. Available from https://doi.org/10.1016/j.electacta.2015.11.133.

Al-Ani, L. A., AlSaadi, M. A., Kadir, F. A., Hashim, N. M., Julkapli, N. M., & Yehye, W. A. (2017). Graphene– gold based nanocomposites applications in cancer diseases: Efficient detection and therapeutic tools. *European Journal of Medicinal Chemistry, 139*, 349–366. Available from https://doi.org/10.1016/j.ejmech.2017.07.036.

Ali, M. A., Mondal, K., Jiao, Y., Oren, S., Xu, Z., Sharma, A., & Dong, L. (2016). Microfluidic Immuno-Biochip for Detection of Breast Cancer Biomarkers Using Hierarchical Composite of Porous Graphene and Titanium Dioxide Nanofibers. *ACS Applied Materials & Interfaces, 8*(32), 20570–20582.

An, Y., Jin, T., Zhu, Y., Zhang, F., & He, P. (2019). An ultrasensitive electrochemical aptasensor for the determination of tumor exosomes based on click chemistry. *Biosensors and Bioelectronics, 142*, 111503.

Antiochia, R., Tortolini, C., Tasca, F., Gorton, L., & Bollella, P. (2018). *Graphene and 2D-like nanomaterials: Different biofunctionalization pathways for electrochemical biosensor development. Graphene Bioelectronics* (pp. 1–35). Elsevier Inc. Available from https://doi.org/10.1016/B978-0-12-813349-1.00001-9.

Amani, J., Khoshroo, A., & Rahimi-Nasrabadi, M. (2017). Electrochemical immunosensor for the breast cancer marker CA 15–3 based on the catalytic activity of a CuS/reduced graphene oxide nanocomposite towards the electrooxidation of catechol. *Microchimica Acta, 185*(1), 79.

Chiu, N. F., Lin, T. L., & Kuo, C. T. (2018). Highly sensitive carboxyl-graphene oxide-based surface plasmon resonance immunosensor for the detection of lung cancer for cytokeratin 19 biomarker in human plasma. *Sensors and Actuators, B: Chemical, 265*, 264–272. Available from https://doi.org/10.1016/j.snb.2018.03.070.

Cui, F., Zhou, Z., & Zhou, H. S. (2020). Review – Measurement and analysis of cancer biomarkers based on electrochemical biosensors. *Journal of the Electrochemical Society, 167*(3). Available from https://doi.org/10.1149/2.0252003JES.

Cruz, S. M. A. (2016). Graphene: The missing piece for cancer diagnosis? *Sensors (Switzerland)*.

Das, P., & Krull, U. J. (2017). Detection of a cancer biomarker protein on modified cellulose paper by fluorescence using aptamer-linked quantum dots. *Analyst, 142*(17), 3132–3135. Available from https://doi.org/10.1039/c7an00624a.

Das, P., Sedighi, A., & Krull, U. J. (2018). Cancer biomarker determination by resonance energy transfer using functional fluorescent nanoprobes. *Analytica Chimica Acta, 1041*, 1–24. Available from https://doi.org/10.1016/j.ac.2018.07.060.

Ding, Y., Ling, J., Wang, H., Zou, J., Wang, K., Xiao, X., & Yang, M. (2015). Fluorescent detection of Mucin 1 protein based on aptamer functionalized biocompatible carbon dots and graphene oxide. *Analytical Methods, 7*(18), 7792–7798. Available from https://doi.org/10.1039/c5ay01680k.

Dong, Y., Zhang, T., Lin, X., Feng, J., Luo, F., Gao, H., ... He, Q. (2020). Graphene/aptamer probes for small molecule detection: from in vitro test to in situ imaging. *Microchimica Acta, 187*(3). Available from https://doi.org/10.1007/s00604-020-4128-8.

Dou, B., Xu, L., Jiang, B., Yuan, R., & Xiang, Y. (2019). Aptamer-functionalized and gold nanoparticle array-decorated magnetic graphene nanosheets enable multiplexed and sensitive electrochemical detection of rare circulating tumor cells in whole blood. *Analytical Chemistry, 91*(16), 10792–10799. Available from https://doi.org/10.1021/acs.analchem.9b02403.

Dreyer, D. R., Park, S., Bielawski, C. W., & Ruoff, R. S. (2010). The chemistry of graphene oxide. *Chemical Society Reviews, 39*(1), 228–240. Available from https://doi.org/10.1039/b917103g.

Duan, F., Zhang, S., Yang, L., Zhang, Z., He, L., & Wang, M. (2018). Bifunctional aptasensor based on novel two-dimensional nanocomposite of MoS2 quantum dots and g-C3N4 nanosheets decorated with chitosan-stabilized Au nanoparticles for selectively detecting prostate specific antigen. *Analytica Chimica Acta*, *1036*, 121–132. Available from https://doi.org/10.1016/j.ac.2018.06.070.

Eskiizmir, G., Baskin, Y., & Yapici, K. (2018). *Graphene-based nanomaterials in cancer treatment and diagnosis. In Fullerenes graphenes and nanotubes: A pharmaceutical approach* (pp. 331–374). Elsevier. Available from https://doi.org/10.1016/B978-0-12-813691-1.00009-9.

Farahavar, G., Abolmaali, S. S., Gholijani, N., & Nejatollahi, F. (2019). Antibody-guided nanomedicines as novel breakthrough therapeutic, diagnostic and theranostic tools. *Biomaterials Science*, *7*(10), 4000–4016. Available from https://doi.org/10.1039/c9bm00931k.

Feagin, T. A., Maganzini, N., & Soh, H. T. (2018). Strategies for Creating Structure-Switching Aptamers. *ACS Sensors*, *3*(9), 1611–1615. Available from https://doi.org/10.1021/acssensors.8b00516.

Feng, J., Li, Y., Li, M., Li, F., Han, J., Dong, Y., … Wei, Q. (2017). A novel sandwich-type electrochemical immunosensor for PSA detection based on PtCu bimetallic hybrid (2D/2D) rGO/g-C3N4. *Biosensors and Bioelectronics*, *91*, 441–448.

Fritea, L., Tertis, M., Sandulescu, R., & Cristea, C. (2018). *Enzyme–graphene platforms for electrochemical biosensor design with biomedical applications*, . *Methods in Enzymology* (Vol. 609, pp. 293–333). Academic Press Inc. Available from https://doi.org/10.1016/bs.mie.2018.05.010.

Furukawa, K., Ueno, Y., Takamura, M., & Hibino, H. (2016). Graphene FRET Aptasensor. *ACS Sensors*, *1*(6), 710–716. Available from https://doi.org/10.1021/acssensors.6b00191.

Ghosal, K., & Sarkar, K. (2018). Biomedical applications of graphene nanomaterials and beyond. *ACS Biomaterials Science and Engineering*, *4*(8), 2653–2703. Available from https://doi.org/10.1021/acsbiomaterials.8b00376.

Gu, H., Tang, H., Xiong, P., & Zhou, Z. (2019). Biomarkers-based biosensing and bioimaging with graphene for cancer diagnosis. *Nanomaterials*, *9* (1). Available from https://doi.org/10.3390/nano9010130.

Gupta, P., Bharti, A., Kaur, N., Singh, S., & Prabhakar, N. (2018). An electrochemical aptasensor based on gold nanoparticles and graphene oxide doped poly(3,4-ethylenedioxythiophene) nanocomposite for detection of MUC1. *Journal of Electroanalytical Chemistry*, *813*, 102–108.

Hassan, E. M., & DeRosa, M. C. (2020). Recent advances in cancer early detection and diagnosis: Role of nucleic acid based aptasensors. *TrAC – Trends in Analytical Chemistry*, 124. Available from https://doi.org/10.1016/j.trac.2020.115806.

Hatami, Z., Jalali, F., Amouzadeh Tabrizi, M., & Shamsipur, M. (2019). Application of metal-organic framework as redox probe in an electrochemical aptasensor for sensitive detection of MUC1. *Biosensors and Bioelectronics*, *141*, 111433.

He, Y., Lin, Y., Tang, H., & Pang, D. (2012). A graphene oxide-based fluorescent aptasensor for the turn-on detection of epithelial tumor marker mucin 1. *Nanoscale*, *4*(6), 2054–2059. Available from https://doi.org/10.1039/c2nr12061e.

Hong, Z., Chen, G., Yu, S., Huang, R., & Fan, C. (2018). A potentiometric aptasensor for carcinoembryonic antigen (CEA) on graphene oxide nanosheets using catalytic recycling of DNase i with signal amplification. *Analytical Methods*, *10*(45), 5364–5371. Available from https://doi.org/10.1039/c8ay02113a.

Hongxia, C., Zaijun, L., Ruiyi, L., Guangli, W., & Zhiguo, G. (2019). Molecular machine and gold/graphene quantum dot hybrid based dual amplification strategy for voltammetric detection of VEGF165. *Microchimica Acta*, *186*(4).

Hu, T., Zhang, L., Wen, W., Zhang, X., & Wang, S. (2016). Enzyme catalytic amplification of miRNA-155 detection with graphene quantum dot-based electrochemical biosensor. *Biosensors and Bioelectronics*, *77*, 451–456. Available from https://doi.org/10.1016/j.bios.2015.09.068.

Huang, J., Tian, J., Zhao, Y., & Zhao, S. (2015). Ag/Au nanoparticles coated graphene electrochemical sensor for ultrasensitive analysis of carcinoembryonic antigen in clinical immunoassay. *Sensors and Actuators, B: Chemical*, *206*, 570–576. Available from https://doi.org/10.1016/j.snb.2014.09.119.

Huang, J.-Y., Zhao, L., Lei, W., Wen, W., Wang, Y.-J., Bao, T., … Wang, S.-F. (2018). A high-sensitivity electrochemical aptasensor of carcinoembryonic antigen based on graphene quantum dots-ionic liquid-nafion nanomatrix and DNAzyme-assisted signal amplification strategy. *Biosensors and Bioelectronics*, *99*, 28–33.

Jang, H. D., Kim, S. K., Chang, H., & Choi, J.-W. (2015). 3D label-free prostate specific antigen (PSA) immunosensor based on graphene–gold composites. *Biosensors and Bioelectronics*, *63*, 546–551.

Jayanthi, V. S. P. K. S. A., Das, A. B., & Saxena, U. (2017). Recent advances in biosensor development for the detection of cancer biomarkers. *Biosensors and Bioelectronics*, *91*, 15–23. Available from https://doi.org/10.1016/j.bios.2016.12.014.

Jeong, H. Y., Baek, S. H., Chang, S. J., Cheon, S. A., & Park, T. J. (2015). Robust fluorescence sensing platform for detection of CD44 cells based on graphene oxide/gold nanoparticles. *Colloids and Surfaces B: Biointerfaces*, *135*, 309–315. Available from https://doi.org/10.1016/j.colsurfb.2015.07.083.

Jeong, S., Barman, S. C., Yoon, H., & Park, J. Y. (2019). A prostate cancer detection immunosensor based on nafion/reduced graphene oxide/aldehyde functionalized methyl pyridine composite electrode. *Journal of the Electrochemical Society*, *166*(12), B920–B926.

Jia, H., Tian, Q., Xu, J., Lu, L., Ma, X., & Yu, Y. (2018). Aerogels prepared from polymeric β-cyclodextrin and graphene aerogels as a novel host-guest system for immobilization of antibodies: a voltammetric immunosensor for the tumor marker CA 15–3. *Microchimica Acta*, *185*(11).

Jian, W., Wang, C., Chen, Z., Yu, Y., Sun, D., Han, L., & Shi, L. (2016). Nonenzymatic electrochemical immunosensor using ferroferric oxide-manganese dioxide-reduced graphene oxide. *Nanocomposite as Label for α-Fetoprotein Detection. Nano*, *11*(10). Available from https://doi.org/10.1142/S1793292016501162.

Jing, A., Xu, Q., Feng, W., & Liang, G. (2020). An electrochemical immunosensor for sensitive detection of the tumor marker carcinoembryonic antigen (CEA) based on three-dimensional porous nanoplatinum/graphene. *Micromachines*, *11*(7). Available from https://doi.org/10.3390/mi11070660.

Karimipour, M., Heydari-Bafrooei, E., Sanjari, M., Johansson, M. B., & Molaei, M. (2019). A glassy carbon electrode modified with TiO2(200)-rGO hybrid nanosheets for aptamer based impedimetric determination of the prostate specific antigen. *Microchimica Acta*, *186*(1).

Kermani, H. A., Hosseini, M., Dadmehr, M., Hosseinkhani, S., & Ganjali, M. R. (2017). DNA methyltransferase activity detection based on graphene quantum dots using fluorescence and fluorescence anisotropy. *Sensors and Actuators, B: Chemical*, *241*, 217–223. Available from https://doi.org/10.1016/j.snb.2016.10.078.

Kim, M. I., Kim, M. S., Woo, M. A., Ye, Y., Kang, K. S., Lee, J., & Park, H. G. (2014). Highly efficient colorimetric detection of target cancer cells utilizing superior catalytic activity of graphene oxide-magnetic-platinum nanohybrids. *Nanoscale*, *6*(3), 1529–1536. Available from https://doi.org/10.1039/c3nr05539f.

Korecká, L., Vytřas, K., & Bílková, Z. (2018). Immunosensors in early cancer diagnostics: From individual to multiple biomarker assays. *Current Medicinal Chemistry*, *25*(33), 3973–3987. Available from https://doi.org/10.2174/0929867324666171121101245.

Kuralay, F. (2019). *Nanomaterials-based enzyme biosensors for electrochemical applications: Recent trends and future prospects. New developments in nanosensors for pharmaceutical analysis* (pp. 381–408). Elsevier. Available from https://doi.org/10.1016/B978-0-12-816144-9.00012-2.

Li, Q., Wang, Q., Yang, X., Wang, K., Zhang, H., & Nie, W. (2017). High sensitivity surface plasmon resonance biosensor for detection of microRNA and small molecule based on graphene oxide-gold nanoparticles composites. *Talanta*, *174*, 521–526. Available from https://doi.org/10.1016/j.talanta.2017.06.048.

Li, Y., Dong, L., Wang, X., Liu, Y., Liu, H., & Xie, M. (2018). Development of graphite carbon nitride based fluorescent immune sensor for detection of alpha fetoprotein. *Spectrochimica Acta – Part A: Molecular and Biomolecular Spectroscopy*, *196*, 103–109. Available from https://doi.org/10.1016/j.saa.2018.02.012.

Li, M., Wang, P., Li, F., Chu, Q., Li, Y., & Dong, Y. (2017). An ultrasensitive sandwich-type electrochemical immunosensor based on the signal amplification strategy of mesoporous core–shell Pd@Pt nanoparticles/amino group functionalized graphene nanocomposite. *Biosensors and Bioelectronics*, *87*, 752–759.

Li, G., Li, S., Wang, Z., Xue, Y., Dong, C., Zeng, J., … Zhou, Z. (2018). Label-free electrochemical aptasensor for detection of alpha-fetoprotein based on AFP-aptamer and thionin/reduced graphene oxide/gold nanoparticles. *Analytical Biochemistry*, *547*, 37–44.

Liang, L., Su, M., Li, L., Lan, F., Yang, G., Ge, S., … Song, X. (2016). Aptamer-based fluorescent and visual biosensor for multiplexed monitoring of cancer cells in microfluidic paper-based analytical devices. *Sensors and Actuators, B: Chemical*, *229*, 347–354. Available from https://doi.org/10.1016/j.snb.2016.01.137.

Mansouri Majd, S., & Salimi, A. (2018). Ultrasensitive flexible FET-type aptasensor for CA 125 cancer marker detection based on carboxylated multiwalled carbon nanotubes immobilized onto reduced graphene oxide film. *Analytica Chimica Acta*, *1000*, 273–282.

Mesgari, F. (2020). Paper-based chemiluminescence and colorimetric detection of cytochrome c by cobalt hydroxide decorated mesoporous carbon. *Micromechanical Journal*.

Miao, L., Jiao, L., Zhang, J., & Li, H. (2017). Amperometric sandwich immunoassay for the carcinoembryonic antigen using a glassy carbon electrode modified with iridium nanoparticles, polydopamine and reduced graphene oxide. *Microchimica Acta*, *184*(1), 169–175.

Mishra, A., & Verma, M. (2010). Cancer biomarkers: Are we ready for the prime time? *Cancers*, *2*(1), 190–208. Available from https://doi.org/10.3390/cancers2010190.

Nemati, F., & Zare-Dorabei, R. (2019). A ratiometric probe based on Ag2S quantum dots and graphitic carbon nitride nanosheets for the fluorescent detection of Cerium. *Talenta*, 249–255.

Nemati, F., Zare-Dorabei, R., Hosseini, M., & Ganjali, M. R. (2018). Fluorescence turn-on sensing of thiamine based on Arginine – functionalized graphene quantum dots (Arg-GQDs): Central composite design for process optimization. *Sensors and Actuators, B: Chemical*, *255*, 2078–2085. Available from https://doi.org/10.1016/j.snb.2017.09.009.

Nie, G., Wang, Y., Tang, Y., Zhao, D., & Guo, Q. (2018). A graphene quantum dots based electrochemiluminescence immunosensor for carcinoembryonic antigen detection using poly(5-formylindole)/reduced graphene oxide nanocomposite. *Biosensors and Bioelectronics*, *101*, 123–128.

Ouyang, L., Hu, Y., Zhu, L., Cheng, G. J., & Irudayaraj, J. (2017). A reusable laser wrapped graphene-Ag array based SERS sensor for trace detection of genomic DNA methylation. *Biosensors and Bioelectronics*, *92*, 755–762. Available from https://doi.org/10.1016/j.bios.2016.09.072.

Pachauri, N., Dave, K., Dinda, A., & Solanki, P. R. (2018). Cubic CeO2 implanted reduced graphene oxide-based highly sensitive biosensor for non-invasive oral cancer biomarker detection. *Journal of Materials Chemistry B*, *6*(19), 3000–3012.

Pal, M., & Khan, R. (2017). Graphene oxide layer decorated gold nanoparticles based immunosensor for the detection of prostate cancer risk factor. *Analytical Biochemistry*, *536*, 51–58.

Parolo, C., & Merkoçi, A. (2013). Paper-based nanobiosensors for diagnostics. *Chemical Society Reviews*, *42*(2), 450–457. Available from https://doi.org/10.1039/c2cs35255a.

Pei, H., Zhu, S., Yang, M., Kong, R., Zheng, Y., & Qu, F. (2015). Graphene oxide quantum dots@silver core-shell nanocrystals as turn-on fluorescent nanoprobe for ultrasensitive detection of prostate specific antigen. *Biosensors and Bioelectronics*, *74*, 909–914. Available from https://doi.org/10.1016/j.bios.2015.07.056.

Peng, X., Hu, T., Bao, T., Zhao, L., Zeng, X., Wen, W., … Wang, S. (2017). A label-free electrochemical biosensor for methyltransferase activity detection and inhibitor screening based on graphene quantum dot and enzyme-catalyzed reaction. *Journal of Electroanalytical Chemistry*, *799*, 327–332. Available from https://doi.org/10.1016/j.jelechem.2017.06.030.

Peng, C., Hua, M. Y., Li, N. S., Hsu, Y. P., Chen, Y. T., Chuang, C. K., … Yang, H. W. (2019). A colorimetric immunosensor based on self-linkable dual-nanozyme for ultrasensitive bladder cancer diagnosis and prognosis monitoring. *Biosensors and Bioelectronics*, *126*, 581–589. Available from https://doi.org/10.1016/j.bios.2018.11.022.

Pumera, M. (2010). Graphene-based nanomaterials and their electrochemistry. *Chemical Society Reviews*, *39*(11), 4146–4157. Available from https://doi.org/10.1039/c002690p.

Qian, L., Li, Q., Baryeh, K., Qiu, W., Li, K., Zhang, J., … Liu, G. (2019). Biosensors for early diagnosis of pancreatic cancer: a review. *Translational Research*, *213*, 67–89. Available from https://doi.org/10.1016/j.trsl.2019.08.002.

Sabet, F. S., Hosseini, M., Khabbaz, H., Dadmehr, M., & Ganjali, M. R. (2017). FRET-based aptamer biosensor for selective and sensitive detection of aflatoxin B1 in peanut and rice. *Food Chemistry*, *220*, 527–532. Available from https://doi.org/10.1016/j.foodchem.2016.10.004.

Salek-Maghsoudi, A., Vakhshiteh, F., Torabi, R., Hassani, S., Ganjali, M. R., Norouzi, P., ... Abdollahi, M. (2018). Recent advances in biosensor technology in assessment of early diabetes biomarkers. *Biosensors and Bioelectronics*, *99*, 122–135. Available from https://doi.org/10.1016/j.bios.2017.07.047.

Şenel, B., Demir, N., Büyükköroğlu, G., & Yıldız, M. (2019). Graphene quantum dots: Synthesis, characterization, cell viability, genotoxicity for biomedical applications. *Saudi Pharmaceutical Journal*, *27*(6), 846–858. Available from https://doi.org/10.1016/j.jsps.2019.05.006.

Sengupta, J., & Hussain, C. M. (2019). Graphene and its derivatives for Analytical Lab on Chip platforms. *TrAC - Trends in Analytical Chemistry*, *114*, 326–337. Available from https://doi.org/10.1016/j.trac.2019.03.015.

Sharifuzzaman, M., Barman, S. C., Rahman, M. T., Zahed, M. A., Xuan, X., & Park, J. Y. (2019). Green synthesis and layer-by-layer assembly of amino-functionalized graphene oxide/carboxylic surface modified trimetallic nanoparticles nanocomposite for label-free electrochemical biosensing. *Journal of the Electrochemical Society*, *166*(12), B983–B993.

Shekari, Z., Zare, H. R., & Falahati, A. (2019). Electrochemical sandwich aptasensor for the carcinoembryonic antigen using graphene quantum dots, gold nanoparticles and nitrogen doped graphene modified electrode and exploiting the peroxidase-mimicking activity of a G-quadruplex DNAzyme. *Microchimica Acta*, *186*(8), 530.

Shi, J., Lyu, J., Tian, F., & Yang, M. (2017). A fluorescence turn-on biosensor based on graphene quantum dots (GQDs) and molybdenum disulfide (MoS2) nanosheets for epithelial cell adhesion molecule (EpCAM) detection. *Biosensors and Bioelectronics*, *93*, 182–188. Available from https://doi.org/10.1016/j.bios.2016.09.012.

Shin, B., Park, J. S., Chun, H. S., Yoon, S., Kim, W. K., & Lee, J. (2020). A fluorescence/colorimetric dual-mode sensing strategy for miRNA based on graphene oxide. *Analytical and Bioanalytical Chemistry*, *412*(1), 233–242. Available from https://doi.org/10.1007/s00216-019-02269-0.

Singh, V. K., Kumar, S., Pandey, S. K., Srivastava, S., Mishra, M., Gupta, G., ... Srivastava, A. (2018). Fabrication of sensitive bioelectrode based on atomically thin CVD grown graphene for cancer biomarker detection. *Biosensors and Bioelectronics*, *105*, 173–181.

Soozanipour, A., & Taheri-Kafrani, A. (2018). *Enzyme Immobilization on Functionalized Graphene Oxide Nanosheets: Efficient and Robust Biocatalysts*, . *Methods in enzymology* (Vol. 609, pp. 371–403). Academic Press Inc. Available from https://doi.org/10.1016/bs.mie.2018.06.010.

Tabasi, A., Noorbakhsh, A., & Sharifi, E. (2017). Reduced graphene oxide-chitosan-aptamer interface as new platform for ultrasensitive detection of human epidermal growth factor receptor 2. *Biosensors and Bioelectronics*, *95*, 117–123.

Tajik, S., Dourandish, Z., Zhang, K., Beitollahi, H., Le, Q. V., Jang, H. W., & Shokouhimehr, M. (2020). Carbon and graphene quantum dots: A review on syntheses, characterization, biological and sensing applications for neurotransmitter determination. *RSC Advances*, *10*(26), 15406–15429. Available from https://doi.org/10.1039/d0ra00799d.

Tombelli, S., Minunni, M., & Mascini, M. (2005). Analytical applications of aptamers. *Biosensors and Bioelectronics*, *20*(12), 2424–2434. Available from https://doi.org/10.1016/j.bios.2004.11.006.

Tsai, H., Lin, W., Chuang, M., Lu, Y., & Fuh, C. B. (2017). Multifunctional nanoparticles for protein detections in thin channels. *Biosensors and Bioelectronics*, *90*, 153–158. Available from https://doi.org/10.1016/j.bios.2016.11.023.

Tuteja, S. K., Chen, R., Kukkar, M., Song, C. K., Mutreja, R., Singh, S., ... Suri, C. R. (2016). A label-free electrochemical immunosensor for the detection of cardiac marker using graphene quantum dots (GQDs). *Biosensors and Bioelectronics*, *86*, 548–556. Available from https://doi.org/10.1016/j.bios.2016.07.052.

Verma, S., Singh, A., Shukla, A., Kaswan, J., Arora, K., Ramirez-Vick, J., ... Singh, S. P. (2017). Anti-IL8/AuNPs-rGO/ITO as an Immunosensing Platform for Noninvasive Electrochemical Detection of Oral Cancer. *ACS Applied Materials and Interfaces*, *9*(33), 27462–27474. Available from https://doi.org/10.1021/acsami.7b06839.

Wang, Huan, et al. (2017). Facile synthesis of cuprous oxide nanowiresdecorated graphene oxide nanosheetsnanocomposites and its application in label-freeelectrochemical immunosensor. *Biosensors and Bioelectronics*, *87*.

Wang, T., Liu, J., Gu, X., Li, D., Wang, J., & Wang, E. (2015). Label-free electrochemical aptasensor constructed by layer-by-layer technology for sensitive and selective detection of cancer cells. *Analytica Chimica Acta*, *882*, 32–37. Available from https://doi.org/10.1016/j.ac.2015.05.008.

Wang, Z., Wu, S., Colombi Ciacchi, L., & Wei, G. (2018). Graphene-based nanoplatforms for surface-enhanced Raman scattering sensing. *Analyst*, *143*(21), 5074–5089. Available from https://doi.org/10.1039/c8an01266k.

Wang, H., Chen, H., Huang, Z., Li, T., Deng, A., & Kong, J. (2018). DNase I enzyme-aided fluorescence signal amplification based on graphene oxide-DNA aptamer interactions for colorectal cancer exosome detection. *Talanta*, *184*, 219–226. Available from https://doi.org/10.1016/j.talanta.2018.02.083.

Wang, Y., Zhang, Y., Wu, D., Ma, H., Pang, X., Fan, D., ... Du, B. (2017). Ultrasensitive Label-free Electrochemical Immunosensor based on Multifunctionalized Graphene Nanocomposites for the Detection of Alpha Fetoprotein. *Scientific Reports*, *7*(1), 42361.

Wang, K., He, M. Q., Zhai, F. H., He, R. H., & Yu, Y. L. (2017). A novel electrochemical biosensor based on polyadenine modified aptamer for label-free and ultrasensitive detection of human breast cancer cells. *Talanta*, *166*, 87–92.

Wang, W., Ge, L., Sun, X., Hou, T., & Li, F. (2015). Graphene-Assisted Label-Free Homogeneous Electrochemical Biosensing Strategy based on Aptamer-Switched Bidirectional DNA Polymerization. *ACS Applied Materials & Interfaces*, *7*(51), 28566–28575.

Wang, Y., Sun, S., Luo, J., Xiong, Y., Ming, T., Liu, J., ... Cai, X. (2020). Low sample volume origami-paper-based graphene-modified aptasensors for label-free electrochemical detection of cancer biomarker-EGFR. *Microsystems and Nanoengineering*, *6*(1).

Wei, B., Mao, K., Liu, N., Zhang, M., & Yang, Z. (2018). Graphene nanocomposites modified electrochemical aptamer sensor for rapid and highly sensitive detection of prostate specific antigen. *Biosensors and Bioelectronics*, *121*, 41–46.

Wu, L., Hu, Y., Sha, Y., Li, W., Yan, T., Wang, S., ... Su, X. (2016). An "in-electrode"-type immunosensing strategy for the detection of squamous cell carcinoma antigen based on electrochemiluminescent AuNPs/g-C3N4 nanocomposites. *Talanta*, *160*, 247–255.

Xu, X., Wei, M., Liu, Y., Liu, X., Wei, W., Zhang, Y., & Liu, S. (2017). A simple, fast, label-free colorimetric method for detection of telomerase activity in urine by using hemin-graphene conjugates. *Biosensors and Bioelectronics*, *87*, 600–606. Available from https://doi.org/10.1016/j.bios.2016.09.005.

Yang, C., Denno, M. E., Pyakurel, P., & Venton, B. J. (2015). Recent trends in carbon nanomaterial-based electrochemical sensors for biomolecules: A review. *Analytica Chimica Acta*, *887*, 17–37. Available from https://doi.org/10.1016/j.ac.2015.05.049.

Yang, L., Zhen, S. J., Li, Y. F., & Huang, C. Z. (2018). Silver nanoparticles deposited on graphene oxide for ultrasensitive surface-enhanced Raman scattering immunoassay of cancer biomarker. *Nanoscale*, *10*(25), 11942–11947. Available from https://doi.org/10.1039/c8nr02820f.

Yang, Y., Yang, X., Yang, Y., & Yuan, Q. (2018). Aptamer-functionalized carbon nanomaterials electrochemical sensors for detecting cancer relevant biomolecules. *Carbon*, *129*, 380–395. Available from https://doi.org/10.1016/j.carbon.2017.12.013.

Zhang, G., Liu, Z., Fan, L., & Guo, Y. (2018). Electrochemical prostate specific antigen aptasensor based on hemin functionalized graphene-conjugated palladium nanocomposites. *Mikrochimica Acta*, *185*(3), 159. Available from https://doi.org/10.1007/s00604-018-2686-9.

Zhang, S., Sun, C., & Zhou, W. (2017). Electrochemical immunoassay determination of a cancer biomarker (CA19-9) by horseradish peroxidase. *International Journal of Electrochemical Science*, *12*(9), 8447–8456. Available from https://doi.org/10.20964/2017.09.31.

Zhang, X., Jing, Q., Ao, S., Schneider, G. F., Kireev, D., Zhang, Z., & Fu, W. (2020). Ultrasensitive field-effect biosensors enabled by the unique electronic properties of graphene. *Small (Weinheim an der Bergstrasse, Germany)*, *16*(15). Available from https://doi.org/10.1002/smll.201902820.

Zou, Y., Huang, S., Liao, Y., Zhu, X., Chen, Y., Chen, L., ... Tan, W. (2018). Isotopic graphene-isolated-Au-nanocrystals with cellular Raman-silent signals for cancer cell pattern recognition. *Chemical Science*, *9*(10), 2842–2849. Available from https://doi.org/10.1039/c7sc05442d.

Chapter 14

# Role of biosensor-based devices for diagnosis of nononcological disorders

Sayali Mukherjee[1] and Surojeet Das[2]
[1]*Amity Institute of Biotechnology, Amity University Uttar Pradesh, Lucknow, India*
[2]*European Molecular Biology Laboratory Australia, Australian Regenerative Medicine Institute, Monash University, Melbourne, VIC, Australia*

## 14.1 Introduction

Biosensors have become very indispensable in the medical arena for the rapid diagnosis of several diseases including cancer, infectious diseases, cardiovascular diseases, neurological diseases, and autoimmune disorders. These devices specifically convert a biological entity into an electrical signal which can be detected and analyzed. Biosensors ensure high sensitivity, portability, stability, selectivity, and real-time analysis for early diagnosis of diseases which is a very important factor in therapeutics. Biosensor-based diagnosis is very important in therapeutics to start early treatment. The detection of pathogenic agents and biological molecules is very essential in clinical diagnosis. The techniques commonly employed in this respect are polymerase chain reaction (PCR), enzyme-linked immunosorbent assay (ELISA), reverse transcriptase PCR, or biosensor-based techniques. Biomolecules have limitations with respect to utilization and stability and therefore, a fast-diagnostic technique that is more selective, stable, and economical is preferred. Biosensors essentially consist of four components: analyte, bioreceptor, transducer, and detector. The target analyte interacts with the bioreceptor, receptor-ligand interaction or chemical reaction takes place and the transducer converts the change in the molecule into a quantifiable signal which is measured by the detector. The detector finally gives the output signal which is directly proportional to the concentration of the analyte. The transduction principles can be widely classified into optical, electrochemical, piezoelectric, or magnetic. Biosensors have been used in clinical diagnostics to detect proteins, peptides, cancer biomarkers (Seda et al., 2016), nucleic acids (Bartold et al., 2018), bacteria, viruses (Matthias et al., 2017), toxins, etc.

Biosensors are ultrasensitive devices that can detect with a high level of precision. It can detect minute biological changes in the cellular microenvironment with minimum reagent consumption and automated monitoring which is exploited for replacing the time-consuming laboratory analysis. New generation biosensors increase the efficacy even more, as for example quantum dots, graphene-based biosensors, carbon nanotubes, microfluidic biosensors, or lab-on-a-chip. In this chapter, the different approaches of developing biosensors for nononcological disorders like viral diseases, bacterial or protozoan diseases, and cardiovascular and neurological diseases will be reviewed for effective diagnosis and management.

## 14.2 Biosensors for infectious diseases

Infectious diseases still pose an omnipresent threat to global and public health, especially in many countries and rural areas of cities. Underlying reasons of such serious maladies can be summarized as the paucity of appropriate analysis methods and subsequent treatment strategies due to the limited access of centralized and equipped healthcare facilities for diagnosis.

The role of the immune system is very important in the pathogenesis of infectious diseases. The immune system is very diverse and specific and it responds differently to different types of pathogens. The clonal selection theory leads to the development of the concept of the production of specific antibodies by B lymphocytes with unique antigen specificity. The T lymphocytes play a central role in the immune system in the activation of B lymphocytes. The pathogen

Biosensor Based Advanced Cancer Diagnostics. DOI: https://doi.org/10.1016/B978-0-12-823424-2.00020-X

when it enters the body is taken up by the antigen-presenting cells (APC), for example, the macrophages and dendritic cells. APC processes the antigen and presents the antigen on a major histocompatibility complex (MHC) to T helper (Th) cells. The T lymphocytes recognize the antigen with its T cell receptor (TCR) and get activated and, in turn, activates the B lymphocytes with the specific antigen specificity. The activated B lymphocytes form a clone of B lymphocytes and some of these mature into plasma cell which secretes antibodies, and some of these B lymphocytes are retained as a memory cell. The degree of response of immune cells to different pathogens in the form of cytokine release, antibodies, or expression of activation receptors on these cells could be monitored and used for diagnostics. The unique markers, cluster of differentiation (CD), present on these immune cells could be tracked and analyzed.

There are several immunological tools available based on the characterization of surface markers and nuclear morphology. Flow cytometry is a very versatile approach of cell analysis and sorting, which is used for various research and clinical purposes. Protein microarray and biosensors for the assay of the serological molecules secreted by immune cells are some of the other techniques based on immunology. Several novel micro-technologies are being developed for the analysis of T lymphocyte and B lymphocyte function and for detecting the markers of inflammation or autoantibodies.

### 14.2.1 Biosensors for pathogenic viruses

Viral infections have caused various episodes of mortality worldwide like the influenza A H1N1 in 2009, the Ebola outbreak in 2014 (Chan & Gack, 2016), and the current pandemic of severe acute respiratory syndrome coronavirus 2 (SARS-CoV-2) causing coronavirus disease 2019 (COVID 19). Viruses infect the host cell and inject their nuclear material into the host cell. The genetic material integrates into the host cell genome, replicates, transcribes, and synthesizes the viral proteins. The viral proteins can manipulate and evade the host immune response. The main challenge in the detection of viruses is that they get modified very quickly and resistant strains might arise. The increased probability of spreading from person to person calls for early intervention. Biosensors hold the promise of applying new diagnostic techniques for the detection of specific biomarkers in diseases.

#### *14.2.1.1 Hepatitis virus*

Hepatitis virus is a potentially life-threatening virus known for long as a virus that causes chronic hepatitis, cirrhosis, and liver cancer. According to the reports of the World Health Organization (WHO), hepatitis B infection accounts for approximately 887,000 deaths (Msomi et al., 2020). Diagnosis of the disease at the initial stage helps in better management of the disease. Several biosensor-based techniques have been devised to detect hepatitis viruses. These devices require the efficient immobilization of the biological molecules over the surface of a transducer, which detects the analyte and transforms it into a detectable signal. The DNA-based hybridization method is used for the diagnosis of the hepatitis B virus using electrochemical impedance spectroscopy (EIS). DNA probes are immobilized on a gold electrode using a magnet. This electrochemical streptavidin-based biosensor can detect 50 picomoles of hepatitis viral DNA in a sample of 20 $\mu$L without PCR amplification with a dynamic range of 2.53–50.6 nmol/mL (Mohamed et al., 2008). Surface plasmon resonance biosensor is also used for the detection of hepatitis B surface antigen which is far more sensitive than detection by ELISA (Joon et al., 2017). Several biosensors are also used with gold and silver nanoparticles for sequence-specific DNA hybridization of hepatitis virus (Adem, Ugur, & Tuncer, 2017). Gold nanorods are also developed for the detection of hepatitis and to study the immobilization of different sequences of DNA (Zahra, Samaneh, Sharmin, Mahdi, & Reza, 2015). DNA-based biosensors with electrodes of silicon oxide are highly sensitive for the diagnosis of hepatitis B virus (Ahangar & Mehrgardi, 2017). A novel fluorescent sensor platform has been developed for nucleic acid detection on low-cost δ-FeOOH nanosheets. Single-stranded DNA probes are immobilized in δ-FeOOH nanosheets. When the analyte is added to these sheets, the target DNA, if present, binds with the probe forming double-stranded DNA which is cleaved by exonuclease III, and the fluorescent signal is analyzed (Wu et al., 2020). This technique has a very high sensitivity (Fig. 14.1).

#### *14.2.1.2 Human immunodeficiency virus*

Human immunodeficiency virus (HIV) causes the communicable disease acquired immune deficiency syndrome (AIDS). These viruses show a slow rate of infection so there is always a window from the start of the infection and the manifestation of the symptoms. HIV targets the T helper cells (CD4 +) T cells and starts to replicate quickly. According to the WHO, HIV has emerged as a major global problem and since its inception has claimed almost 33 million lives globally. There are two subtypes of HIV viruses, HIV-1 and HIV-2; HIV-1 is the most common type to cause disease. Recent studies are approaching disease diagnosis with the help of biosensors. The electro-chemiluminescence biosensor employing aptamer template for HIV-1 gene detection has been developed (Bahareh, Abdollah, & Rahman, 2018).

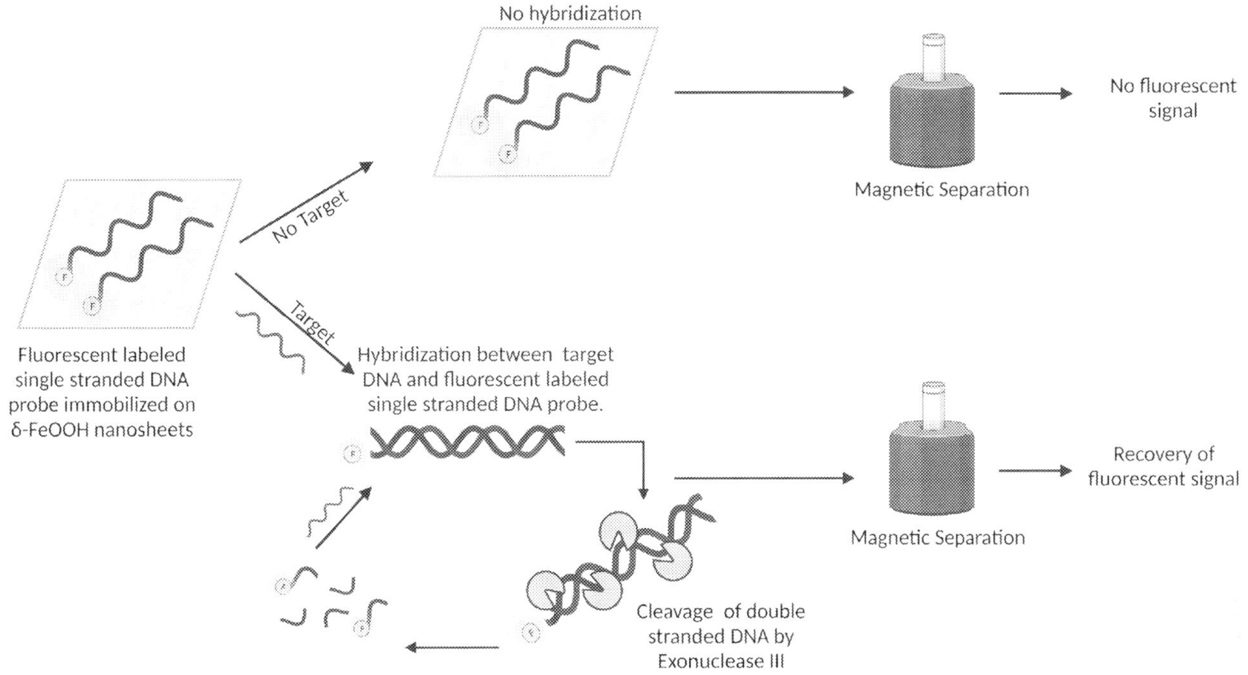


**FIGURE 14.1** Fluorescent sensor platform for detection of hepatitis virus DNA. Single-stranded DNA probes are immobilized in δ-FeOOH nanosheets. The analyte when added to these sheets, the target DNA of the virus, if present binds with the probe forming double-stranded DNA which is cleaved by exonuclease III, and the fluorescent signal is analyzed. This technique has a very high sensitivity. (Figure created with BioRender.com)

Biosensors are also designed based on the transmembrane protein of HIV-1 like glycoprotein 41 (Gp41) which has a major role in the fusion between the virally infected cells and virus. The extent of AIDS progression can be measured by measuring Gp41. A quartz crystal is modified to develop a biosensor by an imprinting method (Chun-Hua et al., 2012). An optical biosensor is created to detect HIV-1 from biological samples by sensing the shift in the resonant peak wavelength during output. This is a very sensitive biosensor where even low concentrations of viruses can be detected (Shafiee et al., 2014). p24 is another HIV-1 capsid protein antigen that can be analyzed before the generation of antibodies against HIV in the host after HIV exposure. An immunosensor-based technique is developed to detect p24. The sensitivity is much increased with a low detection limit of 0.008 ng/mL than that of ELISA for p24 which is about 1 ng/mL. This immunosensor is easy to use, reproducible, noninterfering and can be widely applied to clinical samples.

DNA-based biosensors have also been reported using graphene-polyaniline (G-PANI) modified electrodes. Zinc oxide nanoparticles nanowire has been used to design piezotronic biosensors for the detection of HIV (Cao et al., 2016). Biosensor array for detection of multiplex DNA in microliters of samples. It allows simultaneous detection of the HIV oligonucleotide sequences, HIV-1 and HIV-2, by separate thiolated hairpin-DNA probes. Methylene blue is used as a hybridization redox indicator in voltammetry. This label-free electrochemical biosensor has shown good specificity for the detection of DNA sequences for various clinical applications (Dongdong, Yage, Honglan, Qiang, & Chengxiao, 2010).

### 14.2.1.3 Dengue virus

Dengue fever is a grave disease with a heavy global public health burden caused by a virus transmitted by mosquitoes (Paula et al., 2015). Dengue fever is characterized by shock syndrome which occurs rapidly after viral infections (Lim et al., 2018). The most important biomarker of dengue virus infection was identified as a glycoprotein (GP), nonstructural 1 (NS1), as it accumulates at a high level in dengue patients (Guittet et al., 2011).

The biosensor-based methods include electrochemical detection methods, nanoparticle-based immunosensors (Edwar et al., 2017), surface plasmon resonance methods (Ru, Oleksiy, Devi, Rafiq, & Pierre, 2014), and DNA-based bioassays (Yang et al., 2018). Immittance assay is studied in an electrochemical assay to reduce the detection time to a few seconds. A bifunctional self-assembled monolayer (BSAM) contains polyethylene glycol moieties, and a tethered redox thiol are formed on the electrode (Santos, Bueno, & Davis, 2018). C-type lectin domain family 5, member A

(CLEC5A), a receptor of the dengue virus, is also utilized for designing biosensors. An electrochemical biosensor for the detection of dengue virus RNA is used by immobilizing aminated DNA probes onto the alumina channel walls, which increases the sensitivity of the sensor (Varun, Jiajia, & Chee-Seng, 2012). A silicon nanowire-based biosensor for detection of the target DNA sequence of the dengue virus is created with a detection limit of approximately 2.0 fM and high sensitivity was achieved (Nuzaihan et al., 2016). A biosensor comprised of an immobilized dengue-specific DNA probe on $Cu_2CdSnS_4$ quaternary alloy nanostructures has also been developed for dengue virus (Odeh et al., 2017).

#### 14.2.1.4 Ebola virus

The Ebola virus infection is highly contagious and causes a grave human disease for which there is no specific antiviral treatment. The infection starts manifesting itself with influenza-like symptoms and later affects different organs leading to multiorgan failure. Specific diagnosis of the infection at an early stage is very important. The challenges related to diagnosis are the initial symptoms of Ebola virus infection are nonspecific. An very sensitive electrochemical biosensor has been developed for Ebola virus using a thiolated DNA capture probe and hybridized with biotinylated target strand DNA. EIS is used for detection and a low detection limit value is attained using the fabricated biosensor and the standard deviation of the blank solution (Hoda & Siamak, 2018). Optofluidic biosensors have been developed for detecting whole viruses in biological samples by measuring the change in transmission spectra (Yanik et al., 2010). A modified optical system with detection of single nucleic acid with fluorescence on a silicon chip is devised. This biosensor is much more specific with a low detection value of 0.2pfu/mL (Hawkins et al., 2015).

A field-effect transistor (FET) is also used for fast and specific detection of Ebola virus antigens. Materials with good electronic properties like graphene and black phosphorous are used for this biosensor. Reduced graphene oxide is very useful for FET biosensors for analyzing antibodies against Ebola virus antigens (Mao et al., 2013). Ebola virus surface GP plays a significant role in the disease pathogenesis and causes disruption of endothelial cells and suppresses the host immune system (Mohan et al., 2015). Large amounts of truncated GP, soluble GP, is released after attachment of the Ebola virus which can be detected for diagnosis.

#### 14.2.1.5 Coronavirus

Coronavirus disease 2019 (COVID-19) is a fatal respiratory illness caused by severe acute respiratory syndrome coronavirus 2 (SARS-CoV-2) (Simiao, Juntao, Weizhong, Chen, & Till, 2020; Hu, Matthew, & Joy, 2020). The WHO announced the outbreak of COVID-19 as a global pandemic (Zheng, Ma, Zhang, & Xie, 2020). Nucleic acid tests are commonly applied to diagnose COVID-19 in sputum (or saliva) or nasal secretions (Cormac, 2020; Jiang, Aiqiao, & Tao, 2020). The antibody test is done with a test strip (IgG/IgM test) in blood and antibodies against the virus are detected (Weimin et al., 2020). ELISA IgM and ELISA IgG with merged gold immunochromatography assay (GICA) produced a high detection sensitivity.

Several tools and techniques have been developed for the detection of coronavirus infection and most of the techniques are based on conventional qualitative real-time polymerase chain reaction (qRT-PCR). These techniques are time-consuming and labor-intensive. Chip-based and paper-based biosensors are considered low-cost and user-friendly, which offer tremendous potential for rapid medical diagnosis (Weimin et al., 2020). ELISA-based immunosensor is developed to analyze IgM and IgG against coronavirus in human serum. A biosensor with a single-stranded DNA probe against viral RNA can be designed. A highly sensitive photothermal biosensor to detect SARS-CoV is also configured (Guangyu et al., 2020). Biosensors can also be created using an electrochemical transduction approach. Moreover, DNA hybridization can be considered as a portable electrochemical sensor for point mutation detection of COVID-19-specific viral RNA and cDNA (Suryasnata & Govind, 2020). The SARS-CoV2 is a novel type of virus and much needs to be explored about the pathogenesis of the viral infection. The concept of developing biosensors for COVID 19 is still in its infancy and is dependent on further research findings on the nature of infection of this virus, the mechanism of cytokine storm, and fatality.

### 14.2.2 Biosensors for pathogenic bacteria

#### 14.2.2.1 Tuberculosis

Tuberculosis is a highly infectious disease that mostly affects the lungs. This disease is relevant world-wide and has re-emerged as a grave disease globally. Tuberculosis is caused by gram-negative bacillus, *Mycobacterium tuberculosis*. The bacteria can remain inactive for years in the macrophages of the host immune system without causing any harm, but as soon as the immune system of the host becomes weakened, the bacteria become active and infect mainly the lungs along with other parts of the body. The most common biosensors for the diagnosis of tuberculosis are the electrochemical and electrical biosensors (Maumita, Gajjala, Nagarajan, & Malhotra, 2010). DNA biosensors can be made by immobilizing DNA against specific DNA of *M. tuberculosis* on the electrode. These biosensors are sensitive and flexible but pose certain limitations like maintaining a constant pH and dependence on molecular factors of the analyte. Electronic "nose" biosensors have been designed for the rapid detection of tuberculosis (Fig. 14.2). This biosensor consists of three main parts: a sample delivery chamber, a detection array or pattern recognition system, and a computing system for data interpretation and analysis (Chatterjee, Castro, & Feller, 2013). The distribution pattern of the volatile organic materials on the detector is the characteristic feature of a particular microorganism and differs from other bacteria of the same species. Temperature and humidity must be maintained in the chamber. Electronic nose biosensors are commercially available and are used widely, e.g., "Aeonose" for the detection of tuberculosis. Although this biosensor is cheap and simple, the main limitation of using this biosensor is its low sensitivity. Nanowires or tubes made up of silicon or carbon are also used to build nanowire-based biosensors that work as FETs for the detection of tuberculosis. Nano-biosensors have very high sensitivity, and carbon and silicon provide a better surface functionalization compared to metal electrodes. The major limitation of this biosensor is that technologically it is more challenging.

Optical biosensors for the detection of *M. tuberculosis* include fiber-optic biosensors based on the principle of evanescence. The optical fiber used consists of two components, namely, the core fiber through which light propagates, and the surrounding fiber which is the region of the decaying evanescent field. The evanescent waves interact with the fluorophores on the outer surface of the fibers that are connected to a detector. The light emitted can be detected by a spectrophotometer. The thickness of the fibers can be changed depending on the pathogen type. Niacin is present in high concentrations in the sputum of tuberculosis patients which may react with cyanogen bromide and aniline and give a yellow color in optical biosensors. The surface plasmon resonance (SPR)-based biosensor is another biosensor based on the optical principle. The other side of the noble metal surface is coated with receptors, Surface plasmons are emitted when polarized light is incident on noble metals which interacts with the biological molecules in the analyte. The angle on reflection changes accordingly and can be detected. A different type of optical biosensor used for the detection of pulmonary tuberculosis is the breathalyzer biosensor. The patient is first allowed to cough inside a masked bag-like structure containing a collection tube. The samples are then distributed on the surface of prisms at the bottom of a glass tube coated with *Mycobacterium*-specific antibody and fluorescent peptide epitopes which are analogous to the *M. tuberculosis* T cell epitope, Ag85B. The antibody coated on the prism has a higher affinity for the bacterial antigen than its analog. Therefore, competition assay is detected with a laser detection unit. This biosensor is sensitive, cheap, rapid, and effective for outdoor clinics (McNerney et al., 2010). The breathalyzer is effective for initial screening of patients with tuberculosis, but it is not very effective for relapsed cases of drug-resistant ones.

Piezoelectric quartz crystal biosensors are mechanical biosensors which are developed for detecting either *M. tuberculosis*, or the volatile products or by-products released by the bacteria. Magnetic biosensors like diagnostic magnetic resonance (DMR) are a miniature modification of conventional nuclear magnetic resonance (NMR) to detect *M. tuberculosis*. The magnetic barcode platform is another type of magnetic biosensor for the detection of nucleic acid which is based on a magnetic bar-coding strategy. The mycobacterial sequences can be amplified by PCR and studied on microspheres with magnetic nanoprobes and analyzed by NMR.

### 14.2.2.2 Leprosy

Leprosy is an endemic chronic bacterial disease caused by *Mycobacterium leprae*. Timely diagnosis of the disease is vital to reduce its severity. Electrochemical genosensor is used, where a target single-stranded DNA oligonucleotide sequence against *M. leprae* is immobilized on a graphite electrode. The detection limit ranges from 0.35 to

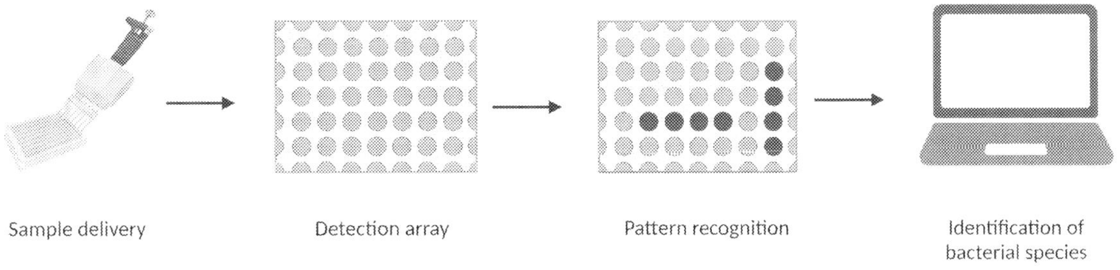


**FIGURE 14.2** An electronic "nose" biosensor. This biosensor consists of three main parts—a sample delivery chamber, a detection array or pattern recognition system, and a computing system for data interpretation and analysis. (Figure created with BioRender.com)

35.0 ng $\mu L^{-1}$. Detection of leprosy by IgM and IgG antibodies reacting with two specific antigens immobilized on nitrocellulose membranes is also a rapid and specific quantitative test (Paula Vaz Cardoso et al., 2013).

A photoelectrochemical sensor has been developed for the detection of leprosy by analyzing the antibodies in patient serum. The system comprises cadmium sulfide and nickel hydroxide layered on fluorine-containing tin oxide-coated glass slides (CdS /Ni $(OH)_2$/ FTO). The detection was done by UV–Vis spectroscopy and EIS and was further evaluated by scanning electrochemical microscopy and scanning electron microscopy (SEM) (Neto et al., 2019). A highly sensitive electrochemical quartz microbalance biosensor is used to detect *M. leprae* with the help of epitope matching. Specific epitopes of this bacterium are imprinted on gold-coated quartz electrodes which can analyze specific binding to blood samples of infected patients (Archana et al., 2019).

#### *14.2.2.3 Meningitis*

Bacterial meningitis is a very serious disease commonly caused by *Streptococcus pneumoniae*, *Haemophilus influenzae*, *Listeria monocytogenes*, or *Neisseria meningitidis* (Bahr & Boulware, 2014). The deadliest causative agent being *N. meningitidis* causing meningococcal meningitis is a worldwide endemic disease (Siqueira et al., 2017). The disease is characterized by severe damage to the brain and spinal cord. Specific bacterial antigens are utilized for designing the biosensors. One such antigen is an outer membrane protein 85 (Omp85) which codes for a virulent membrane protein of *N. meningitidis*. Immunosensors have been developed to detect *Neisseria* by antibodies against the bacterial protein Omp85 (Reddy, Mainwaring, Kobaisi, Zeephongsekul, & Fecondo., 2012). Genosensors are also available for the detection of meningitis which detects the specific virulent gene of the bacterium by immobilizing labeled oligonucleotide probe on gold electrodes and analyzing the electronic signal. Genosensors based on the virulent bacterial gene, rmpM (reduction-modifiable protein M) are more sensitive and can detect *N. meningitidis* in the cerebrospinal fluid sample within 30 min. The detection is done by cyclic voltammetry and differential pulse voltammetry. A modified type of rmpM genosensor was devised using carbon-mercaptooctadecane/carboxylated multi-walled carbon nanotubes for detection of *N. meningitidis* in cerebrospinal fluid with increased specificity and sensitivity (Kumar, Minakshi, Shashi, & Ashok, 2013b). A similar type of biosensor is also designed to target another virulent bacterial gene, Omp85, to detect *N. meningitidis* in the cerebrospinal fluid within 30 min including a 1-min response time (Kumar, Minakshi, Shashi, & Ashok, 2013a). An advanced type of electrochemical genosensor for the detection of *N. meningitidis* is devised in which graphite electrode is fabricated with poly (4-aminophenol). Ethidium bromide is used as an electrochemical indicator. The biosensor is characterized by its specificity, stability, and potential for point-of-care diagnosis (Castro et al., 2018). Immobilization of thiolated DNA probes of *N. meningitidis* on flower-like zinc oxide nanostructures are synthesized on platinized silicon substrates. The detection can be done by cyclic voltammetry and EIS. This DNA biosensor can detect complementary bacterial single-stranded target DNA with good sensitivity and a low detection limit of 5 ng $\mu L^{-1}$ (Manvi, Vinay, & Monika, 2014).

### 14.2.3 Biosensors for pathogenic protozoa

#### *14.2.3.1 Malaria*

Malaria is a deadly disease caused by a protozoan, *Plasmodium*, and transmitted through the bite of female *Anopheles* mosquitoes. The disease can be preventable and curable once detected early. An estimate by WHO reports approximately 228 million cases of malaria have been reported worldwide with 405,000 deaths in 2018 (World Health Organization, 2019) There are five species of *Plasmodium* that have been identified as the causative agent, *Plasmodium falciparum, Plasmodium vivax, Plasmodium ovale* and *Plasmodium knowlesi*. Detecting the disease at an early stage is very important for proper cure. Most of the laboratory tests for malaria are immunological-based methods which are colorimetric, photoelectrochemical, surface plasma resonance, ELISA, or Western Blot. Most of these methods are time-consuming and costly. Several simple, portable biosensors are designed which give more specific, sensitive, and rapid outputs than conventional assay methods. An electrochemical biosensor is the most common type of technique for malaria detection with low cost and increased sensitivity. A modified chronoamperometric ELISA is designed using horseradish peroxidase (HRP) for detecting *P. falciparum* histidine-rich protein 2 (*Pf*HRP 2) in serum samples. A detection range of 1–100 ng/mL is obtained using this biosensor (Aver, Jon, & Ibtisam, 2017). Detection of *Pf*HRP 2 can also be done by magnetic immunoassay (MIA) by utilizing magnetic and nanoparticles. This electrochemical immunosensor fabricated with magnetic nanoparticles has a detection limit of 0.36 ng/mL (De Souza Castilho, Laube, Yamanaka, Alegret, & Pividori, 2011). Aptamers, the sequence of DNA or RNA (12–80 nucleotides) long can bind to proteins, cells, and small molecules with high specificity and sensitivity. The aptamers are advantageous because of their stability, flexibility in the binding affinity, and resistance to the environment. Aptasensor of *P. falciparum* is designed with lactate dehydrogenase (LDH)-specific single-stranded DNA aptamers by SELEX (systematic evolution of ligands by exponential enrichment process) using magnetic beads. This biosensor can effectively detect *P. falciparum* in serum samples with a broad detection range of 100 fM to 100 nM. It has outstanding specificity and can be used for malaria diagnosis (Singh, Arya, Estrela, & Goswami, 2018).

#### 14.2.3.2 *Leishmaniasis*

Leishmaniasis is a parasitic disease caused by the protozoan, *Leishmania*, and transmitted through sandflies. This disease is more common in rural areas than in urban areas. There are 20 species of *Leishmania* that have been identified. Leishmaniasis is mainly of three types: visceral (kala-azar), cutaneous, and mucocutaneous. According to WHO, 700,000 to 1 million new cases are reported annually (Roatt et al., 2020). The rapid and cost-effective detection of the disease is most effective in the management of the disease. Immunosensors have been designed and tested on animal models where anti-*Leishmania infantum* antibodies are deposited on a gold surface, coated with cysteamine and glutaraldehyde, blocked with glycine, and hamster spleens infected with *L. infantum* are added to it. Detection can be done by immunofluorescence or surface plasmon resonance (SPR). Piezoelectric immunosensors are also developed. Specific DNA-based biosensors are created with thiolated sequences of *Leishmania* spp. Genome and detected by electrochemical methods. This technique is novel and ultrasensitive with a very low detection limit. Nanocomposite-based probes are also used in specially designed genosensors developed based on immobilization of a DNA sequence that targeted 18 S rRNA gene sequences from *Leishmania donovani*.

#### 14.2.3.3 *Chagas disease*

Chagas disease is an infectious, inflammatory disease caused by the protozoan parasite *Trypanosoma cruzi*. It is transmitted through the feces of infected blood-sucking triatomine bugs, also known as "kissing bugs." Approximately 8–11 million people are infected with Chagas disease globally; the disease ranks fourth in mortality and morbidity among globally neglected tropical diseases (Tarleton et al., 2014). Chagas disease develops mainly through two phases: acute and chronic. The acute phase starts 6–8 weeks after infection and then a prolonged intermediary phase continues wherein the parasite persists in equilibrium with the host immune system, but the patient is asymptomatic. The disease eventually progresses to the symptomatic chronic phase and affects the digestive system, heart, or peripheral nervous system. The early prevention of the disease requires proper monitoring and diagnosis and initiation of the therapy. However, most of the cases go undiagnosed because of limited access to the medical system and a lack of specificity and sensitivity of the diagnostic tests. Therefore, more accessible and rapid methods of diagnosis must be formulated in areas where this disease is more prevalent. Several immunodiagnostic tests are available for the detection of *T. cruzi* antibodies in the blood of infected patients. The most common techniques used are indirect hemagglutination, immunofluorescence, and ELISA. However, these conventional techniques have certain limitations of cross-reactivity, lack of sensitivity, and time-consumption. Rapid lateral flow tests are commercially available to detect the protozoan parasite in whole blood, plasma, or serum by immunochromatography or immunoblot, but with low sensitivity. Biosensors have been devised for the detection of *T. cruzi*. In electrochemical biosensors, recombinant polypeptide antigens, either cytoplasmic or flagellar, are adsorbed on gold or platinum electrodes. The assay is done in biological samples and analyzed by EIS. Antigenic proteins from epimastigote membranes are used to develop an amperometric immunosensor for the detection of the parasite. Surface plasmon resonance (SPR) transducers also detect the presence of anti-*T. cruzi* antibodies in human serum within 20 min (Luz et al., 2016). This biosensor was found to be more sensitive than ELISA or PCR-based diagnosis of Chagas disease. Surface plasmon resonance-based biosensors are used for commercial purposes, but it is very costly and not portable (Tomazelli, Santis, Jesus, Fracassi da, & Emanuel, 2014). Electrochemical biosensor using magnetic microbeads are also developed for the diagnosis of Chagas disease, which is sensitive, but since it involves electrochemical reactions, indirect steps are involved (Comerci et al., 2016). Electrochemical immunosensors measuring anti-IgM antibodies against *T. cruzi* are designed using gold nanoparticles to improve the limit of the sensor. A highly sensitive biosensor has been developed using nanowires based on FET technology. The antigen used is the chimeric *T. cruzi* antigen, IBMP-8, achieved a detection limit of 6fM (Oliveira et al., 2017).

### 14.2.4 Biosensors for cardiovascular diseases

Cardiovascular diseases are the leading cause of death worldwide demanding 17.9 million lives every year; 85% of these deaths are due to heart attack and stroke (Wilkins et al., 2017). The lifestyle parameters like smoking, alcohol, and junk foods leading to obesity, are some of the risk factors of cardiovascular diseases. Diabetes is also associated with the occurrence of these diseases. Myocardial infarction is the main cause of hospital admissions and death as a timely intervention is the main key behind better prognosis and reduced risk of unfavorable clinical implications. During the episode of heart attack or stroke, most of the biomarkers are released in the blood like lactate dehydrogenase (LDH), triglycerides, creatinine kinase, cardiac troponin, C reactive protein, myoglobin, and brain-type natriuretic peptides (BNP). These biomarkers persist in blood for a short period of time after the heart attack. Rapid and inexpensive diagnosis of these biomarkers is very crucial to minimize death due to cardiovascular diseases. The conventional biochemical tests, nowadays, are replaced by the detection of cardiovascular diseases through biosensors. Electrochemical biosensors are very important and commonly used in cardiovascular diseases. Immunosensors are designed to detect the cardiac biomarkers by immobilizing antibodies against the biomarker on gold surfaces by chemical coupling, using molecules like mercaptoundecanoic acid, dithiobis (succinimidyl propionate), or 3,3′-dithiobis (sulfosuccinimidyl propionate). These linkers allow binding on the gold surface with thiol groups and link the antibody at the other end with carboxylic or ester bonds. Diazonium salts or 3-aminopropyltriethoxysilane (APTS) are also used as linkers (Serafín et al., 2018). Several label-free immunosensors are designed for rapid detection and reduction of complexity but with reduced sensitivity. Therefore, HRP conjugated or fluorescent-tagged secondary antibodies are used in these biosensors which makes it sensitive but time-consuming. Optical biosensors are also available which measure the fluorescent intensity, colorimetric sensing, SPR, and surface-enhanced Raman spectroscopy (SERS). Most of the optical biosensors allow label-free analysis and are easy to perform, but almost all these biosensors are bulky instruments which makes it disadvantageous for application. The diagnosis of cardiovascular diseases by sensing cardiac troponin I (cTnI) using titanium oxide nanotubes (TiO2) can be done with a very low detection limit (Piyush, Archana, Greer, & Karthik, 2012). Silicon nanowire-based biosensors with FETs are also used for the detection of cTnI. DNA aptamers are synthesized to design biosensors to detect cardiac biomarkers. The most common cardiac aptamers are developed for cnTnI, cTnT, BNP, myoglobin. Carbon dots are also used for developing myoglobin-based cardiac biosensors. The myoglobin aptamer is adsorbed on the surface of the carbon dots resulting in quenching of the fluorescence of the carbon dots. When myoglobin is added, the fluorescence is recovered which can be measured with a very low detection limit of 20 pg/mL. This biosensor allows detection of myoglobin concentration in saliva, urine, and serum, and can be effectively used for the detection of myocardial infarction (Jishun et al., 2019).

### 14.2.5 Biosensors for neurological disorders

Neurological disorders involve neuroinflammation which is the basis of several neurological disorders like Parkinson's disease, Alzheimer's disease, and multiple sclerosis. Identification of specific biomarkers is very important in the early diagnosis and management of the disease. However, it is difficult to measure the markers of the central nervous system in these diseases. The inflammatory molecules released by the central nervous system are measured in blood, cerebrospinal fluid, and tears. A biosensor for the detection of multiple cytokines in cerebrospinal fluid has been developed. Microfluidic surface plasmon resonance sensor can detect six cytokines simultaneously with a low detection limit. The sensitivity of these biosensors can further be increased by the addition of nanotubes conjugated with antibodies. Electrochemical immunosensors are designed for the detection of amyloid-beta levels in Alzheimer's disease (Abreu et al., 2018). Although amyloid-beta (Aβ) peptide and tau protein are the main biomarkers in Alzheimer's disease, it is found that sometimes multiple markers may provide a better diagnosis. Biosensors targeting acetylcholine neurotransmitters have also been reported. Optical fluorescent-based biosensors are used for Parkinson's disease for detecting dopamine levels (Leopold, Shcherbakova, & Verkhusha, 2019). In multiple sclerosis, the synthetic antigenic probe is used to detect autoantibodies by SPR.

## 14.3 Recent challenges and future perspectives

In the past few decades, there has been enormous research on the designing of biosensors for the early diagnosis and treatment of nononcological disorders. The use of biosensors is very advantageous in terms of portability, implantability, efficacy, specificity, and sensitivity. The medical and healthcare areas have benefitted enormously with the advent of biosensors. The most widely commercialized biosensor is the "glucometer" for monitoring the blood glucose level in millions of diabetic patients globally. The use of nanomaterials like gold nanoparticles, quantum dots, and carbon nanotubes has led to advancement in the field. However, there are many challenges in the large-scale application of biosensors and that is reflected in the lack of commercialization of biosensor-based products. This can be attributed to problems in production and reproducibility on a large scale. The specificity and predictability of the biosensor poses another challenge and sometimes results in false-positive results in biological samples. The genosensors, which can detect specific sequences of genetic materials of pathogenic viruses or bacteria, are very useful for diagnosing infectious diseases as well as genetic diseases. The major limitation in using genosensors is the sample preparation. Although loop-mediated isothermal amplification (LAMP) has been developed to overcome the limitation of sample preparation, more standardization is required to achieve optimum results in diagnostics.

Implantable and wearable biosensors developed through artificial intelligence should be explored more to get tailored precision treatment for individual patients. Ingestible biosensors are a very important area of research nowadays. The future perspectives in this field would be to devise biosensors for noninfectious diseases with increased accuracy and sensitivity and also to improve the scalability of the biosensors. More incorporation of nanomaterials should be encouraged in the biosensors to increase efficacy. Therefore, more advancement is required in the optimization and designing of low-cost and efficient biosensors for clinical diagnostics on a large scale.

## 14.4 Conclusion

Biosensors are very important and reliable technologies for the detection of nononcological disorders like infectious diseases, cardiovascular diseases, or neurological diseases. Different types of electrochemical, optical, piezoelectric, or genosensors have been developed. Nanoparticles and nanosensor-based strategies further helped in making these technologies more effective. Detection and quantification in a reduced volume of samples with higher specificity and sensitivity is achieved. These technologies overcome the tedious procedure of sample processing and analysis through conventional methods. Miniaturized sizes also lead to increased feasibility in the use of biosensors. Although biosensors are very advantageous for rapid and specific diagnosis of different diseases in the present scenario, several challenges are there which need to be addressed before translating these biosensors from the research laboratory to clinics.

## References


Abreu, C. M., Soares-dos-Reis, R., Melo, P. N., Relvas, J. B., Guimarães, J., Sá, M. J., … Pinto, I. M. (2018). Emerging biosensing technologies for neuroinflammatory and neurodegenerative disease diagnostics. *Frontiers in Molecular Neuroscience*. Available from https://doi.org/10.3389/fnmol.2018.00164.

Adem, Z., Ugur, T., & Tuncer, C. (2017). SERS detection of hepatitis B virus DNA in a temperature-responsive sandwich-hybridization assay. *Journal of Raman Spectroscopy*, 668–672. Available from https://doi.org/10.1002/jrs.5109.

Ahangar, L. E., & Mehrgardi, M. A. (2017). Amplified detection of hepatitis B virus using an electrochemical DNA biosensor on a nanoporous gold platform. *Bioelectrochemistry*, *117*, 83–88. Available from https://doi.org/10.1016/j.bioelechem.2017.06.006.

Archana, K., Juhi, S., Kumar, S. A., Richa, A., Richa, R., Tulika, R., & Meenakshi, S. (2019). Epitope imprinting of *Mycobacterium leprae* bacteria via molecularly imprinted nanoparticles using multiple monomers approach. *Biosensors and Bioelectronics*, 111698. Available from https://doi.org/10.1016/j.bios.2019.111698.

Aver, H., Jon, A., & Ibtisam, T. (2017). Development of an immunosensor for PfHRP 2 as a biomarker for malaria detection. *Biosensors*, 28. Available from https://doi.org/10.3390/bios7030028.

Bahareh, B., Abdollah, S., & Rahman, H. (2018). A molecularly imprinted electrochemiluminescence sensor for ultrasensitive HIV-1 gene detection using EuS nanocrystals as luminophore. *Biosensors and Bioelectronics*, 332–339. Available from https://doi.org/10.1016/j.bios.2018.06.003.

Bahr, N. C., & Boulware, D. R. (2014). Methods of rapid diagnosis for the etiology of meningitis in adults. *Biomarkers in Medicine*, *8*(9), 1085–1103. Available from https://doi.org/10.2217/BMM.14.67.

Bartold, K., Pietrzyk-Le, A., Golebiewska, K., Lisowski, W., Cauteruccio, S., Licandro, E., … Kutner, W. (2018). Oligonucleotide determination via peptide nucleic acid macromolecular imprinting in an electropolymerized CG-rich artificial oligomer analogue. *ACS Applied Materials and Interfaces*, *10*(33), 27562–27569. Available from https://doi.org/10.1021/acsami.8b09296.

Cao, X., Cao, X., Guo, H., Li, T., Jie, Y., Wang, N., & Wang, Z. L. (2016). Piezotronic effect enhanced label-free detection of DNA using a schottky-contacted ZnO nanowire biosensor. *ACS Nano*, *10*(8), 8038–8044. Available from https://doi.org/10.1021/acsnano.6b04121.

Castro, H., Kochi, L. T., Moço, A. C. R., Coimbra, R. S., Oliveira, G. C., Cuadros-Orellana, S., … Brito-Madurro, A. G. (2018). A new genosensor for meningococcal meningitis diagnosis using biological samples. *Journal of Solid State Electrochemistry*, *22*, 2339–2346.

Chan, Y. K., & Gack, M. U. (2016). Viral evasion of intracellular DNA and RNA sensing. *Nature Reviews Microbiology*, 360–373. Available from https://doi.org/10.1038/nrmicro.2016.45.

Chatterjee, S., Castro, M., & Feller, J. F. (2013). An e-nose made of carbon nanotube based quantum resistive sensors for the detection of eighteen polar/nonpolar VOC biomarkers of lung cancer. *Journal of Materials Chemistry B*, *1*(36), 4563–4575. Available from https://doi.org/10.1039/c3tb20819b.

Chun-Hua, L., Yan, Z., Shui-Fen, T., Zhi-Bin, F., Huang-Hao, Y., Xi, C., & Guo-Nan, C. (2012). Sensing HIV related protein using epitope imprinted hydrophilic polymer coated quartz crystal microbalance. *Biosensors and Bioelectronics*, 439–444. Available from https://doi.org/10.1016/j.bios.2011.11.008.

Comerci, D. J., Sabrina, R., Jaime, A., Buscaglia, C. A., Laura, M., Ugalde, J. E., ... Ciocchini, A. E. (2016). Electrochemical magnetic microbeads-based biosensor for point-of-care serodiagnosis of infectious diseases. *Biosensors and Bioelectronics*, 24–33. Available from https://doi.org/10.1016/j.bios.2016.01.021.

Cormac, S. (2020). Fast, portable tests come online to curb coronavirus pandemic. *Nature Biotechnology*, 515–518. Available from https://doi.org/10.1038/d41587-020-00010-2.

De Souza Castilho, M., Laube, T., Yamanaka, H., Alegret, S., & Pividori, M. I. (2011). Magneto immunoassays for plasmodium falciparum histidine-rich protein 2 related to malaria based on magnetic nanoparticles. *Analytical Chemistry*, *83*(14), 5570–5577. Available from https://doi.org/10.1021/ac200573s.

Dongdong, Z., Yage, P., Honglan, Q., Qiang, G., & Chengxiao, Z. (2010). Label-free electrochemical DNA biosensor array for simultaneous detection of the HIV-1 and HIV-2 oligonucleotides incorporating different hairpin-DNA probes and redox indicator. *Biosensors and Bioelectronics*, 1088–1094. Available from https://doi.org/10.1016/j.bios.2009.09.032.

Edwar, I., Tien-Chun, T., I-Fang, C., Tzu-Chuan, H., Chuen, P. G., & Hsien-Chang, C. (2017). A bead-based immunofluorescence-assay on a microfluidic dielectrophoresis platform for rapid dengue virus detection. *Biosensors and Bioelectronics*, 174–180. Available from https://doi.org/10.1016/j.bios.2017.04.011.

Guangyu, Q., Zhibo, G., Yile, T., Jean, S., Gerd, A. K.-U., & Jing, W. (2020). Dual-functional plasmonic photothermal biosensors for highly accurate severe acute respiratory syndrome coronavirus 2 detection. *ACS Nano*, 5268–5277. Available from https://doi.org/10.1021/acsnano.0c02439.

Guittet, E., Rey, F. A., Flamand, M., Salmon, J., D'Alayer, J., Ermonval, M., ... Mégret, F. (2011). Secreted dengue virus nonstructural protein NS1 is an atypical barrel-shaped high-density lipoprotein. *Proceedings of the National Academy of Sciences of the United States of America*, *108*(19), 8003–8008. Available from https://doi.org/10.1073/pnas.1017338108.

Hawkins, A. R., Schmidt, H., Wall, T. A., Stott, M. A., Stambaugh, A., Alfson, K., ... Patterson, J. L. (2015). Optofluidic analysis system for amplification-free, direct detection of Ebola infection. *Scientific Reports*, *5*. Available from https://doi.org/10.1038/srep14494.

Hoda, I., & Siamak, F. (2018). A novel electrochemical DNA biosensor for Ebola virus detection. *Analytical Biochemistry*, 151–155. Available from https://doi.org/10.1016/j.ab.2018.06.010.

Hu, T. Y., Matthew, F., & Joy, W. (2020). Insights from nanomedicine into chloroquine efficacy against COVID-19. *Nature Nanotechnology*, 247–249. Available from https://doi.org/10.1038/s41565-020-0674-9.

Jiang, Z., Aiqiao, F., & Tao, L. (2020). Consistency analysis of COVID-19 nucleic acid tests and the changes of lung CT. *Journal of Clinical Virology*, 104359. Available from https://doi.org/10.1016/j.jcv.2020.104359.

Jishun, C., Fengying, R., Qinhua, C., Dan, L., Weidong, M., Tuo, H., ... Congxia, W. (2019). A fluorescent biosensor for cardiac biomarker myoglobin detection based on carbon dots and deoxyribonuclease I-aided target recycling signal amplification. *RSC Advances*, 4463–4468. Available from https://doi.org/10.1039/c8ra09459d.

Joon, T. Y., Allaudin, Z. N., Akhavan, R. M., Hidayah, M. N., Mohd, A. M. L., Rani, B. A., ... Abdullah, R. (2017). Wide dynamic range of surface-plasmon-resonance-based assay for hepatitis B surface antigen antibody optimal detection in comparison with ELISA. *Biotechnology and Applied Biochemistry*, 735–744. Available from https://doi.org/10.1002/bab.1528.

Kumar, D. S., Minakshi, S., Shashi, K., & Ashok, K. (2013a). Omp85 genosensor for detection of human brain bacterial meningitis. *Biotechnology Letters*, 929–935. Available from https://doi.org/10.1007/s10529-013-1161-2.

Kumar, D. S., Minakshi, S., Shashi, K., & Ashok, K. (2013b). rmpM Genosensor for detection of human brain bacterial meningitis in cerebrospinal fluid. *Applied Biochemistry and Biotechnology*, 198–208. Available from https://doi.org/10.1007/s12010-013-0339-3.

Leopold, A. V., Shcherbakova, D. M., & Verkhusha, V. V. (2019). Fluorescent biosensors for neurotransmission and neuromodulation: Engineering and applications. *Frontiers in Cellular Neuroscience*, *13*, 474. Available from https://doi.org/10.3389/fncel.2019.00474.

Lim, J. M., Kim, J. H., Ryu, M. Y., Cho, C. H., Park, T. J., & Park, J. P. (2018). An electrochemical peptide sensor for detection of dengue fever biomarker NS1. *Analytica Chimica Acta*, *1026*, 109–116. Available from https://doi.org/10.1016/j.ac.2018.04.005.

Luz, J. G. G., Souto, D. E. P., Machado-Assis, G. F., de Lana, M., Luz, R. C. S., Martins-Filho, O. A., ... Martins, H. R. (2016). Applicability of a novel immunoassay based on surface plasmon resonance for the diagnosis of Chagas disease. *Clinica Chimica Acta*, *454*, 39–45. Available from https://doi.org/10.1016/j.cc.2015.12.025.

Manvi, T., Vinay, G., & Monika, T. (2014). Flower-like ZnO nanostructure based electrochemical DNA biosensor for bacterial meningitis detection. *Biosensors and Bioelectronics*, 200–207. Available from https://doi.org/10.1016/j.bios.2014.03.036.

Mao, S., Yu, K., Chang, J., Steeber, D. A., Ocola, L. E., & Chen, J. (2013). Direct growth of vertically-oriented graphene for field-effect transistor biosensor. *Scientific Reports*, 3. Available from https://doi.org/10.1038/srep01696.

Matthias, B., Claudia, K., Sabine, E., Fania, G., Fabian, E., Hartmut, G., ... Schöning, M. J. (2017). Tobacco mosaic virus as enzyme nanocarrier for electrochemical biosensors. *Sensors and Actuators B: Chemical*, 716–722. Available from https://doi.org/10.1016/j.snb.2016.07.096.

Maumita, D., Gajjala, S., Nagarajan, R., & Malhotra, B. D. (2010). Application of nanostructured ZnO films for electrochemical DNA biosensor. *Thin Solid Films*, 1196–1201. Available from https://doi.org/10.1016/j.tsf.2010.08.069.

McNerney, R., Wondafrash, B. A., Amena, K., Tesfaye, A., McCash, E. M., & Murray, N. J. (2010). Field test of a novel detection device for Mycobacterium tuberculosis antigen in cough. *BMC Infectious Diseases*, 10. Available from https://doi.org/10.1186/1471-2334-10-161.

Mohamed, H. W., Carole, C., Adnane, A., François, B., Didier, L., & Nicole, J.-R. (2008). An impedimetric DNA sensor based on functionalized magnetic nanoparticles for HIV and HBV detection. *Sensors and Actuators B: Chemical*, 755–760. Available from https://doi.org/10.1016/j.snb.2008.06.020.

Mohan, G. S., Ling, Y., Wenfang, L., Ana, M., Xiaoqian, L., Bishu, S., ... Lyles, D. S. (2015). Less is more: Ebola virus surface glycoprotein expression levels regulate virus production and infectivity. *Journal of Virology*, 1205–1217. Available from https://doi.org/10.1128/jvi.01810-14.

Msomi, N., Naidoo, K., Yende-Zuma, N., Padayatchi, N., Govender, K., Singh, J. A., ... Mlisana, K. (2020). High incidence and persistence of hepatitis B virus infection in individuals receiving HIV care in KwaZulu-Natal, South Africa. *BMC Infectious Diseases*, *20*(1). Available from https://doi.org/10.1186/s12879-020-05575-6.

Neto, S., Yotsumoto, M. I. S., Lima, S. R. F., Pereira, L. R., Goulart, R., Luz., & Damos, S. (2019). Immunodiagnostic of leprosy exploiting a photoelectrochemical platform based on a recombinant peptide mimetic of a *Mycobacterium leprae* antigen. *Biosensors and Bioelectronics*, *143*.

Nuzaihan, M. M. N., Hashim, U., Md Arshad, M. K., Kasjoo, S. R., Rahman, S. F. A., Ruslinda, A. R., ... Shahimin, M. M. (2016). Electrical detection of dengue virus (DENV) DNA oligomer using silicon nanowire biosensor with novel molecular gate control. *Biosensors and Bioelectronics*, *83*, 106–114. Available from https://doi.org/10.1016/j.bios.2016.04.033.

Odeh, A. A., Al-Douri, Y., Voon, C. H., Mat Ayub, R., Gopinath, S. C. B., Odeh, R. A., ... Bouhemadou, A. (2017). A needle-like Cu2CdSnS4 alloy nanostructure-based integrated electrochemical biosensor for detecting the DNA of Dengue serotype 2. *Microchimica Acta*, *184*(7), 2211–2218. Available from https://doi.org/10.1007/s00604-017-2249-5.

Oliveira, D. S., Kubota, L. T., Cesar, C. L., Souza, A. P. D., Cotta, M. A., Souto, D. E. P., ... Almeida, D. B. (2017). InP nanowire biosensor with tailored biofunctionalization: Ultrasensitive and highly selective disease biomarker detection. *Nano Letters*, *17*(10), 5938–5949. Available from https://doi.org/10.1021/acs.nanolett.7b01803.

Paula Vaz Cardoso, L., Dias, R. F., Freitas, A. A., Hungria, E. M., Oliveira, R. M., Collovati, M., ... Martins Araújo Stefani, M. (2013). Development of a quantitative rapid diagnostic test for multibacillary leprosy using smart phone technology. *BMC Infectious Diseases*, *13*(1). Available from https://doi.org/10.1186/1471-2334-13-497.

Paula, A., Daniel, W., Mattias, P., Robert, B., Anja, B., Paul, Y., ... Marco, D. (2015). Quantification of NS1 dengue biomarker in serum via optomagnetic nanocluster detection. *Scientific Reports*. Available from https://doi.org/10.1038/srep16145.

Piyush, K., Archana, P., Greer, J. J., & Karthik, S. (2012). Ultrahigh sensitivity assays for human cardiac troponin I using TiO2 nanotube arrays. *Lab on a Chip*, 821. Available from https://doi.org/10.1039/c2lc20892j.

Reddy, S. B., Mainwaring, D. E., Kobaisi, M. A., Zeephongsekul, P., & Fecondo, J. V. (2012). Acoustic wave immunosensing of a meningococcal antigen using gold nanoparticle-enhanced mass sensitivity. *Biosensors and Bioelectronics*, 382–387. Available from https://doi.org/10.1016/j.bios.2011.10.051.

Roatt, B. M., de Oliveira Cardoso, J. M., De Brito, R. C. F., Coura-Vital, W., de Oliveira Aguiar-Soares, R. D., & Reis, A. B. (2020). Recent advances and new strategies on leishmaniasis treatment. *Applied Microbiology and Biotechnology*, *104*(21), 8965–8977. Available from https://doi.org/10.1007/s00253-020-10856-w.

Ru, W. W., Oleksiy, K., Devi, S. S., Rafiq, M. A. F., & Pierre, B. (2014). Serological diagnosis of dengue infection in blood plasma using long-range surface plasmon waveguides. *Analytical Chemistry*, 1735–1743. Available from https://doi.org/10.1021/ac403539k.

Santos, A., Bueno, P. R., & Davis, J. J. (2018). A dual marker label free electrochemical assay for Flavivirus dengue diagnosis. *Biosensors and Bioelectronics*, *100*, 519–525. Available from https://doi.org/10.1016/j.bios.2017.09.014.

Seda, A., Kevser, P., Fatma, Y., Canan, Ç., Handan, Y., & Adil, D. (2016). Quartz crystal microbalance based biosensors for detecting highly metastatic breast cancer cells via their transferrin receptors. *Analytical Methods*, 153–161. Available from https://doi.org/10.1039/C5AY02898A.

Serafín, V., Torrente-Rodríguez, R. M., González-Cortés, A., García de Frutos, P., Sabaté, M., Campuzano, S., ... Pingarrón, J. M. (2018). An electrochemical immunosensor for brain natriuretic peptide prepared with screen-printed carbon electrodes nanostructured with gold nanoparticles grafted through aryl diazonium salt chemistry. *Talanta*, *179*, 131–138. Available from https://doi.org/10.1016/j.talanta.2017.10.063.

Shafiee, H., Lidstone, E. A., Jahangir, M., Inci, F., Hanhauser, E., Henrich, T. J., ... Demirci, U. (2014). Nanostructured optical photonic crystal biosensor for HIV viral load measurement. *Scientific Reports*, *4*. Available from https://doi.org/10.1038/srep04116.

Simiao, C., Juntao, Y., Weizhong, Y., Chen, W., & Till, B. (2020). COVID-19 control in China during mass population movements at New Year. *The Lancet*, 764–766. Available from https://doi.org/10.1016/s0140-6736(20)30421-9.

Singh, N. K., Arya, S. K., Estrela, P., & Goswami, P. (2018). Capacitive malaria aptasensor using Plasmodium falciparum glutamate dehydrogenase as target antigen in undiluted human serum. *Biosensors and Bioelectronics*, 246–252. Available from https://doi.org/10.1016/j.bios.2018.06.022.

Siqueira, B. R., Patrícia, G. A., Luiz, D. G. J., Sérgio, B. M. P., Alberto, S. L., Lisa, O., & Mauro, G. (2017). Meningococcal disease, a clinical and epidemiological review. *Asian Pacific Journal of Tropical Medicine*, 1019–1029. Available from https://doi.org/10.1016/j.apjtm.2017.10.004.

Suryasnata, T., & Govind, S. S. (2020). Label-free electrochemical detection of DNA hybridization: A method for COVID-19 diagnosis. *Transactions of the Indian National Academy of Engineering*, 205–209. Available from https://doi.org/10.1007/s41403-020-00103-z.

Tarleton, R. L., Gürtler, R. E., Urbina, J. A., Janine, R., Rodolfo, V., & Eric, D. (2014). Chagas disease and the London declaration on neglected tropical diseases. *PLoS Neglected Tropical Diseases*, e3219. Available from https://doi.org/10.1371/journal.pntd.0003219.

Tomazelli, C. W. K., Santis, N. R. D., Jesus, M. A. D., Fracassi da, S. J. A., & Emanuel, C. (2014). Microfluidic devices with integrated dual-capacitively coupled contactless conductivity detection to monitor binding events in real time. *Sensors and Actuators B: Chemical*, 239–246. Available from https://doi.org/10.1016/j.snb.2013.10.114.

Varun, R., Jiajia, D., & Chee-Seng, T. (2012). Electrochemical nanoporous alumina membrane-based label-free DNA biosensor for the detection of *Legionella* sp. *Talanta*, 112–117. Available from https://doi.org/10.1016/j.talanta.2012.06.055.

Weimin, P., Zhifei, Z., Liyan, C., Feng, Y., Jiasheng, Y., Wensheng, C., ... Kangjun, S. (2020). Development and clinical application of a rapid IgM-IgG combined antibody test for SARS-CoV-2 infection diagnosis. *Journal of Medical Virology*, 1518–1524. Available from https://doi.org/10.1002/jmv.25727.

Wilkins, E., Wilson, L., Wickramasinghe, Bhatnagar, P., Leal, J., Luengo-Fernandez, R., ... Townsend, N. (2017). *European Cardiovascular Disease Statistics*.

World Health Organization. (2019). *World Malaria Report 2019*. Geneva World Health Organization, 232, ISBN: 978-92-4-156572-1, December 4, 2019. https://www.who.int/publications/i/item/world-malaria-report-2019.

Wu, T., Li, X., Fu, Y., Ding, X., Li, Z., Zhu, G., & Fan, J. (2020). A highly sensitive and selective fluorescence biosensor for hepatitis C virus DNA detection based on δ-FeOOH and exonuclease III-assisted signal amplification. *Talanta, 209*.

Yang, C.-F., Chang, S.-F., Hsu, T.-C., Su, C.-L., Wang, T.-C., Lin, S.-H., et al. (2018). Molecular characterization and phylogenetic analysis of dengue viruses imported into Taiwan during 2011–2016. *PLoS Negl Trop Dis*, *12*(9), e0006773. Available from https://doi.org/10.1371/journal.pntd.0006773.

Yanik, A. A., Huang, M., Kamohara, O., Artar, A., Geisbert, T. W., Connor, J. H., & Altug, H. (2010). An optofluidic nanoplasmonic biosensor for direct detection of live viruses from biological media. *Nano Letters*, 4962–4969. Available from https://doi.org/10.1021/nl103025u.

Zahra, S., Samaneh, S., Sharmin, K., Mahdi, A., & Reza, S. (2015). Electrochemical DNA biosensor based on gold nanorods for detecting hepatitis B virus. *Analytical and Bioanalytical Chemistry*, 455–461. Available from https://doi.org/10.1007/s00216-014-8303-9.

Zheng, Y. Y., Ma, Y. T., Zhang, J. Y., & Xie, X. (2020). COVID-19 and the cardiovascular system. *Nature Reviews Cardiology*, *17*(5), 259–260. Available from https://doi.org/10.1038/s41569-020-0360-5.

Chapter 15

# Biosensor-based early diagnosis of gastric cancer

**Saptaka Baruah, Bidyarani Maibam and Sanjeev Kumar**
*Department of Physics, Rajiv Gandhi University, Itanagar, India*

## 15.1 Introduction

Gastric cancer is one of the most commonly found cancers worldwide (Kono, 2016). Gastric adenocarcinomas constitute most of the stomach cancer or gastric cancer, and based on the anatomical location of the tumor, it is sub-divided into cardia (gastro-esophageal junction) and noncardia (true gastric) tumors (Van Cutsem, Sagaert, Topal, Haustermans, & Prenen, 2016). Gastric cancer is uncommon in all populations below the age of 50, and the incidence rate increases with the increase in age, reaching its peak at the age of 55–80 years. The frequency of gastric cancer is two- to three fold higher in men than in women. The age-standardize incidence rate is 15.7 per 1,000,000 men and 7 per 1,000,000 women in 2018 (Thrift & El-Serag, 2020). The highest incidence rate was seen in the high-income Asia Pacific region (29.5 per 100,000 population, age-standardized), especially Japan, South Korea, and East Asia (28.6 per 100,000 population). In East Asia, China contributed about half of the global incident in 2017, followed by Eastern Europe and Andean Latin America. Other than these regions, Mongolia and Afghanistan had the overall highest age-standardized incidence rates. Southern and eastern sub-Saharan Africa and high-income North America experienced the lowest incidence rates. The highest age-standardized death rate is experienced by East Asia, followed by Andean Latin America and central Asia (Etemadi et al., 2020). India falls in the low incidence category in the context of gastric cancer. There is a huge regional difference in gastric cancer occurrence across India. According to the national cancer registries, gastric cancer is the leading problem in the northeastern and southern states of the Indian subcontinent. As per the available report, Mizoram, has the highest recorded incidence of gastric cancer followed by Tamil Nadu. The lowest incidence of gastric cancer in India is reported in Gujrat. Gastric cancer is the fifth most frequent cancer among men and sixth among women in India. It is also the second most common reason for cancer-associated death in Indian men and women among the age group of 15–44. Detection of gastric cancer in the advanced stage in most of the patients leads to a decrease in the 5-year survival rate in comparison with the countries where early diagnosis is made. The treatment standard and protocol in most of the institutions are good as any other country, although it is not observed evenly across the country (Dikshit, Mathur, & Mhatre, 2011; Servarayan Murugesan et al., 2018; Sharma & Radhakrishnan, 2011). The incidence of stomach cancer remarkably decreases in the last half century. Nonetheless, stomach cancer is in the fifth and third positions of cancer incidence and deaths due to cancer, respectively, all over the world (Balakrishnan, George, Sharma, & Graham, 2017).

*Helicobacter pylori* (*H. pylori*) infection is the most important risk factor which causes a prolonged inflammatory reaction of the immune response (Crew & Neugut, 2006; Rawla & Barsouk, 2019). Salt and salt preserved food may also increase the threat of stomach cancer. A decrease in stomach cancer is associated with a reduction of *H. pylori* infection (Cisco, Ford, & Norton, 2008). The decline in infection rate is due to better sanitation, hygienic practice, and better food preservation methods (Sharma & Radhakrishnan, 2011). Stomach cancer epidemiology has significant geographical diversity leading to at least a 10-fold variation of incidence worldwide (Servarayan Murugesan et al., 2018). Part of this variation is related to *H. pylori* infection frequency throughout the population, and environmental factors which are also responsible for stomach cancer (Etemadi et al., 2020). "Cigarette smoking is also a risk factor for both cardia and noncardia type of gastric cancer. Because of the higher occurrence of risk factors such as smoking or hormonal factors, both the cancers are more common in males. Men, for example, have historically been more prone to

Biosensor Based Advanced Cancer Diagnostics. DOI: https://doi.org/10.1016/B978-0-12-823424-2.00023-5

smoke tobacco products, despite the fact that elevated rates in men appear to remain even in countries where men and women smoke in similar ways. Alternatively, physiologic differences may be reflected in sex differences. Estrogens may help to prevent gastric cancer from developing. Delaying menopause and increasing fertility in women may reduce the chance of gastric cancer, however, anti-estrogen medications, such as tamoxifen, may raise the risk. These hormones may protect women from gastric cancer throughout their reproductive years, but their influence fades after menopause, resulting in females developing gastric cancer in the same way as males do, albeit with a 10- to 15-year delay (Karimi et al., 2014)."

The decline in gastric cancer is not universal (Balakrishnan et al., 2017). Reduction in incidence and death cases in East Asia will reduce absolute incidence and death cases as half of cases and deaths occur there. Migrant studies and secular trends in stomach cancer rates reveal that environmental factors play a significant role in the pathogenesis of stomach cancer. In contrast, only about 1–3% are known to be hereditary syndromes (Thrift & El-Serag, 2020; Van Cutsem et al., 2016). Reduction in high salt food consumption in Asian countries is an approach to decrease stomach cancer since lifestyle, particularly high sodium diets in East Asian peoples and smoking in males, plays a significant part in stomach cancer burden. The main focus is on preventing *H. pylori* infection, since it is the most crucial element of danger for stomach cancer.

Gastric cancer is grouped into two: (1) early gastric cancer (EGC, stages I and II) defined as the malignant tumor confined to the mucosa and submucosa irrespective of lymph node metastasis; and (2) advance gastric cancer (AGC, stages III and IV); there is lack of a homogeneous definition of advance gastric cancer. However, gastric cancer is a cancer that has attacked the muscularis propria or gastric wall (Cisco et al., 2008; Ooki et al., 2009; Saragoni, 2015). Surgery can treat EGC, but AGC usually requires multidisciplinary treatment. Early diagnosis and careful staging can reduce mortality. Despite all this, gastric cancer staging is facing difficulties because of the lack of defined risk factors. Thus, late diagnosis and inadequate staging arrangements may cause an increase in mortality. So a fast and noninvasive method is needed for early diagnosis and staging of gastric cancer.

General cancer treatment procedures are related to characterizing the cancer cells at the early stages, like chemotherapy, surgery, and radiation. So the diagnosis of cancer is essential for timely individuating a viable cancer treatment. Existing tumor diagnosis depends on an assortment of complicated clinical settings, which include x-ray, magnetic resonance imaging (MRI), computerized tomography (CT), endoscopy, positron emission tomography (PET), cytology, sonography, thermography, and biopsy. In addition, both genomic- and proteomic-based molecular tools are progressively used, such as polymerase chain reaction (PCR), radioimmunoassay (RIA), enzyme linked immunosorbent assay (ELISA), immunohistochemistry (IHC), and flow cytometry (Altintas & Tothill, 2013; Mittal, Kaur, Gautam, & Mantha, 2017; Prabhakar, Shende, & Augustine, 2018). The current technologies and methods are proficient, but most of them are invasive, costly, time-consuming, and restricted to laboratory centers in big hospitals (Cui, Zhou, & Zhou, 2019). For instance, an invasive method biopsy is a medical process that needs the insertion of the medical tool into the patient's body to deduce specific tissues to be examined to find the presence of cancer cells. Such a procedure is tedious, and further, has numerous constraints. Patients experiencing biopsies complain of weak health, nausea, sleeping disorder with further postbiopsy impacts. Therefore, the requirement for noninvasive detection has come into significance in the present time. Also, rapid detection is needed to give patients instant results to start treatment without wasting any time. So the requirement of rapid noninvasive detection of cancer has driven the researchers to develop instruments that would identify cancer early without an invasive technique. This lead to the development of biosensors for noninvasive early detection of cancer (Devi & Laskar, 2018).

## 15.2 Biomarkers for gastric cancer

Researchers and scientist from all around the world have turned their attention to the noninvasive diagnosis of cancer using cancer biomarkers due to numerous drawbacks of the invasive process of cancer detection (Devi & Laskar, 2018; Grossmann, Avenarius, Mastboom, & Klaase, 2010; Wu & Qu, 2015). Cancer biomarkers are essential indicators of cancer status (Karley, Gupta, & Tiwari, 2011). They are utilized not only to analyze and monitor disease but also to provide a prognostic approach to deal with treatment (Chatterjee & Zetter, 2005; Mayeux, 2004). The National Cancer Institute (NCI) (Park, Ross, Klagholz, & Bevans, 2018) defines a biomarker as "a biological molecule found in blood, other body fluids, or tissues that is a sign of a normal or abnormal process or of a condition or disease." A biomarker may be used to see how well the body responds to a treatment for a disease or condition (Biomarkers Definitions Working Group, 2001). Biomarkers can be of several molecular origins, counting DNA (i.e., specific mutation, translocation, amplification, and loss of heterozygosity), RNA, or protein (i.e., hormone, antibody, oncogene, or tumor suppressor). The existence of biomarkers in blood or some other body fluid confirms the presence of cancer cells in the body (Tothill, 2009). There are different biomarkers for different types of cancers (Meyer & Rustin, 2000; Smith, Humphrey, & Catalona, 1997; Tothill, 2009). The maximum of these biomarkers still has to exhibit adequate sensitivity and specificity for translation into routine clinical use or treatment monitoring. This is an area that biosensor technology can improve upon (Bohunicky & Mousa, 2011).

There are several biomarkers available for the early diagnosis of gastric cancer (Fu, 2016). Fig. 15.1 displays the summary of gastric cancer biomarkers. Serum protein biomarkers of gastric cancer are gastric tissue specific or related to gastric-specific infections and divided into two types: gastric cancer-specific markers, and general tumor markers. Proteins such as pepsinogen I (PGI or PGA), pepsinogen II (PGII or PGC), and gastrin 17 are considered specific markers of gastric cancer because of their gastric specific gene expression (Hallissey, Dunn, & Fielding, 1994; Shiotani et al., 2005). Antibodies linked to gastric specific infections such as *H. Pylori*, CagA, and antiparietal cell antibodies, which reflect current or past gastric infections associated with gastric cancer growth, are useful biomarkers for assessing gastric cancer risk (Kaise et al., 2013; Kikuchi, Crabtree, Forman, & Kurosawa, 1999; Sugiu et al., 2006). Many proteins are regarded as gastric cancer screening markers, although most of them are not gastric cancer specific. These proteins comprise carcinoembryonic antigen (CEA), pyruvate M2 kinase, cancer antigen 125 (CA125), cancer antigen 19-9 (CA19-9), Alpha-fetoprotein (AFP), serum amyloid A, macrophage migration inhibitory factor, leptin, dickkopf (Dkk), olfactomedin 4, VAP-1, UPA, cathepsin B, HMW kininogen, P53 antibody, cytokeratin 18, RegIV, IPO-38, S100A6, thrombin light chain, fibrinopeptide A, angiopoietin-like protein 2 (Capelle et al., 2009; Chan et al., 2007; Ebert et al., 2005, 2006; Gao, Xie, Ren, & Yang, 2012; Ghosh et al., 2013; Hao et al., 2008; Harbeck et al., 2008; Herszenyi et al., 2008; Ick et al., 2004; Kaplan et al., 2014; Kumar, Tapuria, Kirmani, & Davidson, 2007; Lee et al., 2012; Liu, Sheng, & Wang, 2012; Mitani et al., 2007; Suppiah & Greenman, 2013; Tas, Karabulut, Serilmez, Ciftci, & Duranyildiz, 2014; Umemura et al., 2011; Yu, Wang, & Chen, 2011; Zhang, Zhang, Jiang, & Zhang, 2014). Among them, carcinoembryonic antigen (CEA) and cancer antigen 19–9 (CA19–9) are most commonly used. CEA was firstly recognized by Gold and Freedman in 1965 (Gold & Freedman, 1965) and was first used for the diagnosis of early gastric cancer in 1980 (Tatsuta et al., 1980). CEA is currently regarded as the most valuable serum protein marker for identifying patients at risk of developing gastric cancer and for the diagnosis of early-stage gastric cancer (Jin, Jiang, & Wang, 2015). CEA was observed to improve colon carcinoma cells' metastasis with its sialofucosylated glycoforms which function as selecting ligands (Deng et al., 2015; Kikuchi et al., 1999). CEA is produced in a high amount of carcinomas in numerous different organs (Kikuchi et al., 1999; Kumar et al., 2007). CEA significantly affects the tumor prognosis because of its effect on tumor metastasis and may be connected with gastric cancer prognosis. Gastric cancer patients show expanded CEA levels, which are associated with patient survival based on an organized analysis of serum markers for gastric cancer (Sugiu et al., 2006). As per literature, preoperative CEA levels could predict gastric cancer (Ick et al., 2004; Schneider & Schulze, 2003), yet few reports deny this thought (Chan et al., 2007; Kumar et al., 2007; Moshkovskii, 2012). There is still discussion encompassing gastric cancer patients' prognosis with expanded CEA levels (Gao et al., 2012; Lee et al., 2012). Henceforth, it is important to build up a state-of-the-art, highly specific, and sensitive CEA detection technique for clinical examination and diagnostics (Tao, Du, Cheng, & Li, 2018). CA19–9 is a glycoprotein highly associated with malignant tumors and a commonly used marker in gastrointestinal cancer; however, it is present in some cancer types, particularly pancreatic, colorectal, and gastric cancer. The CA 19-9 test combined with the CEA test is a beneficial aide for observing carcinoma of the stomach; though, the sensitivity of performing these tests concurrently is similar to performing the CEA test alone in gastric carcinoma (Szymendera, 1986).

Warburg effect (i.e., cancer cells' dependence on glycolysis for energy and normal cell dependence on oxidative phosphorylation) is the most important difference between cancer cells and normal cells (Vander Heiden, Cantley, &

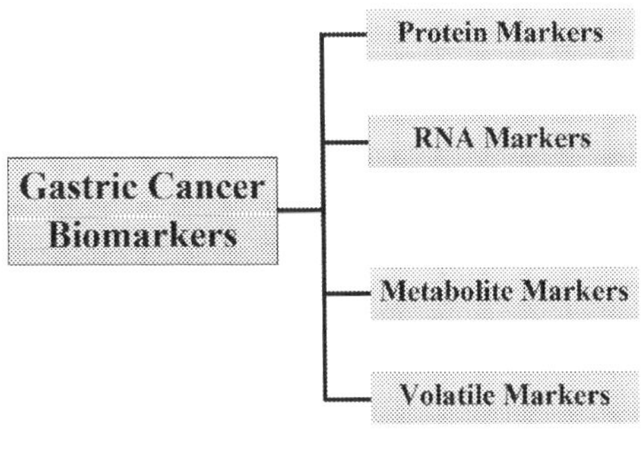


**FIGURE 15.1** Summary of gastric cancer biomarker.

Thompson, 2009; Liberti & Locasale, 2016). In gastric patient's serum or tissue samples, level of lactate which is a result of glucose glycolysis was found to increase constantly (Abbassi-Ghadi et al., 2013; Hirayama et al., 2009). Besides, cancer cells have a high protein synthesis rate. Hence, in gastric cancer patients, numerous metabolic studies showed an increase of amino acids; for example, glycine, asparagine, methionine, tyrosine, and aspartate. Moreover, cancer cells have a high nucleotide synthesis rate for the growing demands of DNA synthesis and DNA repair. Reports also suggested altered nucleotide metabolites in a certain type of cancers. Some of the researchers studied the fatty acid metabolism metabolites in gastric cancer patients. Though both increased fatty acid synthesis (FASN) and fatty acid oxidation (CPT1A) have been related to cancer growth. Fatty acid oxidation metabolites, such as β-hydroxybutyrate and acetone, have been recognized as possible biomarkers of gastric cancer (Fu, 2016).

Usually, RNA is inappropriate for cancer as biomarkers since it is an unsteady species of biomolecules. But current research proposed that certain serum non-coding RNA could also be possible gastric specific markers, for example, RNA HULC and H19 were favorable novel biomarkers in plasma of gastric cancer patients (Abbassi-Ghadi et al., 2013). MicroRNA (miRNA) is a comparatively stable type of RNA in the serum. In gastric cancer, 21 individual miRNAs and six miRNA clusters are consistently upregulated, while miR29c, miR30a5p, miR148a, miR375, and miR638 are usually downregulated (Tatsuta et al., 1980). The most frequently used tumor markers, such as CEA and CA19−9, have limited application in early diagnosis of gastric cancer since they have insufficient sensitivity and specificity. Thus, the foundation of novel robust definite biomarkers with adequate sensitivity is a perfect approach for improving the early detection and the cure rates for gastric cancer patients. Also, these biomarkers should be easy to estimate and consistently linked with clinical results. miRNAs are seen as a desirable cancer biomarker because of the acceptance of their part in tumorigenesis. Discovery of miRNAs and the approval of their role in tumorigenesis and the development of various cancers have presented them as suitable cancer biomarkers. There is also developing evidence that miRNAs exist in cells as well as in an assortment of body fluids, counting blood, saliva, and urine. Those miRNAs that can be found in the circulation system are called circulatory miRNAs. They are generally cancer-specific, and their expression patterns are incredibly comparable among healthy persons and patients. The circulatory miRNAs are remarkably resistant to RNase digestion, non-physiologic pH values, and high temperature. Henceforth, these miRNAs have been considered as a capable biomarker for early detection of cancer (Daneshpour, Omidfar, & Ghanbarian, 2016). But the selection of a high reference gene is an essential element in using miRNA as a tumor biomarker.

Volatile organic compounds (VOCs) released from cancer cell metabolism are considered significant markers for biochemical procedures are happening in cancer cells. The study of VOCs may be capable of predicting and diagnosing early cancer. Volatile metabolites associated with genomics and proteomics represent pathway feedback mechanisms, which positively point out the possible pathophysiological growth in cancer cells. To a certain point, volatile metabolites embody the status of cancer cells. Considering volatile biomarkers from gastric cancer cells and creating an ultrasensitive detection method will help early warning and diagnosis of gastric cancer (Capelle et al., 2009; Chan et al., 2007; Ebert et al., 2005, 2006; Gao et al., 2012; Ghosh et al., 2013; Hao et al., 2008; Harbeck et al., 2008; Herszenyi et al., 2008; Ick et al., 2004; Kaplan et al., 2014; Kumar et al., 2007; Lee et al., 2012; Liu et al., 2012; Mitani et al., 2007; Suppiah & Greenman, 2013; Tas et al., 2014; Umemura et al., 2011; Yu et al., 2011; Zhang et al., 2014).

## 15.3 Biosensor and gastric cancer

Evidence recommends that a growing amount of attention have been focused on developing rapid techniques named "biosensor technology" for the identification, detection, and checking of human health-related conditions (Islam & Uddin, 2017). A biosensor is an analytical device used to identify biological analytes, be it environmental or biological in the source (i.e., inside the human body). A usual biosensor contains a recognition element, a transducer, and a signal-processing unit (Qian et al., 2019). The signal in the form of an analyte is detected by a molecular recognition component converted into an electrical signal by a transducer (Bohunicky & Mousa, 2011). Cammann used the word "biosensor" first (Cammann, 1977), and the International Union of Pure and Applied Chemistry (IUPAC) introduced its definition (Thévenot, Toth, Durst, & Wilson, 2001) and Clark and Lyonsin started biosensor application journey in 1960s (Clark & Lyons, 1962). Biosensors' applications for cancer diagnosis are very promising for conventional methods since it provides better performance in terms of speed, flexibility, automation, and costs (Balaji & Zhang, 2017; Bohunicky & Mousa, 2011; Jainish & Prittesh, 2017; Li, Li, & Yang, 2012; Mittal et al., 2017; Pasinszki, Krebsz, Tung, & Losic, 2017). The recognition of cancer biomarkers present in the blood is the most challenging task because of the low biomarkers' concentration in early-stage patients. A biosensor can measure shallow levels of biomarkers in physiological samples, which can help diagnose cancer at an early stage (Choi, Kwak, & Park, 2010).

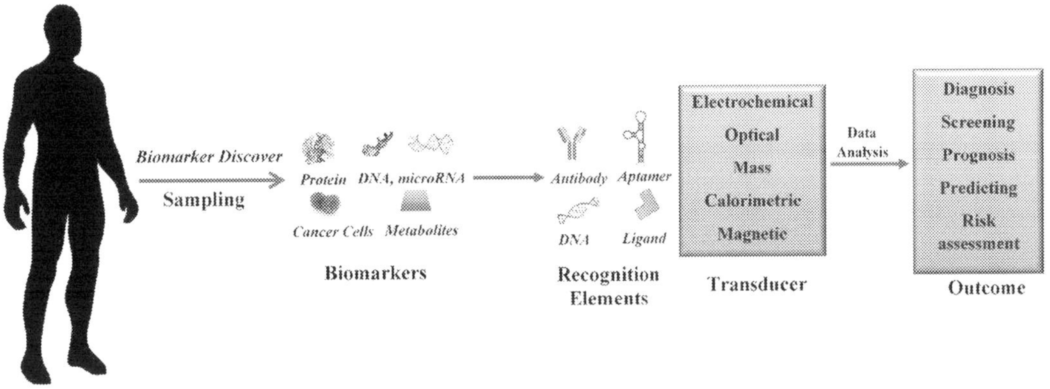


**FIGURE 15.2** Working procedure of biosensors for cancer diagnosis.

Fig. 15.2 demonstrates the working procedure of biosensors for the detection of cancer. The process comprises three key steps: discovery of biomarker, biomarker detection with biosensors, and analysis of data. Every stage plays a vital role and decides the outcomes of the biosensor device (Qian et al., 2019).

## 15.3.1 Role of electrochemical biosensors in early detection of gastric cancer

Among all biosensors, electrochemical sensors have been of great interest, mainly because they are simple, portable, sensitive, inexpensive, and offer a fast response (Topkaya, Azimzadeh, & Ozsoz, 2016). Electrochemical biosensors use electrochemical transducers that transfer a biological entity (i.e., protein, RNA, and DNA) into an electrical signal that can be analyzed and detected (Qian et al., 2019; Wang, 2006). Amperometric and potentiometric transducers are most commonly used in conjunction with electrochemical biosensors. In potentiometric devices, the analytical information is obtained by converting the biorecognition process into a potential signal in connection to the use of ion selective electrodes (ISE). Amperometric biosensors operate by applying a constant potential and monitoring the current associated with the reduction or oxidation of an electroactive species involved in the recognition process. An amperometric biosensor may be more attractive because of its high sensitivity and wide linear range (Wang, 2006). Electrochemical impedance spectroscopy (EIS), differential pulse voltammetry, square wave voltammetry, capacitance measurement, and dielectrophoresis spectroscopy have also been used to measure biosensor response to biomarkers.

Daneshpour et al. (2016) fabricated a novel electrochemical nano biosensor using a double-specific probe approach and a gold-magnetic nanocomposite as tracing tag to detect miR-106a gastric biomarker. EIS and cyclic voltammetry (CV) approaches were used to confirm the electrode's successful modification and hybridization with the target miRNA. For quantifiable estimation of miR-106a, recording the reduction peak current of gold nanoparticles DPV approach was used. The proposed biosensor showed notable selectivity, high specificity, linearity ranging from $1 \times 10^{-3}$ p.m. to $1 \times 10^{3}$ p.m., agreeable storage stability, and great performance in real sample investigations and offered a promising application to be used for medical early detection of gastric cancer. B. Li et al. (2016) carried out a two-stage cyclic enzymatic amplification method (CEAM) to determinate miRNA-21in in the blood serum of gastric cancer patients. The electrochemical biosensor exhibits a low detection limit of 0.36fM with notable specificity. Most importantly, it can be employed to study the expression level of miRNA in the gastric cancer patient blood serum. Tao et al. (2018) developed a selective and sensitive sandwich-type electrochemical aptasensor based on Pt/Au/DN-graphene-CEAapt2-Tb bioconjugate to detect gastric cancer. The proposed method was demonstrated to be sensitive, as indicated by the improved electrochemical response, since the dendritic Pt/Au/DN-graphene showed peroxidase-mimic activity for the reduction of $H_2O_2$ introduced into the electrolytic cell, thereby confirming its desirable catalysis capacity. Since dendritic Pt/Au/ND-graphene is very conductive and possesses peroxidase-mimic activity, the electrochemical response signal and the charge transfer were promoted through catalysis of $H_2O_2$ reduction introduced into the electrolyte cell. Hence, aptasensor was found to enhance analytical capacity and attained desirable sensitivity. Amouzadeh Tabrizi et al. (Amouzadeh Tabrizi, Shamsipur, Saber, Sarkar, & Sherkatkhameneh, 2017) also fabricated a sandwich type electrochemical aptasensor for the sensitive detection of adenocarcinoma gastric cell AGS cancer cells in the presence of $H_2O_2$ by using MWCNT-Aunano as a nanoplatforms and the secondary aptamer-Au@Ag nanoparticles as the labeled aptamers. The aptasensor was also used in the detection of AGS cancer cells in a human serum sample. The developed aptasensor showed a wide linear range and good stability and selectivity. Ilie and Stefan-van Staden (2019) developed a graphite paste modified with 2, 6-bis((E)-2-(furan-2-yl) vinyl)-4-(4,6,8-trimethylazulen-1-yl) pyridine based electrochemical sensor for the detection L-tryptophan gastric cancer biomarker, which is an amino acid in real whole blood samples. The proposed gastric cancer sensor exhibits a high sensitivity with a low limit of detection. Zhang, et al. (Zhang et al., 2014) developed an ultrasensitive electrochemical biosensing interface based on Au-Ag Alloy coated MWCNTs to detect volatile biomarkers of gastric cancer cells. Results displayed that eight various volatile biomarkers were screened out between MGC-803 and GES-1 gastric cancer cells. Fig. 15.3 shows cyclic voltammogram of MWNTs/AU-Ag/GCE was exposed to the head space of MGC-803 gastric cancer cells, GES-1gastric mucosa cells, and cell-free medium. The particular volatile biomarkers of MGC-803 gastric cancer cells and the well-adapted electrochemical system have substantial potential in the near future for applications, for example, screening and warning of early gastric cancer. Rahman et al. fabricated an Ag-Cu bimetallic alloy nanoscale based electrochemical sensor (Rahman et al., 2015) for the monitoring of 2-butanone. The sensor showed the best sensing properties for the detection of 2-butanone with 0.1 μM detection limit. It was expected that the designed sensor could effectively be applied to detect the early stages of gastric and lung cancer caused by 2-butanone. D, Wu et al. (2015) developed a novel and sensitive nonenzymatic sandwich type electrochemical immunosensor for the detection of gastric cancer biomarker CA72−4 using dumbbell-like PtPd-$Fe_3O_4$ nanoparticles (NPs). The immunosensor was fabricated by modifying the glassy carbon electrode by rGO-TEPA for effective immobilization of primary anti-CA72−4 antibody, and the secondary anti-CA72−4 antibody was adsorbed onto the PtPd-$Fe_3O_4$ NPs. The proposed immunosensor showed wide linearity ranging from 0.001−10 U/mL with a low detection limit of 0.0003 U/mL and possessed outstanding clinical value in cancer screening along with suitable point-of-care diagnostics. To meet the clinical demands for early detection of gastric cancer, Yao et al. (Yao et al., 2015) developed a disposable easy-to-use electrochemical microfluidic chip combined with multiple antibodies against six kinds of biomarkers. The electrochemical microfluidic chip showed linearity ranging from 0.37−90 ng $mL^{-1}$, 10.75−172 U $mL^{-1}$, 10−160 U $L^{-1}$, 35−560 ng $mL^{-1}$, 37.5−600 ng $mL^{-1}$, and 2.5−80 ng $mL^{-1}$ for CEA, CA19−9, HP, P53, PG I, and PG II biomarkers, respectively (Fig. 15.4). This method showed improved sensitivity compared with ELISA results of 394 specimens of gastric cancer sera. The electrochemical microfluid chip is a promising candidate for early screening of gastric cancer, therapeutic evaluation, and real-time dynamic review of gastric cancer advancement in the near future. Mohammad Shafiee and Parhizkar (2020) successfully fabricated Au nanoparticles/g-$C_3N_4$ modified electrochemical gastric cancer biosensor for the detection of miRNA. The sensor used a hairpin locked nucleic acids probe and $Zn^{2+}$ functionalized TiP nanospheres labels. The sensor showed linearity ranging from 0.6 nM to 6 nM with a limit of detection to 80 pM. For the detection of miR-100 in the sera gastric cancer patients, Zhuang, Wan, and Zhang (2021) developed a rapid, selective, and sensitive biosensor based on Au electrode (AuE) modified with gold nanoparticle (AuNP) which was attached with DNA capture probes (CPs) (CPs/AuNP-AuE). The range of detection and detection limit of the biosensor for miR-100 was 100 a.m. to 10 p.m. 100 a.m. respectively.

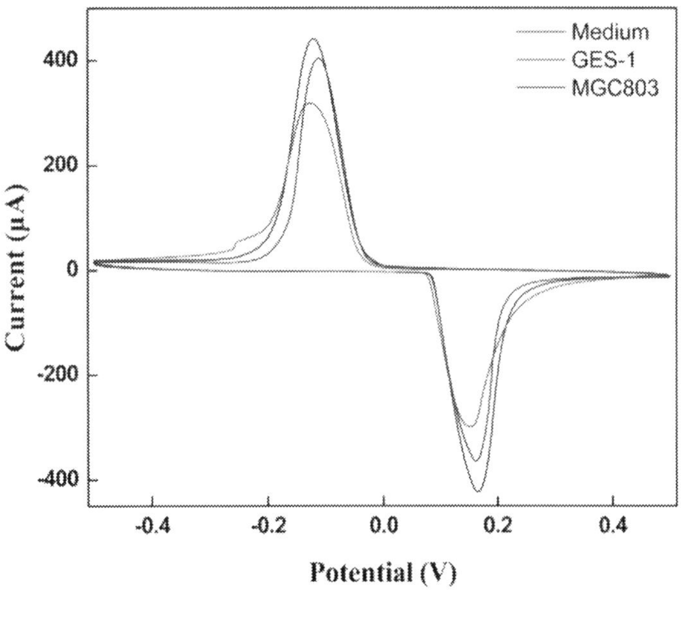


**FIGURE 15.3** CVs of MWNTs/AU-Ag/GCE exposed to the head space of MGC-803 gastric cancer cells, GES-1gastric mucosa cells, and cell-free medium.

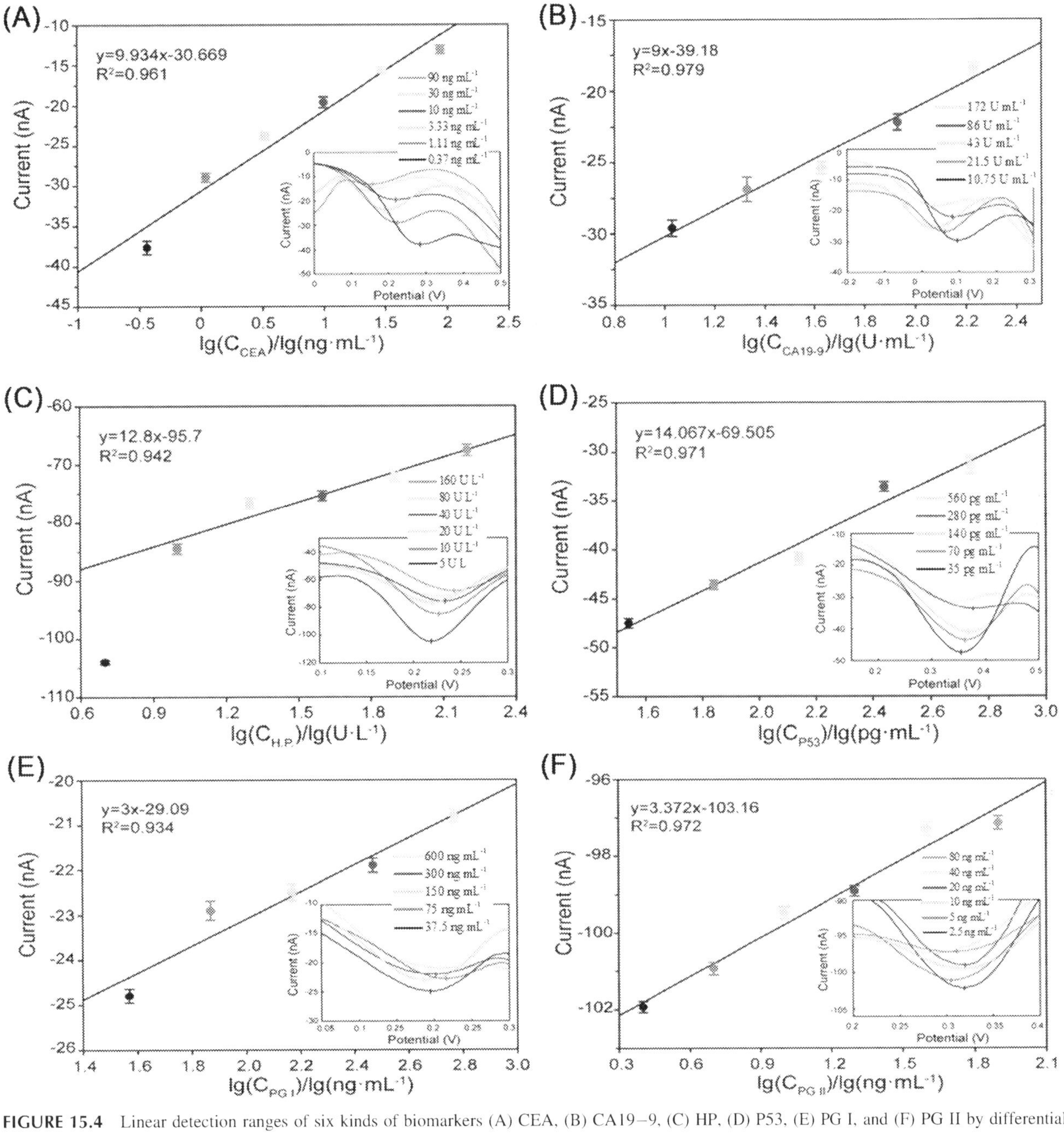


**FIGURE 15.4** Linear detection ranges of six kinds of biomarkers (A) CEA, (B) CA19–9, (C) HP, (D) P53, (E) PG I, and (F) PG II by differential pulse voltammetry.

### 15.3.2 Role of SPR biosensor in early detection of gastric cancer

In recent decades, various optical biosensor approaches have been established, counting surface plasmon resonance (SPR) (Nelson, Grimsrud, Liles, Goodman, & Corn, 2001), ellipsometry (Arwin, Poksinski, & Johansen, 2004), and quartz crystal microbalance (QCM) (Frank, Elke, Neil, Kenichi, & Yoshio, 1997). Amongst them, the SPR-based method is a representative type of label-free procedure for checking biomolecular interactions in a real-time (Nguyen, Park, Kang, & Kim, 2015). SPR is an optical phenomenon take place in the overall internal reflection of light at a metal film-liquid interface (Van Oss & van Regenmortel, 1994; Raether, 1988). At the point when the incident light is completely reflected, a part of the incident light momentum named as evanescent wave penetrates the liquid medium near the metal (generally Au) surface. In the thin metal film surface, the evanescent wave interacts with longitudinally oscillating free electrons termed surface plasmon. During SPR, metal film absorbed the energy of incident light, decreasing the light intensity. While the angle of incidence is fixed, the resonance phenomenon happens only at an accurately defined wavelength, which depends upon the medium's refractive index (RI) near the metal surface. RI changes in a direct extent to the mass and dielectric permittivity of the present medium. Immobilization of antibodies on the metal surface causes the corresponding antigen to bond on the surface when it touches the liquid samples. The binding method can be observed via observing the SPR wavelength which depends on the quantity of antibody-antigen binding. The SPR biosensor is sensitive to refractive index adjustments or thickness of biomaterials at the interface between a metal thin film and a surrounding medium. Therefore, using antibodies peculiar to pathogens of interest can measure the number of pathogenic bacteria existents in a sample by quantifying the change in refractive index and characterize interactions of biomolecules on the surface in real time without labeling (Brockman, Nelson, & Corn, 2000; Fang et al., 2010; Green et al., 2000).

For the early diagnosis of gastric cancer, Fang et al. (2010) fabricated a SPR sensor based on the detection of MG7-Ag, a gastric cancer-specific tumor-associated antigen in human sera. The measurements contained two cases of healthy blood donors, nine cases of gastric cancer patients, and an MKN45 cancer cell lysate sample solution for positive control. Results showed the binding of MG7-Ag onto the sensor surface was observed from SPR spectra. The prepared SPR biosensor showed potential for the early diagnosis of gastric cancer, but the limit of detection and measure for cancer risk assessment in early diagnosis was not confirmed. F. Liu et al. (2011) used surface plasmon resonance phase sensing to detect EGFR (Epidermal growth factor receptor) on active human gastric cancer BGC823 cells. The results showed that the SPR phase sensing is proficient of real-time recognition of molecular interactions and cellular responses on living cells. It also proposed that more studies on the mechanism and method might let SPR sensing become a useful tool for the essential research of cell biology, yet also for medical diagnosis and drug development.

### 15.3.3 Role of surface-enhanced Raman spectroscopy sensor in early detection of gastric cancer

Amongst optical nano biosensors, those established on surface-enhanced Raman scattering (SERS) spectroscopy have been drawing significant attention. It is because of the combination of the intrinsic prerogatives of the technique, such as structural specificity and sensitivity, and the high degree of modification in nano-manufacturing, which translates into consistent and robust real-life applications. In SERS, the excitation of localized surface plasmon resonances (LSPR) at the surface of nanostructured metals with light induces the massive intensification of the Raman scattering from molecules located close to the metallic surface. This effect yields an ultrasensitive plasmon-enhanced spectroscopic technique that retains Raman spectroscopy's intrinsic structural specificity and experimental flexibility. As impressive advances in instrumentation and nanofabrication techniques enabling the engineering of finely tuned plasmonic nanomaterials continue, SERS is progressively expanding into the realm of viable biomedical applications (Guerrini & Alvarez-Puebla, 2019).

There are 14 VOC biomarkers in human breath used for differentiating gastric cancer patients from healthy persons. Chen et al. (2016) fabricated a SERS sensor based on breath analysis to identify VOC biomarkers to distinguish EGC and AGC cancer patients from healthy persons. They prepared a clean SERS sensor using hydrazine vapor adsorbed in graphene oxide (GO) film by in situ formations of gold nanoparticles (AuNPs) on reduced GO (RGO) deprived of any organic stabilizer. The SERS sensor effectively analyzed and distinguished various simulated breath samples and 200 breath samples of medical patients with over 83% and 92% sensitivity and specificity, respectively. Yunsheng Chen et al. (2018) fabricated non-invasive, cheap, fast SERS sensors based on salivary analysis to screen early and advance gastric cancer patients. The developed graphene oxide nanoscrolls wrapped with gold nanoparticle (A/GO NSs)-based SERS sensors detect the biomarkers in 220 clinical liquid saliva. These sensors successfully analyzed and distinguished various stimulated and medical patients' samples with sensitivity and specificity greater than 80% and 87.7%, respectively. For the detection of miR-34a biomarker, Lee et al. (2018) fabricated a uniform, highly robust, and ultra-sensitive surface-enhanced Raman scattering substrate by using silver nanostructures grown in gold nanobowls (SGBs). They were accomplished by consistent and direct detection of miR-34a in human gastric cancer cells by applying the advantages of SGBs in SERS sensing. An essential chemokine named interleukin 8 (IL-8) plays a vital part in tumor growth and angiogenesis and has been found in various human tumors, counting gastric and breast cancer. Zhen-yu Wang et al. (2019) fabricated a double antibody sandwich format-based SERS immunosensor for the determination of IL-8. The immunosensor showed high sensitivity, selectivity, and low detection limits for the detection of IL-8 in PBS (Phosphate-buffered saline) and human serum, hence, providing a great possibility for application in clinical diagnosis.

### 15.3.4 Role of GMI-based biosensing system in early detection of gastric cancer

In recent times, the giant magnetoimpedance (GMI) effect has attracted considerable attention due to its possible application in magnetic field sensing (Wang et al., 2017). The GMI effect is the change of complex impedance of soft magnetic materials conveying alternating current upon the use of the external magnetic field in Beach and Berkowitz (1994), Knobel and Pirota (2002), Phan and Peng (2008), and Panina and Mohri (1994).

Kurlyandskaya et al. (2003) introduced a GMI sensor into the field of biosensors. A GMI-based biosensing system linking with the magnetic labeled technology was used to distinguish gastric cancer cells (Chen et al., 2016). For the recognition of functional nanoparticles-probed gastric cancer cells, Lei Chen et al. (2011) planned, fabricated, and tested a GMI-based biosensing system with a Co-based ribbon sensing element. Functionalized nanoparticles were structured by coating $Fe_3O_4$ with chitosan and conjugating with cyclic RGD peptides. This fabricated system can recognize the dissimilarities among targeted and nontargeted cells.

### 15.3.5 Other types of biosensors in early detection of gastric cancer

Different types of biosensors can also detect gastric cancer related biomarkers. Stefan-van Staden et al. (Stefan-Van Staden, Ilie-Mihai, Pogacean, & Pruneanu, 2019) developed an exfoliated graphene (E-NGr) based high sensitive stochastic sensor used for pattern recognition of CEA, CA19-9, and p53 in whole blood and urine samples of patients found in very early and later gastric cancer stages.

## 15.4 Conclusion and future perspectives

Due to the numerous limitations in conventional detection methods of cancer, scientists and researchers are showing their attention to biosensors' development for effective rapid noninvasive detection of cancer markers. In the body, presence of cancer cells is confirmed by cancer markers. These markers exist in saliva, blood, or some other body fluids. As a complex heterogeneous disease, gastric cancer is one of the most widely recognized malignancies around the world. Gastric malignant growth is the fifth most regular kind of disease and the subsequent driving reason for the third leading malignant growth-related mortality (accounted for 8.2%) overall (Sitarz et al., 2018; Zhou et al., 2018). Early gastric cancer can be cured with surgery. In contrast, advanced gastric cancer often needs combined multidisciplinary therapy, and delayed diagnosis and inadequacies of the staging system may increase mortality. Therefore, it is very demanding to develop a rapid and noninvasive diagnosis technique to realize early detection of gastric cancer and simultaneous staging. Consequently, it is challenging to create a rapid and noninvasive diagnosis technique to realize early detection of gastric cancer and simultaneous staging. Early detection of gastric cancer prominently increases the probabilities for effective treatment and survival rates of cancers. Several types of biosensors have been proposed to detect gastric biomarkers and have shown an excellent opportunity for the early diagnosis of gastric cancer.

## References

Abbassi-Ghadi, N., Kumar, S., Huang, J., Goldin, R., Takats, Z., & Hanna, G. B. (2013). Metabolomic profiling of oesophago-gastric cancer: A systematic review. *European Journal of Cancer, 49*(17), 3625–3637. Available from https://doi.org/10.1016/j.ejc.2013.07.004.

Altintas, Z., & Tothill, I. (2013). Biomarkers and biosensors for the early diagnosis of lung cancer. *Sensors and Actuators, B: Chemical, 188*, 988–998. Available from https://doi.org/10.1016/j.snb.2013.07.078.

Amouzadeh Tabrizi, M., Shamsipur, M., Saber, R., Sarkar, S., & Sherkatkhameneh, N. (2017). Flow injection amperometric sandwich-type electrochemical aptasensor for the determination of adenocarcinoma gastric cancer cell using aptamer-Au@Ag nanoparticles as labeled aptamer. *Electrochimica Acta, 246*, 1147–1154. Available from https://doi.org/10.1016/j.electacta.2017.06.115.

Arwin, H., Poksinski, M., & Johansen, K. (2004). Total internal reflection ellipsometry: Principles and applications. *Applied Optics, 43*(15), 3028–3036. Available from https://doi.org/10.1364/AO.43.003028.

Balaji, A., & Zhang, J. (2017). Electrochemical and optical biosensors for early-stage cancer diagnosis by using graphene and graphene oxide. *Cancer Nanotechnology, 8*(1). Available from https://doi.org/10.1186/s12645-017-0035-z.

Balakrishnan, M., George, R., Sharma, A., & Graham, D. Y. (2017). Changing trends in stomach cancer throughout the world. *Current Gastroenterology Reports, 19*(8). Available from https://doi.org/10.1007/s11894-017-0575-8.

Beach, R. S., & Berkowitz, A. E. (1994). Giant magnetic field dependent impedance of amorphous FeCoSiB wire. *Applied Physics Letters, 64*(26), 3652–3654. Available from https://doi.org/10.1063/1.111170.

Biomarkers Definitions Working Group. (2001). Biomarkers definitions working group. *Biomarkers and surrogate endpoints. Clinical Pharmacology & Therapeutics, 69*(3), 89–95.

Bohunicky, B., & Mousa, S. A. (2011). Biosensors: The new wave in cancer diagnosis. *Nanotechnology, Science and Applications*, 4.

Brockman, J. M., Nelson, B. P., & Corn, R. M. (2000). Surface plasmon resonance imaging measurements of ultrathin organic films. *Annual Review of Physical Chemistry*, *51*, 41–63. Available from https://doi.org/10.1146/annurev.physchem.51.1.41.

Cammann, K. (1977). Bio-sensors based on ion-selective electrodes. *Fresenius' Zeitschrift Für Analytische Chemie*, *287*(1), 1–9. Available from https://doi.org/10.1007/BF00539519.

Capelle, L. G., De Vries, A. C., Haringsma, J., Steyerberg, E. W., Looman, C. W. N., Nagtzaam, N. M. A., ... Kuipers, E. J. (2009). Serum levels of leptin as marker for patients at high risk of gastric cancer. *Helicobacter*, *14*(6), 596–604. Available from https://doi.org/10.1111/j.1523-5378.2009.00728.x.

Chan, D. C., Chen, C. J., Chu, H. C., Chang, W. K., Yu, J. C., Chen, Y. J., ... Chen, J. H. (2007). Evaluation of serum amyloid a as a biomarker for gastric cancer. *Annals of Surgical Oncology*, *14*(1), 84–93. Available from https://doi.org/10.1245/s10434-006-9091-z.

Chatterjee, S. K., & Zetter, B. R. (2005). Cancer biomarkers: Knowing the present and predicting the future. *Future Oncology*, *1*(1), 37–50. Available from https://doi.org/10.1517/14796694.1.1.37.

Chen, L., Bao, C. C., Yang, H., Li, D., Lei, C., Wang, T., ... Cui, D. X. (2011). A prototype of giant magnetoimpedance-based biosensing system for targeted detection of gastric cancer cells. *Biosensors and Bioelectronics*, *26*(7), 3246–3253. Available from https://doi.org/10.1016/j.bios.2010.12.034.

Chen, Y., Cheng, S., Zhang, A., Song, J., Chang, J., Wang, K., ... Cui, D. (2018). Salivary analysis based on surface enhanced Raman scattering sensors distinguishes early and advanced gastric cancer patients from healthy persons. *Journal of Biomedical Nanotechnology*, *14*(10), 1773–1784. Available from https://doi.org/10.1166/jbn.2018.2621.

Chen, Y., Zhang, Y., Pan, F., Liu, J., Wang, K., Zhang, C., ... Cui, D. (2016). Breath analysis based on surface-enhanced raman scattering sensors distinguishes early and advanced gastric cancer patients from healthy persons. *ACS Nano*, *10*(9), 8169–8179. Available from https://doi.org/10.1021/acsnano.6b01441.

Choi, Y. E., Kwak, J. W., & Park, J. W. (2010). Nanotechnology for early cancer detection. *Sensors*, *10*(1), 428–455. Available from https://doi.org/10.3390/s100100428.

Cisco, R. M., Ford, J. M., & Norton, J. A. (2008). Hereditary diffuse gastric cancer implications of genetic testing for screening and prophylactic surgery. *Cancer*, *113*(7), 1850–1856. Available from https://doi.org/10.1002/cncr.23650.

Clark, L. C., & Lyons, C. (1962). Elecyrode systems for continuous monitoring in cardiovascular surgery. *Annals of the New York Academy of Sciences*, *102*(1), 29–45. Available from https://doi.org/10.1111/j.1749-6632.1962.tb13623.x.

Crew, K. D., & Neugut, A. I. (2006). Epidemiology of gastric cancer. *World Journal of Gastroenterology*, *12*(3), 354–362. Available from https://doi.org/10.3748/wjg.v12.i3.354.

Cui, F., Zhou, Z., & Zhou, H. S. (2019). Measurement and analysis of cancer biomarkers based on electrochemical biosensors. *Journal of the Electrochemical Society*, *167*, 3.

Daneshpour, M., Omidfar, K., & Ghanbarian, H. (2016). A novel electrochemical nanobiosensor for the ultrasensitive and specific detection of femtomolar-level gastric cancer biomarker miRNA-106a. *Beilstein Journal of Nanotechnology*, *7*(1), 2023–2036. Available from https://doi.org/10.3762/BJNANO.7.193.

Deng, K., Yang, L., Hu, B., Wu, H., Zhu, H., & Tang, C. (2015). The prognostic significance of pretreatment serum CEA levels in gastric cancer: A metaanalysis including 14651 patients. *PLoS One*, *10*(4), e0124151. Available from https://doi.org/10.1371/journal.pone.0124151.

Devi, N., & Laskar, S. (2018). A review on application of biosensors for cancer detection. *ADBU Journal of Electrical and Electronics Engineering*, *2*(2), 17–21.

Dikshit, R. P., Mathur, G., & Mhatre, S. (2011). Epidemiological review of gastric cancer in India. *Indian Journal of Medical and Paediatric Oncology*, *32*(1), 3–11. Available from https://doi.org/10.4103/0971-5851.81883.

Ebert, M. P. A., Lamer, S., Meuer, J., Malfertheiner, P., Reymond, M., Buschmann, T., ... Seibert, V. (2005). Identification of the thrombin light chain a as the single best mass for differentiation of gastric cancer patients from individuals with dyspepsia by proteome analysis. *Journal of Proteome Research*, *4*(2), 586–590. Available from https://doi.org/10.1021/pr049771i.

Ebert, M. P. A., Niemeyer, D., Deininger, S. O., Wex, T., Knippig, C., Hoffmann, J., ... Röcken, C. (2006). Identification and confirmation of increased fibrinopeptide A serum protein levels in gastric cancer sera by magnet bead assisted MALDI-TOF mass spectrometry. *Journal of Proteome Research*, *5*(9), 2152–2158. Available from https://doi.org/10.1021/pr060011c.

Etemadi, A., Safiri, S., Sepanlou, S. G., Ikuta, K., Bisignano, C., Shakeri, R., ... Almasi-Hashiani, A. (2020). The global, regional, and national burden of stomach cancer in 195 countries, 1990–2017: A systematic analysis for the Global Burden of Disease study 2017. *The Lancet Gastroenterology and Hepatology*, *5*(1), 42–54. Available from https://doi.org/10.1016/S2468-1253(19)30328-0.

Fang, X., Tie, J., Xie, Y., Li, Q., Zhao, Q., & Fan, D. (2010). Detection of gastric carcinoma-associated antigen MG7-Ag in human sera using surface plasmon resonance sensor. *Cancer Epidemiology*, *34*(5), 648–651. Available from https://doi.org/10.1016/j.canep.2010.05.004.

Frank, C., Elke, R., Neil, F. D., Kenichi, N., & Yoshio, O. (1997). Quartz crystal microbalance study of DNA immobilization and hybridization for nucleic acid sensor development. *Analytical Chemistry*, 2043–2049. Available from https://doi.org/10.1021/ac961220r.

Fu, H. (2016). New developments of gastric cancer biomarker research. *Nano Biomedicine and Engineering*, *8*(4), 268–273. Available from https://doi.org/10.5101/nbe.v8i4.p268-273.

Gao, C., Xie, R., Ren, C., & Yang, X. (2012). Dickkopf-1 expression is a novel prognostic marker for gastric cancer. *Journal of Biomedicine and Biotechnology*, 2012. Available from https://doi.org/10.1155/2012/804592.

Ghosh, I., Bhattacharjee, D., Das, A. K., Chakrabarti, G., Dasgupta, A., & Dey, S. K. (2013). Diagnostic role of tumour markers CEA, CA15-3, CA19-9 and CA125 in lung cancer. *Indian Journal of Clinical Biochemistry*, *28*(1), 24–29. Available from https://doi.org/10.1007/s12291-012-0257-0.

Gold, P., & Freedman, S. O. (1965). Specific carcinoembryonic antigens of the human digestive system. *The Journal of Experimental Medicine*, *122*(3), 467–481. Available from https://doi.org/10.1084/jem.122.3.467.

Green, R. J., Frazier, R. A., Shakesheff, K. M., Davies, M. C., Roberts, C. J., & Tendler, S. J. B. (2000). Surface plasmon resonance analysis of dynamic biological interactions with biomaterials. *Biomaterials*, *21*(18), 1823–1835. Available from https://doi.org/10.1016/S0142-9612(00)00077-6.

Grossmann, I., Avenarius, J. K. A., Mastboom, W. J. B., & Klaase, J. M. (2010). Preoperative staging with chest CT in patients with colorectal carcinoma: Not as a routine procedure. *Annals of Surgical Oncology*, *17*(8), 2045–2050. Available from https://doi.org/10.1245/s10434-010-0962-y.

Guerrini, L., & Alvarez-Puebla, R. A. (2019). Surface-enhanced raman spectroscopy in cancer diagnosis, prognosis and monitoring. *Cancers*, *11*(6). Available from https://doi.org/10.3390/cancers11060748.

Hallissey, M. T., Dunn, J. A., & Fielding, J. W. L. (1994). Evaluation of pepsinogen a and gastrin-17 as markers of gastric cancer and high-risk pathologic conditions. *Scandinavian Journal of Gastroenterology*, *29*(12), 1129–1134. Available from https://doi.org/10.3109/00365529409094899.

Hao, Y., Yu, Y., Wang, L., Yan, M., Ji, L., Qu, Y., ... Zhu, Z. (2008). IPO-38 is identified as a novel serum biomarker of gastric cancer based on clinical proteomics technology. *Journal of Proteome Research*, *7*(9), 3668–3677. Available from https://doi.org/10.1021/pr700638k.

Harbeck, N., Schmitt, M., Vetter, M., Krol, J., Paepke, D., Uhlig, M., ... Thomssen, C. (2008). Prospective biomarker trials chemo N0 and NNBC-3 Europe validate the clinical utility of invasion markers uPA and PAI-1 in node-negative breast cancer. In *Breast care* (Vol. 3, Issue 2, pp. 11–15). S. Karger A.G. Available from https://doi.org/10.1159/000151734.

Herszenyi, L., István, G., Cardin, R., De Paoli, M., Plebani, M., Tulassay, Z., & Farinati, F. (2008). Serum cathepsin B and plasma urokinase-type plasminogen activator levels in gastrointestinal tract cancers. *European Journal of Cancer Prevention*, *17*(5), 438–445. Available from https://doi.org/10.1097/CEJ.0b013e328305a130.

Hirayama, A., Kami, K., Sugimoto, M., Sugawara, M., Toki, N., Onozuka, H., ... Soga, T. (2009). Quantitative metabolome profiling of colon and stomach cancer microenvironment by capillary electrophoresis time-of-flight mass spectrometry. *Cancer Research*, *69*(11), 4918–4925. Available from https://doi.org/10.1158/0008-5472.CAN-08-4806.

Ick, H. G., Hak, Y. C., Ho, S. B., Ho, S. J., Lai, P. Y., Dai, K. H., ... Kil, P. W. (2004). Predictive value of preoperative serum CEA, CA19-9 and CA125 levels for peritoneal metastasis in patients with gastric carcinoma. *Cancer Research and Treatment*, 178. Available from https://doi.org/10.4143/crt.2004.36.3.178.

Ilie, R. M., & Stefan-van Staden, R.-I. (2019). Determination of l-tryptophan in whole blood samples using a new electrochemical sensor. *UPB Scientific Bulletin, Series B: Chemistry and Materials Science*, *81*(1), 42–46.

Islam, M. T., & Uddin, M. A. (2017). Biosensors, the emerging tools in the identification and detection of cancer markers. *Journal of Gynecology and Women's Health*, *5*, 4.

Jainish, P., & Prittesh, P. (2017). Biosensors and biomarkers: Promising tools for cancer diagnosis. *Int J Biosen Bioelectron*, *3*, 4.

Jin, Z., Jiang, W., & Wang, L. (2015). Biomarkers for gastric cancer: Progression in early diagnosis and prognosis (review). *Oncology Letters*, *9*(4), 1502–1508. Available from https://doi.org/10.3892/ol.2015.2959.

Kaise, M., Miwa, J., Fujimoto, A., Tashiro, J., Tagami, D., Sano, H., & Ohmoto, Y. (2013). Influence of Helicobacter pylori status and eradication on the serum levels of trefoil factors and pepsinogen test: Serum trefoil factor 3 is a stable biomarker. *Gastric Cancer: Official Journal of the International Gastric Cancer Association and the Japanese Gastric Cancer Association*, *16*(3), 329–337. Available from https://doi.org/10.1007/s10120-012-0185-y.

Kaplan, M. A., Kucukoner, M., Inal, A., Urakci, Z., Evliyaoglu, O., Firat, U., ... Isikdogan, A. (2014). Relationship between serum soluble vascular adhesion protein-1 level and gastric cancer prognosis. *Oncology Research and Treatment*, *37*(6), 340–344. Available from https://doi.org/10.1159/000362626.

Karimi, P., Islami, F., Anandasabapathy, S., Freedman, N. D., & Kamangar, F. (2014). Gastric Cancer: Descriptive Epidemiology, Risk Factors, Screening, and Prevention. *Cancer Epidemiology Biomarkers & Prevention*, *23*(5), 700–713. Available from https://doi.org/10.1158/1055-9965.epi-13-1057.

Karley, D., Gupta, D., & Tiwari, A. (2011). Biomarker for cancer: A great promise for future. *World Journal of Oncology*, *2*, 4.

Kikuchi, S., Crabtree, J. E., Forman, D., & Kurosawa, M. (1999). Association between infections with CagA-positive or-negative strains of *Helicobacter pylori* and risk for gastric cancer in young adults. *American Journal of Gastroenterology*, *94*(12), 3455–3459. Available from https://doi.org/10.1016/S0002-9270(99)00666-8.

Knobel, M., & Pirota, K. R. (2002). Giant magnetoimpedance: Concepts and recent progress. *Journal of Magnetism and Magnetic Materials*, *242–245*(I), 33–40. Available from https://doi.org/10.1016/S0304-8853(01)01180-5.

Kono, S. (2016). *Gastric cancer. International encyclopedia of public health* (pp. 215–222). Elsevier Inc. Available from https://doi.org/10.1016/B978-0-12-803678-5.00167-3.

Kumar, Y., Tapuria, N., Kirmani, N., & Davidson, B. R. (2007). Tumour M2-pyruvate kinase: A gastrointestinal cancer marker. European. *Journal of Gastroenterology and Hepatology*, *19*(3), 265–276. Available from https://doi.org/10.1097/MEG.0b013e3280102f78.

Kurlyandskaya, G. V., Sánchez, M. L., Hernando, B., Prida, V. M., Gorria, P., & Tejedor, M. (2003). Giant-magnetoimpedance-based sensitive element as a model for biosensors. *Applied Physics Letters*, *82*(18), 3053–3055. Available from https://doi.org/10.1063/1.1571957.

Lee, H. S., Lee, H. E., Park, D. J., Kim, H. H., Kim, W. H., & Park, K. U. (2012). Clinical significance of serum and tissue Dickkopf-1 levels in patients with gastric cancer. *Clinica Chimica Acta*, *413*(21–23), 1753–1760. Available from https://doi.org/10.1016/j.cc.2012.07.003.

Lee, T., Wi, J. S., Oh, A., Na, H. K., Lee, J., Lee, K., ... Haam, S. (2018). Highly robust, uniform and ultra-sensitive surface-enhanced Raman scattering substrates for microRNA detection fabricated by using silver nanostructures grown in gold nanobowls. *Nanoscale*, *10*(8), 3680–3687. Available from https://doi.org/10.1039/c7nr08066b.

Li, B., Liu, F., Peng, Y., Zhou, Y., Fan, W., Yin, H., ... Zhang, X. (2016). Two-stage cyclic enzymatic amplification method for ultrasensitive electrochemical assay of microRNA-21 in the blood serum of gastric cancer patients. *Biosensors and Bioelectronics*, *79*, 307–312. Available from https://doi.org/10.1016/j.bios.2015.12.051.

Li, J., Li, S., & Yang, C. F. (2012). Electrochemical biosensors for cancer biomarker detection. *Electroanalysis*, *24*(12), 2213–2229. Available from https://doi.org/10.1002/elan.201200447.

Liberti, M. V., & Locasale, J. W. (2016). The Warburg effect: How does it benefit cancer cells? *Trends in Biochemical Sciences*, *41*(3), 211–218. Available from https://doi.org/10.1016/j.tibs.2015.12.001.

Liu, F., Zhang, J., Deng, Y., Wang, D., Lu, Y., & Yu, X. (2011). Detection of EGFR on living human gastric cancer BGC823 cells using surface plasmon resonance phase sensing. *Sensors and Actuators, B: Chemical*, *153*(2), 398–403. Available from https://doi.org/10.1016/j.snb.2010.11.005.

Liu, X., Sheng, W., & Wang, Y. (2012). An analysis of clinicopathological features and prognosis by comparing hepatoid adenocarcinoma of the stomach with AFP-producing gastric cancer. *Journal of Surgical Oncology*, *106*(3), 299–303. Available from https://doi.org/10.1002/jso.23073.

Mayeux, R. (2004). Biomarkers: Potential uses and limitations. *NeuroRx: the Journal of the American Society for Experimental NeuroTherapeutics*, *1*(2), 182–188. Available from https://doi.org/10.1602/neurorx.1.2.182.

Meyer, T., & Rustin, G. J. S. (2000). Role of tumour markers in monitoring epithelial ovarian cancer. *British Journal of Cancer*, *82*(9), 1535–1538.

Mitani, Y., Oue, N., Matsumura, S., Yoshida, K., Noguchi, T., Ito, M., ... Yasui, W. (2007). Reg IV is a serum biomarker for gastric cancer patients and predicts response to 5-fluorouracil-based chemotherapy. *Oncogene*, *26*(30), 4383–4393. Available from https://doi.org/10.1038/sj.onc.1210215.

Mittal, S., Kaur, H., Gautam, N., & Mantha, A. K. (2017). Biosensors for breast cancer diagnosis: A review of bioreceptors, biotransducers and signal amplification strategies. *Biosensors and Bioelectronics*, *88*, 217–231. Available from https://doi.org/10.1016/j.bios.2016.08.028.

Mohammad Shafiee, M. R., & Parhizkar, J. (2020). Au nanoparticles/g-C3N4 modified biosensor for electrochemical detection of gastric cancer miRNA based on hairpin locked nucleic acids probe. *Nanomedicine Research Journal*, *5*(2), 152–159. Available from https://doi.org/10.22034/NMRJ.2020.02.006.

Moshkovskii, S. A. (2012). Why do cancer cells produce serum amyloid a acute-phase protein? *Biochemistry (Moscow)*, *77*(4), 339–341. Available from https://doi.org/10.1134/S0006297912040037.

Nelson, B. P., Grimsrud, T. E., Liles, M. R., Goodman, R. M., & Corn, R. M. (2001). Surface plasmon resonance imaging measurements of DNA and RNA hybridization adsorption onto DNA microarrays. *Analytical Chemistry*, *73*(1), 1–7. Available from https://doi.org/10.1021/ac0010431.

Nguyen, H. H., Park, J., Kang, S., & Kim, M. (2015). Surface plasmon resonance: A versatile technique for biosensor applications. *Sensors (Switzerland)*, *15*(5), 10481–10510. Available from https://doi.org/10.3390/s150510481.

Ooki, A., Yamashita, K., Kikuchi, S., Sakuramoto, S., Katada, N., & Watanabe, M. (2009). Phosphatase of regenerating liver-3 as a prognostic biomarker in histologically node-negative gastric cancer. *Oncology Reports*, *21*(6), 1467–1475. Available from https://doi.org/10.3892/or_00000376.

Panina, L. V., & Mohri, K. (1994). Magneto-impedance effect in amorphous wires. *Applied Physics Letters*, 1189–1191. Available from https://doi.org/10.1063/1.112104.

Park, J., Ross, A., Klagholz, S. D., & Bevans, M. F. (2018). The role of biomarkers in research on caregivers for cancer patients: A scoping review. *Biological Research for Nursing*, *20*(3), 300–311. Available from https://doi.org/10.1177/1099800417740970.

Pasinszki, T., Krebsz, M., Tung, T. T., & Losic, D. (2017). Carbon nanomaterial based biosensors for non-invasive detection of cancer and disease biomarkers for clinical diagnosis. *Sensors (Switzerland)*, *17*(8). Available from https://doi.org/10.3390/s17081919.

Phan, M. H., & Peng, H. X. (2008). Giant magnetoimpedance materials: Fundamentals and applications. *Progress in Materials Science*, *53*(2), 323–420. Available from https://doi.org/10.1016/j.pmatsci.2007.05.003.

Prabhakar, B., Shende, P., & Augustine, S. (2018). Current trends and emerging diagnostic techniques for lung cancer. *Biomedicine and Pharmacotherapy*, *106*, 1586–1599. Available from https://doi.org/10.1016/j.biopha.2018.07.145.

Qian, L., Li, Q., Baryeh, K., Qiu, W., Li, K., Zhang, J., ... Liu, G. (2019). Biosensors for early diagnosis of pancreatic cancer: A review. *Translational Research*, *213*, 67–89. Available from https://doi.org/10.1016/j.trsl.2019.08.002.

Raether, H. (1988). *Surface plasmons on smooth and rough surfaces and on gratings* (Vol. 111). Springer. Available from https://doi.org/10.1007/BFb0048317.

Rahman, L. U., Shah, A., Lunsford, S. K., Han, C., Nadagouda, M. N., Sahle-Demessie, E., ... Dionysiou, D. D. (2015). Monitoring of 2-butanone using a Ag-Cu bimetallic alloy nanoscale electrochemical sensor. *RSC Advances*, *5*(55), 44427–44434. Available from https://doi.org/10.1039/c5ra03633j.

Rawla, P., & Barsouk, A. (2019). Epidemiology of gastric cancer: Global trends, risk factors and prevention. *Przeglad Gastroenterologiczny*, *14*(1), 26–38. Available from https://doi.org/10.5114/pg.2018.80001.

Saragoni, L. (2015). Upgrading the definition of early gastric cancer: Better staging means more appropriate treatment. *Cancer Biology and Medicine*, *12*(4), 355–361. Available from https://doi.org/10.7497/j.issn.2095-3941.2015.0054.

Schneider, J., & Schulze, G. (2003). Comparison of tumor M2-pyruvate kinase (tumor M2-PK), carcinoembryonic antigen (CEA), carbohydrate antigens CA 19-9 and CA 72-4 in the diagnosis of gastrointestinal cancer. *Anticancer Research*, *23*(6 D), 5089–5093.

Servarayan Murugesan, C., Manickavasagam, K., Chandramohan, A., Jebaraj, A., Jameel, A. R. A., Jain, M. S., & Venkataraman, J. (2018). Gastric cancer in India: Epidemiology and standard of treatment. *Updates in Surgery*, *70*(2), 233–239. Available from https://doi.org/10.1007/s13304-018-0527-3.

Sharma, A., & Radhakrishnan, V. (2011). Gastric cancer in India. *Indian Journal of Medical and Paediatric Oncology*, *32*(1), 12–16. Available from https://doi.org/10.4103/0971-5851.81884.

Shiotani, A., Iishi, H., Uedo, N., Kumamoto, M., Nakae, Y., Ishiguro, S., ... Graham, D. Y. (2005). Histologic and serum risk markers for noncardia early gastric cancer. *International Journal of Cancer*, *115*(3), 463–469. Available from https://doi.org/10.1002/ijc.20852.

Sitarz, R., Skierucha, M., Mielko, J., Offerhaus, G. J. A., Maciejewski, R., & Polkowski, W. P. (2018). Gastric cancer: Epidemiology, prevention, classification, and treatment. *Cancer Management and Research*, *10*, 239–248. Available from https://doi.org/10.2147/CMAR.S149619.

Smith, D. S., Humphrey, P. A., & Catalona, W. J. (1997). The early detection of prostate carcinoma with prostate specific antigen: The Washington University experience. *In Cancer*, *80*(Issue 9), 1852–1856, https://doi.org/10.1002/(SICI)1097-0142(19971101)80:9 < 1852::AID-CNCR25 > 3.0.CO;2-3.

Stefan-Van Staden, R. I., Ilie-Mihai, R. M., Pogacean, F., & Pruneanu, S. (2019). Graphene-based stochastic sensors for pattern recognition of gastric cancer biomarkers in biological fluids. *Journal of Porphyrins and Phthalocyanines*, *23*(11–12), 1365–1370. Available from https://doi.org/10.1142/S1088424619501293.

Sugiu, K., Kamada, T., Ito, M., Kaya, S., Tanaka, A., Kusunoki, H., … Haruma, K. (2006). Anti-parietal cell antibody and serum pepsinogen assessment in screening for gastric carcinoma. *Digestive and Liver Disease*, *38*(5), 303–307. Available from https://doi.org/10.1016/j.dld.2005.10.021.

Suppiah, A., & Greenman, J. (2013). Clinical utility of anti-p53 auto-antibody: Systematic review and focus on colorectal cancer. *World Journal of Gastroenterology*, *19*(29), 4651–4670. Available from https://doi.org/10.3748/wjg.v19.i29.4651.

Szymendera, J. J. (1986). Clinical usefulness of three monoclonal antibody-defined tumor markers: CA 19-9, CA 50, and CA 125. *Tumour Biology*, *7*(5–6), 333–342.

Tao, Z., Du, J., Cheng, Y., & Li, Q. (2018). Electrochemical immune analysis system for gastric cancer biomarker carcinoembryonic antigen (CEA) detection. *International Journal of Electrochemical Science*, *13*(2), 1413–1422. Available from https://doi.org/10.20964/2018.02.21.

Tas, F., Karabulut, S., Serilmez, M., Ciftci, R., & Duranyildiz, D. (2014). Serum levels of macrophage migration-inhibitory factor (MIF) have diagnostic, predictive and prognostic roles in epithelial ovarian cancer patients. *Tumor Biology*, *35*(4), 3327–3331. Available from https://doi.org/10.1007/s13277-013-1438-z.

Tatsuta, M., Itoh, T., Okuda, S., Yamamura, H., Baba, M., & Tamura, H. (1980). Carcinoembryonic antigen in gastric juice as an aid in diagnosis of early gastric cancer. *Cancer*, *46*(12), 2686–2692, https://doi.org/10.1002/1097-0142(19801215)46:12 < 2686::AID-CNCR2820461225 > 3.0.CO;2-E.

Thévenot, D. R., Toth, K., Durst, R. A., & Wilson, G. S. (2001). Electrochemical biosensors: Recommended definitions and classification. *Biosensors and Bioelectronics*, *16*(1–2), 121–131. Available from https://doi.org/10.1016/S0956-5663(01)00115-4.

Thrift, A. P., & El-Serag, H. B. (2020). Burden of gastric cancer. *Clinical Gastroenterology and Hepatology*, *18*(3), 534–542. Available from https://doi.org/10.1016/j.cgh.2019.07.045.

Topkaya, S. N., Azimzadeh, M., & Ozsoz, M. (2016). Electrochemical biosensors for cancer biomarkers detection: Recent advances and challenges. *Electroanalysis*, *28*(7), 1402–1419. Available from https://doi.org/10.1002/elan.201501174.

Tothill, I. E. (2009). Biosensors for cancer markers diagnosis. *Seminars in Cell and Developmental Biology*, *20*(1), 55–62. Available from https://doi.org/10.1016/j.semcdb.2009.01.015.

Umemura, H., Togawa, A., Sogawa, K., Satoh, M., Mogushi, K., Nishimura, M., … Nomura, F. (2011). Identification of a high molecular weight kininogen fragment as a marker for early gastric cancer by serum proteome analysis. *Journal of Gastroenterology*, *46*(5), 577–585. Available from https://doi.org/10.1007/s00535-010-0369-3.

Van Cutsem, E., Sagaert, X., Topal, B., Haustermans, K., & Prenen, H. (2016). Gastric cancer. *The Lancet*, *388*(10060), 2654–2664. Available from https://doi.org/10.1016/S0140-6736(16)30354-3.

Vander Heiden, M. G., Cantley, L. C., & Thompson, C. B. (2009). Understanding the Warburg effect: The metabolic requirements of cell proliferation. *Science*, *324*(5930), 1029–1033. Available from https://doi.org/10.1126/science.1160809.

Van Oss, C.J., & van Regenmortel, M.H.V. (1994). Immunochemistry. Marcel Dekker.

Wang, J. (2006). Electrochemical biosensors: Towards point-of-care cancer diagnostics. *In Biosensors and Bioelectronics*, *21*(Issue 10), 1887–1892. Available from https://doi.org/10.1016/j.bios.2005.10.027.

Wang, T., Zhou, Y., Lei, C., Luo, J., Xie, S., & Pu, H. (2017). Magnetic impedance biosensor: A review. *Biosensors and Bioelectronics*, *90*, 418–435. Available from https://doi.org/10.1016/j.bios.2016.10.031.

Wang, Z. y, Li, W., Gong, Z., Sun, P. r, Zhou, T., & Cao, Xw (2019). Detection of IL-8 in human serum using surface-enhanced Raman scattering coupled with highly-branched gold nanoparticles and gold nanocages. *New Journal of Chemistry*, *43*(4), 1733–1742. Available from https://doi.org/10.1039/C8NJ05353G.

Wu, D., Guo, Z., Liu, Y., Guo, A., Lou, W., Fan, D., & Wei, Q. (2015). Sandwich-type electrochemical immunosensor using dumbbell-like nanoparticles for the determination of gastric cancer biomarker CA72-4. *Talanta*, *134*, 305–309. Available from https://doi.org/10.1016/j.talanta.2014.11.025.

Wu, L., & Qu, X. (2015). Cancer biomarker detection: Recent achievements and challenges. *Chemical Society Reviews*, *44*(10), 2963–2997. Available from https://doi.org/10.1039/c4cs00370e.

Yao, X., Xiao, Z., Su, H., Kan, W., Zhen, Y., He, N., & Cui. (2015). A novel electrochemical microfluidic chip combined with multiple biomarkers for early diagnosis of gastric cancer. *Nanoscale Res Lett*, 10. Available from https://doi.org/10.1186/s11671-015-1153-3.

Yu, L., Wang, L., & Chen, S. (2011). Olfactomedin 4, a novel marker for the differentiation and progression of gastrointestinal cancers. *Neoplasma*, *58*(1), 9–13. Available from https://doi.org/10.4149/neo_2011_01_9.

Zhang, J., Zhang, K., Jiang, X., & Zhang, J. (2014). S100A6 as a potential serum prognostic biomarker and therapeutic target in gastric cancer. *Digestive Diseases and Sciences*, *59*(9), 2136–2144. Available from https://doi.org/10.1007/s10620-014-3137-z.

Zhang, Y., Gao, G., Liu, H., Fu, H., Fan, J., Wang, K., … Cui, D. (2014). Identification of volatile biomarkers of gastric cancer cells and ultrasensitive electrochemical detection based on sensing interface of Au Ag alloy coated MWCNTs. *Theranostics*, *4*(2), 154–162. Available from https://doi.org/10.7150/thno.7560.

Zhou, Z., Lin, Z., Pang, X., Tariq, M. A., Ao, X., Li, P., & Wang, J. (2018). Epigenetic regulation of long non-coding RNAs in gastric cancer. *Oncotarget*, *9*(27), 19443–19458. Available from https://doi.org/10.18632/oncotarget.23821.

Zhuang, J., Wan, H., & Zhang, X. (2021). Electrochemical detection of miRNA-100 in the sera of gastric cancer patients based on DSN-assisted amplification. *Talanta*, *225*, 121981. Available from https://doi.org/10.1016/j.talanta.2020.121981.

Chapter 16

# 3D-printed device with integrated biosensors for biomedical applications

**Shikha Saxena[1] and Deepshikha Pande Katare[2]**
[1]*Amity Institute of Pharmacy, Amity University, Noida, India*
[2]*Proteomic & Translational Research Lab, Centre for Medical Biotechnology, Amity Institute of Biotechnology, Amity University Noida, Noida, India*

## 16.1 Introduction

Biosensors have become an integral part of healthcare system. Sensors are quite a favorite in environment, robotics, automation, as well as the pharmaceutical industry for taking the changes and displaying the results after processing in less time (François, 2004; Lee, 2003; Lynch & Loh, 2006; Sharafeldin, Jones, & Rusling, 2018; Yick, Mukherjee, & Ghosal, 2008). It caters to a variety of material to fabricate the best possible and portable devices. The process of 3D printing is now very well integrated with the biosensors to create the individualization in the health devices. The history of biosensor fabrication dates back to 1984 with the birth of 3D printing by Charles Hull (Matthew, 2014). Since then, this technique is unstoppable and holding its position in various segments of life. Various applications of integrated biosensors using 3D-printing techniques are in the field of education and training, personalized implants, prosthetics, life threatening diseases like cancer, HIV, metabolic disorders such as diabetes, anatomy models, drug target designing, organ printing, dentistry, tissue engineering, etc. (Yana et al., 2018). Its fabrication allows customization based on customer needs, therefore the flexibility of shape and size can be catered wherever possible. The idea can be visualized by computer-aided design and precisely controlled methods of designing are accomplished. There is a wide range of materials that can be used in fabricating like ceramics, plastics, and metal (Yeong, Chua, Leong, Chandrasekaran, & Lee, 2006). The principle is the conversion of biological input into electrochemical or digital response. The construction material is fixed layer by layer to the base using various methods like laser technique, melting, etc. till the 3D like structure is obtained. The low cost, user-friendly platform and customization availability always attracts the researchers and scientists worldwide. In this chapter, the basics and advances of 3D-printing devices with integrated biosensors are discussed along with some case studies and regulatory aspects.

## 16.2 Basics of biosensors

Biosensors are defined as devices that has capability for detection of biological analyte and converting the output into signal which can be enhanced and displayed on the computer. Sample analyte interacts with recognition elements such as receptors, antigens, antibodies, nucleic acid, proteins, enzymes, etc., which when fed into transducer (electrochemical, optical, calorimetric). This is further analyzed by data processor and results are executed (Fig. 16.1). Fig. 16.1 In terms of structure, it has three major parts: an identification element for detection of signal, signal transducer for converting the signal to desired electrical component, and output processor (Figs. 16.2 and 16.3) (Bhalla, Jolly, Formisano, & Estrela, 2016).

## 16.3 Types of biosensors

Biosensors cover the domain of enzymes, antibodies, nucleic acids, hole cells, phages, aptamers, etc. (Ni et al., 2017). Different types can be classified as (Mannoor et al., 2013): They are the combination of physical and chemical sensing techniques. Based on the fact that which transduction principle is to be utilized they can be differentiated into many categories (Ali et al., 2017). Some of the main categories are microbial, cell based, immunosensors, biomolecule-based sensors, DNA sensors, etc. These can be explained as follows:

**Biosensor Based Advanced Cancer Diagnostics. DOI: https://doi.org/10.1016/B978-0-12-823424-2.00018-1**

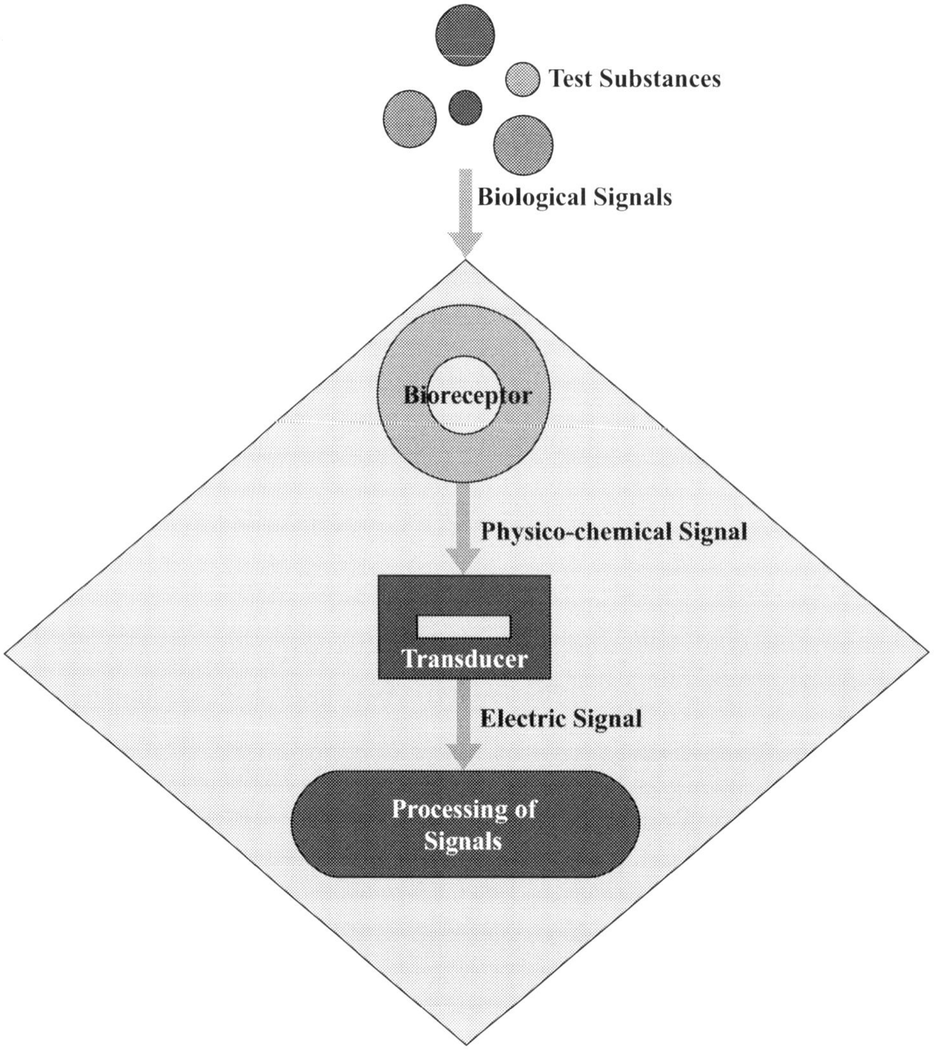


**FIGURE 16.1** Basic structure of biosensor.

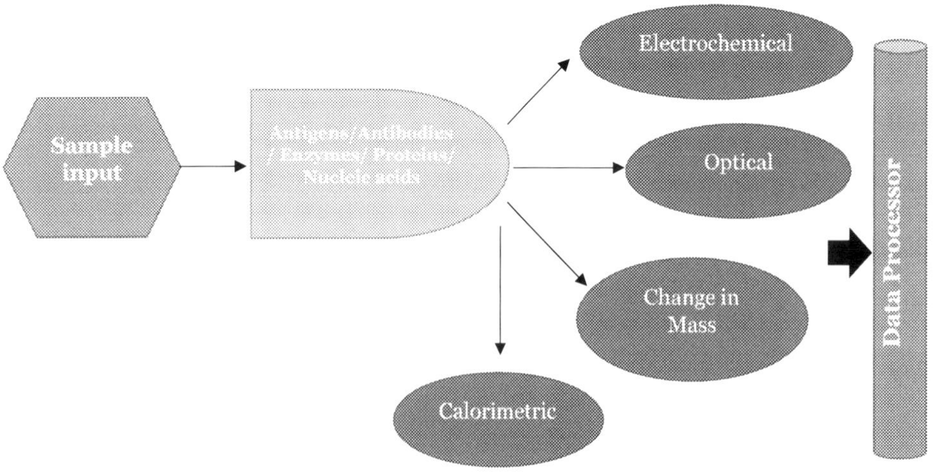


**FIGURE 16.2** Basic layout of biosensor technology.

## 16.3.1 Microbial sensors

In this 3D-printed microfluidic chip (Krejcova et al., 2014) or 3D-printed fluidic device is used as sensor for detection of microbe customization of product is available and detection is based on electro-chemical reaction or change in absorbance intensity.

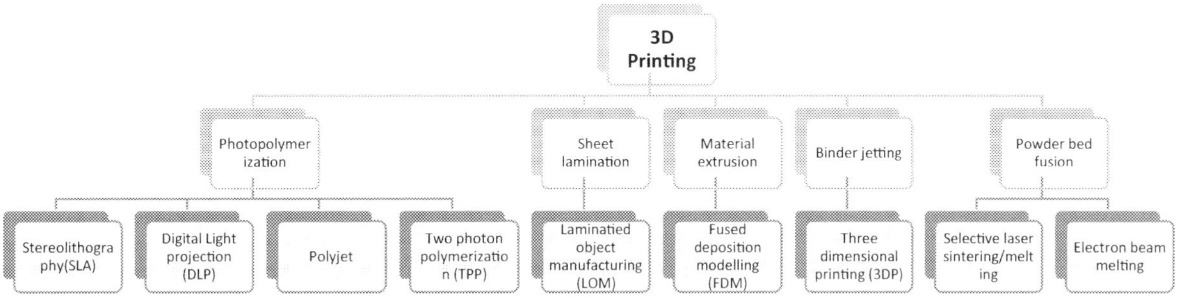


**FIGURE 16.3** 3D-printing types.

### 16.3.2 Cell-based sensors

Based on fluorescence, luminescence potentiometry, amperometry, living cells are utilized as sensing units for evaluating various biological activities. The biological activity includes aging, metabolism, cell toxicity, ATP monitoring, calcium imaging, ALP concentration. Now a days, cell-based sensors gives the option to be attached with smart phones leading to better patient compliance and real time detection (Anderson, Lockwood, Martin, & Spence, 2013; Cevenini, Calabretta, Tarantino, Michelini, & Roda, 2016; Dantism, Takenaga, Wagner, Wagner, & Schöning, 2016; Spivey, Jones, Rybarski, Saifuddin, & Finkelstein, 2017).

### 16.3.3 Immunosensors

In this antigen–antibody, reaction takes place which is both selective and sensitive. The principle of luminescence intensity, fluorescence density or optical density is employed. It is helpful to study biomarker protein, pathogenic bacteria and viruses (Kadimisetty et al., 2016; Salvo et al., 2012; Singh et al., 2015; Wonjae et al., 2015).

### 16.3.4 Biomolecule-based sensors

Biomolecule has two categories (Ni et al., 2017): macromolecules (e.g., protein, nucleic acid, polysaccharides) and micromolecules (e.g., amino acids, nucleic acids, monosaccharide) A biomolecule-based sensor has two types: enzyme based and DNA sensors. The DNA sensors are further classified as three major areas (Bishop, Satterwhite-Warden, Bist, Chen, & Rusling, 2016; Dirkzwager, Liang, & Tanner, 2016; Heger et al., 2016): (1) Fluidic device: This is introduced for ECL measurement and it has customizable/removable parts; (2) Aptamer device: Here malaria can be diagnosed using the principle of colorimetry. It has two prototypes; one is paper based and other is magnetic bead based; and (3) Stratospheric probe: It utilizes principle where DNA damage ca be evaluated (Bishop et al., 2016).

### 16.3.5 Enzyme-based sensors

Enzyme-based sensors have two major parts. One is immobilized enzyme membrane which acts as identification part and other is signal converters. It is based on the principle of detection of concentration when enzyme-based reaction takes place on a 3D-printed device having enzyme membranes. It is usually used for testing glucose or lactose (Cevenini et al., 2016; Gowers et al., 2015; Roda et al., 2014).

### 16.3.6 Bionic sensors

In this type of sensor, living sensors are being cultured using electronic devices and fully developed functional organs are achieved, e.g., bionic ear which can be used as a real ear, heart epicardium in which there is a smart device consisting of heart tissue and it also has pumping unit managed by robotic system, etc. (Mannoor et al., 2013; Xu et al., 2014).

## 16.4 History of 3D-printed biosensors

The history of bioprinting dates back to merging of different fabricating methods for the creation of living cells. The main names which came into the picture were Vladimir Mironov, Gabor Forgacs, and Thomas Boland. A timeline of events can be listed (Zeming, Jianzhong, Hui, & Yong, 2019) as:

1. Birth of 3D printing with Stereolithography (SLA) introduction by Charles Hull (1984);
2. First demonstration pf bioprinting technique through 2D positioning by Klebe (1988);

3. Observation of embryo development by Forgacs and co-workers (1996);
4. 2D pattern of cells were studied through laser technique by Odde and Renn (1999);
5. 3D-printed synthetic scaffold came into picture (2001);
6. Introduction of 3D-bio plotter by Landers and co-workers (2002);
7. Modification of HP inkjet printer to design first inkjet bioprinter (2003);
8. Formation of 3D-tissue devoid of scaffold (2004):
   i. Construction of vasculature related constructs without using scaffold by Norotte (2009);
9. Bioprinting experimented on animals in situ (2012);
10. Coaxial technology introduced to design tubular structure by Gao and co-workers (2015);
11. Digital Light Projection (DLP) technique was used for fabricating optical 3D-printing-based models by Pyo and co-workers (2016);
12. Engineering of collagen human heart using hydrogels technology by Noor and co-workers (2019).

## 16.5 Need of integrated biosensors

Biosensors facilitate the advancement in the field of synthetic biology. They have the power of specific detection to the molecules with a quick sense of analysis. Hence, their usage to various fields like industry, medical, ecological, and scientific technology cannot be questioned. Different approaches pertain to biosensors which make them worthy of their applications (Carpenter, Paulsen, & Williams, 2018). Biosensors are quite relevant to current scientific fields due to their capability to exploit the genomic pool of humans. They allow cost cutting in terms of manufacturing and thus make the technology accessible to all. It helps to incorporate the diversity of materials so that different resources of erratic density, optical rotation, strength, and properties can be used. Further, they help to enhance the customization and control and hence increase the patient acceptability and compliance.

## 16.6 Commercial biosensors in the market

It is reported that there are more than 500 companies fabricating and marketing biosensors. A few of them are involved in the full production while some are providing raw materials. For example, Applied Enzyme Technologies, Bioenzyme Laboratories, Dupont Ltd., Eco Chemie, Ercon Incorporated, Gwent Electronic Materials Ltd., Palm Instruments, and Uniscan Instruments Ltd. are a few names. Some of the available biosensors in the market (Mongra, 2012) are Microvacuum, IBIS, Silion Market, Auto Lab, Graffinity Pharmaceuticals, Biocare, and Biorad (Miroslav, 2019).

## 16.7 Different materials used in 3D-printed biosensors

Many materials can be utilized to fabricate biosensors (Mehrotra, 2016). Some of them are acrylonitrile butadiene styrene (ABS), polylactic acid (PLA), wax blend polymer, nylon, resin (acrylate or epoxy-based with proprietary photo initiator), polymers, metallic powder polyamide, polyvinyl chloride (PVC), photo resin, hydrogels.

## 16.8 Types of 3D-printing techniques

This section covers some of the commonly used 3D-printing methods in the biomedical field (Han, Kundu, Nag, & Xu, 2019) (Fig. 16.3).

### 16.8.1 Fused deposition modeling

In this method, hot mushy flow is present in the heated nozzle of the machine. Thermoplastic material is used to fabricate a thin filament and that filament is feeded and shaped out, one by one cooled down, and deposited (Kumar, Ahuja, & Singh, 2012; Mohan Pandey, Venkata Reddy, & Dhande, 2003). Printing material options include ABS, PLA, blended wax, and nylon. Some of the examples of fused deposition modeling are as follows

- 3D-printed chip for influenza virus using electrochemical detection method;
- Glucose monitoring in artificial serum sample using enzymatic reactors having paper base;
- Toxicity sensor coupled with smart phone.

### 16.8.2 Stereolithography

In this, a stereolithography file is placed in between 2D layers of polymer, and UV light is utilized to perform 3D printing. The full structure is fabricated by layers of resin on a built platform using laser scanning, and the process is on repeat mode till the final structure is produced (Bhargav, Sanjairaj, Rosa, Feng, & Fuh YH, 2018; Mahindru, Mahendru, & Ganj, 2013), Some of the examples of this technique are:

- Alkaline phosphatase (ALP) biomarker detection using portable electrochemical sensor;
- Malaria diagnosis device using principle of capturing enzyme;
- *Escherichia coli* bacterial sensor using water shed resin.

### 16.8.3 Polyjet method

Hardening is achieved using photo multiplier where multiple nozzle is used simultaneously. In printing, nozzles move and eject tiny drops of photopolymer which gets deposited on polymer. The layers are fabricated one by one after hardening of previous layer is achieved (Ionita et al., 2014).

For example,

- Cell viability sensor to check drug transport and cellular levels simultaneously;
- Chemiluminescence-based ATP sensor.

### 16.8.4 Selective laser sintering

This layer by layer fabrication is achieved by binding powder particles using laser power to melt and bind the material. After the formation and scanning of one layer, another layer is fabricated based on CAD design. Material options can be metallic powder, ceramic powder, or polymers. For example, online pH monitoring device for liquid dispensers are desigened by this method. (Frazier, 2014; Olakanmi, Cochrane, & Dalgarno, 2015).

### 16.8.5 3D inkjet printing

In this process, a solid structure is obtained using spreaded powder particles and photocurable resin/hydrogel droplets as printing material. From a fine 3D nozzle, a fixed amount of ink is ejected out to obtain a layer by layer 3D model. The liquid ink is evaporated after deposition leaving a solid structure. However, if wax-based ink is used then it is pushed through a nozzle kept at high temperature and wax is deposited layer by layer (Mannoor et al., 2013; Nakamura et al., 2005).

Example of 3D inkjet printing are

- 3D bionic ear for improvisation with option of stereo audio music;
- Temperature sensor to detect temperature of human body in the range of 20°C –60°C.

### 16.8.6 Digital light processing method

In this method, a digital-based projector is used that flashes an image like layer. This leads to repeated formation of hard 2D layers on exposure of polymer to light projector (Mannoor et al., 2013). For example, glucose sensor using photo resin based on colorimetry uses digital light processing.

## 16.9 Applications of 3D-printed biosensors

3D printing, also known by the name additive printing, was introduced in the 1980s. Here, a new version of a material is created by first taking its blueprint and printing its impression layer-wise (Ghilan et al., 2020).

If forecast to be believed, 3D printing in the biomedical field will be worth US$3.5 billion by the year 2025. Major applications of 3D-printed integrated biosensors are given in this section (Han et al., 2019; Nag, Mukhopadhyay, & Kosel, 2017; Nguyen, Narayan, & Shafiee, 2019)

### 16.9.1 Bioprinting

Here, the printer uses the special ink known by the name bioink that is released by a specially designed computer guided pipette to create an impression on living cells, which leads to the designing of lab-based artificial living tissue. The resultant ones have capabilities to mimic organs. Therefore, they can be employed for research studies as miniatures or can be used as a budget alternative option for organ transplant.

### 16.9.2 As a preparative tool in surgery

Here, replicas which are specific to patients are created and can be used by surgeons to practice before complicated surgery. This increases the speed of the operation as well as decreases the anxiety of the patient.

### 16.9.3 For surgical tools

Various surgical tools can be imprinted by 3D technology like forceps, hemostats, and scalpel instrument clamps.

### 16.9.4 Prosthetics

User friendly and customized prosthetics can be created using 3D printing. Initially, the users have to wait for months, but this technology has reduced the waiting time, with lined up and resultant prosthetics on budget, too.

### 16.9.5 Tissue engineering

It is classified under two domains:

- Scaffold free: According to adhesion hypothesis, cells have a tendency to stick with each other with different strengthening capacities. The behavior of tissue can be considered based on physical parameters like surface tension and viscosity. When cellular principles are amalgamated with the forces, this leads to designing of tissues, e.g., geometrical construction of nerves and blood vessels. Aggregation of bioink particles followed by maturation of bioprinted structures leads to implant designing.
- Scaffold based: Flexibility of designing numerous shapes, cell types, and functional variability can be achieved through scaffold using geometry with different materials. Seeding of scaffold is done with cells and thus human scaled tissues like calvarian bone, cartilage, and skeletal muscle can be constructed.

### 16.9.6 Acellular medical devices

Customizable prosthetics can be fabricated using computer assisted designing (CAD) methods, e.g., child limb prosthetics using their casting. In 3D technology, the cost is minimized and it is also less weight and long lasting, with easily replaceable prosthetics. Dental implants, ortho-deontic aligners, and polymeric casting for broken bones are some of the medical devices utilizing 3D-printing technology.

### 16.9.7 Models and surgical practice

Conventionally before surgery, CT scans and MRI scans were used by surgeons to understand the problematic area before surgery. But as these techniques are 2D in nature, sometimes minute but very important details get missed. 3D-printed technology with integrated sensors gave the opportunity for replica or models of organs to be operated, before and during surgery, thus increasing the success rate of operations and also decreasing the loss of blood during surgical procedures. Thus, it is really useful in orthognathic surgeries in the dentofacial field, liver transplant, neuromodulator treatment, etc. (Dias, Kingsley, & Corr, 2014; Terzaki et al., 2013).

### 16.9.8 Training and education

3D-printed biosensors are widely used in the education field related to medicine especially anatomy and surgery. Students can practice and understand surgical aspects. It further enhances the organs vasculature understanding. The practice is useful in plastic surgery, pediatric surgery, urology, and general surgery.

## 16.10 Advantages of 3D-printed biosensors

3D-printed integrated biosensors have lots of advantages (Didier, Kundu, & Rajaraman, 2020; Xu et al., 2017) Using them allows the estimation and analysis to take place with high speed and, at the same time, a high level of accuracy. In this, durability issues are also quite minimum. Nowadays, biosensors can incorporate a wide range of materials and time for detection output is also less. Many integrated devices have been issued in the market which allows the user to deal with multiple jetting heads, enormous fabrication styles, and colored polymers with high resolution. Surface finishing of the biosensors is equally important. These sensors allow one to use quantitative analysis with limit of detection (LOD), measurement time, and selectivity.

## 16.11 Disadvantages of 3D-printed biosensors

Despite many of the said advantages, the disadvantages associated with the biosensors cannot be ignored. It is reported that in a few cases, fabrication cost becomes high. Sometimes mechanical properties are weak. Biosensors give serious durability challenges due to moisture, heat, and chemical effects. Further, as time passes, the strength of the device becomes an issue and some parts of machinery are fragile.

## 16.12 Some of the case studies of biosensors

1. Wearable medical biosensors are suitable for personalized monitoring of health parameters. In Sung Kyunkwan University, a group of scientists have designed the wearables using inkjet printing technology. The resulting wearable is light in weight and different, rapidly changing signals of the body can be addressed, e.g., electromyography (EMG), electrodermal activity (EDA), and electroencephalogram (EEG) output (Ho et al., 2019).
2. A microfluidic device was fabricated by Erkal et al. having polyester membranes and electrodes (Erkal et al., 2014). The device is reusable and is used for measurement of flowing erythrocytes, ATP and O2 concentration of patient body (Fig. 16.4).
3. A device for detection of ageing in cell was designed by Spivey et al. by the name of FYLM, i.e., fission yeast life span microdessicator. The output is measured on the basis of fluorescence (Fig. 16.5) (Spivey et al., 2017).
4. A low-cost 3D-printed device working as lactate sensor is reported by Roda et al. (2014). It claims to detect the concentration of lactate within 5 min of start. The sample can be in the form of oral fluid or sweat (Cevenini et al., 2016; Spivey et al., 2017).
5. Another biosensor reported is wearable PDMS sweat collectors which consists of collection tubes and channels using this concentration of electrolyte in sweat can be monitored within 7 min of start of exercise (Reeder et al., 2019).

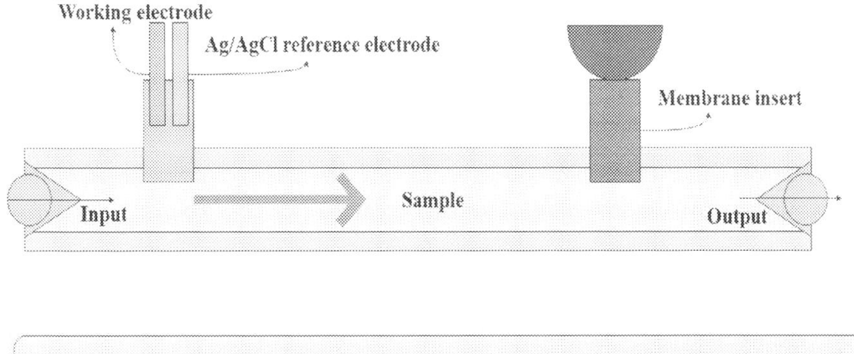


FIGURE 16.4 Microfluidic device layout.

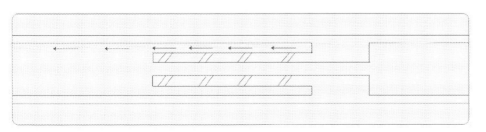

FIGURE 16.5 Fission yeast lifespan micro dissector.

6. A very relevant example of decrease in time of detection is the detection of cancer biomarkers through ECL immunosensors developed by Kadimisetty et al. (2016). In this, three cancer biomarkers can be detected within a span of 35 min in contrast to ELISA which takes 4 h 45 min for one marker (Reeder et al., 2019).
7. Bionic ear is one of the breakthrough examples in the field of 3D biosensors. It has the capacity to respond to frequency modulation and is made up of polyvinylidene fluoride (PVDF) (Park, Yoo, & Hong, 2010). The beauty of this material lies in the fact that it can detect change (Fig. 16.6) in pressure and temperature, too.

   Wearable devices comes in many shapes and sizes. They are generating a lot of interest due to their capability to give real time data. It is persistent to have large cohort studies before putting wearables for clinical acceptance. (Figs. 16.7 and 16.8A–C) (Kim, Campbell, Fernández de Ávila, & Wang, 2019; Zhu et al., 2017).
8. Many people still have phobia from blood test as it causes discomfort and pain to patients. Scientists are always in search of alternative like intestinal fluids, saliva, sweat tears for monitoring health parameters. The idea is to get invasion approach and monitoring parameters in situ.
9. It is reported that one helmet was fabricated by researchers which has ability to combat depression with the ability to send weak signals to brain area. It is claimed that due to interaction of electrical pulses to brain, activation of capillaries take place which causes the depression to treat in the span of 7 days.

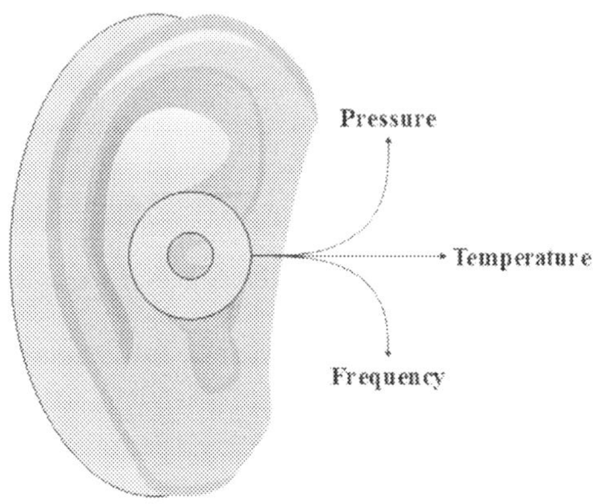


FIGURE 16.6 Bionic ear—basic layout.

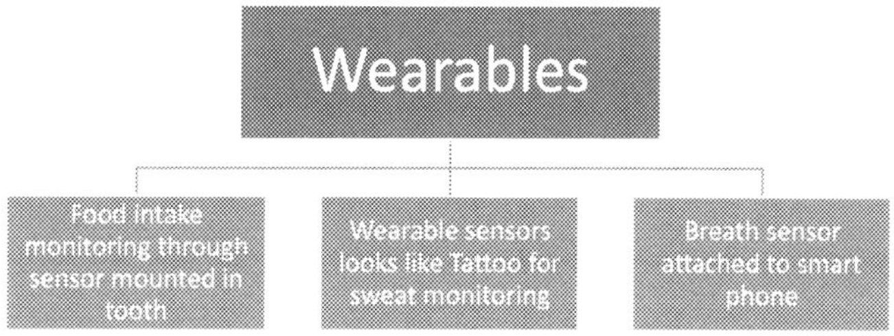


FIGURE 16.7 Wearable devices comes in many shapes and sizes.

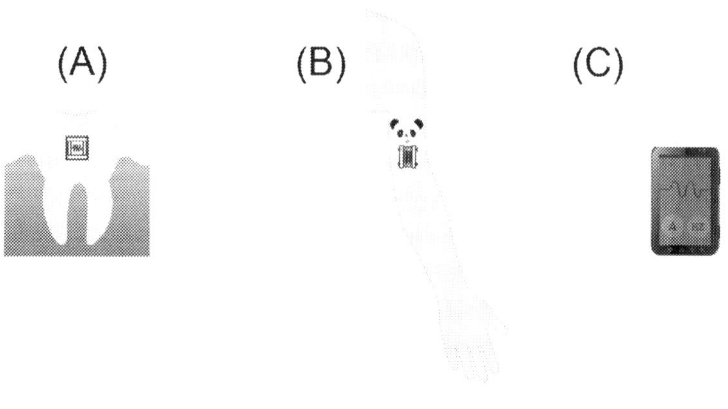


FIGURE 16.8 (A) Biosensor in tooth, (B) tattoo like sensors, (C) breath sensor in smart phone.

## 16.13 Major breakthrough in the field of personalized medicines

Patient centric medicine is whenever chronic and long-term illness like HIV, cancer, hypertension, and diabetes are considered, therapeutic compliance is always difficult to meet especially in the case of elder patients. As per the literature, the noncompliance rate is 50% and annual costing of noncompliance related medication is 100 billion dollars. This noncompliance is the biggest threat in the field of medicine, because it leads to faster disease progression, development of severe hypertension, myocardial infarction, and maybe sudden death. 3D-printed technology opened the gate of personalized medicine where all the prescribed drugs are formulated in a single pill. It also addresses the issues related to dose adjustment and age-related problems of swallowing and chewing, especially in polypharmacy. It also allows the individualization of drugs based on patient references like customized dose strength, size of pill, flavor, color, etc. (Vogenberg, Barash, & Pursel, 2010).

Polypills are the tablets containing several active pharmaceutical ingredient (APIs) to work as polypharmacy. Customization of polypills helps to control the geometry of pill, dissolution of graph, and removal of incompatibility issues.

Transdermal delivery is the personalization of transdermal drug delivery system (TDDS) which leads to better compliance and decreased adverse effects. Personalized microneedles do not cause pain and deliver the drug according to the patients' condition. Similarly, patches can be customized based on the 3D printing. It contains both externally applied patches and internally fixed implants

## 16.14 3D biosensors and cancer

Early detection of cancer is always a challenging task as disease usually gets featured once metastasis takes place. Accuracy and effectiveness of diagnostic methods are equally important. This opens the path to a wide range of opportunities for biosensors. Biosensor devices have the ability to convert a biological response to signal output. Biological responses include antigens, antibodies, enzymes, nucleic acids, and metabolic molecules like glucose (Cheng et al., 2019). They are being employed for the detection of tumor biomarkers and also efficacy of drug targets. The technology allows fast and prior detection, authentic imaging of cancerous cells, and also efficacy profiles of chemotherapeutic agents. It also helps in monitoring angiogenesis, at the metastatic stage. A good biosensor is expected to do multiple detections in single run. Taking the nanoscale occurrence of cancer into consideration, biosensors are designed nowadays using nanotechnology.

## 16.15 Challenges faced by researchers

Despite the huge success and accuracy of 3D biosensor, researchers are still facing some problems in fabricating and dealing with them (Khosravani & Reinicke, 2020). There are many associated problems like biocompatibility issues with material used for printing, implantable printable sensors that are difficult to handle, use of plastics sometimes is unavoidable that leads to threats to living entities, recycling of sensors gives the greatest challenges, and long term usage may lead to a saturation point. Further, liberation of harmful nanoparticles having a tumorigenic nature cannot be ignored in 3D-printing techniques. Machine and instrument cost is also a big issue. Sometimes generating high temperatures cause a distorting effect on material. Keeping all these in mind, if biosensors are portrayed as successful, the statement itself arises many doubts.

## 16.16 Regulatory aspects of biosensors

Regulatory control over 3D-printed biosensors under the domain of medical devices is very important (Sharafeldin et al., 2018; Waheed et al., 2016). Two pathway approach options are followed, one is Conventional 510 (k) pathway and the other is PMA (Premarket approval). In first method, there is no need of clinical trials as the device is following the prototype already present in the market. In second method, the approval process follows the strict guidelines, same as approval of a new drug. In general, approval guidelines depend on their performance and utilization purpose. The quality and regulatory hurdles should be given equal importance to 3D-printed biosensors like drug moieties in the pharmaceutical sector.

## 16.17 3D-printed biosensors in Covid-19

The biggest challenge in coronavirus lies in early stage detection as well as to scrutinize the symptomatic population. Mavrikou et al. demonstrated the portable ultra-rapid and sensitive cell-based biosensor for recognition of the Covid-19 spike protein antigen. The emphasis was on the attachment of protein to the antibodies which are already attached to membrane, leads to bioelectronic changes in cells and, thus, can be evaluated using bioelectric recognition assay. In this process, the fabrication of a biosensor involved the usage of 3D-printed PLA mold designed by 123D Design Software. Fused deposition modeling (FDM) is used for mold printing (M. et al., 2020; Mavrikou, Moschopoulou, Tsekouras, & Kintzios, 2020).

Further, Tarfaoui et al. reviewed the importance of Covid-19 medical devices with 3D printing. Healthcare professionals are working day and night to combat the deadly disorder, but the major issue with the availability of medical devices still persists (Tarfaoui, Nachtane, Goda, Qureshi, & Benyahia, 2020). The demand is overcoming the supply and thus the safety of doctors and nurses is questionable. A mask named NanoHAck2.0 is introduced having a strong mono block structure, with 3D printing and ability to overcome outside environment. The polymer used is PLActive which have recyclable properties. It also has a water resistant effect along with antimicrobial thermoplastic polyurethane (TPU). A recyclable and ecological 3D-printed facemask given by Swennen et al. which is designed by additive manufacturing technique and has polyamide components (Fig. 16.9A).

Reported a N95 mask having a mask adaptor for immediate consumption. The base used is 3D printed and the material is soft and silicon-based (Fig. 16.9B).

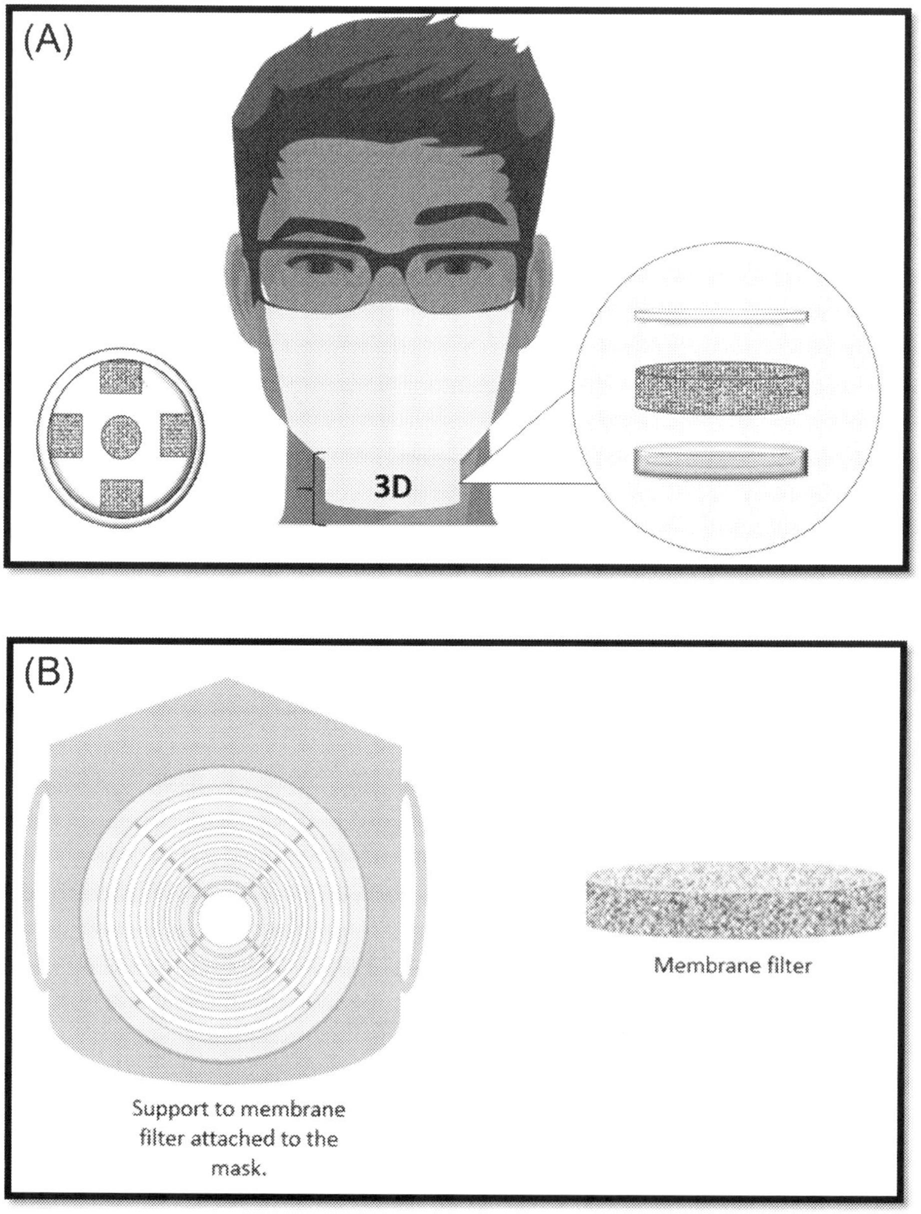


**FIGURE 16.9** (A) 3D-printed NanoHack2.0 mask. (B) 3D-printed facemask73.

## 16.18 Future of 3D-integrated biosensors

Though 3D-integrated biosensors have come a long way, still the destination is far from accepted. The technology has the ability to detect small molecules like glucose, cholesterol, etc., but large molecules, mainly proteins and nucleic acids, possess the problem of detection due to unavailability of specific adsorption techniques. LOD should be small and the usage of nanomaterials is still at premature levels (Chengzhou, Guohai, He, Dan, & Yuehe, 2014; Otero & Magner, 2020). As per the data available in IDTechEx, the possibility of reaching the market level of $44.25 billion by 2021 is there. The sensors used are still not of appropriate quality and need improvement. More understanding and studies need to be done with respect to technology, fabrication, and performance factors. Convenient designing and portability of biosensors are always desirable. Development of better and sophisticated softwares are needed. Scheming and engaging a sensor-based closed-loop control system in metal 3D printing is still in the exploration stage. Recycling is also one of the major criteria to be taken seriously for the future. Maximum use of artificial intelligence is recommended.

## 16.19 Conclusion

The beauty and role of 3D technology has evolved so much that it fascinates biomedical researchers, scientists, and even academician, too. Starting from an industrial platform 3D technology has spread its wings to the medical field in terms of tissue designs, organ printing, diagnostics, anatomical models, drug design, delivery of personalized medicine, etc. It takes the design from the mind to computer screen and finally to reality. Here, fabrication of material is done using powder, metal, alloy, thermoplastics, polymeric, ceramic, wood, paper, or biological material in a specific manner to create a 3D design of choice. As far as traditional methods like molding, machining, forming, casting, and injection laser cutting, they do not give the flexibility of creating nongeometrical and poly-material fabrication. Also, the process is time consuming and fabrication is a cumbersome process. 3D integration with biosensors has gained a lot of popularity, but still the regulatory aspects are not fully addressed. Despite this, the applicative importance and patient compliance with this technology is immensely under accepted.

## References


Ali, J., Najeeb, J., Ali, A., Muhammad., Aslam, F., Muhammad., & Raza, A. (2017). Biosensors: Their fundamentals, designs, types and most recent impactful applications: A review. *Journal of Biosensors and Bioelectronics*, *8*(1), 1–9. Available from https://doi.org/10.4172/2155-6210.1000235.

Anderson, K. B., Lockwood, S. Y., Martin, R. S., & Spence, D. M. (2013). A 3D printed fluidic device that enables integrated features. *Analytical Chemistry*, *85*(12), 5622–5626. Available from https://doi.org/10.1021/ac4009594.

Bhalla, N., Jolly, P., Formisano, N., & Estrela, P. (2016). Introduction to biosensors. *Essays in Biochemistry*, *60*(1), 1–8. Available from https://doi.org/10.1042/EBC20150001.

Bhargav, A., Sanjairaj, V., Rosa, V., Feng, L. W., & Fuh YH, J. (2018). Applications of additive manufacturing in dentistry: A review. *Journal of Biomedical Materials Research - Part B Applied Biomaterials*, *106*(5), 2058–2064. Available from https://doi.org/10.1002/jbm.b.33961.

Bishop, G. W., Satterwhite-Warden, J. E., Bist, I., Chen, E., & Rusling, J. F. (2016). Electrochemiluminescence at bare and DNA-coated graphite electrodes in 3D-printed fluidic devices. *ACS Sensors*, *1*(2), 197–202. Available from https://doi.org/10.1021/acssensors.5b00156.

Carpenter, A. C., Paulsen, I. T., & Williams, T. C. (2018). Blueprints for biosensors: Design, limitations, and applications. *Genes (Basel)*, *9*(8). Available from https://doi.org/10.3390/genes9080375.

Cevenini, L., Calabretta, M. M., Tarantino, G., Michelini, E., & Roda, A. (2016). Smartphone-interfaced 3D printed toxicity biosensor integrating bioluminescent sentinel cells. *Sensors and Actuators, B: Chemical*, *225*, 249–257. Available from https://doi.org/10.1016/j.snb.2015.11.017.

Cheng, N., Du, D., Wang, X., Liu, D., Xu, W., Luo, Y., & Lin, Y. (2019). Recent advances in biosensors for detecting cancer-derived exosomes. *Trends in Biotechnology*, *37*(11), 1236–1254. Available from https://doi.org/10.1016/j.tibtech.2019.04.008.

Chengzhou, Z., Guohai, Y., He, L., Dan, D., & Yuehe, L. (2014). Electrochemical sensors and biosensors based on nanomaterials and nanostructures. *Analytical Chemistry*, 230–249. Available from https://doi.org/10.1021/ac5039863.

Dantism, S., Takenaga, S., Wagner, P., Wagner, T., & Schöning, M. J. (2016). Determination of the extracellular acidification of Escherichia coli K12 with a multi-chamber-based LAPS system. *Physica Status Solidi (A) Applications and Materials Science*, *213*(6), 1479–1485. Available from https://doi.org/10.1002/pssa.201533043.

Dias, A. D., Kingsley, D. M., & Corr, D. T. (2014). Recent advances in bioprinting and applications for biosensing. *Biosensors*, *4*(2), 111–136. Available from https://doi.org/10.3390/bios4020111.

Didier, C., Kundu, A., & Rajaraman, S. (2020). Capabilities and limitations of 3D printed microserpentines and integrated 3D electrodes for stretchable and conformable biosensor applications. *Microsystems and Nanoengineering*, *6*(1). Available from https://doi.org/10.1038/s41378-019-0129-3.

Dirkzwager, R. M., Liang, S. L., & Tanner, J. A. (2016). Development of aptamer-based point-of-care diagnostic devices for malaria using three-dimensional printing rapid prototyping. *Acs Sensors*, *1*(4), 420–426.

Erkal, J. L., Selimovic, A., Gross, B. C., Lockwood, S. Y., Walton, E. L., McNamara, S., ... Spence, D. M. (2014). 3D printed microfluidic devices with integrated versatile and reusable electrodes. *Lab on a Chip, 14*(12), 2023–2032. Available from https://doi.org/10.1039/C4LC00171K.

François, B. (2004). Review of 20 years of range sensor development. *Journal of Electronic Imaging*, 231. Available from https://doi.org/10.1117/1.1631921.

Frazier, W. E. (2014). Metal additive manufacturing: A review. *Journal of Materials Engineering and Performance, 23*(6), 1917–1928. Available from https://doi.org/10.1007/s11665-014-0958-z.

Ghilan, A., Chiriac, A. P., Nita, L. E., Rusu, A. G., Neamtu, I., & Chiriac, V. M. (2020). Trends in 3D printing processes for biomedical field: Opportunities and challenges. *Journal of Polymers and the Environment, 28*(5), 1345–1367. Available from https://doi.org/10.1007/s10924-020-01722-x.

Gowers, S. A. N., Curto, V. F., Seneci, C. A., Wang, C., Anastasova, S., Vadgama, P., ... Boutelle, M. G. (2015). 3D printed microfluidic device with integrated biosensors for online analysis of subcutaneous human microdialysate. *Analytical Chemistry, 87*(15), 7763–7770. Available from https://doi.org/10.1021/acs.analchem.5b01353.

Han, T., Kundu, S., Nag, A., & Xu, Y. (2019). 3D printed sensors for biomedical applications: A review. *Sensors (Switzerland), 19*(7). Available from https://doi.org/10.3390/s19071706.

Heger, Z., Zitka, J., Nejdl, L., Moulick, A., Milosavljevic, V., Kopel, P., ... Kizek, R. (2016). 3D printed stratospheric probe as a platform for determination of DNA damage based on carbon quantum dots/DNA complex fluorescence increase. *Monatshefte Fur Chemie, 147*(5), 873–880. Available from https://doi.org/10.1007/s00706-016-1705-y.

Ho, D. H., Homg, P., Han, J. T., Kim., Kwon, S. J., & Cho, J. H. (2019). 3D printed sugar scaffold for high precision and highly sensitive active and passive wearable sensors. *Advanced Science, 7*(1), 1902521.

Ionita, C.N., Mokin, Varble, N., Bednarek, D.R., Xiang, J., Snyder, K.V., ... Rudin, S. (2014). Challenges and limitations of patient-specific vascular phantom fabrication using 3D Polyjet printing. In *Proceedings of the medical imaging. Biomedical applications in molecular, structural, and functional imaging*.

Kadimisetty, K., Mosa, I. M., Malla, S., Satterwhite-Warden, J. E., Kuhns, T. M., Faria, R. C., ... Rusling, J. F. (2016). 3D-printed supercapacitor-powered electrochemiluminescent protein immunoarray. *Biosensors and Bioelectronics, 77*, 188–193. Available from https://doi.org/10.1016/j.bios.2015.09.017.

Khosravani, M. R., & Reinicke, T. (2020). 3D printed sensors: Current progress and future challenges. *Sensors and Actuators A: Physical*, 305.

Kim, J., Campbell, A. S., Fernández de Ávila, B. E.-, & Wang, J. (2019). Wearable biosensors for healthcare monitoring. *Nature Biotechnology, 37*, 389–406. Available from https://www.nature.com/articles/s41587-019-0045-y.

Krejcova, L., Nejdl, L., Rodrigo, M. A. M., Zurek, M., Matousek, M., Hynek, D., ... Kizek, R. (2014). 3D printed chip for electrochemical detection of influenza virus labeled with CdS quantum dots. *Biosensors and Bioelectronics, 54*, 421–427. Available from https://doi.org/10.1016/j.bios.2013.10.031.

Kumar, P., Ahuja, I. P. S., & Singh, R. (2012). Application of fusion deposition modelling for rapid investment casting – A review. *International Journal of Materials Engineering Innovation, 3*(3–4), 204–227. Available from https://doi.org/10.1504/IJMATEI.2012.049254.

Lee, B. (2003). Review of the present status of optical fiber sensors. *Optical Fiber Technology, 9*(2), 57–79. Available from https://doi.org/10.1016/S1068-5200(02)00527-8.

Lynch, J. P., & Loh, K. J. (2006). A summary review of wireless sensors and sensor networks for structural health monitoring. *Shock and Vibration Digest, 38*, 91–130.

Mahindru, D., Mahendru, S., & Ganj, T. (2013). Review of rapid prototyping-technology for the future. *Global Journal of Computer Science Technology*.

Mannoor, M. S., Jiang, Z., James, T., Kong, Y. L., Malatesta, K. A., Soboyejo, W. O., ... McAlpine, M. C. (2013). 3D printed bionic ears. *Nano Letters, 13*(6), 2634–2639. Available from https://doi.org/10.1021/nl4007744.

Matthew, W. (2014). The history of 3D printing in healthcare. *The Bulletin of the Royal College of Surgeons of England*, 228–229. Available from https://doi.org/10.1308/147363514X13990346756481.

Mavrikou, S., Moschopoulou, G., Tsekouras, V., & Kintzios, S. (2020). Development of a Portable, Ultra-Rapid and Ultra-Sensitive Cell-Based Biosensor for the Direct Detection of the SARS-CoV-2 S1 Spike Protein Antigen. *Sensors (Basel), 20*(11), 3121. Published 2020 May 31.

Mehrotra, P. (2016). Biosensors and their applications – A review. *Journal of Oral Biology and Craniofacial Research, 6*(2), 153–159. Available from https://doi.org/10.1016/j.jobcr.2015.12.002.

Miroslav, P. (2019). Current trends in the biosensors for biological warfare agents assay. *Materials*, 2303. Available from https://doi.org/10.3390/ma12142303.

Mohan Pandey, P., Venkata Reddy, N., & Dhande, S. G. (2003). Slicing procedures in layered manufacturing: A review. *Rapid Prototyping Journal, 9*(5), 274–288. Available from https://doi.org/10.1108/13552540310502185.

Mongra, A. C. (2012). Commercial biosensors: An outlook. *Journal of Academia and Industrial Research, 1*(6), 310–312. Available from http://www.jairjp.com/NOVEMBER%202012/07%20MONGRA.pdf.

Nag, A., Mukhopadhyay, S. C., & Kosel, J. (2017). Sensing system for salinity testing using laser-induced graphene sensors. *Sensors and Actuators, A: Physical, 264*, 107–116. Available from https://doi.org/10.1016/j.sna.2017.08.008.

Nakamura, M., Kobayashi, A., Takagi, F., Watanabe, A., Hiruma, Y., Ohuchi, K., ... Takatani, S. (2005). Biocompatible inkjet printing technique for designed seeding of individual living cells. *Tissue Engineering, 11*(11–12), 1658–1666. Available from https://doi.org/10.1089/ten.2005.11.1658.

Nguyen, A. K., Narayan, R. J., & Shafiee, A. (2019). *3D printing in the biomedical field, . Encyclopedia of Biomedical Engineering* (Vols. 1–3, pp. 275–280). Elsevier, https://doi.org/10.1016/B978-0-12-801238-3.99875-1.

Ni, Y., Ji, R., Long, K., Bu, T., Chen, K., & Zhuang, S. (2017). A review of 3D-printed sensors. *Applied Spectroscopy Reviews, 52*(7), 623–652. Available from https://doi.org/10.1080/05704928.2017.1287082.

Olakanmi, E. O., Cochrane, R., & Dalgarno, K. (2015). A review on selective laser sintering/melting (SLS/SLM) of aluminium alloy powders: Processing, microstructure, and properties. *Progress in Material. Science (New York, N.Y.), 74*, 401–447.

Otero, F., & Magner, E. (2020). *Biosensors – Recent advances and future challenges in electrode materials. Sensors, 20*(12), 3561.

Park, C., Yoo, Y. S., & Hong, S. T. (2010). An update on auricular reconstruction: Three major auricular malformations of microtia, prominent ear and cryptotia. *Current Opinion in Otolaryngology and Head and Neck Surgery, 18*(6), 544–549. Available from https://doi.org/10.1097/MOO.0b013e32833fecb9.

Reeder, J. T., Xue, Y., Franklin, D., Deng, Y., Choi, J., Prado, O., ... Rogers, J. A. (2019). Resettable skin interfaced microfluidic sweat collection devices with chemesthetic hydration feedback. *Nature Communications, 10*(1). Available from https://doi.org/10.1038/s41467-019-13431-8.

Roda, A., Guardigli, M., Calabria, D., Calabretta, M. M., Cevenini, L., & Michelini, E. (2014). A 3D printed device for a smartphone- based chemiluminescence biosensor for lactate in oral fluid and sweat. *Analyst, 139*(24), 6494–6501.

Salvo, P., Raedt, R., Carrette, E., Schaubroeck, D., Vanfleteren, J., & Cardon, L. (2012). A 3D printed dry electrode for ECG/EEG recording. *Sensors and Actuators A: Physical, 174*(1), 96–102. Available from https://doi.org/10.1016/j.sna.2011.12.017.

Sharafeldin, M., Jones, A., & Rusling, J. F. (2018). 3D-printed biosensor arrays for medical diagnostics. *Micromachines, 9*(8). Available from https://doi.org/10.3390/mi9080394.

Singh, H., Shimojima, M., Shiratori, T., Van An, L., Sugamata, M., & Yang, M. (2015). Application of 3D printing technology in increasing the diagnostic performance of enzyme-linked immunosorbent assay (ELISA) for infectious diseases. *Sensors (Switzerland), 15*(7), 16503–16515. Available from https://doi.org/10.3390/s150716503.

Spivey, E. C., Jones, S. K., Rybarski, J. R., Saifuddin, F. A., & Finkelstein, I. J. (2017). An aging-independent replicative lifespan in a symmetrically dividing eukaryote. *ELife*, 6. Available from https://doi.org/10.7554/eLife.20340.

Tarfaoui, M., Nachtane, M., Goda, I., Qureshi, Y., & Benyahia, H. (2020). 3D printing to support the shortage in personal protective equipment caused by COVID-19 pandemic. *Materials, 13*(15). Available from https://doi.org/10.3390/MA13153339.

Terzaki, K., Kalloudi, E., Mossou, E., Mitchell, E. P., Forsyth, V. T., Rosseeva, E., ... Farsari, M. (2013). Mineralized self-assembled peptides on 3D laser-made scaffolds: A new route toward "scaffold on scaffold" hard tissue engineering. *Biofabrication, 5*(4). Available from https://doi.org/10.1088/1758-5082/5/4/045002.

Vogenberg, F. R., Barash, C. I., & Pursel, M. (2010). Personalized medicine – Part 1: Evolution and development into theranostics. *P and T, 35*(10), 560–576. Available from http://www.ptcommunity.com/ptJournal/fulltext/35/10/PTJ3510560.pdf.

Waheed, S., Cabot, J. M., Macdonald, N. P., Lewis, T., Guijt, R. M., Paull, B., & Breadmore, M. C. (2016). 3D printed microfluidic devices: Enablers and barriers. *Lab on a Chip, 16*(11), 1993–2013. Available from https://doi.org/10.1039/c6lc00284f.

Wonjae, L., Donghoon, K., Woong, C., Yeol, J. G., K., A. A., Albert, F., & Sangmin, J. (2015). 3D-printed microfluidic device for the detection of pathogenic bacteria using size-based separation in helical channel with trapezoid cross-section. *Scientific Reports*. Available from https://doi.org/10.1038/srep07717.

Xu, L., Gutbrod, S. R., Bonifas, A. P., Su, Y., Sulkin, M. S., Lu, N., ... Rogers, J. A. (2014). 3D multifunctional integumentary membranes for spatiotemporal cardiac measurements and stimulation across the entire epicardium. *Nature Communications*, 5. Available from https://doi.org/10.1038/ncomms4329.

Xu, Y., Wu, X., Guo, X., Kong, B., Zhang, M., Qian, X., ... Sun, W. (2017). The boom in 3D-printed sensor technology. *Sensors, 17*(5), 1166.

Yana, Q., Dong, H., Su, J., Han, J., Song, B., Wei, Q., & Shi, Y. (2018). A review of 3D printing technology for medical applications. *Engineering, 4*(5), 729–742.

Yeong, W. Y., Chua, C. K., Leong, K. F., Chandrasekaran, M., & Lee, M. W. (2006). Indirect fabrication of collagen scaffold based on inkjet printing technique. *Rapid Prototyping Journal, 12*(4), 229–237. Available from https://doi.org/10.1108/13552540610682741.

Yick, J., Mukherjee, B., & Ghosal, D. (2008). Wireless sensor network survey. *Computer Networks, 52*(12), 2292–2330. Available from https://doi.org/10.1016/j.comnet.2008.04.002.

Zeming, G., Jianzhong, F., Hui, L., & Yong, H. (2019). Development of 3D bioprinting: From printing methods to biomedical applications. *Asian Journal of Pharmaceutical Sciences*. Available from https://doi.org/10.1016/j.ajps.2019.11.003.

## Further reading

Imbrie-Moore, A. M., Park, M. H., Zhu, Y., Paulsen, M. J., Wang, H., & Woo, Y. J. (2020). Quadrupling the N95 Supply during the COVID-19 Crisis with an Innovative 3D-Printed Mask Adaptor. *Healthcare, 8*(3), 225. Available from https://doi.org/10.3390/healthcare8030225, Multidisciplinary Digital Publishing Institute.

Zhu, X., Liu, W., Shuang, S., Nair, M., & Li, C. Z. (2017). *Intelligent tattoos, patches, and other wearable biosensors. Medical biosensors for point of care (POC) applications* (pp. 133–150). Elsevier Inc. Available from https://doi.org/10.1016/B978-0-08-100072-4.00006-X.

Chapter 17

# Novel paper-based diagnostic devices for early detection of cancer

**Maryam Mousavizadegan, Amirreza Roshani and Morteza Hosseini**

*Department of Life Science Engineering, Faculty of New Sciences & Technologies, University of Tehran, Tehran, Iran*

## 17.1 Introduction

The significance of early cancer detection is critically profound as it directly effects disease control and treatment. Existing screening methods for cancer such as computerized topography (CT) scans, magnetic resonance imaging (MRI), x-rays, and ultrasound imaging suffer from being quite costly and unavailable to many people, along with not being strong enough for detection at early stages. Regarding final tumor diagnosis and classification, tissue biopsies are currently the gold standard source (Garcia-Cordero & Maerkl, 2020). However, this method also has many downfalls as it involves an invasive procedure which may not always be feasible depending on the tumor size and location. Furthermore, the challenging nature of this technique hinders its use to assess tumor response to therapeutics and cancer progression (Parikh et al., 2019). Therefore, it is evident that a dire need for reliable, highly specific and accessible point-of-care (POC) diagnosis devices for cancer detection perpetuates.

Biomarkers, defined as indicators of the biological state relating to a disease, are of colossal importance in the development of rapid POC sensors. A vast collection of biomarkers including proteins, protein segments, DNA sequences including methylated DNA and single nucleotide polymorphisms (SNPs) and RNA-based markers such as microRNAs (miRNAs) have been explored for cancer diagnosis. These biomarkers mostly originate from inherited or somatically acquired alterations in regulatory DNA sequences, oncogenes or tumor suppressor genes (Srinivas, Kramer, & Srivastava, 2001). In Chapter 3, Biomarkers associated with different types of cancer as a potential candidate for early diagnosis of oncological disorders, a more detailed explanation on cancer biomarkers has been provided.

Since the advent of paper technology in China during the second century AD, it has become one of the most significant technologies in human history. As it is favorably abundant, lightweight, flexible, available in a variety of thicknesses and environmentally friendly, paper has become one of the forerunners in POC diagnosis development. Furthermore, the hydrophilicity and porosity of paper provides a platform for the flow of fluids through capillary action. Paper-based analytical devices (PADs) were first introduced in the seventeenth century for the detection of uric acid using silver-based filter paper and a colorimetric approach (Schiff, 1866). Since then, PADs have become the focus of attention in the diagnosis domain due to the aforementioned advantages along with the fact that paper can be easily printed and coated with biomolecules in order to develop a portable and ideal platform for the detection of clinically significant markers. PADs have been coupled with numerous optical, electrochemical, and spectroscopy read-out techniques for the detection of a wide range of clinically relevant analytes including tumor markers.

In this chapter, we will introduce the various formats of paper-based sensors namely dipstick assays, lateral flow assays (LAFs) and microfluidic-based devices along with a short description regarding the fabrication techniques for PADs. Later on, various novel sensing approaches based on colorimetric, fluorescence, chemiluminescence (CL), electrochemical, electrochemiluminescence (ECL) and surface-enhanced Raman scattering (SERS) techniques for cancer detection based on different biomarkers will be presented. In the end, the current limitations hindering the commercialization of PADs will be briefly discussed.

**Biosensor Based Advanced Cancer Diagnostics. DOI: https://doi.org/10.1016/B978-0-12-823424-2.00015-6**

## 17.2 Formats of paper-based analytical devices

Since the advent of paper-based sensors, various technologies have been proposed for biochemical analysis starting with dipstick assays, and followed by the more advanced LFAs and the emerging field of microfluidic paper-based devices (μPADs). In this section, an introduction to these technologies has been presented.

### 17.2.1 Paper devices based on dipsticks

Making up the most simplistic paper-based sensors, dipstick assays are commonly known for the detection of glucose in urine (Free, Adams, Kercher, Free, & Cook, 1957). In dipstick assays, the reagents are previously deposited on the paper to develop a device for qualitative assessment of an analyte. These sensors usually adopt a colorimetric approach. Due to their limited capability in handling fluids, dipsticks are inadequate for multistep detection procedures as seen in immunoassays, thus hindering their use in cancer diagnosis.

### 17.2.2 Lateral flow assays

In an attempt to increase the fluid handling capability of PADs, lateral flow test strips were developed. Typically composed of nitrocellulose membranes, these devices consist of four zones: (1) the sample pad which relies on the cellulose fiber matrix for sample filtration and buffer storage; (2) the conjugation pad which is adjacent to the sample pad and contains the dried reagents on glass fibers; (3) the detection pad is where the reagents are captured on the nitrocellulose for signal development; and (4) the absorbent pad with loose mass of nitrocellulose which is responsible for providing the driving force for fluid flow during detection (Fenton, Mascarenas, López, & Sibbett, 2009). Due to the presence of the nitrocellulose matrix, LFAs provide a platform for reactions and efficient multistep detection, which has led to their popularity in bioanalytical sensors. Despite their advanced fluid handling capabilities, limitations regarding precise quantification of biomarkers still persist in this format.

### 17.2.3 Paper devices based on microfluidics

Microfluidic technology, which refers to the control and manipulation of fluids in the range of micro to picoliters, has been coupled with paper-based devices to create advanced sensors with guided fluid flow for the detection of clinical markers. This system minimizes sample loss due to undesired soaking in the paper platform, which enables quantitative and multiplexing detection (Fakhri, Hosseini, & Tavakoli, 2018; Rezk, Qi, Friend, Li, & Yeo, 2012). Exploiting these advantages, numerous μ-PADs have been developed for various biomarkers based on various sensing techniques such as CL, electrochemistry, colorimetric, etc.

## 17.3 Fabrication and development of paper-based analytical devices

Various cellulose derivatives have been used to develop paper devices. The most common materials currently used for paper fabrication and their corresponding attributes have been summarized in Table 17.1. Later in this section, we will briefly discuss the various fabrication methods employed for PAD development. The common techniques used for biomolecule immobilization on paper platforms will also be introduced.

**TABLE 17.1 Common paper substrates and their characteristics.**

| Substrate | Characteristics |
|---|---|
| Whatman paper # 1 | Composed of cotton cellulose: pore size = 11 μm; thickness = 180 μm; medium retention and flow rate |
| Whatman paper # 4 | Composed of cotton cellulose: pore size = 20–25 μm; thickness = 210 μm; very fast filtering and excellent retention rate |
| Nitrocellulose membrane | Very smooth and uniform: pore size = 0.45 μm |
| Bioactive paper | Obtained through the modification of paper matrix with biomolecules; simple without the need for complex equipment to operate |
| Cellulose glossy paper | Composed of cellulose which has been bended with certain inorganic fillers; non-degradable; suitable for surface modifications by nanoparticles |

### 17.3.1 Fabrication methods in paper-based devices

As paper is an extremely flexible material, numerous techniques have been developed for the fabrication of paper sensors. A paper device is made up of hydrophilic channels which act as the assay domain, separated by hydrophobic regions. In general, a paper device should go through two main stages; first the patterning of hydrophobic regions for liquid confinement and then assembly. Amongst the most widely used and established approaches is the use of hydrophobic materials to physically block the pores on the paper. Various water impervious materials have been employed for the implementation of the hydrophobic barriers such as polydimethylsiloxane (PDMS), polystyrene, ethyl cellulose, alkenyl ketene dimer, silicones, rosin, paraffin, printer varnish, cellulose esters, hydrophobic gels, SU-8 and other photoresist materials, which can create regions for the transportation and storage of liquids (Dixit, Kaushik, & Kaushik, 2016; Sharma, Barstis, & Giri, 2018). A brief description of the most common technologies, along with their advantages and limitations has been gathered in Table 17.2.

### 17.3.2 Immobilization of biomolecules on paper

Detection of cancer biomarkers usually requires specific capture molecules such as enzymes, antibodies, complementary nucleotide sequences or aptamers which selectively interact with the target, leading to diagnosis. Thus, in many cases, the paper platform needs to be functionalized with appropriate biomolecules to enable selective detection. Active functionalization of paper can be done at different stages of the fabrication process either during the web formation, during the paper-making conversion process, or after the manufacturing is complete (Kong & Hu, 2012). Generally, biofunctionalization methods of paper can be categorized in two groups: (1) conventional approaches in which the biomolecule is attached to paper through physical adsorption, chemical bonding, affinity-based attachments, entrapment, or surface-modification with reactive molecules; and (2) new approaches that rely on nanoparticles incorporated in the paper which are used as carriers for biomolecule immobilization. Since nanoparticles provide a more active surface, using them as carriers has the obvious merit of enabling more functionalities for biomolecule attachment.

In this section, a brief description of each immobilization technique displayed in Fig. 17.1 for paper biofunctionalization has been provided.

#### *17.3.2.1 Physical adsorption*

A simple approach for paper coating is through the physical adsorption of the biomolecules based on noncovalent interactions such as van der Waals forces, hydrogen bonding, and hydrophobic interactions. The major disadvantage of this method is the low efficiency of attachment since molecules can easily leach from the surface. Furthermore, biomolecules attached via physical adsorption are randomly oriented, which in some cases can lead to the loss of functionality.

As cellulose is hydrophilic with a slightly anionic surface, positively charged molecules can bind to it through electrostatic interactions. Factors such as pH, ionic strength, and specific ion effects can strongly impact the adsorption efficiency of the biomolecules. For instance, a study showed that DNA molecules with high molecular weight are only adsorbed on cellulose at pH 4 (Halder, Chattoraj, & Das, 2005), whereas some reports suggest low molecular weight DNA such as aptamers are more easily adsorbed on cellulose (Su, Nutiu, Filipe, Li, & Pelton, 2007). However, the efficiency of DNA coating via noncovalent interactions is very low as they are easily detached from the surface.

To improve the adsorption of biomolecules, cationic polyamide-epichlorohydrin (PAE) wet-strength resin has been used and it was corroborated that it can influence both the adsorption of DNA aptamers (Su, Ali, Filipe, Li, & Pelton, 2008) and especially antibodies (Wang et al., 2010). Moreover, evidence of the interaction between cellulose and tyrosine groups in proteins can be the basis of the incorporation of tyrosine tags in proteins for more effective adsorption on paper (Lehtiö, Sugiyama, Gustavsson, Fransson, Linder, & Teeri, 2003).

#### *17.3.2.2 Covalent chemical bonding*

This approach provides a more stable, irreversible, and uniform attachment of biomolecules to paper. For chemical bonding, active functional groups, which react under mild conditions, should be present on both the paper surface and the biomolecule. Very few functional groups, mostly backbone hydroxyl groups, are available on pure cellulose, which are too unreactive for direct bioconjugation in aqueous solutions under mild conditions. The only reactive functional groups are some carboxyl groups, which are generated from the oxidation of C-6 hydroxyls, and the active oxidizing end, which are not sufficient for high density biomolecule attachment (Heinze & Liebert, 2001). Thus various activating

**TABLE 17.2 Various fabrication methods for paper-based sensors.**

| Fabrication methods | Description | Advantages | Limitations |
|---|---|---|---|
| Photo lithography | Photo lithography is a noncontact process to pattern small features on paper using light. For paper fabrication, a photoresist polymer like SU-8 is exposed to radiation (mostly UV light) leading to the cross-linking of the polymers and the creation of hydrophobic regions. | High resolution of channels with sharp barriers; suitable for largescale production | Requires expensive instruments and reagents; involves complex steps; fragile while bending; requires organic solvents |
| Inkjet printing | Inkjet printing is a noncontact approach in which through using a small nozzle, reagents are sprayed onto the substrate material. | Rapidly fabricate on a large scale; requires only a desktop printer to produce sensors | Requires a customized inkjet printer and an extra heating step for curing purposes |
| PDMS plotting | PDMS is a hydrophobic compound which can be dispersed onto paper, creating hydrophobic barriers and micro-channels. | Inexpensive technique; cheap patterning agents | Low resolution; modification is required; inconsistent control over the penetration of PDMS due to the ununiform porous nature of paper |
| Laser cutting | Laser cutting employs $CO_2$ to cut or make patterns on the paper. The hydrophilic patterns are then remodeled to create a hydrophobic coating. | Simple and inexpensive technique, sharp defined features | Requires expensive equipment; waste of raw material; yields low mechanical stability; requires a cover tape for preventing pollution |
| Laser printing | This approach enables the fabrication of paper-based microfluidic devices. The laser printer deposits toner layers on transparency sheets. | Simple and inexpensive; can be used to selectively modify the surface structure and property of several papers | Requires special equipment such as laser printer, graphics software, laminator, and paper driller |
| Wax printing | During this process, wax from a printer or a pen in used to create channeling patterns on filter or chromatography paper which control the flow of the fluids. Wax-printed paper devices are usually based on colorimetric or electrochemical techniques for analyte detection. | Fast and simple fabrication technique; Suitable for mass production; Environmentally friendly | Patterned mesh is necessary which makes this method inadequate for prototyping; requires an extra heating step after wax deposition |
| Wax dipping | This method utilizes melted wax to create a hydrophobic barrier. Using a magnetic field, iron mold is temporarily placed on the paper to protect the hydrophilic channels. When the paper attached to the iron mold is dipped into molten wax, hydrophobic barriers are created. | Very cheap and easy; requires solid wax | Inconsistency between batches due to the variation in dipping can be seen; not suitable for mass production |
| Screen printing | In this technique, liquid material is transferred onto paper via a screen. This can be done both manually or automatically, through which pressure and the amount of printed material is regulated. After coating the paper with this material, it is left to dry by heat or other treatments. | Cost-effective; simple process; well-suited for mass production | Low resolution; each pattern requires an individual screen; requires different printing screens for creating different patterns |
| Plasma treatment | Plasma treatment has been used for channel fabrication on paper surfaces and microfluidic chip bonding. In this process, the filter paper is dipped into the alkyl ketene dimer (AKD)-heptane solution and then immediately placed in a fume hood to facilitate evaporation of the heptane. The filter | Inexpensive; the flexibility of paper is maintained | Each pattern requires a specific photomask; the substrate under a mask is often over etched |

(*Continued*)

**TABLE 17.2 (Continued)**

| Fabrication methods | Description | Advantages | Limitations |
|---|---|---|---|
| | paper is then heated in an oven to cure the AKD thus rendering the paper hydrophobic. This paper is then sandwiched between two metal masks with the desired patterns and submitted to plasma treatment to create the hydrophilic regions. | | |
| Flexographic printing | Flexography is a printing process which utilizes a flexible relief plate. During this process, polystyrene is printed on the filter paper by penetrating the depth of the filter paper and forming a hydrophobic wall. The regions of the filter paper without polystyrene are hydrophilic. | Enables fast, commercial roll-to-roll production of paper-based devices | Multi-step process; needs complex reagents and specialized printers; requires frequent cleaning to avoid contamination; roughness of the paper substrate can impact the final quality of printing |

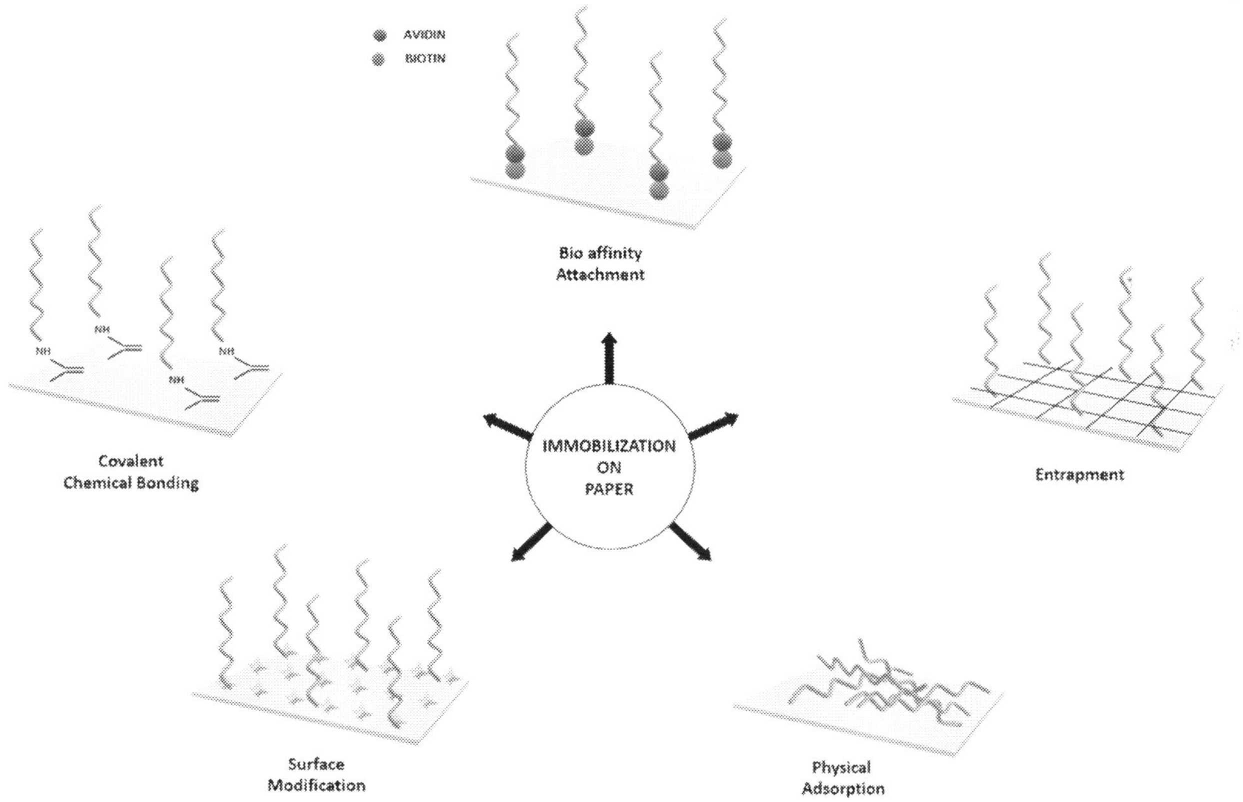


**FIGURE 17.1** Common approaches for biofunctionalization on paper-based analytical devices (PADs).

reactions, some of which are listed below, have been proposed to create a more suitable surface for biomolecule conjugation.

1. Periodate: This reagent leads to the oxidation of the surface hydroxyl groups to aldehyde groups which can react with amine groups in biomolecules. A study used this approach to bind the 3′ amino group of fluorescein-labeled DNA strands of an ATP-binding aptamer to cellulose (Su et al., 2007).

2. Epichlorohydrin: Using this approach, epoxy groups were created from the surface hydroxyl groups (Kong & Hu, 2012).
3. 1-Fluoro-2-nitro-4-azidobenzene: Using this reagent, which reacts with hydroxyl groups in alkaline media, photoreactive cellulose was prepared. In the presence of UV radiation (365 nm) nitrene groups are created from the azido groups on the photoreactive paper, which forms covalent bonds with biomolecules (Bora, Sharma, Kannan, & Nahar, 2006).

#### *17.3.2.3 Bioaffinity attachment*

Bioaffinity attachment is mostly done using the cellulose-binding domain (CBD) of carbohydrate-active enzymes as an affinity tag to create fusion proteins and peptides which can easily be immobilized on the paper surface. CBDs have been identified in more than 13 different protein families and are found to be between 4 and 20 kDa (Terpe, 2003). CBDs can be classified in two groups: those which bind reversibly to cellulose and those which bind irreversibly. For sensing purposes, CBDs which attach to cellulose irreversibly are a more appropriate choice for immobilization of enzymes and antibodies (Cao, Zhang, Wang, Zhu, & Bai, 2007; Xu, Bae, Mulchandani, Mehra, & Chen, 2002).

#### *17.3.2.4 Entrapment*

Entrapment of biomolecules in inorganic materials through the sol–gel process (mostly using silica-based materials) has been widely used as it provides a simple biofunctionalization avenue on solid surfaces and ensures prolonged biomolecule activity. In this process, an inorganic network encapsulates the biomolecule in a tight manner which prevents leaching, but does not limit the local mobility and diffusion required for its function (Luckham & Brennan, 2010; Pankratov & Lev, 1995). Another approach for entrapment is based on polymeric microplates, made from semipermeable polymers such as polyethyleneimine (PEI). Although this method is quite versatile, loss of enzyme activity and biomolecule leaching are amongst its drawbacks which limit the use of this approach (Savolainen et al., 2011).

#### *17.3.2.5 Surface modification with reactive molecules*

A popular approach to generating a more reactive surface is paper-modification. Chitosan, which is a biopolymer attained from the deacetylation of chitin with a similar backbone structure to cellulose, is amongst the most prevalent surface modifying agents for PADs (Koev et al., 2010). In addition to possessing many desirable features such as biocompatibility and biodegradability, chitosan is also rich in amine groups which can actively bind to biomolecule functional groups, such as carboxylic groups through simple reactions such as EDC/NHS coupling (Yi et al., 2005). Chitosan-modified paper has been successfully proven to be effective in biomolecule immobilization in paper-based biosensors.

## 17.4 Diagnostic technologies

Various technologies have been coupled with PADs for the qualitative or quantitative evaluation of tumor markers including colorimetric, fluorescence, CL, electrochemical, ECL, and SERS approaches. In this section, we will provide a brief explanation regarding these technologies along with examples of sensors developed for sensitive cancer detection.

### 17.4.1 Colorimetric

Colorimetric sensing approaches have become the most prevalent in PADs as they require minimum instrumentation and enable qualitative and semiquantitative analysis of the analytes simply done by the naked eye. Through providing a bright, white background with high-contrast, paper substrates ease the reading of color change (Dehghani, Hosseini, Mohammadnejad, & Ganjali, 2019; Nery & Kubota, 2013). The adaptation of various chemical and enzymatic reactions has led to the development of qualitative, semiquantitative and fully-quantitative colorimetric paper-based sensors for the evaluation of one or multiple tumor markers. Various methods which lead to a colorimetric output for PADs include the use of nanoparticles, dyes, redox, or pH indicators (Morbioli, Mazzu-Nascimento, Stockton, & Carrilho, 2017; Sharma et al., 2018).

For quantitative evaluation of colorimetric PADs, the Beer-Lambert Law, in which the intensity of the color signal is directly proportional to the concentration of the analyte, can be used. Using detectors such as CCD, CMOS, scanners, smartphones, and phone cameras, the color produced on the PAD can be captured. Software such as ImageJ can then be used to quantify the intensity of the color (Nery & Kubota, 2013).

An obvious problem with colorimetric readouts is that the naked-eye visual interpretation of the results can be extremely influenced by lighting conditions and each individual's color perceptions. To overcome this complication, a calibrating color code can be used to compare with the results on the paper, thus enabling a more objective approach to the interpretation of the results (Cate, Adkins, Mettakoonpitak, & Henry, 2015).

The most prevalent colorimetric approach in PADs is the use of the enzymatic transformation of chromogenic substrates such as 3,3′,5,5′-tetramethylbenzidine (TMB) in the presence of $H_2O_2$ which leads to the production of colored products. This process is mostly carried out by horseradish peroxidase (HRP) (Busa, Maeki, Ishida, Tani, & Tokeshi, 2016), but recently many nanomaterials have been investigated for enzyme-mimicking activities including peroxidasing and oxidasing activities and used for the detection of various analytes (Dehghani, Hosseini, Mohammadnejad, Bakhshi, & Rezayan, 2018; Kermani et al., n.d.; Mousavizadegan, Azimzadeh Asiabi, Hosseini, & Khoobi, 2020). The implementation of these enzyme-mimicking nanostructures has also been investigated on PADs. In one study, researchers used cysteine-capped gold nanoclusters (Cys-AuNCs) which possess intrinsic peroxidase-mimicking activity to detect citrate ions as an indicator of prostate cancer. Through hydrogen bonding, citrate creates a coat on the surface of the nanoclusters leading to the inhibition of its catalytic activity and a decrease in the blue-colored product (Abarghoei, Fakhri, Borghei, Hosseini, & Ganjali, 2019). In another study, the peroxidase-mimicking activity of DNA-templated Ag/Pt nanoclusters was exploited to develop a paper-based sensor for the detection of miRNA-21. In the presence of miRNA-21, the catalytic activity will be inhibited through hybridization of the target to the DNA template, thus hindering the reaction between $H_2O_2$ and TMB and the creation of the blue-colored product (Fakhri et al., 2020). Nano-based materials composed of β-$Co(OH)_2$ nanoplates (NPls) anchored on ordered mesoporous carbon (carbons mesostructured from Korea or CMK) were also used for their catalytic activity in a colorimetric aptasensor for detection of cytochrome C (Cyt c). As shown in Fig. 17.2, Cyt c will bind to its specific aptamer, thus freeing the β-$Co(OH)_2$ CMK to catalyze the reaction between $H_2O_2$ and TMB, creating a blue signal (Mesgari et al., 2020).

Due to their strong surface plasmon resonance (SPR) absorption from the visible to near-infrared region, silver (Ag) nanostructures are also gaining ample attention in colorimetric sensors. Among the numerous Ag structures, triangular Ag NPls are prevalently being used for the detection of hydrogen peroxide ($H_2O_2$). The reaction between Ag NPls and $H_2O_2$ results in the etching of the NPls to round nanoparticles, creating $Ag^+$ which leads to a color change from blue to mauve (Chen, Zhang, Wu, Li, & Tan, 2015). In one study, researchers took advantage of the outstanding properties of these nanostructures to

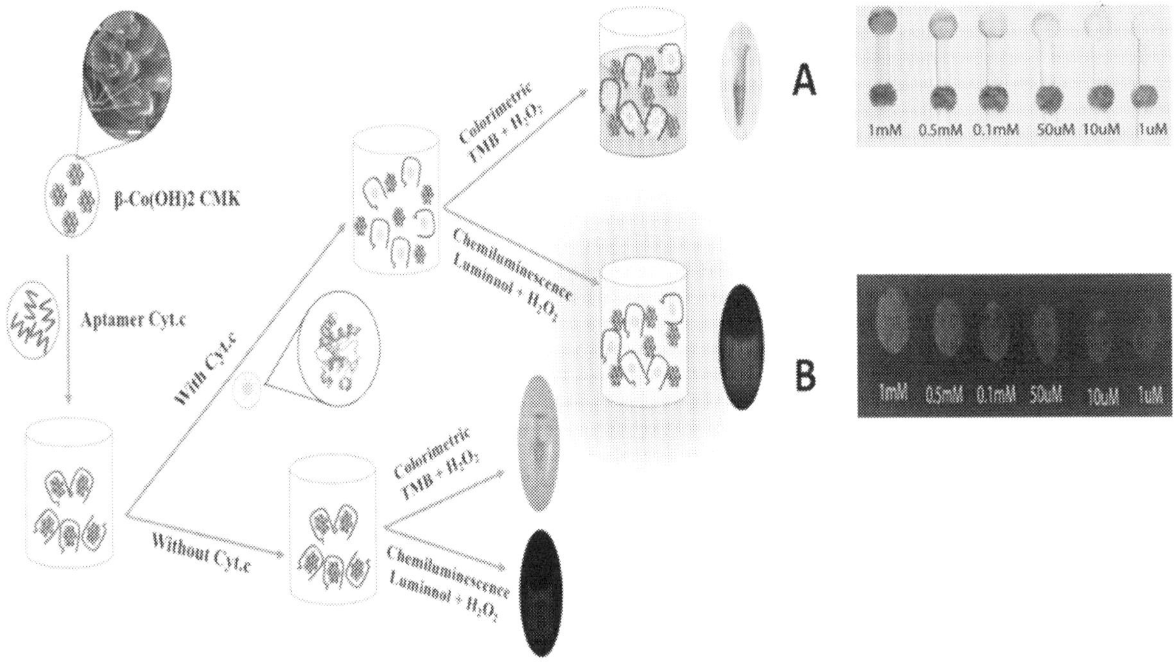


**FIGURE 17.2** General schematic of a paper-based (A) colorimetric and (B) chemiluminescence (CL) aptasensor for the detection of cytochrome C (Cyt c). In the presence of the target, the aptamer will bind to Cyt c, rendering β-Co(OH)2 CMK free, thus increasing the colorimetric and CL signal. *From Mesgari, F., Beigi, S. M., Fakhri, N., Hosseini, M., Aghazadeh, M., & Ganjali, M. R. (2020). Paper-based chemiluminescence (CL) and colorimetric detection of cytochrome C by cobalt hydroxide decorated mesoporous carbon.* Microchemical Journal. *https://doi.org/10.1016/j.microc.2020.104991.*

fabricate Ag NPl-coated paper for the colorimetric detection of $H_2S$ in live prostate cancer cells. In the presence of $H_2S$, brown-colored silver sulfide ($Ag_2S$) is produced, creating a colored signal (Ahn, Gil, Lee, Jang, & Lee, 2020).

Another popular approach for colorimetric output is the use of pH sensitive dyes responsive to certain enzymatic reactions (Feng et al., 2017). In one study, a paper-based litmus test was developed for the colorimetric detection of single nucleotide polymorphisms (SNPs) by the naked eye for the identification of cytosine−cytosine (c-c) mismatches in the *p53* gene. This colorimetric approach was fabricated on the basis of the interaction between c-c mismatched DNA with Ag(I) ions, and the fact that the hydrolyzation of urea with urease is inhibited in the presence of Ag(I). In cell lines with this SNP, the c-c mismatched DNA will bind to the Ag(I) ions resulting in the hydrolyzation of urea. The resulting ammonia will increase the pH leading to the color change of phenol red and a red signal (Wolfe, Ali, & Brennan, 2019).

### 17.4.2 Fluorescence

Due to its advantages, such as high selectivity and sensitivity along with rapid response time, several paper-based platforms based on fluorescence sensing have been developed for the detection of tumor markers. Compared to traditional fluorescent dyes, fluorescent nanotechnologies including quantum dots (QDs), fluorescent micro/nanoparticles, and nanoclusters have become a more prominent choice in PADs as they demonstrate higher stability towards photobleaching and better sensitivity (J. Wang et al., 2010). Although fluorescence detection for PADs has been limited due to some innate features of paper such as its opacity, the fibrous structure which generates back scattering noise thus reducing the sensitivity of the device, and auto-fluorescence (Ulep et al., 2020). As a result, special measures should be taken to enhance the performance of a fluorescent-based PAD.

In addition to many desirable features such as high surface area and rich functionalization strategies, graphene and its derivatives are well-established fluorescence quenchers (Lu et al., 2011). By taking advantage of this feature, a study developed a sensing platform using a combination of NPs and DNA aptamers for the simultaneous detection of various cancer cells. In this approach, aptamers labeled with QDs and mesoporous silica nanoparticles were used to capture cancer cells. Graphene oxide (GO) was doped on the electrodes which, in the presence of DNA-aptamers and through $\pi-\pi$ stacking interactions between the DNA bases and GO, resulted in elevated quenching of the QDs. In the presence of the target cancer cell, and due to the release of the fluorescent probe from the DNA sequence, a fluorescent signal could be recorded leading to specific detection of cancer cells (Fig. 17.3) (Liang et al., 2016).

In another approach, noble metal nanoparticles such as gold nanoparticles, which can also enhance the optical properties of fluorophores due to their SPR (Eustis & El-Sayed, 2006), were used to fabricate a paper-based sensing platform for the detection of urokinase plasminogen activator (uPA), which has been shown to be involved in cancer invasion and metastasis progression. Specific antibodies were used in this approach to detect the presence of uPAs, leading to the release of QDs and a fluorescent signal (Dixit et al., 2016; Sharma et al., 2018).

Smartphone optical sensing strategies due to their portability, user-friendly interface, and feasibility to connect to a network have recently attracted ample attention in POC testing techniques (Ulep & Yoon, 2018). In an attempt to couple smartphone strategies with PADs, a study developed a dual-layer paper microfluidic chip for the detection of receptor tyrosine-like orphan receptor one (ROR1 + ) cancer cells using fluorescent-labeled anti-ROR1 antibodies. In this approach, a primary layer consisting of a glass fiber substrate preloaded with fluorescent-labeled anti-ROR1 was used to capture corresponding cancer cells. A second layer of cellulose chromatography paper enabled the quantification of the flow velocity, thus measuring the antigen concentration in the sample leading to the quantification of cancer cells in a POC manner (Ulep et al., 2020).

### 17.4.3 Chemiluminescence

Owing to its many merits such as extremely high sensitivity and its need of inexpensive reagents and simple instrumentation, CL has been widely employed as an efficient detection approach in bioassays, immunoassays, and in vitro imaging since its introduction in the 1970s (Deo & Roda, 2011). CL detection involves assessing the intensity of light generated from a chemical reaction between luminol and $H_2O_2$, catalyzed by a peroxidase enzyme such as HRP. As there is no requirement for an external light source, the background signal is significantly reduced in CL detection, leading to improved sensitivity and accuracy compared to conventional colorimetric techniques (Cao et al., 2007; Xu et al., 2002).

CL immunoassay, which takes advantage of the selectivity of antibodies along with the sensitivity of CL, has become a key approach in the design of CL-based PADs. Most often, a sandwich immunoassay is used in these sensing systems in which captured antibodies, covalently attached to the paper, interact with their corresponding target antigens.

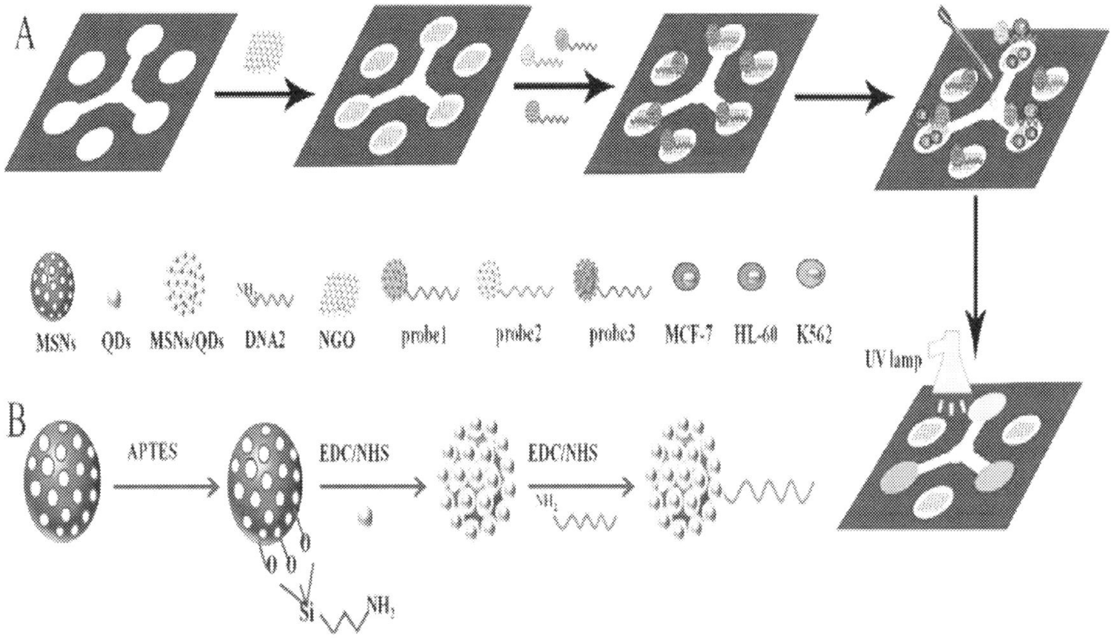


**FIGURE 17.3** Schematic representation of a fluorescent paper-based aptasensor for the multiplexed detection of cancer cells. (A) Shows the fabricaiton process of the sensor on a PAD for which graphene was coated on the paper surface, which through π-π stacking interacts with the DNA probes. (B) Depicts the assembly of the probes through hybridization which results inMSN/QD-modified sequences to adhere to the probes. The presence of the target cells will disrupt the combined quenching effect of DNA and graphene leading to a detectable fluorescent signal from the QDs. *From Liang, L., Su, M., Li, L., Lan, F., Yang, G., Ge, S., Yu, J., & Song, X. (2016). Aptamer-based fluorescent and visual biosensor for multiplexed monitoring of cancer cells in microfluidic paper-based analytical devices.* Sensors and Actuators, B: Chemical, 229, *347–354. https://doi.org/10.1016/j.snb.2016.01.137.*

Signal antibodies labeled with HRP adhere to the tumor biomarker on the capture antibodies and create a signal in the presence of luminol. In these sensors, the stabilization of antibodies on the surface of the paper is of critical importance as it directly affects the outcome of the detection technique (Mazzu-Nascimento et al., 2017). Although pure cellulose has some functional groups, i.e., hydroxyl groups which can be used for the covalent attachment of biomolecules, these groups often do not display sufficient reactivity for efficient functionalization. Thus, many researchers have sought out new strategies to more effectively stabilize antibodies on paper devices. In a study performed by Wang et al., chitosan was employed to stabilize the antibodies on the paper and enhance the wet strength of the paper. Based on this approach, a microfluidic PAD based on sandwich CL-ELISA was developed for the simultaneous detection of several tumor biomarkers (Fig. 17.4) (S. Wang et al., 2012).

As an alternative to antibodies, many PADs have been developed using captured oligonucleotide sequences which can detect complementary DNA or RNA sequences. These paper-based DNA sensors can be used to detect SNPs, circulating cancer DNAs, etc. As with antibodies, the immobilization of captured DNA on the paper surface is key for accurate sensing (Y. Wang et al., 2014). Furthermore, aptamers can also be used for target detection, which significantly lowers cost and complexity compared to dealing with antibodies.

Through acting as a catalyst, metal nanoparticles such as gold and silver nanoparticles can discernibly boost the sensitivity of CL systems (Guo & Cui, 2007). Gold nanoparticles, which enhance the signal in CL systems through facilitating radical generation and electron-transfer processes on their surface after being catalyzed in the luminol − $H_2O_2$ system, have been extensively employed in sensing approaches (Zhang, Cui, Lai, & Liu, 2005). As shown in research carried on by Wang et al., the use of gold nanoparticles can significantly improve the signal leading to augmented sensitivity in the paper-based CL DNA biosensor (Y. Wang et al., 2014). Other nanostructures have also been used for signal amplification in paper-based CL systems. In another study, a paper-based DNA biosensor was developed using carbon dot (C-dot) dotted nanoporous gold (C-dots@NPG) as a signal amplification label which combines the amplification effect of nanoporous gold and the CL properties of the carbon dots which can be attributed to the oxidant injected holes and electrons (Y. Wang et al., 2013).

In addition to gold, other metal-based nanomaterials such as Pt, Pd, Cu, and Co have also been used to enhance the detection performance of CL systems for biological markers (Beigi, Mesgari, Hosseini, Aghazadeh, & Ganjali, 2019).

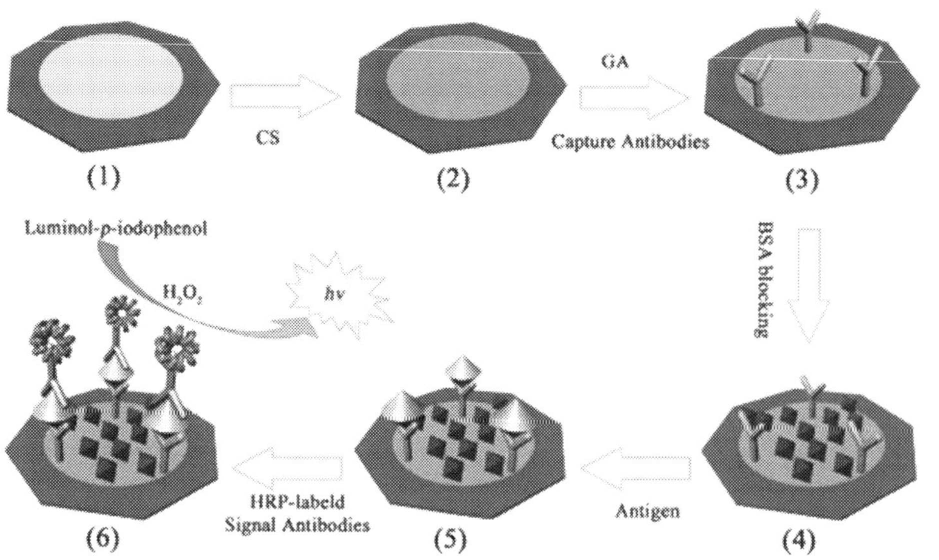


**FIGURE 17.4** Schematic representation of a microfluidic paper-based sandwich CL-ELISA. The paper was generated by wax-screen printing and then modified using chitosan for effective antibody immobilization. Upon the binding of the corresponding antigen to the target antibodies, the HRP-labeled signal antibodies can produce light through a CL reaction in the presence of luminol and $H_2O_2$. *From Wang, S., Ge, L., Song, X., Yu, J., Ge, S., Huang, J., & Zeng, F. (2012). Paper-based chemiluminescence ELISA: Lab-on-paper based on chitosan modified paper device and wax-screen-printing.* Biosensors and Bioelectronics, 31*(1), 212–218. https://doi.org/10.1016/j.bios.2011.10.019.*

In one example, Mesgari et al. took advantage of the ability of Co nanostructures to catalyze oxidation reactions in order to design a CL μPAD for the detection of cytochrome C using specific aptamers for the diagnosis of cancer (Fig. 17.2). They were able to witness a 100-fold enhancement in the CL signal in the presence of the Co nanostructures (Mesgari et al., 2020).

### 17.4.4 Electrochemical

Becoming exceedingly popular since their introduction by Dungchai et al. in 2015 (Dungchai, Chailapakul, & Henry, 2009), electrochemical paper-based analytical devices (ePADs) have become an attractive detection approach in biological sensors. In addition to their small size, these sensing devices show profound selectivity and sensitivity which can be of dire significance especially for quantitative analysis. Many attributes of electrochemical sensing make it an intriguing match for PADs such as the feasibility to fabricate miniature electrodes on paper and its requirement of accessible and portable equipment (Mettakoonpitak et al., 2016). On top of the mentioned advantages of ePADs, they are also listed as one of the best tools for multiplexed detection of tumor markers (Y. Wang et al., 2019).

The performance of an ePAD is highly dependent on the electrodes. As a result, extensive research has been done to assess a variety of materials, e.g., metals, carbon-based materials like graphene and nanoparticles, and fabrication methods for electrode design on ePADs. Included along with many well-established methods are screen-printing, pencil-drawing, and inkjet-printing, which have been widely used and proven effective on ePADs; state-of-the-art techniques such as microwires are also being investigated for electrode fabrication on PADs (Mettakoonpitak et al., 2016).

As the early detection of cancer is critical, it can greatly impact the prognosis and survival rate of the patient, ultrasensitive detection of cancer biomarkers at extremely low levels during the early stages is required. This in turn highlights the need for effective signal amplification strategies which has led to an amplitude of research efforts focused on this matter, including the use of redox-active probes, the assimilation of enzymes, and the integration of nanomaterials (Wu, Xue, Kang, & Hui, 2013). The integration of immunoassays, as they are highly selective and sensitive, with ePADs has attracted ample attention. In order to modify the electrodes in immunoassay-based ePADs for antibody immobilization and also for signal amplifications, nanomaterials are commonly incorporated on the device. Due to high biocompatibility and feasibly accessible surface functionalization strategies, gold nanomaterials have become a prevalent choice for ePADs. Furthermore, the high surface area of gold nanostructures can lead to an increased amount of biomolecule loading capacity along with increased electrochemical signals (Torati, Kasturi, Lim, & Kim, 2017). In one study, researchers developed an immunoassay-based ePAD for the detection of the prostate specific antigen (PSA) with an interconnected layer of gold nanorods (AuNRs) grown on the surface of cellulose paper for the immobilization of capture antibodies. Using specific secondary antibodies labeled with nanocomposites, they were able to detect PSA, as a marker for prostate cancer, in real samples with high accuracy (Sun et al., 2015).

Graphene and its derivatives possess exceptional features which can promote electrical conductivity and boost the mechanical strength of a device. For this reason, many researchers have focused on ways to boost the performance of

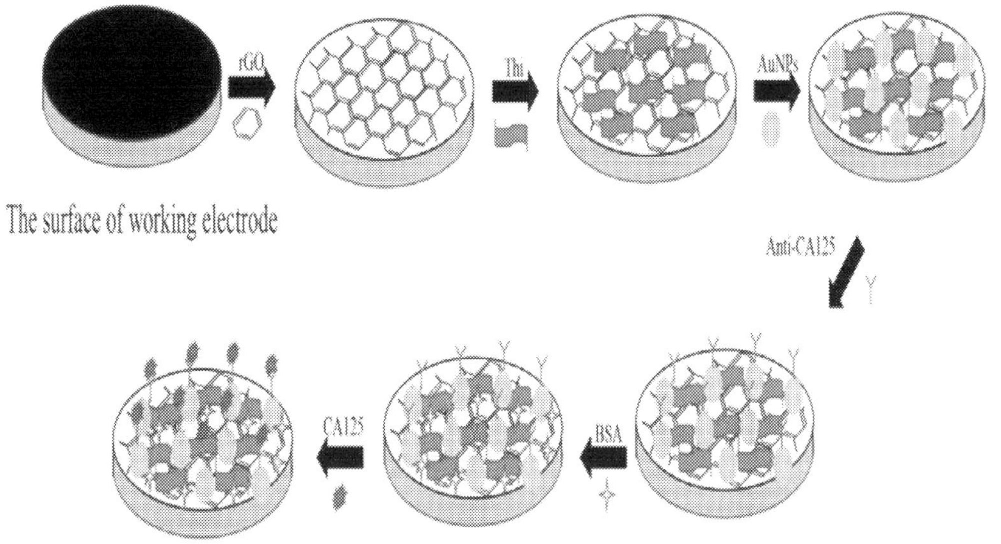


FIGURE 17.5 Schematic representation of an immunoassay-based ePAD. The electrodes were screen-printed on the cellulose paper, which was subsequently modified with rGO/Thi/AuNPs nanocomposites. Anti-CA125 antibodies were immobilized on the AuNPs. The formation of the CA125 antibody and antigen immunocomplex acted as mass-transfer and electron-transfer blocking layer, thus hindering the electron transfer toward the surface of the electrode, which led to a decrease in the current response. *From Fan, Y., Shi, S., Ma, J., & Guo, Y. (2019). A paper-based electrochemical immunosensor with reduced graphene oxide/thionine/gold nanoparticles nanocomposites modification for the detection of cancer antigen 125.* Biosensors and Bioelectronics, 135, *1–7. https://doi.org/10.1016/j.bios.2019.03.063.*

ePADs using graphene (Abergel, Apalkov, Berashevich, Ziegler, & Chakraborty, 2010). In a research, gold and reduced graphene, along with thionine (Thi) which can induce a redox current response during testing, were combined to create rGO/Thi/AuNPs nanocomposites for the development of an electrochemical paper-based immunosensor for the detection of cancer antigen 125 (CA125). Other than significantly enhancing the detection sensitivity of the device, these composites also act as linkers for the immobilization of the capture antibodies on the paper surface, as depicted in Fig. 17.5 (Fan, Shi, Ma, & Guo, 2019).

As with previously mentioned methods, aptamers have become a center of attention in the development of ePADs as they have many merits over antibodies such lower cost, higher stability, lower immunogenicity, and a wider receptor range (Hermann & Patel, 2000). To further improve the performance of aptasensors, various forms of nanomaterials have been employed. In one research, amino functional graphene (NG)-thionin (THI)-gold nanoparticles (AuNPs) and Prussian blue (PB)-poly (3,4-ethylenedioxythiophene) (PEDOT)-AuNPs nanocomposites were coated on the surface of the electrodes on an ePAD to increase the number of immobilized aptamers and also ease the electron transfer process. This ePAD was used to simultaneously detect the carcinoembryonic antigen (CEA) and neuron specific enolase (NSE) in clinical samples with high sensitivity and selectivity (Y. Wang et al., 2019).

As miRNA detection has become highly important as a means of early cancer detection, many researchers have attempted to develop highly specific ePADs for the quantification of miRNA levels in biological samples. In one study, the detection of miRNA-155 was carried out based on hairpin assembly using an ePAD. AuNP-modified Cu-based metal-organic frameworks (Cu-MOFs) were used as nanocarriers for the immobilization of DNA strands. Furthermore, since Cu-MOFs have been shown to possess catalytic activity, they also acted as a catalyst and oxidized glucose leading to the amplification of the electrochemical signal, thus enhancing the detection accuracy of miRNA-155 (Fig. 17.6) (H. Wang et al., 2018).

### 17.4.5 Electrochemiluminescence

Through amalgamating CL and electrochemistry, the ECL technique arises which combines the high sensitivity and wide dynamic range of conventional CL with the portability and stability of electrochemical devices. In ECL light emission occurs from excited species created from an exergonic electron-transfer reaction on the surface of the electrodes. Electrogenerated radicals at the electrode surface can cause bimolecular recombination which leads to the generation of an ECL signal. Regarding the radical source, there are two types of ECL mechanisms: the annihilation pathway, in which the radicals are created from a single emitting species; and the coreact pathway, in which an electrochemical

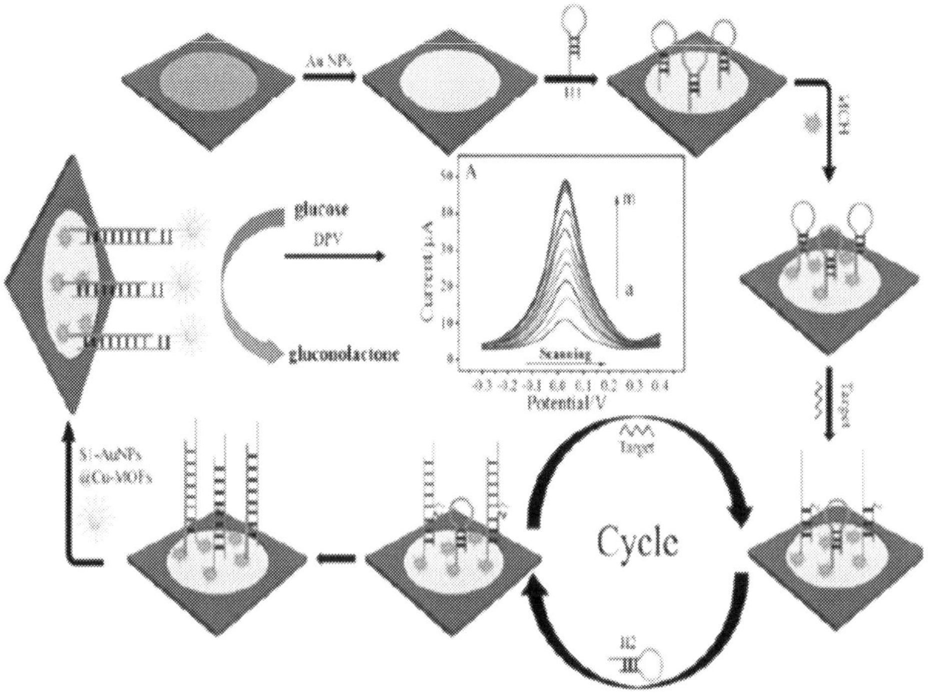


**FIGURE 17.6** A schematic representation of electrochemical biosensor based on hairpin assembly for signal amplification and AuNPs-modified Cu-MOFs for catalysis for highly accurate detection of miRNA-155 in biological samples. *From Wang, H., Jian, Y., Kong, Q., Liu, H., Lan, F., Liang, L., Ge, S., & Yu, J. (2018). Ultrasensitive electrochemical paper-based biosensor for microRNA via strand displacement reaction and metal-organic frameworks.* Sensors and Actuators, B: Chemical, 257, *561–569. https://doi.org/10.1016/j.snb.2017.10.188.*

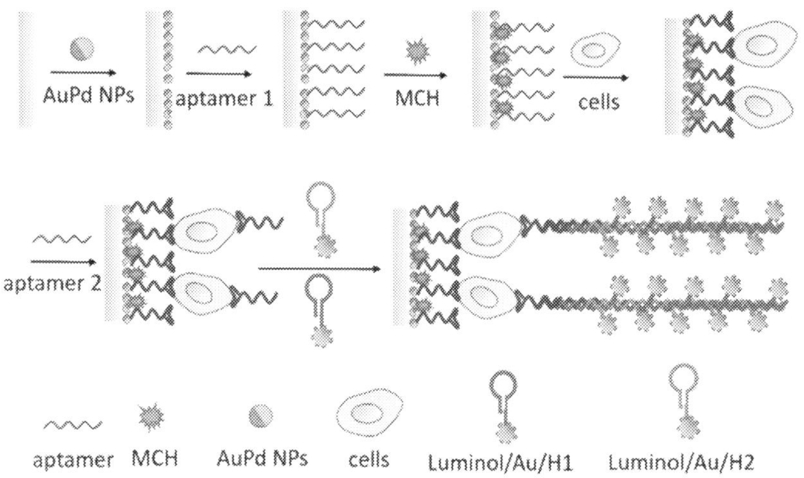


**FIGURE 17.7** Schematic representation of an ECL paper-based aptasensor for the detection of cancer cells. Wax printing was used for the fabrication of reaction zone, and carbon ink-based BPE and driving electrodes were screen-printed into the paper. Two partially complementary hairpins bound to luminol/AuNPs were used for signal amplification via hairpin chain reaction. *From Ge, S., Zhao, J., Wang, S., Lan, F., Yan, M., & Yu, J. (2018). Ultrasensitive electrochemiluminescence assay of tumor cells and evaluation of H2O2 on a paper-based closed-bipolar electrode by in-situ hybridization chain reaction amplification.* Biosensors and Bioelectronics, 102, *411–417. https://doi.org/10.1016/j.bios.2017.11.055.*

reaction occurs between an emitting specie and a coreact leading to the generation of radicals (Chinnadayyala et al., 2019). The emitter is of ample importance and it can be any kind of luminophore such as ruthenium (II) complexes, luminol or QDs (Li et al., 2017).

For the integration of ECL on paper, a bipolar electrode (BPE), which consists of an electronic conductor immersed in an ionic conductive phase without any contact to an external power supply, is commonly used (Zhang et al., 2005). In one study, a paper-based device composed of a BPE was used for the ECL detection of cancer cells. As seen in Fig. 17.7, AuPd nanoparticles were used for the immobilization of the specific aptamers on the carbon ink-based BPE on the paper surface, and also as catalysts for the ECL reaction between luminol and $H_2O_2$. After the corresponding cancer cells were captured, a second aptamer, along with hairpin structures bound to luminol/AuNPs were added to create a sandwich structure on the BPE surface. Cancer cells were stimulated using phorbol myristate acetate to release $H_2O_2$, thus creating an ECL signal, which can be used to quantify the number of cancer cells (Ge et al., 2018). In another study, a paper-based BPE device was coupled with antibodies to create an ECL-based immunosensor for the detection of PSA. In this approach, multiwalled carbon nanotubes were used to modify the BPE and to capture the antibodies (Feng et al., 2017).

Metallic and bimetallic nanostructures have also been commonly used for signal amplification and electrode modification in ECL sensors (Rampazzo et al., 2012). In a recent study, AuPd nanoparticles were chosen for their catalytic activity toward $H_2O_2$ and outstanding conductivity, making them an ideal choice for ECL signal amplification. Specific antibodies against CEA and alpha-fetoprotein (AFP) antigens labeled with CdTe QDs and luminol groups as ECL probes were immobilized on the AuPd modified electrodes to capture and detect MCF-7 cancer cells (Su et al., 2016).

In one study, in order to overcome some of the difficulties relating to fluid control in paper-based devices, a rotational paper-based ECL immunodevice was developed for multiplexed detection of CEA and PSA. Three paper discs were assembled to fabricate the rotational PAD. Tris-(bipyridine)-ruthenium (II)−tripropylamine ECL ($[Ru(bpy)_3]^{2+}$-TPA) detection system was enabled using specific $[Ru(bpy)_3]^{2+}$−labeled antibodies (Sun et al., 2015).

### 17.4.6 Surface-enhanced Raman scattering

SERS possesses many attributes such as outstanding sensitivity and selectivity, along with the ability for multiplexed detection of biomarkers. The characteristic Raman spectra of molecules enable the precise identification and characterization of analytes. Furthermore, due to spatial resolution and its dependence on a single wavelength for excitation, SERS can also be used for multiplex sensing. Thus, many efforts have been devoted to coupling this technique with paper-based devices for clinical detection.

Enhancement is of ample importance in SERS-based detection, which can be achieved when the target is absorbed or attached within a distance of 10−100 nm on roughened noble metal surfaces or metal nanostructures (Doering, Piotti, Natan, & Freeman, 2007). In that regard, recently plasmonic paper, which is paper coated with noble metal nanostructures has been reported for paper-based diagnosis. In one study, AuNRs were selected for SERS enhancement as they have tunable longitudinal SPR and a high extinction coefficient, and plasmonic paper was designed by dipping paper into a solution of AuNRs. This platform was then used to detect cancer cells captured by immunomagnetic particles which contained the specific antibody for the corresponding cells. Via this PAD, colorectal cancer cells were distinguished from other cells based on their distinct SERS pattern (Fig. 17.8) (Reokrungruang, Chatnuntawech, Dharakul, & Bamrungsap, 2019).

Silver (Ag) also displays strong SPR absorption, making it an attractive candidate for SERS enhancement. It has been shown that Ag NPls displayed strong magnetic fields at the corners which resulted in outstanding SERS

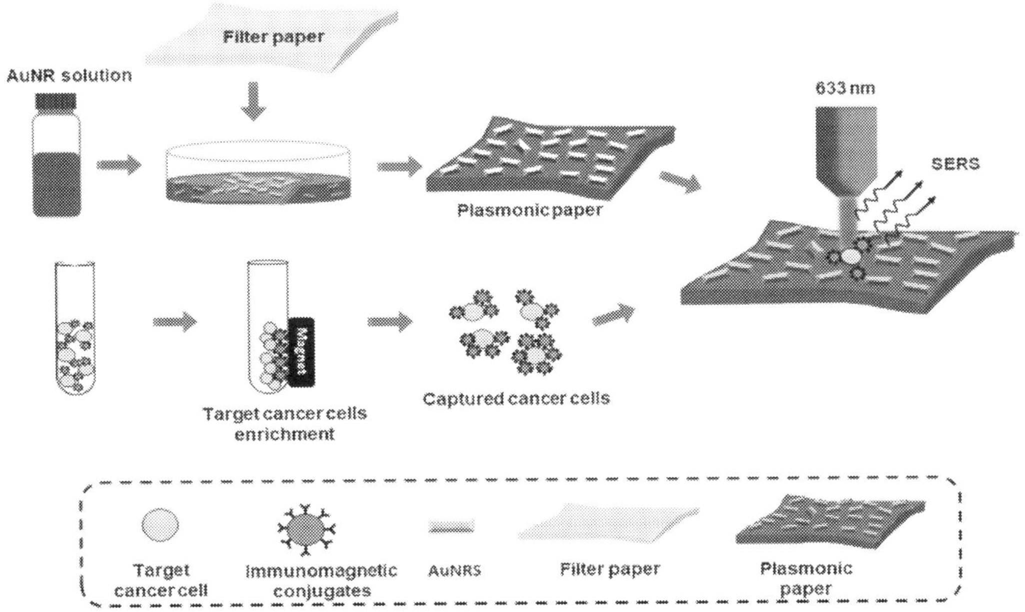


**FIGURE 17.8** Schematic representation of a plasmonic paper-based sensor for the specific detection of cancer cells. Using immunomagnetic conjugates which contained a specific antibody against EpCAM, colorectal cancer cells were captured. The intrinsic and distinctive spectra of cancer cells was used to distinguish them from red blood cells and fibroblasts. *From Reokrungruang, P., Chatnuntawech, I., Dharakul, T., & Bamrungsap, S. (2019). A simple paper-based surface-enhanced Raman scattering (SERS) platform and magnetic separation for cancer screening.* Sensors and Actuators, B: Chemical, 285, *462−469. https://doi.org/10.1016/j.snb.2019.01.090.*

enhancement. Ag NPls were coated on paper to create a PAD for the assessment of $H_2S$ as an indicator of prostate cancer. Rhodamine B was used as the SERS marker for sensitive detection (Ahn et al., 2020).

## 17.5 Current limitations

PADs can be the ideal preliminary screening platform for users, providing cost effective and portable POC diagnosis tools. However, there are still many limitations which need to be addressed. First of all, the cost and fabrication procedure for large scale fabrication of PADs should be optimized. Efficient immobilization of biomolecules also needs to be refined since nonspecific attachment of biomolecules and loss of functionality persist as a challenge. Regarding clinical performance, it has been reported that PADs, especially ones manufactured in large scales for commercial use, can show varying sensitivity and specificity (Pike, Godbert, & Johnson, 2013), which can lead to false negative and false positive results. Various causes have been accounted for this observation: first, various substrates used in the fabrication process or for signal production and amplification can be coupled in different combinations leading to varying results; second, subjective judgment by end-users and different illumination settings can lead to different interpretations of the results especially in optical PADs; third, environmental conditions such as temperature and humidity can impact the robustness of the device, especially regarding the functionality of biomolecules such as enzymes and antibodies, which can ultimately affect the results; and finally, batch-to-batch variations poses a challenge in most POC diagnosis testing devices including PADs. Thus, for large-scale and commercial fabrication and distribution of PADs as POC diagnosis tools, problems pertaining to sensitivity and selectivity, along with challenges regarding biomolecule functionalization need to be addressed.

## 17.6 Conclusion and future perspectives

Early cancer detection based on tumor biomarkers is significantly important as it can directly impact the outlook of the disease. Furthermore, the requirement for perpetual evaluation of the stage and prognosis of the disease during treatment further highlight the need for disease sensing and control. In that regard, the importance of POC diagnosis is most profound in cancer detection. PADs have become one of the forerunners in the attempt to develop POC diagnostics as they are inexpensive, portable, and can be easily up-scaled. The emergence of microfluidics technology and its interfacing with paper technology has greatly impacted PADs, leading to an advancement in the development of more effective POC devices. Furthermore, combining PADs with mobile phone technologies with enable optical signal readouts has eased the progress of POC diagnosis in resource limited environments. However, the full potential of PADs will not be exploited until other aspects of POC diagnosis including sample pretreatment, plasma separation and nucleic acid amplification are effectively achieved on a paper platform.

In this chapter, we have provided an introduction regarding the basic concept of paper-based sensing including various formats of PADs and the commonly used fabrication strategies. Different sensing technologies frequently coupled with PADs for cancer detection were also elaborated. In conclusion, paper-based sensing holds great potential for the development of POC diagnosis tools for rapid and sensitive cancer detection.

## References

Abarghoei, S., Fakhri, N., Borghei, Y. S., Hosseini, M., & Ganjali, M. R. (2019). A colorimetric paper sensor for citrate as biomarker for early stage detection of prostate cancer based on peroxidase-like activity of cysteine-capped gold nanoclusters. *Spectrochimica Acta – Part A: Molecular and Biomolecular Spectroscopy*, *210*, 251–259. Available from https://doi.org/10.1016/j.saa.2018.11.026.

Abergel, D. S. L., Apalkov, V., Berashevich, J., Ziegler, K., & Chakraborty, T. (2010). Properties of graphene: A theoretical perspective. *Advances in Physics*, *59*(4), 261–482. Available from https://doi.org/10.1080/00018732.2010.487978.

Ahn, Y. J., Gil, Y.-G., Lee, Y. J., Jang, H., & Lee, G.-J. (2020). A dual-mode colorimetric and SERS detection of hydrogen sulfide in live prostate cancer cells using a silver nanoplate-coated paper assay. *Microchemical Journal*, 155.

Beigi, S. M., Mesgari, F., Hosseini, M., Aghazadeh, M., & Ganjali, M. R. (2019). An enhancement of luminol chemiluminescence by cobalt hydroxide decorated porous graphene and its application in glucose analysis. *Analytical Methods*, *11*(10), 1346–1352. Available from https://doi.org/10.1039/c8ay02727g.

Bora, U., Sharma, P., Kannan, K., & Nahar, P. (2006). Photoreactive cellulose membrane – A novel matrix for covalent immobilization of biomolecules. *Journal of Biotechnology*, *126*(2), 220–229. Available from https://doi.org/10.1016/j.jbiotec.2006.04.013.

Busa, L. S. A., Maeki, M., Ishida, A., Tani, H., & Tokeshi, M. (2016). Simple and sensitive colorimetric assay system for horseradish peroxidase using microfluidic paper-based devices. *Sensors and Actuators, B: Chemical*, *236*, 433–441. Available from https://doi.org/10.1016/j.snb.2016.06.013.

Cao, Y., Zhang, Q., Wang, C., Zhu, Y., & Bai, G. (2007). Preparation of novel immunomagnetic cellulose microspheres via cellulose binding domain-protein A linkage and its use for the isolation of interferon α-2b. *Journal of Chromatography. A, 1149*(2), 228–235. Available from https://doi.org/10.1016/j.chroma.2007.03.032.

Cate, D. M., Adkins, J. A., Mettakoonpitak, J., & Henry, C. S. (2015). Recent developments in paper-based microfluidic devices. *Analytical Chemistry, 87*(1), 19–41. Available from https://doi.org/10.1021/ac503968p.

Chen, Z., Zhang, C., Wu, Q., Li, K., & Tan, L. (2015). Application of triangular silver nanoplates for colorimetric detection of H2O2. *Sensors and Actuators, B: Chemical, 220*, 314–317. Available from https://doi.org/10.1016/j.snb.2015.05.085.

Chinnadayyala, S. R., Park, J., Le, H. T. N., Santhosh, M., Kadam, A. N., & Cho, S. (2019). Recent advances in microfluidic paper-based electrochemiluminescence analytical devices for point-of-care testing applications. *Biosensors and Bioelectronics, 126*, 68–81. Available from https://doi.org/10.1016/j.bios.2018.10.038.

Dehghani, Z., Hosseini, M., Mohammadnejad, J., Bakhshi, B., & Rezayan, A. H. (2018). Colorimetric aptasensor for *Campylobacter jejuni* cells by exploiting the peroxidase like activity of Au@Pd nanoparticles. *Microchimica Acta, 185*(10). Available from https://doi.org/10.1007/s00604-018-2976-2.

Dehghani, Z., Hosseini, M., Mohammadnejad, J., & Ganjali, M. R. (2019). New colorimetric DNA Sensor for Detection of *Campylobacter jejuni* in milk sample based on peroxidase-like activity of gold/platinium nanocluster. *ChemistrySelect, 4*(40), 11687–11692. Available from https://doi.org/10.1002/slct.201901815.

Deo, S. K., & Roda, A. (2011). Chemiluminescence and bioluminescence: Past, present and future. *Analytical and Bioanalytical Chemistry, 401*, 5.

Dixit, C.K., Kaushik, A.K., & Kaushik, A. (2016). Microfluidics for biologists.

Doering, W. E., Piotti, M. E., Natan, M. J., & Freeman, R. G. (2007). SERS as a foundation for nanoscale, optically detected biological labels. *Advanced Materials, 19*(20), 3100–3108. Available from https://doi.org/10.1002/adma.200701984.

Dungchai, W., Chailapakul, O., & Henry, C. S. (2009). Electrochemical detection for paper-based microfluidics. *Analytical Chemistry, 81*(14), 5821–5826. Available from https://doi.org/10.1021/ac9007573.

Eustis, S., & El-Sayed, M. A. (2006). Why gold nanoparticles are more precious than pretty gold: Noble metal surface plasmon resonance and its enhancement of the radiative and nonradiative properties of nanocrystals of different shapes. *Chemical Society Reviews, 35*(3), 209–217. Available from https://doi.org/10.1039/b514191e.

Fakhri, N., Abarghoei, S., Dadmehr, M., Hosseini, M., Sabahi, H., & Ganjali, M. R. (2020). Paper based colorimetric detection of miRNA-21 using Ag/Pt nanoclusters. *Spectrochimica Acta - Part A: Molecular and Biomolecular Spectroscopy*, 227. Available from https://doi.org/10.1016/j.saa.2019.117529.

Fakhri, N., Hosseini, M., & Tavakoli, O. (2018). Aptamer-based colorimetric determination of Pb2+ using a paper-based microfluidic platform. *Analytical Methods, 10*(36), 4438–4444. Available from https://doi.org/10.1039/c8ay01331d.

Fan, Y., Shi, S., Ma, J., & Guo, Y. (2019). A paper-based electrochemical immunosensor with reduced graphene oxide/thionine/gold nanoparticles nanocomposites modification for the detection of cancer antigen 125. *Biosensors and Bioelectronics, 135*, 1–7. Available from https://doi.org/10.1016/j.bios.2019.03.063.

Feng, C., Mao, X., Shi, H., Bo, B., Chen, X., Chen, T., ... Li, G. (2017). Detection of microRNA: A point-of-care testing method based on a pH-responsive and highly efficient isothermal amplification. *Analytical Chemistry, 89*(12), 6631–6636. Available from https://doi.org/10.1021/acs.analchem.7b00850.

Fenton, E. M., Mascarenas, M. R., López, G. P., & Sibbett, S. S. (2009). Multiplex lateral-flow test strips fabricated by two-dimensional shaping. *ACS Applied Materials and Interfaces, 1*(1), 124–129. Available from https://doi.org/10.1021/am800043z.

Free, A. H., Adams, E. C., Kercher, M. L., Free, H. M., & Cook, M. H. (1957). Simple specific test for urine glucose. *Clinical Chemistry, 3*(3), 163–168. Available from https://doi.org/10.1093/clinchem/3.3.163.

Garcia-Cordero, J. L., & Maerkl, S. J. (2020). Microfluidic systems for cancer diagnostics. *Current Opinion in Biotechnology, 65*, 37–44. Available from https://doi.org/10.1016/j.copbio.2019.11.022.

Ge, S., Zhao, J., Wang, S., Lan, F., Yan, M., & Yu, J. (2018). Ultrasensitive electrochemiluminescence assay of tumor cells and evaluation of H2O2 on a paper-based closed-bipolar electrode by in-situ hybridization chain reaction amplification. *Biosensors and Bioelectronics, 102*, 411–417. Available from https://doi.org/10.1016/j.bios.2017.11.055.

Guo, J. Z., & Cui, H. (2007). Lucigenin chemiluminescence induced by noble metal nanoparticles in the presence of adsorbates. *Journal of Physical Chemistry C, 111*(33), 12254–12259. Available from https://doi.org/10.1021/jp073816w.

Halder, E., Chattoraj, D. K., & Das, K. P. (2005). Adsorption of biopolymers at hydrophilic cellulose-water interface. *Biopolymers, 77*(5), 286–295. Available from https://doi.org/10.1002/bip.20232.

Heinze, T., & Liebert, T. (2001). Unconventional methods in cellulose functionalization. *Progress in Polymer Science, 26*(9), 1689–1762. Available from https://doi.org/10.1016/s0079-6700(01)00022-3.

Hermann, T., & Patel, D. J. (2000). Adaptive recognition by nucleic acid aptamers. *Science (New York, N.Y.), 287*(5454), 820–825. Available from https://doi.org/10.1126/science.287.5454.820.

Kermani, H.A., Hosseini, M., Miti, A., Dadmehr, M., Zuccheri, G., Hosseinkhani, S., & Ganjali, M.R. (n.d.). A colorimetric assay of DNA methyltransferase activity based on peroxidase mimicking of DNA template Ag/Pt bimetallic nanoclusters. Analytical and Bioanalytical Chemistry, 410, 4943–4952.

Koev, S. T., Dykstra, P. H., Luo, X., Rubloff, G. W., Bentley, W. E., Payne, G. F., & Ghodssi, R. (2010). Chitosan: An integrative biomaterial for lab-on-a-chip devices. *Lab on a Chip, 10*(22), 3026–3042. Available from https://doi.org/10.1039/c0lc00047g.

Kong, F., & Hu, Y. F. (2012). Biomolecule immobilization techniques for bioactive paper fabrication. *Analytical and Bioanalytical Chemistry*, *403*(1), 7–13. Available from https://doi.org/10.1007/s00216-012-5821-1.

Lehtiö, J., Sugiyama, J., Gustavsson, M., Fransson, L., Linder, M., & Teeri, T. T. (2003). The binding specificity and affinity determinants of family 1 and family 3 cellulose binding modules. *Proceedings of the National Academy of Sciences*, *100*(2), 484–489. Available from https://doi.org/10.1073/pnas.212651999.

Li, B., Yu, L., Qi, J., Fu, L., Zhang, P., & Chen, L. (2017). Controlling capillary-driven fluid transport in paper-based microfluidic devices using a movable valve. *Analytical Chemistry*, *89*(11), 5707–5712. Available from https://doi.org/10.1021/acs.analchem.7b00726.

Liang, L., Su, M., Li, L., Lan, F., Yang, G., Ge, S., ... Song, X. (2016). Aptamer-based fluorescent and visual biosensor for multiplexed monitoring of cancer cells in microfluidic paper-based analytical devices. *Sensors and Actuators, B: Chemical*, *229*, 347–354. Available from https://doi.org/10.1016/j.snb.2016.01.137.

Lu, C. H., Li, J., Qi, X. J., Song, X. R., Yang, H. H., Chen, X., & Chen, G. N. (2011). Multiplex detection of nucleases by a graphene-based platform. *Journal of Materials Chemistry*, *21*(29), 10915–10919. Available from https://doi.org/10.1039/c1jm11121c.

Luckham, R. E., & Brennan, J. D. (2010). Bioactive paper dipstick sensors for acetylcholinesterase inhibitors based on sol-gel/enzyme/gold nanoparticle composites. *Analyst*, *135*(8), 2028–2035. Available from https://doi.org/10.1039/c0an00283f.

Mazzu-Nascimento, T., Morbioli, G. G., Milan, L. A., Donofrio, F. C., Mestriner, C. A., & Carrilho, E. (2017). Development and statistical assessment of a paper-based immunoassay for detection of tumor markers. *Analytica Chimica Acta*, *950*, 156–161. Available from https://doi.org/10.1016/j.ac.2016.11.011.

Mesgari, F., Beigi, S. M., Fakhri, N., Hosseini, M., Aghazadeh, M., & Ganjali, M. R. (2020). Paper-based chemiluminescence and colorimetric detection of cytochrome C by cobalt hydroxide decorated mesoporous carbon. *Microchemical Journal*. Available from https://doi.org/10.1016/j.microc.2020.104991.

Mettakoonpitak, J., Boehle, K., Nantaphol, S., Teengam, P., Adkins, J. A., Srisa-Art, M., & Henry, C. S. (2016). Electrochemistry on paper-based analytical devices: A review. *Electroanalysis*, *28*(7), 1420–1436. Available from https://doi.org/10.1002/elan.201501143.

Morbioli, G. G., Mazzu-Nascimento, T., Stockton, A. M., & Carrilho, E. (2017). Technical aspects and challenges of colorimetric detection with microfluidic paper-based analytical devices (μPADs) – A review. *Analytica Chimica Acta*, *970*, 1–22. Available from https://doi.org/10.1016/j.ac.2017.03.037.

Mousavizadegan, M., Azimzadeh Asiabi, P., Hosseini, M., & Khoobi, M. (2020). Synthesis of magnetic silk nanostructures with peroxidase-like activity as an approach for the detection of glucose. *ChemistrySelect*, *5*(27), 8093–8098. Available from https://doi.org/10.1002/slct.202002136.

Nery, E. W., & Kubota, L. T. (2013). Sensing approaches on paper-based devices: A review. *Analytical and Bioanalytical Chemistry*, *405*(24), 7573–7595. Available from https://doi.org/10.1007/s00216-013-6911-4.

Pankratov, I., & Lev, O. (1995). Sol-gel derived renewable-surface biosensors. *Journal of Electroanalytical Chemistry*, *393*(1–2), 35–41. Available from https://doi.org/10.1016/0022-0728(95)04020-O.

Parikh, A. R., Leshchiner, I., Elagina, L., Goyal, L., Levovitz, C., Siravegna, G., ... Corcoran, R. B. (2019). Liquid versus tissue biopsy for detecting acquired resistance and tumor heterogeneity in gastrointestinal cancers. *Nature Medicine*, *25*(9), 1415–1421. Available from https://doi.org/10.1038/s41591-019-0561-9.

Pike, J., Godbert, S., & Johnson, S. (2013). Comparison of volunteers' experience of using, and accuracy of reading, different types of home pregnancy test formats. *Expert Opinion on Medical Diagnostics*, *7*(5), 435–441. Available from https://doi.org/10.1517/17530059.2013.830103.

Rampazzo, E., Bonacchi, S., Genovese, D., Juris, R., Marcaccio, M., Montalti, M., ... Prodi, L. (2012). Nanoparticles in metal complexes-based electrogenerated chemiluminescence for highly sensitive applications. *Coordination Chemistry Reviews*, *256*(15–16), 1664–1681. Available from https://doi.org/10.1016/j.ccr.2012.03.021.

Reokrungruang, P., Chatnuntawech, I., Dharakul, T., & Bamrungsap, S. (2019). A simple paper-based surface enhanced Raman scattering (SERS) platform and magnetic separation for cancer screening. *Sensors and Actuators, B: Chemical*, *285*, 462–469. Available from https://doi.org/10.1016/j.snb.2019.01.090.

Rezk, A. R., Qi, A., Friend, J. R., Li, W. H., & Yeo, L. Y. (2012). Uniform mixing in paper-based microfluidic systems using surface acoustic waves. *Lab on a Chip*, *12*(4), 773–779. Available from https://doi.org/10.1039/c2lc21065g.

Savolainen, A., Zhang, Y., Rochefort, D., Holopainen, U., Erho, T., Virtanen, J., & Smolander, M. (2011). Printing of polymer microcapsules for enzyme immobilization on paper substrate. *Biomacromolecules*, *12*(6), 2008–2015. Available from https://doi.org/10.1021/bm2003434.

Schiff, H. (1866). Eine neue Reihe organischer Diamine. *Justus Liebigs Annalen der Chemie*, *140*(1), 92–137. Available from https://doi.org/10.1002/jlac.18661400106.

Sharma, N., Barstis, T., & Giri, B. (2018). Advances in paper-analytical methods for pharmaceutical analysis. *European Journal of Pharmaceutical Sciences*, *111*, 46–56. Available from https://doi.org/10.1016/j.ejps.2017.09.031.

Srinivas, P. R., Kramer, B. S., & Srivastava, S. (2001). Trends in biomarker research for cancer detection. *Lancet Oncology*, *2*(11), 698–704. Available from https://doi.org/10.1016/S1470-2045(01)00560-5.

Su, M., Liu, H., Ge, S., Ren, N., Ding, L., Yu, J., & Song, X. (2016). An electrochemiluminescence lab-on-paper device for sensitive detection of two antigens at the MCF-7 cell surface based on porous bimetallic AuPd nanoparticles. *RSC Advances*, *6*(20), 16500–16506.

Su, S., Ali, M. M., Filipe, C. D. M., Li, Y., & Pelton, R. (2008). Microgel-based inks for paper-supported biosensing applications. *Biomacromolecules*, *9*(3), 935–941. Available from https://doi.org/10.1021/bm7013608.

Su, S., Nutiu, R., Filipe, C. D. M., Li, Y., & Pelton, R. (2007). Adsorption and covalent coupling of ATP-binding DNA aptamers onto cellulose. *Langmuir: The ACS Journal of Surfaces and Colloids*, *23*(3), 1300–1302. Available from https://doi.org/10.1021/la060961c.

Sun, G., Liu, H., Zhang, Y., Yu, J., Yan, M., Song, X., & He, W. (2015). Gold nanorods-paper electrode based enzyme-free electrochemical immunoassay for prostate specific antigen using porous zinc oxide spheres-silver nanoparticles nanocomposites as labels. *New Journal of Chemistry, 39* (8), 6062–6067. Available from https://doi.org/10.1039/c5nj00629e.

Terpe, K. (2003). Overview of tag protein fusions: From molecular and biochemical fundamentals to commercial systems. *Applied Microbiology and Biotechnology, 60*(5), 523–533. Available from https://doi.org/10.1007/s00253-002-1158-6.

Torati, S. R., Kasturi, K. C. S. B., Lim, B., & Kim, C. G. (2017). Hierarchical gold nanostructures modified electrode for electrochemical detection of cancer antigen CA125. *Sensors and Actuators, B: Chemical, 243*, 64–71. Available from https://doi.org/10.1016/j.snb.2016.11.127.

Ulep, T. H., & Yoon, J. Y. (2018). Challenges in paper-based fluorogenic optical sensing with smartphones. *Nano Convergence, 5*(1). Available from https://doi.org/10.1186/s40580-018-0146-1.

Ulep, T. H., Zenhausern, R., Gonzales, A., Knoff, D. S., Lengerke Diaz, P. A., Castro, J. E., & Yoon, J. Y. (2020). Smartphone based on-chip fluorescence imaging and capillary flow velocity measurement for detecting ROR1 + cancer cells from buffy coat blood samples on dual-layer paper microfluidic chip. *Biosensors and Bioelectronics*, 153. Available from https://doi.org/10.1016/j.bios.2020.112042.

Wang, H., Jian, Y., Kong, Q., Liu, H., Lan, F., Liang, L., … Yu, J. (2018). Ultrasensitive electrochemical paper-based biosensor for microRNA via strand displacement reaction and metal-organic frameworks. *Sensors and Actuators, B: Chemical, 257*, 561–569. Available from https://doi.org/10.1016/j.snb.2017.10.188.

Wang, J., Pelton, R., Veldhuis, L. J., Roger Mackenzie, C., Christopher Hall, J., & Filipe, C. D. M. (2010). Wet-strength resins and surface properties affect paper-based antibody assays. *Appita Journal, 63*(1), 32–36.

Wang, S., Ge, L., Song, X., Yu, J., Ge, S., Huang, J., & Zeng, F. (2012). Paper-based chemiluminescence ELISA: Lab-on-paper based on chitosan modified paper device and wax-screen-printing. *Biosensors and Bioelectronics, 31*(1), 212–218. Available from https://doi.org/10.1016/j.bios.2011.10.019.

Wang, Y., Luo, J., Liu, J., Sun, S., Xiong, Y., Ma, Y., … Cai, X. (2019). Label-free microfluidic paper-based electrochemical aptasensor for ultrasensitive and simultaneous multiplexed detection of cancer biomarkers. *Biosensors and Bioelectronics, 136*, 84–90. Available from https://doi.org/10.1016/j.bios.2019.04.032.

Wang, Y., Wang, S., Ge, S., Wang, S., Yan, M., Zang, D., & Yu, J. (2013). Facile and sensitive paper-based chemiluminescence DNA biosensor using carbon dots dotted nanoporous gold signal amplification label. *Analytical Methods, 5*(5), 1328–1336. Available from https://doi.org/10.1039/c2ay26485d.

Wang, Y., Wang, S., Ge, S., Wang, S., Yan, M., Zang, D., & Yu, J. (2014). Ultrasensitive chemiluminescence detection of DNA on a microfluidic paper-based analytical device. *Monatshefte fur Chemie, 145*(1), 129–135. Available from https://doi.org/10.1007/s00706-013-0971-1.

Wolfe, M. G., Ali, M. M., & Brennan, J. D. (2019). Enzymatic litmus test for selective colorimetric detection of C-C single nucleotide polymorphisms. *Analytical Chemistry, 91*(7), 4735–4740. Available from https://doi.org/10.1021/acs.analchem.9b00235.

Wu, Y., Xue, P., Kang, Y., & Hui, K. M. (2013). Paper-based microfluidic electrochemical immunodevice integrated with nanobioprobes onto graphene film for ultrasensitive multiplexed detection of cancer biomarkers. *Analytical Chemistry, 85*(18), 8661–8668. Available from https://doi.org/10.1021/ac401445a.

Xu, Z., Bae, W., Mulchandani, A., Mehra, R. K., & Chen, W. (2002). Heavy metal removal by novel CBD-EC20 sorbents immobilized on cellulose. *Biomacromolecules, 3*(3), 462–465. Available from https://doi.org/10.1021/bm015631f.

Yi, H., Wu, L. Q., Bentley, W. E., Ghodssi, R., Rubloff, G. W., Culver, J. N., & Payne, G. F. (2005). Biofabrication with chitosan. *Biomacromolecules, 6*(6), 2881–2894. Available from https://doi.org/10.1021/bm050410l.

Zhang, Z. F., Cui, H., Lai, C. Z., & Liu, L. J. (2005). Gold nanoparticle-catalyzed luminol chemiluminescence and its analytical applications. *Analytical Chemistry, 77*(10), 3324–3329. Available from https://doi.org/10.1021/ac050036f.

## Further reading

Feng, Q. M., Pan, J. B., Zhang, H. R., Xu, J. J., & Chen, H. Y. (2014). Disposable paper-based bipolar electrode for sensitive electrochemiluminescence detection of a cancer biomarker. *Chemical Communications, 50*(75), 10949–10951. Available from https://doi.org/10.1039/c4cc03102d.

Kong, D., Schuett, W., Dai, J., Kunkel, S., Holtz, M., Yamada, R., … Klinkmann, H. (2002). Development of cellulose-DNA immunoadsorbent. *Artificial Organs, 26*(2), 200–208. Available from https://doi.org/10.1046/j.1525-1594.2002.06721.x.

Sharma, B., Parajuli, P., & Podila, R. (2020). Rapid detection of urokinase plasminogen activator using flexible paper-based graphene-gold platform. *Biointerphases, 15*(1). Available from https://doi.org/10.1116/1.5128889.

Sun, X., Li, B., Tian, C., Yu, F., Zhou, N., Zhan, Y., & Chen, L. (2018). Rotational paper-based electrochemiluminescence immunodevices for sensitive and multiplexed detection of cancer biomarkers. *Analytica Chimica Acta, 1007*, 33–39. Available from https://doi.org/10.1016/j.aca.2017.12.005.

Wang, Y., Zhang, C., Chen, X., Yang, B., Yang, L., Jiang, C., & Zhang, Z. (2016). Ratiometric fluorescent paper sensor utilizing hybrid carbon dots-quantum dots for the visual determination of copper ions. *Nanoscale, 8*(11), 5977–5984. Available from https://doi.org/10.1039/c6nr00430j.

Xu, Y., Liu, M., Kong, N., & Liu, J. (2016). Lab-on-paper micro- and nano-analytical devices: Fabrication, modification, detection and emerging applications. *Microchimica Acta, 183*(5), 1521–1542. Available from https://doi.org/10.1007/s00604-016-1841-4.

Zhang, X., Chen, C., Yin, J., Han, Y., Li, J., & Wang, E. (2015). Portable and visual electrochemical sensor based on the bipolar light emitting diode electrode. *Analytical Chemistry, 87*(9), 4612–4616. Available from https://doi.org/10.1021/acs.analchem.5b01018.

Chapter 18

# Emerging technologies for salivary biomarkers in cancer diagnostics

Ritu Pandey[1], Neha Arya[2,3] and Ashok Kumar[1]

[1]*Department of Biochemistry, All India Institute of Medical Sciences, Bhopal, Bhopal, India*

[2]*Department of Medical Devices, National Institute of Pharmaceutical Education and Research, Ahmedabad, India*

[3]*Department of Translational Medicine Centre, All India Institute of Medical Sciences, Bhopal, Bhopal, India*

## 18.1 Introduction

Globally, cancer is the second most leading cause of death. The average 5-year survival at the early stage is 91%, while the average 5-year survival at the advanced stage is only 26%. Delayed diagnosis of cancer is the major reason for the high mortality of cancer patients since the advanced stage of cancer often lacks effective treatment options. However, one can expect increased survival rates if cancer is detected at early stages. It could be surgically removed or treated with milder chemotherapeutic regimens. Hence, there is a shift towards the development of diagnostic platforms for the early detection of cancer. Conventional detection techniques such as colonoscopy, prostate-specific antigen, mammography, cervical cytology, etc., are available for certain cancer types. However, these methods are invasive, labor-intensive, and expensive, making the detection process quite cumbersome for routine diagnosis. Moreover, mammography may induce radiation-induced mutation, and the results can vary due to breast density (Brooks, Cairns, & Zeleniuch-Jacquotte, 2009; Freer, 2015). To add to this, x-ray scans, magnetic resonance imaging, and positron emission tomography imaging also pose serious concerns of radiation and sensitivity for deep-seated tissues, and require heavy instrumentation. Therefore, the efficacy remains the major question, and most cancer types certainly lack an effective noninvasive early screening. This poses a strong need to develop easy to use, fast, and accurate diagnostic tools. Researchers are working towards the development of diagnostics for the detection of various analytes in a cancer patient's tissue or body fluid. These analytes are termed as biomarkers. More specifically, a "biomarker" is an equitable measured characteristic that describes a normal or abnormal biological state of an organism by analyzing biomolecules such as metabolites, peptides, DNA, RNA, and protein (Strimbu & Tavel, 2010). In terms of clinical utility, cancer biomarkers measure the risk of acquiring cancer at specific tissue or determine cancer progression and its response to therapy. Cancer biomarkers can further be classified as screening, detection, diagnostic, predictive, and prognostic biomarkers (Fig. 18.1) (Goossens, Nakagawa, Sun, & Hoshida, 2015).

An ideal biomarker should fulfill the following four criteria, (1) specificity—it should differentiate diseased from the nondiseased state, (2) noninvasive approach, (3) stability, and (4) should be suitable for automation. Detection of cancer at an early stage remarkably affects recurrence, therapeutic intervention, as well as prognosis. Conventionally, diagnosis, treatment monitoring, recurrence, and detection of residual disease via biomarker detection and analysis in cancer patients involves invasive, painful procedures such as repeated blood draw and biopsies. Thereafter, morphological and histological examination, along with biomarker analysis, is performed. This technique is slow and time-consuming, thereby resulting in delayed diagnosis. A biomarker can also be measured by ELISA, real-time PCR, next-generation sequencing, or mass spectrometry; however, these techniques are tedious, require extensive sample processing, expensive, and require trained personnel. Therefore, the ability to monitor the disease onset, progression, and treatment through non-invasive methods is urgently needed for better healthcare delivery. In this regard, the discovery of saliva-based noninvasive molecular, metabolic, and microbial biomarkers offers distinctive prospects to overcome these limitations for cancer diagnostics.

Saliva is considered an important biofluid that bathes the oral cavity. It contains secretions from the salivary glands and nonsalivary components, including desquamated epithelial linings, blood derivatives, food components,

Biosensor Based Advanced Cancer Diagnostics. DOI: https://doi.org/10.1016/B978-0-12-823424-2.00021-1

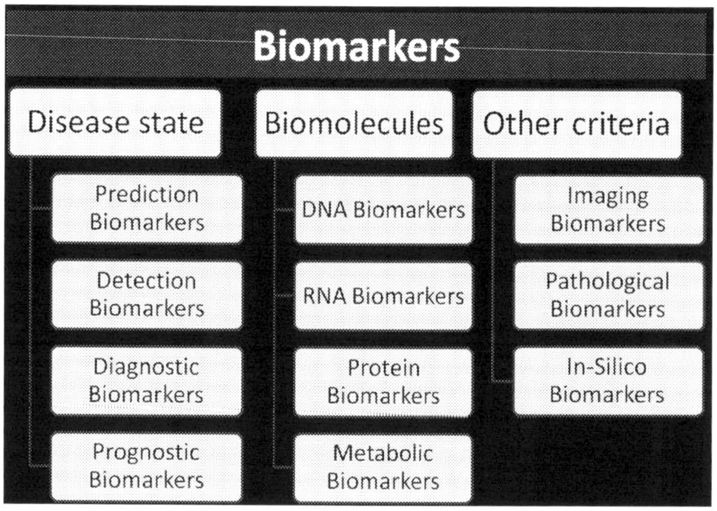


**FIGURE 18.1** Classification of biomarkers based on diseased state, types of biomolecules, and other criteria.

microorganisms, gingival crevicular fluid, and nasal and bronchial secretions. Whole saliva contains 99.5% water, 0.3% proteins (mucins, amylase, and albumin), and 0.2% inorganic (e.g., thiocyanate) substances that reflect the physiological condition of the body (Kaczor-Urbanowicz et al., 2017; Schafer et al., 2014). It includes over 800 identified metabolites (amino acids, carboxylic acids, steroid derivatives, glucose, etc.). It is comparable to human serum metabolomes in terms of chemical complexity and abundance of metabolites (Kaczor-Urbanowicz et al., 2017; Schafer et al., 2014). It has recently been shown that certain proteins or mRNA analogous to a patients' tumor status appear in their saliva; saliva is termed as the "mirror of the body." Therefore, it could be utilized for the diagnosis, prognosis, and post-treatment surveillance of patients with cancer (Kaczor-Urbanowicz et al., 2017). Saliva can provide information about the disease condition of the oral cavity as well as for distant tissues. This chapter will describe the technologies currently used to identify salivary cancer biomarkers and progress toward translating these technologies to point-of-care (POC) systems, followed by a summary of potential biomarkers. Finally, we discuss the challenges to be overcome in the translation of salivary biomarkers-based diagnostics to the clinics.

## 18.2 Technologies for discovery of salivary biomarkers

Currently, the number of biomarkers in clinical use is limited. Thus, identifying new salivary biomarkers that enable better diagnostic tools for cancer with high sensitivity and specificity is imperative. In the quest for new biomarkers, new-omics technologies have provided advances for the development of these markers. These technologies, including transcriptomics, metabolomics, proteomics technologies such as two-dimensional polyacrylamide gel electrophoresis (2D-PAGE) and mass spectrometry, have facilitated the discovery of new biomarkers.

### 18.2.1 Transcriptomics

Exploring transcriptomics is important as both the genome and proteome are dynamically linked to it. It provides information regarding the mRNA transcript generated at a certain point of time (Tuli & Ressom, 2009). Cancer-specific nucleic acids in biological fluids, including saliva, blood, urine, and cerebrospinal fluid, have been employed as biomarkers for cancer diagnosis. Salivary mRNAs are contributed by salivary glands, gingival crevice fluid, and dismantled oral epithelial cells (Zimmermann & Wong, 2008). The association of salivary mRNAs with macromolecules protects their degradation (Zimmermann & Wong, 2008). In this regard, microarray technology is being used for the identification of differentially expressed coding and noncoding RNAs in the biological fluids. In 2004, a team led by D.T. Wong explored the RNA profiling of cell-free saliva (Li et al., 2004). Using high-density oligonucleotide microarrays, the authors showed that cell-free saliva contains thousands of mRNA species. Further, the authors proposed a novel approach to salivary diagnostics, named salivary transcriptome diagnostics (STDs) (Li et al., 2004). More than 1500 genes exhibited significantly different expression levels in saliva from oral squamous cell carcinoma (OSCC) patients and healthy controls, using human genome U133A microarrays (Li et al., 2004). The study further identified a set of seven transcripts: interleukin-8 (IL-8), interleukin-1 beta (IL-1 beta), ornithine decarboxylase antizyme-1 (OAZ1), dual-specificity protein phosphatase 1 (DUSP1), spermidine/spermine N1-acetyltransferase 1 (SAT), H3 histone, family 3 A (HA3), and S100 calcium-binding protein P (S100P) that together yielded a sensitivity and specificity of 91% in discriminating OSCC patients from the healthy controls (Li et al., 2004). As these biomarkers were identified in American patients, a group led by D.T. Wong showed that out of six salivary transcripts (DUSP1, IL-8, IL-1B, OAZ1, SAT1, and S100P), four transcripts (IL-8, IL-1B, SAT1, and S100P) were also significantly elevated in Serbian OSCC patients (Brinkmann et al., 2011). To add to this, Lallemant et al. validated the clinical relevance of nine salivary mRNA biomarkers and showed that IL-1 receptor antagonist (IL1RN), myelin and lymphocyte-associated protein (MAL), and matrix metalloproteinase 1 (MMP1) were the most accurate diagnostic markers of head and neck squamous cell carcinoma (HNSCC), with receiver operating characteristic (ROC) area under the curve (AUC) >0.95 and with both sensitivity and specificity above 91% (Lallemant et al., 2009).

STDs has not only been limited to HNSCC; several transcriptomic-based salivary biomarkers have been discovered in other cancer types as well. Using microarray, a study compared the transcript profiles in the saliva from early-stage resectable pancreatic cancer patients and healthy controls (Zhang et al., 2010). In this study, four mRNA, namely, kirsten rat sarcoma (KRAS), methyl-CpG-binding domain protein 3 like 2 (MBD3L2), acrosomal protein SP-10 (ACRV1), and dolichol phosphate mannose synthase (DPM1) were found to be differentially expressed in the saliva from pancreatic cancer patients when compared to the healthy subjects (Zhang et al., 2010). In yet another study, a set of eight transcripts, S100 calcium-binding protein A8 (S100A8), cystatin A (CSTA), glutamate metabotropic receptor 1 (GRM1), tumor protein, translationally controlled 1 (TPT1), glutamate ionotropic receptor kainate type subunit 1 (GRIK1), hexose-6-phosphate dehydrogenase (H6PD), insulin-like growth factor 2 mRNA binding protein 1 (IGF2BP1), and murine double minute 4 protein (MDM4) showed discriminatory values between saliva of breast cancer patients versus healthy subjects (Zhang et al., 2010).

Standard procedure for Salivary Transcriptome Diagnostic (STD) requires mRNA isolation, which is time-consuming, labor-intensive, and requires maintenance of saliva samples at low temperatures, thereby increasing the logistic complexity during sample handling. In addition to salivary mRNA isolation, particular care is required when working with mRNA owing to its inherent instability and the ubiquitous presence of RNases. In this regard, Wong's group has developed a direct saliva transcriptome analysis (DSTA) protocol for the salivary transcriptomic analysis from cell-free saliva supernatant instead of isolated mRNA. In this protocol, all the steps, including processing, stabilization, and storage of saliva samples, are performed at ambient temperature without a stabilizing reagent (Lee et al., 2011).

### 18.2.2 Cell free microRNAs

MicroRNAs (miRNAs) are RNA molecules of about 18–25 nucleotides in length and act as regulators of gene expression at a post-transcriptional level. Partial complementarity between miRNAs and 3′UTR of the target mRNA transcripts inhibits translation. In contrast, perfect complementarity leads to the cleavage of the targeted mRNAs. Cell-free miRNAs were first discovered in plasma and serum; later, they were also reported in urine, saliva, cerebrospinal fluid, breast milk, colostrum, seminal fluid, and tears (Gablo, Prochazka, Kala, Slaby, & Kiss, 2019). Owing to the stability and the consistency of their expression in primary tumor tissues, miRNAs are believed to be rather selectively and specifically released into the extracellular milieu, conjugated with the structures protecting them against RNase activity (Fujita, Kuwano, Ochiya, & Takeshita, 2014).

In the biological fluids, miRNAs are encapsulated into two distinct classes: the membrane-derived extracellular vesicles (EVs) and exosomes or microparticles. Diameters of exosomes vary in the range of 30–100 nm, whereas microparticles are larger with a diameter of 100–4000 nm Fig. 18.2.

Several circulating miRNAs have demonstrated potential for valuable noninvasive cancer biomarkers (Table 18.1). Quantification of circulating miRNAs is a complex process that requires optimization to reduce pre-analytical and analytical errors (Cristaldi et al., 2019). Several miRNAs, including miR-21, miR-93, miR-125, miR-145, miR-184, miR-200a, and miR-375 have been shown as potential biomarkers to distinguish OSCC patients from healthy subjects (Greither et al., 2017; Park et al., 2009; Wiklund et al., 2011; Zahran, Ghalwash, Shaker, Al-Johani, & Scully, 2015). For example, miR-let-7a-5p ($P < .0001$) and miR-3928 ($P < .01$) were significantly down-regulated in the saliva of HNSCC patients compared to healthy controls (Fadhil, Wei, Nikolarakos, Good, & Nair, 2020). These miRNAs exhibited a good discriminative ability with AUC values of 0.85 and 0.74, respectively (Fadhil et al., 2020). A few miRNAs have been identified as prognostic salivary biomarkers. For example, miR-139-5p has been identified as a valuable biomarker for response to therapy and evaluation of OSCC recurrence (Duz et al., 2016). Salivary miR-15a-5p and miR-15b-5p exhibited differential levels between HNSCC patients with and without complete remission after intensity-

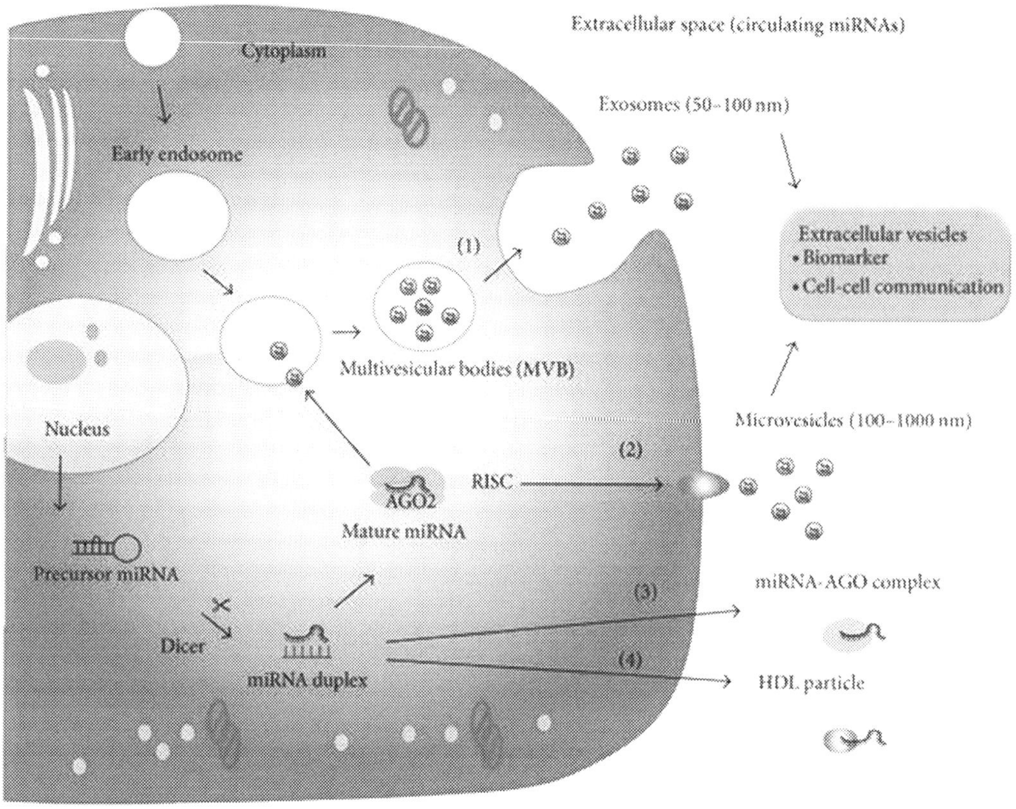


**FIGURE 18.2** Mechanisms depicting release of miRNA into extracellular space. Ribonuclease Dicer processes precursor miRNAs to mature double-stranded miRNAs (miRNA duplex). One strand of the miRNA is recruited into the RNA-induced silencing complex (RISC), containing the Argonaute (AGO) family protein. A fraction of miRNAs are released into the extracellular environment. *From Fujita, Y., Kuwano, K., Ochiya, T., & Takeshita, F. (2014). The impact of extracellular vesicle-encapsulated circulating microRNAs in lung cancer research.* BioMed Research International, 2014. *https://doi.org/10.1155/2014/486413.*

modulated radiation therapy (Ahmad et al., 2020). HNSCC patients with higher levels of miR-15a-5p had a significantly longer locoregional progression-free survival than those with low levels (Fadhil et al., 2020). However, many of these studies lack consistency and reproducibility due to the lack of a standardized miRNA protocol. A list of miRNAs that are differentially expressed in the saliva of healthy subjects and OSCC patients is given in Table 18.1.

## 18.2.3 Proteomics

Proteomics involves large-scale screening of protein expression, protein modification, and interaction utilizing high throughput techniques (Tuli & Ressom, 2009). Proteomics is an indispensable tool for the discovery of cancer biomarkers (Wang, Kaczor-Urbanowicz, & Wong, 2017). It has widely been used to analyze biological fluids, including saliva, blood, synovial fluid, and cerebrospinal fluid. A study identified 56 proteins in the secretions from minor salivary glands (Siqueira, Salih, Wan, Helmerhorst, & Oppenheim, 2008). In another study, approximately 200 differentially expressed proteins and peptides have been identified in the whole saliva of healthy subjects (Castagnola et al., 2017).

Depending on the sample, proteomics platforms are classified as bottom-up and top-down platforms. Top-down proteomics analyses the naturally occurring intact structure of a protein under examination, avoiding any sample alterations. Bottom-up proteomics is centered on pre-digestion of the sample (typical with trypsin) followed by the analysis of peptide fragments by high-throughput analytical methods (Castagnola et al., 2017). Most of the proteins undergo postranslational modification before reaching their mature active state. Therefore a minimalistic approach of bottom-up strategy results in loss of molecular information (Messana et al., 2013). Shot-gun proteomics has the ability to detect the largest range of components, regardless of their mass, because the proteolytic digestion of large proteins generates proteotypic peptides disclosing the presence of the parent protein in a complex mixture. For these reasons, shot-gun approaches can detect components of saliva more than five times greater than the number of components detected by

**TABLE 18.1** Various microRNAs as salivary biomarkers in different types of cancer.

| S. no. | miRNA | Associated cancer | Significance AUC (*P* value) | References |
|---|---|---|---|---|
| 1 | miR-21 | Pancreatic ductal adenocarcinoma CRC | 0.889 (*P* = .001 and 5e-12) | Alemar et al. (2016), Sazanov et al. (2017) |
| 2 | miR-34a | Pancreatic ductal adenocarcinoma | 0.865 (*P* = .002) | Alemar et al. (2016) |
| 3 | miR-93 | HNSCC | *P* = .047 | Greither et al. (2017) |
| 4 | miR-200a | HNSCC | *P* = .036 0.65 (*P* = .01) | Greither et al. (2017), Park et al. (2009) |
| 5 | miR-1246, miR-4644 | Pancreatobiliary tract cancer | 0.814 & 0.763 (*P* < .05) | Machida et al. (2016) |
| 6 | miR-125a | Oral cancer | 0.62 (*P* = .03) | Park et al. (2009) |
| 7 | miR-144 | Esophageal cancer | — | Du and Zhang (2017), Xie et al. (2013) |
| 8 | miR-451 | Esophageal cancer | — | Du and Zhang (2017), Xie et al. (2013) |
| 9 | miR-10b, miR-98, miR-363 | Esophageal cancer | — | Du and Zhang (2017) |
| 10 | miR-let-7a-5p | HNSCC Pancreatic Cancer | *P* = .003 | Fadhil et al. (2020), Torrisani et al. (2009) |
| 11 | miR-3928 | HNSCC | *P* = .049 | Fadhil et al. (2020) |
| 12 | miR-29a-3p, miR-29c-3p, miR-186-5p, miR-491–5p and miR-766-3p | Colorectal cancer | *P* < .05 | Rapado-González et al. (2019) |

any other platform (Castagnola et al., 2017). Top-down platforms are intrinsically limited by the sample treatments necessary for coupling with mass spectrometry (typical treatment with formic acid or trifluoroacetic acid), which inevitably excludes proteins insoluble in acidic solution. Moreover, intact high-molecular-weight proteins and heterogeneous glycosylated proteins are not accessible in their naturally occurring forms, even to the best high-level MS apparatus.

Using the 2D-PAGE platform followed by validation by western blotting, galectine-7 was identified as a good salivary biomarker for OSCC, with a specificity of 90% and a sensitivity of 80.5% (Amenábar, Da Silva, & Punyadeera, 2020). Recent research has suggested that potential biomarkers in other cancer types may be present in human saliva. A study carried out by nano-High Performance Liquid Chromatography (HPLC)-Q-TOF MS investigated the proteome profiles of plasma and saliva of patients with fibroadenoma ($n = 10$), infiltrating ductal carcinoma ($n = 10$), and healthy controls ($n = 8$). In that study, two major differentially expressed proteins in the saliva of patients were $\alpha$2-macroglobulin and ceruloplasmin, as compared to controls. A subtractive proteomics approach identified a set of five differentially expressed protein markers (M2BP, MRP14, CD59, catalase, and profilin) in the saliva of OSCC patients compared to the healthy subjects (Hu et al., 2008). In this approach, pooled saliva samples from 16 subjects were pre-fractionated using C4-reversed-phase liquid chromatography (LC), and subsequent in-solution tryptic digestion followed by quadrupole time-of-flight mass spectrometry (qTOF-MS) (Hu et al., 2008). Platforms based on 2D-electrophoresis are affected by poor reproducibility; therefore, it is often necessary to run multiple replicas of the same sample to avoid bias.

### 18.2.4 Metabolomics

The salivary metabolic profile is called the "mirror of the body" because it can provide a general outlook on change in the metabolites arising from metabolic reprogramming caused due to oncogenic changes in the oral mucosa. To date, more than 100 metabolites have been reported to be dysregulated with malignant progression in the oral cavity. These metabolites include choline, lactate, glutamate, sialic acid, histidine, polyamines, carnitine, pipecolic acid, and trimethylamine N-oxide (Lohavanichbutr et al., 2018; Song et al., 2020; Sridharan, Ramani, Patankar, & Vijayaraghavan, 2019; Yang et al., 2020). A single analytical technique cannot simultaneously analyze all the metabolites in saliva samples.

Therefore, to appropriately identify and analyze various metabolic pathways, suitable analytical platforms shall be chosen. Nuclear magnetic resonance (NMR) is the most common analytical platform used for metabolomics. It has widely been used to discover novel cancer biomarkers (Mikkonen et al., 2018). There are several advantages to using NMR, including the simple and nondestructive preparation of the sample and easy quantification (Sugimoto, Wong, Hirayama, Soga, & Tomita, 2010).

In addition to NMR, mass spectrometry (MS) is a powerful platform for metabolomics. Owing to its in-depth coverage of various molecular species, it has become an important analytical tool for discovering, validating, and diagnosing various diseases. Coupling of various separation techniques with MS has resulted in compelling and versatile MS platforms such as liquid chromatography-mass spectrometry (LC-MS), gas chromatography-mass spectrometry (GC-MS), matrix-assisted laser desorption ionization-time of flight mass spectrometry (MALDI-TOF), capillary electrophoresis mass spectrometry (CE-MS), two-dimensional electrophoresis mass spectrometry (2DE-MS), surface-enhanced laser desorption ionization time-of-flight mass spectrometry (SELDI-TOF), and conductive polymer ionization-mass spectrometry (CPI-MS). LC-MS has widely been used to discover salivary metabolites that could discriminate OSCC patients from healthy subjects and OSCC from premalignant lesions (Sugimoto, 2020). Song et al. analyzed saliva samples from 124 healthy subjects, 124 patients with premalignant lesions, and 125 patients with OSCC, using CPI-MS. This study identified several salivary metabolites that can distinguish healthy individuals from patients with potentially malignant lesions and OSCC with high sensitivity and specificity (Song et al., 2020). In that study, several metabolites, including putrescine, cadaverine, thymidine, adenosine, 5-aminopentoate, hippuric acid, and phosphocholine, were identified as potential salivary biomarkers for OSCC. They showed that the combination of CPI-MS and machine learning provides a simple, fast, and affordable method for OSCC diagnosis with 86.7% accuracy (Song et al., 2020). In addition, a metabolomics approach has also been used for the discovery of salivary biomarkers that could be used for the diagnosis of breast cancer patients. In this study, using the LC-MS method, 18 metabolites, such as citrulline, histidine lyso-phosphatidylcholine, propionyl choline, and palmitic amide were found to be elevated in the saliva of patients with breast cancer compared to those from healthy controls (Zhong, Cheng, Lu, Duan, & Wang, 2016). Furthermore, CE-MS is used to quantify hydrophilic metabolites; it was used for profiling the saliva collected from patients with oral, breast, and pancreatic cancers (Sugimoto, 2020). In that study, the authors identified 57 metabolites that were associated with the susceptibility of developing cancer (Sugimoto, 2020). Taken together, a combination of these analytical methodologies could contribute to increased coverage for metabolite detection towards the discovery of biomarkers.

### 18.2.5 Microbiomics

The oral cavity is inhabited by more than 1,000 different microbes, including bacteria, archaea, viruses, and eukaryotes (Dewhirst et al., 2010). Among all, Streptococci are the dominant class of bacteria, generally accompanied by *Veillonella*, *Gemella*, *Rothia*, *Fusobacterium*, and *Neisseria* species (Aas, Paster, Stokes, Olsen, & Dewhirst, 2005; Bik et al., 2010). Under normal conditions, a balance usually coexists between the human and oral microbiome; however, dysbiosis can lead to various pathologies (Hajishengallis & Lamont, 2012). Besides, identifying the molecular signature of saliva's differential microflora can also predict the disease status. Microbiomics decipher the overall microflora that could potentially alter the molecular constituents of the saliva and therefore emerging as potential screening, diagnostic, and prognostic biomarkers for cancer detection.

Several specific microbial species were found to be associated with various cancers (Mager et al., 2005; Wolf et al., 2017; Zhang et al., 2010). Vesty et al. analyzed the bacterial and fungal community using 16 S rRNA gene and internal transcribed spacer amplicon sequencing, respectively, and revealed that the microbial community of HNSCC were significantly different from healthy individuals; Streptococci were found to be dominant microbial species accounting for 47% of all microbial community (Vesty et al., 2018). Furthermore, the authors showed that the increased levels of interleukin (IL)-1β and IL-8 were positively correlated with the abundance of *Candida albicans* in the saliva from HNSCC patients (Vesty et al., 2018). Several other studies also showed the differential presence of four microbial species, *Dialister* sp., *Selenomonas* sp., *Streptococcus* sp., and *Treponema* sp. in oral cancer patients when compared with healthy individuals (Guerrero-Preston et al., 2016; Pushalkar et al., 2011; Wolf et al., 2017). Biopsy specimens from OSCC patients were relatively abundant (61.2%) with *C. albicans* (Perera et al., 2017). Furthermore, it has been suggested that the ability of *C. albicans* to produce high levels of acetaldehyde might contribute to the process of carcinogenesis (Ramirez-Garcia et al., 2016) and is also attributed to be involved in the transformation of oral lesions to malignant one (Bakri, Hussaini, Holmes, Cannon, & Rich, 2010). Yet another study performed by Wong et al. suggested the two oral microbial markers, *Neisseria elongata* and *Streptococcus mitis*, which yielded a ROC curve of 0.90 with 96.4% sensitivity and 82.1% specificity in determining the patient with early-stage resectable pancreatic cancer from noncancer patients (Zhang et al., 2010). Nevertheless, the microbiota status in the saliva of cancer patients demonstrates the potential for the development of biosensor-based diagnostics for early detection of cancer.

### 18.2.6 Spectroscopy techniques

Optical spectroscopic techniques such as diffuse reflectance spectroscopy, elastic scattering spectroscopy, fluorescence spectroscopy, infrared spectroscopy, Raman spectroscopy, and optical coherence tomography have been employed in detecting molecular changes occurring during malignancy. A few of them will be discussed in this section.

#### *18.2.6.1 Fluorescence spectroscopy*

Fluorescence spectroscopy has been used in distinguishing cancer subjects from normal beings using biofluids and tissues (Huck, Ozaki, & Huck-Pezzei, 2016). Many fluorescence spectroscopic methodologies have been established for nearly all different kinds of cancer. Near-infrared spectroscopy is based on differences of endogenous chromophores between cancer and normal tissues using either oxyhemoglobin or deoxy-hemoglobin, lipid or water bands, or a combination of two or more of these diagnostic markers. Although, fluorescence spectroscopy is sensitive for the detection of a specific biomarker. However, there are several drawbacks of fluorescence spectroscopy, including photobleaching of endogenous fluorophores, and multiple excitation wavelengths are required due to the various fluorophores present in biofluids and tissues. Furthermore, exogenous fluorophores are required to be added in cases of nonfluorescence biomolecules that may affect the native environment of the system (Huck et al., 2016).

#### *18.2.6.2 Vibrational spectroscopy*

Vibrational spectroscopy, Raman, and infrared, provide a detailed salivary molecular fingerprint that can be used for disease biomarker discovery (Derruau et al., 2020). Raman spectra are precise, unique, and require a small sample quantity, and no sample preparation is required. These techniques provide a snapshot of the sample biomolecular composition, and variations can be exploited to identify disease status. In a study by Jaychandran et al., Raman peaks were able to discriminate OSCC, premalignant, and normal subjects on the basis of pyrimidine, amide, mucin, hemocyanin, and carotenoids content in the saliva (Jaychandran, Meenapriya, & Ganesan, 2016). Principal component analysis followed by linear discriminant analysis of saliva samples was able to discriminate spectra from OSCC patients versus healthy controls with an accuracy of 93.1% (Jaychandran et al., 2016).

#### *18.2.6.3 Infrared spectroscopy*

In the last decade, infrared (IR) spectroscopy has attracted attention as a simple and inexpensive method for the biomedical study of several diseases. Mid-IR has been applied for biofluid analysis and microtome tissue-section pathology, which require expensive focal-plane array detectors for imaging preceded by extensive sample preparation in the laboratory. Additionally, near-infrared spectroscopy (NIRS) has gained importance for noninvasive or minimally invasive cancer diagnostic applications. It is based on differences of endogenous chromophores between cancer and normal tissues using either oxy-hemoglobin or deoxy-hemoglobin, lipid or water bands, or a combination of two or more of these diagnostic markers (Huck et al., 2016). Owing to advances in spectrometer hardware, fiber-optic probes and chemometric NIRS are used straightforwardly for noninvasive or minimally invasive diagnostic applications (Huck et al., 2016). Merits and demerits of various spectroscopic techniques are summarized in Table 18.2.

## 18.3 Point-of-care technologies for detection of salivary biomarkers

As mentioned in the previous sections, saliva can be used to monitor both oral and systemic health as it reflects the physiological state of the body. Furthermore, in the previous section, we discussed various technologies, including genomics, MS, HPLC, spectroscopic techniques for the detection of salivary biomarkers. However, most of the techniques or assays are complex, time-consuming, require a large amount of sample volume/reagents as well as require large, bulky, and expensive instruments, thereby limiting their usage in the clinical diagnosis. In the past few decades, saliva-based diagnostics have been developed to monitor oral and systemic diseases that need to be translated from the bench-side to the bed-side. In this regard, POC-diagnostics (POCD) has been developed. By definition, a POC device is a small, rapid, portable device that offers great potential to detect and monitor disease even at resource limiting settings. As per WHO, the list of general characteristics that makes a diagnostic test appropriate for resource-limited sites, abbreviated as ASSURED, and includes, *A*ffordable by those at risk, *S*ensitive, *S*pecific, *U*ser-friendly, *R*apid treatment and

**TABLE 18.2 Merits and demerits of various spectroscopic technologies.**

| Types of spectroscopy | Advantages | Disadvantages |
|---|---|---|
| Infrared Spectroscopy | Can detect the presence and absence of a specific functional group. | Molecular weight of substance cannot be determined. Frequently nonadherence to beer's law of complexity spectra. Does not provide information of the relative position of different functional groups on molecules. Cannot differentiate pure compound and mixture of compound. |
| Raman Spectroscopy | Less change to temperature changes. Minimal sensitivity to water. Suitable for biological samples in native state. Suitable for any substance, since it measures scatter light including opaque substrate. Highly specific. Little or no sample preparation required. Samples can be solid or aqueous. | Specially refrained by optical limits. Long collection time. Fluorescence is a common background issue. Typical detection limits in the parts per thousand range. Requires expensive lasers, detectors and filters. Small sample volume can make it difficult to obtain a representative sample. |
| Fluorescence Spectroscopy | High sensitivity. Unique optical properties of the component make it highly specific. May be immune to the scattering of light. Emitted light is read at the right angle to the exciting light, reducing the background signal. Large range of linearity. Measure analytes concentration in terms of fluorescence intensity and decay times. | All molecules are not fluorescent. Interference due to changes in pH and oxygen levels of the sample. Auto-fluorescence Potential toxicity, due to the foreign material in the biological media. Short lifespan of fluorophores. Photo-stability and loss of recognition capability. |

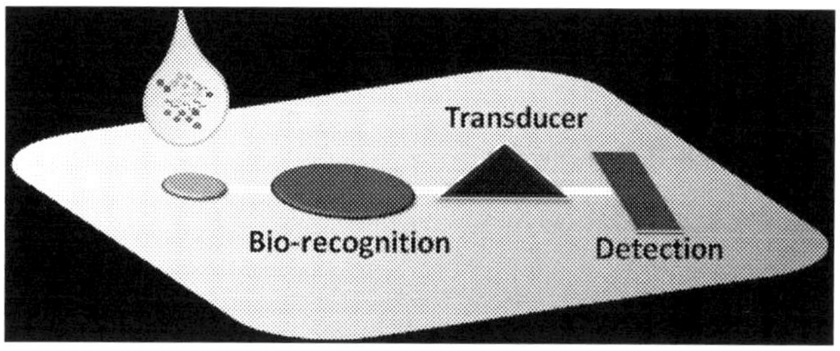


**FIGURE 18.3** Schematic diagram of a point-of-care (POC) system.

robust use, *E*quipment-free and finally, *D*elivered to those who need it (Urdea et al., 2006). A simple schematic of POC device is shown in Fig. 18.3.

In this regard, microfluidics technology possesses remarkable features towards the development of a simple, low-cost, and rapid disease diagnosis; these features include low volumes of reagent consumption, fast analysis, high portability along with integrated processing and analysis of complex biological fluids with high sensitivity and specificity for health care application. Basically, it integrates many processes onto a single chip through a large network of channels where fluid can flow, mix, and react, resulting in analyte analysis. These devices offer chip-based POC diagnosis and real-time monitoring of disease using a small volume of body fluid. Various materials are being used for the fabrication of microfluidic devices, including glass, polydimethylsiloxane, poly(methyl methacrylate) (PMMA), poly(cyclic olefin), and paper-based. Further, the fabrication methods are reactive, ion etching, photolithography, soft lithography, hot embossing, laser ablation, plasma etching, and many more (Fiorini & Chiu, 2005).

Zhou et al. have developed a lung cancer diagnostic kit (LCDK) with the advantage of low-cost, easy-operation, and high-sensitivity for the rapid diagnosis of lung cancer. The proposed LCDK can discriminate lung cancer using noninvasive clinical salivary and urine samples in a short time (Zhou et al., 2020). In this study, the authors designed a strip-based POC testing system to diagnose lung cancer by detecting exosomal miRNA. To increase the detection limit of exosomal miRNAs in the patient's saliva and urine, $Fe_3O_4@SiO_2$-aptamer nanoparticles (FSAs) were used to capture and accumulate lung cancer exosomes and duplex-specific nuclease (DSN) as an amplification tool to amplify miRNA signals. LCDK consisted of three components/steps: (1) screening and concentration of exosomes by FSAs; (2) DSN hydrolysis of reporter DNA triggered by exosomal miRNA; and (3) readout of the data by lateral transverse flow test strips (Zhou et al., 2020). This strategy (Fig. 18.4) was tested for the detection of miRNA-205 in the urine and saliva

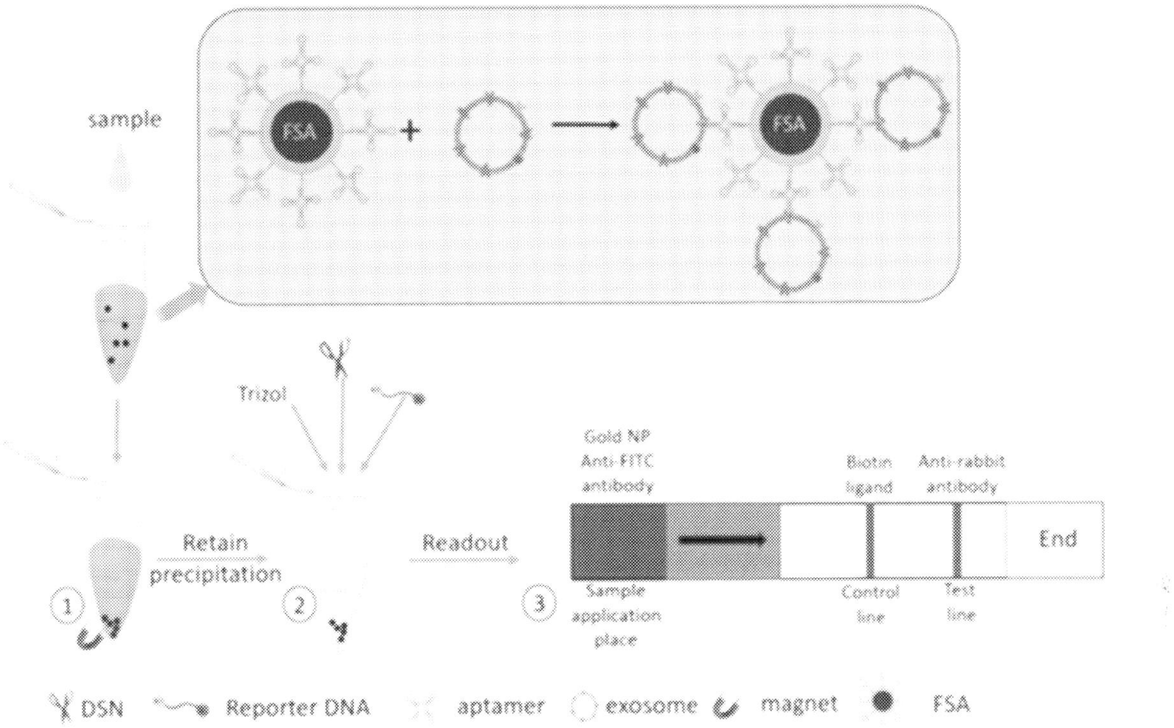


**FIGURE 18.4** Schematics of the detection process, including the collection of exosome by (1) $Fe_3O_4@SiO_2$-aptamer nanoparticles (FSAs), (2) duplex-specific nuclease (DSN)-based reaction, and (3) data readout on lateral flow strips. *From Zhou, P., Lu, F., Wang, J., Wang, K., Liu, B., Li, N., & Tang, B. (2020). A portable point-of-care testing system to diagnose lung cancer through the detection of exosomal miRNA in urine and saliva.* Chemical Communications, *56(63), 8968–8971. https://doi.org/10.1039/d0cc03180a.*

samples of lung cancer patients, and the sensitivity of LCDK for miRNA-205 was found to be approximately 7.76 p.m. Further, the specificity and detection limit of any device (LCDK in this case) depends on the reporter DNA and sequences of aptamer (Zhou et al., 2020).

### 18.3.1 Types of detection system

With the advent of nano- and microfabrication techniques, a wide range of POC diagnostics have been developed. They employ the use of microfluidic devices that require low reaction volumes, less time, are affordable as well as possess portability. A wide range of microfluidic devices have been reported that rely on various nanomaterials towards the generation of detection platforms for various diseases. The upcoming section will discuss a few technologies that have been used for the detection of analytes found in the saliva of cancer patients (Sanjay et al., 2015).

#### *18.3.1.1 Optical detection-based system*

Optical biosensors detect changes in absorption of light (UV/visible/infrared) or the quantity of light emitted following a chemical reaction. These sensors consist of a compact analytical device with biorecognition sensing elements and optical transducers systems; they produce a signal proportional to the concentration of the analyte to be detected (Damborský, Švitel, & Katrlík, 2016). Optical biosensors can be used for label-free or label-based sensing. The label-free mode is based on the detection of the signal following direct interaction between the analyte and the transducer. On the other hand, the label-based mode relies on tagging of an analyte to a reporter system/label. The detection may be based on the surface plasmon resonance, luminescence, fluorescence, colorimetric and interferometric modes.

Various optical biosensors have been reported for the detection of cancer using patient saliva. In this regard, an optical protein sensor was reported for the detection of interleukin-8 (IL-8), a biomarker used for the diagnosis of oral cancer (Tan et al., 2008). The authors could detect IL-8 at a concentration of 1.1 p.m. without enzymatic amplification. Confocal optics further enhanced the limit of detection (LOD) to 4 fM. With this LOD, the optical biosensor was further used to detect the IL-8 levels in saliva samples of cancer patients (IL-8 levels in the control group: 70 p.m., IL-8 levels in cancer patients: 138.5 p.m.), which were well above the range of detection of the optical biosensor. Another study utilized organic photodetectors for optical detection of salivary protein biomarkers. The device comprised of polythiophene-C70 with a calcium-free cathode inner layer; an antibody-functionalized poly(methyl-methacrylate) based microfluidic chip was then used for the detection of IL-8, IL-1β and MMP-8 protein in spiked saliva with detection limit between 80–120 pg/mL leading to the development of attractive optical POC biosensor (Dong & Pires, 2017). It is important to state that the biosensor was capable of detecting IL-8 approximately sixfold lower than the clinically established cut-off, thereby demonstrating its potential as an attractive platform for early detection of oral cancer.

Another interesting study developed an optoelectronic sensor for the detection of clinically relevant levels of carbon dioxide ($CO_2$) and ammonia ($NH_3$) (Fischer, Triggs, & Krauss, 2016). The microfluidic device was comprised of ion-exchange polymer microbeads doped with organic dyes and is sensitive to ppm levels of dissolved $CO_2$ and $NH_3$, usually associated with infection with *Helicobacter pylori* (which is the marker for early identification of stomach cancer). Optical biosensors are advantageous over conventional detection techniques due to their ability to provide direct, real-time, and label-free analytes detection. They also offer high sensitivity and specificity. However, they may demonstrate sensitivity due to interference from light as well as temperature changes (Liao et al., 2019).

### *18.3.1.2 Electrochemical detection systems*

This is based on the detection of electrochemical changes that occurs during the biorecognition process; these could be changed in current, potential, conductance, or impedance, wherein the electrical signal obtained is proportional to the analyte concentration. As a result, they could be further classified as amperometric, potentiometric, conductometric, and impedance-based electrochemical biosensors. In these biosensors, the biomolecules immobilized on electrodes are either receptors or mediators of electron transfer, resulting in oxidation or reduction of the target. In this regard, nanomaterials have been shown to play an important role in providing high surface area for immobilization of biomolecules as well as for the catalysis of redox reactions. Electrochemical sensors have been widely used for the detection of cancer biomarkers using saliva as the biofluid. In one such study, Verma and Singh reported a nanocomposite comprising of zinc oxide-reduced graphene oxide nanocomposite based electrochemical sensor (Verma & Singh, 2019) for detection of IL-8 in the saliva. The biosensor demonstrated a sensitivity of $12.46 \pm 0.82$ μA mL/ng, and a LOD value of $51.53 \pm 0.43$ pg/mL and have been validated using spiked IL8 in saliva samples (Verma & Singh, 2019). Another study reported an electrochemical assay for the detection of oral cancer overexpressed 1 (ORAOV1) by combining nanomaterials such as Pt and Pd-MoS2 and nicking enzyme signaling amplification system (Hu Y et al. 2018). The biosensor demonstrated detection of target DNA in the linear range from 10 fM to 10 nM (Hu Y et al., 2018). The sensor was further employed to detect the target DNA in complex samples such as artificial saliva, and recoveries were in the range of 91.9%–108.6% and demonstrated a potential for diagnosis of ORAOV1 in the clinics. In fact, ORAOV1 has also been used as a target by Chen et al. towards the development of label-free electrochemical sensors based on nuclease-assisted target recycling and amplification of DNAzyme (Chen et al., 2011).

In yet another study, Ma et al. developed an electrochemical DNA biosensor that detects ORAOV1 or related species based on homogeneous exonuclease III-assisted target recycling amplification followed by one-step triggered dual-signal output (Ma et al., 2018). In this biosensor, the target DNA hybridized to a specific ferrocene-labeled hairpin probe, followed by homogeneous exonuclease III-mediated amplification and finally resulting in a decreased concentration of the ferrocene-labeled hairpin probe (Ma et al., 2018). The remaining probe was then measured as a function of its hybridization with the methylene blue-labeled hairpin probe on the gold electrode followed by dual signal ratiometric electrochemical readout (Fig. 18.5). The reported biosensor was able to detect the target DNA at 12.8 fM as well as demonstrate high sensitivity. The biosensor was further applied to detect the target DNA spiked in artificial saliva samples, thereby demonstrating its potential in disease diagnostics.

Tian et al. developed an electrochemical DNA-biosensor using a paper-based microfluidic device to detect epidermal growth factor receptor (EGFR) mutation in the saliva of patients with non-small cell lung cancer (NSCLC) (Tian et al., 2017). The authors immobilized ssDNA on polypyrrole (PPy) modified gold electrodes; the electrochemical signal was obtained by DNA hybridization followed by recognizing the indicator labeled on DNA by horseradish peroxidase (HRP) and further catalysis of redox reaction upon application of differential pulse voltammetry. The developed sensor demonstrated a linear detection range between the current and log value of target DNA in the range of 0.5–500 nM with a detection limit as low as 0.167 nM. The method was also used to detect the target DNA isolated from patient samples and demonstrates potential in real-time monitoring of EGFR mutation status in NSCLC patients.

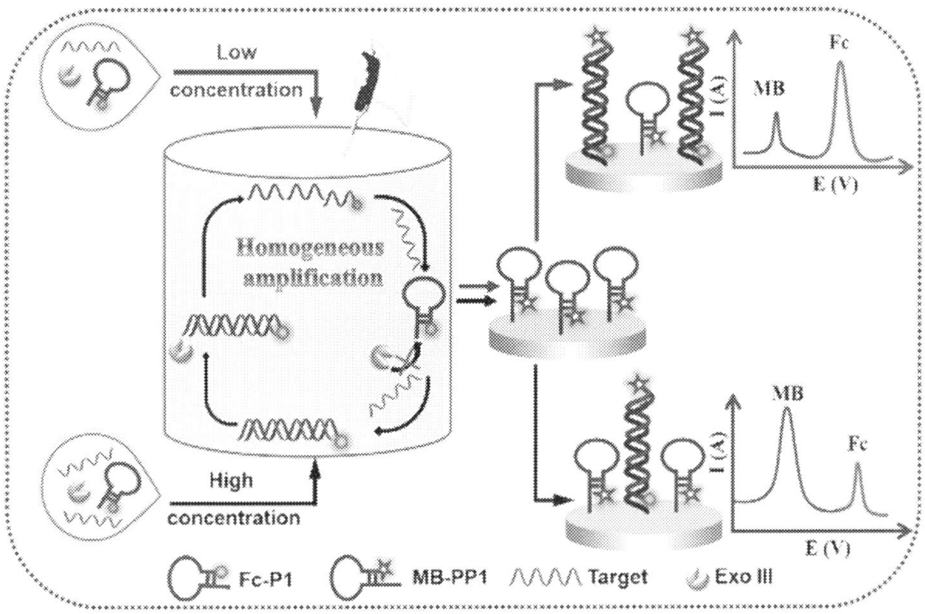


**FIGURE 18.5** Schematic illustration of the homogenous Exo III-assisted target recycling amplification and dual signal ratiometric electrochemical DNA biosensor. *From Ma, R. N., Wang, L. L., Wang, H. F., Jia, L. P., Zhang, W., Shang, L., Xue, Q. W., Jia, W. L., Liu, Q. Y., & Wang, H. S. (2018). Highly sensitive ratiometric electrochemical DNA biosensor based on homogeneous exonuclease III-assisted target recycling amplification and one-step triggered dual-signal output.* Sensors and Actuators, B: Chemical, 269, *173–179. https://doi.org/10.1016/j.snb.2018.04.143.*

David Wong and his team developed an electrochemical sensor for multiplex detection of IL-8 mRNA and IL-8 protein with an LOD of 3.9 fM and 7.4 pg/mL, respectively, and was validated with oral cancer patient saliva (Wei et al., 2009). The sensitivity and specificity were around 90%. It was similar to the results obtained from the same group of saliva samples using traditional assays such as ELISA and PCR. They further developed an automated electrochemical point of care devices termed oral fluid nanosensor test (OFNASET), capable of simultaneous detection of salivary proteins and nucleic acid targets for various disease conditions (Lee & Wong, 2009). This "lab-on-chip" can detect eight biomarkers in just 15 min. Simultaneous detection of IL-8 mRNA and IL-8 protein has been reported by other groups as well for direct determination in raw saliva samples (Torrente-Rodríguez, Campuzano, Ruiz-Valdepeñas Montiel, Gamella, & Pingarrón, 2016).

Another interesting study reported the detection of oral cancer related-miRNA, hsa-miR-200a, in saliva using an electrically magnetic-controllable electrochemical biosensor (Wang et al., 2017). The biosensor was based on an electrically magnetic-controllable gold electrode, "junction-probe" DNA detection technique, and magnetic beads-based enzyme catalysis towards the detection of 0.22 a.m. hsa-mir-200a (Wang ZW et al., 2013), thereby demonstrating potential as a promising POC diagnostic device. Another study by Majidi et al. reported an aptamer-mediated, label-free electrochemical sensor for the detection of L-tryptophan (whose consumption rate can be indicative of a degree of metastasis) in body fluids including saliva (Majidi, Karami, Johari-Ahar, & Omidi, 2016).

Electrochemical sensors offer advantages of easy readout interface, robustness, easy miniaturization, excellent detection limits, and low costs. However, their readouts can be affected by the contamination of the biosensor surface.

### *18.3.1.3 Piezoelectric detection systems*

Piezoelectricity is the property of a material to produce voltage upon mechanical stress or vice versa. Piezoelectric sensors use crystals that undergo elastic deformation upon application of electric potential. Herein, the biosensor is excited following the application of alternating voltage on the surface of the electrode; the alternating voltage then leads to mechanical oscillation, the frequency of which is then recorded. So the binding of an analyte or any mass on the crystal will cause a change in the oscillation frequency, which is converted or read in the form of a signal. While there are no reports of piezoelectric biosensors for detection of cancer biomarkers using patient saliva, a study by Tajima et al. reported a piezoelectric quartz crystal biosensor for the detection of IgA antibodies in diluted human saliva. Further, the study reported detection of antibody concentration as low as 1 μg/mL (Tajima, Asami, & Sugiura, 1998). Piezoelectric biosensors allow direct and label-free detection of the analyte and hold the application in cancer diagnostics using patient saliva.

**TABLE 18.3** Commercially available POC devices for salivary cancer biomarkers.

| S. no. | POC Devices | Name of biomarker (s) | Associated disease | References |
|---|---|---|---|---|
| 1 | Oral Fluid Nanosensor Test (OFNASET) | Protein (thioredoxin and IL-8), mRNA (SAT, ODZ, IL-8, and IL-1b) | Oral cancer | Gau and Wong (2007) |
| 2 | LCDK | Exosomal miRNA-205 | Lung cancer | Zhou et al. (2020) |
| 3 | Integrated microfluidic platform for oral diagnostics (IMPOD) | IL-6 and TNF-α | Chronic periodontitis | Herr et al. (2007) |

*LCDK*, lung cancer diagnostic kit; *POC*, point-of-care.

### 18.3.2 Commercially available POC technologies

Salivary POC devices are divided into two categories: single biomarker detection and multiple biomarker detection. Various groups of scientists are working towards the development of POC systems for the early diagnosis of cancer. The first salivary marker-based POC system was developed by Wong et al. in 2007, termed as OFNASET (Gau & Wong, 2007) which can detect combinations of two proteomic biomarkers (thioredoxin and IL-8) and four mRNA biomarkers (SAT, ODZ, IL-8, and IL-1beta). It combines multiple cutting-edge technologies like self-assembled monolayers (SAM), cyclic enzymatic amplification, bio-nanotechnology, microfluidics, and several other technologies microinjection molding, hybridization-based detection, and molecular purification, allowing it to detect oral cancer with high sensitivity and specificity. Recently, another device, LCDK has been developed to determine miRNA-205 in distinguishing patients with lung cancer. A list of commercially available POC devices for salivary cancer biomarkers is given in Table 18.3.

## 18.4 Challenges in translating salivary biomarkers to the clinics

The presence of differential levels of biomolecules in the saliva serves as an important characteristic that could be exploited for the cancer diagnostics and prognostics. A list of potential biomarkers is given in Table 18.4.

Numerous biomarkers have been studied for the detection of cancer; however, they still lack sensitivity and specificity. Major studies were conducted using the whole saliva that contains the fluid secreted from salivary glands as well as fluids from mucosal and periodontal tissues. Easy collection with less time consumption can be performed by untrained professionals and its cost-effectiveness are some of the key advantages that make use of saliva as an acceptable fluid for diagnostics. However, using salivary biomarkers for cancer diagnostics has certain drawbacks that cannot be ignored; these include the variation in constituents of saliva due to other oral diseases, immune response, using stimulated or unstimulated saliva, sample processing, as well as sample collection time. All these can act as confounding elements, thereby leading to false results. In this regard, certain major challenges associated with the use of salivary biomarkers are described below:

### 18.4.1 Standardization of conditions and methods of saliva sample collection, processing, and storage

Earlier studies were done on both stimulated and unstimulated saliva; the latter does not affect its composition, whereas stimulated saliva provides more accurate detection of cancer biomarkers. No study has ever been done to compare the effect of stimulated and unstimulated saliva on biomarker levels, making it the major challenge in discovering the salivary biomarkers. Usually, saliva is collected in the morning 1 or 1.5 h after eating and drinking to avoid changes after sample collection; variation in saliva collection time, handling processing, and method of analysis influences the biomarker's analysis. Further, the whole saliva consists of epithelial cells, microbes, remnants of food, so it becomes imperative to eliminate solid constituents using centrifugation. Optimal conditions of centrifugation have yet to be standardized for saliva processing. Furthermore, RNA and protein inhibitors were used in certain studies to preserve the RNA and protein present in the saliva. However, most of the studies were performed without the use of any inhibitors. This may result in differential profiles of various biomarkers in the saliva sample. Another parameter is the temperature

**TABLE 18.4** List of potential salivary biomarkers for cancer detection.

| Categories | Biomarkers | Associated cancer | References |
|---|---|---|---|
| Proteins | IL-8, IL-6, TNF-α | Oral | Sahibzada et al. (2017) |
| | IL-8, IL-1B and SAT | Oral | Elashoff et al. (2012) |
| | CFB, C3, C4B, LRG1 and SERPINA1 | OSCC | Kawahara et al. (2016) |
| | Haptoglobin, zinc-a-2-glycoprotein and calprotectin | Lung | Xiao et al. (2012) |
| | Cystatin B (CSTB), triosephosphate isomerase (TPI1), and deleted in malignant brain tumors 1 protein (DMBT1) | Gastric | Xiao et al. (2016) |
| | M2BP, MRP14, CD59, profilin, and catalase | Oral | Hu et al. (2008) |
| | Cytokeratin 19 fragment Cyfra21−1, tissue polypeptide antigen, and cancer antigen 125 | OSCC | Nagler et al. (2006) |
| Polyamines | N1-Ac-SPD, N8-Ac-SPD | Breast | Takayama et al. (2016) |
| mRNA | CCNI, EGFR, FGF19, FRS2 and GREB1 | Lung | Zhang et al. (2012) |
| | IL1b, DUSP1, OAZ1 and HA3, H3F3A | Oral | Elashoff et al. (2012) |
| | S100A8, CSTA, GRM1, TPT1, GRIK1, H6PD, IGF2BP1, MDM4 | Breast | Zhang et al. (2010a) |
| | KRAS, MBD3L2, ACRV1 and DPM1 | Pancreatic | Zhang et al. (2010b) |
| Metabolites | Putrescine, cadaverine, thymidine, adenosine, 5-aminopentoate, hippuric acid and phosphocholine | OSCC | Song et al. (2020) |
| | Leucine, isoleucine, tryptophan, valine, glutamic acid, phenylalanine, glutamine and aspartic acid | Oral, breast and pancreatic | Sugimoto et al. (2010) |
| | Choline, isethionate, cadaverine, N1-acetylspermidine and spermine | Breast | Jinno et al. (2015) |

*OSCC*, oral squamous cell carcinoma.

of storage of saliva samples to avoid any discrepancy in the results. Scientists have used a range of temperature from 4°C to −80°C. Schipper et al. reported that storage at −80°C gives better results as compared to −20°C (Schipper, Silletti, & Vingerhoeds, 2007). The time between the sample collection and analysis should be minimized to avoid altered results. As observed, potential biomarkers in precancer patients were variable in different studies that can be attributed to the fact that the standard method was not used for the sample collection, handling, processing, and analysis. Variable levels of biomarkers in different studies can obstruct the estimation and diagnosis of cancer. Therefore, further studies are required to standardize the saliva collection, processing, and analysis.

## 18.4.2 Variability in the levels of potential salivary biomarkers

This section summarizes the reasons for intra- and intersubject variability that make it difficult to identify standard reference levels. Variable levels of candidate salivary biomarkers have been reported in various studies, making it difficult to determine the reference level. The two most commonly studied OSCC salivary biomarkers are IL-6 and IL-8, which detect OSCC and monitor chronic periodontitis and oral lichen planus. Levels of IL-6 and IL-8 have been shown to vary in OSCC patients in different studies (Rhodus, Ho, Miller, Myers, & Ondrey, 2005; Sahebjamee, Eslami, Atarbashimoghadam, & Sarafnejad, 2008; St. John et al., 2004). When reference levels vary so much, it becomes difficult to determine the range of IL-6 and IL-8 towards potential indication and confirmation of OSCC. Furthermore, intra- and intersubject variability in salivary proteome and transcriptome have been reported in OSCC (Jehmlich et al., 2013; Thomas et al., 2009). This variation in the levels could be attributed to inherent biological variations within different individuals and groups as well as to the variability in the processing methods used (Fraser, 2004). Demographic attributes of the subjects, including age, sex, ethnic background, geographic locations, dietary habits, medications, etc.

contribute to further inherent biological variation, which creates difficulty in determining reference levels of salivary biomarkers in clinical laboratories (Fraser, 2004). Other medical conditions could also serve as confounding factors, thereby making it difficult to identify cancer-associated differential levels of biomarkers. For example, endothein-1, a potential OSCC biomarker, was found to be elevated in chronic heart failure or upper gastrointestinal diseases even in the absence of OSCC (Denver, Tzanidis, Martin, & Krum, 2000; Lam et al., 2004). Besides this, Denny et al. have reported an age-related reduction in the concentration of salivary analytes (Denny et al., 1991).

### 18.4.3 The need for further validation of salivary biomarkers

Ideally, a good clinical test should have the ability to significantly differentiate a diseased from nondiseased state irrespective of other confounding factors, i.e., it should be highly sensitive and specific. In the case of salivary biomarkers, the oral cavity is generally subjected to inflammation from a variety of other causes like dental plaque, infection, or mucocutaneous inflammatory diseases. The contribution of these oral conditions in varying levels of various biomarkers in the saliva is largely unknown, as the previous studies have mostly focused on the levels of biomarkers in cancer patients and healthy control irrespective of other inflammatory conditions that might be present (Denny et al., 1991). Therefore, if increased levels of biomarkers in the presence of inflammation is comparable to cancer patients, then it might lead to false positives and will greatly reduce the clinical value of that biomarker.

Levels of inflammatory interleukins such as IL-6, IL-8, IL-1B are significantly elevated in periodontitis and oral lichen planus patients who did not have cancer. Cheng et al. compared the levels of IL-6 and IL-8 in OSCC as well as chronic periodontitis or lichen planus, and found that the levels of IL-6 were significantly elevated in OSCC than in patients with chronic periodontitis; whereas, IL-8 was marginally higher than in the healthy control. They concluded that salivary IL-6 could be a useful biomarker for OSCC detection (Wu et al., 2010). Indeed, inflammation is the initial step to neoplasia; therefore, there is a great need to validate the potential biomarkers that could differentiate cancer from healthy individual irrespective of other inflammatory condition or there should be a significant difference in the reference levels of the biomarkers in order to establish a panel of reliable salivary biomarkers.

## 18.5 Conclusion

The ultimate goal of molecular diagnostics is to detect the tumors at their earliest stage, which requires high sensitivity and specificity. Monitoring cancer treatment, recurrence, and further detecting residual disease require multiple sampling and/or biopsies, which is sometimes not feasible for various technical reasons. Thus, noninvasive screening methodologies equipped with high accuracy may resolve this problem. In this regard, saliva has emerged as a potential body-fluid for use as a clinical specimen in the development of POCDs. In the last decade, numerous technologies have emerged to evaluate various components as cancer biomarkers and have tremendous potential. As stated previously, cancer salivary biomarkers discovery and validation have made great progress towards clinical application. With the advancement of biosensor detection systems, some of them have even been translated into POC devices. However, to generate the universal reference range of salivary biomarkers and reduce interlaboratory variability, standardization of specimen collection, storage, processing, methodology, and reporting of results is required. Nevertheless, the use of saliva as a potential clinical specimen for the identification of biomarkers holds great potential in cancer diagnostics.

## Acknowledgment

This work was supported by an extramural research grant (EMR/2016/005009) from the Science and Engineering Research Board and from Madhya Pradesh Council of Science & Technology, Bhopal (A/RD/RP-2/2015-16/50) to Ashok Kumar. Ritu Pandey is a recipient of DBT fellowship DBT/2018/AIIMS-B/1153. Neha Arya would like to acknowledge DST INSPIRE Faculty Fellowship (IFA15-LSBM 148).

## References

Aas, J. A., Paster, B. J., Stokes, L. N., Olsen, I., & Dewhirst, F. E. (2005). Defining the normal bacterial flora of the oral cavity. *Journal of Clinical Microbiology*, *43*(11), 5721–5732. Available from https://doi.org/10.1128/JCM.43.11.5721-5732.2005.

Ahmad, P., Slavik, M., Trachtova, K., Gablo, N. A., Kazda, T., Gurin, D., ... Slaby, O. (2020). Salivary microRNAs identified by small RNA sequencing as potential predictors of response to intensity-modulated radiotherapy in head and neck cancer patients. *Cellular Oncology*, *43*(3), 505–511. Available from https://doi.org/10.1007/s13402-020-00507-7.

Alemar, B., Izetti, P., Gregório, C., Macedo, G. S., Castro, M. A., Osvaldt, A. B., ... Ashton-Prolla, P. (2016). miRNA-21 and miRNA-34a Are Potential Minimally Invasive Biomarkers for the Diagnosis of Pancreatic Ductal Adenocarcinoma. *Pancreas*, *45*(1), 84–92. Available from https://doi.org/10.1097/MPA.0000000000000383, PMID: 26262588.

Amenábar, J. M., Da Silva, B. M., & Punyadeera, C. (2020). Salivary protein biomarkers for head and neck cancer. *Expert Review of Molecular Diagnostics*, *20*(3), 305–313. Available from https://doi.org/10.1080/14737159.2020.1722102.

Bakri, M. M., Hussaini, H. M., Holmes, A., Cannon, R. D., & Rich, A. M. (2010). Revisiting the association between candidal infection and carcinoma, particularly oral squamous cell carcinoma. *Journal of Oral Microbiology*, *2*(2010). Available from https://doi.org/10.3402/jom.v2i0.5780.

Bik, E. M., Long, C. D., Armitage, G. C., Loomer, P., Emerson, J., Mongodin, E. F., ... Relman, D. A. (2010). Bacterial diversity in the oral cavity of 10 healthy individuals. *ISME Journal*, *4*(8), 962–974. Available from https://doi.org/10.1038/ismej.2010.30.

Brinkmann, O., Kastratovic, D. A., Dimitrijevic, M. V., Konstantinovic, V. S., Jelovac, D. B., Antic, J., ... Wong, D. T. (2011). Oral squamous cell carcinoma detection by salivary biomarkers in a Serbian population. *Oral Oncology*, *47*(1), 51–55. Available from https://doi.org/10.1016/j.oraloncology.2010.10.009.

Brooks, J., Cairns, P., & Zeleniuch-Jacquotte, A. (2009). Promoter methylation and the detection of breast cancer. *Cancer Causes and Control*, *20*(9), 1539–1550. Available from https://doi.org/10.1007/s10552-009-9415-y.

Castagnola, M., Scarano, E., Passali, G. C., Messana, I., Cabras, T., Iavarone, F., ... Paludetti, G. (2017). Biomarkers e proteomica salivari: Prospettive future cliniche e diagnostiche. *Acta Otorhinolaryngologica Italica*, *37*(2), 94–101. Available from https://doi.org/10.14639/0392-100X-1598.

Chen, J., Zhang, J., Guo, Y., Li, J., Fu, F., Yang, H. H., & Chen, G. (2011). An ultrasensitive electrochemical biosensor for detection of DNA species related to oral cancer based on nuclease-assisted target recycling and amplification of DNAzyme. *Chemical Communications*, *47*(28), 8004–8006. Available from https://doi.org/10.1039/c1cc11929j.

Cristaldi, M., Mauceri, R., Di Fede, O., Giuliana, G., Campisi, G., & Panzarella, V. (2019). Salivary biomarkers for oral squamous cell carcinoma diagnosis and follow-up: Current status and perspectives. *Frontiers in Psychology*, *10*. Available from https://doi.org/10.3389/fphys.2019.01476.

Damborský, P., Švitel, J., & Katrlík, J. (2016). Optical biosensors. *Essays in Biochemistry*, *60*(1), 91–100. Available from https://doi.org/10.1042/EBC20150010.

Denny, P. C., Denny, P. A., Klauser, D. K., Hong, S. H., Navazesh, M., & Tabak, L. A. (1991). Age-related changes in mucins from human whole saliva. *Journal of Dental Research*, *70*(10), 1320–1327. Available from https://doi.org/10.1177/00220345910700100201.

Denver, R., Tzanidis, A., Martin, P., & Krum, H. (2000). Salivary endothelin concentrations in the assessment of chronic heart failure. *Lancet*, *355*(9202), 468–469. Available from https://doi.org/10.1016/S0140-6736(00)82019-X.

Derruau, S., Robinet, J., Untereiner, V., Piot, O., Sockalingum, G. D., & Lorimier, S. (2020). Vibrational spectroscopy saliva profiling as biometric tool for disease diagnostics: A systematic literature. *Molecules (Basel, Switzerland)*, *25*(18). Available from https://doi.org/10.3390/molecules25184142.

Dewhirst, F. E., Chen, T., Izard, J., Paster, B. J., Tanner, A. C. R., Yu, W. H., ... Wade, W. G. (2010). The human oral microbiome. *Journal of Bacteriology*, *192*(19), 5002–5017. Available from https://doi.org/10.1128/JB.00542-10.

Dong, T., & Pires, N. M. M. (2017). Immunodetection of salivary biomarkers by an optical microfluidic biosensor with polyethylenimine-modified polythiophene-C70 organic photodetectors. *Biosensors and Bioelectronics*, *94*, 321–327. Available from https://doi.org/10.1016/j.bios.2017.03.005.

Du, J., & Zhang, L. (2017). Analysis of salivary microrna expression profiles and identification of novel biomarkers in esophageal cancer. *Oncology Letters*, *14*, 1387–1394.

Duz, M. B., Karatas, O. F., Guzel, E., Turgut, N. F., Yilmaz, M., Creighton, C. J., & Ozen, M. (2016). Identification of miR-139–5p as a saliva biomarker for tongue squamous cell carcinoma: A pilot study. *Cellular Oncology*, *39*(2), 187–193. Available from https://doi.org/10.1007/s13402-015-0259-z.

Elashoff, D., Zhou, H., Reiss, J., Wang, J., Xiao, H., Henson, B., ... Le, A. (2012). Prevalidation of Salivary Biomarkers for Oral Cancer Detection. *Cancer Epidemiology, Biomarkers and Prevention*, *21*, 664 LP – 672.

Fadhil, R. S., Wei, M. Q., Nikolarakos, D., Good, D., & Nair, R. G. (2020). Salivary microRNA miR-let-7a-5p and miR-3928 could be used as potential diagnostic bio-markers for head and neck squamous cell carcinoma. *PLoS One*, *15*(3). Available from https://doi.org/10.1371/journal.pone.0221779.

Fiorini, G. S., & Chiu, D. T. (2005). Disposable microfluidic devices: Fabrication, function, and application. *Biotechniques*, *38*(3), 429–446. Available from https://doi.org/10.2144/05383RV02.

Fischer, M., Triggs, G. J., & Krauss, T. F. (2016). Optical sensing of microbial life on surfaces. *Applied and Environmental Microbiology*, *82*(5), 1362–1371. Available from https://doi.org/10.1128/AEM.03001-15.

Fraser, C. G. (2004). Inherent biological variation and reference values. *Clinical Chemistry and Laboratory Medicine*, *42*(7), 758–764. Available from https://doi.org/10.1515/CCLM.2004.128.

Freer, P. E. (2015). Mammographic breast density: Impact on breast cancer risk and implications for screening. *Radiographics: a Review Publication of the Radiological Society of North America, Inc*, *35*(2), 302–315. Available from https://doi.org/10.1148/rg.352140106.

Fujita, Y., Kuwano, K., Ochiya, T., & Takeshita, F. (2014). The impact of extracellular vesicle-encapsulated circulating microRNAs in lung cancer research. *BioMed Research International*, 2014. Available from https://doi.org/10.1155/2014/486413.

Gablo, N. A., Prochazka, V., Kala, Z., Slaby, O., & Kiss, I. (2019). Cell-free microRNAs as non-invasive diagnostic and prognostic biomarkers in pancreatic cancer. *Current Genomics*, *20*(8), 569–580. Available from https://doi.org/10.2174/1389202921666191217095017.

Gau, V., & Wong, D. (2007). Oral fluid nanosensor test (OFNASET) with advanced electrochemical-based molecular analysis platform. *Annals of the New York Academy of Sciences* (Vol. 1098, pp. 401–410). Blackwell Publishing Inc. Available from https://doi.org/10.1196/annals.1384.005.

Goossens, N., Nakagawa, S., Sun, X., & Hoshida, Y. (2015). Cancer biomarker discovery and validation. *Translational Cancer Research*, *4*(3), 256–269. Available from https://doi.org/10.3978/j.issn.2218-676X.2015.06.04.

Greither, T., Vorwerk, F., Kappler, M., Bache, M., Taubert, H., Kuhnt, T., ... Eckert, A. W. (2017). Salivary miR-93 and miR-200a as post-radiotherapy biomarkers in head and neck squamous cell carcinoma. *Oncology Reports*, *38*(2), 1268–1275. Available from https://doi.org/10.3892/or.2017.5764.

Guerrero-Preston, R., Godoy-Vitorino, F., Jedlicka, A., Rodríguez-Hilario, A., González, H., Bondy, J., ... Sidransky, D. (2016). 16S rRNA amplicon sequencing identifies microbiota associated with oral cancer, Human Papilloma Virus infection and surgical treatment. *Oncotarget*, *7*(32), 51320–51334. Available from https://doi.org/10.18632/oncotarget.9710.

Hajishengallis, G., & Lamont, R. J. (2012). Beyond the red complex and into more complexity: The polymicrobial synergy and dysbiosis (PSD) model of periodontal disease etiology. *Molecular Oral Microbiology*, *27*(6), 409–419. Available from https://doi.org/10.1111/j.2041-1014.2012.00663.x.

Herr, A. E., Hatch, A. V., Giannobile, W. V., Throckmorton, D. J., Tran, H. M., Brennan, J. S., & Singh, A. K. (2007). Integrated microfluidic platform for oral diagnostics. *Annals of the New York Academy of Sciences*, *1098*, 362–374.

Hu, S., Arellano, M., Boontheung, P., Wang, J., Zhou, H., Jiang, J., ... Wong, D. T. (2008). Salivary proteomics for oral cancer biomarker discovery. *Clinical Cancer Research*, *14*(19), 6246–6252. Available from https://doi.org/10.1158/1078-0432.CCR-07-5037.

Hu, Y., Chang, Y., Chai, Y., & Yuan, R. (2018). An electrochemical biosensor for detection of DNA species related to oral cancer based on a particular host-guest recognition-assisted strategy for signal tag in situ. *Journal of the Electrochemical Society*, *165*(7), B289–B295. Available from https://doi.org/10.1149/2.0851807jes.

Huck, C. W., Ozaki, Y., & Huck-Pezzei, V. A. (2016). Critical review upon the role and potential of fluorescence and near-infrared imaging and absorption spectroscopy in cancer related cells, serum, saliva, urine and tissue analysis. *Current Medicinal Chemistry*, *23*(27), 3052–3077. Available from https://doi.org/10.2174/0929867323666160607110507.

Jaychandran, S., Meenapriya, P., & Ganesan, S. (2016). Raman spectroscopic analysis of blood, urine, saliva and tissue of oral potentially malignant disorders and malignancy – A diagnostic study. *International Journal of Oral and Craniofacial Science*, *2*(1), 011–014. Available from https://doi.org/10.17352/2455-4634.000013.

Jehmlich, N., Dinh, K. H. D., Gesell-Salazar, M., Hammer, E., Steil, L., Dhople, V. M., ... Völker, U. (2013). Quantitative analysis of the intra- and inter-subject variability of the whole salivary proteome. *Journal of Periodontal Research*, *48*(3), 392–403. Available from https://doi.org/10.1111/jre.12025.

Jinno, H., Murata, T., Sunamura, M., & Sugimoto, M. (2015). Investigation of potential salivary biomarkers for the diagnosis of breast cancer. *Journal of Clinical Oncology*, *33*, 145.

Kaczor-Urbanowicz, K. E., Martin Carreras-Presas, C., Aro, K., Tu, M., Garcia-Godoy, F., & Wong, D. T. W. (2017). Saliva diagnostics – Current views and directions. *Experimental Biology and Medicine*, *242*(5), 459–472. Available from https://doi.org/10.1177/1535370216681550.

Kawahara, R., Bollinger, J. G., Rivera, C., Ribeiro, A. C. P., Brandão, T. B., Leme, A. F. P., & Maccoss, M. J. (2016). A targeted proteomic strategy for the measurement of oral cancer candidate biomarkers in human saliva. *Proteomics*, *16*, 159–173.

Lallemant, B., Evrard, A., Combescure, C., Chapuis, H., Chambon, G., Raynal, C., ... Brouillet, J. P. (2009). Clinical relevance of nine transcriptional molecular markers for the diagnosis of head and neck squamous cell carcinoma in tissue and saliva rinse. *BMC Cancer*, *9*, 370. Available from https://doi.org/10.1186/1471-2407-9-370.

Lam, H. C., Lo, G. H., Lee, J. K., Lu, C. C., Chu, C. H., Sun, C. C., ... Wang, M. C. (2004). Salivary immunoreactive endothelin in patients with upper gastrointestinal diseases. *Journal of Cardiovascular Pharmacology*, *44*(Issue 1), S413–S417. Available from https://doi.org/10.1097/01.fjc.0000166288.87571.ae.

Lee, Y. H., & Wong, D. T. (2009). Saliva: An emerging biofluid for early detection of diseases. *American Journal of Dentistry*, *22*(4), 241–248.

Lee, Y. H., Zhou, H., Reiss, J. K., Yan, X., Zhang, L., Chia, D., & Wong, D. T. W. (2011). Direct saliva transcriptome analysis. *Clinical Chemistry*, *57*(9), 1295–1302. Available from https://doi.org/10.1373/clinchem.2010.159210.

Li, Y., St., John, M. A. R., Zhou, X., Kim, Y., Sinha, U., Jordan, R. C. K., ... Wong, D. T. (2004). Salivary transcriptome diagnostics for oral cancer detection. *Clinical Cancer Research*, *10*(24), 8442–8450. Available from https://doi.org/10.1158/1078-0432.CCR-04-1167.

Liao, Z., Zhang, Y., Li, Y., Miao, Y., Gao, S., Lin, F., ... Geng, L. (2019). Microfluidic chip coupled with optical biosensors for simultaneous detection of multiple analytes: A review. *Biosensors and Bioelectronics*, *126*, 697–706. Available from https://doi.org/10.1016/j.bios.2018.11.032.

Lohavanichbutr, P., Zhang, Y., Wang, P., Gu, H., Nagana Gowda, G. A., Djukovic, D., ... Chen, C. (2018). Salivary metabolite profiling distinguishes patients with oral cavity squamous cell carcinoma from normal controls. *PLoS One*, *13*(9). Available from https://doi.org/10.1371/journal.pone.0204249.

Ma, R. N., Wang, L. L., Wang, H. F., Jia, L. P., Zhang, W., Shang, L., ... Wang, H. S. (2018). Highly sensitive ratiometric electrochemical DNA biosensor based on homogeneous exonuclease III-assisted target recycling amplification and one-step triggered dual-signal output. *Sensors and Actuators, B: Chemical*, *269*, 173–179. Available from https://doi.org/10.1016/j.snb.2018.04.143.

Machida, T., Tomofuji, T., Maruyama, T., Yoneda, T., Ekuni, D., Azuma, T., ... Tsutsumi, K. (2016). MIR 1246 and MIR-4644 in salivary exosome as potential biomarkers for pancreatobiliary tract cancer. *Oncology Reports*, *36*, 2375–2381.

Mager, D. L., Haffajee, A. D., Delvin, P. M., Norris, C. M., Posner, M. R., & Goodson, J. M. (2005). The salivary microbiota as a diagnostic indicator of oral cancer: A descriptive, non-randomized study of cancer-free and oral squamous cell carcinoma subjects. *Journal of Translational Medicine*, *3*. Available from https://doi.org/10.1186/1479-5876-3-27.

Majidi, M. R., Karami, P., Johari-Ahar, M., & Omidi, Y. (2016). Direct detection of tryptophan for rapid diagnosis of cancer cell metastasis competence by an ultra-sensitive and highly selective electrochemical biosensor. *Analytical Methods*, *8*(44), 7910–7919. Available from https://doi.org/10.1039/c6ay02103d.

Messana, I., Cabras, T., Iavarone, F., Vincenzoni, F., Urbani, A., & Castagnola, M. (2013). Unraveling the different proteomic platforms. *Journal of Separation Science*, *36*(1), 128–139. Available from https://doi.org/10.1002/jssc.201200830.

Mikkonen, J. J. W., Singh, S. P., Akhi, R., Salo, T., Lappalainen, R., González-Arriagada, W. A., ... Myllymaa, S. (2018). Potential role of nuclear magnetic resonance spectroscopy to identify salivary metabolite alterations in patients with head and neck cancer. *Oncology Letters, 16*(5), 6795–6800. Available from https://doi.org/10.3892/ol.2018.9419.

Nagler, R., Bahar, G., Shpitzer, T., & Feinmesser, R. (2006). Concomitant analysis of salivary tumor markers - A new diagnostic tool for oral cancer. *Clinical Cancer Research, 12*, 3979–3984.

Park, N. J., Zhou, H., Elashoff, D., Henson, B. S., Kastratovic, D. A., Abemayor, E., & Wong, D. T. (2009). Salivary microRNA: Discovery, characterization, and clinical utility for oral cancer detection. *Clinical Cancer Research, 15*(17), 5473–5477. Available from https://doi.org/10.1158/1078-0432.CCR-09-0736.

Perera, M., Al-Hebshi, N. N., Perera, I., Ipe, D., Ulett, G. C., Speicher, D. J., ... Johnson, N. W. (2017). A dysbiotic mycobiome dominated by candida albicans is identified within oral squamous-cell carcinomas. *Journal of Oral Microbiology, 9*. Available from https://doi.org/10.1080/20002297.2017.1385369.

Pushalkar, S., Mane, S. P., Ji, X., Li, Y., Evans, C., Crasta, O. R., ... Saxena, D. (2011). Microbial diversity in saliva of oral squamous cell carcinoma. *FEMS Immunology and Medical Microbiology, 61*(3), 269–277. Available from https://doi.org/10.1111/j.1574-695X.2010.00773.x.

Ramirez-Garcia, A., Rementeria, A., Aguirre-Urizar, J. M., Moragues, M. D., Antoran, A., Pellon, A., ... Hernando, F. L. (2016). Candida albicans and cancer: Can this yeast induce cancer development or progression? *Critical Reviews in Microbiology, 42*(2), 181–193. Available from https://doi.org/10.3109/1040841X.2014.913004.

Rapado-González., Majem., Álvarez-Castro., Díaz-Peña., Abalo., Suárez-Cabrera., ... Muinelo-Romay. (2019). A Novel Saliva-Based miRNA Signature for Colorectal Cancer Diagnosis. *Journal of Clinical Medicine, 8*, 2029.

Rhodus, N. L., Ho, V., Miller, C. S., Myers, S., & Ondrey, F. (2005). NF-κB dependent cytokine levels in saliva of patients with oral preneoplastic lesions and oral squamous cell carcinoma. *Cancer Detection and Prevention, 29*(1), 42–45. Available from https://doi.org/10.1016/j.cdp.2004.10.003.

Sahibzada, H. A., Khurshid, Z., Khan, R. S., Naseem, M., Siddique, K. M., Mali, M., & Zafar, M. S. (2017). Salivary IL-8, IL-6 and TNF-α as Potential Diagnostic Biomarkers for Oral Cancer. *Diagnostics, 7*, 21.

Sahebjamee, M., Eslami, M., Atarbashimoghadam, F., & Sarafnejad, A. (2008). Salivary concentration of TNFα, IL1α, IL6, and IL8 in oral squamous cell carcinoma. *Medicina Oral, Patologia Oral y Cirugia Bucal, 13*(5), 292–295.

Sanjay, S. T., Fu, G., Dou, M., Xu, F., Liu, R., Qi, H., & Li, X. (2015). Biomarker detection for disease diagnosis using cost-effective microfluidic platforms. *Analyst, 140*(21), 7062–7081. Available from https://doi.org/10.1039/c5an00780a.

Sazanov, A. A., Kiselyova, E. V., Zakharenko, A. A., Romanov, M. N., & Zaraysky, M. I. (2017). Plasma and saliva miR-21 expression in colorectal cancer patients. *Journal of Applied Genetics, 58*, 231–237.

Schafer, C. A., Schafer, J. J., Yakob, M., Lima, P., Camargo, P., & Wong, D. T. W. (2014). Saliva diagnostics: Utilizing oral fluids to determine health status. *Monographs in Oral Science, 24*, 88–98. Available from https://doi.org/10.1159/000358791.

Schipper, R. G., Silletti, E., & Vingerhoeds, M. H. (2007). Saliva as research material: Biochemical, physicochemical and practical aspects. *Archives of Oral Biology, 52*(12), 1114–1135. Available from https://doi.org/10.1016/j.archoralbio.2007.06.009.

Siqueira, W. L., Salih, E., Wan, D. L., Helmerhorst, E. J., & Oppenheim, F. G. (2008). Proteome of human minor salivary gland secretion. *Journal of Dental Research, 87*(5), 445–450. Available from https://doi.org/10.1177/154405910808700508.

Song, X., Yang, X., Narayanan, R., Shankar, V., Ethiraj, S., Wang, X., ... Zare, R. N. (2020). Oral squamous cell carcinoma diagnosed from saliva metabolic profiling. *Proceedings of the National Academy of Sciences of the United States of America, 117*(28), 16167–16173. Available from https://doi.org/10.1073/pnas.2001395117.

Sridharan, G., Ramani, P., Patankar, S., & Vijayaraghavan, R. (2019). Evaluation of salivary metabolomics in oral leukoplakia and oral squamous cell carcinoma. *Journal of Oral Pathology and Medicine, 48*(4), 299–306. Available from https://doi.org/10.1111/jop.12835.

St. John, M. A. R., Li, Y., Zhou, X., Denny, P., Ho, C. M., Montemagno, C., ... Wong, D. T. W. (2004). Interleukin 6 and interleukin 8 as potential biomarkers for oral cavity and oropharyngeal squamous cell carcinoma. *Archives of Otolaryngology—Head and Neck Surgery, 130*(8), 929–935. Available from https://doi.org/10.1001/archotol.130.8.929.

Strimbu, K., & Tavel, J. A. (2010). What are biomarkers? *Current Opinion in HIV and AIDS, 5*(6), 463–466. Available from https://doi.org/10.1097/COH.0b013e32833ed177.

Sugimoto, M. (2020). Salivary metabolomics for cancer detection. *Expert Review of Proteomics*. Available from https://doi.org/10.1080/14789450.2020.1846524.

Sugimoto, M., Wong, D. T., Hirayama, A., Soga, T., & Tomita, M. (2010). Capillary electrophoresis mass spectrometry-based saliva metabolomics identified oral, breast and pancreatic cancer-specific profiles. *Metabolomics: Official Journal of the Metabolomic Society, 6*(1), 78–95. Available from https://doi.org/10.1007/s11306-009-0178-y.

Tajima, I., Asami, O., & Sugiura, E. (1998). Monitor of antibodies in human saliva using a piezoelectric quartz crystal biosensor. *Analytica Chimica Acta, 365*(1–3), 147–149. Available from https://doi.org/10.1016/S0003-2670(97)00596-5.

Takayama, T., Tsutsui, H., Shimizu, I., Toyama, T., Yoshimoto, N., Endo, Y., ... Mizuno, H. (2016). Diagnostic approach to breast cancer patients based on target metabolomics in saliva by liquid chromatography with tandem mass spectrometry. *Clinica Chimica Acta; International Journal of Clinical Chemistry, 452*, 18–26.

Tan, W., Sabet, L., Li, Y., Yu, T., Klokkevold, P. R., Wong, D. T., & Ho, C. M. (2008). Optical protein sensor for detecting cancer markers in saliva. *Biosensors and Bioelectronics, 24*(2), 266–271. Available from https://doi.org/10.1016/j.bios.2008.03.037.

Thomas, M. V., Branscum, A., Miller, C. S., Ebersole, J., Al-Sabbagh, M., & Schuster, J. L. (2009). Within-subject variability in repeated measures of salivary analytes in healthy adults. *Journal of Periodontology, 80*(7), 1146–1153. Available from https://doi.org/10.1902/jop.2009.080654.

Tian, T., Liu, H., Li, L., Yu, J., Ge, S., Song, X., & Yan, M. (2017). Paper-based biosensor for noninvasive detection of epidermal growth factor receptor mutations in non-small cell lung cancer patients. *Sensors and Actuators, B: Chemical*, *251*, 440–445. Available from https://doi.org/10.1016/j.snb.2017.05.082.

Torrente-Rodríguez, R. M., Campuzano, S., Ruiz-Valdepeñas Montiel, V., Gamella, M., & Pingarrón, J. M. (2016). Electrochemical bioplatforms for the simultaneous determination of interleukin (IL)-8 mRNA and IL-8 protein oral cancer biomarkers in raw saliva. *Biosensors and Bioelectronics*, *77*, 543–548. Available from https://doi.org/10.1016/j.bios.2015.10.016.

Torrisani, J., Bournet, B., Du Rieu, M. C., Bouisson, M., Souque, A., Escourrou, J., ... Cordelier, P. (2009). Let-7 microRNA transfer in pancreatic cancer-derived cells inhibits in vitro cell proliferation but fails to alter tumor progression. *Human Gene Therapy*, *20*, 831–844.

Tuli, L., & Ressom, H. W. (2009). LC-MS based detection of differential protein expression. *Journal of Proteomics and Bioinformatics*, *2*(10), 416–438. Available from https://doi.org/10.4172/jpb.1000102.

Urdea, M., Penny, L. A., Olmsted, S. S., Giovanni, M. Y., Kaspar, P., Shepherd, A., ... Hay Burgess, D. C. (2006). Requirements for high impact diagnostics in the developing world. *Nature*, *444*, 73–79. Available from https://doi.org/10.1038/nature05448.

Verma, S., & Singh, S. P. (2019). Non-invasive oral cancer detection from saliva using zinc oxide-reduced graphene oxide nanocomposite based bioelectrode. *MRS Communications*, *9*(4), 1227–1234. Available from https://doi.org/10.1557/mrc.2019.138.

Vesty, A., Gear, K., Biswas, K., Radcliff, F. J., Taylor, M. W., & Douglas, R. G. (2018). Microbial and inflammatory-based salivary biomarkers of head and neck squamous cell carcinoma. *Clinical and Experimental Dental Research*, *4*(6), 255–262. Available from https://doi.org/10.1002/cre2.139.

Wang, X., Kaczor-Urbanowicz, K. E., & Wong, D. T. W. (2017). Salivary biomarkers in cancer detection. *Medical Oncology*, *34*(1). Available from https://doi.org/10.1007/s12032-016-0863-4.

Wang, Z. W., Zhang, J., Guo, Y., Wu, X. Y., Yang, W. J., Xu, L. J., ... Fu, F. F. (2013). A novel electrically magnetic-controllable electrochemical biosensor for the ultra sensitive and specific detection of attomolar level oral cancer-related microRNA. *Biosensors and Bioelectronics*, *45*(1), 108–113. Available from https://doi.org/10.1016/j.bios.2013.02.007.

Wei, F., Patel, P., Liao, W., Chaudhry, K., Zhang, L., Arellano-Garcia, M., ... Wong, D. T. (2009). Electrochemical sensor for multiplex biomarkers detection. *Clinical Cancer Research*, *15*(13), 4446–4452. Available from https://doi.org/10.1158/1078-0432.CCR-09-0050.

Wiklund, E. D., Gao, S., Hulf, T., Sibbritt, T., Nair, S., Costea, D. E., ... Kjems, J. (2011). MicroRNA alterations and associated aberrant DNA methylation patterns across multiple sample types in oral squamous cell carcinoma. *PLoS One*, *6*(11). Available from https://doi.org/10.1371/journal.pone.0027840.

Wolf, A., Moissl-Eichinger, C., Perras, A., Koskinen, K., Tomazic, P. V., & Thurnher, D. (2017). The salivary microbiome as an indicator of carcinogenesis in patients with oropharyngeal squamous cell carcinoma: A pilot study. *Scientific Reports*, *7*(1). Available from https://doi.org/10.1038/s41598-017-06361-2.

Wu, J. Y., Yi, C., Chung, H. R., Wang, D. J., Chang, W. C., Lee, S. Y., ... Yang, W. C. V. (2010). Potential biomarkers in saliva for oral squamous cell carcinoma. *Oral Oncology*, *46*(4), 226–231. Available from https://doi.org/10.1016/j.oraloncology.2010.01.007.

Xiao, H., Zhang, L., Zhou, H., Lee, J. M., Garon, E. B., & Wong, D. T. W. (2012). Proteomic analysis of human saliva from lung cancer patients using two-dimensional difference gel electrophoresis and mass spectrometry. *Molecular & Cellular Proteomics*, *11*, M111.012112.

Xiao, H., Zhang, Y., Kim, Y., Kim, S., Kim, J. J., Kim, K. M., ... Wong, D. T. W. (2016). Differential Proteomic Analysis of Human Saliva using Tandem Mass Tags Quantification for Gastric Cancer Detection. *Scientific Reports*, *6*, 22165.

Xie, Z., Chen, G., Zhang, X., Li, D., Huang, J., Yang, C., ... Gong, B. (2013). Salivary MicroRNAs as Promising Biomarkers for Detection of Esophageal Cancer. *PLoS ONE*, *8*, e57502.

Yang, Y., Zhang, R., Li, Z., Mei, L., Wan, S., Ding, H., ... Zhou, B. (2020). Discovery of highly potent, selective, and orally efficacious p300/CBP histone acetyltransferases inhibitors. *Journal of Medicinal Chemistry*, *63*(3), 1337–1360. Available from https://doi.org/10.1021/acs.jmedchem.9b01721.

Zahran, F., Ghalwash, D., Shaker, O., Al-Johani, K., & Scully, C. (2015). Salivary microRNAs in oral cancer. *Oral Diseases*, *21*(6), 739–747. Available from https://doi.org/10.1111/odi.12340.

Zhang, L., Xiao, H., Karlan, S., Zhou, H., Gross, J., Elashoff, D., ... Karlan, B. (2010a). Discovery and preclinical validation of salivary transcriptomic and proteomic biomarkers for the non- invasive detection of breast cancer. *PLoS ONE*, *5*, e15573.

Zhang, L., Farrell, J. J., Zhou, H., Elashoff, D., Akin, D., Park, N. H., ... Wong, D. T. (2010a). Salivary Transcriptomic Biomarkers for Detection of Resectable Pancreatic Cancer. *Gastroenterology*, *138*, 949–957.

Zhang, L., Xiao, H., Zhou, H., Santiago, S., Lee, J. M., Garon, E. B., ... Akin, D. (2012). Development of transcriptomic biomarker signature in human saliva to detect lung cancer. *Cellular and Molecular Life Sciences*, *69*, 3341–3350.

Zhong, L., Cheng, F., Lu, X., Duan, Y., & Wang, X. (2016). Untargeted saliva metabonomics study of breast cancer based on ultra performance liquid chromatography coupled to mass spectrometry with HILIC and RPLC separations. *Talanta*, *158*, 351–360. Available from https://doi.org/10.1016/j.talanta.2016.04.049.

Zhou, P., Lu, F., Wang, J., Wang, K., Liu, B., Li, N., & Tang, B. (2020). A portable point-of-care testing system to diagnose lung cancer through the detection of exosomal miRNA in urine and saliva. *Chemical Communications*, *56*(63), 8968–8971. Available from https://doi.org/10.1039/d0cc03180a.

Zimmermann, B. G., & Wong, D. T. (2008). Salivary mRNA targets for cancer diagnostics. *Oral Oncology*, *44*(5), 425–429. Available from https://doi.org/10.1016/j.oraloncology.2007.09.009.

Chapter 19

# Two-dimensional nanomaterials for cancer application

Tripti Rimza[1], Shiv Singh[2] and Pradip Kumar[1]
[1]Integrated Approach for Design and Product Development Division, CSIR-Advanced Materials and Processes Research Institute (CSIR-AMPRI), Bhopal, India
[2]Lightweight Metallic Materials Division, CSIR-Advanced Materials and Processes Research Institute (CSIR-AMPRI), Bhopal, India

## 19.1 Introduction

2D materials are defined to be 1-atom thick sheets isolated from bulk as free-standing layers and do not need a substrate to exist. Often, layers with up to 10 atoms thick are also considered as 2D sheets (Mas-Ballesté, Gómez-Navarro, Gómez-Herrero, & Zamora, 2011). 2D-layered materials received a significant research interest after the identification of graphene in 2004. Graphene, a single layer of carbon atoms, was first identified by Geim and Novoselov using the simple adhesive tape method, for which they won the Nobel Prize in 2010 (Geim & Novoselov, 2007). Graphene is a single layer of $sp^2$-hybridized carbon atoms arranged in a hexagonal structure. In graphite, carbon layers are stacked through weak van der Waals forces. A single layer of graphite (graphene) isolation was possible due to weak interlayer bonding. Initial graphene identification gained interest due to linear energy-momentum relation for low energies, half-integer quantum hall effect, and ultrahigh carrier mobility. The ultimate properties are thermal conductivity ($\sim$5000 W/mK), high Young's modulus ($\sim$1100 GPa), electron mobility (200,000 $cm^2/V \cdot s$), surface area (2630 $m^2/g$), and optically transparent (Ferrari et al., 2015). These unique properties of graphene opened a lot of opportunities in opto-electronics, thermal management, EMI shielding, biosensor, drug delivery, energy storage, and aerospace applications (Cho et al., 2015, 2017; Kumar et al., 2015, 2016).

2D materials basically possess neutral atoms, covalently or ionically bonded with their neighbors in a plane/layer while the layers are held together by van der Waals coupling along the third axis. The weak van der Waals energies ($\sim$40–70 meV) allow the facile exfoliation of these layers from their parent bulk material (Nicolosi, Chhowalla, Kanatzidis, Strano, & Coleman, 2013). Since the re-discovery of graphene in 2004, many other 2D nanomaterials like 2D dichalcogenide, MXenes, borophene, and oxides, were explored for many potential applications. These 2D nanomaterials included insulators (e.g., h-BN), semiconductors (e.g., MoS2), and conductors (e.g., graphene) (Butler et al., 2013). In a layered crystal, the electronic wave function exceeds three dimensions while in the 2D layers, there is strong quantum confinement of electrons with their wave functions constrained to adopt a 2D form. This imparts the 2D materials with new electronic band structures and with unique and fascinating physical/electrical/electronic properties. The above-mentioned interesting properties of graphene has been usefully exploited for technological development in the next generation devices in the areas of electronics, optoelectronics, energy storage applications, etc. The disadvantage of graphene lies in the fact that it lacks an intrinsic band gap making it unsuitable for certain logic applications like transistors requiring a high ON/OFF ratio. Bandgap engineering, like functionalization, doping, introducing defects, etc., has been adapted to open a bandgap in graphene without considerable modification of the pristine sheet. Unfortunately, during such processes, the conductivity of graphene reduces drastically.

## 19.2 Synthesis of two-dimensional nanomaterials

Similar to other nanomaterials, 2D nanomaterials can be synthesized using both top-down and bottom-up approaches. The process of thinning-down the bulk materials to a few layers or single layer is known as the top-down method. The process

Biosensor Based Advanced Cancer Diagnostics. DOI: https://doi.org/10.1016/B978-0-12-823424-2.00025-9

of thinning down the bulk to a few layers is called exfoliation. The weak van der Waals energies ( ~40–70 meV) allows the facile exfoliation of these layers from their parent bulk material (Nicolosi et al., 2013). The exfoliated single-layer sheets will have a high surface area, which is necessary for active and catalytic activity. Additionally, the strong layer thickness-dependent electronic properties emphasize the necessity for developing strategies to produce reliable and high-quality ultrathin sheets of these 2D materials from their bulk counterpart. The success of exfoliation is gauged by the layer thickness, lateral dimensions, size dispersity, crystal quality, etc. The synthesis of 2D nanosheets is mainly based on two major approaches, namely, top-down and bottom-up strategies. The nanosheets are exfoliated from the 3D bulk by mechanical or chemical methods in the former. In the latter case, they are prepared to start from the basic building blocks, which upon chemical reactions complete the formation of covalently linked large area 2D sheets. Some of the common methods relevant to the bottom-up approach are chemical vapor deposition (CVD), pulsed laser deposition (PLD), a sputtering method, atomic layer deposition (ALD), and physical vapor deposition (PVD).

### 19.2.1 Mechanical exfoliation

Over the past decade, many methods have been attempted to exfoliate layered solids to monolayer thickness. Such exfoliation resulted in high aspect ratio flakes with large surface areas, enough to process them for device applications and other surface-related activities.

Mechanical exfoliation of graphene from graphite using adhesive tape, was the first one to be attempted has allowed the production of flakes in which many novel properties were discovered (Mas-Ballesté et al., 2011; Novoselov et al., 2004). The success observed in graphene has allowed this method to continue to be the primary route for producing ultra-thin flakes. Due to the strong in-plane and weak out of plane bonding, layered materials can be sheared parallel or expanded normal to the in-plane direction. This can overcome the van der Waals forces and results in a process called exfoliation, producing nanometer or even atomically thin sheets or nanosheets. In mechanical exfoliation, the competition between the inter sheet van der Waals forces in a layered solid and that between the substrate and the outermost sheet plays a crucial role in efficient thinning of the solids (Huang et al., 2015). Typically, a well cleaned SiO2/Si substrate is used for transferring the flakes. The adhesive tape is contacted with the bulk crystal and then removed such that multilayered flakes are transferred onto the tape. During the few times of the repetitive process of contacting the flake-loaded tape with fresh tape, the flake thickness reduces with each contact. After considerable thinning of the flakes, the adhesive tape is brought in contact with the substrate and peeled off, leaving the high-quality ultra-thin single-crystal flakes on the substrate, suitable for fundamental studies. Many materials like MoS2, ReS2, WSe2, TaS2, TaSe2, and BP have been successfully exfoliated into thin sheets using this method and have been employed for developing FETs, phototransistors, etc. (Du, Liu, Deng, & Ye, 2014; Radisavljevic, Radenovic, Brivio, Giacometti, & Kis, 2011; Rahman, Davey, & Qiao, 2017). One main disadvantage of this method is that it leaves contaminants from the adhesive tape. Some studies have introduced an additional step where the substrates with flakes are dipped in acetone for about 5 min, to remove the residues of the tape (Castellanos-Gomez et al., 2012)

### 19.2.2 Liquid phase exfoliation

Liquid phase exfoliation (LPE) of layered materials, where the bulk layered solids are exfoliated in liquids mostly under ultrasonication, is a highly promising route for large scale production of 2D nanosheets. LPE involves three steps: (1) dispersion of the bulk the material in a liquid medium, (2) exfoliation of the nanosheets by the use of ultrasonic waves, and (3) purification and extraction of the exfoliated sheets. The LPE method permits hybrids and composites by simple mixing of the dispersions of the two materials (Yang et al., 2017). Large-area thin films can also be formed by coating the resultant dispersions by spray coating, inkjet or screen printing, spin coating, doctor blading, etc.

There are various ways to carry out LPE-involving oxidation, surface passivation by solvents, and use of intercalation compounds. One of the methods applied much earlier was to oxidize the layered crystals and then disperse them in suitable solvents. For example, graphene was exfoliated from graphite by ultrasonication using Brodie's, Staudenmaier's, or Hummer's method (Hernandez et al., 2008; Nicolosi et al., 2013). Here, graphite was treated with sulfuric acid and potassium permanganate resulting in the addition of hydroxyl and epoxy groups. This caused graphene sheets to turn hydrophilic allowing easy intercalation of water, leading to well-exfoliated graphene oxide (GO) layers. The sheets were stabilized against reaggregation due to the negative surface charge, yielding a concentration of up to 1 mg/mL. But this procedure disrupts the electronic structure of graphene where the GO produced an exhibited loss of conductivity. The functional groups were then removed by reducing the GO to produce reduced graphene oxide (rGO). The disadvantage of this method is that a large number of holes and sp3 defects are produced during the oxidation and the removal of functional groups. Apart from the tedious multiple steps involved, toxic gases like $NO_2$, $N_2O_4$, etc. are generated, posing serious safety issues. Additionally, interfering hetero-atomic species like sulfur (S) and metal-containing impurities like K, Mn, etc. can get covalently attached to the nanosheets, altering their properties (Shahzad, Kumar, Kim, Hong, & Koo, 2016; Shahzad et al., 2015).

LPE was modified by changing the liquid from high boiling point solvents like NMP (N-Methyl-2-pyrrolidone), DMF (Dimethylformamide), etc. to low boiling point solvents like methanol, ethanol, isopropyl alcohol, acetone, etc. This was because (1) solvents like NMP were toxic and could pose safety issues and (2) as they are high boiling points, they did not evaporate easily, thus remaining on the surface of the exfoliated flakes and hindering the use of these flakes in practical applications. Coleman group have also demonstrated exfoliation of graphene in low boiling point solvents like chloroform, isopropanol, acetone along with cyclohexanone, NMP and DMF. Concentrations of graphene for various solvents under different centrifugation rates are shown in Fig. 19.1 (O'Neill, Khan, Nirmalraj, Boland, & Coleman, 2011).

The electrochemical route is an efficient and mild method to intercalate and exfoliate layered materials into nanosheets and has been well explored by many groups (Ejigu, Kinloch, Prestat, & Dryfe, 2017; Liu et al., 2008; Parvez et al., 2014). The bulk material to be exfoliated is exposed to positive and negative cycles of voltage and, hence, subjecting it to oxidation and reduction reactions. This cycle of opposite polarity would facilitate intercalation of oppositely charged electrolyte species into bulk materials. The basic setup required for this kind of exfoliation is a two-electrode system consisting of the bulk-layered material acting as the anode, a counter electrode (mostly platinum) acting as the cathode, the electrolyte, and a suitable power supply. Depending on the potential applied, electrochemical exfoliation can be divided into two categories, i.e., anodic and cathodic exfoliation. In anodic exfoliation, a positive potential is applied to the electrode with the layered solid; following that, the anions of the electrolyte or that were created during the electrolysis would intercalate into the layered material under the applied electric field. In the case of cathodic exfoliation, negative potential is applied to the electrode with the bulk material, causing cations to intercalate into them. In the former, aqueous solutions of acids like H2SO4, H3PO4, etc., and some ionic liquid–water mixtures are used as electrolytes. In the case of the latter, organic solvents like DMSO (Dimethylsulfoxide) and propylene carbonate, containing lithium, alkylammonium salts, etc. are generally used as electrolytes. Use of aqueous electrolytes have made the anodic exfoliation more preferable than the cathodic exfoliation. Different groups have adopted different architectures where bulk materials are used as electrodes in some studies, while they are added to the electrolyte solution with platinum acting as the electrodes in other studies. Electrochemical exfoliation of graphite into graphene has been successfully demonstrated using ionic liquids and aqueous acids of $H_2SO_4$ or $H_3PO_4$ (Zeng et al., 2012).

A ball mill is a type of grinder used to grind and blend materials in industries, laboratory research on materials science, etc. Ball milling is a mechano-chemical exfoliation method combining mechanical forces followed by ultrasonication, which has been successfully employed to isolate nanolayers of graphene, h-BN, MoS2, WS2, etc. (Tao et al., 2017). Ball milling imparts two kinds of forces: shear, as well as compressive forces on the layered solids. Shear forces cleave off the layers from the top/bottom surface, whereas the compression forces peel off the layers from the edges. This method employs either high-energy or low-energy ball milling where the former is dominantly caused through plane fracture than the delamination of the flakes, and hence causing severe defects in the exfoliated layers. Control of the defect production is attempted by modifying the milling conditions and the interactions of correct guest molecules. Ball milling can be of different types: planetary, wet (liquid), etc. In liquid ball milling, the ball to bulk powder ratio, the size of the milling balls, milling duration, the type of liquid chosen, etc. affect the efficiency of exfoliation.

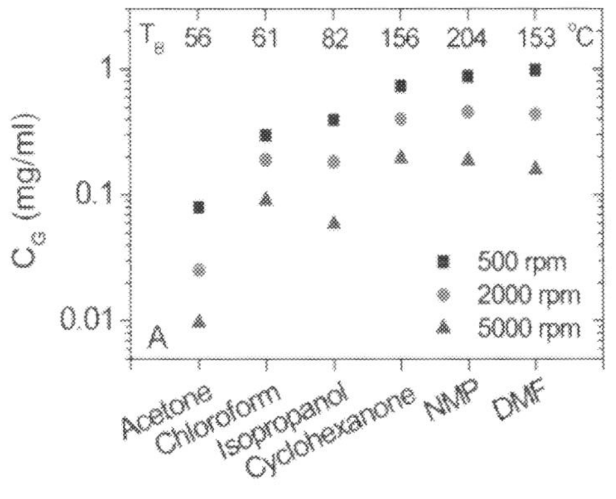


**FIGURE 19.1** Concentrations of exfoliated graphene for different solvents and centrifugation rate. *From O'Neill, A., Khan, U., Nirmalraj, P.N., Boland, J., Coleman, J.N. (2011). Graphene dispersion and exfoliation in low boiling point solvents.* The Journal of Physical Chemistry C, 115, *5422–5428.*

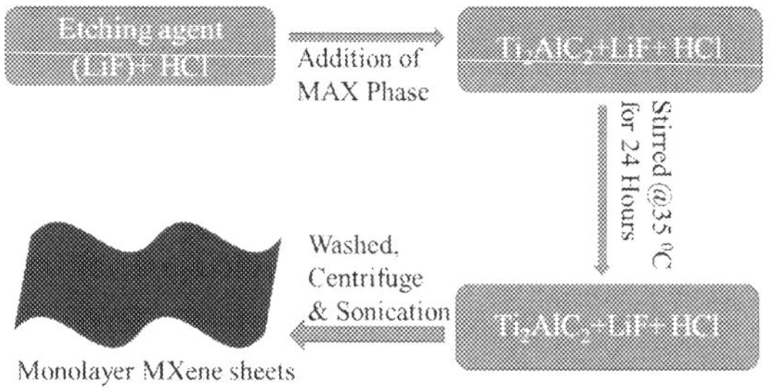


**FIGURE 19.2** A proposed fabrication method for 2D MXenes sheets.

Titanium carbide MXene was prepared in a typical procedure as follows (Rasid, Omar, Nazeri, A'ziz, & Szota, 2017): a mixture of titanium hydrate, aluminum, and graphite powder were thoroughly mixed in a required stoichiometric ratio using ball milling for about 20–80 h, and pelletized. The pellet was then sintered at a high temperature of around 1350oC in an argon atmosphere. The resulting pellet in MAX phase was pulverized, immersed, and stirred in hydrofluoric acid (HF). The mixture was then centrifuged and washed with DI (Deionised) water for about 20 h. Ding et al. has used different raw materials like TiC, Ti, and Al to synthesize T3AlC2 precursors (Ding et al., 2017). There are reports of synthesizing Ti3AlC2 using Ti, Al, and C; Ti, Al4C3, and C; or even TiO2, Al, and C. Though in the initial studies, pressure was applied on MAX powders to get fully dense samples, this step was removed in the later studies as the powder samples were even advantageous for further milling processes. In some studies, the A layer from the MAX phase were also etched using LiF/HCl combination. Under this condition, delamination of the MXene sheets can occur simultaneously due to the intercalation of Li in between the MXene layers. MXenes can also be intercalated with different organic compounds such as dimethyl sulfoxide (DMSO, tetraalkylammonium hydroxides, etc.). or inorganic salts, to exfoliate and produce the MXene sheets. A schematic showing the synthesis of MXene sheets from MAX phase is shown in Fig. 19.2.

## 19.3 Two-dimensional nanomaterials for cancer applications

### 19.3.1 Black phosphorous nanosheets

Black phosphorous is found bulk in nature, its single planar sheet is known as phosphorene. It has a strong utility in biomedicine due to the presence of a high surface area, tunable energy band gap, weak van der Waals force, and high P–P interaction (Wang, Yang et al., 2015). The use of phosphorene in biomedical refers to their main function as drug-loading vehicles for the tumored body parts. These drugs include biomolecules such as nucleic acids, genes, and antibodies (Wang, Yang et al., 2015; Wang and Yu, 2018). Apart from this, BP nanosheets have moderate carrier mobility, intrinsic photoacoustic properties, and excellent light-absorbing ability (Li et al., 2016).

#### *19.3.1.1 Photodynamic therapy*

Photodynamic therapy (PDT) is a kind of treatment that uses a photosensitizing drug, molecular oxygen, as well as suitable electromagnetic radiation to perform its operation. The whole process works in three steps; firstly, the photosensitizing substance is applied to the tumor-affected body through cream, liquid, or via intravenous injection. Afterwards, photosensitizing drugs need to be activated. But prior to this, there an incubation period of minutes, hours, or even days. After completing the incubation period, the targeted tumor area is exposed to electromagnetic radiation which leads to tumor tissue destruction. But the proper functioning of the mechanism needs the presence of properly activated molecular oxygen inside the body. Having treatment with suitable light photosensitizers generates various molecular oxygen species, which leads to cell death (necrosis) and cell suicide (apoptosis) inside the tumor. Thus, PDT is a good biomedical technique that is less painful to patients who are unable to tolerate chemotherapy, surgery, etc. But there is some limitation with PDT, i.e., the depth of penetration of applied radiation; if the tumor is near to the skin, the skin is healed soon compared to the tumor being deep inside the body. Presently some researchers have proven the use of BP nanosheets as an efficient photosensitizer (Zheng et al., 2015). In this direction, some test was conducted on tumor-bearing mice after injecting very thin BP nanosheets as a photosensitizing agent into mice. After completing the incubation period when laser spotted to the tumor affected area, the tumor growth was significantly altered. The size of the tumor was reduced, which truly indicates the qualities of phosphorene in PDT of cancer.

### 19.3.1.2 Photothermal therapy

The photothermal therapy (PTT) technique is an extension of PDT. Like PDT, this technique also uses a photosensitizer which is excited with EM radiation having a certain bandgap. This radiative excitation moves the sensitizing agent to a higher-level, consequently it emits thermal vibrations that kill the tumor cells. The influence of infrared radiation on the black phosphorous quantum dots (BPQDs) have shown the end of various kinds of cancer cells such as C6 and MCF7; it is very less toxic to other cells also. In recent years, some scientific groups have done a detailed study on novel hydrogel nanostructure BP@hydrogel stating their importance in PTT and PDT therapies. This gel consists of phosphorene, hydrogel, and an anticancer drug (Fig. 19.3) (Qiu et al., 2018). When BP@hydrogel is spotted with IR light, it releases a therapeutic drug that could break the DNA chains inside the cancer cells, which leads to the development of normal cells along with the death of cancer cells (this process is termed as apoptosis induction) (Qiu et al., 2018).

Further, it is important to study about PEGylated BP nanoparticles, which are synthesized from red phosphorus and polyethylene glycol by the ball-milling approach. To test their utility in biomedical an experiment is done taking tumor-bearing mice. First, the PEGylated BP nanosheets are injected into the body of mice. After injecting, the PEGylated BP nanoparticles are spotted with infrared radiation for 5 min. It is found that the tumor temperature abruptly rose to 59°C; consequently, a complete removal of the tumor is recognized without any further recurrence mark. This proves that the PEGylated BP nanoparticles act as both heat mediator and heat enhancer. Moreover, the effect of PTT on the health of the normal tissue of mice is found null which is an extraordinary result highlighted among the researcher community. Fig. 19.3 illustrates the potential applications of PEGylated BP nanoparticles in biomedicine in diagnosis, imaging, and treatment of tumors (Sun et al., 2016).

### 19.3.1.3 Therapeutic agent delivery

Since phosphorene nanosheets have a large surface area, they can be used to load biomedical drugs in bulk; hence, they are a good therapeutic agent. Further, due to the presence of phosphoric acid on the surface of sheets, they can be easily combine with positive charges.

### 19.3.1.4 Bioimaging

It is a biological technique used to study living cells and body parts of various living beings. It uses different kinds of radiations to produce images of the body to examine diseases that can't be viewed by the naked eye. There are many bioimaging techniques used nowadays such as bioluminescence, gallium, molecular, optical, optoacoustic, photoacoustic, and ultrasound imaging. There is an important application of phosphorene in the photoacoustic imaging of cancer tumors because of its photothermal properties. Its working principle is such that when phosphorene nanosheets are injected into the tumored body, the tumor sensitive area is irradiated with a suitable wavelength; then, as a result some thermal vibrations will be generated that are carried by an ultrasonic wave. These ultrasonic waves then convert into the form of signals and later carried in the form of the in vivo imaging of nano drugs. It is found that the tumored area produces a strong signal as compared to other body parts. Furthermore, BP can also be used for photothermal imaging by means of infrared thermography.

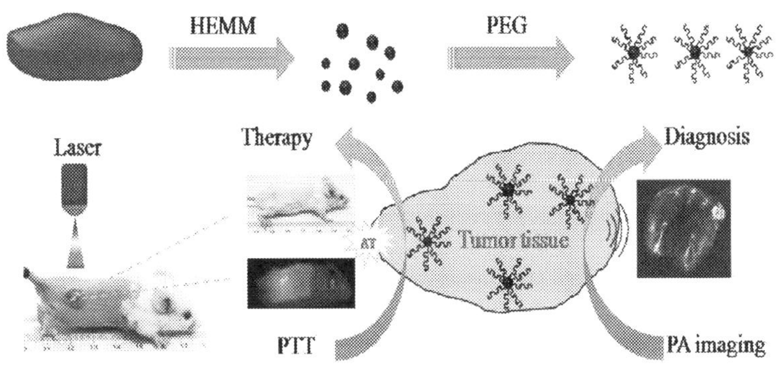


**FIGURE 19.3** Figure shows various applications of working scheme of PEGylated BP nanoparticles (Sun et al., 2016).

#### 19.3.1.5 Theranostics

A theranostic nano vehicle is a combination of imaging as well as therapeutic therapy that enables combining diagnosis as well as therapy of diseases (Fig. 19.4). PEGylated BPQDs presented an active theranostic approach against cancer tumor treatment and revealed vast potential in biomedical as well.

### 19.3.2 Graphene-based materials

In recent years, the utility of GO in biomedicine, especially in the treatment of cancer, has attracted a lot of researchers to explore more. GO has a sheet-like structure, and various types of oxygen functional groups are also present on its surface as shown in Fig. 19.5. Such types of functional group enhanced GO to hydroxyl and epoxide-enriched GO and carboxyl-rich GO (Yang, Feng, Dougherty et al., 2016). These oxidation groups further promote GO to combine with other biocompatible species such as anticancer drugs, proteins, etc. (Zhu et al., 2017). It has proved useful in drug delivery for different chemotherapy agents. Therefore, all such things make GO a better candidate for biomedical purpose.

In GO, the carboxyl groups are located at the edge sites while hydroxyl and epoxide groups are on the plane parallel to the horizontal axis, which shows solid $\pi-\pi$ stacking and hydrogen bonding that is useful in drug loading (Prabakaran, Jeyaraj, Nagaraj, Sadasivuni, & Rajan, 2019). Although GO is good for drug delivery, still there is some issue of poor target grasping ability and lack of cancer cell internalization. To resolve this issue, a nanovehicle based on hyaluronic acid-modified GO (HSG) has been developed which shows a high loading capacity to adriamycin or doxorubicin (DOX) (Yin, Gu et al., 2017; Yin, Liu et al., 2017). DOX is the most prominent anticancer drugs ever invented. After injecting it into the affected body, it is carried by nanovehicles. Its specialty is it can kill cancer cells at every stage of their life cycle, and is widely used in biomedicine. Further, GO can also be used as a photothermal agent for efficient PTT due to its high NIR absorbance with suitable therapeutic results. The modification of GO with magnetic nanomaterials will prominently boost the photothermal performance of GO with various multifunctional groups, and surface coating will prominently boost the photothermal performance of GO, hence, making it more appropriate for targeted drug delivery and bioimaging purposes without toxicity to treated cells/animals (Fig. 19.6) (Yang, Feng, & Liu, 2016).

Apart from GO, the rGO prepared from GO has vast applications in cancer therapy. rGO is a fast nanovehicle compared to GO and also has better electrical conductivity (Cheng, Wang, Gong, Liu, & Liu, 2019). In recent times, some nanocapsules based on rGO were fabricated by some researchers that are showing excellent photo-triggered capability

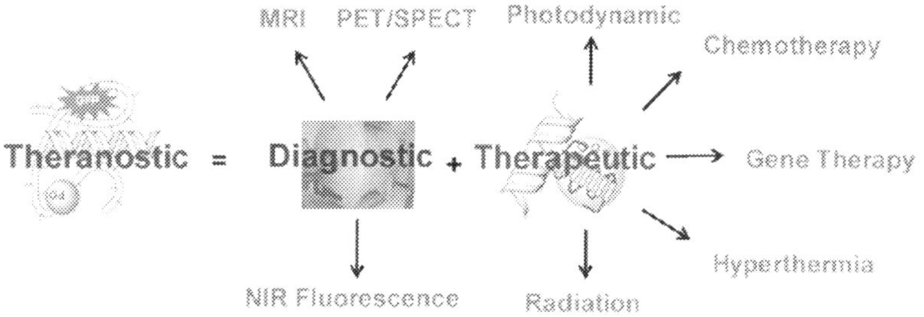


**FIGURE 19.4** Schematic diagram of theranostic treatment strategy (Kelkar & Reineke, 2011).

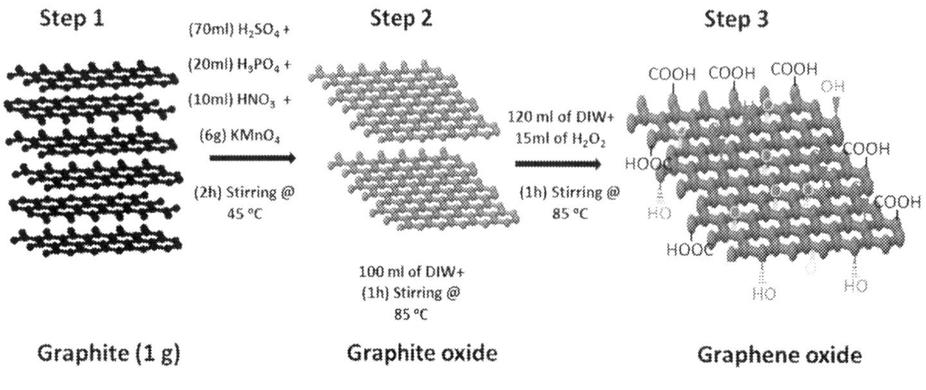


**FIGURE 19.5** Diagram represents the formation of graphene oxide (GO) from graphite (Gaashani, Najjar, Zakaria, Mansour, & Atieh, 2019).

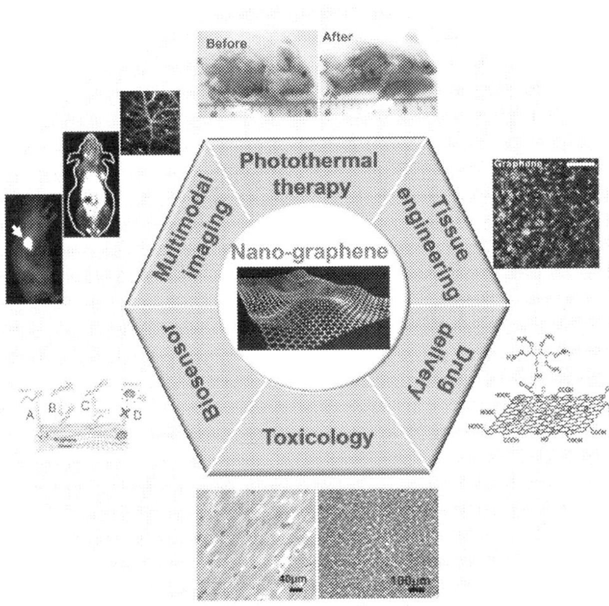


FIGURE 19.6 Figure shows the utility of graphene in various biomedical applications in cancer treatment such as bioimaging, biosensing, drug delivery, etc. (Yang, Feng, & Liu, 2016).

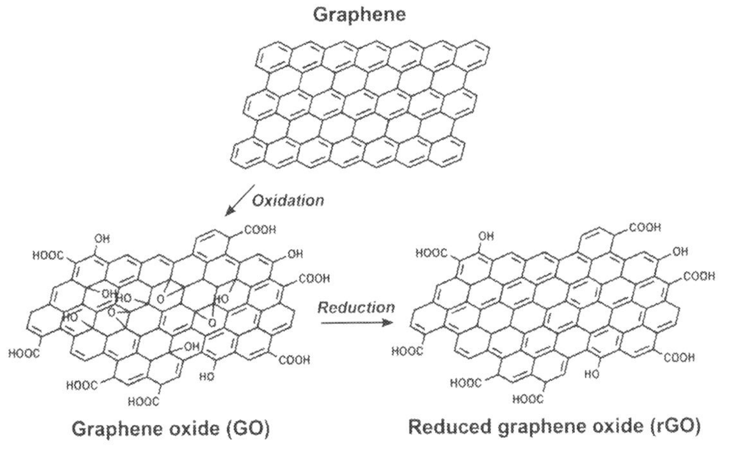


FIGURE 19.7 Diagram represents the preparation of rGO and GO from graphene (McCoy, Urpin, Teo, & Tabor, 2019).

with an appropriate drug loading capacity of 9.43 mmol/g toward DOX (Hu et al., 2014). Furthermore, in using rGOs, a careful drug delivery and PTT were attained by integrating a single-strand with polydopamine rGO nanosheets (Robinson et al., 2011; Singh, Kumar, & Singh, 2016). Thus, most of the authors verified the role of rGO in combination with IR light as an efficient PTT agent against cancer (Fig. 19.7) (McCoy et al., 2019).

Further it is important to study about PEGylated nano-GO (nGO-PEG). To test its behavior, a tumor-bearing mouse is injected with nGO-PEG and after careful observations by scientists, it is established that nGO-PEG has a high deposit rate due to greater permeability and retention effects. It is seen that using a 808 nm laser, the tumors of mice were fully eradicated (Yang et al., 2010).

### 19.3.3 Layered double hydroxides

The 2D sheet-like structures, layered double hydroxides (LDHs), were first developed as a nanovehicle in 1999 (Mei et al., 2018). The variation's in its chemical composition and crystal structure lead to a variety of application of LDHs in biomedicine (Asif et al., 2017; Cheng et al., 2019). They react easily with various types of organic anions, and due to high charge density and interlayer anion exchange performance, it contributes to pH-triggered drug release actions and also helps in quick delivery of the drug to the tumor-affected area (Gao and Yan, 2018). This works in such a way that the presence of a positive charge on LDH surfaces attracts the oppositely charged drug molecules and is then absorbed by LDHs nanosheets which helps for effective delivery of biomolecules (Asif et al., 2017). In addition, it is suitable for LDHs to drive through cell membranes without any kind of surface modification, because such membranes show negative charge, hence, enhancing the healing performance (Zhu et al., 2016). LDHs are a useful drug delivery agent with predictable and systematic performances for drug molecules, like DOX, 5-FU, vancomycin, and sodium fusidate (Jin, Liu, Sun, Ni, & Wei, 2010; Sun, Li, Fan, & Ai, 2013). Further, the combination of intercalation chemistry with rare earth elements along with magnetic and fluorescent properties in the layers of LDHs may lead to their uses in nanoplatforms, especially in cancer imaging. For example, LDH codoped with Mn and Fe (called as Mn−Fe LDH) is used for pH-responsive T1 Mr bioimaging. This Mn−Fe LDH rapidly responds to the tumor sites and activates a fast and immense release of paramagnetic $Fe^{3+}$ and $Mn^{2+}$ ions due to its alkaline behavior that results in substantial T1 Mr imaging improvement (Huang et al., 2017). Similarly, some researchers have also reported that Gd-LDH and Mn-LDH nanoparticles are excellent contrast agents for MRI and are having wonderful longitudinal relaxivity also (Li, Gu, Kurniawan, Chen, & Xu, 2017).

### 19.3.4 Transition metal carbides and nitrides (MXenes)

MXenes are layered inorganic compounds that are composed of early transition metal carbides, nitrides, or carbon nitrides such as $Nb_2C$, $Zr_3C_2$, $Ti_3C_2$, $Mo_2C$, $V_2C$, and $Ta_4C_3$ (Naguib et al., 2011). MXenes are having three typical formulas $M_2X$, $M_3X_2$, and $M_4X_3$ (here M is metals and X is for carbon or nitrogen). The surface of the MXenes is normally depleted with oxygen, fluorine, and other hydroxide functional groups, which allows many surface modifications that result in several practical uses as reported by many authors (Naguib, Mochalin, Barsoum, & Gogotsi, 2014). Surface modification assists rich anchoring sites with various functional groups and high specific surface area that enable MXenes to have vast applications as a drug or protein carrier (Lin, Chen, & Shi, 2018). Some scientists have developed a thin mesoporous-silica shell coated $Ti_3C_2$ MXene ($Ti_3C_2$@mMSNs), which fruitfully upgraded the original interfacial properties of $Ti_3C_2$ and the photothermal-conversion capacity for PTT (Li et al., 2018). The ultrathin-layered structure of MXene has unique physicochemical properties, for instance photothermal conversion (Li, Zhang, Shi, & Wang, 2017), enzyme-triggered biodegradation (Lin, Gao, Dai, Chen, & Shi, 2017), localized surface plasmon resonance (LSPR), cellular endocytosis, etc. (Bolotsky et al., 2019). Nowadays, multifunctional MXenes are also designed and used in various cancer applications. In this direction, some scientists have worked on $MnO_x/Ti_3C_2$ nanocomposites by growing $MnO_x$ onto the surface of $Ti_3C_2$ MXene. It is established that this composite presented a specific T1-weighted Mr imaging ability. Moreover, the IR absorbance of this nanocomposite gives good a PA imaging function and also confirmed effective tumor excisions. Thus, 2D material MXenes have attracted a lot of researchers to explore more about its utilities in biomedicine such as bioimaging, biosensing, etc.

### 19.3.5 Transition metal dichalcogenides

The most common formula for transition metal dichalcogenides (TMDs) is $MA_2$, here M denotes transition metals and A represents the groups 16 elements of the periodic table. Having a huge surface area, 2D TMDCs nanosheets can react with different biomolecules including anticancer drugs, thereby presenting its applications in tissue engineering, drug delivery, gene transfection, and biosensors (Mei et al., 2020). Different from other 2D materials, TMDs possess good potential in biomedical areas like PTT, photoacoustic (PA) imaging, and computer tomography (CT) imaging. Presently some authors have utilized the low toxicity as well as high NIR absorbance capacity of 2D TMDCs for PTT of cancer cells with promising therapeutic conclusions. The large surface area as well as multipurpose surface chemistries of 2D TMDCs have a significant role to add PTT with other therapeutic moieties, which further improves them with various functionalities for the treatment of cancer. Due to the presence of metal atoms, TMDs also exhibit good flexibility towards drug loading. Furthermore, some researchers have specified a methodical and deep study on the applications of 2D TMDCs in bioimaging, biosensing, and tissue engineering.

### 19.3.6 Molybdenum disulfide

The 2D molybdenum disulfide ($MoS_2$) nanosheets have attracted a lot of researchers in past years. It is a type of TMDCs with good chemical and physical properties. It shows strong photoluminescence when transiting from bulk to nanoscale; this is due to the quantum confinement effect (Ellis, Lucero, & Scuseria, 2011). Hence, owing to its photoluminescence property, it has strong applications in making biosensors, doing PTT, as well as bioimaging (Cheng, Wang,

Feng, Yang, & Liu, 2014). Further study was carried taking PEGylated $MoS_2$, consequently, proven a useful tool for drug delivery as well as PTT (using NIR radiation) of cancer patients (Chen, Roy, Yang, & Prasad, 2016).

Further, the processing of radar-like $MoS_2$ nanovehicles ($MoS_2$ RNPs) for NIR PTT has also been reported (Huang, Qi, Yu, & Zhan, 2016). The photothermal conversion efficiency (PCE) of radar-like $MoS_2$ nanovehicles has been reported by some authors, including the synthesis process. It is established that the PCE of such nanovehicles ($MoS_2$ RNPs) is 53.3%, which is far greater than $MoS_2$ nanosheets (24.4%), and $MoS_2$ nanoflakes (27.6%). Further, some research has been done taking multilayered $MoS_2$ nanosheets linked together by DNA, which is useful in drug delivery applications. $MoS_2$-based nano combinations and MoS2 QDs are quite useful in bioimaging applications to perform intracellular positioning of bio cells (Li, Setyawati et al., 2017; Wang, Nan et al., 2015). In recent years, various types of biosensors have been developed using $MoS_2$ and its composites. These include electrochemical, optical, and FET-based biosensors. Some biosensors which use electrodes for their functionalization are electrochemical biosensors, while optical biosensor uses fluorescent material; yet another FET-based sensor uses the ion-conducting liquid solution for their operation. The electrochemical biosensors are used for the detection of DNA (Wang, Nan et al., 2015). The presence of electrodes provides higher conductivity and a high electron transfer rate, therefore, is used to detect biomolecules such as dopamine (DA), glucose, DNA, and many more.

## Conclusion

A new trend is running to study 2D nanomaterials and find their utility in almost every branch of science. Recently, after the discovery of graphene many other 2D nanomaterials are also studied for their extraordinary features. Such nanomaterials are graphene based materials such as graphene, graphene oxide, reduced graphene oxide, other materials are black phosphorous quantum dots, PEGylated BP nanoparticles, 2D dichalcogenide, layered double hydroxides, Transition metal dichalcogenides, MXenes, Molybdenum disulphide etc. were explored for many potential applications. These 2D nanosheets were exfoliated from its bulk counterpart either by mechanical exfoliation or liquid phase exfoliation that are discussed in the chapter. Due to the huge surface area and outstanding mechanical and electronic properties, these 2D nanomaterials are studied flawlessly nowadays. In this chapter we have concisely discussed the importance of these nanomaterials in the field of biomedical. Particular focus has been given to study the application of these nanomaterials for cancer treatment. A brief method/mechanism is given regarding the working of these 2D nanomaterials in theranostics, therapeutic application. The usefulness of these nanoparticles in photo thermal therapy, Photodynamic therapy, how they act as chemotherapy agents and obviously how they hinder the cancer tumor growth? all these questions have been answered in this chapter. Apart from this what are the limitations of these nano-tools have also mentioned. In this chapter some useful figures are also given to show the utility of such nanomaterials in various types of biomedical applications in cancer treatment. However, it is seen that these nanomaterials can act as very fast nano-vehicles in supplying the drug and taking fast action against the tumor. Some of them has good flexibility towards drug loading, others are useful in bioimaging, biosensing, and tissue engineering etc. Furthermore, a lot of research is still to be done in future. Thus, there is more to explore about these nanomaterials for biological application point of view.

## References

Asif, M., Liu, H., Aziz, A., Wang, H., Wang, Z., Ajmal, M., ... Liu, H. (2017). Core-shell iron oxide-layered double hydroxide: High electrochemical sensing performance of H2O2 biomarker in live cancer cells with plasma therapeutics. *Biosensors and Bioelectronics*, *97*, 352–359.

Bolotsky, A., Butler, D., Dong, C., Gerace, K., Glavin, N. R., Muratore, C., ... Ebrahimi, A. (2019). Two-dimensional materials in biosensing and healthcare: From in vitro diagnostics to optogenetics and beyond. *ACS Nano*, *13*, 9781–9810.

Butler, S. Z., Hollen, S. M., Cao, L., Cui, Y., Gupta, J. A., Gutiérrez, H. R., ... Goldberger, J. E. (2013). Progress, challenges, and opportunities in two-dimensional materials beyond graphene. *ACS Nano*, *7*, 2898–2926.

Castellanos-Gomez, A., Poot, M., Steele, G. A., van der Zant, H. S. J., Agraït, N., & Rubio-Bollinger, G. (2012). Elastic properties of freely suspended MoS2 nanosheets. *Advanced Materials*, *24*, 772–775.

Chen, G., Roy, I., Yang, C., & Prasad, P. N. (2016). Nanochemistry and nanomedicine for nanoparticle-based diagnostics and therapy. *Chemical Reviews*, *116*, 2826–2885.

Cheng, L., Wang, C., Feng, L., Yang, K., & Liu, Z. (2014). Functional nanomaterials for phototherapies of cancer. *Chemical Reviews*, *114*, 10869–10939.

Cheng, L., Wang, X., Gong, F., Liu, T., & Liu, Z. (2019). 2D nanomaterials for cancer theranostic applications. *Advanced Materials*, *1902333*.

Cho, K. Y., Yeom, Y. S., Seo, H. Y., Kumar, P., Lee, A. S., Baek, K.-Y., & Yoon, H. G. (2015). Ionic block copolymer doped reduced graphene oxide supports with ultra-fine Pd nanoparticles: Strategic realization of ultra-accelerated nanocatalysis. *Journal of Materials Chemistry A*, *3*, 20471–20476.

Cho, K. Y., Yeom, Y. S., Seo, H. Y., Kumar, P., Lee, A. S., Baek, K.-Y., & Yoon, H. G. (2017). Molybdenum-doped PdPt@Pt core−shell octahedra supported by ionic block copolymer-functionalized graphene as a highly active and durable oxygen reduction electrocatalyst. *ACS Applied Materials & Interfaces*, *9*, 1524−1535.

Ding, L., Wei, Y., Wang, Y., Chen, H., Caro, J., Wang, H., & Two-Dimensional, A. (2017). Lamellar membrane: MXene nanosheet stacks. *Angewandte Chemie International Edition*, *56*, 1825−1829.

Du, Y., Liu, H., Deng, Y., & Ye, P. D. (2014). Device perspective for black phosphorus field-effect transistors: Contact resistance, ambipolar behavior, and scaling. *ACS Nano*, *8*, 10035−10042.

Ejigu, A., Kinloch, I. A., Prestat, E., & Dryfe, R. A. W. (2017). A simple electrochemical route to metallic phase trilayer MoS2: Evaluation as electrocatalysts and supercapacitors. *Journal of Materials Chemistry A*, *5*, 11316−11330.

Ellis, J. K., Lucero, M. J., & Scuseria, G. E. (2011). The indirect to direct band gap transition in multilayered MoS2 as predicted by screened hybrid density functional theory. *Applied Physics Letters*, *99*, 261908.

Ferrari, A. C., Bonaccorso, F., Fal'ko, V., Novoselov, K. S., Roche, S., Bøggild, P., … Kinaret, J. (2015). Science and technology roadmap for graphene, related two-dimensional crystals, and hybrid systems. *Nanoscale*, *7*, 4598−4810.

Gaashani, R. A., Najjar, A., Zakaria, Y., Mansour, S., & Atieh, M. A. (2019). XPS and structural studies of high qualitygraphene oxide and reduced graphene oxide prepared by different chemical oxidation methods. *Ceramics International*, *45*(11), 14439−14448.

Gao, R., & Yan, D. (2018). Fast formation of single-unit-cell-thick and defect-rich layered double hydroxide nanosheets with highly enhanced oxygen evolution reaction for water splitting. *Nano Research*, *11*, 1883−1894.

Geim, A. K., & Novoselov, K. S. (2007). The rise of graphene. *Nature Materials*, *6*, 183−191.

Hernandez, Y., Nicolosi, V., Lotya, M., Blighe, F. M., Sun, Z., De, S., … Coleman, J. N. (2008). *High-yield production of graphene by liquid-phase exfoliation of graphite*, . *Nature Nanotechnology* (3, pp. 563−568). .

Hu, S.-H., Fang, R.-H., Chen, Y.-W., Liao, B.-J., Chen, I.-W., & Chen, S.-Y. (2014). Photoresponsive protein−graphene−protein hybrid capsules with dual targeted heat-triggered drug delivery approach for enhanced tumor therapy. *Advanced Functional Materials*, *24*, 4144−4155.

Huang, G., Zhang, K.-L., Chen, S., Li, S.-H., Wang, L.-L., Wang, L.-P., … Yang, H.-H. (2017). Manganese-iron layered double hydroxide: A theranostic nanoplatform with pH-responsive MRI contrast enhancement and drug release. *Journal of Materials Chemistry B*, *5*, 3629−3633.

Huang, Y., Sutter, E., Shi, N. N., Zheng, J., Yang, T., Englund, D., … Sutter, P. (2015). Reliable exfoliation of large-area high-quality flakes of graphene and other two-dimensional materials. *ACS Nano*, *9*, 10612−10620.

Huang, Z., Qi, Y., Yu, D., & Zhan, J. (2016). Radar-like MoS2 nanoparticles as a highly efficient 808 nm laser-induced photothermal agent for cancer therapy. *RSC Advances*, *6*, 31031−31036.

Jin, L., Liu, Q., Sun, Z., Ni, X., & Wei, M. (2010). Preparation of 5-fluorouracil/β-cyclodextrin complex intercalated in layered double hydroxide and the controlled drug release properties. *Industrial & Engineering Chemistry Research*, *49*, 11176−11181.

Kelkar, S. S., & Reineke, T. M. (2011). Theranostics: Combining imaging and therapy. *Bioconjugate Chemistry*, *22*, 1879−1903.

Kumar, P., Shahzad, F., Yu, S., Hong, S. M., Kim, Y.-H., & Koo, C. M. (2015). Large-area reduced graphene oxide thin film with excellent thermal conductivity and electromagnetic interference shielding effectiveness. *Carbon*, *94*, 494−500.

Kumar, P., Yu, S., Shahzad, F., Hong, S. M., Kim, Y.-H., & Koo, C. M. (2016). Ultrahigh electrically and thermally conductive self-aligned graphene/polymer composites using large-area reduced graphene oxides. *Carbon*, *101*, 120−128.

Li, B., Gu, Z., Kurniawan, N., Chen, W., & Xu, Z. P. (2017). Manganese-based layered double hydroxide nanoparticles as a T1-MRI contrast agent with ultrasensitive pH response and high relaxivity. *Advanced Materials*, *29*, 1700373.

Li, B. L., Setyawati, M. I., Chen, L., Xie, J., Ariga, K., Lim, C.-T., … Leong, D. T. (2017). Directing assembly and disassembly of 2D MoS2 nanosheets with DNA for drug delivery. *ACS Applied Materials & Interfaces*, *9*, 15286−15296.

Li, L., Yang, F., Ye, G. J., Zhang, Z., Zhu, Z., Lou, W., … Zhang, Y. (2016). Quantum Hall effect in black phosphorus two-dimensional electron system. *Nature Nanotechnology*, *11*, 593−597.

Li, R., Zhang, L., Shi, L., & Wang, P. (2017). MXene Ti3C2: An effective 2D light-to-heat conversion material. *ACS Nano*, *11*, 3752−3759.

Li, Z., Zhang, H., Han, J., Chen, Y., Lin, H., & Yang, T. (2018). Surface nanopore engineering of 2D MXenes for targeted and synergistic multitherapies of hepatocellular carcinoma. *Advanced Materials*, *30*, 1706981.

Lin, H., Chen, Y., & Shi, J. (2018). Insights into 2D MXenes for versatile biomedical applications: Current advances and challenges ahead. *Advanced Science*, *5*, 1800518.

Lin, H., Gao, S., Dai, C., Chen, Y., & Shi, J. (2017). A two-dimensional biodegradable niobium carbide (MXene) for photothermal tumor eradication in NIR-I and NIR-II biowindows. *Journal of the American Chemical Society*, *139*, 16235−16247.

Liu, N., Luo, F., Wu, H., Liu, Y., Zhang, C., & Chen, J. (2008). One-step ionic-liquid-assisted electrochemical synthesis of ionic-liquid-functionalized graphene sheets directly from graphite. *Advanced Functional Materials*, *18*, 1518−1525.

Mas-Ballesté, R., Gómez-Navarro, C., Gómez-Herrero, J., & Zamora, F. (2011). 2D materials: To graphene and beyond. *Nanoscale*, *3*, 20−30.

McCoy, T. M., Urpin, G. T., Teo, B. M., & Tabor, R. F. (2019). Graphene oxide: A surfactant or particle? *Current Opinion in Colloid & Interface Science*, *39*, 98−109.

Mei, X., Hu, T., Wang, Y., Weng, X., Liang, R., & Wei, M. (2020). Recent advancements in two-dimensional nanomaterials for drug delivery. *WIREs Nanomedicine and Nanobiotechnology*, *12*, e1596.

Mei, X., Ma, J., Bai, X., Zhang, X., Zhang, S., Liang, R., … Duan, X. (2018). A bottom-up synthesis of rare-earth-hydrotalcite monolayer nanosheets toward multimode imaging and synergetic therapy. *Chemical Science*, *9*, 5630−5639.

Naguib, M., Kurtoglu, M., Presser, V., Lu, J., Niu, J., Heon, M., … Barsoum, M. W. (2011). Two-dimensional nanocrystals produced by exfoliation of Ti3AlC2. *Advanced Materials*, *23*, 4248−4253.

Naguib, M., Mochalin, V. N., Barsoum, M. W., & Gogotsi, Y. (2014). 25th Anniversary article: MXenes: A new family of two-dimensional materials. *Advanced Materials*, *26*, 992–1005.

Nicolosi, V., Chhowalla, M., Kanatzidis, M. G., Strano, M. S., & Coleman, J. N. (2013). Liquid exfoliation of layered materials. *Science (New York, N.Y.)*, *340*, 1226419.

Novoselov, K. S., Geim, A. K., Morozov, S. V., Jiang, D., Zhang, Y., Dubonos, S. V., … Firsov, A. A. (2004). Electric field effect in atomically thin carbon films. *Science (New York, N.Y.)*, *306*, 666.

O'Neill, A., Khan, U., Nirmalraj, P. N., Boland, J., & Coleman, J. N. (2011). Graphene dispersion and exfoliation in low boiling point solvents. *The Journal of Physical Chemistry C*, *115*, 5422–5428.

Parvez, K., Wu, Z.-S., Li, R., Liu, X., Graf, R., Feng, X., & Müllen, K. (2014). Exfoliation of graphite into graphene in aqueous solutions of inorganic salts. *Journal of the American Chemical Society*, *136*, 6083–6091.

Prabakaran, S., Jeyaraj, M., Nagaraj, A., Sadasivuni, K. K., & Rajan, M. (2019). Polymethyl methacrylate–ovalbumin @ graphene oxide drug carrier system for high anti-proliferative cancer drug delivery. *Applied Nanoscience*, *9*, 1487–1500.

Qiu M., Wang D., Liang W., Liu L., Zhang Y., Chen X., … Cao Y. (2018). Novel concept of the smart NIR-light–controlled drug release of black phosphorus nanostructure for cancer therapy. *Proceedings of the National Academy of Sciences*, *115*, 501–506.

Radisavljevic, B., Radenovic, A., Brivio, J., Giacometti, V., & Kis, A. (2011). Single-layer MoS2 transistors. *Nature Nanotechnology*, *6*, 147–150.

Rahman, M., Davey, K., & Qiao, S.-Z. (2017). Advent of 2D rhenium disulfide (ReS2): Fundamentals to applications. *Advanced Functional Materials*, *27*, 1606129.

Rasid Z.A.M., Omar M.F., Nazeri M.F.M., A'ziz M.A.A., Szota M. (2017). Low cost synthesis method of two-dimensional titanium carbide Mxene. *IOP Conference Series: Materials Science and Engineering*, *209*, 012001.

Robinson, J. T., Tabakman, S. M., Liang, Y., Wang, H., Sanchez Casalongue, H., Vinh, D., & Dai, H. (2011). Ultrasmall reduced graphene oxide with high near-infrared absorbance for photothermal therapy. *Journal of the American Chemical Society*, *133*, 6825–6831.

Shahzad, F., Kumar, P., Kim, Y.-H., Hong, S. M., & Koo, C. M. (2016). Biomass-derived thermally annealed interconnected sulfur-doped graphene as a shield against electromagnetic interference. *ACS Applied Materials & Interfaces*, *8*, 9361–9369.

Shahzad, F., Kumar, P., Yu, S., Lee, S., Kim, Y.-H., Hong, S. M., & Koo, C. M. (2015). Sulfur-doped graphene laminates for EMI shielding applications. *Journal of Materials Chemistry C*, *3*, 9802–9810.

Singh, R. K., Kumar, R., & Singh, D. P. (2016). Graphene oxide: Strategies for synthesis, reduction and frontier applications. *RSC Advances*, *6*, 64993.

Sun, C., Wen, L., Zeng, J., Wang, Y., Sun, Q., Deng, L., … Li, Z. (2016). One-pot solventless preparation of PEGylated black phosphorus nanoparticles for photoacoustic imaging and photothermal therapy of cancer. *Biomaterials*, *91*, 81–89.

Sun, J., Li, J., Fan, H., & Ai, S. (2013). Ag nanoparticles and vancomycin comodified layered double hydroxides for simultaneous capture and disinfection of bacteria. *Journal of Materials Chemistry B*, *1*, 5436–5442.

Tao, H., Zhang, Y., Gao, Y., Sun, Z., Yan, C., & Texter, J. (2017). Scalable exfoliation and dispersion of two-dimensional materials – An update. *Physical Chemistry Chemical Physics*, *19*, 921–960.

Wang, H., Yang, X., Shao, W., Chen, S., Xie, J., Zhang, X., … Xie, Y. (2015). Ultrathin Black Phosphorus Nanosheets for Efficient Singlet Oxygen Generation. *Journal of the American Chemical Society*, *137*, 11376–11382.

Wang, H., & Yu, X.-F. (2018). Few-layered black phosphorus: From fabrication and customization to biomedical applications. *Small (Weinheim an der Bergstrasse, Germany)*, *14*, 1702830.

Wang, X., Nan, F., Zhao, J., Yang, T., Ge, T., & Jiao, K. (2015). A label-free ultrasensitive electrochemical DNA sensor based on thin-layer MoS2 nanosheets with high electrochemical activity. *Biosensors and Bioelectronics*, *64*, 386–391.

Yang, D., Feng, L., Dougherty, C. A., Luker, K. E., Chen, D., Cauble, M. A., … Hong, H. (2016). In vivo targeting of metastatic breast cancer via tumor vasculature-specific nano-graphene oxide. *Biomaterials*, *104*, 361–371.

Yang, K., Feng, L., & Liu, Z. (2016). Stimuli responsive drug delivery systems based on nano-graphene for cancer therapy. *Advanced Drug Delivery Reviews*, *105*, 228–241.

Yang, K., Zhang, S., Zhang, G., Sun, X., Lee, S.-T., & Liu, Z. (2010). Graphene in mice: Ultrahigh in vivo tumor uptake and efficient photothermal therapy. *Nano Letters*, *10*, 3318–3323.

Yang, Q., Xu, Z., Fang, B., Huang, T., Cai, S., Chen, H., … Gao, C. (2017). MXene/graphene hybrid fibers for high performance flexible supercapacitors. *Journal of Materials Chemistry A*, *5*, 22113–22119.

Yin, F., Gu, B., Lin, Y., Panwar, N., Tjin, S. C., Qu, J., … Yong, K.-T. (2017). Functionalized 2D nanomaterials for gene delivery applications. *Coordination Chemistry Reviews*, *347*, 77–97.

Yin, T., Liu, J., Zhao, Z., Zhao, Y., Dong, L., Yang, M., … Huo, M. (2017). Redox sensitive hyaluronic acid-decorated graphene oxide for photothermally controlled tumor-cytoplasm-selective rapid drug delivery. *Advanced Functional Materials*, *27*, 1604620.

Zeng, Z., Sun, T., Zhu, J., Huang, X., Yin, Z., Lu, G., … Zhang, H. (2012). An effective method for the fabrication of few-layer-thick inorganic nanosheets. *Angewandte Chemie International Edition*, *51*, 9052–9056.

Zheng, X., Liu, M., Hui, J., Fan, D., Ma, H., Zhang, X., … Wei, Y. (2015). Ln3 + -doped hydroxyapatite nanocrystals: Controllable synthesis and cell imaging. *Physical Chemistry Chemical Physics*, *17*, 20301–20307.

Zhu, J., Xu, M., Gao, M., Zhang, Z., Xu, Y., Xia, T., & Liu, S. (2017). Graphene oxide induced perturbation to plasma membrane and cytoskeletal meshwork sensitize cancer cells to chemotherapeutic agents. *ACS Nano*, *11*, 2637–2651.

Zhu, R., Wang, Q., Zhu, Y., Wang, Z., Zhang, H., Wu, B., … Wang, S. (2016). pH sensitive nano layered double hydroxides reduce the hematotoxicity and enhance the anticancer efficacy of etoposide on non-small cell lung cancer. *Acta Biomaterialia*, *29*, 320–332.

Chapter 20

# Challenges and future prospects and commercial viability of biosensor-based devices for disease diagnosis

Niloy Chatterjee[1,*], Krishnendu Manna[2,3,*], Niladri Mukherjee[2,*] and Krishna Das Saha[2]

[1]*Food and Nutrition Division, University of Calcutta, Kolkata, India*

[2]*Cancer Biology and Inflammatory Disorder Division, CSIR-Indian Institute of Chemical Biology, Kolkata, India*

[3]*Department of Food and Nutrition, University of Kalyani, Kalyani, India*

## 20.1 Introduction

Sensors can be generally referred to as instruments, module, device or subsystems electrical or bioelectronic in nature that can respond to an observable stimulus, such as temperature, brightness, chemicals or pressure relevant to their presence in different scenarios (Han et al., 2017). They are used for bio-analysis providing real-time, precise and consistent data and results about different target molecules qualitatively as well as quantitatively. Sensors can be of three types depending on the type of analyte measured: physical sensors (quantify physical quantities), chemical sensors (quantify chemical entities), and biosensors (quantify analytes using biological foundations) (Loock & Wentzell, 2012; Mehrotra, 2016; Stetter et al., 2003). Biosensors are analytical instruments with a physicochemical transducer and biological detection elements. Such technologies retort to any biological stimulus by generating an indication or signal that can be easily inferred or determined (Perumal & Hashim, 2014b). All kinds of sensors including biosensors detect various measurable quantities by converting them into voltage with the help of a transducer (Abid et al., 2021). Biosensors can be defined as scientific analytical instruments which are proficient and competent in translating a wide variety of biotic impulses or signs into an easy electrical response (Song et al., 2006; Turner, 2013b) utilizing the principles of electronics to medicine as well as biology. They not only sense but also recognize biological components. A propitious and superior biosensor must possess several attributes such as high sensitivity, rapid response time, multimode sensing, reusability, disposability, cost-effective, long shelf life, easy to use, self-regulating, and self-sufficient without dependency on physical constraints like pH, temperature, etc. (Metkar & Girigoswami, 2019; Pravin & Sadaf, 2020). The field of biosensors is not something new but complex and multifaceted, involving a trans-disciplinary knowledge of diverse domains of science such as chemistry, physics, material science, biology, and engineering (Simranjeet et al., 2020). The human body is the best example of a sensor with various sensory parts: the ear, nose, eye, tongue, ability to touch (Hanson et al., 2009). The biosensing industry is a booming and prosperous sector recently, possessing huge future potential. Based on the bio-recognition method of these biosensors they can be classified by the following parameters: biocatalyst (enzyme, chemicals, etc.), bio-affinity molecule (antibodies, nucleic acids), and target for recognition (biomolecule, contaminants, microbe, etc.) (Alhadrami, 2018; Fang et al., 2020; Pravin & Sadaf, 2020). A biosensor consists of several parts or constituents and is illustrated in Fig. 20.1. They generate signals depending on the concentration of the analytes which can be easily visualized and interpreted. Bioreceptors possess the ligands which actually detect the target, then the biological signal is converted to electrical by the transducer, and after amplification by the microcontroller, they are finally sent to a data processor which converts the signal to visualizable data (Sang et al., 2016). Biosensors are often termed "biological sensor" owing to their involvement in biological elements (Morales & Halpern, 2018). The present chapter aims at summarizing the development, types, pros and cons, most recent findings and challenges associated with them, and future challenges regarding the use of biosensors in the domain of disease detection and diagnosis. The overall structure of a biosensor and its general components is given in figure Fig. 20.1 (Kissinger, 2005; Korotkaya, 2014; Koyun et al., n.d.)

*Niladri Mukherjee, Krishnendu Manna, and Niloy Chatterjee contributed equally.

**Biosensor Based Advanced Cancer Diagnostics. DOI: https://doi.org/10.1016/B978-0-12-823424-2.00013-2**

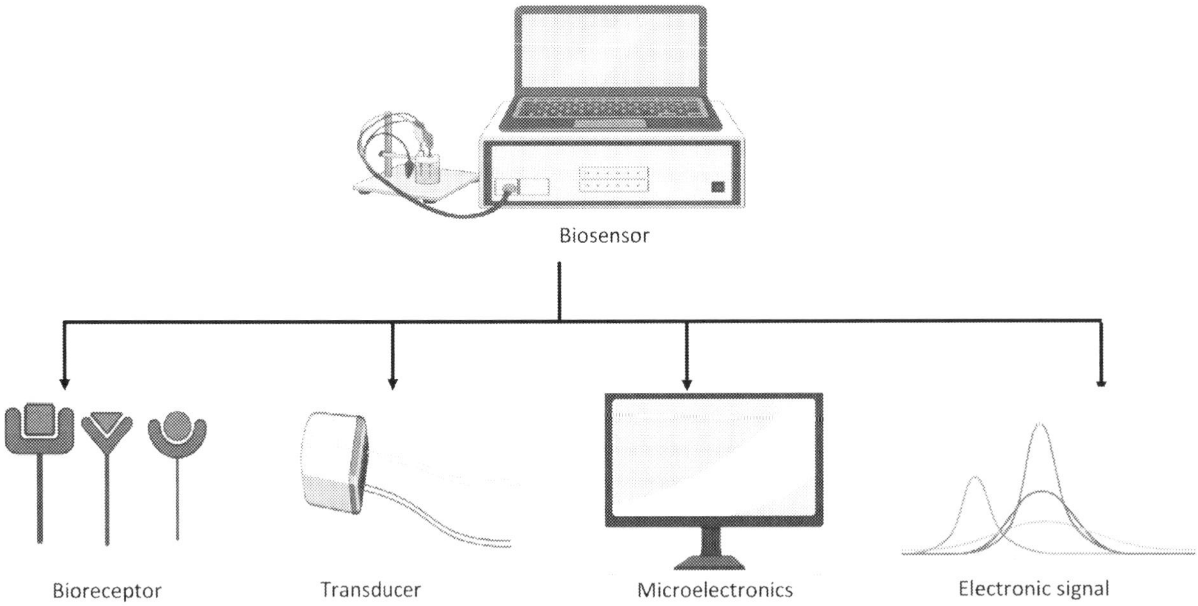


**FIGURE 20.1** Biosensor and their general components.

The latest developments in biomarker development and bioengineering, as well as for biotechnology, are shedding light on the complexities of different biological as well as physiological processes in healthy as well as diseased states, revealing novel diagnostic as well as therapeutic goals (Kussmann et al., 2006). Such scientific and technological advancements hold great promise and are particularly significant in the areas of diseases, infections, and maladies, where actual as well as expected mortality is still very high due to poor medical infrastructure, environment deterioration by humans, urbanization, the emergence of pandemics, epidemics, and the rise of a wide variety of antibiotic-resistant pathogens along with various emerging and reemerging diseases (Majumder & Deen, 2019; Steinhubl et al., 2015). As a consequence, providing and discovering accurate, reliable, and sensitive diagnostic methods that can early and prematurely detect diseases is important and the need of the hour.

## 20.2 Biosensor classification for disease diagnosis

Biosensors for the detection of diseases can be divided into different classification groups, such as the type of ligand or the biocomponent probe being used for the detection of diseases, type of transducer used in the instrument, and detection strategy utilized in the functioning of the biosensor (Rastislav et al., 2012). A wide variety of biological ligands and probes are currently being employed for prognosis as well as detection: protein and enzymes, antigens, antibodies, nucleotides, aptamers, cells and tissues, receptors, microbes, etc. (Karunakaran et al., 2015; Morales & Halpern, 2018). Enzyme-based biosensors work on catalysis mechanism and are, therefore, very selective and specific, offering huge variations in term of analytes (Ispas et al., 2012; Songa & Okonkwo, 2016). The first biosensors were developed by Clark et al., based on the usage of the enzyme glucose oxidase (GO) as a probe linked to an electrode for detecting glucose and also providing the uptake of $O_2$ (Wu et al., 2010). They are generally employed to detect different disease protein or similar markers at various stages (Leca-Bouvier & Blum, 2005; Rocchitta et al., 2016). Again, DNA or aptamer-based sensors are used to detect genes, oligonucleotides, and their mutations in diseases and pathogenic microbe identification based on genetics (Abolhasan et al., 2019; Chao et al., 2016). In the past few years, specific immunosensors have been developed that possess antibodies as the probe and can specifically detect antigens (Asal et al., 2018b). Certain antigens or similar analytes vary greatly between normal and unhealthy states and can be very important. Based on the type of transducer or signal used in the detection mechanism of diseases, they can be calorimetric (detects the change of enthalpy) (Wignarajah et al., 2015); optical (detects light or SPR or related phenomena) (Yoo & Lee, 2016); piezoelectric (detects changes in mass) (do Nascimento et al., 2017); cell-based (detect various types of cells and their amounts) (Gopinath, Anitha, & Mastani, 2015); and electrochemical (formation or loss of electro-active molecules and analytes) (Bellagambi et al., 2017; Esteves-Villanueva et al., 2014).

## 20.3 Biomarkers

A biomarker is "an objectively calculated and assessed feature that serves as an indication of normal biological processes, pathogenic processes, or pharmacologic responses to a therapeutic intervention" (Herrera-Espejo et al., 2019). The biomarker can include different types of biomolecules, such as proteins, cells, genes, and metabolites, that are linked with a specific or a fraction of people. The overall health status of an individual can be assessed upon the amount of specific biomarker(s) (Alexandra et al., 2019; Ayesha et al., 2020; Kelley, 2017). These are widely used in the field of transcriptomics, genomics, microbiology, metabolomics, as well as proteomics along with drug discovery (Koushki et al., 2018; Riedmaier & Pfaffl, 2013). Biomarkers are of immense importance for prediction, diagnosis, progression, and even therapeutic outcome of diseases after treatment with different medications and drugs. Biomarkers are accurately evaluated and measured to indicate abnormality and pathogenicity as well as normal biological processes and pharmacological responses (Mayeux, 2004; Strimbu & Tavel, 2010). Based on the application, biomarkers can be of several types: (1) biomarkers that can detect disease risk, (2) biomarkers screening illnesses in subclinical forms, (3) biomarkers that help in the diagnosis of disease, (4) biomarkers that stage or classify severe diseases, and (5) biomarkers that predict future diseases, their relapse and even therapeutic response (Attur et al., 2013; Cova & Priori, 2018; Valenti, 2013). A commonly used biomarker used widely in the field of diseases is C-reactive protein (CRP), applied for severe infection, sepsis, heart disease risk, and rheumatologic conditions. This protein is usually given to test by the doctor and are generally regarded as the primitive or foremost mark of any infection in the body. The importance of biomarkers to determine the state of health or illness of an individual as well as their multifaceted roles is deciphered in Fig. 20.2 (Bonassi et al., 2001; Kraus et al., 2011; Parikh & Mansour, 2017)

The desired biomarker must possess a broad range of favorable features, such as reusability, specificity, robust, and high sensitivity that can be easily measured in real-time. It must be capable of independent use during clinical dispositions; and their application ought to be applied in various economical noninvasive testing methods with high duplicability and constancy for processing a large number of samples (Bennett & Devarajan, 2011).

Simple statistical studies, the creation and analysis of classification models and subset-selection optimization are some widely used options for identifying biomarkers. Due to advances in biomolecule screening techniques in silico studies and bioinformatics research, many biomarkers for different diseases have been reported in recent decades. However, their clinical application is still limited (Lin et al., 2017). Table 20.1 shows examples of some widely used diagnostic biomarkers and their diseases and frequency with which they are assessed.

## 20.4 Application of biosensors in disease detection

Biosensors provide a wide range of rewards in comparison to other detection or analysis technologies, instruments, etc., including high discrimination power and sensitiveness, miniaturization and compactness, economical, real-time recognition, small sample volumes, and fast response (Singh et al., 2014; Song et al., 2006).

Various kinds of biosensors are being currently developed successfully by researchers globally and are being functional in the field of diagnosis as well as detection and screening of maladies such as such as autoimmune disorders,

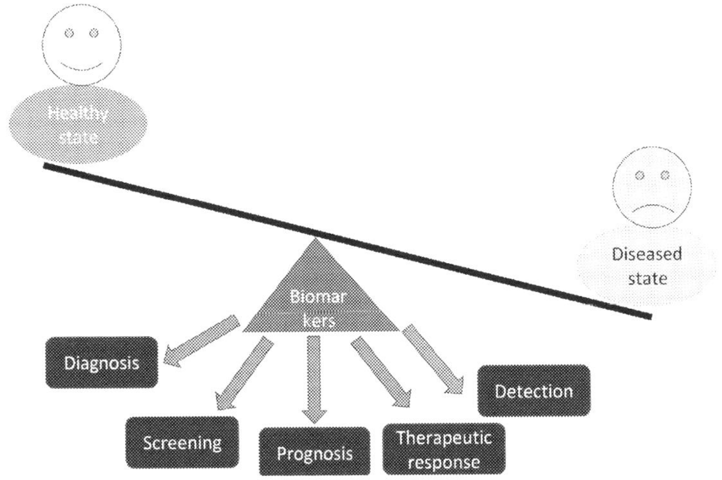


**FIGURE 20.2** Importance of biomarkers in diseased conditions of individuals.

**TABLE 20.1 Some common biomarkers applied in the field of disease diagnosis.**

| Disease diagnosed | Important biomarkers | Currently widely used or not |
|---|---|---|
| Diabetes mellitus | RBS, FBS, HbA1c | Widely used. |
| Cardiac problems | PRO-BNP | Moderately used |
| Cancer | EGFR, PSA, Her2/Neu | Widely used |
| Antioxidative conditions | Catalase, SOD, Glutathione | Widely used |
| Oxidative stress | $H_2O_2$, MDA, α-1 antiproteinase | Moderately used |
| Hypertension | Blood pressure, Aldosterone, Angiotensin, Plasma Renin | Widely used |
| COPD/Asthma | Leukotrienes | Widely used |

*EGFR*, Epidermal growth factor receptor; *FBS*, Fasting blood glucose; *HbA1c*, hemoglobin A1c; *Her2/Neu*, Receptor tyrosine-protein kinase erbB-2; *MDA*, Malondialdehyde; *PRO-BNP*, ProB-type natriuretic peptide; *PSA*, Prostate-specific antigen; *RBS*, Random blood sugar; *SOD*, Superoxide dismutase.

cancer, neurodegenerative diseases, and cardiovascular diseases. But still they are in their initial phase and has a long path to travel to be used extensively (Gu & Liu, 2020).

Most relevant or useful biosensors nowadays are affinity-based, which relies or depends on an immobilized detection probe with a highly selective attachment only towards a particular molecular entity, also known as the target or analyte (Arugula & Simonian, 2014). Therefore, the complex task to detect or sense a miniature chemical, molecule, or entity in an environment or a matrix physically is gradually shifted to the detection of any physical phenomenon or action which can be easily and straightforwardly perceived (Perumal & Hashim, 2014a; Karunakaran et al., 2015). The wide variation of detection procedures, such as detection of light (e.g., SPR, fluorescence, chemiluminescence, etc.) (Singh et al., 2014; Song et al., 2006), mechanical or power-driven signal (e.g., quartz crystal microbalance (QCM) or resonant cantilever) (Montagut et al., 2011; Singh & Yadava, 2020; Tamayo et al., 2013), or magnetic elements (Nabaei et al., 2018), can be calculated with the aid of biosensors to detect wide variety of molecules of interest. The electrical or physical arm of implementing bio-sensors is commonly utilized in the design of "mark-free" biosensors from the variety of techniques available to the consumer, thereby averting the need of any identifier or label to confirm the recognition of particular molecules or bio-markers (Sang et al., 2016; Zanchetta et al., 2017). For binding detection, electrochemical biosensors depend exclusively on measuring electrical currents as well as voltages depending indirectly on various physics parameters (Hammond et al., 2016). Because of their low expense, minimal power consumption, and ease of diminishment, electrochemical biosensors are highly promising for a wide variety of applications (Feiyun et al., 2020; Ronkainen et al., 2010) such as "point-of-care" (POC) disease diagnostics along with early detection, where diminishing dimensions and price is of utmost importance (Anik, 2017; da Silva et al., 2017). The detailed mechanism of action of biosensors is given in Fig. 20.3 (Kirsch et al., 2013; Gilchrist et al., 2001; Shi et al., 2018)

Numerous bio-based sensors are already present in the market and are highly successful commercially. They utilize various time-saving, robust, as well as sensitive analytical approaches for the identification of several analytes together as well as sometimes individually (Koyun et al., n.d.). Similarly, in the field of disease or pathogenic biology, they have facilitated the detection of pathogens or disease detection in early stages, targeting specific biomolecular markers which is possible only in recent times (Cesewski & Johnson, 2020; Chen et al., 2018; Huang et al., 2017; J et al., 2018). The initial illustration of an electrochemical biosensor based on enzyme detection was first studied by Professor Leland C. Clark (1962), which trapped GO over a Clark-type electrode within a dialysis sheath, based on oxygen (Palchetti & Mascini, 2010). In addition, Guilbault and Montalvo reported urease-coupled glass electrodes for the measurement of urea concentration through potentiometry. In addition to these primary sensors, electrochemical transducers were combined as biochemical recognition components with enzymes, antibodies, and DNA (Asal et al., 2018a). Currently, they constitute the largest class of food, clinical, and environmental sensing biosensors. An extensive array of analytes can therefore be detected by these modern electrochemical instruments which can be effortlessly altered and also integrated into strong, transportable, low-priced, miniature devices or systems that can be customized for specific usages. Electrochemical biosensors can also particularly distinguish a wide range of target chemical species or analytes, possessing the above advantages as given previously, along with the integration of highly precise biotic recognition molecules (enzymes, nucleic acids, cells, tissues, etc.) (Darsanaki, Azizzadeh, & Nourbakhsh, 2013).

The mounting figures of biosensor-based scientific studies suggest an increased interest in these devices among the larger scientific community. The various biomolecular structure used in biosensors (e.g., biocatalytic enzymes, DNAzymes, biocatalytic abiotic nanospecies, bioreceptors, DNA/RNA, aptasensors, etc.) (Buddhadev et al., 2020) as well as mechanisms of physical signal transduction (e.g., electrochemical, electrical, magnetic, etc.), from the sensors

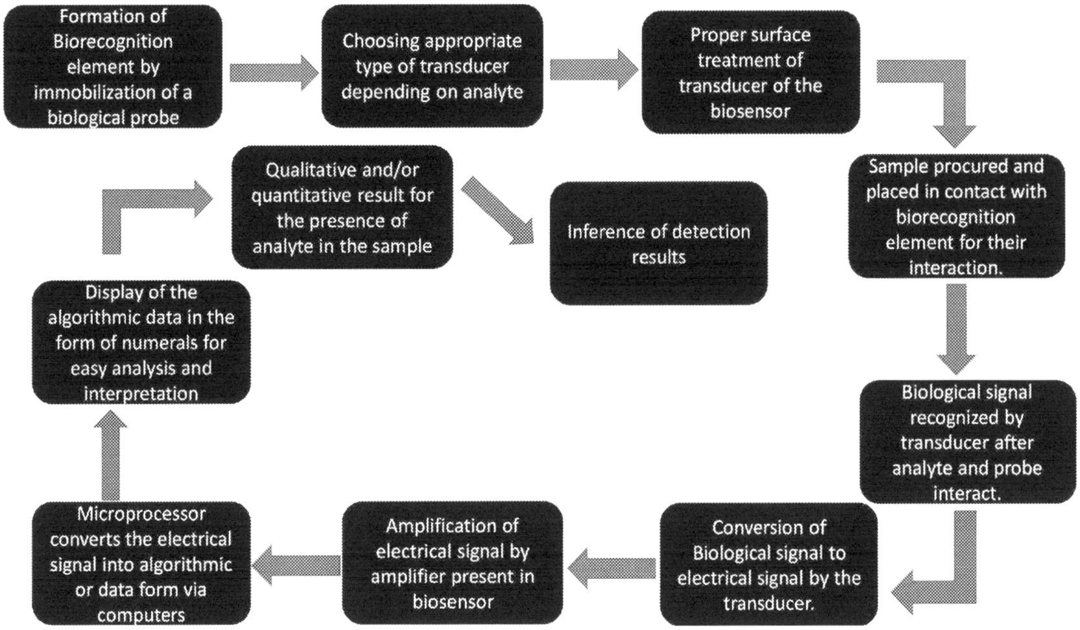


**FIGURE 20.3** Mechanism of action of biosensors.

constitute important part of the whole system. The instruments can either operate in an analog manner, accurately measuring the concentrations of various analytes in various solutions, matrices, etc., or binary devices operating in the +/− format, showing just the presence or absence of analytes, likely through logical processing of input signals.

## 20.5 The market trend of biosensors in disease detection

Three decades back, biochemical sensors generated global revenues of about US$5 million per annum which increased several folds nearing about US$20 billion in 2019 with expectations to generate revenue of around US$50 billion by 2022 (Ferreira et al., 2017). The bio-electrochemical sensing instruments currently have been hugely commercialized, successfully moving from a lab to the fields of applications and POC. In the market, several sensing systems are currently available from various manufacturers for the detection of various metabolites or bio-molecules (lactose, uric acid, cholesterol, lactic acid, sugars, proteins, fats, etc.) (Touhami, 2014), but blood glucose sensors dominate the market, capping more than 85% of the marketplace linked with modern glucose detecting devices (Cesewski & Johnson, 2020; Chen et al., 2018; Huang et al., 2017; J et al., 2018). Advancements in nanomaterials triggered the development of more advanced bio-sensors. Nanomaterial-based biosensors typically use a restricted biologically sensitive probe that is extremely discriminatory for the analyte fragments under examination. The technologically advanced nanomaterial is further receptive to analytes providing very delicate biological as well as chemical sensors that have distinguishing characteristics not found in their bulk counterparts (Girigoswami & Akhtar, 2019; McKeating et al., 2016).

## 20.6 Research trends of novel biosensors in disease detection

The future of biosensors is optimistic in various areas or domains. The range of biosensors available, built on different artificial recognition principles of the machinery, permits biosensors to be used not only against a broader variety of analytes but also using various modern transduction approaches. They generally possess ultra-low limits of discrimination and sensitivity or selectivity. Effective shrinkage of instrumentation often allows biosensors to be assembled as arrays. In the near future, biosensors will provide analytical approaches as an influential device with the development being a radically alterable and gradual evolving process. The production of swift and accurate laboratory sample analysis was inspired by different contemporary medicinal and pharmacological commerce. Some key objectives of research priorities for the use of biosensors are in the fields of environmental monitoring, environmental assessment, response time optimization for chemical detection, choosiness, and durability. Along with being economical, biosensors have recently been sought by various manufacturers resulting in future effective and accurate measurement systems (Choi et al., 2017).

Various biosensor ranges have been developed and others are under investigation (Jazib et al., 2017). Different antibodies, important enzymes, including artificial cell matrices, dynamic membranes, and even whole animal cells are important constituents of these sensors. Accessible literature revealed that ground-breaking research in the field of electrical engineering, biophysics, and nanoscience are now revolutionizing the field of biosensor operation as well as their basic mechanisms of different disease and ailment studies, which have been previously impossible. These consist of biochemical and chemical analytics that in the coming era will offer new theranostic instruments to clinicians (Acharya et al., 2017; Liu et al., 2018; Sokolov et al., 2017).

For the development of efficient biosensors, there lies two main hindrances to be solved. The first of them is to diminish the detection limit of the analytes. Often the challaenge is that, very minuscule or little amounts of test molecules are present. For their detection, very sensitive probes or molecules possessing high affinity to analytes are required (Wang et al., 2014). The other hinderance is that they must be able to detect multiple analytes in a sample, also termed as multiplexing. Multiplexing is very crucial since the consistent and reliable analysis of several diseases needs simultaneous identification of a wide variety of biomolecular indicators (Bottazzi et al., 2014). The impediment for effective reduction of detection limits lies in the thermodynamics of affinity between probe and target (Idili et al., 2013). Classically, dissociation constants ($K_d$) lie between $10^8-10^{12}$ M, and within this range, the association is not able to detect miniature amounts of molecules. The known strongest affinity for any known biomolecule is the reaction between biotin and streptavidin of about $10^{15}$ M. However, using orthodox transduction systems, the detection limits lie within $10^9-10^{14}$ M. The detection limits of biosensors can be greatly lowered using nanoscience or by manufacturing various nanobiosensors (Wu et al., 2010). Various nanomaterials like nanotubes, nanowires, nanoparticles, etc. can be utilized. Amplification by these nanobiosensors however requires rigorous sample preparation and proper labeling which limits the use of such advanced technologies. They are often time-consuming and require skilled operators. In recent times, cantilevers and field-effect transistors are utilized in biorecognition processes for extremely delicate and label-free detection.

## 20.7 Advantages of use of biosensors in the field of disease detection

For the maintenance of a normal and healthy lifestyle, the early unearthing of illness or health disorders is very important. Various biomolecules are can help in the determination of the diseased condition of individuals. These are known as biomarkers, and are very important in disease diagnosis. Various food and environmental contaminants can also be detected in different food matrices or soils, water bodies, etc., thereby helping in monitoring human health against harmful chemicals (Morales & Halpern, 2018; Senveli & Tigli, 2013; Soldatkin et al., 2013). There is an urgent requirement for fast analysis, dynamic continuous real-time, accurate and sensitive monitoring approaches (Meshram et al., 2018; Hamm et al., 2019). A real-time biosensor can be one of the best alternatives to detect molecules of interest in a continuous manner with time playing a significant part in achieving actual information genesis along with data processing, thereby aiding in actual result analysis, taking out inferences, and swiftness in handling and operation (Mauriz et al., 2019; Rogers et al., 2015). One of the most important inventions in this field of disease-based biosensors is hybrid analytical systems in a wide variety of settings and milieus at very minute amounts, which can detect various analytes simultaneously (Kim et al., 2019; Sun et al., 2018). These are currently widely used as standard methods of choice in various POC settings. Modern instruments of these principles currently employ nanomaterials as well as microfluidic approaches similar to that of semiconductor devices. Much of the contemporary investigations, therefore, attempt to combine similar complicated clinical and complex measurement requirements with on-chip detection techniques to create multipurpose nanobiosensors for early detection of various maladies (Sanjay et al., 2016). The basic perceptions of nanobiosensors are therefore important in the current scenario. Nanotechnology-based diagnostic instruments will become operational during the next years and will be capable of performing hundreds of analyses very quickly and cheaply. Future diagnostic developments will intensify as biochip design is miniaturized to the nanometer scale (Acharya et al., 2017; Liu et al., 2018; Sokolov et al., 2017). Examination of serum proteins would be the most common clinical medical procedure. The overall health or illness in human beings is generally embodied easily by blood or liquid connective tissue in the systemic circulation. Blood molecular fingerprint identification, therefore, would offer a much more delicate measure of healthiness and sickness of individuals.

Usage of nanobiosensor technology may provide several advantages such as decreased time for results of lab tests in the coming years (Sagadevan & Periasamy, 2014, 2016). For instance, when they first arrive at the hospital, individuals with infectious diseases should have urine samples, and the findings could be available by the time they see the doctor with simultaneous detection of several disease markers (Jarockyte et al., 2020). A medication could perhaps be given

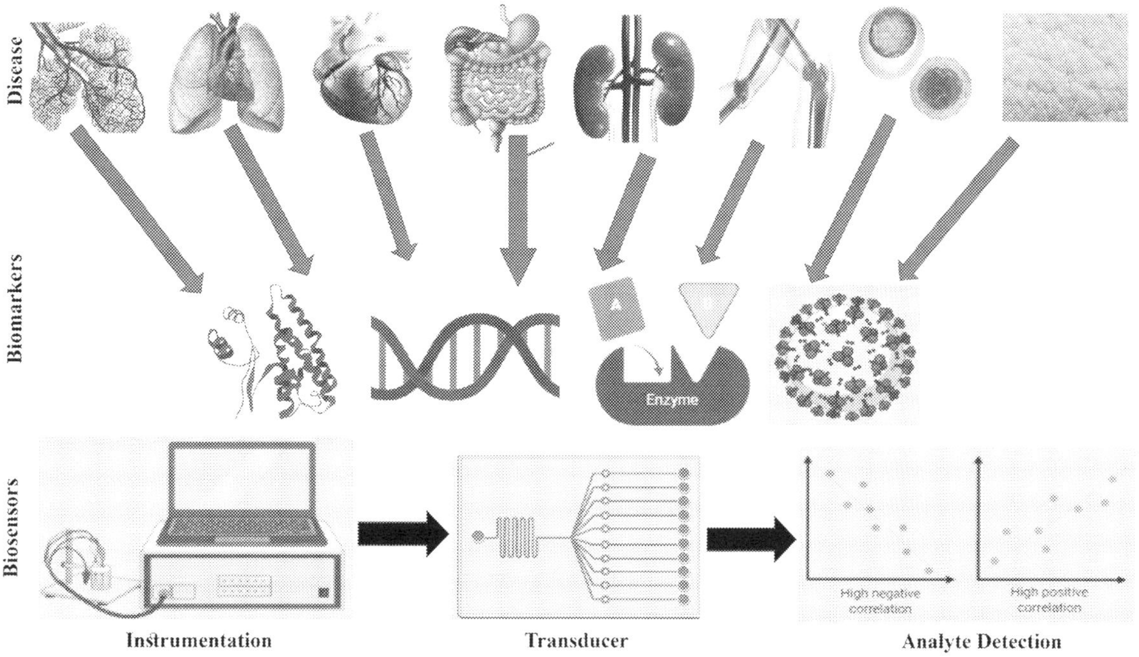


FIGURE 20.4 Schematic diagram of disease detection using biosensors to detect specific biomarkers.

directly to patient, minimizing the time span the patient has to wait for outcomes, thereby reducing stress, relieving anxiety, increasing acceptance, and making the entire procedure much less costly.

Quantum circuitry as well as nanostructural advancements aid in chemical biosensors engineering, making it possible to create tiny sensors that are sensitive enough to detect biochemical signatures in a liquid (Zhang et al., 2017). Data from a huge proportion of such instruments that continuously flow through the bloodstream or are used for analysis enables the characterization and detection of small chemical entities in external matrices, body fluids, cells, as well as tissues, to be measured. Predictions of different modern system capacities were used to examine their efficiency in reaction to specific injuries or infection for the identification of standard chemical entities released in the body systems. These outcomes suggest that the devices can easily differentiate very unique chemicals or slight quantitative changes in their amounts from the background chemical concentration or noise in vivo, offering enhanced detection in body systems (Kassal et al., 2018; Yao et al., 2014). The various current procedures generally used for the analysis of blood samples, after extraction of various chemical analytes, are generally difficult to discriminate from such a context. The schematic picture of the roadmap from disease to their diagnosis by biosensors is given in Fig. 20.4 (Metkar & Girigoswami, 2019; Rebelo et al., 2019; Vigneshvar et al., 2016)

## 20.8 Designing and advancements of biosensor design

The basic designing and construction of a biosensor is multifaceted and depends on various factors. Beginning from one of the most basic components, the strategy is to fabricate or create diagnostic devices utilizing the bottom-up approach or cumulative growing technique (Song et al., 2006; Turner, 2013a) (Papadimitrakopoulos, Vaddiraju, Jain, & Tomazos, 2013). Such a long-term outlook for nanobiosensors in diagnostic techniques is challenging because there have been no significant accomplishments or developments in this domain which can transform these principles to instrumentation and then to large-volume sales, thereby leading to widespread acceptance (Sagadevan & Periasamy, 2014). The recent phenomenon of shifting from fluorescent labeling techniques, since miniaturization and more advancements in technology, decreases the threshold of analyte detection. These scientific advancements could help the adoption of nanobiosensors in near future, but there have been some developments that make fluorescent labeling techniques practical with nanomaterials (Rizwan et al., 2018). The production of nonamplification yet modernized diagnostic technologies for biomolecules detection will also be enabled by nanobiosensors. Nanomedicine can theoretically be included as a mere enhancement to examine a single cell or organelle to diagnose a genetic disease (Debnath & Das, 2020).

## 20.9 Biosensor ligands used for disease diagnosis

A wide variety of ligands are employed as detection agents by biosensors. These ligands generally include various biomolecules, whole cells, and receptors.

### 20.9.1 Nucleic acid ligands

Fluorescence in situ hybridization (FISH) and polymerase chain reaction (PCR) are two of the most commonly used techniques that are generally employed for the detection of DNA, nucleic acids, genes, oligonucleotides, etc. (Wang et al., 2014). These technologies are widely employed for pathogen or microbe identification which are responsible for causing various human diseases (Wang et al., 2017). FISH relies on tagging oligonucleotides using a fluorescent dye and these oligonucleotides selectively hybridize with complementary sequences increasing specificity and sensitivity many-folds. The other common technique, PCR, involves selective in vitro amplification of certain specific DNA sequences. Both these approaches depend on the DNA sequences and, hence, are to be known for the identification of corresponding disease-causing pathogenic organisms. To develop such novel diagnostics based on oligonucleotides ligands and biomarkers generally, they are compared with a reference or standard DNA sequences of known pathogens (Zhang et al., 2017). This strategy uses such genes or sequences which are present invariably and in high amounts in the cells of the targeted species. These methods are therefore very specific given very fewer false-positives. Due to their inherent specificity, they are widely employed in hybridization assays (Kellis et al., 2004). Some specific conditions are criteria to be met if nucleic acids or oligonucleotides are to be used as ligands to target specific gene sequences (Wang et al., 2019). Firstly, they must be complementary and highly specific for a particular target sequence. Secondly, the nucleotide sequence must not give rise to any secondary structures like hairpin, loop, dimer, etc. Thirdly, the melting temperature (Tm) of the sequence must be accurately calculated so that they can bind to the complementary sequences stringently only and no other nearby complementary ones. Probes based on nucleic acid targets and their design are widely used in microarray techniques. FISH, as well as PCR technologies, can also be used easily on a wide range of samples detecting multiple sequences simultaneously. Some studies showed (Kim et al., 2019) the detection of *Vibrio cholerae* based on 23-mer primers detection by a quantitative PCR. These primer sequences of the microbes use the lipoprotein lolB gene present in the microbial membrane as a target leading to the precise identification and isolation of various strains, species, subspecies. *Vibro cholerae* is a very important pathogen causing severe diarrhea and dehydration. They can be present in a wide variety of samples and hence their detection can be of much importance. Ogura et al. described a similar technique based on a probe of microarray (Ogura et al., 2001). This strategy compared the edited distances among various sequences of a single organism thereby evading cross hybridization with similar probes on the array. Researchers used the *Mycobacterium tuberculosis pili* (MTP) gene and its corresponding sequence of proteins to detect tuberculosis along with their specific potential biomarkers. They are found to be highly conserved and are very specific strains of the mycobacterium tuberculosis complex (MTBC) as evident using various alignment and BLAST (Basic Local Alignment Search Tool) tools or similar databases (Naidoo et al., 2014). Such studies demonstrate and illustrate that the emphasis of nucleotide-based diagnostic approaches and the selection of suitable biomarker ligands has been on sequence comparison. However, through the use of next-generation sequencing (NGS), which does not rely on prior knowledge of the pathogen's genetic sequence, this paradigm is shifting (Deneke et al., 2017).

### 20.9.2 Protein and peptide ligands

For protein biomarker discovery and ligand discovery, a range of methodologies are used, including mass spectrometry, gel electrophoresis, and protein microarrays; the latter of which allows for the analysis of entire proteomes. After computational preselection, peptide microarrays can also be used to pick peptide ligands (Amir et al., 2015). Another notable approach is phage display of proteins, which involves the expression and appearance of a large variety of peptides or proteins on the surfaces of bacteriophages, allowing them to be selected against a target biomolecule (Ch'ng et al., 2012). As a result, this technique can be used to discover new immunological assay reagents, such as phage-displayed peptides that mimic pathogen antigens with uses such as leprosy diagnosis and the advancement of visceral leishmaniasis vaccines. Studies showed that three single-chain variable fragments (scFv) for detecting influenza virus strains could help with outbreak detection and control. Antibodies may also be presented on the surface of the phage for possible therapeutic roles and can be used against various targets or antigens with easier and accurate detection strategies (Chaisri & Chaicumpa, 2018).

### 20.9.3 Other ligands

This group includes unconventional ligands like aptamers and possesses a very selective affinity being chosen by different in vitro methods. They are highly specific giving much more sensitive results in respect to other ligands. Similar to antibodies, unique protein isoforms or conformations (Zhou et al., 2014) could be their targets. Systematic evolution of ligands by exponential enrichment (SELEX) is used to pick aptamers from libraries against a variety of biomolecular targets, including proteins and carbohydrates, as well as inorganic molecules (Wu & Kwon, 2016) and whole cells. Studies conducted (Aimaiti et al., 2015) used the whole-cell SELEX method to select species-specific aptamers for discriminating *Mycobacterium tuberculosis* strains. In other studies (Shiratori et al., 2014) DNA aptamers specific for different subtypes of influenza A proteins from viruses are used as a sandwich detection system that showed good results. Besides, aptamers were applied as therapeutic agents, one example being the newly recognized S15 aptamer, which attaches to the dengue envelope protein and also their several serotypes and neutralizes these infections (Engelberg et al., 2019). These illustrations demonstrate how combining biochemistry, physiology, bioinformatics, genetics, and other allied research fields can help with the discovery of biomarkers and also choosing ligands, permitting a deeper understanding of various illnesses and diseases along with the creation of successful diagnostic methods like biosensors (Cho, 2007).

## 20.10 Detection of pathogenic organisms in diseases by biosensors

Microorganisms or germs are wide known to be the cause of various diseases. They cannot be seen with the naked eye but exist ubiquitously. They are the major causes of infections, illnesses, or diseases of organisms including plants and animals. These microorganisms can be detected by using biosensors. Although a wide variety of microbes are generally present, among them bacteria and viruses are the most important ones. They are the causative agents of several dangerous diseases and some of them don't even have any cure. Therefore, detection and screening technologies of such microbial cells, their protein, or any extracellular part may greatly avoid disease. Moreover, it may decrease the economic burden related to therapy and drastically decrease the mortality of people.

### 20.10.1 Virus detecting biosensors

The world is currently suffering from a pandemic of COVID-19. Although vaccine and some medications are available, these are not always effective. Therefore, any detection technique that can indicate the presence of this virus would be highly beneficial. A field-effect transistor-based biosensor has been developed by coating sheets of graphene with a specific antibody that may target an antigen of coronavirus. The instrument was found to be highly sensitive, detecting as low as 1–100 fg/mL of the virus in samples. Similar methods or strategies can be devised and widely used (Afzal et al., 2017). Again, gravimetric diagnosis methods based on QCM transducers along with synthetic as well as natural receptors are minute sensing instruments that can specifically identify and enumerate detrimental virus species. In another study, researchers developed an economical and portable biosensor for the Zika virus. The probe was made of graphene and studded with a monoclonal antibody specific for the Zika virus. It can quantitatively detect the virus and in real-time. It is highly sensitive at detecting about 450 p.m. of antigens in clinical samples (Afsahi et al., 2018). Some novel biosensors are even developed that can detect viruses in food samples thereby preventing the entry of virus from the food matrix to the human system (Neethirajan et al., 2017). Similarly, another electrochemical biosensor based on DNA tetrahedral nanostructure was established for the detection of avian influenza A (H7N9) virus by identifying and comparison of gene sequences of hemagglutinin fragment (Dong et al., 2015). This technique showed a sensitivity limit of 100 fm. Norouzi et al., developed a cadmium-tellurium quantum dot to detect human T-lymphotropic virus-1. Two probes such as biotin-labeled acceptor and NH2-reporter probes with target DNA were hybridized. The resulted sandwich complex was immobilized on a well containing streptavidin and quantum dots were applied for detection. This technique has a sensitivity limit of around 20 pg/μL and can be widely used for the identification of nucleic acids (Norouzi et al., 2017). Biosensors specific for the Zika virus have been developed, being coated with antibodies; and biosensors also for the Dengue virus that can detect it with high accuracy (Cabral-Miranda et al., 2018; Eivazzadeh-Keihan et al., 2019). Some other biosensors specific for the detection of viruses are given in Table 20.2.

### 20.10.2 Bacteria detecting biosensors

In many areas, such as medicine and food safety, pathogenic bacteria are important targets for identification. Different methods for detecting pathogenic bacteria have been established since these microorganisms play a role in global

**TABLE 20.2** Some pathogens detected by biosensor technology along with their detection limits.

| Organisms | Type of probe or biomarker used | Biosensor type | Detection limit |
|---|---|---|---|
| HPV | DNA oligonucleotide | Electrochemical | 4.03 × 10−14 M |
| Hepatitis B virus | DNA oligonucleotide | Electrochemical | 2.61 nM |
| Dengue virus | DNA oligonucleotide | Electrochemical | Not detected |
| HPV 16 | 5051 mAb | Electrochemical | Not detected |
| Dengue virus | NS1 protein/anti-NS1 antibody | Electrochemical | 3 ng $mL^{-1}$ (PBS) 30 ng $mL^{-1}$ (Neat blood) |
| *Mycobacterium tuberculosis* | Genomic DNA/DNA oligonucleotide | Electrochemical | 6 ng $\mu L^{-1}$ |
| *Mycobacterium tuberculosis, Mycobacterium avium* | ITS gene/DNA oligonucleotide | Optical | 4.2 × 104 CFU $mL^{-1}$ and 3.7 × 104 CFU $mL^{-1}$, respectively |
| *Mycobacterium leprae* | IgM and IgG antibodies/ND-O and LID-1 antigens | Immunosensor | Not detected |
| *Neisseria meningitidis* | CtrA gene/DNA oligonucleotide | Electrochemical | Not detected |
| *Plasmodium falciparum* | PfHRP-2/specific antibody | Electrochemical immunosensor | 8 ng $mL^{-1}$ |
| *Plasmodium falciparum and Plasmodium vivax* | pLDH/aptame | Electrochemical aptasensor | 1 p.m. |
| *Toxoplasma gondii* | IgG antibodies/specific antigen | Piezoelectric immunosensor | ~1:5500 dilution |
| *Leishmania infantum* | *L. infantum* antibodies/*L. infantum antigens* | Optical immunosensor | 1:6400 dilution |
| *Leishmania donovani* | 18 S rRNA gene/DNA oligonucleotide | Electrochemical | 0.02 ± 0.002 ng/μL |

diseases such as tuberculosis, leprosy, and meningitis. Bacteria are the most disease-causing organisms in humans. They can inhabit any environment and survive. They are known to cause a huge variety of diseases some of which are curable, unlike viruses. Medications and drugs are available to cure bacterial infections. While in the case of viruses as discussed previously, cures are not so much available. Bacteria are also found in water samples and food samples which causes diseases in healthy people. In the past few years innovative biosensors have been developed that can detect the whole bacterial cell qualitatively as well as differentiating them from other bacterial species (Ahmed et al., 2014). They generally use some specific proteins present, particularly in the target bacteria and not in any other bacteria. Optical as well as electrochemical biosensors are widely used for diagnosis of bacterial pathogens. Impedimetric biosensors are also coming in the market but the nanobiosensors are the most sought ones (Mobed et al., 2019; Yoo & Lee, 2016). They possess various advantages not present in hefty and bulky sensors. Some bacterial species detected by biosensors are given in Table 20.2 along with their detection phenomenon and detection limit.

### 20.10.3 Protozoan-detecting biosensors

Protozoa are one of the most common parasitic groups that infect humans. Biosensors for the diagnosis of protozoan-caused diseases such as malaria, leishmaniasis, American trypanosomiasis (Chagas disease), and toxoplasmosis have been developed using a range of approaches (Mukherjee & Mukherjee, 2021).

Malaria has been well known among people and since early times many peoples are known to be infected. However, there is still no full-proof method of diagnosis of this disease currently available. Therefore, biosensors can be an exciting and promising alternative that can indicate the presence of any antigen-specific for protozoa to be detected at a very minimal amount. Real-time monitoring can be also possible to see the progression of such diseases in an individual. Late treatment can prove highly fatal in some cases. So biosensors are the need of the hour to efficiently track the cause as well as the progression of diseases. Some common protozoans which are being detected using biosensors in contemporary times are tabulated below.

The following table shows some microorganisms which were studied to evaluate the sensing ability of biosensors employing various probes and mechanisms along with their sensitivity (Table 20.2).

## 20.11 Nanoscience and disease biosensor

Nanotechnology and nanoscience are a strong-growing domain and hold a major effect on knowledge and development of biosensors as well as disease detection, monitoring, screening, and diagnosis (Ramos et al., 2017; Rupak et al., 2021). Most diseases are usually diagnosed after it has already progressed to a great extent, affecting the whole physiological system and making patients' therapy sometimes very problematic and unsuccessful as at that stage no drugs or medications can cure those disease. Therefore, premature detection of diseases is very necessary. However, unlike for cancer, there are medicines available for several human diseases and they work on the human system efficiently. Utilizing nanotechnology in the production of biosensors increases the probabilities of detecting cancer earlier, leading to the improvement of overall treatment efficacy as well as reduction of patient mortality (Bayford et al., 2017). A wide variety of diagnostic procedures are currently used in the screening of diseases possessing pros and cons and none of these technologies is perfect. Clinicians physically examine the patients primitively, then if any abnormality is noticed they give some tests of biomarkers which usually indicate whether the whole system is in a state of inflammation or not, or if any immune responses are generated. The presence of a diseased state or pathogenic organisms inside the body is indicated by some physical markers as well as the immunological response of the individual (Moussavi et al., 2007). These diagnostic techniques have a major disadvantage as these biomarkers cannot be observed in a few cases and go undetected and unidentified (Stern et al., 2010) Nanomaterials can be used as imaging compounds enabling diseased states to be differentiated and diagnose more sensitively and reliably. The nanostructures widely exploited in such techniques are dendrimers, liposomes, carbon nanotubes, buckyballs, etc. and can hugely improve as well as accelerate the imaging procedures of diseases (Swierczewska et al., 2012). In addition, nanoscience applications can create minuscule sensors, resulting in cheap technology and improved recognition of disease markers, more effective and precise signal detection and analysis, as well as high-performance detection (Girigoswami & Akhtar, 2019; Hammond et al., 2016).

Nanomaterials usually encompass particles or structures having dimensions in the range of 1−100 nm. The small size of nanoparticles enables a higher ratio of surface to volume. This ratio increase enables better methods of diagnosis, imaging, prognosis, and enhanced delivery of drugs to diseased states and environment that were earlier unavailable and of no significance (Moradi et al., 2018). In various novel instrumentations nanowires, nanocantilevers, and nanochannels have been exploited for improved transduction of signals as well as identification of disease-specific events. To detect micro-RNAs, researchers have developed a biosensor which depends on nanowire technology. MiRNAs are major regulators that control the expression level of genes and are connected to disease origin, their inherent physiology, and progression. Outdated miRNA detection approaches such as northern blotting are time-consuming, expensive, and possess low sensitivity. The development of an easy-to-use, low-cost, noninvasive biosensor to detect miRNAs associated with specific diseases might result in a huge advancement in the field of disease screening, diagnosis, and prognosis (Chao et al., 2016).

Innovations in single-walled carbon nanotubes (SWCNTs) have led to their increased usage and also dramatically improved their electrochemical biosensor detection capabilities, and greatly increasing sensitivity (Yang et al., 2015). They possess amplified action against $H_2O_2$ and NADH and are being widely used to boost signal detection and transduction in immunosensors and oligonucleotide sensors for cancer biomarkers. Surface-enhanced Raman scattering (SERS), in association with nanotechnology has led to huge advancements in techniques of an optical biosensor. With SERS, more than existing techniques can be done with a degree of multiplicity. Without any intervention, SERS can estimate and distinguish simultaneously up to 25 disease markers.

A typical instance as how nanotechnology can revolutionize disease diagnosis is the development of microfluidic lab-on-a-chip (LOC) devices (Luka et al., 2015). A laboratory's complexity is taken up by LOC technology and streamlined into an affordable, compact, easy-to-use device that can be easily handled by patients, clinicians, and diagnostic personnel. LOC approaches using immunological assays and arrays of DNA hybridization have been tested for their ability to identify people who may be at high risk of some genetic and rare diseases. The use of quantum dots is yet another significant application of nanotechnology (Hildebrandt, 2011). Nanocrystalline quantum dots possess luminescence, phosphorescence, and fluorescence showing many novel properties and are widely applied in optical biosensors for screening cancer. Photons of various intensity, wavelength, and spectra are released by quantum dots, allowing several unique molecular elements to be diagnosed and detected (Wegner & Hildebrandt, 2015). As they travel through an environment, they interact with various molecular entities, molecules, cells, and analytes. The pattern of interaction varies which leads to the identification of matter. As such, by detecting cellular abnormality, presence or absence of certain markers, their selective differentiation and drug therapy efficacy, these QDs holds great promise and advantage in monitoring the growth and development of several diseases (Faridbod & Sanati, 2019). Their high stability, multidimensionality, and miniature size make such quantum dots very alluring for application in these fields. Quantum dots are also able to supply particular target areas with therapeutic agents to enhance pharmaceutical efficacy while reducing toxic effects (Nguyen et al., 2015). Nanobiosensors have turned out to be an emerging area of interdisciplinary research in recent years. Nanomaterials have allowed the advancement of normal biosensors of ultrasensitive biosensors in the field of detection of diseases. Their efficacy has increased further owing to their good biocompatibility, high surface area, electrocatalytic activity, superior electronic characteristics. Within the next decade, nanodiagnostics will be hugely accessible, and will be able to detect hundreds of biomarkers simultaneously, quickly, and economically (Wang et al., 2017). The most common clinical diagnostic application will be analysis of biofluids by nanobiosensors. Nanobiosensors not only offer ultra-sensitiveness in cancer biomarker detection but may open new avenues to detect diseases in their primitive stages along with treatment. Nanodevices for such applications are still in the stage of feasibility. Nanotheranostics is an emerging area of nanotechnology where nanodevices useful for therapy will be rooted as a prophylaxis target in persons without any noticeable appearances or symptoms of illness or diseases and monitoring can be easily done by smartphones. Such intense levels of screening and monitoring can perceive such maladies in the preliminary stages thereby enabling healthcare professionals for suitable therapeutic intervention. Early detection and treatment might upsurge the likelihoods of survival and therapy. Forthcoming inclinations in disease diagnostics will endure in shrinking of biochip technology to the nanoscale array (Park et al., 2018).

## 20.12 Conclusion

Since their invention half a century ago, biosensors have shown huge promise and potential to revolutionize detection as well as prevention of dangerous diseases. Rapidity, specificity, and sensitivity are all important characteristics for early detection and also the administration of medications and drugs, so their impact on clinical management is well understood. Novel technologies such as microfluidics and nanotechnology, as well as the unearthing of efficient biomarkers, can greatly increase their efficacy. Some diseases are highly communicable and possess the potential to spread and result in epidemics and pandemics amongst the human population. They can affect a large section of the population and be lethal with huge chances of recurrence. Therefore, keeping such points in mind, the convenience of effective diagnostic technologies is vital.

The future seems to be optimistic as well. Genetic as well as epigenetic regulations, with knowledge of physiology, biochemistry, chemistry, and bioinformatics can greatly enrich this field of disease diagnosis by biosensors to focus and thereby rectify the complexities of different biological paradigms and developments in healthy as well as diseased states. Innovative research goals emerge, and others become clear due to the unraveling of facts, detailed studies, etc., making physiology more understandable, especially in the areas of therapeutics and diagnosis, or even both (i.e., theranostics). As a result, integrating such optimistic findings with hopeful advancements such as biosensors has the potential to completely change the clinical diagnosis landscape.

For effective therapy as well as recovery of patients suffering from a disease, a precise diagnosis is needed. Diagnostic techniques ought to be easy, penetrating, and capable of detecting numerous biomarkers in noninvasive way within biological fluids at low concentrations. These specifications can be met with biosensors. Again, these instruments must be more technologically advanced and enhanced to meet the new challenges posed by discovery of various pathogens as well as diseases in recent times, such as multiplex analysis of multiple biomarkers that necessitates the development of arrays of sensors on a single platform.

Biosensors are well-known for their use in clinical chemical analysis. Biosensors for measuring blood metabolites such as glucose, lactate, urea, and creatinine, using both electrochemical and optical modes of transduction, have been developed commercially and are routinely used in labs, POC environments, and, in the case of glucose, for self-testing. Although immunosensors struggle to compete with conventional immunoassays based solely on sensitivity criteria, they show promise in tests where some sensitivity can be sacrificed in exchange for enhanced ease of use and faster turnaround times, such as near-patient testing for cardiac and cancer markers. While biosensors are used in a variety of clinical applications, only a few biosensors for cardiovascular and cancer-related clinical testing have been created. The evolution of genomic and proteomic molecular tools to profile tumors and create molecular signatures based on genetic and epigenetic signatures, changes in gene expression and protein profiles, and protein post-translational modifications has opened up new avenues for using biosensors in cancer research. Because of the complexity and diversity of cancer, harnessing the ability of biosensors is difficult. Continued development of biomarkers and ligands for those biomarkers, as well as sample preparation methods and multichannel biosensors capable of analyzing multiple cancer markers simultaneously, will be needed for the successful development of biosensor-based cancer testing. Biosensors for cancer clinical testing can improve assay speed and versatility, allow for multitarget analyses and automation, and lower

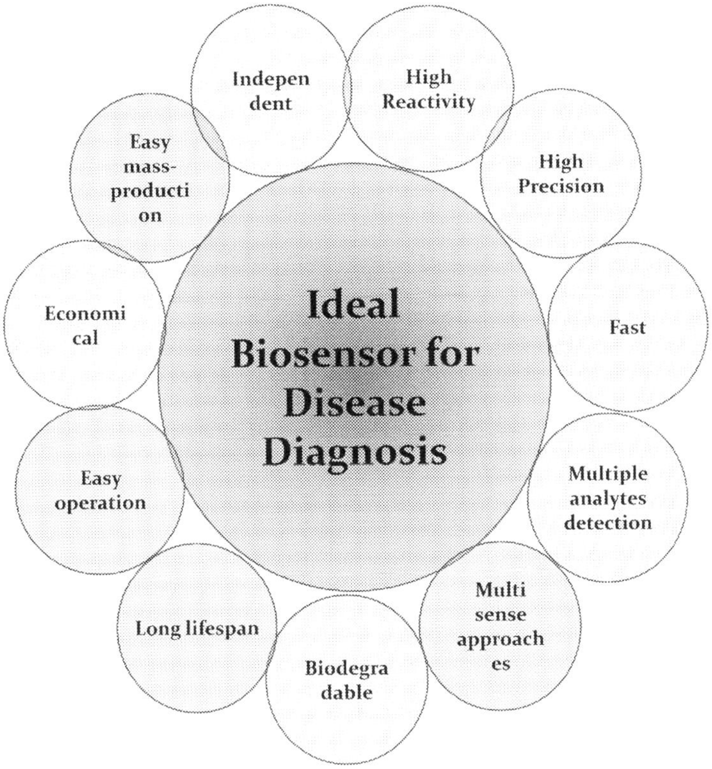


**FIGURE 20.5** Characteristics of ideal biosensor for disease diagnosis.

diagnostic testing costs. Biosensors have the ability to carry molecular research to underserved communities and the community healthcare environment. Validation of cancer biomarkers discovered through basic and clinical research, as well as genomic and proteomic analyses, is needed. Biosensors for cancer-related clinical research can be made by combining ligands and probes for these markers with detectors. Integration and automation of technology, as well as the creation of effective sample preparation methods, are needed for POC cancer testing. An ideal biosensor must possess some characteristics or properties which are depicted in Fig. 20.5.

## 20.13 Future aspects

The implementation of biosensors is growing tremendously in several domains of environmental monitoring and assessment. Though, for investigators, certain drawbacks such as specificity, responsiveness, reaction time, and operational lifespan are of critical significance and are crucial to the field of analysis (Castillo et al., 2004; Brazaca et al., 2020). In all aspects, there is an urgent requirement to advance existing biosensors and to render them more attractive investigative apparatuses. The utmost likely zones of improvements that are anticipated to evolve significantly and increase the use of biosensor technology for the identification of multiple pollutants may be control and detection systems, nanodevelopment, compression, and multisensory array strengths (Liao et al., 2018; Nikitin et al., 2003). In addition, it is important to build innovative biosensors based on the generalized base components and signaling components that must be robust, simple to make, and worthy of extending the specificity array.

Studying diverse individual specimens is possible with the advent of sophisticated sensors, which will aid in field observations, reducing contributions from nonspecific background results. This research has shown that the potential of biosensors is exciting, but it may be focused on the successful enforcement of evolving different scientific technologies, chemistry, biochemistry, doping physics, and electronics. Biosensor advancement is an evolutionary area possessing multiple areas for progress and development in the strategy and architectures of evolving biosensors. Unearthing and inclusion of different chemical substances which were previously unexplored and unknown, can become a threat to humans and the natural ecosystems, and may require novel, quicker and convenient biosensors for detection. Bearing in mind that the implementation of new technologies is the best indicator of success and achievement for an emerging technology, further advancements are very much important to gain the approval of prospective consumers.

Molecular diagnostics is a straightforward path for future biosensor studies. A significant aim is to improve the sensitivity of DNA biosensors for single-molecule detection in unamplified samples. To achieve this goal, the signal-to-noise ratio must be improved, the signal provided by the biochemical reaction must be improved, or the transducer's sensitivity must be increased while background noise is reduced. Technologies such as ultrasensitive transducers will be needed. Microcantilevers for detecting mass changes upon detection of a binding event and QCMs capable of tracking formation are two recent examples of enhanced sensitivity transduction modes by detecting acoustic emissions and rupturing chemical bonds. The latter has shown the ability to detect a single virus particle with high sensitivity. Increasing the amplitude of the arrays for more complete and rapid DNA sequencing information is another area of focus, and progress in this area may be restricted in the end by the detection transducer's resolution. DNA chips are being used in complete analysis systems that combine microfluidics and a biosensor into a single structure. In the future, these systems could provide sample preparation capabilities, a user-friendly handling system, chemical analysis, and signal acquisition capabilities. Homogeneous sensing formats and microfabrication technologies for DNA analysis will be critical to the advancement of LOC analysis systems. The creation of synthetic polymeric probes that emit fluorescence only after hybridization to native DNA targets, allowing monitoring of hybridization in real time without the need for separation measures, is one recent step toward a homogeneous assay. Nanotechnologies would need to be further developed and improved in order to manufacture nanoscale devices with larger array sizes by using less sample volume. The future of such devices for rapid disease diagnosis may be particularly useful in POC settings. However, in order for reliable devices to achieve widespread acceptance, their cost and quality control must be strictly adjusted. To cut costs, homogeneous assay formats, which remove the need for sample preparation and amplification steps, as well as mass fabrication, will be crucial.

Molecular biology will play a key role in biosensor production in the future, for example, in improving biocomponent stability and evolving aptamers. The highly repeatable synthetic approach and ease of immobilization of aptamers bode well for the custom design of potential molecular diagnostics biosensors. Future advancements in biosensor technology, such as biomarker patterns, software, and microfluidics, may make these devices extremely useful in health-related applications. The use of nanomaterials in the production of biomarker detection sensors would make these instruments more sensitive and useful for POC early diagnosis. Early diagnosis will help patients live longer, and the effective implementation of biosensors for disease diagnosis and monitoring will necessitate adequate funding to take the technology from testing to commercialization.

Biosensor research and development over the past few decades has shown that the technology is only in its infancy. Cost factors and some key technical obstacles may have led to the slow and restricted technology transfer. Many of the most recent big breakthroughs have had to wait for miniaturization technologies to emerge from research in the electronic and optical solid-state circuit industries. Automation, miniaturization, and system integration with high throughput for multiple tasks have all influenced analytical chemistry. Biosensor technology, which is often designed to detect one or a few target analytes, faces a significant challenge in meeting such requirements. Effective biosensors must be flexible enough to support interchangeable biorecognition components, as well as miniaturized enough to allow for simultaneous sensing automation and ease of operation at a reasonable cost. Biosensors have a promising future. These advances, however, would necessitate a concerted multidisciplinary approach if sensor systems are to successfully make the transition from the research and development lab to the marketplace. The combination of many new techniques derived from physical chemistry, molecular biology, biochemistry, thick and thin film physics, materials science, and electronics, as well as the necessary skills, has revealed the potential for developing a viable and clinically useful biosensor.

## References


Abid, S. A., Ahmed Muneer, A., Al-Kadmy, I. M. S., Sattar, A. A., Beshbishy, A. M., Batiha, G. E. S., & Hetta, H. F. (2021). Biosensors as a future diagnostic approach for COVID-19. *Life Sciences*, 273. Available from https://doi.org/10.1016/j.lfs.2021.119117.

Abolhasan, R., Mehdizadeh, A., Rashidi, M. R., Aghebati-Maleki, L., & Yousefi, M. (2019). Application of hairpin DNA-based biosensors with various signal amplification strategies in clinical diagnosis. *Biosensors and Bioelectronics*, *129*, 164–174. Available from https://doi.org/10.1016/j.bios.2019.01.008.

Acharya, G., Mitra, A. K., & Cholkar, K. (2017). *Nanosystems for Diagnostic Imaging, Biodetectors, and Biosensors. Emerging Nanotechnologies for Diagnostics, Drug Delivery and Medical Devices* (pp. 217–248). Elsevier Inc. Available from https://doi.org/10.1016/B978-0-323-42978-8.00010-3.

Afsahi, S., Lerner, M. B., Goldstein, J. M., Lee, J., Tang, X., Bagarozzi, D. A., ... Goldsmith, B. R. (2018). Novel graphene-based biosensor for early detection of Zika virus infection. *Biosensors and Bioelectronics*, *100*, 85–88. Available from https://doi.org/10.1016/j.bios.2017.08.051.

Afzal, A., Mujahid, A., Schirhagl, R., Bajwa, S. Z., Latif, U., & Feroz, S. (2017). Gravimetric viral diagnostics: QCM based biosensors for early detection of viruses. *Chemosensors*, *5*(1). Available from https://doi.org/10.3390/chemosensors5010007.

Ahmed, A., Rushworth, J. V., Hirst, N. A., & Millner, P. A. (2014). Biosensors for whole-cell bacterial detection. *Clinical Microbiology Reviews*, *27* (3), 631–646. Available from https://doi.org/10.1128/CMR.00120-13.

Aimaiti, R., Qin, L., Cao, T., Yang, H., Wang, J., Lu, J., ... Hu, Z. (2015). Identification and application of ssDNA aptamers against H37Rv in the detection of Mycobacterium tuberculosis. *Applied Microbiology and Biotechnology*, *99*(21), 9073–9083. Available from https://doi.org/10.1007/s00253-015-6815-7.

Alexandra, R., C., R. L., I., R. C., E., L. R., Simina, B., & I., M. R. (2019). A new approach for the diagnosis of systemic and oral diseases based on salivary biomolecules. *Disease Markers*, 1–11. Available from https://doi.org/10.1155/2019/8761860.

Alhadrami, H. A. (2018). Biosensors: Classifications, medical applications, and future prospective. *Biotechnology and Applied Biochemistry*, *65*(3), 497–508. Available from https://doi.org/10.1002/bab.1621.

Amir, S., Kenji, U., Kin-ya, T., Kotaro, K., & Hisakazu, M. (2015). Label and label-free detection techniques for protein microarrays. *Microarrays*, 228–244. Available from https://doi.org/10.3390/microarrays4020228.

Anik, U. (2017). *Electrochemical medical biosensors for POC applications. Medical Biosensors for Point of Care (POC) Applications* (pp. 275–292). Elsevier Inc. Available from https://doi.org/10.1016/B978-0-08-100072-4.00012-5.

Arugula, M. A., & Simonian, A. (2014). Novel trends in affinity biosensors: Current challenges and perspectives. *Measurement Science and Technology*, *25*(3). Available from https://doi.org/10.1088/0957-0233/25/3/032001.

Asal, M., Özen, Ö., Şahinler, M., & Polatoğlu, İ. (2018a). Recent developments in enzyme, DNA and immuno-based biosensors. *Sensors (Switzerland)*, *18*(6). Available from https://doi.org/10.3390/s18061924.

Asal, M., Özen, Ö., Şahinler, M., & Polatoğlu, İ. (2018b). Recent developments in enzyme, DNA and immuno-based biosensors. *Sensors (Switzerland)*, *18*(6). Available from https://doi.org/10.3390/s18061924.

Attur, M., Krasnokutsky-Samuels, S., Samuels, J., & Abramson, S. B. (2013). Prognostic biomarkers in osteoarthritis. *Current Opinion in Rheumatology*, *25*(1), 136–144. Available from https://doi.org/10.1097/BOR.0b013e32835a9381.

Ayesha, T., Abdul, R., & Z, B. S. (2020). *Biomarkers and Their Role in Detection of Biomolecules* (pp. 73–94). Wiley. Available from https://doi.org/10.1002/9783527345137.ch4.

Bayford, R., Rademacher, T., Roitt, I., & Wang, S. X. (2017). Emerging applications of nanotechnology for diagnosis and therapy of disease: A review. *Physiological Measurement*, *38*(8), R183–R203. Available from https://doi.org/10.1088/1361-6579/aa7182.

Bellagambi, F. G., Baraket, A., Longo, A., Vatteroni, M., Zine, N., Bausells, J., ... Errachid, A. (2017). Electrochemical biosensor platform for TNF-α cytokines detection in both artificial and human saliva: Heart failure. *Sensors and Actuators, B: Chemical*, *251*, 1026–1033. Available from https://doi.org/10.1016/j.snb.2017.05.169.

Bennett, M. R., & Devarajan, P. (2011). *Characteristics of an ideal biomarker of kidney diseases. Biomarkers of Kidney Disease* (pp. 1–24). Elsevier Inc. Available from https://doi.org/10.1016/B978-0-12-375672-5.10001-5.

Bonassi, S., Neri, M., & Puntoni, R. (2001). Validation of biomarkers as early predictors of disease. Mutation. *Research - Fundamental and Molecular Mechanisms of Mutagenesis*, *480–481*, 349–358. Available from https://doi.org/10.1016/S0027-5107(01)00194-4.

Bottazzi, B., Fornasari, L., Frangolho, A., Giudicatti, S., Mantovani, A., Marabelli, F., ... Valsesia, A. (2014). Multiplexed label-free optical biosensor for medical diagnostics. *Journal of Biomedical Optics*, *19*(1). Available from https://doi.org/10.1117/1.JBO.19.1.017006.

Brazaca, L. C., Sampaio, I., Zucolotto, V., & Janegitz, B. C. (2020). Applications of biosensors in Alzheimer's disease diagnosis. *Talanta*, 210. Available from https://doi.org/10.1016/j.talanta.2019.120644.

Buddhadev, P., R., V. P., P., S. N., & Pranjal, C. (2020). Biosensor nanoengineering: Design, operation, and implementation for biomolecular analysis. *Sensors International*, *1*, 100040. Available from https://doi.org/10.1016/j.sintl.2020.100040.

Cabral-Miranda, G., Cardoso, A. R., Ferreira, L. C. S., Sales, M. G. F., & Bachmann, M. F. (2018). Biosensor-based selective detection of Zika virus specific antibodies in infected individuals. *Biosensors and Bioelectronics*, *113*, 101–107. Available from https://doi.org/10.1016/j.bios.2018.04.058.

Castillo, J., Gáspár, S., Leth, S., Niculescu, M., Mortari, A., Bontidean, I., ... Csöregi, E. (2004). Biosensors for life quality – Design, development and applications. *Sensors and Actuators, B: Chemical*, *102*(2), 179–194. Available from https://doi.org/10.1016/j.snb.2004.04.084.

Cesewski, E., & Johnson, B. N. (2020). Electrochemical biosensors for pathogen detection. *Biosensors and Bioelectronics*, 159. Available from https://doi.org/10.1016/j.bios.2020.112214.

Chaisri, U., & Chaicumpa, W. (2018). Evolution of therapeutic antibodies, influenza virus biology, influenza, and influenza immunotherapy. *BioMed Research International*, 2018. Available from https://doi.org/10.1155/2018/9747549.

Chao, J., Zhu, D., Zhang, Y., Wang, L., & Fan, C. (2016). DNA nanotechnology-enabled biosensors. *Biosensors and Bioelectronics*, *76*, 68–79. Available from https://doi.org/10.1016/j.bios.2015.07.007.

Chen, Y., Wang, Z., Liu, Y., Wang, X., Li, Y., Ma, P., ... Li, H. (2018). Recent advances in rapid pathogen detection method based on biosensors. *European Journal of Clinical Microbiology and Infectious Diseases*, *37*(6), 1021–1037. Available from https://doi.org/10.1007/s10096-018-3230-x.

Ch'ng, A. C. W., Choong, Y. S., & Lim, T. S. (2016). Phage display-derived antibodies: Application of recombinant antibodies for diagnostics. In S. K. Saxena (Ed.), Proof and concepts in rapid diagnostic tests and technologies (pp. 107–135). London, UK: IntechOpen.

Cho, W. C. S. (2007). Contribution of oncoproteomics to cancer biomarker discovery. *Molecular Cancer*, 6. Available from https://doi.org/10.1186/1476-4598-6-25.

Choi, J., Seong, T. W., Jeun, M., & Lee, K. H. (2017). Field-effect biosensors for on-site detection: Recent advances and promising targets. *Advanced Healthcare Materials*, *6*(20). Available from https://doi.org/10.1002/adhm.201700796.

Cova, I., & Priori, A. (2018). Diagnostic biomarkers for Parkinson's disease at a glance: Where are we? *Journal of Neural Transmission, 125*(10), 1417–1432. Available from https://doi.org/10.1007/s00702-018-1910-4.

da Silva, E. T. S. G., Souto, D. E. P., Barragan, J. T. C., de F. Giarola, J., de Moraes, A. C. M., & Kubota, L. T. (2017). Electrochemical biosensors in point-of-care devices: Recent advances and future trends. *ChemElectroChem, 4*(4), 778–794. Available from https://doi.org/10.1002/celc.201600758.

Darsanaki, RK, Azizzadeh, A, Nourbakhsh, M, et al. (2013). Biosensors: functions and applications. *Journal of Biology and Today's World, 2*(1), 53–61. Available from https://doi.org/10.15412/J.JBTW.01020105.

Debnath, N., & Das, S. (2020). *Nanobiosensor: Current trends and applications. NanoBioMedicine* (pp. 389–409). Singapore: Springer. Available from https://doi.org/10.1007/978-981-32-9898-9_16.

Deneke, C., Rentzsch, R., & Renard, B. Y. (2017). PaPrBaG: A machine learning approach for the detection of novel pathogens from NGS data. *Scientific Reports*, 7. Available from https://doi.org/10.1038/srep39194.

do Nascimento, N. M., Juste-Dolz, A., Grau-García, E., Román-Ivorra, J. A., Puchades, R., Maquieira, A., . . . Gimenez-Romero, D. (2017). Label-free piezoelectric biosensor for prognosis and diagnosis of Systemic Lupus Erythematosus. *Biosensors and Bioelectronics, 90*, 166–173. Available from https://doi.org/10.1016/j.bios.2016.11.004.

Dong, S., Zhao, R., Zhu, J., Lu, X., Li, Y., Qiu, S., . . . Song, H. B. (2015). Electrochemical DNA biosensor based on a tetrahedral nanostructure probe for the detection of avian influenza A (H7N9) virus. *ACS Applied Materials and Interfaces, 7*(16), 8834–8842. Available from https://doi.org/10.1021/acsami.5b01438.

Eivazzadeh-Keihan, R., Pashazadeh-Panahi, P., Mahmoudi, T., Chenab, K. K., Baradaran, B., Hashemzaei, M., . . . Maleki, A. (2019). Dengue virus: a review on advances in detection and trends – From conventional methods to novel biosensors. *Microchimica Acta, 186*(6). Available from https://doi.org/10.1007/s00604-019-3420-y.

Engelberg, S., Netzer, E., Assaraf, Y. G., & Livney, Y. D. (2019). Selective eradication of human non-small cell lung cancer cells using aptamer-decorated nanoparticles harboring a cytotoxic drug cargo. *Cell Death and Disease, 10*(10). Available from https://doi.org/10.1038/s41419-019-1870-0.

Esteves-Villanueva, J. O., Trzeciakiewicz, H., & Martic, S. (2014). A protein-based electrochemical biosensor for detection of tau protein, a neurodegenerative disease biomarker. *Analyst, 139*(11), 2823–2831. Available from https://doi.org/10.1039/c4an00204k.

Fang, Y., Yuanyuan, M., G., S. S., & Aiguo, W. (2020). *Transduction Process-Based Classification of Biosensors* (pp. 23–44). Wiley. Available from https://doi.org/10.1002/9783527345137.ch2.

Faridbod, F., & Sanati, A. L. (2019). Graphene quantum dots in electrochemical sensors/biosensors. *Current Analytical Chemistry, 15*(2), 103–123. Available from https://doi.org/10.2174/1573411014666180319145506.

Feiyun, C., Zhiru, Z., & Susan, Z. H. (2020). Review—Measurement and analysis of cancer biomarkers based on electrochemical biosensors. *Journal of the Electrochemical Society*, 037525. Available from https://doi.org/10.1149/2.0252003jes.

Ferreira, M. F. S., Castro-Camus, E., Ottaway, D. J., López-Higuera, J. M., Feng, X., Jin, W., . . . Quan, Q. (2017). Roadmap on optical sensors. *Journal of Optics (United Kingdom), 19*(8). Available from https://doi.org/10.1088/2040-8986/aa7419.

Gilchrist, K. H., Barker, V. N., Fletcher, L. E., DeBusschere, B. D., Ghanouni, P., Giovangrandi, L., & Kovacs, G. T. A. (2001). General purpose, field-portable cell-based biosensor platform. *Biosensors and Bioelectronics, 16*(7–8), 557–564. Available from https://doi.org/10.1016/S0956-5663(01)00169-5.

Girigoswami, K., & Akhtar, N. (2019). Nanobiosensors and fluorescence based biosensors: An overview. *International Journal of Nano Dimension, 10*(1), 1–17.

Gopinath, P, Anitha, V, & Mastani, SA (2015). Microcantilever based biosensor for disease detection applications. *Journal of Medical and Bioengineering, 4*(34). Available from https://doi.org/10.12720/jomb.4.4.307-311.

Gu, N., & Liu, S. (2020). Introduction to biosensors. *Journal of Materials Chemistry B, 8*(16), 3168–3170. Available from https://doi.org/10.1039/d0tb90051f.

Hamm, L., Gee, A., & Indrasekara, A. S. D. S. (2019). Recent advancement in the surface-enhanced raman spectroscopy-based biosensors for infectious disease diagnosis. *Applied Sciences (Switzerland), 9*(7). Available from https://doi.org/10.3390/app9071448.

Hammond, J. L., Formisano, N., Estrela, P., Carrara, S., & Tkac, J. (2016). Electrochemical biosensors and nanobiosensors. *Essays in Biochemistry, 60*(1), 69–80. Available from https://doi.org/10.1042/EBC20150008.

Han, S. T., Peng, H., Sun, Q., Venkatesh, S., Chung, K. S., Lau, S. C., . . . Roy, V. A. L. (2017). An overview of the development of flexible sensors. *Advanced Materials, 29*(33). Available from https://doi.org/10.1002/adma.201700375.

Hanson, M. A., Powell, H. C., Barth, A. T., Ringgenberg, K., Calhoun, B. H., Aylor, J. H., & Lach, J. (2009). Body area sensor networks: Challenges and opportunities. *Computer, 42*(1), 58–65. Available from https://doi.org/10.1109/MC.2009.5.

Herrera-Espejo, S., Santos-Zorrozua, B., Álvarez-González, P., Lopez-Lopez, E., & Garcia-Orad, Á. (2019). A systematic review of microRNA expression as biomarker of late-onset Alzheimer's disease. *Molecular Neurobiology, 56*(12), 8376–8391. Available from https://doi.org/10.1007/s12035-019-01676-9.

Hildebrandt, N. (2011). Biofunctional quantum dots: Controlled conjugation for multiplexed biosensors. In *ACS Nano* (Vol. 5, Issue 7, pp. 5286–5290). American Chemical Society. https://doi.org/10.1021/nn2023123

Huang, Y., Xu, J., Liu, J., Wang, X., & Chen, B. (2017). Disease-related detection with electrochemical biosensors: A review. *Sensors (Switzerland), 17*(10). Available from https://doi.org/10.3390/s17102375.

Idili, A., Plaxco, K. W., Vallée-Bélisle, A., & Ricci, F. (2013). Thermodynamic basis for engineering high-affinity, high-specificity binding-induced DNA Clamp Nanoswitches. *ACS Nano, 7*(12), 10863–10869. Available from https://doi.org/10.1021/nn404305e.

Ispas, C. R., Crivat, G., & Andreescu, S. (2012). Review: Recent developments in enzyme-based biosensors for biomedical analysis. *Analytical Letters*, *45*(2–3), 168–186. Available from https://doi.org/10.1080/00032719.2011.633188.

J, B., Chanda, K., & Balamurali, M. M. (2018). Biosensors for pathogen surveillance. *Environmental Chemistry Letters*, *16*(4), 1325–1337. Available from https://doi.org/10.1007/s10311-018-0759-y.

Jarockyte, G., Karabanovas, V., Rotomskis, R., & Mobasheri, A. (2020). Multiplexed nanobiosensors: Current trends in early diagnostics. *Sensors (Switzerland)*, *20*(23), 1–23. Available from https://doi.org/10.3390/s20236890.

Jazib, A., Jawayria, N., Muhammad, A. A., Muhammad, F. A., & Ali, R. (2017). Biosensors: Their fundamentals, designs, types and most recent impactful applications: A review. *Journal of Biosensors & Bioelectronics*. Available from https://doi.org/10.4172/2155-6210.1000235.

Karunakaran, C., Rajkumar, R., & Bhargava, K. (2015). *Introduction to biosensors. Biosensors and Bioelectronics* (pp. 1–68). Elsevier Inc. Available from https://doi.org/10.1016/B978-0-12-803100-1.00001-3.

Kassal, P., Steinberg, M. D., & Steinberg, I. M. (2018). Wireless chemical sensors and biosensors: A review. *Sensors and Actuators, B: Chemical*, *266*, 228–245. Available from https://doi.org/10.1016/j.snb.2018.03.074.

Kelley, S. O. (2017). What are clinically relevant levels of cellular and biomolecular analytes? *ACS Sensors*, *2*(2), 193–197. Available from https://doi.org/10.1021/acssensors0.6b00691.

Kellis, M., Patterson, N., Birren, B., Berger, B., & Lander, E. S. (2004). Methods in comparative genomics: Genome correspondence, gene identification and regulatory motif discovery. *In Journal of Computational Biology, Vol. 11*(Issues 2–3), 319–355. Available from https://doi.org/10.1089/1066527041410319.

Kim, J., Campbell, A. S., de Ávila, B. E. F., & Wang, J. (2019). Wearable biosensors for healthcare monitoring. *Nature Biotechnology*, *37*(4), 389–406. Available from https://doi.org/10.1038/s41587-019-0045-y.

Kirsch, J., Siltanen, C., Zhou, Q., Revzin, A., & Simonian, A. (2013). Biosensor technology: Recent advances in threat agent detection and medicine. *Chemical Society Reviews*, *42*(22), 8733–8768. Available from https://doi.org/10.1039/c3cs60141b.

Kissinger, P. T. (2005). Biosensors – A perspective. *Biosensors and Bioelectronics*, *20*(12), 2512–2516. Available from https://doi.org/10.1016/j.bios.2004.10.004.

Korotkaya, E. V. (2014). Biosensors: Design, classification, and applications in the food industry. *Foods and Raw Materials*, *2*(2), 161–171. Available from https://doi.org/10.12737/5476.

Koushki, M., Amiri-Dashatan, N., Ahmadi, N., Abbaszadeh, H. A., & Rezaei-Tavirani, M. (2018). Resveratrol: A miraculous natural compound for diseases treatment. *Food Science and Nutrition*, *6*(8), 2473–2490. Available from https://doi.org/10.1002/fsn3.855.

Koyun, A., Ahlatcolu, E., & Koca, Y. (n.d.). Biosensors and their principles. *A Roadmap of Biomedical Engineers and Milestones* (Vol. 2012, pp. 117–142). https://doi.org/10.5772/48824

Kraus, V. B., Burnett, B., Coindreau, J., Cottrell, S., Eyre, D., Gendreau, M., ... Todman, M. (2011). Application of biomarkers in the development of drugs intended for the treatment of osteoarthritis. *Osteoarthritis and Cartilage*, *19*(5), 515–542. Available from https://doi.org/10.1016/j.joc.2010.08.019.

Kussmann, M., Raymond, F., & Affolter, M. (2006). OMICS-driven biomarker discovery in nutrition and health. *Journal of Biotechnology*, *124*(4), 758–787. Available from https://doi.org/10.1016/j.jbiotec.2006.02.014.

Leca-Bouvier, B., & Blum, L. J. (2005). Biosensors for protein detection: A review. *Analytical Letters*, *38*(10), 1491–1517. Available from https://doi.org/10.1081/AL-200065780.

Liao, Z., Wang, J., Zhang, P., Zhang, Y., Miao, Y., Gao, S., ... Geng, L. (2018). Recent advances in microfluidic chip integrated electronic biosensors for multiplexed detection. *Biosensors and Bioelectronics*, *121*, 272–280. Available from https://doi.org/10.1016/j.bios.2018.08.061.

Lin, Y., Qian, F., Shen, L., Chen, F., Chen, J., & Shen, B. (2017). Computer-aided biomarker discovery for precision medicine: Data resources, models and applications. *Briefings in Bioinformatics*, *20*(3), 952–975. Available from https://doi.org/10.1093/bib/bbx158.

Liu, H., Ge, J., Ma, E., & Yang, L. (2018). *Advanced biomaterials for biosensor and theranostics. Biomaterials in translational medicine: a biomaterials approach* (pp. 213–255). Elsevier. Available from https://doi.org/10.1016/B978-0-12-813477-1.00010-4.

Loock, H. P., & Wentzell, P. D. (2012). Detection limits of chemical sensors: Applications and misapplications. *Sensors and Actuators, B: Chemical*, *173*, 157–163. Available from https://doi.org/10.1016/j.snb.2012.06.071.

Luka, G., Ahmadi, A., Najjaran, H., Alocilja, E., Derosa, M., Wolthers, K., ... Hoorfar, M. (2015). Microfluidics integrated biosensors: A leading technology towards lab-on-A-chip and sensing applications. *Sensors (Switzerland)*, *15*(12), 30011–30031. Available from https://doi.org/10.3390/s151229783.

Majumder, S., & Deen, M. J. (2019). Smartphone sensors for health monitoring and diagnosis. *Sensors (Switzerland)*, *19*(9). Available from https://doi.org/10.3390/s19092164.

Mauriz, E., Dey, P., & Lechuga, L. M. (2019). Advances in nanoplasmonic biosensors for clinical applications. *Analyst*, *144*(24), 7105–7129. Available from https://doi.org/10.1039/c9an00701f.

Mayeux, R. (2004). Biomarkers: Potential uses and limitations. *NeuroRx: The Journal of the American Society for Experimental NeuroTherapeutics*, *1*(2), 182–188. Available from https://doi.org/10.1602/neurorx.1.2.182.

McKeating, K. S., Aubé, A., & Masson, J. F. (2016). Biosensors and nanobiosensors for therapeutic drug and response monitoring. *Analyst*, *141*(2), 429–449. Available from https://doi.org/10.1039/c5an01861g.

Mehrotra, P. (2016). Biosensors and their applications – A review. *Journal of Oral Biology and Craniofacial Research*, *6*(2), 153–159. Available from https://doi.org/10.1016/j.jobcr.2015.12.002.

Meshram, B. D., Agrawal, A. K., Adil, S., Ranvir, S., & Sande, K. K. (2018). Biosensor and its application in food and dairy industry: A review. *International Journal of Current Microbiology and Applied Sciences*, 3305–3324. Available from https://doi.org/10.20546/ijcmas.2018.702.397.

Metkar, S. K., & Girigoswami, K. (2019). Diagnostic biosensors in medicine – A review. *Biocatalysis and Agricultural Biotechnology*, *17*, 271–283. Available from https://doi.org/10.1016/j.bcab.2018.11.029.

Mobed, A., Baradaran, B., Guardia, Mdl, Agazadeh, M., Hasanzadeh, M., Rezaee, M. A., ... Hamblin, M. R. (2019). Advances in detection of fastidious bacteria: From microscopic observation to molecular biosensors. *TrAC - Trends in Analytical Chemistry*, *113*, 157–171. Available from https://doi.org/10.1016/j.trac.2019.02.012.

Montagut, Y., Narbon, Jiménez, Y., March, C., Montoya, A., & Arnau. (2011). QCM Technology in Biosensors.

Moradi, S., Khaledian, S., Abdoli, M., Shahlaei, M., & Kahrizi, D. (2018). Nano-biosensors in cellular and molecular biology. *Cellular and Molecular Biology*, *64*(5), 85–90. Available from https://doi.org/10.14715/cmb/2018.64.5.14.

Morales, M. A., & Halpern, J. M. (2018). Guide to selecting a biorecognition element for biosensors. *Bioconjugate Chemistry*, *29*(10), 3231–3239. Available from https://doi.org/10.1021/acs.bioconjchem.8b00592.

Moussavi, S., Chatterji, S., Verdes, E., Tandon, A., Patel, V., & Ustun, B. (2007). Depression, chronic diseases, and decrements in health: Results from the World Health Surveys. *Lancet*, *370*(9590), 851–858. Available from https://doi.org/10.1016/S0140-6736(07)61415-9.

Mukherjee, Suprabhat, & Mukherjee, Niladri (2021). Current Developments in Diagnostic Biosensor Technology: Relevance to Therapeutic Intervention of Infectious and Inflammatory Diseases of Human. In G Dutta, et al. (Eds.), *Modern Techniques in Biosensors. Studies in Systems, Decision and Control* (pp. 1–36). Singapore: Springer, Singapore. Available from https://doi.org/10.1007/978-981-15-9612-4_1.

Nabaei, V., Chandrawati, R., & Heidari, H. (2018). Magnetic biosensors: Modelling and simulation. *Biosensors and Bioelectronics*, *103*, 69–86. Available from https://doi.org/10.1016/j.bios.2017.12.023.

Naidoo, N., Ramsugit, S., & Pillay, M. (2014). Mycobacterium tuberculosis pili (MTP), a putative biomarker for a tuberculosis diagnostic test. *Tuberculosis*, *94*(3), 338–345. Available from https://doi.org/10.1016/j.tube.2014.03.004.

Neethirajan, S., Ahmed, S. R., Chand, R., Buozis, J., & Nagy, É. (2017). Recent advances in biosensor development for foodborne virus detection. *Nanotheranostics*, *1*(3), 272–295. Available from https://doi.org/10.7150/ntno.20301.

Nguyen, N. H., Duong, T. G., Hoang, V. N., Pham, N. T., Dao, T. C., & Pham, T. N. (2015). Synthesis and application of quantum dots-based biosensor. *Advances in Natural Sciences: Nanoscience and Nanotechnology*, *6*(1). Available from https://doi.org/10.1088/2043-6262/6/1/015015.

Nikitin, P. I., Valeiko, M. V., & Gorshkov, B. G. (2003). New direct optical biosensors for multi-analyte detection. *In Sensors and Actuators, B: Chemical, Vol. 90*(Issues 1–3), 46–51. Available from https://doi.org/10.1016/S0925-4005(03)00020-0.

Norouzi, M., Zarei Ghobadi, M., Golmimi, M., Mozhgani, S. H., Ghourchian, H., & Rezaee, S. A. (2017). Quantum dot-based biosensor for the detection of human T-lymphotropic virus-1. *Analytical Letters*, *50*(15), 2402–2411. Available from https://doi.org/10.1080/00032719.2017.1287714.

Ogura, M., Yamaguchi, H., Yoshida, K. I., Fujita, Y., & Tanaka, T. (2001). DNA microarray analysis of Bacillus subtilis DegU, ComA and PhoP regulons: An Approach to comprehensive analysis of *B.subtilis* two-component regulatory systems. *Nucleic Acids Research*, *29*(18), 3804–3813. Available from https://doi.org/10.1093/nar/29.18.3804.

Palchetti, I., & Mascini, M. (2010). Biosensor technology: A brief history. In *Lecture notes in electrical engineering* (Vol. 54, pp. 15–23). https://doi.org/10.1007/978-90-481-3606-3_2

Papadimitrakopoulos, F., Vaddiraju, S., Jain, F.C., & Tomazos, I.C. (2013). United States. Patent 608.

Parikh, C. R., & Mansour, S. G. (2017). Perspective on clinical application of biomarkers in AKI. *Journal of the American Society of Nephrology*, *28* (6), 1677–1685. Available from https://doi.org/10.1681/ASN.2016101127.

Park, M., Kang, B. H., & Jeong, K. H. (2018). Paper-based biochip assays and recent developments: A review. *Biochip Journal*, *12*(1). Available from https://doi.org/10.1007/s13206-017-2101-3.

Perumal, V., & Hashim, U. (2014a). Advances in biosensors: Principle, architecture and applications. *Journal of Applied Biomedicine*, *12*(1), 1–15. Available from https://doi.org/10.1016/j.jab.2013.02.001.

Perumal, V., & Hashim, U. (2014b). Advances in biosensors: Principle, architecture and applications. *Journal of Applied Biomedicine*, *12*(1), 1–15. Available from https://doi.org/10.1016/j.jab.2013.02.001.

Pravin, B., & Sadaf, H. (2020). *Basics of biosensors and nanobiosensors* (pp. 1–22). Wiley. Available from https://doi.org/10.1002/9783527345137.ch1.

Ramos, A. P., Cruz, M. A. E., Tovani, C. B., & Ciancaglini, P. (2017). Biomedical applications of nanotechnology. *Biophysical Reviews*, *9*(2), 79–89. Available from https://doi.org/10.1007/s12551-016-0246-2.

Rastislav, M., Miroslav, S., & Ernest, Š. (2012). Biosensors — Classification, characterization and new trends. *Acta Chimica Slovaca*, 109–120. Available from https://doi.org/10.2478/v10188-012-0017-z.

Rebelo, R., Barbosa, A. I., Caballero, D., Kwon, I. K., Oliveira, J. M., Kundu, S. C., ... Correlo, V. M. (2019). 3D biosensors in advanced medical diagnostics of high mortality diseases. *Biosensors and Bioelectronics*, *130*, 20–39. Available from https://doi.org/10.1016/j.bios.2018.12.057.

Riedmaier, I., & Pfaffl, M. W. (2013). Transcriptional biomarkers – High throughput screening, quantitative verification, and bioinformatical validation methods. *Methods (San Diego, Calif.)*, *59*(1), 3–9. Available from https://doi.org/10.1016/j.ymeth.2012.08.012.

Rizwan, M., Mohd-Naim, N. F., & Ahmed, M. U. (2018). Trends and advances in electrochemiluminescence nanobiosensors. *Sensors (Switzerland)*, *18*(1). Available from https://doi.org/10.3390/s18010166.

Rocchitta, G., Spanu, A., Babudieri, S., Latte, G., Madeddu, G., Galleri, G., ... Serra, P. A. (2016). Enzyme biosensors for biomedical applications: Strategies for safeguarding analytical performances in biological fluids. *Sensors (Switzerland)*, *16*(6). Available from https://doi.org/10.3390/s16060780.

Rogers, J. K., Guzman, C. D., Taylor, N. D., Raman, S., Anderson, K., & Church, G. M. (2015). Synthetic biosensors for precise gene control and real-time monitoring of metabolites. *Nucleic Acids Research*, *43*(15), 7648–7660. Available from https://doi.org/10.1093/nar/gkv616.

Ronkainen, N. J., Halsall, H. B., & Heineman, W. R. (2010). Electrochemical biosensors. *Chemical Society Reviews*, *39*(5), 1747–1763. Available from https://doi.org/10.1039/b714449k.

Rupak, N., Avinash, S., Deepak, K., Soham, M., Fatih, S., & Praveen, K. A. (2021). Amalgamation of biosensors and nanotechnology in disease diagnosis: Mini-review. *Sensors International*, *2*, 100089. Available from https://doi.org/10.1016/j.sintl.2021.100089.

Sagadevan, S., & Periasamy, M. (2014). Recent trends in nanobiosensors and their applications – A review. *Reviews on Advanced Materials Science*, *36*(1), 62–69. Available from http://www.ipme.ru/e-journals/RAMS/no_13614/06_13614_suresh.pdf.

Sang, S., Wang, Y., Feng, Q., Wei, Y., Ji, J., & Zhang, W. (2016). Progress of new label-free techniques for biosensors: A review. *Critical Reviews in Biotechnology*, *36*(3), 465–481. Available from https://doi.org/10.3109/07388551.2014.991270.

Sanjay, S. T., Dou, M., Sun, J., & Li, X. (2016). A paper/polymer hybrid microfluidic microplate for rapid quantitative detection of multiple disease biomarkers. *Scientific Reports*, 6. Available from https://doi.org/10.1038/srep30474.

Senveli, S. U., & Tigli, O. (2013). Biosensors in the small scale: Methods and technology trends. *IET Nanobiotechnology*, *7*(1), 7–21. Available from https://doi.org/10.1049/iet-nbt.2012.0005.

Shi, S., Ang, E. L., & Zhao, H. (2018). In vivo biosensors: mechanisms, development, and applications. *Journal of Industrial Microbiology and Biotechnology*, *45*(7), 491–516. Available from https://doi.org/10.1007/s10295-018-2004-x.

Shiratori, I., Akitomi, J., Boltz, D. A., Horii, K., Furuichi, M., & Waga, I. (2014). Selection of DNA aptamers that bind to influenza A viruses with high affinity and broad subtype specificity. *Biochemical and Biophysical Research Communications*, *443*(1), 37–41. Available from https://doi.org/10.1016/j.bbrc.2013.11.041.

Simranjeet, S., Vijay, K., Singh, D. D., Shivika, D., Ram, P., & Joginder, S. (2020). *Biological biosensors for monitoring and diagnosis* (pp. 317–335). Springer Science and Business Media LLC. Available from https://doi.org/10.1007/978-981-15-2817-0_14.

Singh, P., & Yadava, R. D. S. (2020). Stochastic resonance induced performance enhancement of MEMS cantilever biosensors. *Journal of Physics D: Applied Physics*, *53*(46). Available from https://doi.org/10.1088/1361-6463/ab98c4.

Singh, R., Mukherjee, M. D., Sumana, G., Gupta, R. K., Sood, S., & Malhotra, B. D. (2014). Biosensors for pathogen detection: A smart approach towards clinical diagnosis. *Sensors and Actuators, B: Chemical*, *197*, 385–404. Available from https://doi.org/10.1016/j.snb.2014.03.005.

Sokolov, I. L., Cherkasov, V. R., Tregubov, A. A., Buiucli, S. R., & Nikitin, M. P. (2017). Smart materials on the way to theranostic nanorobots: Molecular machines and nanomotors, advanced biosensors, and intelligent vehicles for drug delivery. *Biochimica et Biophysica Acta - General Subjects*, *1861*(6), 1530–1544. Available from https://doi.org/10.1016/j.bbagen.2017.01.027.

Soldatkin, A. P., Dzyadevych, S. V., Korpan, Y. I., Sergeyeva, T. A., Arkhypova, V. N., Biloivan, O. A., . . . El'skaya, A. V. (2013). Biosensors. A quarter of a century of R&D experience. *Biopolymers and Cell*, *29*(3), 188–206. Available from https://doi.org/10.7124/bc.000819.

Song, S., Xu, H., & Fan, C. (2006). Potential diagnostic applications of biosensors: Current and future directions. *International Journal of Nanomedicine*, *1*(4), 433–440. Available from https://doi.org/10.2147/nano.2006.1.4.433.

Songa, E. A., & Okonkwo, J. O. (2016). Recent approaches to improving selectivity and sensitivity of enzyme-based biosensors for organophosphorus pesticides: A review. *Talanta*, *155*, 289–304. Available from https://doi.org/10.1016/j.talanta.2016.04.046.

Steinhubl, S. R., Muse, E. D., & Topol, E. J. (2015). The emerging field of mobile health. *Science Translational Medicine*, *7*(283). Available from https://doi.org/10.1126/scitranslmed.aaa3487.

Stern, E., Vacic, A., Rajan, N. K., Criscione, J. M., Park, J., Ilic, B. R., . . . Fahmy, T. M. (2010). Label-free biomarker detection from whole blood. *Nature Nanotechnology*, *5*(2), 138–142. Available from https://doi.org/10.1038/nnano.2009.353.

Stetter, J. R., Penrose, W. R., & Yao, S. (2003). Sensors chemical sensors, electrochemical sensors, and ECS. *Journal of the Electrochemical Society*, *150*(2), S11–S16. Available from https://doi.org/10.1149/1.1539051.

Strimbu, K., & Tavel, J. A. (2010). What are biomarkers? *Current Opinion in HIV and AIDS*, *5*(6), 463–466. Available from https://doi.org/10.1097/COH.0b013e32833ed177.

Sun, L., Zhang, Y., Wang, Y., Yang, Y., Zhang, C., Weng, X., . . . Yuan, X. (2018). Real-time subcellular imaging based on graphene biosensors. *Nanoscale*, *10*(4), 1759–1765. Available from https://doi.org/10.1039/c7nr07479d.

Swierczewska, M., Liu, G., Lee, S., & Chen, X. (2012). High-sensitivity nanosensors for biomarker detection. *Chemical Society Reviews*, *41*(7), 2641–2655. Available from https://doi.org/10.1039/c1cs15238f.

Tamayo, J., Kosaka, P. M., Ruz, J. J., Paulo, Á. S., & Calleja, M. (2013). Biosensors based on nanomechanical systems. *Chemical Society Reviews*, *42* (3), 1287–1311. Available from https://doi.org/10.1039/c2cs35293a.

Touhami, A. (2014). Biosensors and nanobiosensors: Design and applications. *Nanomedicine: Nanotechnology, Biology, and Medicine*, *15*, 374–403.

Turner, A. P. F. (2013a). Biosensors: Sense and sensibility. *Chemical Society Reviews*, *42*(8), 3184–3196. Available from https://doi.org/10.1039/c3cs35528d.

Turner, A. P. F. (2013b). Biosensors: Sense and sensibility. *Chemical Society Reviews*, *42*(8), 3184–3196. Available from https://doi.org/10.1039/c3cs35528d.

Valenti, D. A. (2013). Alzheimer's disease: Screening biomarkers using frequency doubling technology visual field. *Alzheimer*, *9*, 31–34. Available from https://doi.org/10.1016/j.jalz.2013.05.1588.

Vigneshvar, S., Sudhakumari, C. C., Senthilkumaran, B., & Prakash, H. (2016). Recent advances in biosensor technology for potential applications – An overview. *Frontiers in Bioengineering and Biotechnology*, 4. Available from https://doi.org/10.3389/fbioe.2016.00011.

Wang, Q., Zhang, B., Xu, X., Long, F., & Wang, J. (2017). CRISPR-typing PCR (ctPCR), a new Cas9-based DNA detection method. *BioRxiv*. Available from https://doi.org/10.1101/236588.

Wang, T., Chen, C., Larcher, L. M., Barrero, R. A., & Veedu, R. N. (2019). Three decades of nucleic acid aptamer technologies: Lessons learned, progress and opportunities on aptamer development. *Biotechnology Advances*, *37*(1), 28–50. Available from https://doi.org/10.1016/j.biotechadv.2018.11.001.

Wang, X., Lu, X., & Chen, J. (2014). Development of biosensor technologies for analysis of environmental contaminants. *Trends in Environmental Analytical Chemistry*, *2*, 25–32. Available from https://doi.org/10.1016/j.teac.2014.04.001.

Wang, Y., Yu, L., Kong, X., & Sun, L. (2017). Application of nanodiagnostics in point-of-care tests for infectious diseases. *International Journal of Nanomedicine*, *12*, 4789–4803. Available from https://doi.org/10.2147/IJN.S137338.

Wegner, K. D., & Hildebrandt, N. (2015). Quantum dots: Bright and versatile in vitro and in vivo fluorescence imaging biosensors. *Chemical Society Reviews*, *44*(14), 4792–4834. Available from https://doi.org/10.1039/c4cs00532e.

Wignarajah, S., Suaifan, G. A. R. Y., Bizzarro, S., Bikker, F. J., Kaman, W. E., & Zourob, M. (2015). Colorimetric assay for the detection of typical biomarkers for periodontitis using a magnetic nanoparticle biosensor. *Analytical Chemistry*, *87*(24), 12161–12168. Available from https://doi.org/10.1021/acs.analchem.5b03018.

Wu, C. C., Luk, H. N., Lin, Y. T. T., & Yuan, C. Y. (2010). A Clark-type oxygen chip for in situ estimation of the respiratory activity of adhering cells. *Talanta*, *81*(1–2), 228–234. Available from https://doi.org/10.1016/j.talanta.2009.11.062.

Wu, Y. X., & Kwon, Y. J. (2016). Aptamers: The "evolution" of SELEX. *Methods (San Diego, Calif.)*, *106*, 21–28. Available from https://doi.org/10.1016/j.ymeth.2016.04.020.

Yang, N., Chen, X., Ren, T., Zhang, P., & Yang, D. (2015). Carbon nanotube based biosensors. *Sensors and Actuators, B: Chemical*, *207*, 690–715. Available from https://doi.org/10.1016/j.snb.2014.10.040.

Yao, J., Yang, M., & Duan, Y. (2014). Chemistry, biology, and medicine of fluorescent nanomaterials and related systems: New insights into biosensing, bioimaging, genomics, diagnostics, and therapy. *Chemical Reviews*, *114*(12), 6130–6178. Available from https://doi.org/10.1021/cr200359p.

Yoo, S. M., & Lee, S. Y. (2016). Optical biosensors for the detection of pathogenic microorganisms. *Trends in Biotechnology*, *34*(1), 7–25. Available from https://doi.org/10.1016/j.tibtech.2015.09.012.

Zanchetta, G., Lanfranco, R., Giavazzi, F., Bellini, T., & Buscaglia, M. (2017). Emerging applications of label-free optical biosensors. *Nanophotonics*, *6*(4), 627–645. Available from https://doi.org/10.1515/nanoph-2016-0158.

Zhang, L., Wan, S., Jiang, Y., Wang, Y., Fu, T., Liu, Q., . . . Tan, W. (2017). Molecular elucidation of disease biomarkers at the interface of chemistry and biology. *Journal of the American Chemical Society*, *139*(7), 2532–2540. Available from https://doi.org/10.1021/jacs.6b10646.

Zhang, S., Geryak, R., Geldmeier, J., Kim, S., & Tsukruk, V. V. (2017). Synthesis, assembly, and applications of hybrid nanostructures for biosensing. *Chemical Reviews*, *117*(20), 12942–13038. Available from https://doi.org/10.1021/acs.chemrev.7b00088.

Zhou, W., Jimmy Huang, P. J., Ding, J., & Liu, J. (2014). Aptamer-based biosensors for biomedical diagnostics. *Analyst*, *139*(11), 2627–2640. Available from https://doi.org/10.1039/c4an00132j.

Chapter 21

# Cancer diagnosis by biosensor-based devices: types and challenges

Krishnendu Manna[1,2,*], Niladri Mukherjee[1,*], Niloy Chatterjee[3,*] and Krishna Das Saha[1]
[1]*Cancer Biology and Inflammatory Disorder Division, CSIR-Indian Institute of Chemical Biology, Kolkata, India*
[2]*Department of Food and Nutrition, University of Kalyani, Kalyani, India*
[3]*Food and Nutrition Division, University of Calcutta, Kolkata, India*

## 21.1 Introduction

Cancer is generally conceived as a disease with no treatment or an incurable, insufferably debilitating condition aptly termed as the "silent killer" (Blank, Haanen, Ribas, & Schumacher, 2016; Lujambio & Lowe, 2012). Such perception of cancer, although very common, is often misunderstood and over-comprehensive (Klein et al., 2014; Liora, 1999). The results of several cutting-edge studies and research complement the mounting body of validation, suggesting that cancer is a common word that encompasses a broad cluster of contagium involving the human body system as a whole (David & Zimmerman, 2010). Regrettably, it is a tissue-level variation disorder, and their unpredictability or dearth of detection is a significant obstacle for its medical diagnosis (Smith, Pope, & Botha, 2005), followed by therapeutic response (Padma, 2015). Cancer is the world's second-biggest reason for mortality, accounting for approximately 10 million fatalities in 2020 (Sung et al., 2021). Approximately 17% of deaths worldwide were attributed to cancer. About 70% of cancer cases occur in regions with low and average income (Shah, Kayamba, Peek, & Heimburger, 2019). The economic burden of cancer is wide-ranging and is rising year by year. The gross annual monetary cost of cancer was projected to be around US$158 billion in 2020 (Glode & May, 2017; Nils et al., 2017). Inaccessible diagnosis, care, and treatment, and delayed appearance are prevalent and cause concerns to researchers and doctors (Price, Ndom, Atenguena, Mambou Nouemssi, & Ryder, 2012). In 2017, only 26% of poverty-stricken countries informed possessing cancer pathology services normally accessible to common people (Sullivan, Pramesh, & Booth, 2017). Over 90% of the hospital services have been identified in affluent countries to maintain treatment options and choices and modern prognosis technology for cancers, in contrast to less than 30% in poor countries (De Souza, Hunt, Asirwa, Adebamowo, & Lopes, 2016; Gelband et al., 2016). Lung, colorectal, stomach, liver, and breast cancers are the foremost reasons for mortality from cancer worldwide. Studies suggest that lung and breast cancer are the most dominant forms of cancer in males and females, being followed by prostate, colorectal, stomach, and liver in men, colorectal, lung, cervical, and thyroid women, respectively (Miller et al., 2019; Yuan et al., 2016).

Cancer can be characterized as unusual and unregulated cell growth and development caused by the build-up of particular hereditary and epigenetic abnormalities triggered by environmental or genetic origins (Blanpain, 2013). Uncontrolled growth of cells results in the development of a tumor mass, which becomes free of usual homeostatic safeguards and checkpoints across time. In turn, tumor cells become immune within the body to apoptosis and other antigrowth defenses, growing profusely and uncontrollably (Hanselmann & Welter, 2016; Ichim & Tait, 2016). With the gradual development and evolution of cancer, the tumor continues to spread far beyond the source site and metastasize or spread vigorously to many other areas and parts of the body. During this stage, the cancer is essentially incurable and irresponsive to any medication or treatment (Seyfried & Huysentruyt, 2013; Zheng et al., 2018). The leading underlying causes or mechanisms that lead to cancer are stimulation of oncogene and inhibition of tumor suppressor genes (TSGs) (Kontomanolis et al., 2020; Vicente-Dueñas, Romero-Camarero, Cobaleda, & Sánchez-García, 2013). The activity of oncogenes increases is dysregulated due to mutation or duplication of proto-oncogenes, which are generally

*Krishnendu Manna, Niladri Mukherjee, and Niloy Chatterjee contributed equally to this chapter.

Biosensor Based Advanced Cancer Diagnostics. DOI: https://doi.org/10.1016/B978-0-12-823424-2.00017-X

responsible for cell maturation, proliferation, and differentiation (Gil-Bazo, 2020; Imran et al., 2017). This aberration leads to intrinsic stimulation or additional production of some gene products, which results in dysregulation of cell growth, and the anomalous cell cycle finally paving the way to tumorigenesis and henceforth cancer (Ponder, 2001; Shelton & Jaiswal, 2013). As growth is intrinsically related to cancer, hence growth factors, as well as their receptors, have been widely studied and explored as therapeutic targets and bioindicators or markers for cancer diagnosis (Hursting & Berger, 2010; Witsch, Sela, & Yarden, 2010). About one third of patients diagnosed with breast cancer show a proportionate increase in human epidermal growth factor receptor (Her-2), and cancers with augmented Her-2 receptors in their cells appear to grow and metastasize in a hostile manner and develop and disseminate more ruthlessly. Thus, in deciding the suitable treatment regime, science, familiarity, and knowledge of any such underlying growth factors is vital (Lower, Glass, Blau, & Harman, 2009; Ross et al., 2004). For patients with enhanced expression of the Her-2 gene, a recombinant humanized MAb engaged against this protein as a targeted treatment for breast cancer is now a routine adjunct treatment (Sun et al., 2015). By retardation or halting cell division, TSGs play a critical role as the regulator of excessive cell growth and proliferation (Liu et al., 2015). Retinoblastoma (Rb) protein, BRCA1/2, and p53 are some of the most widely studied TSGs exploited in cancer. Rb is a central transcription factor that regulates the division of cells, and therefore alteration in gene sequence plays a significant role in most human cancers (Giacinti & Giordano, 2006; Sherr & McCormick, 2002). The most likely reasons for downregulation of the Rb1 gene are single nucleotide mutations and deletions (Dick & Rubin, 2013). BRCA1 is an enzyme for DNA repair that is engaged in the "proofreading" of new DNA replicated during the cell cycle for the presence of mutation tolerance besides location (Narod & Foulkes, 2004; Venkitaraman, 2002). DNA repair enzymes usually act until the cell divides to eliminate replication errors. Around one half of inherited breast malignancies and about 40% of inherited ovarian cancers account for BRCA1 gene mutations (Welcsh & King, 2001). Lastly, a principal supervisor of programmed cell death or apoptosis is the p53 protein. Mutations in p53 often accompany carcinomas of the brain, lung, colon, breast, leukemia, as well as hepatocellular carcinomas. A further fundamental issue about p53 loss is that it can act as an inhibitor to chemotherapeutic drugs (Karin, André, Dipita, & Thomas, 2017; Muller & Vousden, 2014; Parrales & Iwakuma, 2015) preventing them from proper functioning. The invention and emergence of novel biosensors that can indicate the occurrence of abnormalities or changes in such vital genes seem to be of considerable significance in precise diagnostic and therapeutic regimens of recent times, which can accurately determine cancer early and prevent them.

The global prevalence of cancer tends to surge continually, imposing tremendous physical, expressive, and economic stress on individuals and their families, communities, societies, and healthcare systems (Fenn et al., 2014; Girgis, Lambert, Johnson, Waller, & Currow, 2013). Many healthcare facilities are far less prepared to handle this burden or responsibility in low- and middle-income nations. Significant numbers of people with cancer worldwide do not even have access to a diplomatic quality of diagnosis and care. Today, many cancers may be treated successfully to remove, minimize, or slow the effects of the disease in the lives of patients (Fitzmaurice et al., 2015; Sankaranarayanan, Ramadas, & Qiao, 2014). Although a cancer diagnosis can always leave patients feeling powerless and out of control, there is the reason for hope instead of hopelessness in many instances today (Li, Guo, Tang, & Yang, 2018; Li, & Zeng, et al., 2018). Cancer mortality may decline if early diagnosis and treatment are possible (Weller et al., 2012). Cancer is expected to retort to effective care when diagnosed early, leading to higher survival chances, reduced mortality and complications, and fewer treatment expenses, thereby leading to significant differences in the lives of cancer patients (Bohunicky & Mousa, 2011; Hamilton, Walter, Rubin, & Neal, 2016; Vedsted & Olesen, 2015). In all conditions and the majority of cancers, early detection is critical. Patients are diagnosed at late stages, in the absence of an early diagnosis, where curative care may no longer be an option. There lie several barriers that need to be solved effectively if cancer diagnosis and treatment are done early; this is highlighted in Fig. 21.1

## 21.2 Disadvantages of conventional methods of cancer detection

There is no standard test to detect cancer appropriately. Detailed patient history and corporal investigation in consort with diagnostic inquiry are typically required for a patient's total assessment. Several tests are required (Buchen, 2011; Lyratzopoulos, Vedsted, & Singh, 2015) to confirm or determine whether a person has cancer or another infection or disease showing symptoms similar to cancer. Cancer may result in deregulated levels of particular constituents in the human body (Li et al., 2018) thereby aiding in early cancer diagnosis. So, laboratory diagnostic tests that measure the anomaly or changes of these certain substances in blood plasma, excretory materials, and other body fluids can be facilitative and of utmost importance. Some lab tests involve testing blood or tissue samples for tumor markers (Amann et al., 2014; Hüttenhain et al., 2012). Other conventional diagnostic methods include diagnostic imaging methods like computed tomography scan, which shows computer image using X-rays, nuclear scan, PET scan, bone scan, ultrasound,

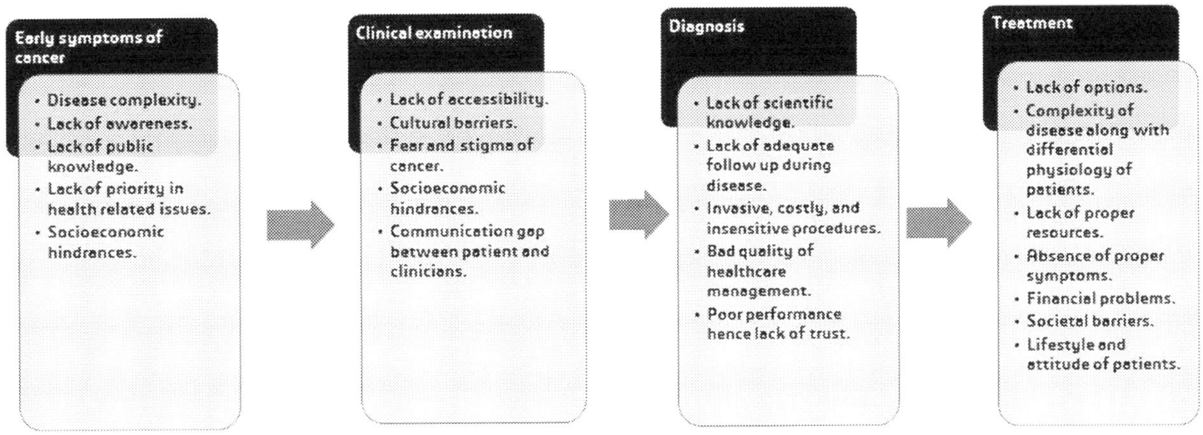


**FIGURE 21.1** Barriers in various stages of early cancer disease to treatment.

magnetic resonance imaging (MRI), to detect abnormalities by using magnetic fields and X-rays, endoscopies (cystoscopy, endoscopic retrograde cholangiopancreatography, colonoscopy, sigmoidoscopy and esophagogastroduodenoscopy) (Hussain & Nguyen, 2014), isotopic techniques by the introduction of some radioactive tracers in patient's body, various genetic tests (testing for mutations) (Lynch, Venne, & Berse, 2015), and tumor biopsy (bone marrow biopsy, endoscopic biopsy, punch biopsy, fine needle aspiration biopsy, excisional or incisional biopsy, shave biopsy, skin biopsy, etc.) (Francesc et al., 2018; Vaidyanathan, Soon, Zhang, Jiang, & Lim, 2019). Biopsy techniques are invasive in nature, costly, and impractical for testing again and again (Schlange & Pantel, 2016). Widespread limitations in the biopsy of tissues have motivated the advancement of liquid biopsy, by means of which cancer biomarkers can be detected in a wide variety of body liquids or fluids such as blood, serum, saliva, urine, etc., that supports easier and early inspection of cancer, along with their diagnosis and treatment (Barlebo Ahlborn & Østrup, 2019). Biopsy of liquid samples is comparatively unaggressive, low-cost, convenient, and easy-going for huge populaces. It can be performed simultaneously and permits repetitive testing in real-time (Siravegna, Marsoni, Siena, & Bardelli, 2017). Such innovative procedures can monitor the disease's progression and the effectiveness of treatment or drug regimens. One of the significant advantages of bio fluids is that it contains details and holistic data from tumor cells as a whole in the patient rather than from a small section or segment of the tumor as in biopsy of tissue samples (Mattox et al., 2019). The traditional methods used for bio fluids detection are the conventional enzyme-linked immunosorbent assay (ELISA) or Polymerase chain reaction (PCR)-based approaches. These technological limitations require costly reagents and chemicals in every test, slow rate of detection, complexity of machinery, and provision of skilled personnel for carrying out the assay (Miranda et al., 2010; Wang, Wang, Tu, & Wong, 2016).

Furthermore, since these procedures are manual, they are incapable of continuous monitoring of disease throughout treatment. Additionally, all cancers generally involve more than one event or molecule, ultimately leading to cancer. These hindrances prevent an early and correct diagnosis of this catastrophic disease. Therefore novel and robust technologies for simultaneous detection of multiple biomarkers are needed for accurate diagnosis and prognosis. Some additional traditional methods such as Northern blot, Southern blot analysis, and other blotting techniques suffer low sensitivity, time consumption, and high cost (Laurie et al., 2012). Therefore there is a vital and crucial need for fast and quick, consistent, precise, and delicately modified alternate methods for cancer detection. These methods suffer from several drawbacks: lack of sensitivity, expensiveness, high risk of false positives, inability to detect fine entities, absence of precision, and early detection, and are clearly depicted in Fig. 21.2 (Surbone, Zwitter, & Rajer, 2012). These disadvantages can be greatly resolved using biosensors, which allows for much accurate, sensitive, specific and early detection of cancer.

## 21.3 Cancer biomarkers

The National Cancer Institute (NCI) explains or describes biomarker as "a biological molecule found in blood, other body fluids, or tissues that is a sign of a normal or abnormal process, or of a disease or condition" (Prensner, Chinnaiyan, & Srivastava, 2012). A biomarker or similar molecules can help doctors or clinicians differentiate how an individual responds to the treatment of a disease or ailment (Bailey et al., 2014; Brown, Harhay, & Harhay, 2015).

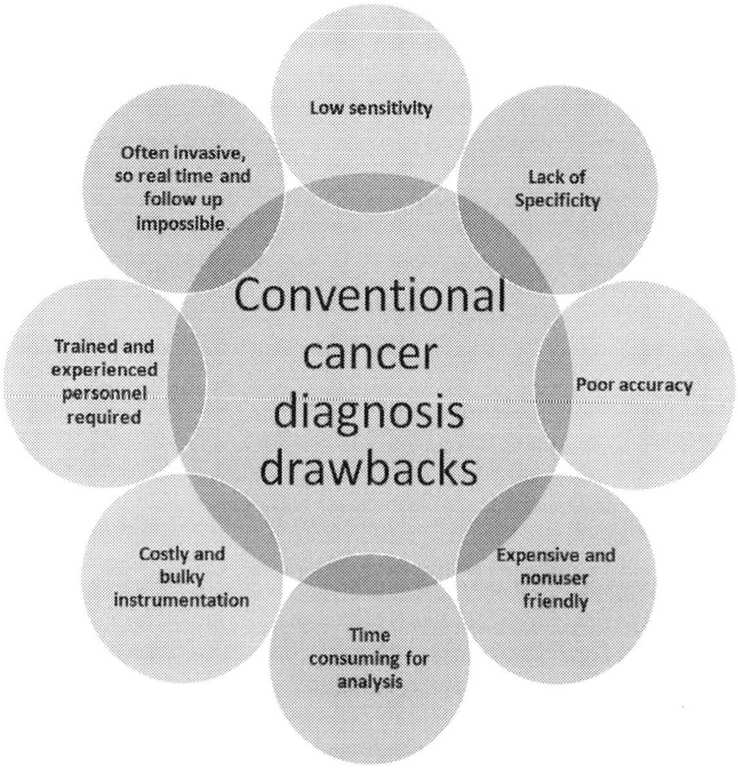


**FIGURE 21.2** Drawbacks or pitfalls of conventional methods of cancer diagnosis.

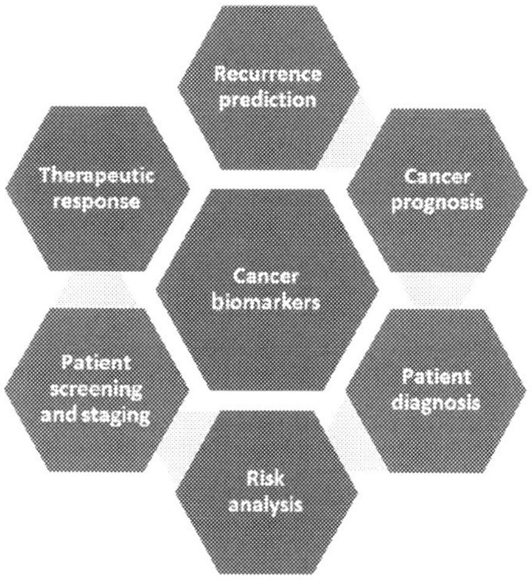


**FIGURE 21.3** Application of cancer biomarkers in the field of cancer diagnosis.

Biomarkers may belong to different cellular origins or sources, such as DNA (i.e., translocation, deletion, point mutation, amplification, insertion and loss of heterozygosity), RNA (change in structure, conformation, etc.), or proteins, enzymes, antigens as well as antibody, hormone, oncogene, proto-oncogene or TSG (changes in structure, function, activation or repression). Biomarkers for cancer are possibly among the most valuable and preliminary methods used for pretreatment staging for early cancer diagnosis, assessing cancer's response to chemotherapy, other drugs, and medicaments, and surveying the progression of the disease (Bhatt, Mathur, Farooque, Verma, & Dwarakanath, 2010; Falanga & Marchetti, 2018). Biomarkers are usually found in human cells and fluids such as blood, urine, serum, plasma, or cerebral spinal fluid (CSF), but they may also be present in or on tumor cells (Li et al., 2018). Functionally, they can be divided into diagnostic biomarkers (help in disease detection), prognostic biomarkers (provides information on the stage of disease), and treatment biomarkers (indicate the response of treatment modalities). Biomarkers are multifaceted, playing a very versatile role in cancer, and are shown in Fig. 21.3.

**TABLE 21.1** List of conventional cancer biomarkers.

| Biomarker | Cancer types | Where present | Functions |
|---|---|---|---|
| Calcitonin | Medullary thyroid cancer | Blood | Treatment response, recurrence and prognosis |
| CA-125<br>(Cancer antigen-125) | Epithelial ovarian cancer | Blood | Treatment efficacy, recurrence |
| ß-2 Microglobulin | Colorectal | Blood, urine, cerebral spinal fluid | Prognosis and treatment response |
| α-Fetoprotein | Liver, breast, gastric, rectal, prostate and ovarian | Blood | Staging and prognosis of cancer, treatment response |
| Estrogen receptor | Breast | Tumor tissue | Appropriateness of therapy |
| Progesterone receptor | Breast | Tumor tissue | Appropriateness of therapy |
| EGFR gene mutation | Lung | Tumor tissue | Determine treatment and prognosis |
| PD-L1<br>(Programmed death-ligand 1) | Nonsmall cell lung cancer | Tumor tissue | Appropriateness of therapy |
| Her2<br>(Human epidermal growth factor receptor 2) | Breast | Tumor tissue | Therapy choice |
| CA19−9<br>(Carbohydrate antigen 19−9) | Pancreatic, breast | Blood | Monitoring therapy |
| PSA, Pro2PSA<br>(Prostate-specific antigen) | Prostate | Blood | Screening, discriminating cancer from benign disease |

The cancer biomarker detection theory is ancient and of considerable significance in cancer diagnosis and can be traced back to Dr. Joseph Gold in 1965 when he discovered carcinoembryonic antigen (CEA) (Gold, Banjo, Freedman, & Gold, 1973). Nowadays, a wide variety of potential testing markers in human body fluids are routinely used to detect tumor evolution, monitor, and prognosis of the disease. In Table 21.1, a list of tumor biomarkers is given. However, the maximum of such available biomarkers is to show delicate sensitiveness and selectivity for transformation into regular clinical use or assess the response of treatment. This domain or field can be theoretically as well as experimentally enhanced by biosensor technology.

## 21.3.1 Proteomics-based cancer biomarker detection

Advances in proteomic technology (i.e., MALDI-Ms, LC-Ms, and Ms/MS), greater genomic annotations, sequencing of large genomes, along with progress in innovative and cutting-edge algorithms, computational techniques, and software, for comparison and analysis of large data sets obtained from mass spectrometry results and data, have given rise to rapid evolution and development in identification of novel and unexplored cancer biomarkers (Srivastava & Creek, 2019). Various *in silico* and bioinformatic approaches have aided and supported these techniques creating new research avenues and solving massive problems that cannot be solved previously. The supremacy of merging these approaches may prove beneficial to produce detailed proteomic profiles (Hanash, Pitteri, & Faca, 2008) when biopsies of tumor tissue are used for test or proteins present in blood plasma are used as a source for differential diagnosis of patients. The protein data may aid in specific and much earlier detection of cancer in individuals presently impossible by the advanced diagnostic technologies available. The use of biosensors and such sensitive, robust detection technologies may significantly revolutionize this field, providing a better recovery and less mortality. From a given protein sample, present Ms methods may produce tons of millenary of individual mass spectra. To allow correct staging and progression of cancer disease in individuals, sometimes qualitative and quantitative assessment of proteins from diverse sources can provide huge information (Shukla, 2017). Publicly accessible software tools and databases can sort and simplify these large amounts of data or spectra along with their comparison to other reference patients and also annotated genome databases and point out the differences. Tumor and normal tissues and body fluids can be used for such proteomic analysis giving robust and accurate results in a short time (Ling, Chen, Liu, & Yang, 2014; Rodland, 2004). In this manner, whether proteins are overexpressed or downregulated, in a tumor sample they can be recognized as reputed cancer biomarkers. These proteins can be also used as therapeutic targets for treatment accordingly. Several recent studies in the area of proteomics and tumor biomarkers appeal to readers (Ludwig & Weinstein, 2005; Sallam, 2015).

## 21.4 Need of biosensors for cancer diagnosis

A cancer diagnosis is currently based on the development of investigative and analytical techniques that are evidently proficient and competent in the responsive and simultaneous identification of this fatal disease (Le, Rana, & Rotello, 2013). The current tests used to detect or confirm cancer are not so advanced and up to the mark. So early and correct detection technologies are the need of the hour. Biosensors can be good alternatives against various contemporary diagnostic approaches. They can provide more accuracy, sensitivity, and robustness. This technology possesses the potential to deliver a rapid and precise diagnosis, with vivid and accurate images of cancer cells. On the other hand, the technology can differentiate the cancer cells from normal ones and assess angiogenesis progression and detect the metastatic stage. This technology can be an excellent option to detect the therapeutic improvement of any cancer. Generally, biosensors employ a device called a "transducer" that transforms the recognition/biological signal to an electric signal. The transducer can be multifaceted: optical (detecting either luminescent, color, interference, or fluorescence), heat-sensitive (thermistor), electrochemical (utilizing amperemeter, conductometer, impedimeter, or potentiometer), or grounded on mass changes (using piezoelectric effect or sound waves). It gives high accuracy and excellent signal-to-noise ratios, ideal resolution, high performance, economical and light-weight instrumentation, and reliable and dependable results (Ferhan, Jackman, Park, & Cho, 2018). The most commonly used biosensors are electrochemical, and the least is calorimetric sensors possessing complex detection mechanisms (Sheikhpour, Golbabaie, & Kasaeian, 2017). They can calculate suboptimal points of biomarkers qualitatively as well as quantitatively within diverse physiological and biological samples, including the breath of patients (Queralto et al., 2014; Shehada et al., 2016) that can aid in early-stage cancer diagnosis owed to their lower minute limits of detection. Furthermore, they encourage recycling the various molecules used in biorecognition and avoiding lapse and waste of time between sample collection, preparation, and sample analysis, thereby paving ways for much better real-time analysis. Besides, biosensors possess a considerable possibility for rapid and synchronized detection of numerous biomarkers simultaneously. One of the most novel and currently used technology are antigens and aptamers that can specifically bind to miRNAs and finally to the corresponding single-stranded DNAs to detect cancer monoclonal antibodies (ssDNA) (Wu, Chen, Wu, & Zhao, 2015).

## 21.5 Fabrication strategies for cancer biosensors

The working procedure of biosensing analysis depends on analyte detection and can be of two types: label-free or labeled. Label-free identification is based on the direct binding of the original and unmodified analyte molecule to the biorecognition element leading to their recognition (Chen et al., 2016; Kumar, Kumar, Kumar, & Panda, 2017). In contrast, in labeled techniques, only analyte molecules marked with a label or tag can be recognized by the biorecognition element (e.g., amperometric, voltammetric or fluorescent experiments) for obtaining an electroactive signal (Freitas, Nouws, & Delerue-Matos, 2018; Pasinszki, Krebsz, Tung, & Losic, 2017). Some commonly used labels in the domain of cancer molecule detection are fluorophores (especially for fluorescent microscopy) (Seyfried & Huysentruyt, 2013; Zheng et al., 2018), enzymes (often for Western blot analysis) (Metkar & Girigoswami, 2019) and nanoparticles or nanomaterials (essential for MRI) (Viswambari Devi, Doble, & Verma, 2015) as well as magnetic particles (Rocha-Santos, 2014). For medical applications, the usage of magnetic nanoparticles (MNPs), magnetic beads (MBs), semiconductor quantum dots (QDs) with combined biosensor techniques are increasingly justified (Krishna et al., 2018; Moro, Turemis, Marini, Ippodrino, & Giardi, 2017). Metal nanoparticles (mainly gold or silver) have a substantial affinity for cancer cells, which is why they are frequently used in cancer research. Using chemical synthesis or genetic engineering techniques, unique markers are also implemented into the tested compound (Zhang, Chen, He, Wang, & Hu, 2016). However, the labeling method can involve additional sample preparation or a second molecule binding must be followed. Unfortunately, label attachment may alter the properties of the tested molecule significantly; substances used as markers may bind to molecules other than the target and may interfere with their metabolism when living cells are used. Despite all the above, more accuracy, less errors, and quick detection time is leading to more acceptance of label-free approaches. Quartz crystal microbalance (QCM) (Lim, Saha, Tey, Tan, & Ooi, 2020), surface plasmon resonance (SPR) (Pothipor et al., 2019), and microelectronic mechanical resonance (MEM) cantilevers (Eivazzadeh-Keihan et al., 2018), where cantilever sensors emerged from the atomic force microscopy (AFM), are standard label-free techniques today. These methods allow the kinetic/thermodynamic analysis of two complementary molecules' interaction phenomenon to be monitored and calculated in real-time, where one molecule is immobilized on the surface and the second is in flux. In addition, to observe the mass shifts on the sensor, QCM and MEM-cantilevers use resonant frequency changes, but SPR uses changes in the refractive index of thin metal layers (like gold surfaces) to measure the biomolecule's binding mechanism to the sensor surface.

Current cancer diagnosis techniques rely heavily on morphological as well as physiological parameters of the cell, using intrusive methods of microscopy, staining, and imaging (Glunde, Pathak, & Bhujwalla, 2007). In addition, the removal of tissues can lead to the missing of essential cells and markers at the early stages of the disease. Biosensor-based detections are far more realistic and beneficial for cancer clinical studies, extra user-friendly, quicker, less costly, and technically much less challenging than other hefty and bulky proteomic or microarray analyses (Bohunicky & Mousa, 2011). Noteworthy technological progress, particularly for protein-based biosensors, is still required. However, multiarray sensors for multimarker detection will be advantageous for cancer diagnosis (Emami Nejad, Mir, & Farmani, 2019). For biomarker identification, various analyte or marker recognition fragments or structures are in use, and antibodies are the most predominant and frequent ones. Lately, molecular recognition components like aptamers, nanostructures, phage proteins, synthetic peptides, binding peptides, and metal oxide materials are currently widely prepared and manufactured to find analytes of cancer but also their analysis (Perumal & Hashim, 2014). In cancer diagnostic and testing experiments for aiming cancer biomarkers, cells, antigens, and Abs (monoclonal and polyclonal) are occasionally used (Justino, Freitas, Pereira, Duarte, & Rocha Santos, 2015). Polyclonal antibodies can be generated towards any biomarker or cell, and high throughput technologies have allowed such substances to be successfully applied to sensors. Monoclonal antibodies, on the other hand, result in more accurate studies. Monoclonal antibodies have the drawback of being more challenging to maintain and are much more expensive than polyclonal antibodies.

Consequently, replacing natural biological molecules with synthetic receptors or biomimics has become common in the industry. Such molecules have the advantages of being tougher, more stable, less expensive to make, and easily adjustable along with adding labels as the detection producer, to aid immobilization on the sensor surface (Gharatape & Khosroushahi, 2019). Following a selection from combinatorial libraries, such molecules can be synthesized with higher specificity and sensitivity than antibody molecules.

## 21.6 Biosensors for cancer detection

A biosensor is a term for any device or technology that can be applied to distinguish a biological molecule or compound or analyte (i.e., inside the human body) or abnormality in terms of chemical entity to help the detection of cancer. Qualitative, as well as quantitative assessment of analytes, is possible by using Biosensors (Prittesh, 2017). Therefore diagnosis and assay methods are currently trying to incorporate biosensor elements where they can analyze the presence or absence of cancer or tumor and the progression stage. Details such as whether an analyte is present or absent, can be easily detected without any strenuous or time-consuming approaches. The instrument can transform various degrees of biological signal to an electric signal and can be easily amplified, viewed, and analyzed (Ahmed, Saaem, Wu, & Brown, 2014). Different proteins such as enzymes, antigens, antibodies, nucleic acid, other biological entities (e.g., glucose), or metabolites are examples of some of the most widely used analytes. Biosensors are commonly used in translational or contemporary medicine to observe the changes or levels of specific and essential molecules in various diseases, identify bacteria or other pathogenic microorganisms, and diagnose and monitor incurable and hard-to-detect diseases like cancer (Metkar & Girigoswami, 2019). Recently biosensors have been developed to detect exosomes in patients, which may help in effective cancer detection in poor and unresourceful countries (Cheng et al., 2019). For biosensors, the vision of the future also involves minuscule instruments based on chips and integrated circuits which can be attached to the skin surface, various parts of the human body, or such biodegradable and nontoxic technologies that can be consumed or inhaled to monitor dynamic changes, difference in parameters, correct anomalies, or convey signal during worse or disastrous physiological conditions. The applications of biosensors, in theory, are infinite.

In the domain or field of cancer, a tumor biomarker is an analyte being identified by the biosensor. Some of the most common examples that are clinically used were provided earlier in the study. Thus, biosensors can detect the presence or absence of a tumor in their primitive stage, distinguish them depending on characteristics if they are benign or changed to malignant, and whether the treatment has been positive or fruitful in reducing or eradicating cancer by measuring levels of specific proteins expressed and secreted by cancerous cells. In cancer diagnosis and monitoring,

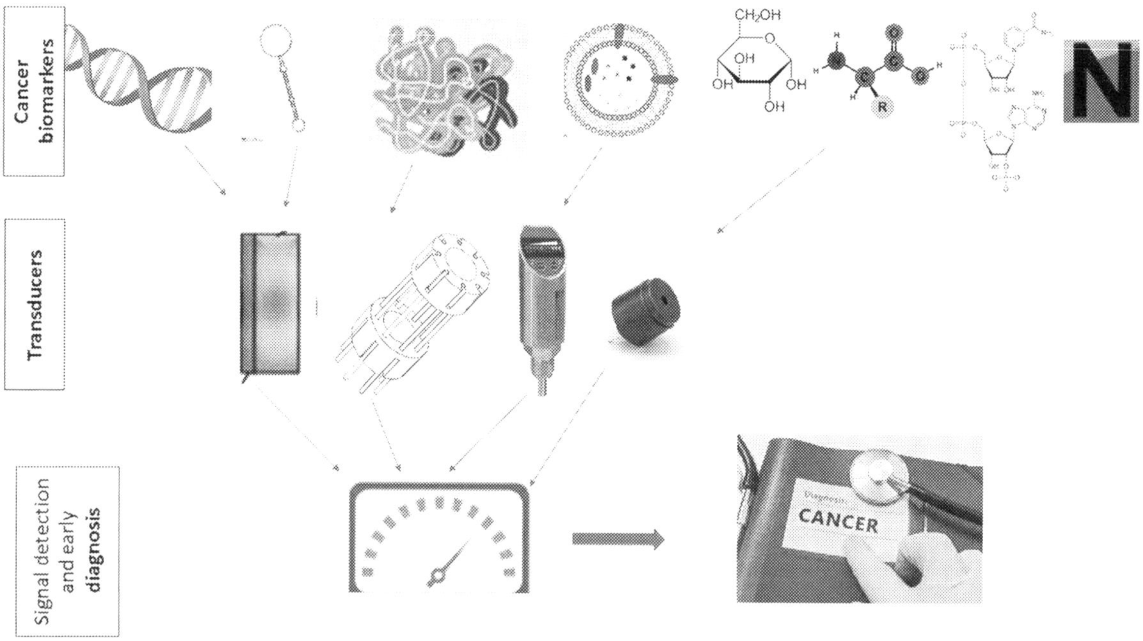


FIGURE 21.4 Structure of a cancer biosensor.

biosensors can sense and distinguish multiple analytes which can be especially advantageous because multiple biomarkers are involved in most forms of cancer. A biosensor's ability to screen for various markers simultaneously besides early diagnosis is less detailed, and saves time and money.

## 21.7 Structure of cancer biosensor

A biosensor is composed of three parts: a part suitable for detection or identification of analytes or molecules, a transducing system that helps transmit the biosignal into another easily detectable signal, and a signal processor that transmits the results and presents them (Vigneshvar, Sudhakumari, Senthilkumaran, & Prakash, 2016). The molecular recognition identifies a "signal" from the physiological or biological sample as the analyte; after that, the transducer transforms the biotic sign or parameter into an electric signal, which the user finally sees in the form of simple data. Fig. 21.4 reveals the elementary construction and function of a biosensor.

### 21.7.1 Biosensor recognition element

The recognition mechanism is an essential part of the biosensor technology used in cancer, as the success or failure of the whole procedure depends on this crucial step. Early biosensors employed natural or inherent recognition structures of biological or environmental origins. They were easier to prepare and use. There has been a shift in the paradigm, and fabricated functional ligands are being prepared or formulated according to need. They show higher cancer detection efficacy and also prevent false-negative results. Numerous biosensor recognition components are now synthesized in the laboratory thanks to technological advances and synthetic chemistry, allowing for improved biosensor function durability and robustness. Recognition elements or ligands include antigens, receptor proteins, antibodies, nucleic acids, enzymes, substrates, etc. They can also be modified by adding fluorochromes and chromogens luminescent substances, increasing their detection limits in cancer diagnosis (Cieplak & Kutner, 2016; Morales & Halpern, 2018).

### 21.7.2 Receptors

Receptors present on the surface of cells are therapeutic targets that are utilized to deliver and target cancer or tumor cells specifically and are highly useful to track cancer therapeutics' efficacy in individuals. They are generally highly selective and aptly suited to recognize patients in cancer diagnosis and prognosis. The types and number of various receptors can change progressively depending on the extent to which an individual is affected by cancer, how they are responding to therapeutics, etc. (Osman et al., 2019). After binding specific ligands, extracellular activation or repression of receptor molecules causes numerous signaling cascades to initiate, which may later lead to various events or physiological processes in cells leading to either promotion or inhibition of cancer. Changes in receptor structure, as well as function, may sometimes result in the abnormal functioning of ion channels and also variations in permeability of membrane structure, stimulation, or inhibition of messenger molecules such as adenyl cyclase and even secondary transmitters, secondary messengers, and stimulation of other kinases, minor G-proteins, phosphatases, and also a wide variety of transcription factors (Uings & Farrow, 2000). These proteins can regulate gene expression and their translation in a multidimensional way, sometimes resulting in abnormal cell growth or development, dysregulated cell division, abnormality in apoptosis and angiogenesis, and activation of metastasis (Martin, 2003). Even though receptor function regulates several dimensions of cellular activity and cellular processes, these receptors may also be helpful as recognition components, but their application in real scenarios in the case of biosensors is challenging and complex. Since receptors can only operate suitably in a particular cellular-simulated environment and the presence of lipid bilayer constituents, thus making it very problematic to acquire purified receptor proteins in high amounts with their full functionality, stability, and activity.

### 21.7.3 Antigen/antibody

Antigen- and antibody-based recognition components are the most common as well as the rapid features used in cancer detection. The instruments utilizing such features can help in a much easier detection technique of cancers. The binding as well as interaction by utilizing such antigen or antibody components are highly specific and, therefore there is no need to be purified before being used as ligand (Khanmohammadi et al., 2020). Besides, the affinity between ligand and cancer target is sometimes reversible, which may lead to reusability of the material. This decreases the cost as well as the time of analysis. These crucial advantages make them stand-out from other similar constituents. For example, an anti-PSA antibody is the recognition factor for detecting PSA and is very frequently used for clinical diagnosis of prostate cancer detection (Najeeb, Ahmad, Shakoor, Mohamed, & Kahraman, 2017). In modern methods, transducers based on microcantilever and sensors that depend on SPR phenomena have been connected to anti-PSA recognition elements and are being widely used. These instruments are very sensitive and advanced as they detect the vibration-induced due to binding, leading to much easier detection.

### 21.7.4 Enzymes

As recognition elements, allosteric enzymes demonstrate colossal promise and potential. The controlling subunit performs the recognizing component's function in most situations, while the catalytic subunit or fragment may function as the transducer (Gruhl, Rapp, & Länge, 2013). One of the most innovative sensors in this group includes the glucose sensor, in which the recognition element used is enzyme glucose oxidase. In the presence of oxygen, the enzyme catalyzes glucose oxidation to give the products hydrogen peroxide and gluconolactone. Usually, the rate of oxygen removal or hydrogen peroxide formation rate will be determined by an amperometric transducer and converted into results that are later analyzed in terms of glucose concentration present in sample (Fu, Qi, Lin, & Huang, 2018).

### 21.7.5 Nucleic acid

Some oligonucleotides, commonly known as aptamers, are chosen and widely used for their specificity and very high affinities for their selected analytes (i.e., DNA/gene from within a pool of thousands of different sequences). Consequently, they are increasingly evolving as elements of biosensor recognition (Liang et al., 2016). Nucleic acid ligands are generally produced from a combinatorial oligonucleotide library composed of DNA and RNA. A chemistry-based approach termed SELEX (systematic ligand evolution by exponential enrichment) forms the backbone of the cancer biomolecule or analyte detection using nucleotides. SELEX applications are almost infinite since the possible ligands number is countless and such molecules are chosen that possess a powerful attraction as well as the binding capacity to targeted ligands (Lee et al., 2008). Biosensors based on this form of the device have proven to be helpful for the discovery or unearthing of new biomarkers that are crucial to early cancer detection. The CT antigens, CAGE-1, NY-BR-1, NY-ESO-1, and Ing-1, have been identified using SELEX knowledge. Aptamers have also been used for the identification of cancer proteins in small sensing arrays. Another feature of aptamers is the allosteric properties they possess. Samples of allosteric aptamers are RNAzymes and DNAzymes (Tothill, 2009).

### 21.7.6 Biosensor transducer

It is possible to interface a particular biomolecular cellular or recognition component to various signal transducers. The transducer converts the molecular signal into an electrical or digital signal that can be quantified, displayed, and analyzed (Su et al., 2013). Electrochemical (i.e., amperometric and potentiometric), optical (i.e., colorimetric, fluorescent, luminescent, and interferometric), mass-based (i.e., piezoelectric and acoustic wave), and calorimetric transducers fall into four general groups (Kazemi-Darsanaki, Azizzadeh, Nourbakhsh, Raeisi, & Aliabadi, 2013).

Due to their portability, cost-effectiveness, small size, and ease of use, electrochemical biosensors are the most common type of biosensor in use today (An et al., 2018). Electrochemical biosensors can be used as point-of-care (POC) instruments at home or the doctor's office. The glucose sensor, which has revolutionized how blood glucose readings are taken and registered, is an electrochemical biosensor, as described above (Cui, Zhou, & Zhou, 2020). The two most common types of electrochemical biosensors are potentiometric and amperometric. In order to detect an electrical response in the molecular recognition portion, potentiometric biosensors use ion-selective electrodes. A light-addressable potentiometric sensor (LAPS) that is coupled to a phage recognition factor is not yet in clinical use, but shows great promise in the field of cancer detection (Bohunicky & Mousa, 2011). The phage-LAPS was able to detect with high sensitivity the cancer biomarker hPRL-3 and cancer cells (MDA/MB231 breast cancer cell line). This new potentiometric-based biosensor has been suggested for use in cancer clinical diagnosis and anticancer drug assessment (Mavrič et al., n.d.; Sadighbayan, Sadighbayan, Tohid-kia, Khosroushahi, & Hasanzadeh, 2019).

Amperometric transducers calculate the current produced when a potential between two electrodes is mounted. A current is produced by oxidation or reduction reactions, which can then be calculated. Amperometric-based cancer detection biosensors that use sequence-specific DNA as the factor of recognition have the potential to be extremely useful in cancer diagnosis (Ferreira, Uliana, & Castilho, 2013; Zhao, Yan, Zhu, Li, & Li, 2013). Via identification of and hybridization to unique DNA sequences present in the cancer cell genomes, these sensors can detect the existence of gene mutations associated with cancer. Wang and Kawde, who used chronopotentiometric biosensors for transduction, made a breakthrough using this form of technology to establish the mutations of BRCA1 and BRCA2 are associated with hereditary breast cancer (Miranda et al., 2010; Wang et al., 2016).

GlucoWatch is a glucose monitor designed for the noninvasive, ongoing measurement of blood glucose levels in patients with diabetes and can also be applied in the case of cancer (Tierney, Tamada, Potts, Jovanovic, & Garg, 2001). The GlucoWatch is worn like a wristwatch and, through reverse iontophoresis, takes blood glucose readings through the skin, a mechanism whereby a tiny electrical signal brings glucose to the surface of the skin so that it can be measured. Via an amperometer transducer, glucose signals are converted into electrical signals. This type of device allows patients without the discomfort associated with conventional injections to control their blood glucose readings (Wysocki, 2005). The GlucoWatch also has a warning to the patient when blood glucose measurements are out of the normal range and record multiple glucose readings (Dunn, Eastman, & Tamada, 2004). The capacity to detect damaged DNA as well as the carcinogens that caused the damage is supported by electrochemical biosensors. In immunoassays and protein arrays, electrochemical transducer technology is also commonly used. For cancer biomarkers' measurements, immunosensors in which an antibody is coupled to an electrochemical transducer have also been used (Keshavarz, Behpour, & Rafiee-Pour, 2015; Mollarasouli, Kurbanoglu, & Ozkan, 2019).

A crucial development in biosensor technology is the realization of multiplex biosensors capable of detecting multiple CAs, thereby providing more precise diagnosis and monitoring (Bohunicky & Mousa, 2011). This type of biosensor must contain several transducers targeted for detection by unique proteins or antigens. With the use of semiconductor materials, electrochemical-based biosensors have been key players in the field of multiplex biosensors since the technology is accurate, low-cost, and easily scaled down. The most popular multiplex instruments used for cancer protein analysis are multifunctional antibody arrays based on ELISA methodology (Anik & Timur, 2016).

Asphahani and Zhang (Sun et al., 2015) have shown the feasibility of a cell-based electrochemical biosensor that measures changes in cell impedance in response to analytes. In order to track changes caused by different stimuli, these cell-based biosensors are commonly referred to as cytosensors and use live cells as the biological sensing element (Sun, Lu, Zhang, & Chen, 2019). This type of sensor can be used to track the effects of anticancer agents on their target molecules when it comes to cancer therapy. By killing cancer cells via p53-mediated apoptosis, most anticancer drugs acts. The cell undergoes dramatic shifts in membrane integrity, ion channel permeability, and overall morphology during apoptosis. Such cellular changes could be identified more easily and efficiently by cytosensors than by a sensor that uses a purified component (i.e., receptor or enzyme), thereby more reliably assessing the anticancer agent's pharmaceutical efficacy in question (Wang, Xiong, Xiao, & Duan, 2017; Xu, Hu, Wang, Ma, & Guo, 2020).

## 21.8 Novel biosensors

Electrochemical biosensors will be most widely chosen because of their minimal price, flexibility, compact size, and simple usability. They may be worked on in or out of a doctor's office. The most frequently used instance of an electrochemical biosensor centered on a screen-printed disposable amperometric electrode is the glucose biosensor. These kinds of the biosensor are currently widely used also for uncovering cancers in patients in various clinics all over the world, offering diagnosis for on-site study (El Aamri, Yammouri, Mohammadi, Amine, & Korri-Youssoufi, 2020; Wang et al., 2017). Another invention that integrates several electrochemical detection technologies in a solitary multifaceted instrument is a single chip that is used in the analysis of clinical cancer analytes for numerous compounds, and electrolyte levels are the handheld STAT clinical analyzer. Initial discovery of malignancy is established by antibody-based electrochemical affinity sensors with excellent vision and responsiveness. A wide assortment of electrochemical transducers and potentiometric, impedimetric/conductivity, and amperometric devices are included among such techniques (Sadighbayan et al., 2019). Some of the most popularly employed electrochemical biosensors are amperometric and potentiometer biosensors.

The potentiometric sensor uses electrodes susceptible to various ions and analytes that perceive electric response every time the molecular recognition of different analytes or element occurs. Such biosensors possess a high potential for cancer diagnosis (Cui et al., 2020; Topkaya, Azimzadeh, & Ozsoz, 2016). The MDA / MB231 marker hPRL-3 has been reported for highly selective breast tumor cell detection (Prittesh, 2017). Amperometric transducers can measure the current generated owing to the potential difference generated between electrodes. The current formed by oxidation-reducing reactions can be calculated. For cancer detection by using a nucleotide sequence, the use of amperometric biosensors has been documented. These sensors can detect interlinked tumors with genetic variations; detection occurs by hybridization of tumor-specific DNA sequences within the tumor cell genome (Povedano et al., 2020). This kind of biosensors test for mutations such as BRCA1 and BRCA2 is associated with inherited breast cancer. Electrochemical biosensors accurately identify damaged DNA together with carcinogens that cause the damage. Protein and immunoassay scans and electrochemical transducers are also included. In immunosensors, which are often used for cancer detection, antibodies are connected to an electrochemical transducer. Biosensors that are capable of detecting several types of tumors are the safest way to diagnose and avoid cancer. In order to detect particular antigens or proteins, specific biosensors with multiple transducers are independently designed and synthesized. These biosensors can indeed be constructed with semiconductor materials to be cheap and robust.

## 21.9 Cell and tissue-based biosensors

Information that is more agreeable than the techniques using samples extracted from cells or tissues (for example, DNA/RNA, proteins) is given by methods in which you could add whole cells. This is due to the phase of isolation, which could harm or modify the biomolecules' conformations, affect the analyte concentration or the stability of the sample required. Moreover, measurements carried out on entire cell-based biosensors will mimic processes under physiological conditions (Gupta, Renugopalakrishnan, Liepmann, Paulmurugan, & Malhotra, 2019). For experiments with living cells (for compound absorption tests) or fixed cells, cells seeded on sensors can be allocated (like compound binding ability tests). Usually, in vitro cells are grown on polystyrene or glass plates, often on precoated extracellular matrix (ECM) protein surfaces such as fibronectin, collagen, laminin, vitronectin, but seldom on metals such as titanium, zirconium, or gold. Due to the nonspecific adsorption of several biological molecules such as DNA and proteins that can be used as analytes, bare gold sensors, popular for QCM and SPR methods, are not desirable surfaces for cell culture experiments. That is why polystyrene or silicon dioxide are commonly coated with gold sensors. However, surface modification plays an important role in controlling the up and down of cell physiological processes such as adhesion, proliferation, and differentiation (Edmondson, Broglie, Adcock, & Yang, 2014). For the design of devices that regulate living cells' activities, cell attachment monitoring and spreading may be crucial. At the same time, it could bring more comprehensive information about the phase of metastasis. Cell adhesion is a complex process that starts with cell sedimentation, followed by the development of nonspecific interactions between cells and substrates, and then the establishment of specific receptor-ligand type molecular binding. Owing to the ongoing physiological processes, the adhesion of time cells to the surface varies (Katrine & Noemi, 2012). The QCMD method can be extended to calculate cell-type-specific interactions (Wang et al., 2012). Such examples of applications for cell-/tissue-based biosensors are compound absorption tests on cells. It is possible to achieve a combination of whole-cell sensing and real-time label-free monitoring of the absorption of nanoparticles by cells via the SPR technique. The kinetic absorption of selected nanoparticles at μg/mL concentrations has already been checked on HeLa cells. This form, however, is temperature-dependent; the uptake is greater at about 20°C, whereas it is lower for 37°C. Two phases of human colorectal cancer cells were collected from the same patient (primary and metastases), seeded on a polystyrene-coated gold QCM sensor, and lectin-carbohydrate interaction was calculated with lectin Helix pomatia agglutinin (HPA). At the end, a higher affinity of HPA to metastatic cells was obtained. The extent of glycosylation of melanocytes and melanoma cells (cultured on polystyrene coated QCM-D gold sensors) was also investigated by lectin Con A[150]. The study revealed that long and branched structures consist of mannose and glucose forms of oligosaccharides present on metastatic melanoma cells. In contrast, primary tumor cells and normal cells have short and less ramified oligosaccharides. In addition, Con A's affinity with oligosaccharides was ten times higher for metastatic melanoma cells than for primary tumor cells and melanocytes. Cancer drug experiments can also use cell-based biosensors. The antibody-conjugated medication Herceptin detects the receptor of human epidermal growth factor 2 (HER2) protein, which is overexpressed in 25%–30% of breast cancers. It induces a cytostatic effect and antibody-dependent cell-mediated cytotoxicity associated with the G1 phase cell cycle arrest. The G protein-coupled receptors (GPCRs), on the other hand, are essential drug targets that can be regulated by histamine. A triphasic response of HeLa cells to histamine interaction was seen in the SPR study: 1-GPCRs activated calcium release, 2-alternations of cell-matrix adhesion following activation of Protein Kinase C (PKC), 3-dynamic redistribution of mass in cells. Surprisingly, there are only a few tissue-based biosensors listed so far. Specimens of tonsil, prostate, and breast tumors were collected and immobilized on the QCM gold sensor surface. Next, most cancer cell interaction between the rVAR2 protein and placental-like chondroitin sulfate was analyzed, and the measured affinity was within the nanomolar range. Cell adhesion test methods can be used to characterize the function of cell membrane receptors in cancer cells and to scan for other cell-specific ligands. Cell attachment to the surface, for example, is regulated mainly by the integrin receptor of the cell transmembrane that binds to the sequence of Arg-Gly-Asp (RDG). In order to determine the time point of presentation of the adhesive ligand from human umbilical vein endothelial cells (HUVEC), the QCM-D sensor was updated with a photo-activable RGD peptide. The HeLa cells propagating kinetics on the ligand RGD tripeptide were also calculated using a novel high-throughput label-free resonant waveguide grating (RWG) imager. On the other hand, for the binding of suspended melanoma, cervix, and ovarian cancer cells, vitronectin protein and antibody (CA-125)-based QCM biosensors were used (Atay et al., 2016; Gupta et al., 2019).

## 21.10 Biosensors and nanotechnology

Nanotechnology and nanoscience is a fast-growing area and has a significant effect on biosensor technology as well as cancer detection, monitoring, screening, and diagnosis (Hosnedlova, Sochor, Baron, Bjørklund, & Kizek, 2020; Yata, Tiwari, & Ahmad, 2018). Most cancer can be usually diagnosed after it has metastasized or spread, not in the benign stage, making their treatment much tricky and futile as at that stage, no drugs or medications can cure this disease. After the tumor has metastasized, approximately 60% of cancer cases are diagnosed in patients. Applying nanotechnology in the production of biosensors increases the probabilities of detecting cancer earlier, leading to improvement of overall treatment efficacy and reducing mortality of patients (Guerrini et al., 2018). Many diagnostic procedures are currently used in cancer, possessing pros and cons, and none of these technologies are perfect. A similar technique used is MRI, one of the most popular image technologies. This expensive technology has gained much importance and holds much promise in monitoring, staging, and diagnosing cancer (Menezes, Sekar, & Cylma, 2011; Schoots & Padhani, 2020). MRI suffers from a significant disadvantageous minute particle having dimensions in the lower end of centimeters or below, which cannot be observed and unidentified (Perfézou, Turner, & Merkoçi, 2012). Nanomaterials can be used as imaging compounds that enable cancerous tissues to be differentiated and measured more sensitively and reliably. The nanostructures widely exploited in such techniques are dendrimers, liposomes, carbon nanotubes, buckyballs, etc., and can hugely improve as well as accelerate the imaging procedures of cancers (Menezes et al., 2011; Schoots & Padhani, 2020). In addition, nanoscience applications can result in minuscule sensors, resulting in cheap technology and enhanced recognition of cancer markers, more effective and precise signal detection and analysis, as well as high-performance detection (Grodzinski, Silver, & Molnar, 2006; Perfézou et al., 2012).

Nanomaterials usually encompass particles or structures having dimensions in the range of 1–100 nm. The small size of nanoparticles enables a higher ratio of surface to volume. This ratio increase allows better diagnosis, imaging, prognosis, and enhanced delivery of drugs to cancer microenvironments that earlier were unavailable and of no significance (Bayda, Adeel, Tuccinardi, Cordani, & Rizzolio, 2020). In various novel instrumentations, nanowires, nanocantilevers, and nanochannels have been exploited for improved transduction of signals as well as identification of cancer-specific events (Erramilli, 2008). To detect micro-RNAs, researchers have developed a biosensor that depends on nanowire technology. MiRNAs are major regulators that control genes' expression levels and are connected to cancer development and progression. Outdated miRNA detection approaches such as Northern blotting are time-consuming, expensive, and possess low sensitivity. The development of an easy-to-use, low-cost, noninvasive biosensor to detect miRNAs associated with cancer might result in massive advancement in the field of using miRNAs for cancer detection (Lyberopoulou, Efstathopoulos, & Gazouli, 2016).

Innovations in single-walled carbon nanotubes (SWCNTs) have led to increased usage and dramatically improved their electrochemical biosensor detection capabilities, increasing sensitivity many folds (Claussen, Franklin, Haque, Marshall Porterfield, & Fisher, 2009). They possess amplified action against $H_2O_2$and NADH and are being widely used to boost signal detection and transduction in immunosensors and oligonucleotide sensors for cancer biomarkers. Surface-enhanced Raman scattering (SERS), in association with nanotechnology, has led to considerable advancements in optical biosensor techniques. With SERS, more than existing designs can be done with a degree of diversity. Without any intervention, SERS can estimate and distinguish simultaneously up to 25 cancer markers.

A typical instance of how nanotechnology can revolutionize cancer diagnosis is the development of microfluidic lab-on-a-chip (LOC) devices (Gupta et al., 2019). A laboratory's complexity is taken up by LOC technology and streamlined into an affordable, compact, easy-to-use device that can be easily handled by patients, clinicians, and diagnostic personnel. LOC approaches using immunological assays and arrays of DNA hybridization have been tested for their ability to identify people who may be at high risk of cancer. The use of QDs is yet another significant application of nanotechnology (Erickson et al., 2014). Nanocrystalline QDs possess luminescence, phosphorescence, and fluorescence showing many novel properties and are widely applied in optical biosensors for screening cancer (Jiandong, Meili, Claudio, Yong, & Francis, 2018). Photons of various intensity, wavelengths, and spectra are released by QDs, allowing several unique molecular elements to be diagnosed and detected (Fang, Peng, Pang, & Li, 2012). As they travel through an environment, they interact with various molecular entities: molecules, cells, and analytes. The pattern of interaction varies, which leads to the identification of matter. By tracking cancer cell migration, cancer metastasis, and drug therapy efficacy, these QDs hold promise as an advantage in monitoring the growth and development of cancer (Fang, Chen, Liu, & Li, 2017). Their high stability, multidimensionality, and miniature size make such QDs very alluring for application in these fields. QDs can also supply particular target areas with therapeutic agents to enhance pharmaceutical efficacy while reducing toxic effects (McHugh et al., 2018).

Nanobiosensors have turned out to be an emerging area of interdisciplinary research in recent years. Nanomaterials have allowed the advancement of normal biosensors of ultrasensitive biosensors in the field of cancer detection (Gdowski, Ranjan, Mukerjee, & Vishwanatha, 2014). Their efficacy has increased further owing to their excellent biocompatibility, high surface area, electrocatalytic activity, superior electronic characteristics. Within the next decade, nanodiagnostics will be hugely accessible and will be able to detect hundreds of biomarkers simultaneously, quickly, and economically (Sharifi et al., 2019). The most common clinical diagnostic application will be an analysis of biofluids by nanobiosensors. Nanobiosensors not only offer ultra-sensitiveness in cancer biomarkers detection but may open new avenues to detect cancer in their primitive stages along with cancer treatment. Nanodevices for such applications are still in the stage of feasibility (Shandilya et al., 2019). Nanotheranostics is an emerging area of nanotechnology where nanodevices are helpful for therapy. It will be rooted as a prophylaxis target in persons without any noticeable appearances or symptoms of cancer, and smartphones can quickly do the monitoring. Such an intense level of screening and monitoring can perceive cancers in the preliminary stages, thereby enabling healthcare professionals for suitable therapeutic intervention. Early detection and treatment might upsurge the likelihood of survival and therapy. Forthcoming inclinations in cancer diagnostics will endure in shrinking of biochip technology to the nanoscale array (Franier & Thompson, 2019; Sharifi, Hasan, Attar, Taghizadeh, & Falahati, 2020).

## 21.11 Challenges

To a large degree, in vivo testing and experimentation have not yet been carried out to discover biosensors suitable for cancer diagnosis. These studies are generally time-consuming, strenuous, economical, and convenient models that need to be developed. Sometimes cancer transitions do not occur, and as a whole, the whole study fails (Franier & Thompson, 2019). Again, they also suffer from drawbacks of similar hallmarks, markers during the progression of cancer. However, unique analyte molecules and specific proteins which are only expressed during cancers are being gradually unearthed in cancer are targeted by molecular studies. The discovery of such molecules requires good knowledge of physiology as well as chemistry, modern instrumentations, and trained personnel (Rasooly & Jacobson, 2006). Powering is a critical problem in in vivo systems. Inductive connections for remotely operated devices have already entered the market. At the same time, size reductions in inductors remain an open subject for in vivo applications. The use of microfabricated inductors demonstrates the most remarkable ability. There were studies of less power usage and autonomous platforms (Johnson & Mutharasan, 2014). Nanoparticle magnetic stimulation switches for in vivo sensing, for example, have been created. The sensor is covered by a semipermeable membrane which allows cancer biomarkers or drug molecules to be selectively diffused into the sensor surface. In addition, biocompatibility concerns have not been discussed sufficiently. Their position in system engineering is inevitably dual: preventing the reaction of the foreign body and fouling of the sensor. Several polymeric materials have been proposed as coating materials, such as polyallylamines, horseradish peroxidase, or polyethylene glycol derivatives, but have been shown to be unsuccessful. The use of poly(lactic-co-glycolic) acid microsphere hydrogels dispersed in poly(vinyl alcohol) has recently been proposed by Wang et al. (Chen et al., 2016; Kumar et al., 2017). The results of preliminary in vivo testing were very positive, but further research is needed to assess its efficiency with different biosensor systems. Nonetheless, a fascinating field has been launched recently: nanoelectronics (Daniels & Pourmand, 2007). In brief, to address current problems in biosensors, nanomaterials are combined with biology and electronics. The downsizing of electronic transducers gives them a more biocompatible and nature-relevant character that is supposed to offer near-nature level sensitivity. The study of neural circuits at the cellular and subcellular level uses nanobioelectronic devices. For minimally invasive recordings, nanowire-nanotube heterostructures can penetrate cell membranes; these nanoprobes can allow spontaneous membrane penetration and a strong membrane seal of high resistance when combined with phospholipid functionalization. Intracellular sensing is possible, opening new avenues for the diagnosis of cancer (Liu et al., 2015).

Modern technologies need to be developed that can detect cancer in its presymptomatic stages. Also, cancer precursors and a precancerous lesion screening technique need to be given importance. Further, technologies need to be developed where patients can overcome the doctor's dependence and be relieved from the invasive methods of on-site screening. Cheaper alternatives for cancer prognosis need to be developed, allowing more responsive and reliable patient outcomes in real-time and remotely reducing follow-up. More and more clinical trials are to be conducted with governmental interventions.

## 21.12 Future aspects

Necessary measures of tumor growth are cancer biomarkers. They are used to diagnose and control illness and provide care with a predictive approach. POC cancer detection is one of biosensor technology's ultimate targets. Point-of-care-testing (POCT), or on-site diagnostic testing, is an area in which biosensors could significantly affect patients and medical staff to quickly and easily get results. POCT facilitates quicker diagnosis and can reduce costs dramatically. However, multitarget detection of multiple biomarkers is essential to transfer biosensors to POC devices. Furthermore, while taking biosensor technology to the bedside of the patient, POCT and multibiosensors must preserve the accuracy and reliability of the laboratory. LOC biosensors are a relatively new technology that shows promise in this field.

Nanomaterials, especially QDs, are the significant influences of nanotechnology on biosensor growth, as they can not only facilitate the diagnosis and tracking of cancer cells, but can also accurately deliver drugs to target sites and allow more sensitive imaging systems that can detect cancer at an earlier stage. Nanotechnology will undoubtedly revolutionize cancer detection and therapy in the next 5–10 years. Integrating nanomaterials and biosensors would make it possible for us to detect disease sooner, enhance cancer imaging, and help diagnose/prognosis and advance drug delivery while mitigating adverse reactions. Cancer exists at the nanoscale stage; it only makes sense that the battle against this disease often takes place at the nanoscale. While the complexity and diversity of cancer have raised many challenges in the medical field, biosensor technology can provide quick, precise results while preserving cost-effectiveness. In cancer diagnosis, clinical biosensors certainly have plenty to offer. Recent developments in the development of multiplexed platforms are encouraging, although nanosensors' sensitivity and selectivity in early diagnosis and therapy monitoring may prove very useful for novel approaches. LOC systems demonstrate a steady ability to commercialize POC and implantable systems quickly. Nanomaterials, especially QDs, can facilitate the tracking of cancer cells or drug molecules. Integrating nanomaterials and biosensors could enhance the imaging of cancer and the delivery of drugs. Soon, customized healthcare services could be a reality. In addition to serving the existing cancer diagnostic strategy, biosensor technology offers the ability to propose and endorse new, more effective schemes. For instance, cancer is typically expressed with a collection of biomarkers at the molecular level; multiplexed platforms could be built to provide accurate information at ultra-low detectability for an extensive dynamic range of several different biomarkers. In addition, the creation of a diagnostic tool for informing pre- or perioperatively about tumor borders may increase therapeutic success rates. Notwithstanding, when developing biosensor platforms, many problems require careful consideration. The assembly of the biosensor components into a fully integrated system could autonomously conduct the analysis phase, despite advances in microfluidics, miniaturized transducers, and materials that have not yet been realized. The evolving nanobioelectronics technology could probably support this objective since the single-cell analysis has only

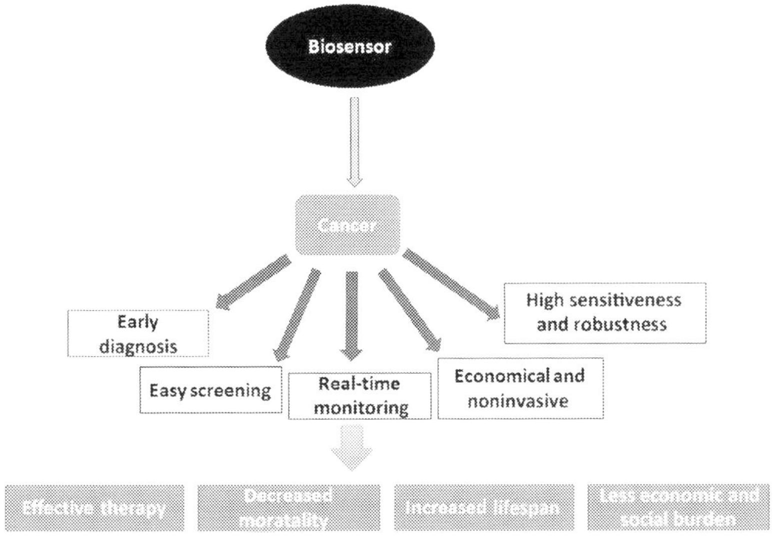


**FIGURE 21.5** Cancer diagnosis by biosensor.

begun to appear as a prerequisite for early cancer diagnosis. Reports demonstrated that evolved-nanoplatforms had not shown the ability to accurately detect only a small number of biomolecules within a given cell. Also, personalized medicine goes beyond the diagnosis of the disease; further, clinical knowledge is needed for comprehensive molecular profiling, particularly for the tumor genesis level, the appropriate treatment regimen, or for disease recurrence monitoring. Therefore biosensors need to be established that could quickly test for DNA mutations and gene products. Another area where biosensing may prove beneficial and successful could be drug development and delivery. The active ingredient properties and how these properties are changed in vivo by transport, binding, or metabolism are influenced by in vivo drug kinetics. This approach requires new strategies for reliably predicting drug delivery properties early in pharmaceutical development so that the most efficient and suitable compounds move to clinical studies. This is especially true for the new therapeutic classes of gene-based drugs. However, the proteomic information is now available from gene expression data which offers new prospects in cancer management and biosensor development (Fig. 21.5).

## References

Ahmed, M. U., Saaem, I., Wu, P. C., & Brown, A. S. (2014). Personalized diagnostics and biosensors: A review of the biology and technology needed for personalized medicine. *Critical Reviews in Biotechnology*, *34*(2), 180–196. Available from https://doi.org/10.3109/07388551.2013.778228.

Amann, A., Costello, B. D. L., Miekisch, W., Schubert, J., Buszewski, B., Pleil, J., ... Risby, T. (2014). The human volatilome: Volatile organic compounds (VOCs) in exhaled breath, skin emanations, urine, feces and saliva. *Journal of Breath Research*, *8*(3). Available from https://doi.org/10.1088/1752-7155/8/3/034001.

An, L., Wang, G., Han, Y., Li, T., Jin, P., & Liu, S. (2018). Electrochemical biosensor for cancer cell detection based on a surface 3D micro-array. *Lab on a Chip*, *18*(2), 335–342. Available from https://doi.org/10.1039/c7lc01117b.

Anik, Ü., & Timur, S. (2016). Towards the electrochemical diagnosis of cancer: Nanomaterial-based immunosensors and cytosensors. *RSC Advances*, *6*(113), 111831–111841. Available from https://doi.org/10.1039/C6RA23686C.

Atay, S., Pişkin, K., Yilmaz, F., Çakir, C., Yavuz, H., & Denizli, A. (2016). Quartz crystal microbalance based biosensors for detecting highly metastatic breast cancer cells via their transferrin receptors. *Analytical Methods*, *8*(1), 153–161. Available from https://doi.org/10.1039/c5ay02898a.

Bailey, A. M., Mao, Y., Zeng, J., Holla, V., Johnson, A., Brusco, L., ... Meric-Bernstam, F. (2014). Implementation of biomarker-driven cancer therapy: Existing tools and remaining gaps. *Discovery Medicine*, *17*(92), 101–114. Available from http://www.discoverymedicine.com/Prakash-Jayabalan/2014/02/03/the-development-of-biomarkers-for-degenerative-musculoskeletal-conditions/.

Barlebo Ahlborn, L., & Østrup, O. (2019). Toward liquid biopsies in cancer treatment: Application of circulating tumor DNA. *APMIS: Acta Pathologica, Microbiologica, et Immunologica Scandinavica*, *127*(5), 329–336. Available from https://doi.org/10.1111/apm.12912.

Bayda, S., Adeel, M., Tuccinardi, T., Cordani, M., & Rizzolio, F. (2020). The history of nanoscience and nanotechnology: From chemical–physical applications to nanomedicine. *Molecules (Basel, Switzerland)*, *25*(1). Available from https://doi.org/10.3390/molecules25010112.

Bhatt, A. N., Mathur, R., Farooque, A., Verma, A., & Dwarakanath, B. S. (2010). Cancer biomarkers – Current perspectives. *Indian Journal of Medical Research*, *132*(8), 129–149. Available from http://icmr.nic.in/ijmr/2010/august/0803.pdf.

Blank, C. U., Haanen, J. B., Ribas, A., & Schumacher, T. N. (2016). The cancer immunogram. *Science (New York, N.Y.)*, *352*(6286), 658–660. Available from https://doi.org/10.1126/science.aaf2834.

Blanpain, C. (2013). Tracing the cellular origin of cancer. *Nature Cell Biology*, *15*(2), 126–134. Available from https://doi.org/10.1038/ncb2657.

Bohunicky, B., & Mousa, S. A. (2011). Biosensors: The new wave in cancer diagnosis. *Nanotechnology, Science and Applications*, *4*(1), 1–10. Available from https://doi.org/10.2147/NSA.S13465.

Brown, J. C., Harhay, M. O., & Harhay, M. N. (2015). Physical function as a prognostic biomarker among cancer survivors. *British Journal of Cancer*, *112*(1), 194–198. Available from https://doi.org/10.1038/bjc.2014.568.

Buchen, L. (2011). Cancer: Missing the mark. *Nature*, *471*(7339), 428–432. Available from https://doi.org/10.1038/471428a.

Chen, X., Pan, Y., Liu, H., Bai, X., Wang, N., & Zhang, B. (2016). Label-free detection of liver cancer cells by aptamer-based microcantilever biosensor. *Biosensors and Bioelectronics*, *79*, 353–358. Available from https://doi.org/10.1016/j.bios.2015.12.060.

Cheng, N., Du, D., Wang, X., Liu, D., Xu, W., Luo, Y., & Lin, Y. (2019). Recent advances in biosensors for detecting cancer-derived exosomes. *Trends in Biotechnology*, *37*(11), 1236–1254. Available from https://doi.org/10.1016/j.tibtech.2019.04.008.

Cieplak, M., & Kutner, W. (2016). Artificial biosensors: How can molecular imprinting mimic biorecognition? *Trends in Biotechnology*, *34*(11), 922–941. Available from https://doi.org/10.1016/j.tibtech.2016.05.011.

Claussen, J. C., Franklin, A. D., Haque, A. U., Marshall Porterfield, D., & Fisher, T. S. (2009). Electrochemical Biosensor of nanocube-augmented carbon nanotube networks. *ACS Nano*, *3*(1), 37–44. Available from https://doi.org/10.1021/nn800682m.

Cui, F., Zhou, Z., & Zhou, H. S. (2020). Review – Measurement and analysis of cancer biomarkers based on electrochemical biosensors. *Journal of the Electrochemical Society*, *167*(3). Available from https://doi.org/10.1149/2.0252003JES.

Daniels, J. S., & Pourmand, N. (2007). Label-free impedance biosensors: Opportunities and challenges. *Electroanalysis*, *19*(12), 1239–1257. Available from https://doi.org/10.1002/elan.200603855.

David, A. R., & Zimmerman, M. R. (2010). Cancer: An old disease, a new disease or something in between? *Nature Reviews. Cancer*, *10*(10), 728–733. Available from https://doi.org/10.1038/nrc2914.

De Souza, J. A., Hunt, B., Asirwa, F. C., Adebamowo, C., & Lopes, G. (2016). Global health equity: Cancer care outcome disparities in high-, middle-, and low-income countries. *Journal of Clinical Oncology*, *34*(1), 6–13. Available from https://doi.org/10.1200/JCO.2015.62.2860.

Dick, F. A., & Rubin, S. M. (2013). Molecular mechanisms underlying RB protein function. *Nature Reviews. Molecular Cell Biology*, *14*(5), 297–306. Available from https://doi.org/10.1038/nrm3567.

Dunn, T. C., Eastman, R. C., & Tamada, J. A. (2004). Rates of glucose change measured by blood glucose meter and the gluco watch biographer during day, night, and around mealtimes. *Diabetes Care*, *27*(9), 2161–2165. Available from https://doi.org/10.2337/diacare.27.9.2161.

Edmondson, R., Broglie, J. J., Adcock, A. F., & Yang, L. (2014). Three-dimensional cell culture systems and their applications in drug discovery and cell-based biosensors. *Assay and Drug Development Technologies*, *12*(4), 207–218. Available from https://doi.org/10.1089/adt.2014.573.

Eivazzadeh-Keihan, R., Pashazadeh-Panahi, P., Baradaran, B., Maleki, A., Hejazi, M., Mokhtarzadeh, A., & de la Guardia, M. (2018). Recent advances on nanomaterial based electrochemical and optical aptasensors for detection of cancer biomarkers. *TrAC – Trends in Analytical Chemistry*, *100*, 103–115. Available from https://doi.org/10.1016/j.trac.2017.12.019.

El Aamri, M., Yammouri, G., Mohammadi, H., Amine, A., & Korri-Youssoufi, H. (2020). Electrochemical biosensors for detection of microRNA as a cancer biomarker: Pros and cons. *Biosensors*, *10*(11). Available from https://doi.org/10.3390/bios10110186.

Emami Nejad, H., Mir, A., & Farmani, A. (2019). Supersensitive and tunable nano-biosensor for cancer detection. *IEEE Sensors Journal*, *19*(13), 4874–4881. Available from https://doi.org/10.1109/JSEN.2019.2899886.

Erickson, D., O'Dell, D., Jiang, L., Oncescu, V., Gumus, A., Lee, S., … Mehta, S. (2014). Smartphone technology can be transformative to the deployment of lab-on-chip diagnostics. *Lab on a Chip*, *14*(17), 3159–3164. Available from https://doi.org/10.1039/c4lc00142g.

Erramilli, S. (2008). *Development of nanomechanical sensors for breast cancer biomarkers*. Boston University Press, MA.

Falanga, A., & Marchetti, M. (2018). Hemostatic biomarkers in cancer progression. *Thrombosis Research*, *164*, S54–S61. Available from https://doi.org/10.1016/j.thromres.2018.01.017.

Fang, M., Chen, M., Liu, L., & Li, Y. (2017). Applications of quantum dots in cancer detection and diagnosis: A review. *Journal of Biomedical Nanotechnology*, *13*(1), 1–16. Available from https://doi.org/10.1166/jbn.2017.2334.

Fang, M., Peng, C. W., Pang, D. W., & Li, Y. (2012). Quantum dots for cancer research: Current status, remaining issues, and future perspectives. *Cancer Biology and Medicine*, *9*(3), 151–163. Available from https://doi.org/10.7497/j.issn.2095-3941.2012.03.001.

Fenn, K. M., Evans, S. B., McCorkle, R., DiGiovanna, M. P., Pusztai, L., Sanft, T., … Chagpar, A. B. (2014). Impact of financial burden of cancer on survivors' quality of life. *Journal of Oncology Practice*, *10*(5), 332–338. Available from https://doi.org/10.1200/JOP.2013.001322.

Ferhan, A. R., Jackman, J. A., Park, J. H., & Cho, N. J. (2018). Nanoplasmonic sensors for detecting circulating cancer biomarkers. *Advanced Drug Delivery Reviews*, *125*, 48–77. Available from https://doi.org/10.1016/j.addr.2017.12.004.

Ferreira, A, Uliana, C., & Castilho, M. (2013). Amperometric biosensor for diagnosis of disease. In *State of the art in biosensors – Environmental and medical applications* (pp. 253–286).

Fitzmaurice, C., Dicker, D., Pain, A., Hamavid, H., Moradi-Lakeh, M., MacIntyre, M. F., … Cooke, G. S. (2015). The global burden of cancer 2013. *JAMA Oncology*, *1*(4), 505–527. Available from https://doi.org/10.1001/jamaoncol.2015.0735.

Francesc, C.-G., Sofia, G., Cinzia, D., Ilaria, A., Luca, Q., Charlotte, N., … Nicola, A. (2018). Cancer diagnosis using a liquid biopsy: Challenges and expectations. *Diagnostics*, *8*, 31. Available from https://doi.org/10.3390/diagnostics8020031.

Franier, B. D. L., & Thompson, M. (2019). Early stage detection and screening of ovarian cancer: A research opportunity and significant challenge for biosensor technology. *Biosensors and Bioelectronics*, *135*, 71–81. Available from https://doi.org/10.1016/j.bios.2019.03.041.

Freitas, M., Nouws, H. P. A., & Delerue-Matos, C. (2018). Electrochemical biosensing in cancer diagnostics and follow-up. *Electroanalysis*, *30*(8), 1576–1595. Available from https://doi.org/10.1002/elan.201800193.

Fu, L. H., Qi, C., Lin, J., & Huang, P. (2018). Catalytic chemistry of glucose oxidase in cancer diagnosis and treatment. *Chemical Society Reviews*, *47*(17), 6454–6472. Available from https://doi.org/10.1039/c7cs00891k.

Gdowski, A., Ranjan, A. P., Mukerjee, A., & Vishwanatha, J. K. (2014). Nanobiosensors: Role in cancer detection and diagnosis. *Advances in Experimental Medicine and Biology*, *807*, 33–58. Available from https://doi.org/10.1007/978-81-322-1777-0_4.

Gelband, H., Sankaranarayanan, R., Gauvreau, C. L., Horton, S., Anderson, B. O., Bray, F., … Trimble, E. L. (2016). Costs, affordability, and feasibility of an essential package of cancer control interventions in low-income and middle-income countries: Key messages from disease control priorities, 3rd edition. *The Lancet*, *387*(10033), 2133–2144. Available from https://doi.org/10.1016/S0140-6736(15)00755-2.

Gharatape, A., & Khosroushahi, A. Y. (2019). Optical biomarker-based biosensors for cancer/infectious disease medical diagnoses. *Applied Immunohistochemistry and Molecular Morphology*, *27*(4), 278–286. Available from https://doi.org/10.1097/PAI.0000000000000586.

Giacinti, C., & Giordano, A. (2006). RB and cell cycle progression. *Oncogene*, *25*(38), 5220–5227. Available from https://doi.org/10.1038/sj.onc.1209615.

Gil-Bazo, I. (2020). Oncogenes in cancer: Using the problem as part of the solution. *Cancers*, *12*(11), 1–3. Available from https://doi.org/10.3390/cancers12113373.

Girgis, A., Lambert, S., Johnson, C., Waller, A., & Currow, D. (2013). Physical, psychosocial, relationship, and economic burden of caring for people with cancer: A review. *Journal of Oncology Practice*, *9*(4), 197–202. Available from https://doi.org/10.1200/JOP.2012.000690.

Glode, A. E., & May, M. B. (2017). Rising cost of cancer pharmaceuticals: Cost issues and interventions to control costs. *Pharmacotherapy*, *37*(1), 85–93. Available from https://doi.org/10.1002/phar.1867.

Glunde, K., Pathak, A. P., & Bhujwalla, Z. M. (2007). Molecular-functional imaging of cancer: To image and imagine. *Trends in Molecular Medicine*, *13*(7), 287–297. Available from https://doi.org/10.1016/j.molmed.2007.05.002.

Gold, J. M., Banjo, C., Freedman, S. O., & Gold, P. (1973). Immunochemical studies of the intramolecular heterogeneity of the carcinoembryonic antigen (CEA) of the human digestive system. *Journal of Immunology*, *111*(6), 1872–1879.

Grodzinski, P., Silver, M., & Molnar, L. K. (2006). Nanotechnology for cancer diagnostics: Promises and challenges. *Expert Review of Molecular Diagnostics*, *6*(3), 307–318. Available from https://doi.org/10.1586/14737159.6.3.307.

Gruhl, F. J., Rapp, B. E., & Länge, K. (2013). Biosensors for diagnostic applications. *Advances in Biochemical Engineering/Biotechnology*, *133*, 115–148. Available from https://doi.org/10.1007/10_2011_130.

Guerrini, L., Garcia-Rico, E., & Alvarez-Puebla, R. (2018). Chapter 7—The role of nanoscience in cancer diagnosis. In *Handbook of nanomaterials for cancer theranostics* (pp. 177–197). <https://doi.org/10.1016/B978-0-12-813339-2.00007-4>.

Gupta, N., Renugopalakrishnan, V., Liepmann, D., Paulmurugan, R., & Malhotra, B. D. (2019). Cell-based biosensors: Recent trends, challenges and future perspectives. *Biosensors and Bioelectronics*, *141*. Available from https://doi.org/10.1016/j.bios.2019.111435.

Hamilton, W., Walter, F. M., Rubin, G., & Neal, R. D. (2016). Improving early diagnosis of symptomatic cancer. *Nature Reviews Clinical Oncology*, *13*(12), 740–749. Available from https://doi.org/10.1038/nrclinonc.2016.109.

Hanash, S. M., Pitteri, S. J., & Faca, V. M. (2008). Mining the plasma proteome for cancer biomarkers. *Nature*, *452*(7187), 571–579. Available from https://doi.org/10.1038/nature06916.

Hanselmann, R. G., & Welter, C. (2016). Origin of cancer: An information, energy, and matter disease. *Frontiers in Cell and Developmental Biology*, *4*. Available from https://doi.org/10.3389/fcell.2016.00121.

Hosnedlova, B., Sochor, J., Baron, M., Bjørklund, G., & Kizek, R. (2020). Application of nanotechnology based-biosensors in analysis of wine compounds and control of wine quality and safety: A critical review. *Critical Reviews in Food Science and Nutrition*, *60*(19), 3271–3289. Available from https://doi.org/10.1080/10408398.2019.1682965.

Hursting, S. D., & Berger, N. A. (2010). Energy balance, host-related factors, and cancer progression. *Journal of Clinical Oncology*, *28*, 4058–4065. Available from https://doi.org/10.1200/jco.2010.27.9935.

Hussain, T., & Nguyen, Q. T. (2014). Molecular imaging for cancer diagnosis and surgery. *Advanced Drug Delivery Reviews*, *66*, 90–100. Available from https://doi.org/10.1016/j.addr.2013.09.007.

Hüttenhain, R., Soste, M., Selevsek, N., Röst, H., Sethi, A., Carapito, C., … Aebersold, R. (2012). Reproducible quantification of cancer-associated proteins in body fluids using targeted proteomics. *Science Translational Medicine*, *4*(142). Available from https://doi.org/10.1126/scitranslmed.3003989.

Ichim, G., & Tait, S. W. G. (2016). A fate worse than death: Apoptosis as an oncogenic process. *Nature Reviews. Cancer*, *16*(8), 539–548. Available from https://doi.org/10.1038/nrc.2016.58.

Imran, A., Qamar, H. Y., Ali, Q., Naeem, H., Riaz, M., Amin, S., … Nasir, I. A. (2017). Role of molecular biology in cancer treatment: A review article. *Iranian Journal of Public Health*, *46*(11), 1475–1485. Available from http://ijph.tums.ac.ir/index.php/ijph/article/download/11475/5814.

Jiandong, W., Meili, D., Claudio, R., Yong, L., & Francis, L. (2018). Lab-on-chip technology for chronic disease diagnosis. *NPJ Digital Medicine*, *1*. Available from https://doi.org/10.1038/s41746-017-0014-0.

Johnson, B. N., & Mutharasan, R. (2014). Biosensor-based microRNA detection: Techniques, design, performance, and challenges. *Analyst*, *139*(7), 1576–1588. Available from https://doi.org/10.1039/c3an01677c.

Justino, C. I. L., Freitas, A. C., Pereira, R., Duarte, A. C., & Rocha Santos, T. A. P. (2015). Recent developments in recognition elements for chemical sensors and biosensors. *TrAC — Trends in Analytical Chemistry*, *68*, 2–17. Available from https://doi.org/10.1016/j.trac.2015.03.006.

Karin, H., André, M., Dipita, B.-G., & Thomas, E. (2017). The role of p53 in cancer drug resistance and targeted chemotherapy. *Oncotarget*, *8*, 8921–8946. Available from https://doi.org/10.18632/oncotarget.13475.

Katrine, K.-P., & Noemi, R. (2012). Cell-based biosensors: Electrical sensing in microfluidic devices. *Diagnostics*, *2*, 83–96. Available from https://doi.org/10.3390/diagnostics2040083.

Kazemi-Darsanaki, R., Azizzadeh, A., Nourbakhsh, M., Raeisi, G., & Aliabadi, M.A. (2013). Biosensors: Functions and applications. *Journal of Biology and Today's World*. <https://doi.org/10.15412/J.JBTW.01020105>.

Keshavarz, M., Behpour, M., & Rafiee-Pour, H. A. (2015). Recent trends in electrochemical microRNA biosensors for early detection of cancer. *RSC Advances*, *5*(45), 35651–35660. Available from https://doi.org/10.1039/c5ra01726b.

Khanmohammadi, A., Aghaie, A., Vahedi, E., Qazvini, A., Ghanei, M., Afkhami, A., ... Bagheri, H. (2020). Electrochemical biosensors for the detection of lung cancer biomarkers: A review. *Talanta*, *206*. Available from https://doi.org/10.1016/j.talanta.2019.120251.

Klein, W. M. P., Bloch, M., Hesse, B. W., McDonald, P. G., Nebeling, L., O'Connell, M. E., ... Tesauro, G. (2014). Behavioral research in cancer prevention and control: A look to the future. *American Journal of Preventive Medicine*, *46*(3), 303–311. Available from https://doi.org/10.1016/j.amepre.2013.10.004.

Kontomanolis, E. N., Koutras, A., Syllaios, A., Schizas, D., Mastoraki, A., Garmpis, N., ... Fasoulakis, Z. (2020). Role of oncogenes and tumor-suppressor genes in carcinogenesis: A review. *Anticancer Research*, *40*(11), 6009–6015. Available from https://doi.org/10.21873/anticanres.14622.

Krishna, V. D., Wu, K., Su, D., Cheeran, M. C. J., Wang, J. P., & Perez, A. (2018). Nanotechnology: Review of concepts and potential application of sensing platforms in food safety. *Food Microbiology*, *75*, 47–54. Available from https://doi.org/10.1016/j.fm.2018.01.025.

Kumar, N., Kumar, S., Kumar, J., & Panda, S. (2017). Investigation of mechanisms involved in the enhanced label free detection of prostate cancer biomarkers using field effect devices. *Journal of the Electrochemical Society*, *164*(9), B409–B416. Available from https://doi.org/10.1149/2.0541709jes.

Laurie, C. C., Laurie, C. A., Rice, K., Doheny, K. F., Zelnick, L. R., McHugh, C. P., ... Weir, B. S. (2012). Detectable clonal mosaicism from birth to old age and its relationship to cancer. *Nature Genetics*, *44*(6), 642–650. Available from https://doi.org/10.1038/ng.2271.

Le, N. D., Rana, S., & Rotello, V. M. (2013). Chemical nose sensors: An alternative strategy for cancer diagnosis. *Expert Review of Molecular Diagnostics*, *13*(2), 111–113. Available from https://doi.org/10.1586/erm.12.143.

Lee, J. O., So, H. M., Jeon, E. K., Chang, H., Won, K., & Kim, Y. H. (2008). Aptamers as molecular recognition elements for electrical nanobiosensors. *Analytical and Bioanalytical Chemistry*, *390*(4), 1023–1032. Available from https://doi.org/10.1007/s00216-007. Available from 1643-y.

Li, P., Guo, Y. J., Tang, Q., & Yang, L. (2018). Effectiveness of nursing intervention for increasing hope in patients with cancer: A *meta*-analysis. *Revista Latino-Americana de Enfermagem*, *26*. Available from https://doi.org/10.1590/1518-8345.1920.2937.

Li, Y., Zeng, X., He, J., Gui, Y., Zhao, S., Chen, H., ... Yuan, H. (2018). Circular RNA as a biomarker for cancer: A systematic meta analysis. *Oncology Letters*, *16*(3), 4078–4084. Available from https://doi.org/10.3892/ol.2018.9125.

Liang, L., Su, M., Li, L., Lan, F., Yang, G., Ge, S., ... Song, X. (2016). Aptamer-based fluorescent and visual biosensor for multiplexed monitoring of cancer cells in microfluidic paper-based analytical devices. *Sensors and Actuators, B: Chemical*, *229*, 347–354. Available from https://doi.org/10.1016/j.snb.2016.01.137.

Lim, H. J., Saha, T., Tey, B. T., Tan, W. S., & Ooi, C. W. (2020). Quartz crystal microbalance-based biosensors as rapid diagnostic devices for infectious diseases. *Biosensors and Bioelectronics*, *168*. Available from https://doi.org/10.1016/j.bios.2020.112513.

Ling, B., Chen, L., Liu, Q., & Yang, J. (2014). Gene expression correlation for cancer diagnosis: A pilot study. *BioMed Research International*, *2014*. Available from https://doi.org/10.1155/2014/253804.

Liora, N. (1999). Cultural views of cancer around the world. *Cancer Nursing*, *22*, 39–45. Available from https://doi.org/10.1097/00002820-199902000-00008.

Liu, Y., Hu, X., Han, C., Wang, L., Zhang, X., He, X., & Lu, X. (2015). Targeting tumor suppressor genes for cancer therapy. *Bioessays: News and Reviews in Molecular, Cellular and Developmental Biology*, *37*(12), 1277–1286. Available from https://doi.org/10.1002/bies.201500093.

Lower, E. E., Glass, E., Blau, R., & Harman, S. (2009). HER-2/neu expression in primary and metastatic breast cancer. *Breast Cancer Research and Treatment*, *113*(2), 301–306. Available from https://doi.org/10.1007/s10549-008. Available from 9931-6.

Ludwig, J. A., & Weinstein, J. N. (2005). Biomarkers in cancer staging, prognosis and treatment selection. *Nature Reviews. Cancer*, *5*(11), 845–856. Available from https://doi.org/10.1038/nrc1739.

Lujambio, A., & Lowe, S. W. (2012). The microcosmos of cancer. *Nature*, *482*(7385), 347–355. Available from https://doi.org/10.1038/nature10888.

Lyberopoulou, A., Efstathopoulos, E., & Gazouli, M. (2016). *Nanotechnology-based rapid diagnostic tests*. <https://doi:10.5772/63908>.

Lynch, J. A., Venne, V., & Berse, B. (2015). Genetic tests to identify risk for breast cancer. *Seminars in Oncology Nursing*, *31*(2), 100–107. Available from https://doi.org/10.1016/j.soncn.2015.02.007.

Lyratzopoulos, G., Vedsted, P., & Singh, H. (2015). Understanding missed opportunities for more timely diagnosis of cancer in symptomatic patients after presentation. *British Journal of Cancer*, *112*, S84–S91. Available from https://doi.org/10.1038/bjc.2015.47.

Martin, G. S. (2003). Cell signaling and cancer. *Cancer Cell*, *4*(3), 167–174, Cell Press. Available from https://doi.org/10.1016/S1535-6108(03)00216-2.

Mattox, A. K., Bettegowda, C., Zhou, S., Papadopoulos, N., Kinzler, K. W., & Vogelstein, B. (2019). Applications of liquid biopsies for cancer. *Science Translational Medicine*, *11*(507). Available from https://doi.org/10.1126/scitranslmed.aay1984.

Mavrič, T., Benčina, M., & Imani, R. (2018). Electrochemical biosensor based on $TiO_2$ nanomaterials for cancer diagnostics. In *Advances in biomembranes and lipid self-assembly* (pp. 63–105). Academic Press.

McHugh, K. J., Jing, L., Behrens, A. M., Jayawardena, S., Tang, W., Gao, M., ... Jaklenec, A. (2018). Biocompatible semiconductor quantum dots as cancer imaging agents. *Advanced Materials*, *30*(18). Available from https://doi.org/10.1002/adma.201706356.

Menezes, D., Sekar, P., & Cylma, M. (2011). Nanoscience in diagnostics: A short review. *Internet Journal of Medical Update*, *6*.

Metkar, S. K., & Girigoswami, K. (2019). Diagnostic biosensors in medicine — A review. *Biocatalysis and Agricultural Biotechnology*, *17*, 271–283. Available from https://doi.org/10.1016/j.bcab.2018.11.029.

Miller, K. D., Nogueira, L., Mariotto, A. B., Rowland, J. H., Yabroff, K. R., Alfano, C. M., ... Siegel, R. L. (2019). Cancer treatment and survivorship statistics, 2019. *CA Cancer Journal for Clinicians*, *69*(5), 363–385. Available from https://doi.org/10.3322/caac.21565.

Miranda, O. R., Chen, H. T., You, C. C., Mortenson, D. E., Yang, X. C., Bunz, U. H. F., & Rotello, V. M. (2010). Enzyme-amplified array sensing of proteins in solution and in biofluids. *Journal of the American Chemical Society*, *132*(14), 5285–5289. Available from https://doi.org/10.1021/ja1006756.

Mollarasouli, F., Kurbanoglu, S., & Ozkan, S. A. (2019). The role of electrochemical immunosensors in clinical analysis. *Biosensors*, *9*(3). Available from https://doi.org/10.3390/bios9030086.

Morales, M. A., & Halpern, J. M. (2018). Guide to selecting a biorecognition element for biosensors. *Bioconjugate Chemistry*, *29*(10), 3231–3239. Available from https://doi.org/10.1021/acs.bioconjchem.8b00592.

Moro, L., Turemis, M., Marini, B., Ippodrino, R., & Giardi, M. T. (2017). Better together: Strategies based on magnetic particles and quantum dots for improved biosensing. *Biotechnology Advances*, *35*(1), 51–63. Available from https://doi.org/10.1016/j.biotechadv.2016.11.007.

Muller, P. A. J., & Vousden, K. H. (2014). Mutant p53 in cancer: New functions and therapeutic opportunities. *Cancer Cell*, *25*(3), 304–317. Available from https://doi.org/10.1016/j.ccr.2014.01.021.

Najeeb, M. A., Ahmad, Z., Shakoor, R. A., Mohamed, A. M. A., & Kahraman, R. (2017). A novel classification of prostate specific antigen (PSA) biosensors based on transducing elements. *Talanta*, *168*, 52–61. Available from https://doi.org/10.1016/j.talanta.2017.03.022.

Narod, S. A., & Foulkes, W. D. (2004). BRCA1 and BRCA2: 1994 and beyond. *Nature Reviews. Cancer*, *4*(9), 665–676. Available from https://doi.org/10.1038/nrc1431.

Nils, W., Gilberto, L., Klaus, M., Steven, S., Wim., van, H., & Arnold, V. (2017). Can we continue to afford access to cancer treatment? *European Oncology & Haematology*, *13*, 114. Available from https://doi.org/10.17925/EOH.2017.13.02.114.

Osman, D. I., El-Sheikh, S. M., Sheta, S. M., Ali, O. I., Salem, A. M., Shousha, W. G., ... Shawky, S. M. (2019). Nucleic acids biosensors based on metal-organic framework (MOF): Paving the way to clinical laboratory diagnosis. *Biosensors and Bioelectronics*, *141*. Available from https://doi.org/10.1016/j.bios.2019.111451.

Padma, V. V. (2015). An overview of targeted cancer therapy. *BioMedicine (Netherlands)*, *5*(4), 1–6. Available from https://doi.org/10.7603/s40681-015-0019-4.

Parrales, A., & Iwakuma, T. (2015). Targeting oncogenic mutant p53 for cancer therapy. *Frontiers in Oncology*, *5*. Available from https://doi.org/10.3389/fonc.2015.00288.

Pasinszki, T., Krebsz, M., Tung, T. T., & Losic, D. (2017). Carbon nanomaterial based biosensors for non-invasive detection of cancer and disease biomarkers for clinical diagnosis. *Sensors (Switzerland)*, *17*(8). Available from https://doi.org/10.3390/s17081919.

Perfézou, M., Turner, A., & Merkoçi, A. (2012). Cancer detection using nanoparticle-based sensors. *Chemical Society Reviews*, *41*(7), 2606–2622. Available from https://doi.org/10.1039/c1cs15134g.

Perumal, V., & Hashim, U. (2014). Advances in biosensors: Principle, architecture and applications. *Journal of Applied Biomedicine*, *12*(1), 1–15. Available from https://doi.org/10.1016/j.jab.2013.02.001.

Ponder, B. A. J. (2001). Cancer genetics. *Nature*, *411*(6835), 336–341. Available from https://doi.org/10.1038/35077207.

Pothipor, C., Wiriyakun, N., Putnin, T., Ngamaroonchote, A., Jakmunee, J., Ounnunkad, K., ... Aroonyadet, N. (2019). Highly sensitive biosensor based on graphene–poly (3-aminobenzoic acid) modified electrodes and porous-hollowed-silver-gold nanoparticle labelling for prostate cancer detection. *Sensors and Actuators, B: Chemical*, *296*. Available from https://doi.org/10.1016/j.snb.2019.126657.

Povedano, E., Ruiz-Valdepenas Montiel, V., Gamella, M., Pedrero, M., Barderas, R., Pelaez-Garcia, A., ... Pingarron, J. M. (2020). Amperometric bioplatforms to detect regional DNA methylation with single-base sensitivity. *Analytical Chemistry*, *92*(7), 5604–5612. Available from https://doi.org/10.1021/acs.analchem.0c00628.

Prensner, J. R., Chinnaiyan, A. M., & Srivastava, S. (2012). Systematic, evidence-based discovery of biomarkers at the NCI. *Clinical and Experimental Metastasis*, *29*(7), 645–652. Available from https://doi.org/10.1007/s10585-012-9507-z.

Price, A. J., Ndom, P., Atenguena, E., Mambou Nouemssi, J. P., & Ryder, R. W. (2012). Cancer care challenges in developing countries. *Cancer*, *118*(14), 3627–3635. Available from https://doi.org/10.1002/cncr.26681.

Prittesh, P. (2017). Biosensors and biomarkers: Promising tools for cancer diagnosis. *International Journal of Biosensors & Bioelectronics*. Available from https://doi.org/10.15406/ijbsbe.2017.03.00072.

Queralto, N., Berliner, A. N., Goldsmith, B., Martino, R., Rhodes, P., & Lim, S. H. (2014). Detecting cancer by breath volatile organic compound analysis: A review of array-based sensors. *Journal of Breath Research*, *8*(2). Available from https://doi.org/10.1088/1752-7155/8/2/027112.

Rasooly, A., & Jacobson, J. (2006). Development of biosensors for cancer clinical testing. *Biosensors and Bioelectronics*, *21*(10), 1851–1858. Available from https://doi.org/10.1016/j.bios.2006.01.003.

Rocha-Santos, T. A. P. (2014). Sensors and biosensors based on magnetic nanoparticles. *TrAC – Trends in Analytical Chemistry*, *62*, 28–36. Available from https://doi.org/10.1016/j.trac.2014.06.016.

Rodland, K. D. (2004). Proteomics and cancer diagnosis: The potential of mass spectrometry. *Clinical Biochemistry*, *37*(7), 579–583. Available from https://doi.org/10.1016/j.clinbiochem.2004.05.011.

Ross, J. S., Fletcher, J. A., Bloom, K. J., Linette, G. P., Stec, J., Symmans, W. F., ... Hortobagyi, G. N. (2004). Targeted therapy in breast cancer: The HER-2/neu gene and protein. *Molecular and Cellular Proteomics*, *3*(4), 379–398. Available from https://doi.org/10.1074/mcp.R400001-MCP200.

Sadighbayan, D., Sadighbayan, K., Tohid-kia, M. R., Khosroushahi, A. Y., & Hasanzadeh, M. (2019). Development of electrochemical biosensors for tumor marker determination towards cancer diagnosis: Recent progress. *TrAC – Trends in Analytical Chemistry*, *118*, 73–88. Available from https://doi.org/10.1016/j.trac.2019.05.014.

Sallam, R. M. (2015). Proteomics in cancer biomarkers discovery: Challenges and applications. *Disease Markers*, *2015*. Available from https://doi.org/10.1155/2015/321370.

Sankaranarayanan, R., Ramadas, K., & Qiao, Y. L. (2014). Managing the changing burden of cancer in Asia. *BMC Medicine*, *12*(1). Available from https://doi.org/10.1186/1741-7015-12-3.

Schlange, T., & Pantel, K. (2016). Potential of circulating tumor cells as blood-based biomarkers in cancer liquid biopsy. *Pharmacogenomics*, *17*(3), 183–186. Available from https://doi.org/10.2217/pgs.15.163.

Schoots, I. G., & Padhani, A. R. (2020). Personalizing prostate cancer diagnosis with multivariate risk prediction tools: How should prostate MRI be incorporated? *World Journal of Urology*, *38*(3), 531–545. Available from https://doi.org/10.1007/s00345-019-02899-0.

Seyfried, T. N., & Huysentruyt, L. C. (2013). On the origin of cancer metastasis. *Critical Reviews in Oncogenesis*, *18*(1–2), 43–73. Available from https://doi.org/10.1615/CritRevOncog.v18.i1-2.40.

Shah, S. C., Kayamba, V., Peek, R. M., & Heimburger, D. (2019). Cancer control in low- and middle-income countries: Is it time to consider screening? *Journal of Global Oncology*, *2019*(5). Available from https://doi.org/10.1200/JGO.18.00200.

Shandilya, R., Bhargava, A., Bunkar, N., Tiwari, R., Goryacheva, I. Y., & Mishra, P. K. (2019). Nanobiosensors: Point-of-care approaches for cancer diagnostics. *Biosensors and Bioelectronics*, *130*, 147–165. Available from https://doi.org/10.1016/j.bios.2019.01.034.

Sharifi, M., Avadi, M. R., Attar, F., Dashtestani, F., Ghorchian, H., Rezayat, S. M., … Falahati, M. (2019). Cancer diagnosis using nanomaterials based electrochemical nanobiosensors. *Biosensors and Bioelectronics*, *126*, 773–784. Available from https://doi.org/10.1016/j.bios.2018.11.026.

Sharifi, M., Hasan, A., Attar, F., Taghizadeh, A., & Falahati, M. (2020). Development of point-of-care nanobiosensors for breast cancers diagnosis. *Talanta*, *217*. Available from https://doi.org/10.1016/j.talanta.2020.121091.

Shehada, N., Cancilla, J. C., Torrecilla, J. S., Pariente, E. S., Brönstrup, G., Christiansen, S., … Haick, H. (2016). Silicon nanowire sensors enable diagnosis of patients via exhaled breath. *ACS Nano*, *10*(7), 7047–7057. Available from https://doi.org/10.1021/acsnano.6b03127.

Sheikhpour, M., Golbabaie, A., & Kasaeian, A. (2017). Carbon nanotubes: A review of novel strategies for cancer diagnosis and treatment. *Materials Science and Engineering C*, *76*, 1289–1304. Available from https://doi.org/10.1016/j.msec.2017.02.132.

Shelton, P., & Jaiswal, A. K. (2013). The transcription factor NF-E2-related factor 2 (Nrf2): A protooncogene? *FASEB Journal*, *27*(2), 414–423. Available from https://doi.org/10.1096/fj.12-217257.

Sherr, C. J., & McCormick, F. (2002). The RB and p53 pathways in cancer. *Cancer Cell*, *2*(2), 103–112. Available from https://doi.org/10.1016/S1535-6108(02)00102-2.

Shukla, H. D. (2017). Comprehensive analysis of cancer-proteogenome to identify biomarkers for the early diagnosis and prognosis of cancer. *Proteomes*, *5*(4). Available from https://doi.org/10.3390/proteomes5040028.

Siravegna, G., Marsoni, S., Siena, S., & Bardelli, A. (2017). Integrating liquid biopsies into the management of cancer. *Nature Reviews Clinical Oncology*, *14*(9), 531–548. Available from https://doi.org/10.1038/nrclinonc.2017.14.

Smith, L. K., Pope, C., & Botha, J. L. (2005). Patients' help-seeking experiences and delay in cancer presentation: A qualitative synthesis. *Lancet*, *366* (9488), 825–831. Available from https://doi.org/10.1016/S0140-6736(05)67030-4.

Srivastava, A., & Creek, D. J. (2019). Discovery and validation of clinical biomarkers of cancer: A review combining metabolomics and proteomics. *Proteomics*, *19*(10). Available from https://doi.org/10.1002/pmic.201700448.

Su, L., Zou, L., Fong, C. C., Wong, W. L., Wei, F., Wong, K. Y., … Yang, M. (2013). Detection of cancer biomarkers by piezoelectric biosensor using PZT ceramic resonator as the transducer. *Biosensors and Bioelectronics*, *46*, 155–161. Available from https://doi.org/10.1016/j.bios.2013.01.074.

Sullivan, R., Pramesh, C. S., & Booth, C. M. (2017). Cancer patients need better care, not just more technology. *Nature*, *549*(7672), 325–328. Available from https://doi.org/10.1038/549325a.

Sun, D., Lu, J., Zhang, L., & Chen, Z. (2019). Aptamer-based electrochemical cytosensors for tumor cell detection in cancer diagnosis: A review. *Analytica Chimica Acta*, *1082*, 1–17. Available from https://doi.org/10.1016/j.ac.2019.07.054.

Sun, Z., Shi, Y., Shen, Y., Cao, L., Zhang, W., & Guan, X. (2015). Analysis of different HER-2 mutations in breast cancer progression and drug resistance. *Journal of Cellular and Molecular Medicine*, *19*(12), 2691–2701. Available from https://doi.org/10.1111/jcmm.12662.

Sung, H., Ferlay, J., Siegel, R. L., Laversanne, M., Soerjomataram, I., Jemal, A., & Bray, F. (2021). Global cancer statistics 2020: GLOBOCAN estimates of incidence and mortality worldwide for 36 cancers in 185 countries. *CA Cancer Journal for Clinicians*. Available from https://doi.org/10.3322/caac.21660.

Surbone, A, Zwitter, M., & Rajer, M. (2012). *New challenges in communication with cancer patients*. ISBN: 978-1-4614-3368-2.

Tierney, M. J., Tamada, J. A., Potts, R. O., Jovanovic, L., & Garg, S. (2001). Clinical evaluation of the GlucoWatch® biographer: A continual, non-invasive glucose monitor for patients with diabetes. *Biosensors and Bioelectronics*, *16*(9–12), 621–629. Available from https://doi.org/10.1016/S0956-5663(01)00189-0.

Topkaya, S. N., Azimzadeh, M., & Ozsoz, M. (2016). Electrochemical biosensors for cancer biomarkers detection: Recent advances and challenges. *Electroanalysis*, *28*(7), 1402–1419. Available from https://doi.org/10.1002/elan.201501174.

Tothill, I. E. (2009). Biosensors for cancer markers diagnosis. *Seminars in Cell and Developmental Biology*, *20*(1), 55–62. Available from https://doi.org/10.1016/j.semcdb.2009.01.015.

Uings, I. J., & Farrow, S. N. (2000). Cell receptors and cell signalling. *Journal of Clinical Pathology – Molecular Pathology*, *53*(6), 295–299. Available from https://doi.org/10.1136/mp.53.6.295.

Vaidyanathan, R., Soon, R. H., Zhang, P., Jiang, K., & Lim, C. T. (2019). Cancer diagnosis: From tumor to liquid biopsy and beyond. *Lab on a Chip*, *19*(1), 11–34. Available from https://doi.org/10.1039/c8lc00684a.

Vedsted, P., & Olesen, F. (2015). A differentiated approach to referrals from general practice to support early cancer diagnosis – The Danish three-legged strategy. *British Journal of Cancer*, *112*, S65–S69. Available from https://doi.org/10.1038/bjc.2015.44.

Venkitaraman, A. R. (2002). Cancer susceptibility and the functions of BRCA1 and BRCA2. *Cell*, *108*(2), 171–182. Available from https://doi.org/10.1016/S0092-8674(02)00615-3.

Vicente-Dueñas, C., Romero-Camarero, I., Cobaleda, C., & Sánchez-García, I. (2013). Function of oncogenes in cancer development: A changing paradigm. *EMBO Journal*, *32*(11), 1502–1513. Available from https://doi.org/10.1038/emboj.2013.97.

Vigneshvar, S., Sudhakumari, C. C., Senthilkumaran, B., & Prakash, H. (2016). Recent advances in biosensor technology for potential applications - An overview. *Frontiers in Bioengineering and Biotechnology*, *4*. Available from https://doi.org/10.3389/fbioe.2016.00011.

Viswambari Devi, R., Doble, M., & Verma, R. S. (2015). Nanomaterials for early detection of cancer biomarker with special emphasis on gold nanoparticles in immunoassays/sensors. *Biosensors and Bioelectronics*, *68*, 688–698. Available from https://doi.org/10.1016/j.bios.2015.01.066.

Wang, A., Wang, C. P., Tu, M., & Wong, D. T. W. (2016). Oral biofluid biomarker research: Current status and emerging frontiers. *Diagnostics*, *6*(4). Available from https://doi.org/10.3390/diagnostics6040045.

Wang, J., Wu, C., Hu, N., Zhou, J., Du, L., & Wang, P. (2012). Microfabricated electrochemical cell-based biosensors for analysis of living cells in vitro. *Biosensors*, *2*(2), 127–170. Available from https://doi.org/10.3390/bios2020127.

Wang, L., Xiong, Q., Xiao, F., & Duan, H. (2017). 2D nanomaterials based electrochemical biosensors for cancer diagnosis. *Biosensors and Bioelectronics*, *89*, 136–151. Available from https://doi.org/10.1016/j.bios.2016.06.011.

Welcsh, P. L., & King, M. C. (2001). BRCA1 and BRCA2 and the genetics of breast and ovarian cancer. *Human Molecular Genetics*, *10*(7), 705–713. Available from https://doi.org/10.1093/hmg/10.7.705.

Weller, D., Vedsted, P., Rubin, G., Walter, F. M., Emery, J., Scott, S., ... Neal, R. D. (2012). The Aarhus statement: Improving design and reporting of studies on early cancer diagnosis. *British Journal of Cancer*, *106*(7), 1262–1267. Available from https://doi.org/10.1038/bjc.2012.68.

Witsch, E., Sela, M., & Yarden, Y. (2010). Roles for growth factors in cancer progression. *Physiology*, *25*(2), 85–101. Available from https://doi.org/10.1152/physiol.00045.2009.

Wu, X., Chen, J., Wu, M., & Zhao, J. X. (2015). Aptamers: Active targeting ligands for cancer diagnosis and therapy. *Theranostics*, *5*(4), 322–344. Available from https://doi.org/10.7150/thno.10257.

Wysocki, T. (2005). Youth and parent satisfaction with clinical use of the glucoWatch G2 Biographer in the management of pediatric type 1 diabetes. *Diabetes Care*, *28*(8), 1929–1935. Available from https://doi.org/10.2337/diacare.28.8.1929.

Xu, J., Hu, Y., Wang, S., Ma, X., & Guo, J. (2020). Nanomaterials in electrochemical cytosensors. *Analyst*, *145*(6), 2058–2069. Available from https://doi.org/10.1039/c9an01895f.

Yata, V. K., Tiwari, B. C., & Ahmad, I. (2018). Nanoscience in food and agriculture: Research, industries and patents. *Environmental Chemistry Letters*, *16*(1), 79–84. Available from https://doi.org/10.1007/s10311-017-0666-7.

Yuan, Y., Liu, L., Chen, H., Wang, Y., Xu, Y., Mao, H., ... Liang, H. (2016). Comprehensive characterization of molecular differences in cancer between male and female patients. *Cancer Cell*, *29*(5), 711–722. Available from https://doi.org/10.1016/j.ccell.2016.04.001.

Zhang, X., Chen, B., He, M., Wang, H., & Hu, B. (2016). Gold nanoparticles labeling with hybridization chain reaction amplification strategy for the sensitive detection of HepG2 cells by inductively coupled plasma mass spectrometry. *Biosensors and Bioelectronics*, *86*, 736–740. Available from https://doi.org/10.1016/j.bios.2016.07.073.

Zhao, J., Yan, Y., Zhu, L., Li, X., & Li, G. (2013). An amperometric biosensor for the detection of hydrogen peroxide released from human breast cancer cells. *Biosensors and Bioelectronics*, *41*(1), 815–819. Available from https://doi.org/10.1016/j.bios.2012.10.019.

Zheng, G., Ma, Y., Zou, Y., Yin, A., Li, W., & Dong, D. (2018). HCMDB: The human cancer metastasis database. *Nucleic Acids Research*, *46*(1), D950–D955. Available from https://doi.org/10.1093/nar/gkx1008.

Chapter 22

# Miniaturized devices for point-of-care testing/miniaturization and integration with microfluidic systems

Ankur Kaushal[1], Amit Seth[2], Deepak Kala[1], Shagun Gupta[3], Lucky Krishnia[1] and Vivek Verma[3]
[1]Amity Centre of Nanotechnology, Amity University, Gurugram, India
[2]School of Life Science, Manipur University, Imphal, India
[3]Shoolini University, Solan, India

## 22.1 Introduction

Miniaturized devices are remarkably operative in plummeting the period of scrutiny, besides easing diminished usage of diagnostic chemicals (Fig. 22.1). Current progressions have further enabled the conservation of sample integrity b before evading contamination. Moreover, proficiency in test precision has improved. A range of strategies, formulations, and innovative consolidation of high-end technologies is additionally being conceived. These new systems even though holding slight dimensions have the ability to carefully analyze minute samples with inadequate precision. The assemblage of these devices by integrating novel constituents and exceedingly sensitive sensors has amplified their involvement in high throughput investigations. One of the foremost criteria for the prolific suitability of these devices is the categorical identification of the target molecule. It is also anticipated that their worth can be boosted by promoting additional diversification apt for broad-spectrum analysis. The advent of microfluidic technology has provided it with the vital impetus for rapid growth and applicability. The efficacious amalgamation of assorted designs has unlocked new prospects for successful commercialization. The swift analysis is of utmost significance in the judicious detection of grave diseases like cancer and HIV. Moreover, the development of low-cost analytical kits is indispensable for affordable and inclusive healthcare and can become effective point-of-care (POC) technology.

Integration of advanced microfluidic technology with the miniaturized POC technology has revolutionized the diagnostic platform many fold. So microfluidics is the manipulation of different fluids at a microscale level and having dimensions at around less than 1 mm. This integration offers the ability to work with less volume, short reaction times, perform simultaneous operations, and can act as an entire laboratory on a single chip

## 22.2 Detection of infectious and chronic diseases

Devices have been designed for the detection of pathogenic bacteria, viruses, and fungi. These devices can have wide ramifications in the healthcare, pharma, agriculture, and biotech industries. A number of model organisms along with subcellular components have been incorporated and successfully applied in these ventures. The outer covering of bacterial or fungal cell and virus coat proteins have been targeted for appropriate sensor detection. These devices are known for their accuracy and precision in being successful even for infinitesimal test samples. These on-site transportable gadgets are very effective in the timely estimation of pathogenic microbes at the very early stages of infection. Furthermore, these devices can estimate microbial pathogens in diverse environmental samples ranging from patient samples to water and soil samples (Lee, Sun, Ham, & Weissleder, 2008). Infectious diseases tend to spread quickly over vast regions thus giving rise to pandemics. The safety of public health is a major concern for governments around the globe, especially in the current post-COVID-19 challenging times. Reasonably priced detection tests are the need of the hour. A number of electronic gadgets have been successfully integrated to form amalgamated biosensors capable of carrying out rapid trials. Similar

Biosensor Based Advanced Cancer Diagnostics. DOI: https://doi.org/10.1016/B978-0-12-823424-2.00005-3

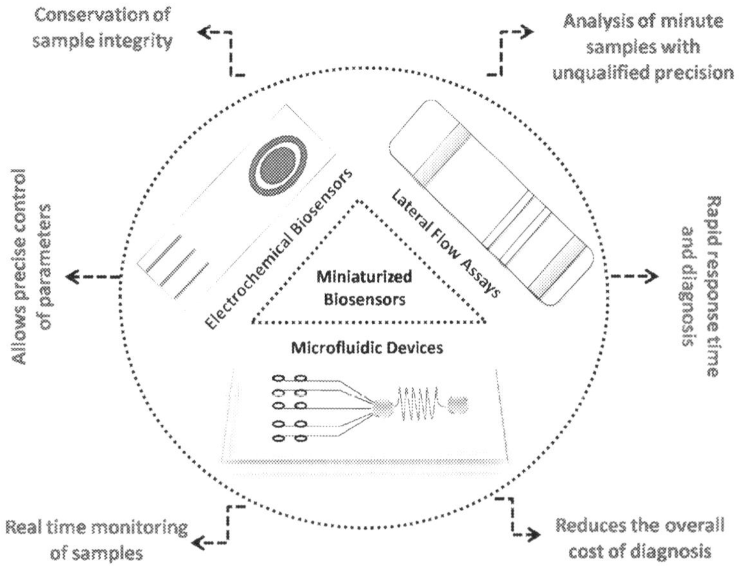


**FIGURE 22.1** The miniaturized devices and their advantages.

devices have also been developed for fungal pathogens particularly appropriate for the agricultural sector. These fungal detectors are either based on nucleotide sequences derived from the microbial genome or rely on explicit antigen-antibody interactions. PCR (Polymerase chain reaction) and ELISA (Enzyme linked immunosorbent assay) are the fundamental technologies that form the source for designing these miniaturized instruments. The ability to quantitatively estimate microbial population and diversity has devices with improved efficiency and applicability. These devices typically target surface markers or mycotoxins released by the pathogen. The analysis based on antibody detection seems to have emerged as the preferable choice against DNA-based detection. This is due to rapid analysis timing and precise detection accompanied with a high level of accuracy thus serving as an effective tool for high throughput screening. Both monoclonal and polyclonal antibodies have been successfully applied for pathogen detection. However, recent advancements in recombinant antibody fragment production have further elevated the precision power of these devices. Immobilization of antibodies on a suitable sensor surface through diverse physical and chemical interventions has concurred with the advent of highly sensitive sensors. The sensor surface used as a substratum for attachment is either of the inorganic origins or composed of high molecular weight organic hydrogels. The efficiency of antibody attachment has been enhanced by the use of various cross-linkers. It is highly imperative that the conformation and orientation of biomolecules should be appropriately maintained so as to ensure their viability during such interactions (Skottrup, Nicolaisen, & Justesen, 2008). The usage of impedance-based biosensors for pathogen detection has been extensively used in the past. The analysis of membrane potential and relative entry and exit of targeted metabolites can rapidly detect desired pathogenic microorganisms. Major progress has been attained in the development of impedance-based microchips. These microchips have enabled accurate estimation of growth patterns of various pathogenic microbes. Miniaturized impedance biosensors can readily discriminate dead cells from living cells (Yang & Bashir, 2008). Han, Koo, Ki, and Kim (2020) developed a low-cost biosensing system capable of integration with smartphone devices that facilitated sample analysis within 20 min. A similar point-of care-testing (POCT) device was fabricated for detection of the COVID-19 virus which involved the usage of gold nanoparticles. Some other recent advancements clearly indicate that POCT device development is constantly evolving and successfully used for various other pathogens like meningitis pathogens, *Escherichia coli* (He et al., 2020), Pseudorabies virus (Huang et al., 2020) and COVID-19 virus (Yakoh et al., 2021). The continuous evolution of microbes in the face of increasing usage of antibiotics is posing new challenges in the field of diagnostic research. The advent of new hyperpathogenic strains is stimulating the neo-fabrication and refabrication of sensitive devices.

## 22.3 Role of nanotechnology in the development of miniaturized devices

POCT is known for its on-site investigations. A variety of biosensors are involved in POCT sample scrutiny. The target is usually an enzyme, DNA segment, or antibody molecule. Nanotechnology has emerged as a promising field having growing pertinence in therapeutic analysis. The nanomaterials range from metallic nanoparticles to complex nanocomposites. They have served as building blocks for the development of a variety of exceedingly sensitive probes. These nanomaterials own supplementary features like magnetism and fluorescence consequently making them enormously apt for analytical procedures. Traditional testing aids typically suffer from dual drawbacks of low sensitivity and selectivity. Recent trends suggest that the electrodes used in biosensors are increasingly incorporating nanomaterials, thus imparting them with improved sensitivity. The usage of small nanoparticle-derived substances has provided an added impetus towards the development of miniaturized devices. However, the applicability of these nanomaterials is found wanting when the test sample volume is present in minuscule amounts far below the detection limit of these advanced biosensors. The efficacy of nanomaterial-based POCT kits can be further enhanced by guaranteeing concurrent scrutiny of miscellaneous biological components for a better holistic analysis. Multidisciplinary and collaborative research is the need of the hour for the development of nano platform-based sensors.

### 22.3.1 Magnetic nanoparticles

Magnetic nanoparticles (MNPs) can be used for tagging desired biological components. A large number of MNPs including that of iron have been proven to be biologically compatible. A variety of techniques have been standardized to develop MNPs. These MNPs have been successfully integrated with diverse biomolecules. High molecular weight polymers are being increasingly used in conjunction with MNPs for improved relevance. Recently a variety of MNPs has been developed that exhibit fluorescence which has eventually facilitated quicker recognition of targeted biomolecules. Magnetic labeling of analyte molecule also ensures its rapid separation under the effect of the applied magnetic field. Owing to their large surface area these MNPs can facilitate simultaneous analysis of a large population of desired biomolecules. Similarly, MNPs labeled antibodies are progressively being used in numerous assays. These MNPs tend to remain stable during the entire process thus exhibiting a higher degree of reliability and robustness. MNPs have been used for testosterone detection (Sanli, Moulahoum, Ghorbanizamani, Gumus, & Timur, 2020). MNPs are known to increase the sensitivity of biosensor electrodes. MNPs also form the core components of colorimetric detectors for identifying pathogenic strains of microorganisms (Eissa & Zourob, 2020). MNPs have also been successfully used in capturing antibodies for rapid detection (Cao, Xiao, Fang, Zhao, & Chen, 2020). These nanoparticles are also finding increasing relevance in the development of sensors for detecting enzyme activity (Adem, Jain, & Sveiven, 2020). MNPs have the ability to initiate the enrichment of various biological molecules including nucleic acids and proteins. DNA molecules can be assembled and readily detected by the use of MNPs. The changes in nucleotide sequences can also be readily detected with the aid of these nanoparticles. The DNA purification at the end of the PCR cycle is also facilitated by MNPs. A careful and concerted selection of biocompatible nanoparticles is imperative for successful application in medical diagnostics.

### 22.3.2 Carbon nanotubes

Akin to MNPs, carbon nanotubes (CNTs) are demonstrating their tremendous relevance in the field of medical diagnostics. CNTs possess the unique ability to serve as versatile components of modern biosensors. CNTs have the ability to magnify signal transduction thus imparting heightened sensitivity to diagnostic applications. Their small size is a critical feature in their efficacious involvement in the assembly of miniaturized systems. The incorporation of CNTs has seen a marked effect on speedy detection and precise prediction for minuscule samples. Successful integration of CNTs and polymeric substances has produced inimitable nanocomposites. CNTs are characterized by their high electrical conductivity, high surface-to-volume ratio, and structural stability. The high sensitivity of CNTs is owed to the presence of carbon atoms at their core and is making them enormously appropriate for multiplexed operations. CNTs are further envisioned to serve as critical components of implantable devices and thus are in the midst of various immunological trials to determine their biocompatibility. CNTs are being applied in the development of highly sensitive paper-based sensors (Shen, Tran, Modha, Tsutsui, & Mulchandani, 2019; Wang et al., 2018). CNTs have also proved to be successful in Alzheimer's disease detection wherein fluorescent sensors are being fabricated. CNT-based composites display higher selectivity and conductivity as compared to unaggregated CNTs. CNTs have also proved to be critical towards the development of enzyme-based biosensors. These CNTs are instrumental in generating unique functional groups on the biosensor surface for improved detection. CNT-equipped biosensors are highly selective and are characterized by minimum deviation in signal transmission due to the presence of any contaminating molecule. CNTs provide a favorable surface for the attachment of MNPs. These CNTs also facilitate stable attachment of sensitive biomolecules like antibodies. CNTs-based biosensors are marked by their low cost and easy portability (Weng, Ahmed, & Neethirajan, 2018).

### 22.3.3 Graphene

Graphene is a 2D-nanomaterial revealing a high level of biocompatibility and economic viability for microfabrication of miniaturized POCTs. The production of graphene has increased many folds in recent years due to its wide applicability. Graphene has the ability to be easily modified to form multilayered structures, which can eventually be transformed into nanosheets. Graphene possesses a large surface area due to its high degree of porosity. Graphene-based biosensors are known for ready responsivity towards diverse biological signals. Graphene also displays appreciable compatibility with DNA. Graphene-based nanosheets can easily be generated by the application of high temperature and sonication. Various studies have clearly demonstrated the stability of graphene during the fabrication process. Graphene incorporation amplifies signal transmission thus leading to increased sensitivity. Graphene is also known to form stable nanocomposites for varied applications. It is also known to enhance electron transfer rate which is a characteristic feature of highly sensitive electrodes (Chaiyo et al., 2018). Graphene-based biosensors are also used in the detection of contraband drugs in biological fluids. Graphene amplifies the surface area of the electrode thus imparting heightened sensitivity. Graphene impregnated paper strips also display amicable shelf life and long-term stability (Narang et al., 2017). Graphene-paper-based biosensors are practically relevant due to the stable and uniform attachment of graphene onto the filter paper surface (Khan et al., 2018).

## 22.4 Integration of microfluidics with miniaturized point-of-care systems

POC testing is an indispensable factor for public healthcare and allows rapid, accurate diagnosis of different diseases in less time. It is becoming a very cheap technology especially in remote areas with the integration of advanced technologies like microfluidics. Microfluidics is a multidisciplinary technology that includes nanotechnology, biotechnology, biochemistry, physics, and engineering (Manz et al., 1992). Microfluidics is an advanced field that pertains to the manipulation of fluids on the microscale level, with critical dimensions of less than 1 mm. By understanding microscale phenomena, it can be used to perform experiments that are not possible on the macro scale and take the biosensing platform to the next level by performing simultaneous experimentations. As advancements in the field have progressed, different methods have emerged for fabricating channels with the required dimensions and integration with electrochemical devices (Fig. 22.2).

While designing a microfluidic device some of the factors should be considered, such as the fabrication material, the compatibility of materials with the solvent used, the number of inlets, and the types of synthesizers and mixers. Soft lithography is the most preferred technique used in microfluidics, especially in polydimethylsiloxane (PDMS) and poly (methyl methacrylate) (PMMA) (Whitesides, Ostuni, Takayama, Jiang, & Ingber, 2001). Other techniques that have

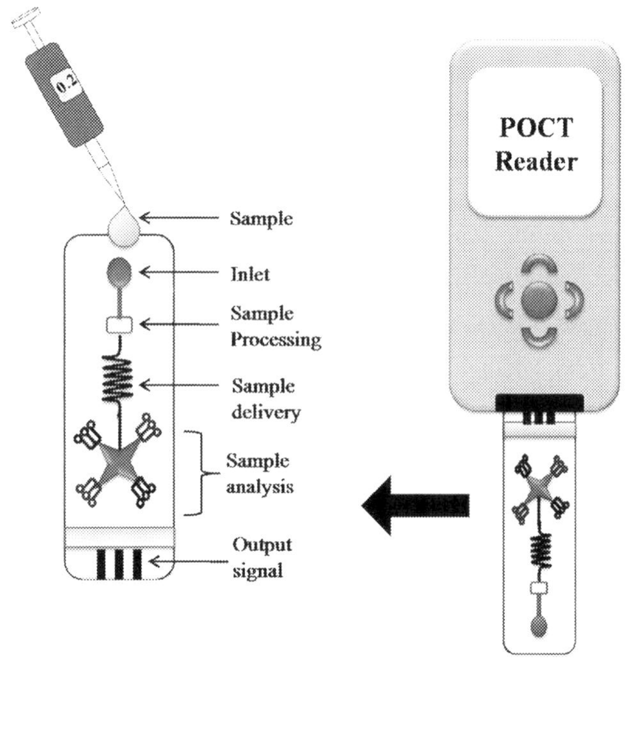


FIGURE 22.2 Schematic representation of microfluidics.

been used in the fabrication of microfluidics are micromachining, injection molding, and laser ablation. Each technique used for the fabrication of microfluidics has its own advantages and disadvantages and can be used depending on the application of the device.

### 22.4.1 Fabrication of microfluidics

Different materials have been used for the fabrication of microfluidics like silicon, glass, and the most common, polymer. The application of polymeric materials to microfluidic devices has influenced greatly the commercialization of POC-based devices, due to their low cost and easy fabrication steps in comparison to glass and silicon. Microfluidic devices can be fabricated by conventional fabrication techniques such as chemical vapor deposition (Lam, Devadhasan, Howse, & Kim, 2017), lithography (Manz et al., 1992), soft-lithography (Kim et al., 2008; McDonald & Whitesides, 2002; Qin, Xia, & Whitesides, 2010), micromachining, plasma treatment (Li, Tian, Nguyen, & Shen, 2008), in situ construction (Beebe et al., 2000; Beebe, Mensing, & Walker, 2002; Khoury, Mensing, & Beebe, 2002; Yu, Bauer, Moore, & Beebe, 2001), injection and polymer molding (Ugolini, Visone, Redaelli, Moretti, & Rasponi, 2017), laser ablation (Gale et al., 2018), polymer laminates and 3D-printing approaches. A cleanroom is required/mandatory for the fabrication of microfluidic devices. The materials which are used in microfluidic devices need to have some specific properties such as optical transparency, mechanically strong, high thermal stability, adaptable to various changes, and mass production that can be easily done. All these properties would not be of a single material, so, according to the type of application the material is selected and so is the fabrication process. The first material used for microfluidic applications were silicon and glass. However, with the new requirements and advancements in technology, there has been the addition of many other new materials, for example, polymer substrates, composites, or paper. The choice of materials that will be used for the fabrication of microfluidic devices varies at the research level and at the commercial level, while versatility and performance are key parameters for the research-level, but at the commercial level, production cost, reliability, and ease of usability are the top priorities (McDonald & Whitesides, 2002)

#### *22.4.1.1 Micromachining*

The technique by which microfluidic devices were fabricated initially was through micromachining on silicon and/or glass (Becker & Heim, 1999a, 1999b; Jacobson, Hergenröder, Koutny, & Ramsey, 1994; Manz et al., 1992; Terry, Jerman, & Angell, 1979), which was very similar to the traditional method used in the semiconductor industry. There is the involvement of various techniques such as wet and/or dry etching, photolithography, e-beam lithography, heat treatment, and many others which require the usage of cleanroom facilities and equipment. Although micromachining techniques generated the best quality devices and are still widely used, the silicon is replaced as it is not an ideal material for microfluidic applications as it is optically opaque, costly, component integration is difficult, and above all, the surface characteristics are not suitable for biological applications. Silicon and glass-based devices are well suited for chemistry-based applications where there are strong solvents used, with the requirement of high temperatures and chemically stable surfaces (Becker & Heim, 1999a, 1999b; Jacobson et al., 1994; Kameoka, Craighead, Zhang, & Henion, 2001; Manz et al., 1992; Terry et al., 1979; Xu, Locascio, Gaitan, & Lee, 2000). Therefore, these materials have been replaced by disposable plastic-based devices, although the master device is first generated by the micromachining steps and from that the plastic devices are being replicated.

#### *22.4.1.2 Soft lithography*

The term soft lithography was first coined and used by Xia and Whitesides in 1998 (Faustino, Catarino, Lima, & Minas, 2016; Whitesides et al., 2001). This is a combination of a diverse set of techniques which includes replica molding using PDMS for the fabrication of the microfluidic device as well as microcontact printing. Mainly, this technology relies on the usage of photolithography for the generation of silicon and photoresist molds; on top of it PDMS is being poured and cured. Photolithography requires the usage of a photoresist and in this technique, SU-8 is commonly used due to its unique properties such as mold durability, capacity to handle high aspect ratios, and high resolution (Faustino et al., 2016). The basic technique utilized for making PDMS microchannels includes the manufacture of a silicon first master imprint with designed highlights made out of a photoresist (a photograph dynamic polymer regularly utilized in photolithography, e.g., SU-8). In this cycle, silicon wafers are covered with photoresist and afterward presented to bright UV light through a veil (Duffy, McDonald, Schueller, & Whitesides, 1998; McDonald & Whitesides, 2002; McDonald et al., 2000). High-goal transparencies can be utilized as the photomask to fast model SU-8 designed experts, from which PDMS molds can be imitated. When utilizing straightforward photomasks, the component sizes are generally restricted to 8 μm and bigger. For more modest element estimates, a chrome cover, which can be more costly (roughly multiple times more costly) than transparencies, is normally required. The photolithography interaction is effortlessly rehashed on wafers to deliver various layers of designed photoresist (i.e., multi-facet), and slight films of designed (Anderson et al., 2000) PDMS can be stacked together to make multi-facet 3D microfluidic frameworks (Jo, Van Lerberghe, Motsegood, & Beebe, 2000). From the time this technology has been invented, its usage has always grown and with the advent of time, it became one of the most promising and standard techniques for microfluidic device fabrication. In general, this technique became the standard technique for microfluidic devices in biomedical applications due to the incorporation of a high-resolution, flexible, optically transparent, biocompatible polymer, namely PDMS.

#### 22.4.1.3 Embossing

Embossing is a technique that includes the utilization of thermoplastic materials, normally as level sheets, which are designed against an expert (stamp) utilizing pressure and warmth. Thermoplastics can be reshaped when warmed close to the glass change temperature (Tg) of the material. Instances of thermoplastics utilized in hot decorating incorporate polymethylmethacrylate (PMMA) (Fig. 22.1 polycarbonate (PC), cyclic olefin copolymer (COC), polystyrene (PS), polyvinylchloride (PVC), and polyethylene terephthalate glycol (PETG) (Becker & Heim, 2000; Becker & Locascio, 2002). Commonly, aces are made in one or the other silicon or metal. Silicon wafers are prepared to utilize micromachining strategies to make a silicon stamp (Becker & Heim, 1999a; Kameoka et al., 2001; Martynova et al., 1997; Xu et al., 2000), while metal stamps are either electroplated against micromachined silicon aces or are electroformed created utilizing the LIGA cycle (a German abbreviation for lithography, electroplating, and shaping) (Galloway et al., 2002; McCormick, Nelson, Alonso-Amigo, Benvegnu, & Hooper, 1997; Qi et al., 2002). When made, the stamp and the chosen thermoplastic are set into a water-powered press, and afterward warmth and pressure are applied to embellish the plastic against the stamp. Decorating can likewise be refined without heat by utilizing more noteworthy measures of applied pressure (room temperature engraving) (Xu et al., 2000). Embossing is genuinely direct, just as quick and modest if there is admittance to the essential water-driven press gear and a designed stamp. Manufacture of the emblazoning stamp can be a tedious interaction, and accordingly, decorating is ideal for circumstances including routine microfluidic plans (i.e., not ideal for prototyping a couple of gadgets for testing). Particular vacuum presses are frequently utilized in emblazoning to dispose of air bubbles caught between the substrate and stamp and accommodate exact replications. The replication ability to embellish is restricted by the interaction (e.g., micromachining or LIGA) utilized for the manufacture of the decorating stamp. Most embellished channel frameworks are one-layer planar constructions.

#### 22.4.1.4 Polymer laminates

Microfluidic devices which are fabricated using laminates act like chips. For fabricating a microfluidic device using the laminated method, there are four principal steps involved: (1) selection of appropriate material, (2) defining features in each layer, (3) cutting the defined features of each layer, and (4) assembling the already fabricated independent layers to form one functional device.

## 22.5 Microfluidics as an emerging platform for point-of-care diagnosis

Over the past two decades, microfluidic technologies have experienced significant growth in POC diagnostic systems. Advanced instrumentation and miniaturization of the device have shown accurate results in blood-borne illness, immunoassays, amplification assays, and infectious disease diagnosis. It offers several advantages over conventional methods of diagnosis that include reagent storage, multiple steps for reagent addition, the potential for incorporation of centrifugal steps at various speeds, and the application of a range of detection strategies leading to greater sensitivity and specificity. Various microfluidic-based POCs have been developed till now for the detection of different analytes. Recently, a microfluidic-based POC device for the determination of prostate-specific antigen (PSA), present in cases of prostate cancer, has been developed and further approved by the US Food and Drug Administration (FDA). It consists of a microfluidic-based immuno sandwich assay using neogold-labeled anti-PSA monoclonal antibodies. Different colorimetric, chemiluminescence, and electrochemical-based automated sensing mechanisms have been used for the detection of analytes. Integration of these parameters with microfluidics plays a major role in the development of miniaturized POC diagnostic systems.

## 22.6 Conclusion

This chapter summarized technological advancements in the area of miniaturized microfluidics and its applications in the development of POC devices. A miniaturized POC diagnosis system is a need for early diagnosis with more sensitivity and specificity. Due to the presence of high disease burdens and overpopulation, there are relatively very few diagnostic tools present having excellent sensitivity and specificity. New categories of POC devices, with significant progress in sensitivity and specificity, have been introduced during the past years. Numerous materials such as silicon, polymer, and paper are being used for microfluidic device development, but the cost of these devices is a major concern. Different technologies are present for the fabrication of these devices and nanotechnology plays an important factor to increase their effectiveness. The nanomaterials range from metallic nanoparticles to complex nanocomposites. Further, efforts should be made for the fabrication of polymer or paper-based MF devices, which can be a promising alternative to these substrates. Its low cost, lightweight, portability, flexibility, self-driven fluidic properties, simple fabrication method without cleanroom facilities, and small sample requirement is continuously attracting many researchers to this highly potential field. Integration of advanced microfluidic technology with the miniaturized POC technology has revolutionized the diagnostic platform many fold. So miniaturized POC microfluidics plays an important role in early-stage diagnosis with more sensitivity and specificity.

## References


Adem, S., Jain, S., & Sveiven, M. (2020). Giant magneto resistive biosensors for real-time quantitative detection of protease activity. *Scientific Reports*, *10*.

Anderson, J. R., Chiu, D. T., Jackman, R. J., Chemiavskaya, O., McDonald, J. C., Wu, H., ... Whitesides, G. M. (2000). Fabrication of topologically complex three-dimensional microfluidic systems in PDMS by rapid prototyping. *Analytical Chemistry*, *72*(14), 3158–3164. Available from https://doi.org/10.1021/ac9912294.

Becker, H., & Heim, U. (1999a). Polymer hot embossing with silicon master structures. *Sensors and Materials*, *11*(5), 297–304.

Becker, H., & Heim, U. (1999b). Polymer hot embossing with silicon master structures. *Sensors and Materials*, *11*(5), 297–304.

Becker, H., & Heim, U. (2000). Hot embossing as a method for the fabrication of polymer high aspect ratio structures. *Sensors and Actuators, A: Physical*, *83*(1), 130–135. Available from https://doi.org/10.1016/S0924-4247(00)00296-X.

Becker, H., & Locascio, L. E. (2002). Polymer microfluidic devices. *Talanta*, *56*(2), 267–287. Available from https://doi.org/10.1016/S0039-9140(01)00594-X.

Beebe, D. J., Mensing, G. A., & Walker, G. M. (2002). Physics and applications of microfluidics in biology. *Annual Review of Biomedical Engineering*, *4*, 261–286. Available from https://doi.org/10.1146/annurev.bioeng.4.112601.125916.

Beebe, D. J., Moore, J. S., Yu, Q., Liu, R. H., Kraft, M. L., Jo, B. H., & Devadoss, C. (2000). Microfluidic tectonics: A comprehensive construction platform for microfluidic systems. *Proceedings of the National Academy of Sciences of the United States of America*, *97*(25), 13488–13493. Available from https://doi.org/10.1073/pnas.250273097.

Cao, L., Xiao, H., Fang, C., Zhao, F., & Chen, Z. (2020). Electrochemical immunosensor based on binary nanoparticles decorated rGO-TEPA as magnetic capture and Au@PtNPs as probe for CEA detection. *Microchimica Acta*, *187*(10). Available from https://doi.org/10.1007/s00604-020-04559-2.

Chaiyo, S., Mehmeti, E., Siangproh, W., Hoang, T. L., Nguyen, H. P., Chailapakul, O., & Kalcher, K. (2018). Non-enzymatic electrochemical detection of glucose with a disposable paper-based sensor using a cobalt phthalocyanine–ionic liquid–graphene composite. *Biosensors and Bioelectronics*, *102*, 113–120. Available from https://doi.org/10.1016/j.bios.2017.11.015.

Duffy, D. C., McDonald, J. C., Schueller, O. J. A., & Whitesides, G. M. (1998). Rapid prototyping of microfluidic systems in poly(dimethylsiloxane). *Analytical Chemistry*, *70*(23), 4974–4984. Available from https://doi.org/10.1021/ac980656z.

Eissa, S., & Zourob, M. (2020). A dual electrochemical/colorimetric magnetic nanoparticle/peptide-based platform for the detection of Staphylococcus aureus. *Analyst*, *145*(13), 4606–4614. Available from https://doi.org/10.1039/d0an00673d.

Faustino, V., Catarino, S. O., Lima, R., & Minas, G. (2016). Biomedical microfluidic devices by using low-cost fabrication techniques: A review. *Journal of Biomechanics*, *49*(11), 2280–2292. Available from https://doi.org/10.1016/j.jbiomech.2015.11.031.

Gale, B. K., Jafek, A. R., Lambert, C. J., Goenner, B. L., Moghimifam, H., Nze, U. C., & Kamarapu, S. K. (2018). A review of current methods in microfluidic device fabrication and future commercialization prospects. *Inventions*, *3*(3). Available from https://doi.org/10.3390/inventions3030060.

Galloway, M., Stryjewski, W., Henry, A., Ford, S. M., Llopis, S., McCarley, R. L., & Soper, S. A. (2002). Contact conductivity detection in poly (methyl methacylate)-based microfluidic devices for analysis of mono- and polyanionic molecules. *Analytical Chemistry*, *74*(10), 2407–2415. Available from https://doi.org/10.1021/ac011058e.

Han, G. R., Koo, H. J., Ki, H., & Kim, M. G. (2020). Paper/soluble polymer hybrid-based lateral flow biosensing platform for high-performance point-of-care testing. *ACS Applied Materials and Interfaces*, *12*(31), 34564–34575. Available from https://doi.org/10.1021/acsami.0c07893.

He, P. J. W., Katis, I. N., Kumar, A. J. U., Bryant, C. A., Keevil, C. W., Somani, B. K., ... Sones, C. L. (2020). Laser-patterned paper-based sensors for rapid point-of-care detection and antibiotic-resistance testing of bacterial infections. *Biosensors and Bioelectronics*, *152*. Available from https://doi.org/10.1016/j.bios.2020.112008.

Huang, L., Xiao, W., Xu, T., Chen, H., Jin, Z., Zhang, Z., & Tang, Y. (2020). Miniaturized paper-based smartphone biosensor for differential diagnosis of wild-type pseudo rabies virus infection vs vaccination immunization. *Sensors and Actuators B: Chemical, 327*.

Jacobson, S. C., Hergenröder, R., Koutny, L. B., & Ramsey, J. M. (1994). High-speed separations on a microchip. *Analytical Chemistry, 66*(7), 1114–1118. Available from https://doi.org/10.1021/ac00079a029.

Jo, B. H., Van Lerberghe, L. M., Motsegood, K. M., & Beebe, D. J. (2000). Three-dimensional micro-channel fabrication in polydimethylsiloxane (PDMS) elastomer. *Journal of Microelectromechanical Systems, 9*(1), 76–81. Available from https://doi.org/10.1109/84.825780.

Kameoka, J., Craighead, H. G., Zhang, H., & Henion, J. (2001). A polymeric microfluidic chip for CE/MS determination of small molecules. *Analytical Chemistry, 73*(9), 1935–1941. Available from https://doi.org/10.1021/ac001533t.

Khan, M. S., Dighe, K., Wang, Z., Srivastava, I., Daza, E., Schwartz-Dual, A. S., … Pan, D. (2018). Detection of prostate specific antigen (PSA) in human saliva using an ultra-sensitive nanocomposite of graphene nanoplatelets with diblock: Co-polymers and Au electrodes. *Analyst, 143*(5), 1094–1103. Available from https://doi.org/10.1039/c7an01932g.

Khoury, C., Mensing, G. A., & Beebe, D. J. (2002). Ultra rapid prototyping of microfluidic systems using liquid phase photopolymerization. *Lab on a Chip, 2*(1), 50–55. Available from https://doi.org/10.1039/b109344d.

Kim, P., Kwon, K. W., Park, M. C., Lee, S. H., Kim, S. M., & Suh, K. Y. (2008). Soft lithography for microfluidics: A Review. *Biochip Journal, 2* (1), 1–11, http://biochips.or.kr/website/in_journal/download.php?u = KDItMSk0Mi0yMDA4MDMzMTE2MjI0NS5wZGY = lpdfl%5B%C1%A62%B1%C71%C8%A3%5D-%282–1%2942–20080331162245.pdf.

Lam, T., Devadhasan, J. P., Howse, R., & Kim, J. (2017). A chemically patterned microfluidic paper-based analytical device (C-μPAD) for point-of-care diagnostics. *Scientific Reports, 7*.

Lee, H., Sun, E., Ham, D., & Weissleder, R. (2008). Chip-NMR biosensor for detection and molecular analysis of cells. *Nature Medicine, 14*(8), 869–874. Available from https://doi.org/10.1038/nm.1711.

Li, X., Tian, J., Nguyen, T., & Shen, W. (2008). Paper-based microfluidic devices by plasma treatment. *Analytical Chemistry, 80*(23), 9131–9134. Available from https://doi.org/10.1021/ac801729t.

Manz, A., Harrison, D. J., Verpoorte, E. M. J., Fettinger, J. C., Paulus, A., Lüdi, H., & Widmer, H. M. (1992). Planar chips technology for miniaturization and integration of separation techniques into monitoring systems. *Capillary electrophoresis on a chip. Journal of Chromatography A, 593* (1–2), 253–258. Available from https://doi.org/10.1016/0021-9673(92)80293-4.

Martynova, L., Locascio, L. E., Gaitan, M., Kramer, G. W., Christensen, R. G., & MacCrehan, W. A. (1997). Fabrication of plastic microfluid channels by imprinting methods. *Analytical Chemistry, 69*(23), 4783–4789. Available from https://doi.org/10.1021/ac970558y.

McCormick, R. M., Nelson, R. J., Alonso-Amigo, M. G., Benvegnu, D. J., & Hooper, H. H. (1997). Microchannel electrophoretic separations of DNA in injection-molded plastic substrates. *Analytical Chemistry, 69*(14), 2626–2630. Available from https://doi.org/10.1021/ac9701997.

McDonald, J. C., & Whitesides, G. M. (2002). Poly(dimethylsiloxane) as a material for fabricating microfluidic devices. *Accounts of Chemical Research, 35*(7), 491–499. Available from https://doi.org/10.1021/ar010110q.

McDonald, J. C., Duffy, D. C., Anderson, J. R., Chiu, D. T., Wu, H., Schueller, O. J. A., & Whitesides, G. M. (2000). Fabrication of microfluidic systems in poly(dimethylsiloxane). *Electrophoresis, 21*(1), 27–40, https://doi.org/10.1002/(SICI)1522–2683(20000101)21:1 < 27::AID-ELPS27 > 3.0.CO;2-C.

Narang, J., Malhotra, N., Singhal, C., Mathur, A., Chakraborty, D., Anil, A., … Pundir, C. S. (2017). Point of care with micro fluidic paper based device integrated with nano zeolite–graphene oxide nanoflakes for electrochemical sensing of ketamine. *Biosensors and Bioelectronics, 88*, 249–257. Available from https://doi.org/10.1016/j.bios.2016.08.043.

Qi, S., Liu, X., Ford, S., Barrows, J., Thomas, G., Kelly, K., … Soper, S. A. (2002). Microfluidic devices fabricated in poly(methyl methacrylate) using hot-embossing with integrated sampling capillary and fiber optics for fluorescence detection. *Lab on a Chip, 2*(2), 88–95. Available from https://doi.org/10.1039/b200370h.

Qin, D., Xia, Y., & Whitesides, G. M. (2010). Soft lithography for micro- and nanoscale patterning. *Nature Protocols, 5*(3), 491–502. Available from https://doi.org/10.1038/nprot.2009.234.

Sanli, S., Moulahoum, H., Ghorbanizamani, F., Gumus, Z. P., & Timur, S. (2020). On-site testosterone biosensing for doping detection: Electrochemical immunosensing via functionalized magnetic nanoparticles and screen-printed electrodes. *ChemistrySelect, 5*(47), 14911–14916. Available from https://doi.org/10.1002/slct.202004204.

Shen, Y., Tran, T. T., Modha, S., Tsutsui, H., & Mulchandani, A. (2019). A paper-based chemiresistive biosensor employing single-walled carbon nanotubes for low-cost, point-of-care detection. *Biosensors and Bioelectronics, 130*, 367–373. Available from https://doi.org/10.1016/j.bios.2018.09.041.

Skottrup, P. D., Nicolaisen, M., & Justesen, A. F. (2008). Towards on-site pathogen detection using antibody-based sensors. *Biosensors and Bioelectronics, 24*(3), 339–348. Available from https://doi.org/10.1016/j.bios.2008.06.045.

Terry, S. C., Jerman, J. H., & Angell, J. B. (1979). A gas chromatographic air analyzer fabricated on a silicon wafer. *IEEE Transactions on Electron Devices*, 1880–1886. Available from https://doi.org/10.1109/T-ED.1979.19791.

Ugolini, G. S., Visone, R., Redaelli, A., Moretti, M., & Rasponi, M. (2017). Generating multicompartmental 3D biological constructs interfaced through sequential injections in microfluidic devices. *Advanced Healthcare Materials, 6*(10). Available from https://doi.org/10.1002/adhm.201601170.

Wang, Y., Luo, J., Liu, J., Li, X., Kong, Z., Jin, H., & Cai, X. (2018). Electrochemical integrated paper-based immunosensor modified with multi-walled carbon nanotubes nanocomposites for point-of-care testing of 17β-estradiol. *Biosensors and Bioelectronics, 107*, 47–53. Available from https://doi.org/10.1016/j.bios.2018.02.012.

Weng, X., Ahmed, S. R., & Neethirajan, S. (2018). A nanocomposite-based biosensor for bovine haptoglobin on a 3D paper-based analytical device. *Sensors and Actuators, B: Chemical*, *265*, 242–248. Available from https://doi.org/10.1016/j.snb.2018.03.061.

Whitesides, G. M., Ostuni, E., Takayama, S., Jiang, X., & Ingber, D. E. (2001). Soft lithography in biology and biochemistry. *Annual Review of Biomedical Engineering*, *3*, 335–373. Available from https://doi.org/10.1146/annurev.bioeng.3.1.335.

Xu, J., Locascio, L., Gaitan, M., & Lee, C. S. (2000). Room-temperature imprinting method for plastic microchannel fabrication. *Analytical Chemistry*, *72*(8), 1930–1933. Available from https://doi.org/10.1021/ac991216q.

Yakoh, A., Pimpitak, U., Rengpipat, S., Hirankarn, N., Chailapakul, O., & Chaiyo, S. (2021). Paper-based electrochemical biosensor for diagnosing COVID-19: Detection of SARS-CoV-2 antibodies and antigen. *Biosensors and Bioelectronics*, *176*. Available from https://doi.org/10.1016/j.bios.2020.112912.

Yang, L., & Bashir, R. (2008). Electrical/electrochemical impedance for rapid detection of foodborne pathogenic bacteria. *Biotechnology Advances*, *26*(2), 135–150. Available from https://doi.org/10.1016/j.biotechadv.2007.10.003.

Yu, Q., Bauer, J. M., Moore, J. S., & Beebe, D. J. (2001). Responsive biomimetic hydrogel valve for microfluidics. *Applied Physics Letters*, *78*(17), 2589–2591, https://doi.org/10.106.1/1.1367010.

Chapter 23

# Integrated low-cost biosensor for rapid and point-of-care cancer diagnosis

Ankur Kaushal[1], Deepak Kala[1], Vivek Verma[2] and Shagun Gupta[2]
[1]*Amity Centre of Nanotechnology, Amity University, Gurugram, India*
[2]*Shoolini University, Solan, India*

## 23.1 Introduction

Cancer is a generic word used for a group of diseases that involves uncontrolled or abnormal cell growth with the potential to influence other parts of the body. According to World Health Organization (WHO), cancer is the second largest cause of death globally with estimated casualties of 9.6 million people in 2018 (Ferlay et al., 2019). Most of the deaths (70%) from cancer were reported from low- and middle-income countries, where the proper diagnosis and prognosis facilities are not available to cure the disease (Wild & Stewart, 2014). The early cancer diagnosis has been reported with significant improvements in survival rate of cancer patients. The disease is curable if diagnosed at early stages. Delayed diagnosis or diagnosis at an advanced stage reduces the patient's survival rate (Ferlay, 2010; WHO, 2020). The occurrence of cancer in underdeveloped countries accounts for 53%, with a 56% mortality rate, and by 2020 it is expected that the total number of new cases in developing countries will increase up to 73% (Kanavos, 2006). According to WHO, in 2017 only 26% of low-income countries had the pathological facilities available in the public sector for the diagnosis of cancer. Owing to the limited resources for cancer diagnosis, there is a need for affordable point-of-care (POC) systems for the early diagnosis of cancer especially in middle- and low-income countries, where most of the causalities are reported due to the unaffordable financial burden of cancer. The conventional methods for cancer diagnosis include digital mammography, ultrasound, magnetic resonance imaging (MRI), computed tomography (CT) scan, and positron emission tomography (PET). Other diagnosis methods such as immunohistochemistry, in situ hybridization (FISH, CSH), polymerase chain reaction (conventional PCR, RT-PCR), flow cytometry, and microarray are used nowadays (Kumar & Pawaiya, 2010). The conventional methods discussed above are time-consuming, costly, and mostly not available in rural areas for early-stage diagnosis of cancer, thereby lack of availability becomes a cause of a higher mortality rate. Researchers across the globe are working on the development of an affordable POC diagnosis system for early-stage screening and diagnosis of cancer that can combat the limitations of the available methods of diagnosis. Therefore, the advanced diagnostic techniques such as lateral flow assays (LFAs), microfluidic devices, and biosensors are gaining prominence in the field of cancer diagnosis due to their short response time, the requirement of low sample volumes, portability, high sensitivity, and selectivity. This chapter focuses on the introduction of the cancer biomarkers, followed by the advanced low-cost POC diagnostics systems developed by researchers across the globe in the last decade with their merits, demerits, and future applicability.

## 23.2 Cancer biomarkers

Cancer biomarkers are the biomolecules expressed by the tumor cells or other cells of the body to respond against the tumor cells. They are essential for the early diagnosis and prognosis of cancer. The cancer biomarkers are specific and used for the detection of different types of cancers. These can be categorized as nucleic acid biomarkers, proteins/enzyme biomarkers, small molecules, extracellular vesicles, and the circulating tumor cells (Mahmoudi, de la Guardia, & Baradaran, 2020). The most common cancer biomarkers used in clinical setups for diagnosis of cancer are prostate-specific antigen (PSA), alpha-fetoprotein (AFP), cancer antigen 125 (CA125), and carcinoembryonic antigen (CEA) (Kamel & Al-Amodi, 2016). The cancer biomarkers act like a signature for cancer diagnosis, such as PSA is used for

Biosensor Based Advanced Cancer Diagnostics. DOI: https://doi.org/10.1016/B978-0-12-823424-2.00002-8

assessment of prostate cancer, CA125 is a marker of ovarian cancer, and CEA is elevated in patients with colorectal, breast, lung, or pancreatic cancer (Lokshin & Nolen, 2011). The other biomarkers used in clinical setups for diagnosis of cancers are AFP for hepatocellular carcinoma, human chorionic gonadotropin-β (HCG-β) for testicular cancer, and calcitonin for medullary carcinoma of the thyroid (Kamel & Al-Amodi, 2016). One of the major challenges in cancer diagnosis is to develop a promising cancer biomarker with higher diagnostic sensitivity and specificity.

## 23.3 New low-cost point-of-care diagnostics for cancer detection

### 23.3.1 Low-cost disposable material for the construction of biosensors

The selection of a sensing platform is an important step towards the development of low-cost disposable biosensors. The success of biosensors depends on the interaction and compatibility of bioreceptors with the materials (biocomposite, ion-selective membranes) on the sensing platform surface. It prevents the denaturation of biological receptors (Ag, Ab, DNA probe) and provides a reliable interface for good adhesion and biological interaction (Ahmed et al., 2016). Several advancements have been done so far for improving the performance of transducers and to reduce the overall cost of the biosensor. One such approach is screen-printing technology that has emerged as a promising approach for the development of simple, rapid, and low-cost biosensors. In screen-printing technology (Fig. 23.1) the ink or paste of the desired material (nanomaterials, nanocomposites) is printed on a substrate using a squeeze blade through a mesh-screen that contains the pattern of electrodes (Ahmed et al., 2016). The paste connects the bioreceptor to the substrate. Several substrates have been used for designing the screen printed electrodes such as alumina, ceramics, plastic [polyvinyl chloride (PVC), polycarbonate], and fiberglass (Ghosale et al., 2018; Moreira, Ferreira, Puga, & Sales, 2016).

The paper-based substrate is an emerging platform for the development of disposable sensors due to their low cost and high flexibility in comparison to the other substrate materials (Cao, Han, Xiao, Chen, & Fang, 2020; Mohanraj et al., 2020; Yáñez-Sedeño, Campuzano, & Pingarrón, 2020). Screen-printing technology has revolutionized the field of electrochemical biosensing. The screen-printed electrode's (SPE) smaller size and its ability to work with less sample volumes enables the development of low-cost POC biosensors. Several paper-based SPEs have been used for the development of electrochemical biosensors (Farshchi, Saadati, & Hasanzadeh, 2020; Gutiérrez-capitán, Baldi, & Fernández-sánchez, 2020; Loo & Pui, 2020; Sakata et al., 2020). The paper electrode is a better substitute for the development of disposable, inexpensive, and eco-friendly electrochemical biosensors. Its lightweight and high flexibility provide additional benefits to the sensors that confers unique exploitable properties for application in electroanalysis (Yáñez-Sedeño et al., 2020). SPE-based sensors are in tune with the growing need for performing rapid and accurate in situ analyses, and for the development of portable devices (Taleat, Khoshroo, & Mazloum-Ardakani, 2008).

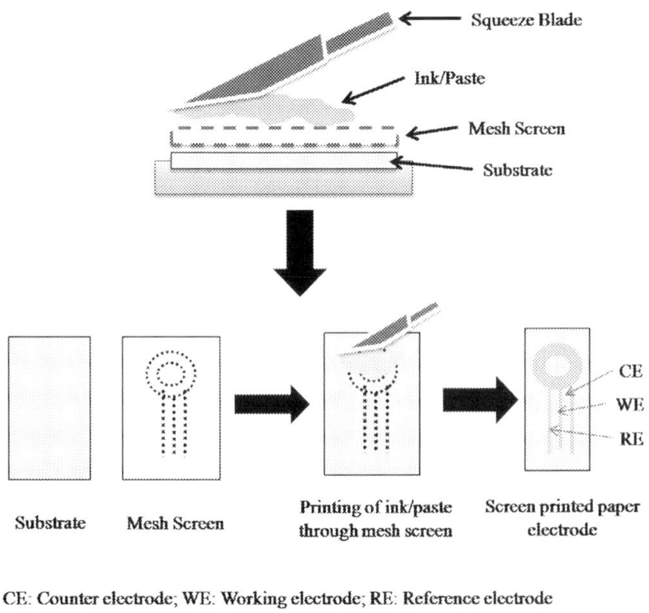


FIGURE 23.1 Illustration of the basic processes involved in the screen-printing technology for fabrication of screen-printed electrodes.

The LFAs are another example of a low-cost, paper-based POC diagnosis that uses a nitrocellulose membrane as a functional material. LFAs consists of a labeled probe used to capture and detect an analyte in a sample using fluorescent and colorimetric readouts. The results were analyzed, by the naked eye or optical devices, by colored bands on strips that offered qualitative and/or semi-quantitative readouts (LFAs). In addition to cellulose and nitrocellulose papers, there are other low-cost cellulose-based materials available such as cellophane and nanocellulose-based materials (Dincer et al., 2019). Further, the analytical performance of the paper-based platforms can improve by introducing nanomaterials, including various nanoparticles (López-Marzo & Merkoçi, 2016; Parolo & Merkoçi, 2013; Sánchez-Calvo, Blanco-López, & Costa-García, 2020), graphene materials (Figueredo, Garcia, Cortón, & Coltro, 2016; Morales-Narváez, Baptista-Pires, Zamora-Gálvez, & Merkoçi, 2017), or coatings using biopolymers such as chitosan (Gabriel et al., 2016). The use of microfluidic paper-based analytical devices (μPADs) has increased exponentially in recent years. The μPADs resolve the problems of poor sensitivity of the devices made from cellulose and modified cellulose. The paper-based microfluidic systems consist of channels that can direct the flow of fluid in different regions of the device without any need for external forces. The potential of μPADs such as lower cost, high sensitivity, and multiplexed analysis ability makes them ideal for POC analytical systems (Carrell et al., 2019).

### 23.3.2 Paper electrode-based electrochemical biosensors for cancer assessment

Over the past decade, various advancements have been done in the field of electrochemical biosensing for the development of a rapid and low-cost POC system for the detection of life-threatening diseases (Kala, Gupta et al., 2020; Kala, Sharma, et al., 2020; Kaushal, Singh, Kala, Kumar, & Kumar, 2016; Kaushal, Singh, Kumar, & Kumar, 2017; Verma et al., 2020). So the diagnosis at an early stage has been a challenge over decades and is attracting the focus of researchers for improvement of diagnosis methods. The paper-based electrode (Fig. 23.2) has been exploited as an option for the development of electrochemical assays due to their low cost, portability, biodegradability, and biocompatibility (Ge et al., 2015). A 3D microfluidic paper-based electrochemical immunodevice (m-PEID) for simultaneous sensitive detection of two tumor biomarkers AFP and CEA was fabricated. The paper working electrode (WE) was modified with nanoporous Pt (platinum) particle (NPPt-PWE), and metal ions loaded on the L-cysteine-capped flowerlike Au (gold) nanoparticles (Au−Cys) were used as a signal amplification label. Further, the cancer biomarker's specific antibodies (Ab1) were immobilized by chemical adsorption onto the nanoparticle-modified paper electrode (NPPt-PWE) surface.

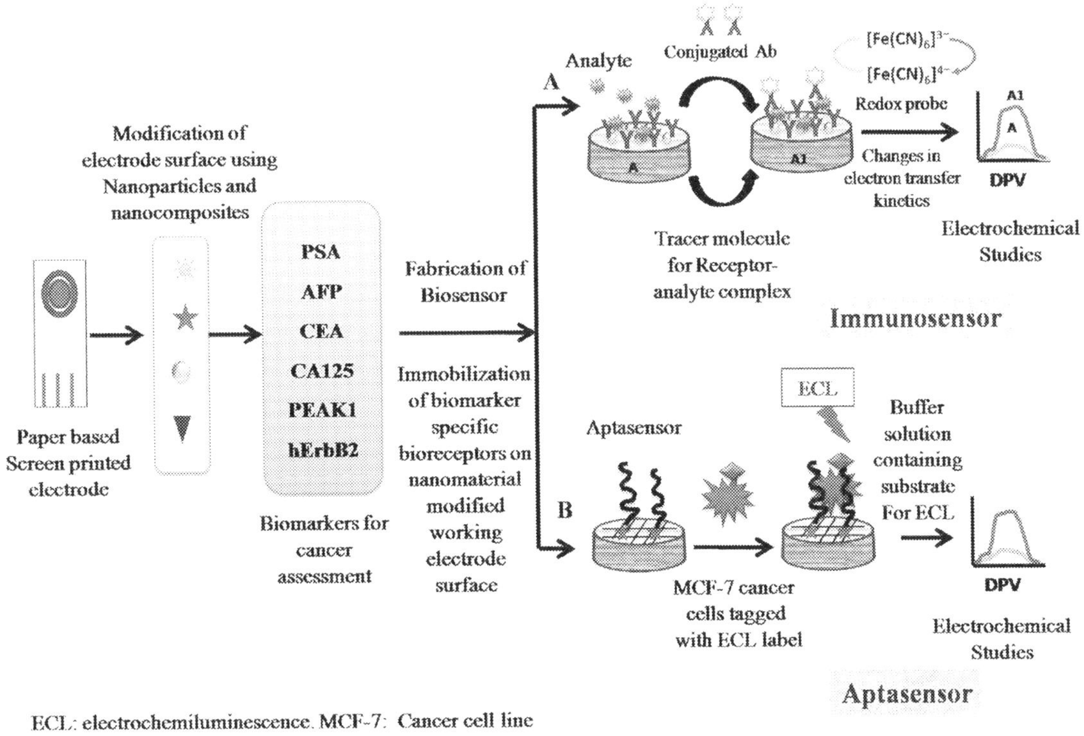


**FIGURE 23.2** Illustration of paper-based electrochemical biosensors developed for cancer assessment.

The modified electrode was then incubated with analytes (AFP, CEA) and subsequently with the bioconjugate (Ab2) for monitoring the Ag-Ab interaction on the WE surface. The electrochemical measurements were carried out using differential pulse voltammetry (DPV) in pH 4.6 HAc/NaA (acetic acid/ sodium acetate) buffer solution. The limit of detection of the developed sensor for the cancer biomarkers AFP and CEA was 1.0 and 1.3 pg/mL, respectively. The good stability, reproducibility, and accuracy of the m-PEID indicate that it has great potential application in clinical diagnostics (Li et al., 2015). The paper-based immunosensor was also developed for the detection of CEA cancer biomarkers. The sensor was constructed using filter paper modified with poly (3, 4-ethylene dioxythiophene) a conducting polymer, doped with polys (styrene sulfonate) anions and iron oxide nanoparticles (nFe2O3@PEDOT:PSS). The anti-CEA monoclonal antibody was immobilized on the modified filter paper surface to construct an immunosensing platform. The linearity of the sensor for the detection of CEA is ranged from 4 to 25 ng/mL with a sensitivity of 10.2 $\mu A\ ng^{-1}\ mL\ cm^{-2}$. The proposed biosensor offered good long-term stability (Kumar et al., 2019). A low-cost paper-based electrochemical immunosensor has been developed for the detection of pancreatic cancer using a new biomarker, i.e., pseudopodium-enriched atypical Kinase one, SGK269 (PEAK1) (Prasad et al., 2020). The immunosensor was fabricated using a paper-based WE and modified with graphene oxide nanoparticles. The nanomaterial was used as a matrix for the immobilization of antibodies specific to the cancer biomarker (PEAK1). The paper-based biosensor has been constructed in such a way so that extra complicated surface modifications can be avoided for immobilization of the bioreceptors. The developed immunosensor was a sandwich-type assay in which graphene oxide layers were immobilized with anti-PEAK1, and the secondary antibody was conjugated with gold nanoparticles (AuNPs-tagged-anti-PEAK1) and used for monitoring the Ag—Ab interactions. Incubation of the electrode with cancer biomarker PEAK1 changed the electrochemical property of the electrode surface due to the AuNPs tagged conjugate antibody and provided a detectable signal for electrochemical measurements. The electrochemical measurement was carried out using DPV in potassium ferricyanide (redox probe) solution. The immunosensor showed a lower limit of detection of 10 pg/mL PEAK1; that confirmed its higher sensitivity for early cancer diagnosis.

Apart from the immunosensor, paper-based DNA sensors have also been developed for the assessment of cancer biomarkers. One of the promising approaches involves the use of aptamers, the oligonucleotide sequences synthesized by the sequential evolution of ligands by exponential enrichment (SELEX) method. Aptamers can bind with a wide range of target molecules from micro to macromolecules with higher affinity (Dalirirad & Steckl, 2019). Wu et al. (2016) developed a disposable paper-based aptasensor for the detection of cancerous cells. This electrochemiluminescence-based sensor device was constructed using Au@Pd nanoparticles modified cellulose surface, an aptamer against MCF-7 cells, and nanoporous PtNi alloy particles loaded with carbon dots (PtNi@C-dot) and conjugated with concanavalin A (Con A). The electrochemiluminescence signal was recorded in the presence of peroxodisulfate ions in the measurement buffer. The sensor was able to detect cancer cells within a linear range from 480 to $2.0 \times 107$ cells/mL with a detection limit of 300 cells/mL (Wu et al., 2016).

The early-stage diagnosis of cancer biomarkers using advanced analytical technique viz biosensor enables rapid POC diagnostics of cancer with higher sensitivity, specificity, and lower cost. A biosensor is a bioanalytical device, which converts a biological response into a measurable signal. Biosensors detect an analyte (DNA, Ag, Ab, and substrate) in the sample and convert it into a measurable signal, i.e., change in current, pH, color, heat, etc. A biosensor consists of two major components: biological (bioreceptor) and physical transducer, for the conversion of biochemical response into a measurable signal (Malhotra, Kumar, & Pandey, 2016). Based on the transducers, biosensors can be classified as optical, piezoelectric, electrochemical, magnetic, and thermal sensors. Among all of the transducers, the electrochemical is generally used due to their high sensitivity, the lower limit of detection, lower cost, and applicability in the development of disposable devices and methodologies capable of working with low sample volumes (Kala, Gupta et al., 2020; Kala, Sharma, et al., 2020; Kaushal et al., 2016, 2017; Verma et al., 2020).

### 23.3.3 Low-cost optical biosensors

An optical biosensor consists of a biorecognition (bioreceptors) layer with an optical transducer. Optical detection is performed by measuring the changes on the sensor surface due to the interaction of the optical field with a biorecognition element. Optical biosensors enable real-time monitoring and label-free detection of a wide range of biomolecules and chemical entities (Damborský, Švitel, & Katrlík, 2016). A label-free sensor was developed for the detection of the human tumor marker CEA using plasmonic metasurface sensors with gold nanobump arrays (Zhu et al., 2020). The sensor was constructed by modification of plasmonic metasurface with covalently attached anti-CEA antibodies. A spectrometer with an integrated optical fiber probe was used as a detector to measure the reflectance spectra of all plasmonic metasurface samples. Detection of analyte was based on differences in the optical activity of bare metasurface, with anti-CEA antibodies, and after interaction with the analyte (CEA). The developed sensor was reported with a linear detection range of CEA in human serum samples from less than 10 ng/mL to more than 87 ng/mL. The assay was able to detect the threshold tumor marker concentration of 20 ng/mL for early cancer prediction.

The low-cost paper-based optical sensor has also been developed for the detection of cancer biomarker CEA (Damborský et al., 2016). A chemiluminescence-based sensor has been developed using a paper chip treated with plasma and modified with a captured antibody specific to the cancer biomarker (CEA). In this sensor, carbon nanospheres functionalized with HRP (horseradish peroxidase) conjugate with the secondary antibody and used as a signaling label. The analytical chemiluminescence signal was obtained using luminol and hydrogen peroxide onto the paper surface. The developed sensor had a linear dynamic range from 0.01 to 20.0 ng/mL and a detection limit of 3 pg/mL (Chen et al., 2018).

The other optical sensors developed using the paper-based chips are based on fluorescence spectroscopy for the detection of CEA and PSA biomarkers. The cancer biomarkers CEA and PSA specific antibodies modified with NH2 groups were immobilized on the paper surface and CdTe (cadmium telluride) quantum dots modified antibodies were used as a signaling label for monitoring Ag–Ab (antigen/antibody) interaction on the paper surface. The emission peak was recorded at 525 and 605 nm, respectively for CEA-Ab and PSA-Ab. The LOD was reported as 0.3 and 0.4 ng/mL for detection of CEA and PSA, respectively. The developed method showed a good agreement with the standard ELISA method (Chen et al., 2019).

An optomagnetic detection method was developed for cancer assessment using human epidermal growth factor receptor-2 (hErbB2) as a cancer biomarker in serum. The sensor was developed using a bioactivated borofloat PoC-chip (Ab1) and immuno-optomagnetic quantum dots (MQDs-Ab2). The Ab1 and Ab2 were both designed to bind with the hErbB2 biomarker but at different epitopes. The hErbB2 biomarker was captured initially by the MQDs-Ab2 to form a hErbB2-MQDs-Ab2 complex hErbB2which later on presented to the PoC-chip (Ab1). The optomagnetic signal from the chip was directly detected by the naked eye using a UV lamp and a smartphone camera in the range of 620 pg/mL to 5 ng/mL of hErbB2 biomarker in serum. The developed assay provided a better alternative to the existing methods due to its high speed, sensitivity, portability, and lower cost (Qureshi, Tufani, Corapcioglu, & Niazi, 2020).

### 23.3.4 Lateral flow assays

The lateral flow assays are becoming a better POC diagnosis system as they are user friendly, less in cost, highly sensitive, selective, and doesn't require sophisticated instrument facility and expertise person to conduct the test. Lateral flow kits are constructed using a paper strip and consist of major parts, i.e., sample pad, conjugate pad, biorecognition pad (test and control line), and adsorption pad (Fig. 23.3).

The sample pad is used to load the samples and ensures the binding of the analyte with the reagent of conjugate and on the membrane (Ratajczak & Stobiecka, 2020). The sample migrates along the strip without any external forces using capillary action. The sample moves to the conjugate pad that contains the target-specific antibodies tagged with colored or fluorescent particles as a tracer for an analyte. The sample bound with the tracer antibody moves to the test pad which is made up of porous membrane (nitrocellulose) and contains captured antibodies immobilized in a line and react with the analyte tracer antibody complex. The positive and negative samples are assessed by comparing the test and control line on the paper strip with different intensities by the naked eye or using a dedicated reader (Koczula & Gallotta, 2016).

The traditional colorimetric LFA has the limitations of the lower limit of detection, lower sensitivity, and semi-quantitative detection. A new approach, surface-enhanced Raman scattering (SERS) has been developed as an alternative technique due to its better sensitivity and antiinterference ability. The assay (Fig. 23.4) was developed by targeting prostate cancer antigen3 mimic DNA. The assay was constructed using a competitive approach where the competition

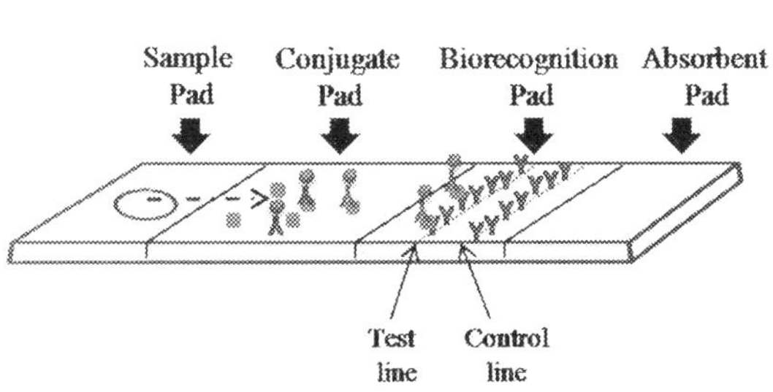


FIGURE 23.3 Schematic representation of the lateral flow assay (LFA) strip composition.

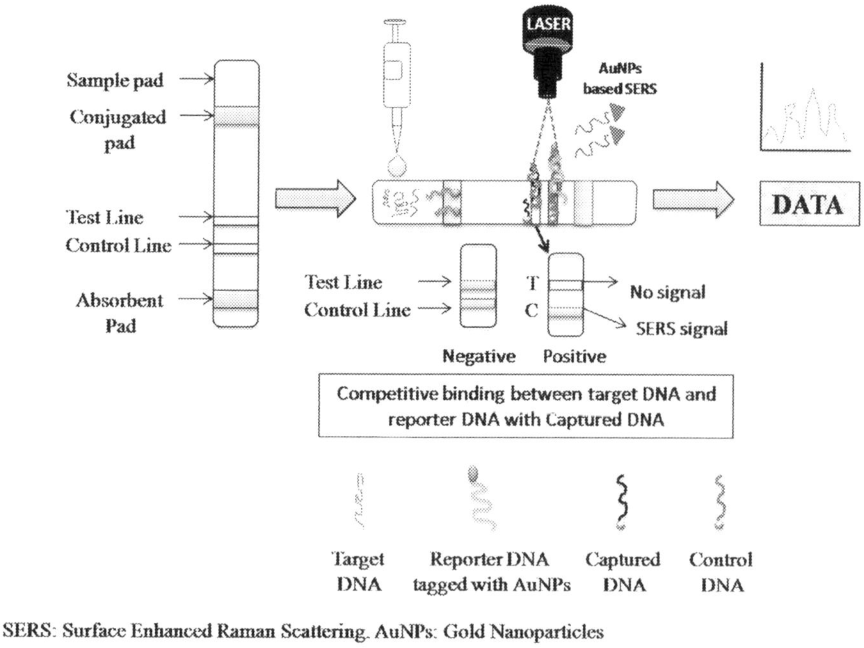


**FIGURE 23.4** Schematic representation of surface-enhanced Raman scattering (SERS)-based LFA for cancer assessment.

occurs in between reporter DNA (SERS nanotags) and cancer antigen3 mimic DNA for the hybridization with captured probe (streptavidin-biotinylated capture DNA) on the test line. The color and Raman intensity of SERS nanotags on the test line was inversely proportional to the concentration of cancer antigen3 mimic DNA. The LOD of the assay was reported as 3 fm and linearity was in the range of 0.01 p.m. to 50,000 p.m. The developed assay had the potential to become a valuable tool for the detection of genetic disorders at early stages (Fu et al., 2019).

A lateral flow immunoassay (LFIA) has been developed for ovarian cancer associated with serum CA125 biomarker. The assay was developed using a lateral flow strip in which the anti-Stn CA125 protein antibodies were captured in the test line and the anti-CA125 mab on UCNPs (upconverting nanoparticles) used as a tracer for the target antigen, i.e., CA125 antigen. The abnormal glycosylation of ovarian cancer cells used as a unique feature for enhancing the sensitivity of the diagnosis method. The developed LFIA was found superior to the conventional CA125 immunoassay (CA125IA) with 72% sensitivity, 98% specificity, which was 4.5 times higher than the sensitivity of the conventional CA125 ELISA method (Bayoumy et al., 2020). A DNA-based approach was developed for the detection of lung cancer using urine and saliva samples. The developed assay was based on the detection of exosomal miRNA. The main limitation of the assay was the lower concentration of exosomal miRNA in saliva and urine samples. To overcome the limitation new approach was developed using $Fe_3O_4@SiO_2$-aptamer nanoparticles (FSAs) as a bioreceptor for lung cancer exosomes and Duplex-Specific Nuclease (DSN) as a tool to amplify the signals. The developed lung cancer diagnosis kit consists of three sections: (1) The screening of exosomes by FASs, (2) hydrolysis of reporter DNA by exosomal miRNA, and (3) Analysis of results by lateral transverse flow test strips. The reported sensitivity of the developed detection kit was 7.76 p.m. The developed kit proved to be a noninvasive method for the detection of lung cancer with higher sensitivity and selectivity. The method was also found to be consistent with the standard real-time PCR method (Zhou et al., 2020).

## 23.4 Conclusion

The POC diagnosis system is a need of the hour for cancer assessment at early stages. The biosensor is a solution to overcome the limitations of current diagnosis methods, viz. requirement of sophisticated instruments. The short response time, portability, lower cost, higher sensitivity, and selectivity makes it an alternative for the current approaches in practice for cancer diagnosis. Biosensors provide real-time monitoring of cancer which is required for improving the survival rate of the patients suffering from this life-threatening disease. The lower cost of the biosensor is very helpful for the low- to middle-income countries where most of the deaths occur due to delay in proper diagnosis. In these resource-limited settings, low-cost biosensors and LFAs can play an important role in the rapid detection of the disease. The studies suggested that the nanomaterial-modified paper electrode has a great potential for the development of low-cost cancer detection kits due to their enhanced sensitivity, lower detection limit, and ability of early cancer detection. The development of new cancer biomarkers also has an importance in enhancing the accuracy of diagnosis methods. The development of new approaches based on the detection of cancer biomarkers in body fluids like urine and saliva will make these methods noninvasive and user friendly. The LFAs are very rapid, user friendly, and highly specific, but their limitations are their semiquantitative detection ability. But new approaches have been developed to overcome these limitations like the use of Raman-SERS and integration of the LFA with other optical sensing devices to record the data rather than just visual interpretations. Reduction in the hardware costs for the POC techniques will make it a realistic option in resource-poor settings. The use of new labels such as nanomaterials like quantum dots and the upconverting phosphor technology can improve the sensitivity of the method and allow the use of more diluted samples to minimize the matrix interference or use of matrices with lower concentrations, such as tears or saliva.

## References

Ahmed, M. U., Hossain, M. M., Safavieh, M., Wong, Y. L., Rahman, I. A., Zourob, M., & Tamiya, E. (2016). Toward the development of smart and low cost point-of-care biosensors based on screen printed electrodes. *Critical Reviews in Biotechnology*, *36*(3), 495–505. Available from https://doi.org/10.3109/07388551.2014.992387.

Bayoumy, S., Hyytiä, H., Leivo, J., Talha, S. M., Huhtinen, K., Poutanen, M., ... Pettersson, K. (2020). Glycovariant-based lateral flow immunoassay to detect ovarian cancer–associated serum CA125. *Communications Biology*, *3*(1). Available from https://doi.org/10.1038/s42003-020-01191-x.

Cao, L., Han, G. C., Xiao, H., Chen, Z., & Fang, C. (2020). A novel 3D paper-based microfluidic electrochemical glucose biosensor based on rGO-TEPA/PB sensitive film. *Analytica Chimica Acta*, *1096*, 34–43. Available from https://doi.org/10.1016/j.ac.2019.10.049.

Carrell, C., Kava, A., Nguyen, M., Menger, R., Munshi, Z., Call, Z., ... Henry, C. (2019). Beyond the lateral flow assay: A review of paper-based microfluidics. *Microelectronic Engineering*, *206*, 45–54. Available from https://doi.org/10.1016/j.mee.2018.12.002.

Chen, Y., Chu, W., Liu, W., Guo, X., Jin, Y., & Li, B. (2018). Based chemiluminescence immunodevice for the carcinoembryonic antigen by employing multi-enzyme carbon nanosphere signal enhancement. *Microchimica Acta*, *185*, 3.

Chen, Ying, Guo, X., Liu, W., & Zhang, L. (2019). Paper-based fluorometric immunodevice with quantum-dot labeled antibodies for simultaneous detection of carcinoembryonic antigen and prostate specific antigen. *Microchimica Acta*, *186*(2), 112. Available from https://doi.org/10.1007/s00604-019-3232-0.

Dalirirad, S., & Steckl, A. J. (2019). Aptamer-based lateral flow assay for point of care cortisol detection in sweat. *Sensors and Actuators, B: Chemical*, *283*, 79–86. Available from https://doi.org/10.1016/j.snb.2018.11.161.

Damborský, P., Švitel, J., & Katrlík, J. (2016). Optical biosensors. *Essays in Biochemistry*, *60*(1), 91–100. Available from https://doi.org/10.1042/EBC20150010.

Dincer, C., Bruch, R., Costa-Rama, E., Fernández-Abedul, M. T., Merkoçi, A., Manz, A., ... Güder, F. (2019). Disposable sensors in diagnostics, food, and environmental monitoring. *Advanced Materials*, *31*(30). Available from https://doi.org/10.1002/adma.201806739.

Farshchi, F., Saadati, A., & Hasanzadeh, M. (2020). Optimized DNA-based biosensor for monitoring: Leishmania infantum in human plasma samples using biomacromolecular interaction: A novel platform for infectious disease diagnosis. *Analytical Methods*, *12*(39), 4759–4768. Available from https://doi.org/10.1039/d0ay01516d.

Ferlay, J. (2010). Cancer incidence and mortality world-wide. IARC Cancer Base No (Vol. 2).

Ferlay, J., Colombet, M., Soerjomataram, I., Mathers, C., Parkin, D. M., Piñeros, M., ... Bray, F. (2019). Estimating the global cancer incidence and mortality in 2018: GLOBOCAN sources and methods. *International Journal of Cancer*, *144*(8), 1941–1953. Available from https://doi.org/10.1002/ijc.31937.

Figueredo, F., Garcia, P. T., Cortón, E., & Coltro, W. K. T. (2016). Enhanced analytical performance of paper microfluidic devices by using $Fe_3O_4$ nanoparticles, MWCNT, and graphene oxide. *ACS Applied Materials and Interfaces*, *8*(1), 11–15. Available from https://doi.org/10.1021/acsami.5b10027.

Fu, X., Wen, J., Li, J., Lin, H., Liu, Y., Zhuang, X., ... Chen, L. (2019). Highly sensitive detection of prostate cancer specific PCA3 mimic DNA using SERS-based competitive lateral flow assay. *Nanoscale*, *11*(33), 15530–15536. Available from https://doi.org/10.1039/c9nr04864b.

Gabriel, E. F. M., Garcia, P. T., Cardoso, T. M. G., Lopes, F. M., Martins, F. T., & Coltro, W. K. T. (2016). Highly sensitive colorimetric detection of glucose and uric acid in biological fluids using chitosan-modified paper microfluidic devices. *Analyst*, *141*(15), 4749–4756. Available from https://doi.org/10.1039/c6an00430j.

Ge, S., Zhang, L., Zhang, Y., Liu, H., Huang, J., Yan, M., & Yu, J. (2015). Electrochemical K-562 cells sensor based on origami paper device for point-of-care testing. *Talanta*, *145*, 12–19. Available from https://doi.org/10.1016/j.talanta.2015.05.008.

Ghosale, A., Shrivas, K., Deb, M. K., Ganesan, V., Karbhal, I., Bajpai, P. K., & Shankar, R. (2018). A low-cost screen printed glass electrode with silver nano-ink for electrochemical detection of H2O2. *Analytical Methods*, *10*(26), 3248–3255. Available from https://doi.org/10.1039/c8ay00652k.

Gutiérrez-capitán, M., Baldi, A., & Fernández-sánchez, C. (2020). Electrochemical paper-based biosensor devices for rapid detection of biomarkers. *Sensors (Switzerland)*, *20*(4). Available from https://doi.org/10.3390/s20040967.

Kala, D., Gupta, S., Nagraik, R., Verma, V., Thakur, A., & Kaushal, A. (2020). Diagnosis of scrub typhus: recent advancements and challenges. *3 Biotech*, *10*(9). Available from https://doi.org/10.1007/s13205-020-02389-w.

Kala, D., Sharma, T. K., Gupta, S., Nagraik, R., Verma, V., Thakur, A., & Kaushal, A. (2020). AuNPs/CNF-modified DNA biosensor for early and quick detection of O. tsutsugamushi in patients suffering from scrub typhus. *3 Biotech*, *10*(10), 446. Available from https://doi.org/10.1007/s13205-020-02432-w.

Kamel, H. F. M., & Al-Amodi, H. S. B. (2016). Cancer biomarkers. In *Role of biomarkers in medicine*. *IntechOpen*.

Kanavos, P. (2006). The rising burden of cancer in the developing world. *Annals of Oncology*, *17*(8), viii15−viii23. Available from https://doi.org/10.1093/annonc/mdl983.

Kaushal, A., Singh, S., Kala, D., Kumar, D., & Kumar, A. (2016). speB genosensor for rapid detection of Streptococcus pyogenes causing damage of heart valves in human. *Cellular and Molecular Biology*, 62.

Kaushal, A., Singh, S., Kumar, A., & Kumar, D. (2017). Nano-Au/cMWCNT modified speB gene specific amperometric sensor for rapidly detecting streptococcus pyogenes causing rheumatic heart disease. *Indian Journal of Microbiology*, *57*(1), 121−124. Available from https://doi.org/10.1007/s12088-016-0636-y.

Koczula, K. M., & Gallotta, A. (2016). Lateral flow assays. *Essays in Biochemistry*, *60*(1), 111−120. Available from https://doi.org/10.1042/EBC20150012.

Kumar, P., & Pawaiya, R. V. S. (2010). Advances in cancer diagnostics. *Brazilian Journal of Veterinary Pathology*, *3*(2), 142−153. Available from http://www.abpv.vet.br/upload/documentos/DOWNLOAD-FULL-ARTICLE-27-20881_2011_1_7_16_45.pdf.

Kumar, S., Umar, M., Saifi, A., Kumar, S., Augustine, S., Srivastava, S., & Malhotra, B. D. (2019). Electrochemical paper based cancer biosensor using iron oxide nanoparticles decorated PEDOT:PSS. *Analytica Chimica Acta*, *1056*, 135−145. Available from https://doi.org/10.1016/j.ac.2018.12.053.

Li, L., Kong, Q., Zhang, Y., Dong, C., Ge, S., & Yu, J. (2015). A 3D electrochemical immunodevice based on a porous Pt-paper electrode and metal ion functionalized flower-like Au nanoparticles. *Journal of Materials Chemistry B*, *3*(14), 2764−2769. Available from https://doi.org/10.1039/c4tb01946f.

Lokshin, A. E., & Nolen, B. (2011). The expansion and advancement of cancer biomarkers. *Cancer Biomarkers*, *10*(2), 61−62. Available from https://doi.org/10.3233/CBM-2012-0244.

Loo, S. W., & Pui, T. S. (2020). Cytokine and cancer biomarkers detection: The dawn of electrochemical paper-based biosensor. *Sensors (Switzerland)*, *20*(7). Available from https://doi.org/10.3390/s20071854.

López-Marzo, A. M., & Merkoçi, A. (2016). Paper-based sensors and assays: A success of the engineering design and the convergence of knowledge areas. *Lab on a Chip*, *16*(17), 3150−3176. Available from https://doi.org/10.1039/c6lc00737f.

Mahmoudi, T., de la Guardia, M., & Baradaran, B. (2020). Lateral flow assays towards point-of-care cancer detection: A review of current progress and future trends. *TrAC - Trends in Analytical Chemistry*, 125. Available from https://doi.org/10.1016/j.trac.2020.115842.

Malhotra, B.D., Kumar, S., & Pandey, C.M. (2016). Nanomaterials based biosensors for cancer biomarker detection. *Journal of Physics: Conference Series* (Vol. 704, Issue 1). https://doi.org/10.1088/1742-6596/704/1/012011

Mohanraj, J., Durgalakshmi, D., Rakkesh, R. A., Balakumar, S., Rajendran, S., & Karimi-Maleh, H. (2020). Facile synthesis of paper based graphene electrodes for point of care devices: A double stranded DNA (dsDNA) biosensor. *Journal of Colloid and Interface Science*, *566*, 463−472. Available from https://doi.org/10.1016/j.jcis.2020.01.089.

Morales-Narváez, E., Baptista-Pires, L., Zamora-Gálvez, A., & Merkoçi, A. (2017). Graphene-Based Biosensors: Going Simple. *Advanced Materials*, *29*(7). Available from https://doi.org/10.1002/adma.201604905.

Moreira, F. T. C., Ferreira, M. J. M. S., Puga, J. R. T., & Sales, M. G. F. (2016). Screen-printed electrode produced by printed-circuit board technology. Application to cancer biomarker detection by means of plastic antibody as sensing material. *Sensors and Actuators, B: Chemical*, *223*, 927−935. Available from https://doi.org/10.1016/j.snb.2015.09.157.

Parolo, C., & Merkoçi, A. (2013). Paper-based nanobiosensors for diagnostics. *Chemical Society Reviews*, *42*(2), 450−457. Available from https://doi.org/10.1039/c2cs35255a.

Prasad, K. S., Cao, X., Gao, N., Jin, Q., Sanjay, S. T., Henao-Pabon, G., & Li, X. (2020). A low-cost nanomaterial-based electrochemical immunosensor on paper for high-sensitivity early detection of pancreatic cancer. *Sensors and Actuators B: Chemical*.

Qureshi, A., Tufani, A., Corapcioglu, G., & Niazi, J. H. (2020). CdSe/CdS/ZnS nanocrystals decorated with Fe3O4 nanoparticles for point-of-care optomagnetic detection of cancer biomarker in serum. *Sensors and Actuators, B: Chemical*, 321. Available from https://doi.org/10.1016/j.snb.2020.128431.

Ratajczak, K., & Stobiecka, M. (2020). High-performance modified cellulose paper-based biosensors for medical diagnostics and early cancer screening: A concise review. *Carbohydrate Polymers*, 229. Available from https://doi.org/10.1016/j.carbpol.2019.115463.

Sakata, T., Hagio, M., Saito, A., Mori, Y., Nakao, M., & Nishi, K. (2020). Biocompatible and flexible paper-based metal electrode for potentiometric wearable wireless biosensing. *Science and Technology of Advanced Materials*, *21*(1), 379−387. Available from https://doi.org/10.1080/14686996.2020.1777463.

Sánchez-Calvo, A., Blanco-López, M. C., & Costa-García, A. (2020). Based working electrodes coated with mercury or bismuth films for heavy metals determination. *Biosensors*, *10*(5).

Taleat, Z., Khoshroo, A., & Mazloum-Ardakani, M. (2008). Screen-printed electrodes for biosensing: A review (Vol. 181, pp. 865−891).

Verma, V., Goyal, M., Kala, D., Gupta, S., Kumar, D., & Kaushal, A. (2020). Recent advances in the diagnosis of leptospirosis. *Frontiers in Bioscience - Landmark*, *25*(9), 1655−1681. Available from https://doi.org/10.2741/4872.

WHO (2020). Report on cancer: Setting priorities, investing wisely and providing care for all.

Wild, C. P., & Stewart, B. W. (2014). *World cancer report 2014* (pp. 482−494). Geneva, Switzerland: World Health Organization.

Wu, L., Zhang, Y., Wang, Y., Ge, S., Liu, H., Yan, M., & Yu, J. (2016). A paper-based electrochemiluminescence electrode as an aptamer-based cytosensor using PtNi@carbon dots as nanolabels for detection of cancer cells and for in-situ screening of anticancer drugs. *Microchimica Acta*, *183*(6), 1873–1880. Available from https://doi.org/10.1007/s00604-016-1827-2.

Yáñez-Sedeño, P., Campuzano, S., & Pingarrón, J. M. (2020). Screen-printed electrodes: Promising paper and wearable transducers for (bio)sensing. *Biosensors*, *10*(7). Available from https://doi.org/10.3390/BIOS10070076.

Zhou, P., Lu, F., Wang, J., Wang, K., Liu, B., Li, N., & Tang, B. (2020). A portable point-of-care testing system to diagnose lung cancer through the detection of exosomal miRNA in urine and saliva. *Chemical Communications*, *56*(63), 8968–8971. Available from https://doi.org/10.1039/d0cc03180a.

Zhu, J., Wang, Z., Lin, S., Jiang, S., Liu, X., & Guo, S. (2020). Low-cost flexible plasmonic nanobump metasurfaces for label-free sensing of serum tumor marker. *Biosensors and Bioelectronics*, 150. Available from https://doi.org/10.1016/j.bios.2019.111905.

Chapter 24

# Scope of biosensors, commercial aspects, and miniaturized devices for point-of-care testing from lab to clinics applications

Pushpesh Ranjan[1,2], Ayushi Singhal[1,2], Mohd Abubakar Sadique[1], Shalu Yadav[1,2], Arpana Parihar[1,3] and Raju Khan[1,2]

[1]*Microfluidics & MEMS Centre, CSIR-Advanced Materials and Processes Research Institute (AMPRI), Bhopal, India*

[2]*Academy of Scientific and Innovative Research (AcSIR), Ghaziabad, India*

[3]*Department of Biochemistry and Genetics, Barkatullah University, Bhopal, India*

## 24.1 Introduction

The concept of biosensors was started by Dr. Leland C. Clark Jr. who developed the first biosensor for detection of glucose based on oxidase enzyme electrode in 1962. He is also recognized as the "father of biosensors." Later on, in 1972 the term biosensor was coined by Karl Cammann. A few years later, this glucose biosensor was successfully commercialized in 1975 named as "Model 23 A YSI analyzer" by Yellow Springs Instrument Co., Inc. (Yoo & Lee, 2010). International Union of Pure and Applied Chemistry (IUPAC) define biosensor as "a device that uses specific biochemical reactions mediated by isolated enzymes, immunosystems, tissues, organelles or whole cells to detect chemical compounds usually by electrical, thermal or optical signals." As the name says, the word biosensor is composed of two words, "bio" and "sensor." "Bio" represents the biological component, the analyte under investigation such as living cells, antibodies, enzymes, proteins, nucleic acid, and "sensor" means the physical components which may include transducer and amplifier (Pravin & Sadaf, 2020; Soleymani & Li, 2017). Biosensor is utilized for the investigation of target bio-component either by qualitative or quantitative manner. It consists of a bio-recognition element, amplifier, transducer, and detector. In a biosensor array, the bioreceptor selectively targets the analyte and generates the signal which is transduced and gets a result (Khan, Mohammad, & Asiri, 2019). Biosensors are categorized on the basis of analytes and also on the nature of biorecognition element or bioreceptor, which are aptasensor, immunosensor, genosensor, (enzyme, antibody, cell, DNA, etc.). Moreover, on the basis of physical transducer nature, they are electrochemical (voltammetry, potentiometry, amperometry, electrochemical impedance, FET, conductometry), optical (Surface Plasmon Resonance (SPR), Surface-enhanced Raman spectroscopy (SERS), fluorescence, fiber optics), lateral flow immunoassay (LFIA) and mass (piezoelectric, magnetoelastic) based biosensor. They offer the detection of target analyte at ultra-low levels of concentration in small sample volumes. Further, lots of improvements were done on technical aspects regarding to enhance the sensitivity, selectivity, lower the detection limit, etc., the schematic representation of different biosensors are shown in Fig. 24.1 (Chikkaveeraiah, Bhirde, Morgan, Eden, & Chen, 2012; Jayanthi, Das, & Saxena, 2017; Mahmoudi, de la Guardia, & Baradaran, 2020; Shandilya et al., 2019).

## 24.2 Scope of biosensors

Nowadays, several different types of biosensors are reported and available for the detection of different varieties of analyte. They are widely applicable in clinical diagnosis such as cancer biomarkers, cardio vascular biomarker, uric acid, glucose, lactates, etc. detection. In addition, they estimate the different kinds of electrolyte ($Na^+$, $K^+$, $Mg^{+2}$, $Fe^{+2}$, etc.

Biosensor Based Advanced Cancer Diagnostics. DOI: https://doi.org/10.1016/B978-0-12-823424-2.00004-1

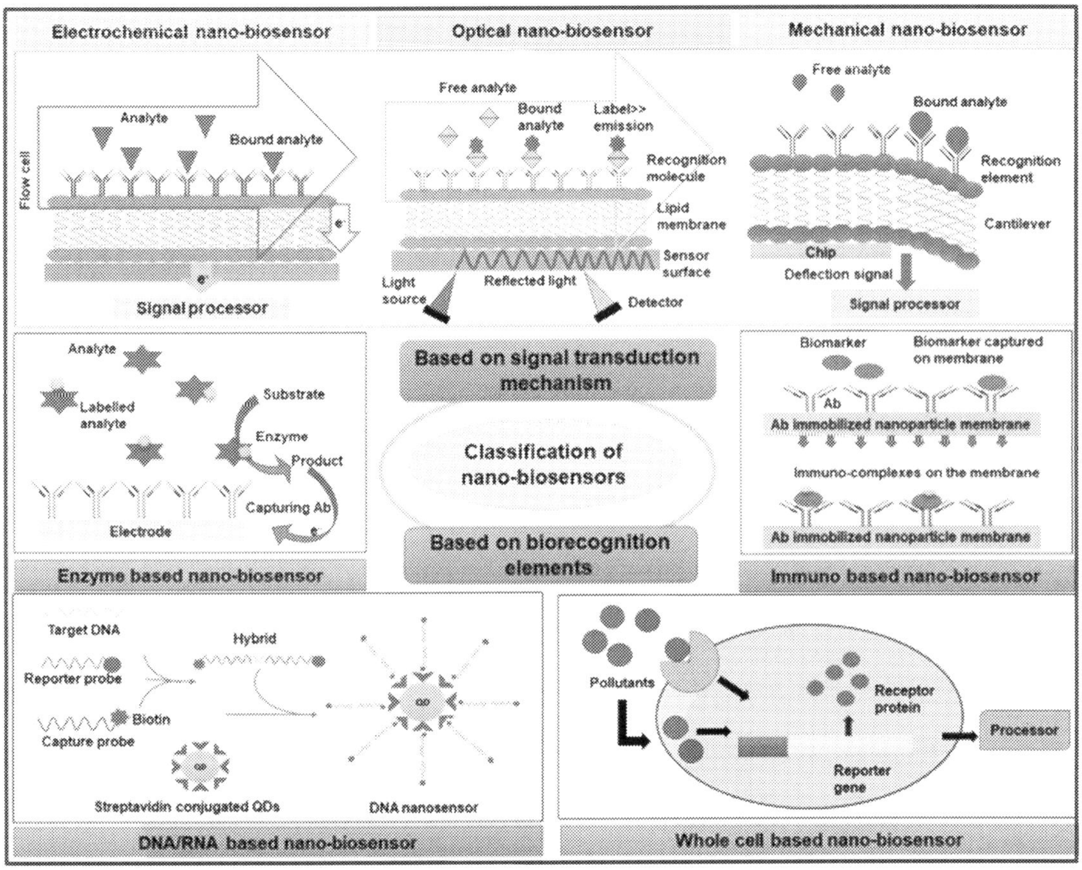


**FIGURE 24.1** Nanobiosensors: point-of-care (POC) approaches for cancer diagnostics. *Reprinted with permission from Shandilya, R., Bhargava, A., Bunkar, N., Tiwari, R., Goryacheva, I. Y., & Mishra, P. K. (2019). Nanobiosensors: Point-of-care approaches for cancer diagnostics. Biosensors and Bioelectronics, 130, 147–165. https://doi.org/10.1016/j.bios.2019.01.034.*

ions) in biological samples such as blood, tear, serum, swab, sweat, etc. Moreover, the estimation of pollutants of environment, food contaminations and biothreats (algae, fungi, bacteria, virus, etc.) is also done through the biosensor. The qualitative or quantitative measurement of concentration of the target analyte provides valuable information for the diagnostic stages. In the past decade, biosensors have become an emerging device which has been utilized for the rapid detection of biological analyte. They offer the real-time, on-site diagnostic platform. Moreover, the potential for the development of biosensors from laboratory scale to clinical application. They provide a simple platform at the clinical scale which is reliable, cost-effective, highly sensitive, specific, and rapid. At this stage, the uses of biosensors are not restricted and further innovation and improvements are on the way (Bahadir & Sezgintürk, 2015; Mehrotra, 2016). With the technical improvements, advanced, smart, and portable biosensors would open the new era in biosensor platforms. Along with being flexible and wearable, biosensors play a key role in the diagnosis of disease and management of healthcare. Advanced smart-care management helps to control and minimize the disease progression at an early stage and could diminish the mortality rate (Salim & Lim, 2019). Several application of biosensor are shown in Fig. 24.2.

## 24.3 Cancer biomarker detection

Nowadays, global facing various serious challenges, where health related concern get top priority. Millions of peoples are affected due to numerous types of serious illness, where cancer attracted major serious health concerns. According to the World Health Organization (WHO), cancer is the major cause of death globally and responsible for about 9.6 million deaths in 2018 which increase to 10 million in 2020 (Cancer, n.d.). Where, approximately 70% of deaths occur in low and middle socio-economic countries. Due to cancer, approximately US$1.16 trillion of economic burdens are faced. To reduce the mortality rate and economic burden of cancer, it needs to be diagnosed at a very early stage (Cancer, n.d.). There are several diagnostic platforms that are utilized for detection of cancer. In conventional diagnostic

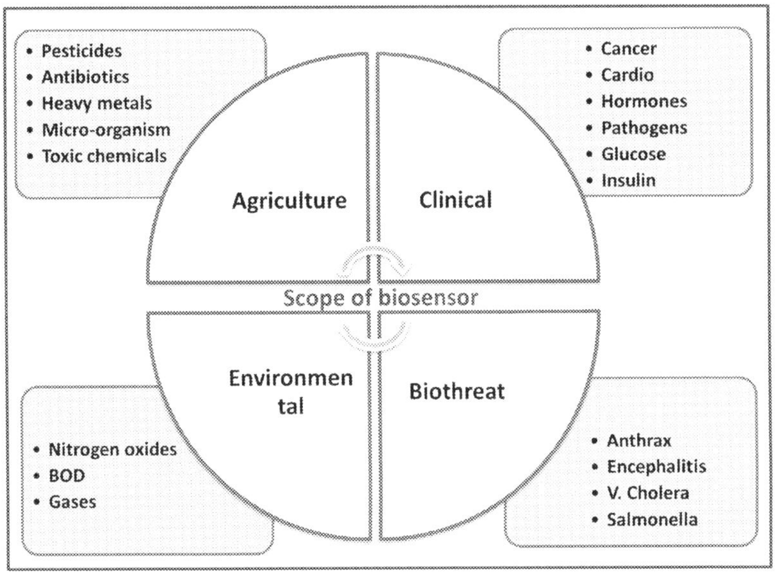


FIGURE 24.2 Represents the application of biosensors in various fields.

platforms such as biopsy, ultrasound imaging, MRI, positron emission tomography (PET), CT scan, etc., the diagnosis of cancer biomarker are done through the sophisticated instruments, usually equipped in centralized laboratories. However, the sensitivity and specificity of conventional techniques are limited. Moreover, the molecular and microbiology-based diagnosis is most commonly used in this regard. On the other hand, the biosensors-based platform is now in almost all the fields that aim to improve the quality of life including clinical application, drug discovery, environmental monitoring, food analysis, security defense, etc. Although, the emerging lab-on-a-chip-based biosensor has opportunities and demands more work to further develop new biosensors with better performances. The miniaturized and ease-to-use characteristics make biosensor ideal for POC applications (Anneng & Feng, 2020; Liu et al., 2020).

There is a vast scope of biosensors in the field of advanced cancer diagnosis. Several conventional methods are available for the diagnosis of cancer, but biosensor-based cancer diagnosis is attracting more due to their several advantages over the conventional techniques. They offers several advantages such as inexpensive testing, high specificity, sensitivity, rapid test, low volume of sample requirements, and easy to operate. In addition, it doesn't require a sophisticated instrumentation setup and trained personnel. There are several biosensors reported for the diagnosis of cancer biomarkers even at the early stage, which helps to monitor and control the progression of disease (Mahato, Maurya, & Chandra, 2018). Detection and monitoring of cancer diseases at early stages need a great deal of effort to perform and record the routine test either by the test of a clinical sample or the examination through conventional tests such as biopsy, CT-scanning, etc. Since cancer does not express significant symptoms at early stage, it is difficult to diagnose and maintain the routine test at a very early stage (Bossmann & Troyer, 2013).

### 24.3.1 Breast cancer

Breast cancer is the most common cancer in women. The number of breast cancer cases is growing in a rapid manner all over the world. Breast cancer caused about 6.6% of cancer deaths in 2018, and it was the fourth leading cause of cancer deaths in the world. It is anticipated that the number of breast cancer cases will reach approximately 3.2 million by 2050. The survival rate can be enhanced if the cancer could be diagnosed at an early stage. Most cases of death are occurring due to the late diagnosis. For cancer cases that were diagnosed at an early stage (stage-1), the survival rate was found to be more than 90%. The role of early diagnosis has high importance; the diagnosis should be performed in such a way that it would detect even the cellular or molecular level at initial changes. These changes can be identified with the changing number of biomarkers. Biomarkers are biological molecules found in blood, other body fluids, or tissues that are a sign of a normal or abnormal process or of a condition or disease. The number of these biomarkers varies with the changing stages of disease. Therefore, these biomarkers are considered an important tool for the diagnosis of cancer at an early stage. The early detection of biomarkers would be possible with the help of biosensors (Ranjan et al., 2020).

### 24.3.2 Lung cancer

Lung cancer is the second leading cause of death globally. In 2020, approximately 2.2 million new cancer cases and 1.8 million cancer deaths were estimated to occur in world wide (GLOBOCAN, 2020: New Global Cancer Data, n.d.). Each year, lung cancer causes approximately 1.5 million deaths globally. The major causes for lung cancer include smoking, use of tobacco products, exposure to various carcinogens present as environmental pollutants, and numerous genetic level changes. The most common symptoms include coughing up blood, loss of appetite or unexplained weight loss, etc. There are two types of lung cancer: small cell lung carcinomas (SCLC) and nonsmall cell lung carcinomas (NSCLC). It was found that the NSCLC accounts for 90% of all cases of lung cancer. In comparison to other types of cancers, the survival rate of lung cancer is lower (about 18%) and its prognosis is also very poor (Saijo, 2012). For increasing the chances of rate of survival, early diagnosis of such a fatal disease is of utmost priority. Most conventional methods are available but they have some drawbacks which are required to overcome, because they can lead to a delay in diagnosis and even can cause mistreatment and worsen the health conditions. There is an urgent demand for highly sensitive biosensing devices to identify and diagnose patients with premalignant and premetastatic malignant tumors (Khanmohammadi et al., 2020b).

### 24.3.3 Oral cancer

Globally, oral cancer (OC) is the sixth most happening disease and approximately 3.0 million new patients are diagnosed with OC annually, and in turn over 1.4 million deaths. Usually, early stages of OCs have no symptoms, and hence being disregarded at the primary stage results in high death rate. The efficient detection of biomarkers concentration with the help of biosensors at an early stage can be utilized for the early detection of malignancy (Mishra et al., 2016).

### 24.3.4 Pancreatic cancer

Pancreatic cancer is the fourth most happening cause of death in the United States. For the cancer cases, of all the stages, the survival rate was found to be approximately 8%. Among the various subtypes of pancreatic cancer, the majority of pancreatic tumors are represented by the pancreatic ductal adenocarcinoma (PDAC). The only cure from PDAC is the surgery pancreatic intraepithelial neoplasia (PanIN), which is the most common precursor of PDAC. Main diagnostic approaches are imaging diagnosis, histopathology diagnosis, genetic diagnosis (detection of gene mutation), detection of serum tumor markers, biopsy of circulation tumor DNA (ctDNA), and circulation tumor cells in a noninvasive manner. Although PCR and ELISA are sensitive approaches to pancreatic cancer detection, they have some drawbacks like it is expensive, time consuming, and requires well-trained personnel as well as restrictive experimental conditions (Qian et al., 2019).

## 24.4 Biomarkers for predicting the outcome of various cancer immunotherapies

Nowadays, cancer immunotherapies are used to treat various types of cancer. Although immunotherapy is an efficient technique, it can also result in adverse reactions. Biomarkers can help in predicting the outcomes of various immunotherapies and can identify the group of people getting benefited from immunotherapies. Use of biomarkers can improve the health line and can also avoid extra health-related expenditures. The use of biomarkers can give insight into the mechanism of action of immunotherapies. The various cellular and molecular predictive biomarkers such as serum proteins, tumor-specific factors, and their related microenvironment components as well as host genetic background can be used for various immunotherapies (Jafarzadeh, Khakpoor-Koosheh, Mirzaei, & Mirzaei, 2020).

## 24.5 Miniaturized devices for point-of-care testing from lab to clinical applications

The most conventional technologies such as magnetic resonance, biopsy, ultrasound imaging, computed and PET, and CT-scan are currently utilized for the diagnosis of cancer at the clinical level. They have high accuracy and much preferred techniques. But despite the advantages, they have some limitations such as they require costly instruments, well-trained personnel, limited sensitivity, and time-consuming process. In addition, some technique are painful, such as a biopsy (Karen, Pushpa, Rao, Houston, & C, 2017; Tuğba, 2020). Over the conventional techniques, the genomic and proteomic detection techniques like PCR, ELISA, colorimetric, and DNA quenching gain much attention in cancer diagnosis because they are highly specific, simple to use, and cost-effective (Tuğba, 2020). Moreover, they face serious problems like tedious process, complicated instrumentation handling, and limited sensitivity along with high false negative and false positive results, making them a limitation in POC application in cancer diagnosis. The monitoring and diagnosis of cancer progress at an early stage suffer due to lack of insufficient POC technical devices. There is an urgent need for a highly sensitive and specific POC platforms with high accuracy for early diagnosis of cancer biomarkers (Sandbhor Gaikwad & Banerjee, 2018). A fast and reliable personalized diagnostic platform is suitable for at home screening of biomarkers. For instance, the nanobiosensor and microfluidic-based platform have the capability to miniaturize in small diagnostic devices which have potential to be a personal cancer diagnostic platform. These types of devices not only provide the result in a small volume of samples but they are also rapid and cost-effective. Along with, they have capability to point out the early diagnosis, classify the tumor progression, and risk indication (Khondakar, Dey, Wuethrich, Sina, & Trau, 2019; Syedmoradi, Norton, & Omidfar, 2020). On the other hand, an in vivo POC biosensor is potentially suitable for cancer diagnosis at the clinical stage (Huang, Liu, Yung, Xiong, & Chen, 2017). The schematic potential of POC cancer diagnostics over conventional techniques, and the unmet clinical needs in healthcare systems are shown in Fig. 24.3.

## 24.6 Miniaturized point-of-care biosensor for cancer diagnosis

There are numerous electrochemical, optical, and microfluidic biosensor platforms reported for the diagnosis of cancer protein/antigen/exosome in different clinical samples (Sharifi, Hasan, Attar, Taghizadeh, & Falahati, 2020; Xu et al., 2020). Miniaturized POC technology offers detection of trace levels of cancer biomarkers in biological specimens. They have easy handling, sample preparation, evaluation, and diagnosis. However, the presence of several other biological components in the specimen could hamper the result. Therefore a highly sensitive and specific biosensor is recommended in POC applications. In this regard, several biosensors such as electrochemical, optical, and mass-based biosensors have been reported (Wongkaew, 2019).

### 24.6.1 Electrochemical biosensor for cancer diagnosis

Electrochemical-based biosensors monitor the change in current, conductance, potential, or impedance of electroactive material. Miniaturized electrochemical POC biosensors monitor the risk assessment of cancer disease. They are cost-effective and have high specificity, sensitivity, wide linearity, and reproducibility. In addition, they have long-term stability and could monitor the specific biomarkers without affecting their performance. Hence, they offer the on-site

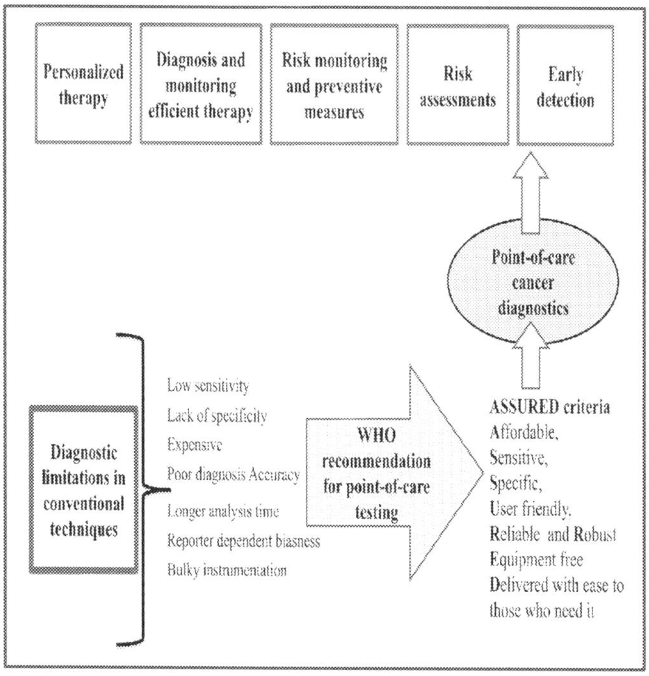


**FIGURE 24.3** Schematic of potential of POC cancer diagnostics over conventional techniques and the unmet clinical need in healthcare systems. *From Sandbhor Gaikwad, P., & Banerjee, R. (2018). Advances in point-of-care diagnostic devices in cancers. Analyst, 143(6), 1326–1348. https://doi.org/10.1039/c7an01771e.*

detection of cancer biomarkers and efficiently replace the conventional technique in the near future (Siontorou et al., 2017; Tuğba, 2020; Pal and Khan, 2017). In a recently published review, Nur et al. describes the several electrochemical biosensors for the detection of different kinds of cancer biomarkers such as lung, breast, ovarian, pancreatic, prostate, cervical, etc. in the numerous clinical specimens. In addition, the aptamer-based biosensors are also reported which are highly specific and sensitive towards cancer biomarker detection (Nur, Mostafa, & Mehmet, 2016). Moreover, several other electrochemical-based biosensors for the detection of lung, breast, and prostate cancer are reported and described in a well manner (Khanmohammadi et al., 2020a). Recently, Kumar et al. reported a label free paper-based electrochemical biosensor of conducting polymer of poly(3,4-ethylenedioxythiophene):poly(styrenesulfonate) and nanostructured iron oxide ($nFe_2O_3$@PEDOT:PSS) nanocomposite for the diagnosis of cancer biomarker carcinoembryonic antigen (CEA). This biosensor has good sensitivity, a low detection limit, and a shelf-life up to one month (Fig. 24.4) (Kumar et al., 2019). In another report, Shahrokhian et al. reported the nanocomposite of conductive polymer and reduced graphene oxide for efficiently detection of BRCA1 gene of breast cancer as low as femtomolar concentration (Shahrokhian & Salimian, 2018). Recently, an electrochemical bio-barcode assay termed as "e-biobarcode" in which the electrode surface is modified by nanomaterial and biorecognition elements which integrate with the transducer to signal readout. Moreover, this assay has advantages; it uses the single step for sample analysis and eliminates the multistep process and offers a POC diagnostics platform for detection of PSA in unprocessed and undiluted human plasma samples. This device has an ultra-low LOD and detected nanogram concentration of PSA in samples (Traynor, Wang, Pandey, Li, & Soleymani, 2020). Since electrochemical biosensor offers several advantages such as high throughput, excellent sensitivity, ultra-low LOD, long shelf-life, and easy miniaturization makes them a POC biosensor in cancer diagnosis.

## 24.6.2 Optical biosensor for cancer diagnosis

For a long time, optical-based biosensors have been utilized for the detection of cancer biomarkers. They offer the on-site detection, along with the strip/paper-based colorimetric visual observation through naked eye offer promising in biosensor application. For instance, Pan et al. reported a portable fluorescence device for detection of skin cancer through imaging of squamous cell carcinoma, noninvasion tumor. Through this device, the spot of the cancer cell is easily found on the basis of imaging of the cell. Moreover, it has the ability to integrate with other devices such as ultrasound probes which are further capable to detect multiple biomarkers of cancer cells. This device is light weight and operates at low power (1.5 W) (Pan et al., 2020). Recently, Tian et al. demonstrated hydrogel microparticle-coated chips having great fluorescence properties were utilized for the diagnosis of miRNA of cancer cells. They offer the easy platform through smartphone-based detection of miRNA (Tian et al., 2019). In another approach, a fluorescent-enabled smartphone-based portable device platform was reported for the detection of cancer biomarkers.

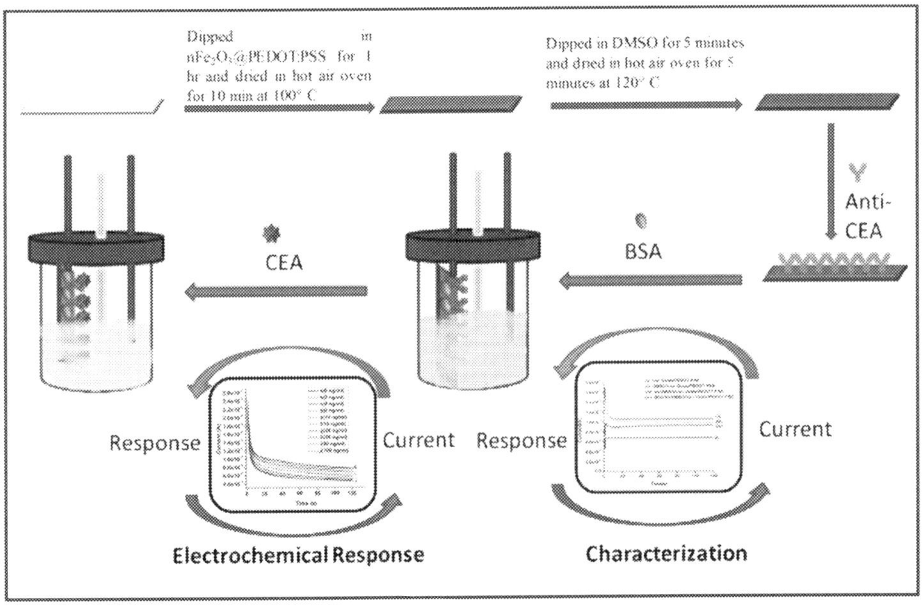


**FIGURE 24.4** Schematic representation of fabrication of electrochemical paper based cancer biosensor. *From Kumar, S., Umar, M., Saifi, A., Kumar, S., Augustine, S., Srivastava, S., & Malhotra, B. D. (2019). Electrochemical paper based cancer biosensor using iron oxide nanoparticles decorated PEDOT:PSS. Analytica Chimica Acta, 1056, 135–145. https://doi.org/10.1016/j.ac.2018.12.053.*

Herein, they fabricated the sensor chip by phenylboronic acid conjugated gold nanocluster, which efficiently detected the mucin through the imaging and further applicable in therapy of cancer cells (Dutta, Sailapu, Chattopadhyay, & Ghosh, 2018). In a recent study, Kim et al. reported a SERS-based biosensor for detection of breast cancer in human tear samples. Herein, they fabricated the plasmonic SERS chip by a single layer of gold nanoparticles functionalized with hexagonal close packed polystyrene nanosphere. These sensors have excellent LOD which detects up to femtomolar concentration of analyte. In addition, they show good reproducibility, sensitivity and provide real-time diagnosis (Fig. 24.5) (Kim et al., 2020).

### 24.6.3 Microfluidics biosensor for cancer diagnosis

Microfluidics is an emerging diagnostics platform for rapid and accurate diagnosis of target analytes. It requires minimal volume (μL) to perform detection. In addition, it is easily integrated with electrochemical and optical techniques to further enhance the sensitivity and achieve a low detection limit. However, the detection of multiple biomarkers offers the great platform in biosensor application. For instance, Prasad et al. reported a paper-based microfluidics device integrated with a smartphone for the diagnosis of BRCA-1, breast cancer biomarker. This device, FluoreZen, is being utilized for imaging of cancer genes (Prasad, Hasan, Grouchy, & Gartia, n.d.). In another report, Zheng et al. reported a SERS-integrated microfluidics chip for the multiplex detection of breast cancer biomarkers. Herein, they fabricated a microfluidic chip by surface coating of microfluidic channel by self-assembled silver nanoparticles followed by its tagging through Raman reporter. They detected the CA15-3, CA125, and CEA, breast cancer biomarkers in real samples with high sensitivity and selectivity. Further, they validated and confirmed the same result through the ELISA kit (Fig. 24.6) (Zheng et al., 2018). In another report, Uliana et al. demonstrated an electrochemical integrated disposable microfluidic chip (μFED) for diagnosis of breast cancer biomarker estrogen receptor-α at femtogram concentration in clinical sample. The fabrication cost of the device is low (US$0.20) which could turn into a low-cost device and efficiently benefit a large group of the population (Uliana, Peverari, Afonso, Cominetti, & Faria, 2018). Chen et al. reported a microfluidic-based biosensor for simultaneous detection of prostate specific antigen (PSA) and CEA at picomolar concentration within 20 min. In this sensor, during the acoustic mixing, trapping of magnetic beads with the biomarkers of sampled leads to detection and further result the improvement of LOD of sensor (Chen et al., 2018).

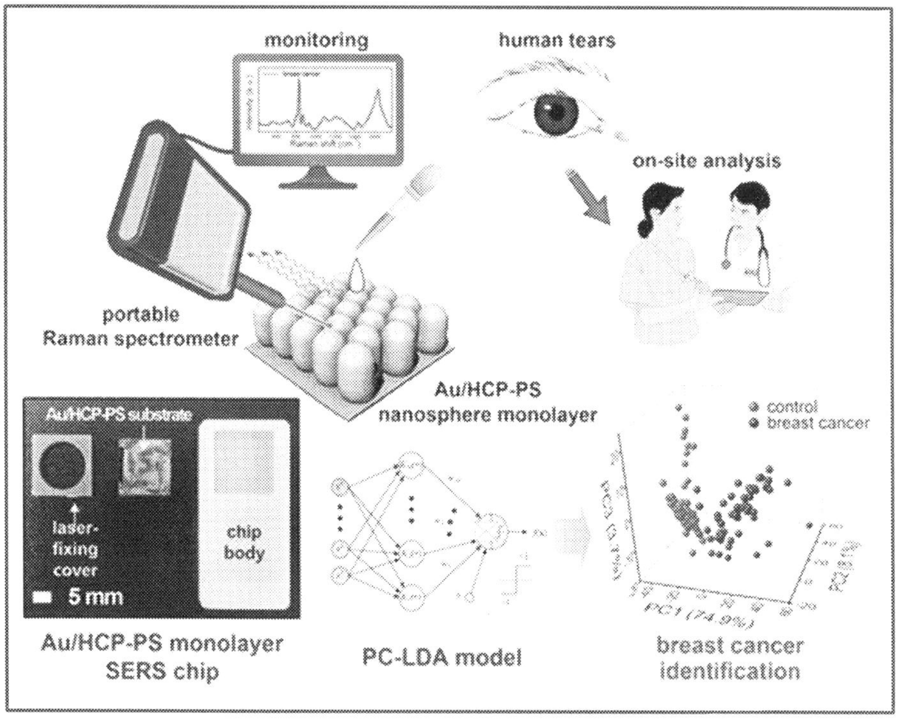


**FIGURE 24.5** Portable SERS-based biosensor for detection of breast cancer in human tear sample. *From Kim, S., Kim, T. G., Lee, S. H., Kim, W., Bang, A., Moon, S. W., Song, J., Shin, J. H., Yu, J. S., & Choi, S. (2020). Label-free surface-enhanced raman spectroscopy biosensor for on-site breast cancer detection using human tears. ACS Applied Materials and Interfaces, 12(7), 7897–7904. https://doi.org/10.1021/acsami.9b19421.*

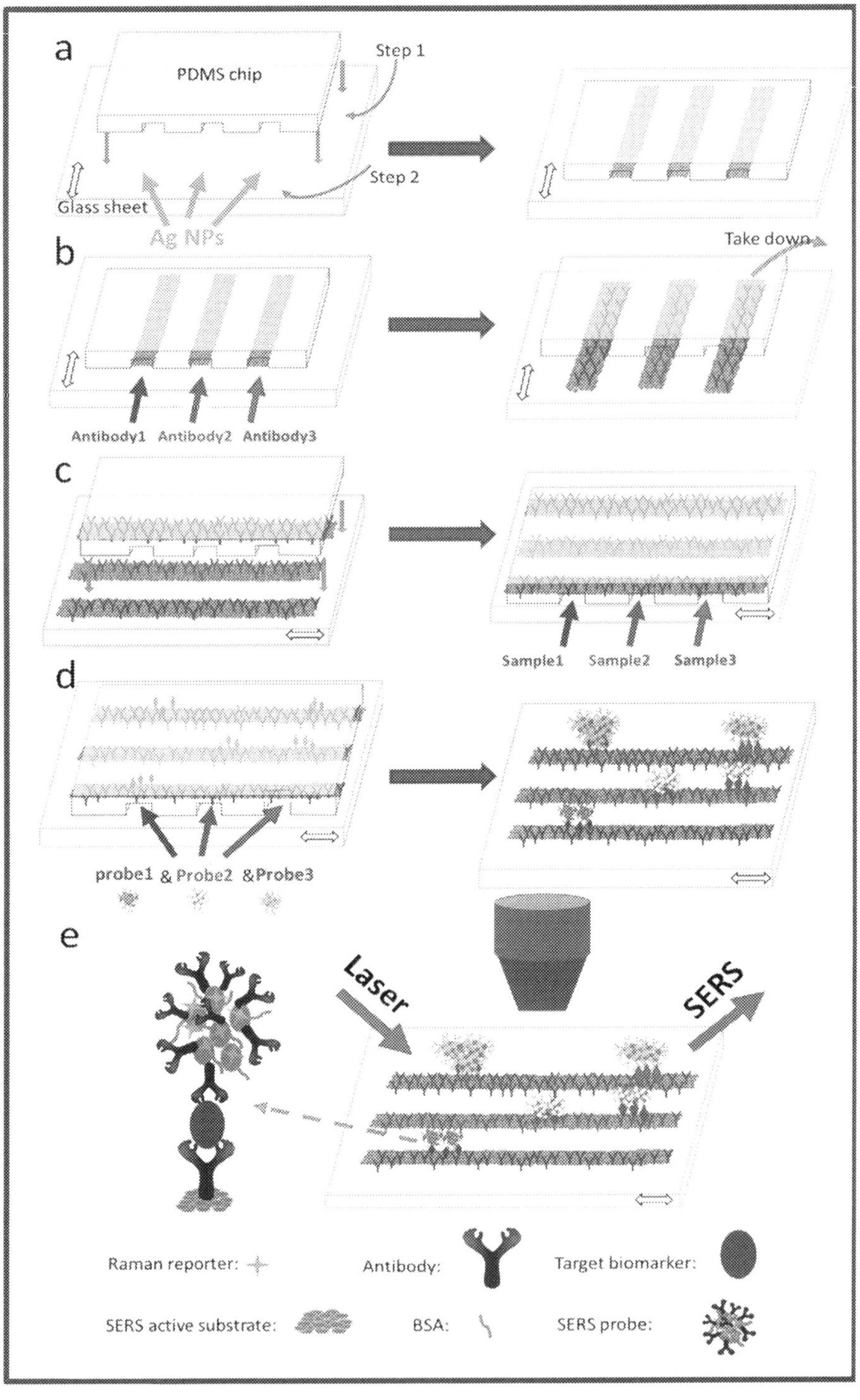


**FIGURE 24.6** Schematic diagram of (A) Illustration of the microfluidic system. Polydimethylsiloxane (PDMS) chip was bonded to the glass sheet. Ag NPs were injected into the microfluidic channels by syringe pumps. (B) Different capture antibodies were injected into the Ag NPs modified microfluidic channels via syringe pumps. (C) The previous PDMS chip was removed and a new PDMS chip was bonded to the glass sheet perpendicular to the previous placement. Different samples were injected into the microfluidic channels. (D) The mixture of three kinds of SERS probes was injected into the microfluidic channels. (E) The sandwich structure formed in the microfluidic channels and SERS signal of each crossing region (i.e., the reaction spot) was measured with a confocal microscope. *From Zheng, Z., Wu, L., Li, L., Zong, S., Wang, Z., & Cui, Y. (2018). Simultaneous and highly sensitive detection of multiple breast cancer biomarkers in real samples using a SERS microfluidic chip. Talanta, 188, 507–515. https://doi.org/10.1016/j.talanta.2018.06.013.*

## 24.7 Current status of point-of-care cancer diagnostic devices

Numerous POC biosensors were reported and available in the market for the diagnosis of specific cancer biomarkers. They get approval from different government agencies such as the Food and Drug Administration (FDA), Indian Council of Medical Research (ICMR), etc. after complete analysis and observation of the biosensor product. For instance, the latex agglutination test as bladder tumor antigen (BTA) test is performed for the diagnosis of basement membrane antigen, a cancer biomarker in the urine sample. Moreover, the FDA approved BTAstat and the BTA TRAK POC device replaced these conventional tests for the diagnosis of target analyte. The BTAstat test is performed with chromatographic tools which are not only cost-effective, but they are also rapid and could be performed by a layman. Moreover, the BTA TRAK test is a sandwich assay and could take a few more times. In another way, BTAstat and BTA TRAK have sensitivity of 57%–83% and 66% while the specificity is 60%–92% and 65%, respectively (Xylinas et al., 2014). The nuclear matrix protein-22 bladder cancer (NMP22 BC) FDA-approved test kit (Matritec Inc., Newton, MA) and NMP22 Bladder Chek (Alere) are quantitative and qualitative tests for the diagnosis of NMP22 in bladder cancer, respectively. The NMP22 test assay has 75.0% and 55.7% of sensitivity and specificity, respectively which were recorded for this test. Whereas 75.0% and 85.7% of sensitivity and specificity, respectively for the NMP22 Bladder Chek test. The two tests such as UBC-rapid kit and UBC-ELISA (IDL Biotech, Sweden) are promising in POC application for diagnosis of bladder cancer. UBC-ELISA is a sandwich assay; however, UBC-rapid is made of gold-labeled antibodies and utilized for the quantitative estimation of cytokeratin-8 and 18 in urine samples. Further, it has improved to detect tumor markers of bladder cancer. The UBC-rapid test kit has 59% and 86% of sensitivity and specificity, respectively. The result of the UBC-rapid test kit produced inconsistent results. So the present analysis was performed to examine the overall accuracy of the UBC-rapid test kit for diagnosis of bladder cancer. Moreover, the accuracy of UBC-rapid test kit was estimated by calculating the several parameter pooled sensitivity, specificity, positive likelihood ratio (PLR), negative likelihood ratio (NLR), diagnostic odds ratio (DOR) and the area under the curve (AUC). In current analysis, it suggests that the performance of UBC-rapid test kit is superior as comparison to cytology and cystoscopy (Lu et al., 2018; Tian et al., 2019). Another test kit, BioDoct (dot-blot assay), Fujirebio Diagnostics, Inc. is utilizing it for the diagnosis of survival (antiapoptotic protein) in bladder cancer (Horstmann et al., 2010; Liou et al., 2006). Similarly, The uCyT+/Immonocyt is used for the diagnosis of cellular biomarkers of bladder cancer. The combination of cytology and immunofluorescence which generate the signal are utilized for the diagnosis of bladder cancer biomarkers. In another diagnostic platform, Cha et al. reported the prognostics potential application of Immunocyt on 1182 patient with painless disease hematuria and resulting to improvement of diagnostic precision up to 90% (Cha, 2012). A fluorescence-based device UroVysion (Abbott Molecular, InC., Des Plaines, IL) was reported for the monitoring of bladder cancer. This device was approved by the FDA for clinical application. An another FDA-approved POC biosensor, Accu-Dx (Intracel Corp., Rockville, IL) is reported for the diagnosis of high levels of Vascular endothelial growth factor (VEGF) and other serum proteins like plasminogen and fibrin in bladder cancer. However, another immunosensor, UBC II-ELISA (IDL Biotech) is approved for the detection of cytokeratins-8 and 18 protein biomarkers of bladder cancer. In this context, OSTEOMARkNTx point-of-care immunosensor device is reported with laboratory-based ELISA assay for the diagnosis of N-terminal type-I collagen protein biomarker in metastatic bone cancer. Moreover, Todenhofer et al. documented a quantitative diagnostic POC UBC rapid test kit incorporated with/combined with photometric observer Concile Ω100 for the detection of bladder cancer. Further, Ritter et al. had done a comparative test of photometric reader Concile Ω100 integrated UBC rapid test kit with customary visual analysis of POC test along with laboratory ELISA kit for UBC, NMP22 bladder Chek and cytology (Ritter, 2014; Todenhofer, 2013). The POC devices septin-9 (ColoVantage) and vimentin (ColoSure) have satisfactorily performed the testing of DNA-methylation biomarkers of colorectal cancer. Another FDA-approved Cologuard Exact Science Corporation (Medison), which has the capability to target and detect multiple biomarkers of colorectal cancer. More recently, Elsawy et al. reported the Xpert monitor for the detection of multiple mRNA biomarkers (ABL1, ANXA10, UPK1B, CRH and IGF2) of bladder cancer in urine samples. ABL1 serves as a sample adequacy control. It indicates that the sample contains human cells and RNA. For a valid test result, it has to give a positive ABL1 signal. Xpert monitor gives the positive and negative test result based on the outcome of the linear regression algorithm. Moreover, to achieve the positive test result it doesn't require detecting all the mRNA biomarkers. The sensitivity of Xpert monitor is 73.7% with a negative predictive value of 96.3% (Elsawy, Awadalla, Elsayed, Abdullateef, & Abol-Enein, 2020; Sandbhor Gaikwad & Banerjee, 2018). Several commercially available lateral flow test strips for cancer biomarker detection are listing in Table 24.1 (Mahmoudi et al., 2020). In addition, the FDA approved the numerous POC diagnostic devices for the diagnosis of several cancer biomarkers in vitro testing, that are listed in Tables 24.1 and 24.2 (List of Cleared or Approved Companion Diagnostic Devices In Vitro and Imaging Tools).

**TABLE 24.1** Commercially available lateral flow test strips for cancer biomarker detection.

| Company | Product name | Cancer type | Analyte | Required sample | Detection time (min) | Sensitivity | Specificity |
|---|---|---|---|---|---|---|---|
| CTK Biotech | On-Site PSA Semi-quantitative Rapid Test | Prostate | PSA | 60–90 μL Blood/Serum/ Plasma | 10 | 100% | 99.0% |
| | On-Site FOB Rapid Test | Colorectal | hHB | Human fecal specimens | 10 | Not Reported | Not Reported |
| Alere | Alere NMP22 BladderChek | Bladder | Nuclear matrix protein (NMP2 2) | 4 drops of urine | 30 | 99% when Combined with cystoscopy | 99.0% NPV along with cystoscopy |
| | Clearview iFOBT | Colon | Fecal Occult Blood | Feces | 5 | 93.60 | 99.10% |
| Arbor Vita Corp. | OncoE6 Cervical Test | Cervical | E6 oncoproteins | Cervical swab | 150 | 84.6% | 98.5% |
| | OncoE6 Oral Test | Oral | E6 oncopr oteins | NR | 150 | Not Reported | Not Reported |
| Quicking Biotech Co., Ltd | CA125 Rapid Screen Test kit | Overian | CA125 Ag | 100.0 μ L of Serum | 5–10 | 40.0 U/mL | Not Reported |
| Innovation Biotech | AFP Test | Hepatocellular | AFP | Serum, Plasma | 10 | 25.0 ng/mL | Not Reported |
| Diagnostic Automation /Cortez Diagnostics Inc. | CEA Serum Rapid Test | Adenocarcinoma of the colon | CEA | Serum | 10 | 5.0 ng/mL | Not Reported |
| LifeSign, LLC | Status BTA | Bladder | Bladder tumor associated antigen | 3 drops of urine | 5 | 67.0% | 70.0% |
| ALFA SCIENTIFIC DESIGNS | INSTANT- VIEW | Prostate | PSA | Serum | 4–7 | 4.0 ng/mL | Not Reported |
| AccuBioTech Co., Ltd | AFP Whole Blood Test Cassette | Hepatocellular | AFP | Blood/Serum/ Plasma | <10 | 99.3% | 99.0% |
| | PSA Whole Blood Test Cassette | Prostate | PSA | Blood | <10 | 98.9% | 98.6% |
| | CEA Whole Blood Test Cassette | Hepatocellular | CEA | Blood | <10 | 98.7% | 99.3% |
| | FOB Feces Test Strip | Colon | Human Occult Blood in feces | Blood | <10 | 94.6% | 99.3% |

| | | | | | | | |
|---|---|---|---|---|---|---|---|
| TÜRKLAB | CEA (CARCIN O- EMBRYONIC ANTIGEN) TEST | Diagnosis of primary carcinomas | CEA | Blood/Serum/ Plasma | <10 | 99.3% | 99.3% |
| | AFP (ALFA FETA PROTEIN) TEST | Detection of many tumors at an earlier stage/detection of fetal open neural tube defects | AFP | Blood/Serum/ Plasma | <10 | 99.3% | 99.0% |
| | FOB (FECAL OCCULT BLOOD) TEST | Colorectal cancer | human hemoglobin (hHb) | Human feces | <10 | 99.9% | 97.0% |
| Ulti med | FOB CASSETT E | Colorectal cancer | Human Occult Blood | Human feces | <10 | Not Reported | Not Reported |

*Source*: Mahmoudi, T., de la Guardia, M., & Baradaran, B. (2020). Lateral flow assays towards point-of-care cancer detection: A review of current progress and future trends. *TrAC—Trends in Analytical Chemistry*, 125. https://doi.org/10.1016/j.trac.2020.115842.

**TABLE 24.2** List of cleared or FDA-approved companion diagnostic devices (in vitro and imaging tools).

| Diagnostic name | Diagnostic manufacturer | Trade name (generic) |
|---|---|---|
| Abbott Real*Time* IDH1 | Abbott Molecular, Inc. | Acute myeloid leukemia |
| Abbott RealTime IDH2 | Abbott Molecular, Inc. | Acute myeloid leukemia |
| BRACAnalysis CDx | Myriad Genetic Laboratories, Inc. | Breast cancer |
| Bond Oracle HER2 IHC System | Leica Biosystems | Breast cancer |
| Cobas EGFR Mutation Test v2 | Roche Molecular Systems, Inc. | Non-small cell lung cancer |
| Cobas KRAS Mutation Test | Roche Molecular Systems, Inc. | Colorectal cancer |
| Cobas 4800 BRAF V600 Mutation Test | Roche Molecular Systems, Inc. | Melanoma |
| Cobas EZH2 Mutation Test | Roche Molecular Systems, Inc. | Follicular lymphoma tumor |
| Dako EGFR pharmDx Kit | Dako North America, Inc. | Colorectal cancer |
| Dako c-KIT pharmDx | Dako North America, Inc. | Gastrointestinal stromal tumors |
| FerriScan | Resonance Health Analysis Services Pvt Ltd | Non-transfusion-dependent thalassemia |
| FoundationOne Liquid CDx | Foundation Medicine, Inc. | Non-small cell lung cancer, Metastatic castrate resistant Prostate cancer (mCRPC) (plasma) Ovarian cancer |
| FoundationOne CDx | Foundation Medicine, Inc. | Non-small cell lung cancer, Melanoma, Breast cancer, Colorectal cancer, Ovarian cancer |
| FoundationFocus CDxBRCA Assay | Foundation Medicine, Inc. | Ovarian cancer |
| Guardant360 CDx | Guardant Health, Inc. | On-small cell lung cancer |
| HercepTest | Dako Denmark A/S | Breast cancer, Gastric cancer, Gastroesophageal cancer |
| HER2 CISH pharmDx Kit | Dako Denmark A/S | Breast cancer |
| HER2 FISH pharmDx Kit | Dako Denmark A/S | Breast cancer, Gastric cancer, Gastroesophageal cancer |
| INFORM HER-2/neu | Ventana Medical Systems, Inc. | Breast cancer |
| INFORM HER2 Dual ISH DNA Probe Cocktail | Ventana Medical Systems, Inc. | Breast cancer |
| InSite Her-2/neu KIT | Biogenex Laboratories, Inc. | Breast cancer |
| LeukoStrat CDx FLT3 Mutation Assay | Invivoscribe Technologies, Inc. | Acute myelogenous leukemia |
| Myriad MyChoice | Myriad Genetic Laboratories, Inc. | Ovarian cancer |
| MRDx BCR-ABL Test | MolecularMD Corporation | Chronic myeloid leukemia |

*(Continued)*

**TABLE 24.2 (Continued)**

| Diagnostic name | Diagnostic manufacturer | Trade name (generic) |
|---|---|---|
| Oncomine Dx Target Test | Life Technologies Corporation | Non-small cell lung cancer |
| PathVysion HER-2 DNA Probe Kit | Abbott Molecular Inc. | Breast cancer |
| PATHWAY anti- Her2/neu (4B5) Rabbit Monoclonal Primary Antibody | Ventana Medical Systems, Inc. | Breast cancer |
| PD-L1 IHC 28–8 pharmDx | Dako North America, Inc. | Non-small cell lung cancer |
| PD-L1 IHC 22C3 pharmDx | Dako North America, Inc. | ~~Non-small cell lung cancer,~~ gastric or gastroesophageal junction adenocarcinoma, cervical cancer, urothelial carcinoma, head and neck squamous cell carcinoma, esophageal squamous cell carcinoma, triple-negative breast cancer |
| PDGFRB FISH for Gleevec Eligibility in Myelodysplastic Syndrome/Myeloproliferative Disease (MDS/MPD) | ARUP Laboratories, Inc. | Myelodysplastic syndrome/myeloproliferative disease |
| Praxis Extended RAS Panel | Illumina, Inc. | Colorectal cancer |
| SPOT-LIGHT HER2 CISH Kit | Life Technologies Corporation | Breast cancer |
| Therascreen BRAF V600E RGQ PCR Kit | QIAGEN GmbH | Colorectal cancer |
| Therascreen PIK3CA RGQ PCR Kit | QIAGEN GmbH | Breast cancer |
| Therascreen FGFR RGQ RT-PCR Kit | QIAGEN Manchester Ltd. | Urothelial cancer |
| Therascreen EGFR RGQ PCR Kit | Qiagen Manchester, Ltd. | Non-small cell lung cancer |
| Therascreen KRAS RGQ PCR Kit | Qiagen Manchester, Ltd. | Colorectal cancer |
| THXID BRAF Kit | BioMérieux Inc. | Melanoma |
| VENTANA PD-L1 (SP142) Assay | Ventana Medical Systems, Inc. | Urothelial carcinoma, Triple- Negative Breast Carcinoma, and Non-small cell lung cancer |
| VENTANA HER2 Dual ISH DNA Probe Cocktail | Ventana Medical Systems, Inc. | Breast cancer |
| VENTANA ALK (D5F3) CDx Assay | Ventana Medical Systems, Inc. | Non-small cell lung cancer |
| Vysis ALK Break Apart FISH Probe Kit | Abbott Molecular Inc. | Non-small cell lung cancer |
| Vysis CLL FISH Probe Kit | Abbott Molecular, Inc. | B-cell chronic lymphocytic leukemia |

## 24.8 Global market of point-of-care devices

The global POC diagnostics market rising from US$34.49 billion in 2020 to US$81.37 billion by 2028 at a compound annual growth rate (CAGR) of 9.4%. Moreover, the market of cancer diagnostics was estimated to be US$156.27 billion in 2020. The cancer diagnostic market is one of the fastest growing fields with the (CAGR) of 8.9%. It is predicted to enhance the US$239.23 billion in 2025. This is because of the rapid growth of POC devices that play an important role to improve the healthcare system. POC miniaturized devices provide the high sensitivity, specificity, and ultralow detection limit against target analyte. In addition, they provide on-site detection, ease of operation, are compact, reliable, and have minimum false positive results. However, they don't require clinical sample storage, tedious process, bulky instruments or more space. As the need for healthcare awareness, the demands of POC application are rapidly growing in the 21st century (By End User Hospital Bedside, Physician's Office Lab, Urgent Care & Retail Clinics, and Homecare/Self-Testing, and Regional Forecast, n.d.). The Business Research Company (N.d.).

## 24.9 Limitations and challenges in cancer diagnostics

Conventional techniques such as MRI, x-ray, CT-imaging, biopsy, and mammography are being used for the diagnosis of tumor or single biomarkers associated to cancer, for example, prostate specific antigen for prostate cancer. Moreover, biopsy is a painful process. MRI and x-ray are a concern for safety purpose due to radiation exposure. In addition, they are low sensitive tools and face the challenges of real-time monitoring. Such techniques are slow, time consuming, labor intensive, and require well sophisticated instrument facilities. The longer time consumption delays the diagnosis of patients, which in turn enhances the cost of testing. However, some techniques such as mammography have poor sensitivity to imaging of dense breast cancer tissue and also have the risk of radiation-induced mutation. Enzyme-linked immunosorbent assay (ELISA) is a well-known platform for diagnosis of cancer biomarker. But it takes more steps and the possibility to give false positive and false negative results. Further, chemiluminescence, SERS, SPR, and mass-spectrometry are enough to detect cancer biomarkers with high sensitivity, but the bulkiness and costly instruments limit their usage. So there are lots of challenges in the biosensor field for in vivo cancer diagnostics application. Powering of devices, miniaturization and cost of biosensor chips are still challenging (Caracciolo, Vali, Moore, & Mahmoudi, 2019; Sandbhor Gaikwad & Banerjee, 2018).

## 24.10 Conclusions and future prospects

Clinical diagnosis offers a suitable platform for cancer diagnosis. The conventional techniques are utilizing a broad range for cancer cell diagnosis, but they are lacking in sensitivity and are costly. The development and progress is needed of newly developed biosensors that not only provide high sensitivity and specific but also offer multiplex diagnostics platform. In addition, microfluidics and lab-on-a-chip platforms have the potential to offer POC applications for multiple detection of cancer cells. They have the ability to convert into portable devices which will be further commercialized. However, nanomaterials such as graphene, quantum dots, metal nanoparticles, etc., provide low-cost materials for fabrication of sensor surfaces with high sensitivity and long-term stability (Dincer, Bruch, Kling, Dittrich, & Urban, 2017; Siontorou et al., 2017). The miniaturized device platform for cancer diagnostics could be promising in POC applications to provide early diagnosis of cancer, which helps the doctor to diagnosis and therapeutic of cancer at early stage. Several FDA-approved POCT miniaturized devices were reported for cancer detection, which have capability to perform a wide range of POC testing to achieve rapid results. But there is a further need for several technical improvements for integration and miniaturization into compact devices that can easily perform for diagnostic applications at the clinical level. They effectively perform multitasks simultaneously detecting the multiplex biomarkers in a single device platform. Moreover, they have potential to minimize the false positive and false negative result to gain the high sensitivity and specificity against target analysis. Several POC devices were reported in this regard, but further improvements are still required to improve their testing ability to achieve the excellent results. Nowadays, highly specific, sensitive, and rapid diagnostic devices are in demand, which results more accurately for on-site detection. They have the potential to replace the conventional diagnostics platforms which are painful, costly, and time consuming. So the biosensor-based diagnostics platform will be highly demandable in future days.

## Acknowledgments

The authors would like to thank Council of Scientific and Industrial Research (CSIR), India for SRF fellowship and Science and Engineering Research Board (SERB) for providing funds in the form of IPA/2020/000130 project for the financial support.

## References

Anneng, Y., & Feng, Y. (2020). Flexible electrochemical biosensors for health monitoring. *ACS Applied Electronic Materials*. Available from https://doi.org/10.1021/acsaelm.0c00534.

Bahadir, E. B., & Sezgintürk, M. K. (2015). Applications of commercial biosensors in clinical, food, environmental, and biothreat/biowarfare analyses. *Analytical Biochemistry*, *478*, 107–120. Available from https://doi.org/10.1016/j.ab.2015.03.011.

Bossmann, S. H., & Troyer, D. L. (2013). Point-of-care routine rapid screening: The future of cancer diagnosis? *Expert Review of Molecular Diagnostics*, *13*(2), 107–109. Available from https://doi.org/10.1586/erm.13.3.

By End User (Hospital Bedside, Physician's Office Lab, Urgent Care & Retail Clinics, and Homecare/Self-Testing), and Regional Forecast. (n.d.). In Point-of-care (POC) Diagnostics Market Size, Share & COVID-19 Impact Analysis, By Product (Blood Glucose Monitoring, Infectious Diseases, Cardiometabolic Diseases, Pregnancy & Infertility Testing, Hematology Testing, and Others) (pp. 2020–2027). <https://www.fortunebusinessinsights.com/amp/industry-reports/point-of-care-diagnostics-market-101072>.

Cancer. (n.d.). Cancer. <https://www.who.int/news-room/fact-sheets/detail/cancer>.

Caracciolo, G., Vali, H., Moore, A., & Mahmoudi, M. (2019). Challenges in molecular diagnostic research in cancer nanotechnology. *Nano Today*, *27*, 6–10. Available from https://doi.org/10.1016/j.nantod.2019.06.001.

Cha, E, K (2012). Immunocytology Is Strong Predictor of Bladder CancerPresence in Patients With Painless Hematuria: A MulticentreStudy. *Eur Urol.*, *61*(1), 185–192. Available from https://doi.org/10.1016/j.eururo.2011.08.073.

Chen, H., Chen, C., Bai, S., Gao, Y., Metcalfe, G., Cheng, W., & Zhu, Y. (2018). Multiplexed detection of cancer biomarkers using a microfluidic platform integrating single bead trapping and acoustic mixing techniques. *Nanoscale*, *10*(43), 20196–20206. Available from https://doi.org/10.1039/c8nr06367b.

Chikkaveeraiah, B. V., Bhirde, A. A., Morgan, N. Y., Eden, H. S., & Chen, X. (2012). Electrochemical immunosensors for detection of cancer protein biomarkers. *ACS Nano*, *6*(8), 6546–6561. Available from https://doi.org/10.1021/nn3023969.

Dincer, C., Bruch, R., Kling, A., Dittrich, P. S., & Urban, G. A. (2017). Multiplexed point-of-care testing – xPOCT. *Trends in Biotechnology*, *35*(8), 728–742. Available from https://doi.org/10.1016/j.tibtech.2017.03.013.

Dutta, D., Sailapu, S. K., Chattopadhyay, A., & Ghosh, S. S. (2018). Phenylboronic acid templated gold nanoclusters for mucin detection using a smartphone-based device and targeted cancer cell theranostics. *ACS Applied Materials and Interfaces*, *10*(4), 3210–3218. Available from https://doi.org/10.1021/acsami.7b13782.

Elsawy, A. A., Awadalla, A., Elsayed, A., Abdullateef, M., & Abol-Enein, H. (2020). Prospective validation of clinical usefulness of a novel mRNA-based urine test (Xpert® Bladder Cancer Monitor) for surveillance in non muscle invasive bladder cancer. *Urologic Oncology: Seminars and Original Investigations*, *000*, 8.

GLOBOCAN (2020). GLOBOCAN 2020: New Global Cancer Data. <https://www.uicc.org/news/globocan-2020-new-global-cancer-data>.

Horstmann, M., Bontrup, H., Hennenlotter, J., Taeger, D., Weber, A., Pesch, B., ... Brüning, T. (2010). Clinical experience with survivin as a biomarker for urothelial bladder cancer. *World Journal of Urology*, *28*(3), 399–404. Available from https://doi.org/10.1007/s00345-010-0538-2.

<https://www.fda.gov/medical-devices/vitro-diagnostics/list-cleared-or-approved-companion-diagnostic-devices-vitro-and-imaging-tools>.

Huang, X., Liu, Y., Yung, B., Xiong, Y., & Chen, X. (2017). Nanotechnology-enhanced no-wash biosensors for in vitro diagnostics of cancer. *ACS Nano*, *11*(6), 5238–5292. Available from https://doi.org/10.1021/acsnano.7b02618.

Jafarzadeh, L., Khakpoor-Koosheh, M., Mirzaei, H., & Mirzaei, H. R. (2020). Biomarkers for predicting the outcome of various cancer immunotherapies. *Critical Reviews in Oncology*.

Jayanthi, V. S. P. K. S. A., Das, A. B., & Saxena, U. (2017). Recent advances in biosensor development for the detection of cancer biomarkers. *Biosensors and Bioelectronics*, *91*, 15–23. Available from https://doi.org/10.1016/j.bios.2016.12.014.

Karen, H., Pushpa, T., Rao, D. R. O. M., Houston, B., & C, P. P. (2017). The role of affordable, point-of-care technologies for cancer care in low- and middle-income countries: A review and commentary. *IEEE Journal of Translational Engineering in Health and Medicine*, 1–14. Available from https://doi.org/10.1109/JTEHM.2017.2761764.

Khan, R., Mohammad, A., & Asiri, A.M. (2019). Advanced biosensors for health care applications. Available from https://doi.org/10.1016/C2017-0-02661-7.

Khanmohammadi, A., Aghaie, A., Vahedi, E., Qazvini, A., Ghanei, M., Afkhami, A., ... Bagheri, H. (2020a). Electrochemical biosensors for the detection of lung cancer biomarkers: A review. *Talanta*, 206. Available from https://doi.org/10.1016/j.talanta.2019.120251.

Khanmohammadi, A., Aghaie, A., Vahedi, E., Qazvini, A., Ghanei, M., Afkhami, A., ... Bagheri, H. (2020b). Electrochemical biosensors for the detection of lung cancer biomarkers: A review. *Talanta*, 206. Available from https://doi.org/10.1016/j.talanta.2019.120251.

Khondakar, K. R., Dey, S., Wuethrich, A., Sina, A. A. I., & Trau, M. (2019). Toward personalized cancer treatment: From diagnostics to therapy monitoring in miniaturized electrohydrodynamic systems. *Accounts of Chemical Research*, *52*(8), 2113–2123. Available from https://doi.org/10.1021/acs.accounts.9b00192.

Kim, S., Kim, T. G., Lee, S. H., Kim, W., Bang, A., Moon, S. W., ... Choi, S. (2020). Label-free surface-enhanced raman spectroscopy biosensor for on-site breast cancer detection using human tears. *ACS Applied Materials and Interfaces*, *12*(7), 7897–7904. Available from https://doi.org/10.1021/acsami.9b19421.

Kumar, S., Umar, M., Saifi, A., Kumar, S., Augustine, S., Srivastava, S., & Malhotra, B. D. (2019). Electrochemical paper based cancer biosensor using iron oxide nanoparticles decorated PEDOT:PSS. *Analytica Chimica Acta*, *1056*, 135–145. Available from https://doi.org/10.1016/j.ac.2018.12.053.

Liou, L. S., Grossman, H. B., Dinney, C. P. N., Blute, M. L., Reuter, V. E., & Jones, J. S. (2006). Urothelial cancer biomarkers for detection and surveillance. *Urology*, *67*(3), 25–33. Available from https://doi.org/10.1016/j.urology.2006.01.034.

Liu, D., Wang, J., Wu, L., Huang, Y., Zhang, Y., Zhu, M., ... Yang, C. (2020). Trends in miniaturized biosensors for point-of-care testing. *TrAC – Trends in Analytical Chemistry*, 122. Available from https://doi.org/10.1016/j.trac.2019.115701.

Lu, P., Cui, J., Chen, K., Lu, Q., Zhang, J., Tao, J., ... Gu, M. (2018). Diagnostic accuracy of the UBC® rapid test for bladder cancer: A *meta*-analysis. *Oncology Letters*, *16*(3), 3770–3778. Available from https://doi.org/10.3892/ol.2018.9089.

Mahato, K., Maurya, P. K., & Chandra, P. (2018). Fundamentals and commercial aspects of nanobiosensors in point-of-care clinical diagnostics. *3 Biotech*, *8*(3). Available from https://doi.org/10.1007/s13205-018-1148-8.

Mahmoudi, T., de la Guardia, M., & Baradaran, B. (2020). Lateral flow assays towards point-of-care cancer detection: A review of current progress and future trends. *TrAC – Trends in Analytical Chemistry*, 125. Available from https://doi.org/10.1016/j.trac.2020.115842.

Mehrotra, P. (2016). Biosensors and their applications – A review. *Journal of Oral Biology and Craniofacial Research*, *6*(2), 153–159. Available from https://doi.org/10.1016/j.jobcr.2015.12.002.

Mishra, S., Saadat, D., Kwon, O., Lee, Y., Choi, W. S., Kim, J. H., & Yeo, W. H. (2016). Recent advances in salivary cancer diagnostics enabled by biosensors and bioelectronics. *Biosensors and Bioelectronics*, *81*, 181–197. Available from https://doi.org/10.1016/j.bios.2016.02.040.

Nur, T. S., Mostafa, A., & Mehmet, O. (2016). Electrochemical biosensors for cancer biomarkers detection: Recent advances and challenges. *Electroanalysis*, 1402–1419. Available from https://doi.org/10.1002/elan.201501174.

Pal, M., & Khan, R. (2017). Graphene oxide layer decorated gold nanoparticles based immunosensor for the detection of prostate cancer risk factor. *Analytical Biochemistry*, *536*, 51–58. Available from https://doi.org/10.1016/j.ab.2017.08.001.

Pan, J., Liu, Q., Sun, H., Zheng, W., Wang, P., Wen, L., ... Lu, W. (2020). A miniaturized fuorescence imaging device for rapid early skin cancer detection. *Journal of Innovative Optical Health Sciences*.

Prasad, A., Hasan, S.M.A., Grouchy, S., & Gartia, M.R. (n.d.). DNA microarray analysis using a smartphone to detect the BRCA-1 gene. Analyst.

Pravin, B., & Sadaf, H. (2020). *Basics of biosensors and nanobiosensors* (pp. 1–22). Wiley. Available from https://doi.org/10.1002/9783527345137.ch1.

Qian, L., Li, Q., Baryeh, K., Qiu, W., Li, K., Zhang, J., ... Liu, G. (2019). Biosensors for early diagnosis of pancreatic cancer: A review. *Translational Research*, *213*, 67–89. Available from https://doi.org/10.1016/j.trsl.2019.08.002.

Ranjan, P., Parihar, A., Jain, S., Kumar, N., Dhand, C., Murali, S., ... Khan, R. (2020). Biosensor-based diagnostic approaches for various cellular biomarkers of breast cancer: A comprehensive review. *Analytical Biochemistry*, 610. Available from https://doi.org/10.1016/j.ab.2020.113996.

Ritter, R (2014). *Urologic Oncology: Seminars and Original Investigations*, *32*(3), 337–344. Available from https://doi.org/10.1016/j.urolonc.2013.09.024.

Saijo, N. (2012). Problems involved in the clinical trials for non-small cell lung carcinoma. *Cancer Treatment Reviews*, *38*(3), 194–202. Available from https://doi.org/10.1016/j.ctrv.2011.06.001.

Salim, A., & Lim, S. (2019). Recent advances in noninvasive flexible and wearable wireless biosensors. *Biosensors and Bioelectronics*, 141. Available from https://doi.org/10.1016/j.bios.2019.111422.

Sandbhor Gaikwad, P., & Banerjee, R. (2018). Advances in point-of-care diagnostic devices in cancers. *Analyst*, *143*(6), 1326–1348. Available from https://doi.org/10.1039/c7an01771e.

Shahrokhian, S., & Salimian, R. (2018). Ultrasensitive detection of cancer biomarkers using conducting polymer/electrochemically reduced graphene oxide-based biosensor: Application toward BRCA1 sensing. *Sensors and Actuators, B: Chemical*, *266*, 160–169. Available from https://doi.org/10.1016/j.snb.2018.03.120.

Shandilya, R., Bhargava, A., Bunkar, N., Tiwari, R., Goryacheva, I. Y., & Mishra, P. K. (2019). Nanobiosensors: Point-of-care approaches for cancer diagnostics. *Biosensors and Bioelectronics*, *130*, 147–165. Available from https://doi.org/10.1016/j.bios.2019.01.034.

Sharifi, M., Hasan, A., Attar, F., Taghizadeh, A., & Falahati, M. (2020). Development of point-of-care nanobiosensors for breast cancers diagnosis. *Talanta*, 217. Available from https://doi.org/10.1016/j.talanta.2020.121091.

Siontorou, C. G., Nikoleli, G. P. D., Nikolelis, D. P., Karapetis, S., Tzamtzis, N., & Bratakou, S. (2017). Point-of-care and implantable biosensors in cancer research and diagnosis. In *Next generation point-of-care biomedical sensors technologies for cancer diagnosis* (pp. 115–132). Singapore: Springer. Available from https://doi.org/10.1007/978–981-10–4726-8_5.

Soleymani, L., & Li, F. (2017). Mechanistic challenges and advantages of biosensor miniaturization into the nanoscale. *ACS Sensors*, *2*(4), 458–467. Available from https://doi.org/10.1021/acssensors.7b00069.

Syedmoradi, L., Norton, M. L., & Omidfar, K. (2020). Point-of-care cancer diagnostic devices: From academic research to clinical translation. *Talanta*.

The Business Research Company (N.d.). Cancer Diagnostics Global Market Report 2021: COVID-19 Growth And Change To 2030. <https://www.thebusinessresearchcompany.com/report/cancer-diagnostics-global-market-report>.

Tian, Y., Zhang, L., Wang, H., Ji, W., Zhang, Z., Zhang, Y., ... Chang, J. (2019). Intelligent detection platform for simultaneous detection of multiple miRNAs based on smartphone. *ACS Sensors*, *4*(7), 1873–1880. Available from https://doi.org/10.1021/acssensors.9b00752.

Todenhofer, T (2013). *European Urology Supplements*, *12*(1). Available from https://doi.org/10.1016/S1569-9056(13)60849-0.

Traynor, S. M., Wang, G. A., Pandey, R., Li, F., & Soleymani, L. (2020). Dynamic bio-barcode assay enables electrochemical detection of a cancer biomarker in undiluted human plasma: A sample-in-answer-out approach. *Angewandte Chemie - International Edition*, *59*(50), 22617–22622. Available from https://doi.org/10.1002/anie.202009664.

Tuğba, Ö. V. (2020). *Electrochemical sensors and biosensors for the detection of cancer biomarkers and drugs* (pp. 15–43). Springer Science and Business Media LLC. Available from https://doi.org/10.1007/978–981-15–7586-0_2.

Uliana, C. V., Peverari, C. R., Afonso, A. S., Cominetti, M. R., & Faria, R. C. (2018). Fully disposable microfluidic electrochemical device for detection of estrogen receptor alpha breast cancer biomarker. *Biosensors and Bioelectronics*, *99*, 156–162. Available from https://doi.org/10.1016/j.bios.2017.07.043.

Wongkaew, N. (2019). Nanofiber-integrated miniaturized systems: An intelligent platform for cancer diagnosis. *Analytical and Bioanalytical Chemistry*, *411*(19), 4251–4264. Available from https://doi.org/10.1007/s00216-019-01589-5.

Xu, L., Shoaie, N., Jahanpeyma, F., Zhao, J., Azimzadeh, M., & Al – Jamal, K. T. (2020). Optical, electrochemical and electrical (nano)biosensors for detection of exosomes: A comprehensive overview. *Biosensors and Bioelectronics*, 161. Available from https://doi.org/10.1016/j.bios.2020.112222.

Xylinas, E., Kluth, L. A., Rieken, M., Karakiewicz, P. I., Lotan, Y., & Shariat, S. F. (2014). Urine markers for detection and surveillance of bladder cancer. *Urologic Oncology: Seminars and Original Investigations*, *32*(3), 222–229. Available from https://doi.org/10.1016/j.urolonc.2013.06.001.

Yoo, E. H., & Lee, S. Y. (2010). Glucose biosensors: An overview of use in clinical practice. *Sensors*, *10*(5), 4558–4576. Available from https://doi.org/10.3390/s100504558.

Zheng, Z., Wu, L., Li, L., Zong, S., Wang, Z., & Cui, Y. (2018). Simultaneous and highly sensitive detection of multiple breast cancer biomarkers in real samples using a SERS microfluidic chip. *Talanta*, *188*, 507–515. Available from https://doi.org/10.1016/j.talanta.2018.06.013.

# Index

*Note*: Page numbers followed by "*f*" and "*t*" refer to figures and tables, respectively.

## Q



## R

## S



## T

Printed in the United States
by Baker & Taylor Publisher Services